PROPERTIES AND APPLICATIONS OF SILICON CARBIDE

Edited by **Rosario Gerhardt**

Properties and Applications of Silicon Carbide
Edited by Rosario Gerhardt

Published by InTech
Janeza Trdine 9, 51000 Rijeka, Croatia

Publishing Process Manager Katarina Lovrecic

Technical Editor Teodora Smiljanic

Cover Designer InTech Design Team

Additional hard copies can be obtained from orders@intechopen.com

Properties and Applications of Silicon Carbide, Edited by Rosario Gerhardt
p. cm.
ISBN 978-953-307-201-2

For EU product safety concerns:
IN TECH d.o.o., Prolaz Marije Krucifikse Kozulić 3, 51000 Rijeka, Croatia,
info@intechopen.com or visit our website at intechopen.com.

Meet the editor

Dr. Rosario A. Gerhardt is a professor of materials science and engineering at the Georgia Institute of Technology in Atlanta, GA, USA, where she has taught for the last 20 years. She was previously with the Center for Ceramics Research at Rutgers University in Piscataway, NJ and has also served as a visiting scientist to the NASA Marshall Space Flight Center in Hunstville, Alabama and the Center for Nanophase Materials Sciences at Oak Ridge National Laboratory in Oak Ridge, Tennessee. She received her D.Eng.Sc. in metallurgy and materials science from Columbia University in New York City under the guidance of Prof. A.S. Nowick. Prof. Gerhardt's research interests include developing structure-processing-property relationships in all classes of materials. Over the years she has worked with ceramics, metals, polymers as well as composites in bulk, thin film and in nanostructure form. Most of her work has dealt with the electrical and microstructural characterization of materials using impedance and dielectric spectroscopy, microscopy via optical, SEM, TEM and AFM as well as x-ray and neutron scattering methods.

Contents

Preface

Silicon carbide (SiC) is an interesting material that has found application in a variety of industries. The two best known applications of this material are its use as an abrasive material and its more recent use as a wide band gap semiconductor for high power, high temperature electronic devices. The high hardness of this material, known for many years, led to its use in machining tools and in other structural applications. Usage of SiC in semiconductor devices only became possible in the last twenty years, when commercially available SiC single crystals became available. Thin films and nanoparticles of SiC are still rare, but monolithic SiC and composites containing SiC have been available much longer. One of the challenges of working with this material is that it can crystallize into many different polymorphs, the most common being the 3C (β-SiC), and the hexagonal (α-SiC): 2H, 4H and 6H phases. Because of its high melting point, achieving reasonable bulk densities in polycrystalline materials is difficult. In addition, the semiconducting material forms Schottky barriers with most metals, while the formation of its native oxide, SiO2, can pose additional issues when used in oxidizing atmospheres. However, the scientific community has shown ingenuity in turning some of the pitfalls into decided advantages for a variety of applications.

In this book, we explore an eclectic mix of articles that highlight some new potential applications of SiC and different ways to achieve specific properties. Some articles describe well-established processing methods, while others highlight phase equilibria or machining methods. A resurgence of interest in the structural arena is evident, while new ways to utilize the interesting electromagnetic properties of SiC continue to increase. The reader is encouraged to explore the vast literature in this field, ranging from 40,000 up to 150,000 articles depending on which database one chooses to search in, but several gems may be found among the chapters in this book.

The articles have been grouped according to the following three categories: Part A: Thin Films and Electromagnetic Applications. Part B: Other Applications: Electrical, Structural and Biomedical. Part C: Bulk Processing Methods, Phase Stability and Machining.

Katarina Lovrecic, the publishing process manager, deserves much credit for this work. She initiated contact with the various authors and kept everyone on task throughout this process. I would like to also acknowledge the assistance of my graduate students, Brian D. Bertram and Timothy L. Pruyn, who helped proofread the chapters and make suggestions to the authors. The final editing of all materials was conducted by In-Tech publications.

Rosario A. Gerhardt
Professor of Materials Science and Engineering
Georgia Institute of Technology
Atlanta, USA

Part 1

Thin Films and Electromagnetic Applications

1

Identification and Kinetic Properties of the Photosensitive Impurities and Defects in High-Purity Semi-Insulating Silicon Carbide

D. V. Savchenko, B. D. Shanina and E. N. Kalabukhova
V.E. Lashkaryov Institute of Semiconductor Physics,
National Academy of Science of Ukraine
Ukraine

1. Introduction

Semi-insulating (SI) silicon carbide substrates are required for high power microwave devices and circuits based on SiC and GaN. SI properties of SiC can be achieved by introducing deep levels from either impurities or intrinsic defects into the material to compensate shallow donors and acceptors and pin the Fermi level near the middle of the bandgap. Intrinsic defects with deep levels are believed to be responsible for the SI properties of undoped material. Most of the intrinsic defects studied in SI 4H-SiC grown by physical vapour transport (PVT) and high temperature chemical vapour deposition (HTCVD), referred to as high purity semi-insulating (HPSI) material, have energies ranging from 0.85 eV to 1.8 eV below the conduction band (Müller et al., 2003; Son et al., 2004). However, not all of these defects are temperature stable and desirable for SI SiC. Among the defects, which are stable after annealing at 1600 ^{0}C – 1800 ^{0}C and believed to be responsible for SI properties of HPSI 4H-SiC are the ID and SI-5 defects. ID defect with energy level of 1.79 eV below the conduction band which occupies quasi-cubic (ID1) and hexagonal (ID2) lattice sites were observed in both HTCVD and PVT p-type HPSI wafers and identified as the carbon vacancy in the single positive charge state, V_C^+ (Konovalov et al., 2003). The parameters of the ID1 and ID2 defects are coincided with those of E15 and E16 centers, respectively, which were observed in electron-irradiated p-type 4H-SiC and originate from V_C^+ at quasi-cubic (c) and hexagonal (h) lattice sites (a. Umeda et al., 2004; b. Umeda et al., 2004).

SI-5 defect with energy level of 1.2 ÷ 1.5 eV below the conduction band (E_C) is found to be a dominating defect among a series of other defects (SI1-SI9) observed in HTCVD and PVT HPSI 4H-SiC substrates in the dark and under light illumination with different photon energies (Son et al. 2004). The HEI4 center with the g-values similar to those found for SI-5 defect was observed in electron irradiated n-type 4H-SiC samples but with the concentration higher than that in HPSI samples which was sufficient to observe the hyperfine (HF) structure of the defect (a. Umeda et al., 2006). As a result, based on the analysis of the HF structures of ^{29}Si and ^{13}C and first principles calculations, the SI-5 center was identified to be originated from the negatively charged carbon antisite-vacancy (AV) pair ($C_{Si}V_C^-$) (b. Umeda et al., 2006). The energy level of the SI-5 center obtained from the photo-EPR studies of the electron irradiated n-type 4H-SiC samples amounts to 1.1 eV below the conduction band and coincides with the ionization levels (E_C – 1.0 eV and E_C – 0.9 eV) calculated from the first principles for the (0|–) and (–|2–) charge states of the $C_{Si}V_C$, respectively (Bockstedte et al., 2006). According to theory the Fermi level E_F is pinned at ~ E_C – 1.1 eV or lower due to $C_{Si}V_C$ in SiC. Thus, the single vacancies as well as the carbon AV pairs are found to be the dominant defects responsible for the SI property of SiC.

The presented review indicates that the investigation of the SI SiC material mainly was focused on the identification of the intrinsic defects responsible for the SI properties of SiC. At the same time, investigation of the role, which they play in the trapping, recombination, and ionization of non-equilibrium charge carriers, processes of paramount importance for semiconducting materials, has not received proper attention. HPSI 4H-SiC samples have a specific feature of the so-called persistent relaxation (PR) of the photo-response and persistent photoconductivity (PPC), which originate from a very long (over 30 h) low-temperature lifetime of photo-carriers trapped into defect and impurity levels (Kalabukhova et al., 2006).

In this chapter we present the results of identification of the intrinsic defects observed in EPR spectrum of HPSI 4H and 6H-SiC wafers in the dark and under photo-excitation and investigation of the PR and PPC which form in HPSI 4H-SiC samples at low temperatures after termination of photo-excitation.

The existence of very long lifetime of photo-carriers (PR of the photo-response) in semiconductors is reported since the 1930s (Sheinkman & Shik, 1976). The phenomenon of the PPC has been observed for a wide set of binary (Queisser & Theodorou, 1986; Dissanayake, et al., 1991; Evwaraye et al., 1995) and amorphous semiconductor materials (Kakalios & Fritzsche, 1984).

The main specifications of PPC are the following: 1) the type of semiconductor and it's state (mono- or poly-crystal, or amorphous, or powder) is not important; 2) the wavelength of photo-excitation does not mean; 3) the photo-response time has a temperature dependence proportional to $\exp(E_{rec}/kT)$, where E_{rec} is the recombination barrier, which depends on a time; 4) residual photoconductivity can achieve a large value $\sigma_{PPC} >> \sigma_0$, where σ_0 is a conductivity before photo-excitation. Two main models have been proposed for the explanation of the PPC. In the first model, the reason of PPC existence is a significant concentration of traps, which serve as the recombination centres for electrons and holes and have enough high activation energy for ionization (Litton & Reynolds, 1964). The second model is the so called a 'barrier model', which supposes the space separation of photo-carriers due to an appearance of the electrical barriers, which are potential barriers for the recombination of carriers. Macroscopic potential barriers may arise at surfaces, interfaces, junctions, around doping inhomogeneities. Microscopic barriers against recombination may arise due to impurity atoms with large lattice relaxations (Lang & Logan, 1977; Dissanayake & Jiang, 1992). The second model was considered in (Shik A.Ya., 1976; Ryvkin & Shlimak, 1973) as the main mechanism of PPC in CdS. The height of the barriers was determined to be equal to 10^5 V/cm.

To distinguish between the models responsible for the PR and PPC in HPSI 4H-SiC we have studied kinetic properties of the photosensitive paramagnetic impurities (nitrogen and boron centers) and deep defects observed in HPSI 4H SiC, using EPR, photo EPR methods and optical admittance spectroscopy. It was expected that in the case of 'barrier model' electrostatic potential fluctuations will cause the noticeable shift of the g-factor of nitrogen donor centers with respect to that measured in the dark or broadening of their spectral lines as long as 4HSiC does not have the center of inversion symmetry. Otherwise, we have to reject this model and consider the role played in PR and PPC by deep traps.

2. Samples and experimental technique

The nature of the intrinsic defects in HPSI 4H and 6H-SiC was studied on the samples grown up by the PVT method at the Cree Research Inc. and Bandgap Technologies Inc., respectively, by EPR and photo-EPR methods. The nature of PR of the photo-response and of PPC in HPSI 4H-SiC material was studied on the samples grown by PVT method at the Cree Research Inc. by photo-EPR and optical admittance spectroscopy. The HPSI material was purposefully undoped SiC with a residual impurity concentration of the order of 10^{15} cm^{-3}, which has a high room-temperature electrical resistivity (on the order of 10^{10} Ω·cm). Before carrying out experiments, the samples were annealed in an inert atmosphere at T = 1800 ^{0}C to remove surface intrinsic defects which are known to be always present in a SI SiC material before its annealing and characterized by an EPR signal with isotropic g-factor $g_{\|} = g_{\perp} = 2.0025$ (Macfarlane & Zvanut, 1999; Kalabukhova et al., 2001).
A study of the temperature dependence of charge carrier concentration performed on the same HPSI 4H-SiC samples on which the EPR measurements were carried out revealed that the samples were n-type, and that the Fermi level was localized in the upper half of the band gap. The activation energy, derived from the slope of the dependence of charge carrier concentration on 1/T, turned out to be 1.1 eV. The charge carrier concentration determined at the highest temperature of the experiment was about 1×10^{15} cm^{-3} (Kalabukhova et al., 2004).
The EPR and photo-EPR spectra were measured in an X-band (9 GHz) and Q-band (37 GHz) EPR spectrometers in the temperature range of 4.2 K – 140 K. Photo-excitation of samples by interband light was provided by a 250-W high-pressure mercury vapor lamp equipped with interference filters for wavelengths from 365 nm to 380 nm. To illuminate a sample with impurity light, a 100-W xenon and halogen incandescent lamps were used in combination with either an UM-2 prism monochromator or the interference or glass filters, which enabled us to carry out photo experiments in the wavelength range from 380 nm to 1000 nm. Light focused by a short-focus doubleconvex quartz lens was coupled into the resonator through a light guide, with the sample of about 1.7 x 4 mm^2 in size fixed to its end face oriented such that the c axis of the crystal was perpendicular to the direction of the external magnetic field. The thickness of the illuminated sample was about 35 μm, which was thin enough for the light to illuminate the total sample.
The PPC data obtained by optical admittance spectroscopy at T = 300 K were taken from (Kalabukhova et al., 2006). The technique employed in optical admittance spectroscopy measurements was described in considerable detail in (Evwaraye et al., 1995).
The EPR spectra were simulated with the help of an Easyspin toolbox (Stoll & Schweiger, 2005). The EPR lineshape was Gaussian. The determination error of the g-factor was ± 0.0002. The determination error for the defect and impurity energy levels was approximately ± 0.06 eV.

3. Identification of the photosensitive impurities and defects in HPSI SiC

3.1. Hydrogenated carbon vacancy in HPSI 4H-SiC

Fig. 1 presents a typical EPR spectrum of the HPSI 4H-SiC samples under study annealed at T = 1800°C, which were measured in the dark and under excitation with the different photon energies at T = 77 K when magnetic field $\mathbf{B_0}$ is parallel to the **c**-axis.

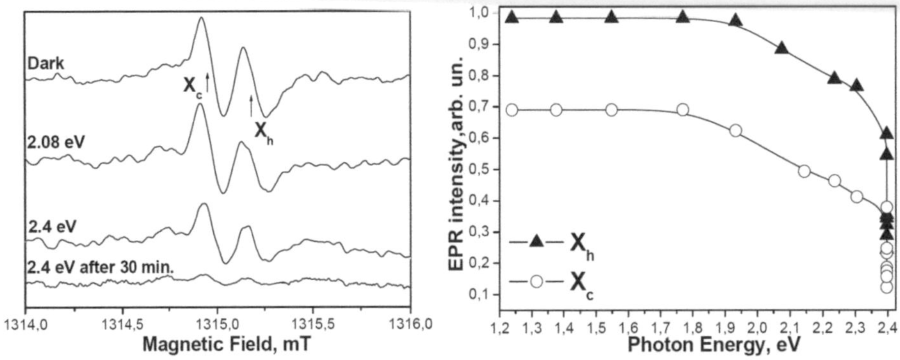

Fig. 1. Spectral dependence of the Q-band photo-EPR spectra of the X-defect measured in HPSI 4H-SiC at T = 77 K. $\mathbf{B_0}||\mathbf{c}$. After (b. Kalabukhova et al., 2006).

The spectrum obtained in the dark reveals two EPR signals with an intensity ratio of 1:1.6 deriving from a thermally stable X defect with S = 1/2 which occupies the quasi-cubic (X_c) and hexagonal (X_h) positions in the 4H-SiC lattice. At 77 K, the EPR lines of X_h and X_c are both characterized by axial g-tensors: $g_{||}$ = 2.0025, $g_{\perp}$ = 2.0044 and $g_{||}$ = 2.0028, $g_{\perp}$ = 2.0043, respectively (Kalabukhova et al., 2004).

The energy level of the X-defect was determined from the spectral dependences of the intensities of the X defect EPR lines. As was shown in Fig. 1, the intensity of the X-defect photo-EPR lines remains constant until the photon energy hν reaches to 1.90 eV. Above that photon energy, the intensity decreases. With a threshold photon energy of hν = 1.90 eV for the X_h defect the energy level of the X_h defect can be determined as $\Delta E = E_C - 1.26$ eV and for X_c as $\Delta E = E_C - 1.36$ eV. When the photon energy approaches hν = 2.4 eV the EPR signals due to the X-defect gradually vanished. The transition of the single X_c EPR line into six lines is observed at the temperature below 40 K, showing that the symmetry of the EPR spectrum for the defect residing on a c site is lowered from axial C_{3v} to C_{1h}, while the axial symmetry of the EPR spectrum due to the X-defect residing on the h site remains.

The ligand HF structure of the X_h and X_c defect was analyzed at X-band frequency. As was seen from Fig. 2, at X-band frequency, the X_h and X_c EPR signals are superimposed and the EPR spectrum consists of a single line with a slightly asymmetric line shape at $\mathbf{B_0}||\mathbf{c}$. Besides the strong central line, six weaker satellite lines (labeled 1 to 6) belonging to the X_h, and X_c defect are found symmetrically around the central line. The HF interaction constants determined from comprehensive analysis of the complete angular dependence of all satellites are in excellent agreement with those found for the Si-nearest neighbors of the carbon vacancy $V_C^{+/0}$, both at the c and h lattice sites known as EI5 and EI6 (a. Umeda et al., 2004; b. Umeda et al., 2004) or ID and ID2 (Konovalov et al., 2003) centers. The results of this analysis are summarized in Table 1.

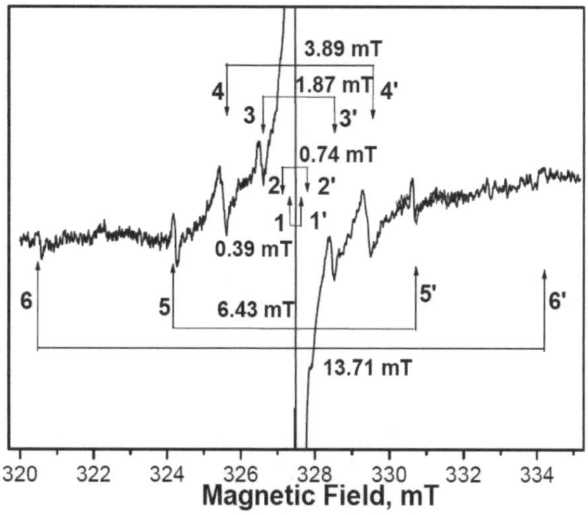

Fig. 2. X-band EPR spectrum of the X-defect measured at 77 K. $\mathbf{B}_0 \| \mathbf{c}$. After (b. Kalabukhova et al., 2006).

Satellite line number[1)]	$A_{\|}$, mT	$A_{\perp}$, mT	Relative intensities and assignments of X_c/X_h ligand structure	Assignments of ID1/ID2 ligand structure	Assignments of E15/E16 ligand structure
(5-5')	6.43	4.46	X_C-1: 0.060 1 x ^{29}Si	ID1-1: 1 x ^{29}Si	E15-1: 1 x ^{29}Si
(4-4')	3.89	4.08	X_C-2: 0.146 3 x ^{29}Si	ID1-2: 3 x ^{29}Si	E15-2: 3 x ^{29}Si
(1-1')	0.39	0.34	X_C-3: 0.543 12 x ^{13}C +8 x ^{29}Si	ID1-3: 12 x ^{13}C	E15(*)-3: 5-11 x ^{13}C + 3 x ^{29}Si
(6-6')	13.71	9.52	X_h-1: 0.062 1 x ^{29}Si	ID2-1: 1 x ^{29}Si	E16-1: 1 x ^{29}Si
(3-3')	1.87	2.76	X_h-2: 0.146 3 x ^{29}Si	ID2-2: 3 x ^{29}Si	E16-2: 3 x ^{29}Si
(2-2')	0.74	0.66	X_h-3: 0.156 3 x ^{29}Si	ID2-3: 3 x ^{29}Si	E16-3: 3 x ^{29}Si

Table 1. EPR parameters of the X_c and X_h defects with S = 1/2 in HPSI 4H-SiC samples measured at T = 77 K. $A_{\|}$ and $A_{\perp}$ are the HF splitting as determined from angular dependent EPR measurements. The number of magnetic nuclei determined from the relative intensities of the HF satellites is given after (Kalabukhova et al., 2006) for X_c/X_h, after (Konovalov et al., 2003) for ID1/ID2 and after (a, b. Umeda, et al., 2004; Bockstedte et al., 2003) for E15/E16 centers residing c/h positions. 1) The lines are labeled after Fig. 2. (*) The number of magnetic nuclei was determined by pulsed ENDOR (b. Umeda et al., 2004).

However, there are significant discrepancies in the intensity ratio of line (1-1') with HF splitting of $A_{\|}$ = 0.39 mT and $A_{\perp}$ = 0.34 mT in all four references given in Table 1. Since all other intensity ratios coincide within experimental error, the mistakes in the measurements

can be excluded. Thus varying sample preparations might be responsible, and therefore probably slightly different defects might occur. Nevertheless, the disturbance is rather small and in particular cannot be explained by another charge state of the carbon vacancy, e.g. V_C^- is known to have distinctly different HF interaction constants (Umeda et al., 2005).
Since the HF splitting is the same in all cases, but the number of nuclei involved is different, one has to discuss a common additional rather small disturbance of the carbon vacancies. The twofold HF splitting suggests that a nucleus with nuclear spin $I = 1/2$ and high (probably 100%) natural abundance is involved. There are only a few plausible nuclei in question. The most probable candidate is hydrogen, which e.g. also in silicon provokes a small disturbance of defects, introduced by sample preparation and only hardly observable with EPR (Langhanki et al., 2001).
This conclusion is supported by the ionization energy of X-defect which is close to that calculated for V_C with adjacent hydrogen (V_C+H) (Aradi et al., 2001; Gali et al., 2003) and has significant higher ionization energy than that known for EI5/EI6 center, see Table 2.

Experimental E_V - E_i	Theory (A. Gali et al., 2003)	Model
1.47 eV (Son et al., 2002)	1.57 eV	$V_C^{+/0}$ acceptor
1.90 eV, 2.00 eV (b. Kalabukhova et al., 2006)	2.05 eV	$(V_C$+H$)^{0/-}$ donor

Table 2. Ionization energies (E_i) of the isolated carbon vacancy V_C and hydrogenated carbon vacancies (V_C+H) in respect to the valence band (E_v) from experiment and theory.

As was seen from Table 2, the energy level of the $V_C^{+/0}$ in contrast to the X-defect is pinned in the lower half of the band gap and shows acceptor-like behavior. Therefore, the X-defect which shows the donor-like behavior was assigned to the hydrogenated carbon vacancy $(V_C$+H$)^{0/-}$ which occupies the c and h positions in the 4H-SiC lattice (Kalabukhova et al., 2006).
In accordance with calculations performed in (Aradi et al., 2001; Gali et al., 2003), the most stable configuration of the $(V_C$+H$)^{0/-}$ is that where hydrogen is built in bond-bridging between two Si-ligands of V_C as was shown in Fig. 3.

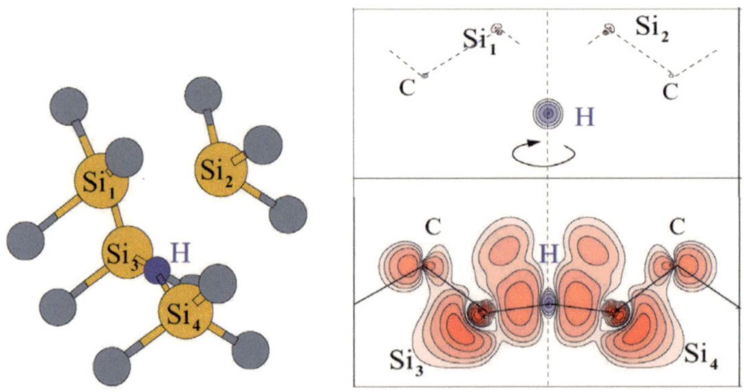

Fig. 3. Calculated ground state geometry and corresponding spin-density (in two perpendicular planes) of the hydrogenated carbon vacancy (V_C+H) at the quasi-cubic site in 4H-SiC. After (b. Kalabukhova et al., 2006).

On the other hand accounting the C_{3v} symmetry of the high-temperature (T = 77 K) EPR spectrum of X_C defect caused by a dynamic Jahn-Teller effect the motional averaged configuration with the H-atom in the center of three ligands Si_3, Si_4 and Si_2 has been proposed in (b. Kalabukhova et al., 2006) which provides nearly isotropic HF parameters for X_C defect and are in reasonable agreement with the experimental values of the HF splitting given in Table 1.

According to our analysis, in the present sample about 20% of the carbon vacancies would be contaminated with hydrogen. This is compatible with the lower signal-to-noise ratio observed in our study compared to that in the literature.

The proposed model for X defect is in agreement with the temperature dependent Hall effect measurements indicating that the Fermi level is pinned in the upper half of the band gap at an energy close to that of the X-defect, suggesting it is donor-like.

3.2. Photosensitive impurities in HPSI 4H-SiC

Fig. 4 shows the EPR spectrum of the HPSI 4H-SiC samples measured in the dark and under interband light at T = 77 K. The EPR spectrum consists of a single line due to the X_h and X_c defects which EPR lines are superimposed when magnetic field $\mathbf{B_0}$ is perpendicular to the **c**-axis.

The EPR spectra of nitrogen and boron are not seen in the dark, which suggests that shallow donor and acceptor centers of the nitrogen and boron impurities reside in the ionized state because of the mutual compensation of their charge or partial compensation by deep-level defects. In our particular case, such a defect is the X defect lying 1.36–1.26 eV below the conduction band bottom.

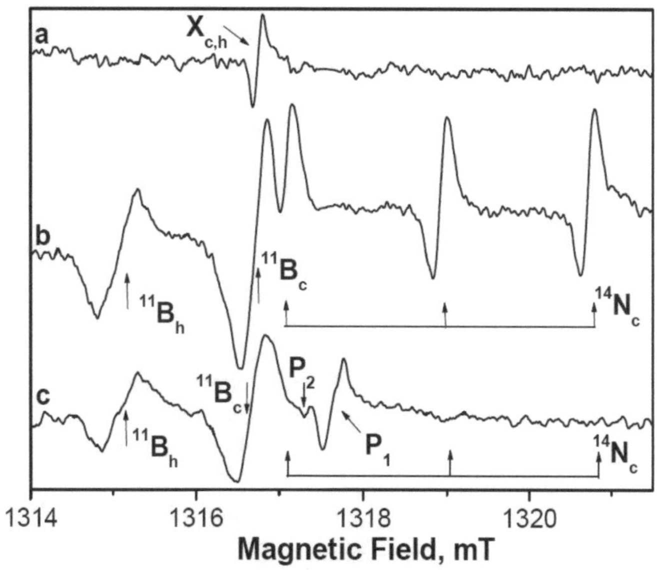

Fig. 4. Photo-response of the Q-band EPR spectrum in HPSI 4H-SiC samples at T = 77 K and $\mathbf{B_0} \perp \mathbf{c}$: (a) – in the dark, (b) – excitation with interband light, and (c) – EPR spectrum measured in the dark 21 h after termination of photo-excitation. After (a. Savchenko et al., 2009).

Illumination of the samples with interband light of wavelength 365 nm gives rise to trapping of non-equilibrium charge carriers into the donor and acceptor levels of nitrogen and boron, which generates in the EPR spectrum simultaneously a triplet of EPR lines due to nitrogen in the quasi-cubic position (N_c) (Greulich-Weber, 1997) and two EPR signals of boron occupying quasi-cubic (B_c) and hexagonal (B_h) positions of C_{3v} symmetry above 50 K (Greulich-Weber, et al., 1998). At the same time trapping of nonequilibrium charge carriers into the levels of the X defect changes its charge state and initiates its transition to the nonparamagnetic state.

As will be shown in Section 4.1, the EPR signal due to nitrogen in hexagonal position (N_h) was observed in EPR spectrum of the HPSI 4H-SiC at 50 K under illumination with interband light.

The EPR parameters of the nitrogen and boron centers measured in HPSI 4H-SiC in the presence of photo-excitation are listed in Table 3. In addition, the energy ionizations of the nitrogen and boron centers are given in Table 3 after (Evwaraye et al., 1996) and (Sridhara et al., 1998), respectively.

Impurity	N_c	N_h	B_c	B_h
$g_{\parallel}$	2.0043	2.0063	2.0063	2.0019
$g_{\perp}$	2.0013	2.0006	2.0046	2.0070
$A_{\parallel}$, mT	1.82	0.10	0.20	0.20
$A_{\perp}$, mT	1.82	0.10	0.11	0.12
$E_C - E_i$, eV	0.10	0.053		
$E_V + E_i$, eV			0.628	

Table 3. EPR parameters and energy ionizations of nitrogen (N_c and N_h) and boron (B_c and B_h) centers measured in HPSI 4H-SiC under photo-excitation with interband light in the temperature interval from 50 K to 80 K. $A_{\parallel}$ and $A_{\perp}$ are the HF splitting.

Comparing the obtained data with the literature data has shown that there is no difference between g-tensor of nitrogen EPR spectrum measured in HPSI 4H-SiC in the presence of the photo-excitation and in n-type 4H SiC crystals with ($N_D - N_A$) = $10^{17} - 10^{18}$ cm^{-3} measured in the dark (Kalabukhova et al., 2007).

Therefore one can exclude the presence of the electrical barriers in the HPSI 4H-SiC sample after its photo-excitation which may disturb the local environment of the donors and give rise to the shift of the g-factor of nitrogen EPR spectrum with respect to that observed in the dark (Kalabukhova et al., 1990).

It was found that the lifetime of the nonequilibrium charge carriers trapped into the donor and acceptor levels of nitrogen and boron is very large (on the order of 30 h and longer). This PR of the photo-response after termination of photo-excitation is accompanied by the PPC phenomenon (a. Kalabukhova et al., 2006).

Examining Fig. 4, we see that the EPR line intensities of nitrogen and boron signals decay very slowly after termination of the pump light. Within 21 h after switching off the light, the EPR signal intensities of boron, B_c and B_h, are 0.4 and 0.7 of those under illumination, respectively, while the N_c EPR line intensity drops in this time to 0.07 of that observed with the light on.

3.3. Carbon antisite-vacancy pair $C_{Si}V_C$ and silicon vacancy in HPSI 4H-SiC

As evident from Fig. 4, when the EPR nitrogen triplet line intensities decay, EPR signals due to another defect center, labeled as the P_1 and P_2 defects, appear in the EPR spectrum. This suggests that electrons detrapping from the nitrogen level become trapped into the level of the P_1 and P_2 centers. On the other hand, as can be seen in Fig. 5, the transformation of the P_1 and P_2 centers into an EPR-active charge state can be achieved by the photo-excitation of the HPSI 4H-SiC sample with a threshold photon energy of approximately $h\nu = 2.11$ eV (598 nm) making it possible to determine the position of the energy level of P_1 and P_2 defects as: $\Delta E = \Delta E_g - 2.11$ eV $= E_C - 1.15$ eV (Kalabukhova et al., 2004).

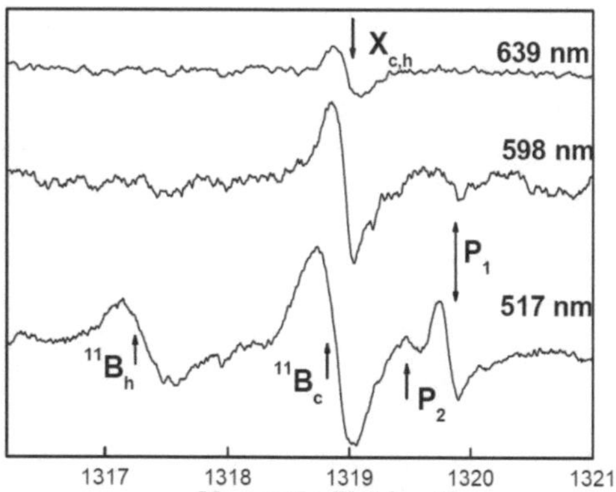

Fig. 5. Behavior of the EPR spectrum of HPSI 4H-SiC sample measured at 37 GHz and 77 K under illumination with the light of different photon energies. $\mathbf{B_0}\perp\mathbf{c}$. After (Kalabukhova et al., 2004).

The EPR line due to P_1 defect has axially symmetric g-tensor ($g_{\parallel} = 2.0048$, $g_{\perp} = 2.0030$) but was too weak in intensity to be identified using its HF structure due to the small concentration of the defect, estimated to be about $5\cdot10^{14}$cm^{-3}. It should be noted that the second EPR line of small intensity with isotropic g-value $g = 2.0037$, labeled P_2 defect, always appears in the EPR spectrum of HPSI 4H-SiC along with the P_1 defect signal having the axially symmetric g-tensor.

The high symmetry of the this line gave us argument to compare the symmetry, energy ionization and spin state of the P_2 defect with electronic and structural properties of silicon vacancies calculated in (Torpo et al., 2001) which are generally have T_d symmetry group and less frequently a lower symmetry like C_{3V}. Among all possible configuration the silicon vacancy in the (3–) charge state can provide the characteristics (low spin state, $S = 1/2$, energy ionization level of about 2.25 eV above the valence band maximum) which are suitable for the P_2 defect. Therefore, it was suggested that P_2 defect is due to the V_{Si}^{3-}. But at the same time the EPR parameters of the P_1 defect with axially symmetric g-tensor agree well with those of the SI-5 center observed in HPSI 4H SiC (Son et al., 2004) and in electron irradiated n-type 4H SiC (a. Umeda et al., 2006) at 77 K. In addition, energy position of the P_1 defect coincides with that of SI-5 center which amounts to 1.1 eV below the conduction band and coincides with the ionization levels (E_C – 1.0 eV and E_C – 0.9 eV) calculated from the first principles for the (0|–) and (–|2–) charge states of the carbon antisite-vacancy (AV) pair $C_{Si}V_C$, respectively (Bockstedte et al., 2006). Therefore, $C_{Si}V_C$ pair could be suggested as the most possible model for the P_1 defect. This suggestion is supported by the fact that the transformation of $C_{Si}V_C$ from nonparamagnetic (–2) into paramagnetic (–1) state similar to the P_1 defect is caused by capture of the electrons from nitrogen donors into the $C_{Si}V_C$ defect level under light illumination (b. Umeda et al., 2006).

It should be noted that the similar isotropic line with g = 2.0037, labeled as SI-11, has also been observed together with SI-5 defect signal in HPSI 4H-SiC in (Son et al., 2004, Carlsson P. et al., 2007) but was not assigned with the inequivalent position of the SI-5 center due its small intensity. Theoretically predicted bistabillity between carbon AV complex $C_{Si}V_C$ and the isolated Si vacancy depending on the position of the Fermi level (E_F) make it possible to suggest that this weak EPR line, with isotropic g = 2.0037 is due to the presence of small concentration of the Si vacancy which was predicted to be more stable in n-type than in p-type HPSI 4H-SiC samples (Rauls et al., 2003; b Bockstedte et al., 2003; Bockstedte et al., 2004). Spin-Hamiltonian parameters and energy levels of the thermally stable deep intrinsic defects with S = 1/2 measured in HPSI 4H-SiC samples at T = 77 K were listed in Table 4.

Defects	P_1	P_2	X_C	X_h
$g_{\parallel}$	2.0048	2.0037	2.0028	2.0025
$g_{\perp}$	2.0030	2.0037	2.0043	2.0044
$E_C - E_i$, eV	1.15	1.15	1.36	1.26
Model	$C_{Si}\, V_C^{0/-}$	V_{Si}^{3-}	$(V_C+H)^{0/-}$	

Table 4. Spin-Hamiltonian parameters and energy levels of the deep intrinsic defects with S = 1/2 measured in HPSI 4H-SiC samples at T = 77 K along with their identification.

3.4. Carbon vacancy and carbon antisite-vacancy pair $C_{Si}V_C$ in HPSI 6H-SiC

As was already mentioned above, most of the study of intrinsic defects in SI SiC has been made on the 4H polytype due to its relatively wide availability but very little reported on SI 6H-SiC. Fig. 6 shows the EPR spectrum observed in annealed HPSI 6H-SiC sample at 77 K in the dark and under illumination with the light of different photon energy. As seen from Fig. 6 HPSI 6H-SiC reveals a series of photosensitive paramagnetic centers including boron ($B_{c1,c2}$ and B_h), nitrogen ($N_{c1,c2}$) and two thermally stable deep intrinsic defects labeled as XX and PP (Savchenko et al., 2006; Savchenko & Kalabukhova, 2009). The EPR parameters of the nitrogen and boron centers measured in HPSI 6H-SiC in the presence of photo-excitation are coincided with those measured in n-type 6H SiC (b. Savchenko et al., 2009) and p-type 6H SiC (Greulich-Weber et al., 1998) in the dark and are listed in Table 5. The notation for the XX and PP defects was selected by analogy with that adopted for two thermally stable intrinsic defects X and P, which are observed in HPSI 4H-SiC with very similar EPR parameters.

The photo EPR data placed the energy level of the defects in the region 1.24 ÷ 1.29 eV above the valence band (Savchenko & Kalabukhova, 2009). The EPR parameters and symmetry features of the XX defect which substitutes three inequivalent positions (XX_h, XX_{c1} and XX_{c2}) in the 6H-SiC lattice was found to be similar to those of the Ky3(h), Ky1(c) and Ky2(c) defect observed in p-type electron irradiated 6H-SiC samples and assigned to the carbon vacancy in the single positive charge state V_C^+ at three inequivalent positions (Bratus et al., 2005).

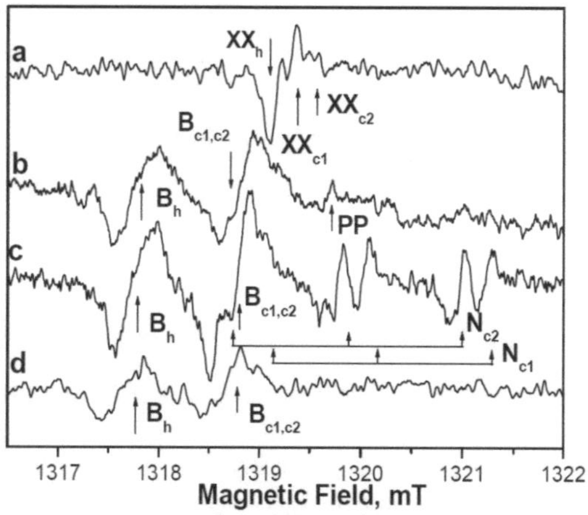

Fig. 6. Q-band EPR spectrum measured in HPSI 6H-SiC sample in the dark (a) and under illumination with the light of different photon energies: (b) – 2.79 eV; (c) – 3.06 eV, (d) - EPR spectrum measured in the dark 17 h after termination of photo-excitation. $\mathbf{B_0 \perp c}$. T = 77 K. After (Savchenko & Kalabukhova, 2009).

Impurity	N_{c1}	N_{c2}	B_{c1}	B_{c2}	B_h
$g_{\parallel}$	2.0040	2.0037	2.0055	2.0062	2.0020
$g_{\perp}$	2.0026	2.0030	2.0045	2.0045	2.0068
$A_{\parallel}$, mT	1.20	1.19	0.22	0.19	0.19
$A_{\perp}$, mT	1.20	1.19	0.13	0.12	0.15
$E_C - E_i$, eV	0.138	0.142			
$E_V + E_i$, eV			0.31-0.38		0.27

Table 5. EPR parameters of nitrogen and boron centers measured in HPSI 6H-SiC under photo-excitation with interband light in the temperature interval from 50 K to 80 K. The energy ionizations of the nitrogen and boron centers are given after (Suttrop et al., 1992) and (Evwaraye et al., 1997), respectively. $A_{\parallel}$ and $A_{\perp}$ are the HF splitting

Therefore, XX defect was also attributed to the carbon vacancy in the single positive charge state $V_C^{0/+}$ that substitutes three inequivalent positions. The spin-Hamiltonian parameters of the thermally stable deep intrinsic defects observed in HPSI 6H-SiC samples at T = 77 K along with their identification and energy levels are summarized in Table 6.

Sample	HPSI 6H-SiC			e-irradiated p-type 6H-SiC			HPSI 6H-SiC
Defects	XX_h	XX_{c1}	XX_{c2}	Ky3(h)	Ky2(c)	Ky1(c)	PP
$g_{\parallel}$	2.0024	2.0027	2.0035	2.0025	2.0028	2.0036	2.0047
$g_{\perp}$	2.0045	2.0043	2.0040	2.0045	2.0043	2.0045	2.0028
$E_C - E_i$, eV	1.84						1.79
Model	$V_C^{0/+}$			$V_C^{0/+}$			$C_{Si} V_C^{0/+}$

Table 6. Spin-Hamiltonian parameters, energy levels and electronic models of the thermally stable deep intrinsic defects with S = 1/2 measured in HPSI 6H-SiC samples at T = 77 K. For comparison, the literature data for Ky center were taken after (Bratus et al., 2005).

As seen from Table 6 the EPR parameters of the PP defect are consistent with those of the P_1 defect (see Table 4) and SI-5 center observed in HPSI 4H-SiC (Son et al., 2004) and electron irradiated n-type 4H-SiC at 77 K (a. Umeda, et al., 2004; a. Umeda et al., 2006). Therefore, similar to the P_1 defect, the PP defect was attributed to the pair ($C_{Si}\,V_C^{0/+}$) in the positive charge state, which was previously found in p-type 4H SiC with the energy level closed to the V_C^+ (Umeda et al., 2007).

4. Kinetic properties of the photosensitive impurities and defects in HPSI SiC

4.1. Thermally stimulated charge carrier trapping and transfer process in HPSI 4H-SiC

As already mentioned in Sect. 3.2, the lifetime of the nonequilibrium charge carriers trapped into the donor and acceptor levels of nitrogen and boron in HPSI 4H-SiC is very large (on the order of 30 h and longer) and the recombination rate of the photo-excited carriers is very small. The recombination between nonequilibrium charge carriers is impeded by the intercenter charge transfer process occurred in HPSI 4H-SiC in the dark after termination of photo excitation.

As was shown in Fig. 4, in the dark, after termination of photo excitation, charge carrier transfer from the shallow nitrogen donor to a deep P_1 and P_2 defect centers lying 1.15 eV below the conduction band bottom in HPSI 4H-SiC samples and identified as ($C_{Si}\,V_C^{0/-}$), V_{Si}^{3-}, respectively. The efficiency of this electron transfer does not depend on the concentration of the nitrogen donor but is activated with increasing concentration of the deep donor centers, as the Fermi level approaches the band gap center. Studies of the thermally stimulated evolution of the photo-EPR spectra in the dark showed that, as the temperature increases, the nonequilibrium charge carriers, rather than becoming excited into the conduction band, are trapped into deep levels near which the Fermi quasi-level is localized at this temperature. As seen from Fig. 7, within 26 h after photo excitation by interband light has been terminated, EPR signals of boron and P_1, P_2 centers appear in the EPR spectrum obtained at T = 77 K. As the temperature increases to 106 K, the EPR signal intensity from boron and the P_1, P_2 centers decays, which suggests that nonequilibrium charge carriers are released from the trapping levels. Now, as the temperature decreases, a single signal of a trapping center of unknown nature with g = 2.0048, which is labeled in Fig. 7 by the letter L, appears in the EPR spectrum.

This means that nonequilibrium charge carriers detrapped from leaving the level of the P_1, P_2 centers, rather than being ionized into the conduction band, become trapped with decreasing temperature into another, deeper level, close to which the Fermi quasi-level at this temperature is located. If we heat the sample again to 140 K, and reduce the temperature subsequently to 85 K, the charge carriers will be released from the L center at the high temperature, and then trapped again at a low temperature into the level of the $X_{c,h}$ defect, which lies deeper in the band gap than the L center.

Because the nonequilibrium charge carrier transfer occurs within a narrow temperature interval, it appears only natural to assume that the levels of the P_1, P_2, L, and $X_{c,h}$ trapping centers are spaced from one another by an energy on the order of $\Delta T = 140 - 107\text{ K} = 33\text{ K} = 2.85\text{ meV}$.

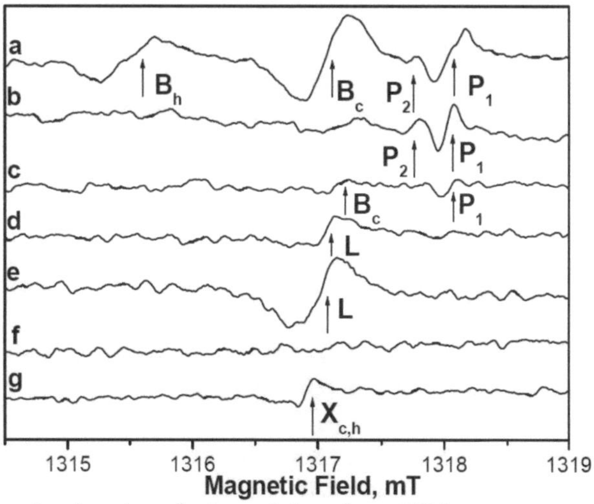

Fig. 7. Temperature-stimulated evolution of the photo-EPR spectra measured on HPSI 4H-SiC samples in the dark at 37 GHz. $\mathbf{B_0} \perp \mathbf{c}$. The spectrum was obtained 26 h (a) after termination of photo-excitation by interband light at T = 77 K and (b-g) after a change in the temperature to (b) T = 98.5 K, (c) T = 106 K, (d) T = 86 K, (e) T = 77 K, (f) T = 140 K and (g) T = 88.5 K. After (a. Savchenko et al., 2009).

Thus, within the temperature interval from 77 K to 140 K, nonequilibrium charge carriers undergo thermally stimulated cascade transfer from the nitrogen donor level into three closely lying trapping centers in the band gap.

As the temperature decreases, the Fermi quasi-levels will approach the band gap and valence band edges, thus increasing the probability of trapping of photo-induced electrons and holes by shallow donors and acceptors. This pattern of the behavior of the Fermi quasi-levels is corroborated by the low-temperature transformation of the photo-EPR spectra in HPSI 4H-SiC samples shown in Fig. 8.

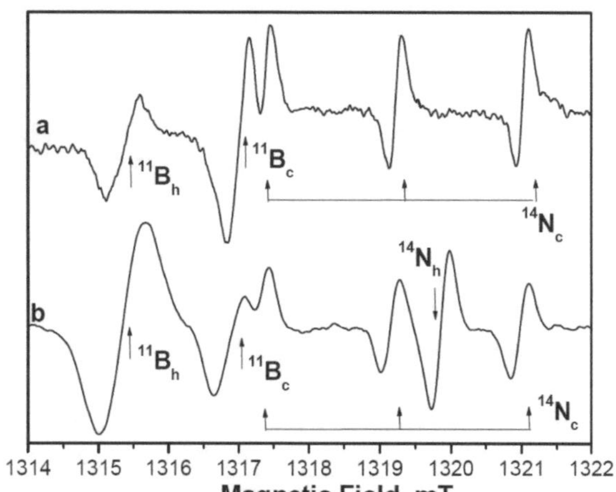

Fig. 8. EPR spectra observed in HPSI 4H-SiC samples measured at 37 GHz following by interband light. $\mathbf{B_0} \perp \mathbf{c}$. a - T = 77, b - T = 50 K. After (a. Savchenko et al., 2009).

At T = 50 K, a signal appears in the photo-EPR spectrum due to nitrogen occupying the hexagonal (N_h) position in the 4H-SiC lattice and localized in a more shallow level in the band gap as compared to the N_c position, see Table 3. Now, the ratio of EPR signals from boron in the c and h positions, 0.4, becomes opposite to that observed at T = 77 K.
The latter observation suggests that, under the conditions where the Fermi hole quasi-level lies above the boron levels (0,628 eV, see Table 3) and their occupation is equally probable, the level of the boron in the h position is more shallow than that in the c position in the 4H-SiC lattice and the concentration of paramagnetic centers in the h lattice positions is higher than that in the c positions.

4.2. Decay kinetics of the PR of the nonequilibrium charge carriers trapped by the paramagnetic centers in HPSI 4H-SiC

Figures 9 a,b display the decay with time of EPR line intensities of nitrogen and boron after termination of photo-excitation, which translate the variation with time of the concentration of nonequilibrium charge carriers trapped in the nitrogen and boron levels under photo-excitation.

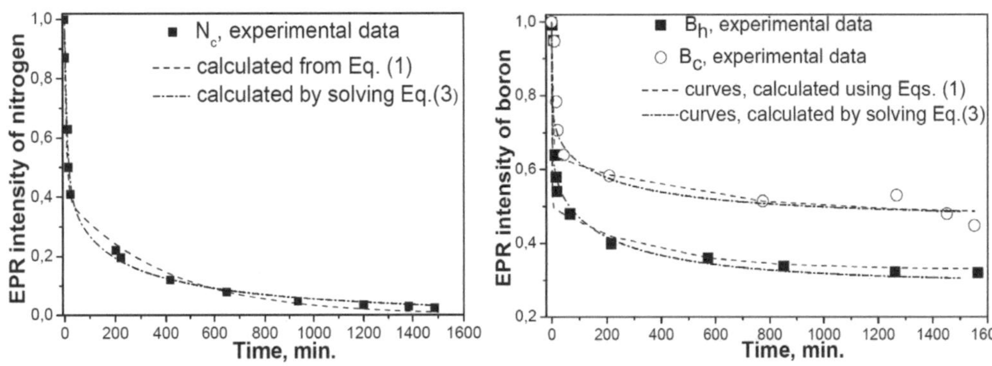

a b

Fig. 9. Decay with time of the nitrogen (a) and boron (b) EPR signal intensity at T = 77 K measured after termination of photo-excitation of the sample by interband light. After (a. Savchenko et al., 2009).

An analysis of the experimental curves yields the following relation for the decay of EPR signal intensities or concentrations of nonequilibrium charge carriers trapped by the paramagnetic centers:

$$\begin{aligned} I_D &= 0.6\exp(-\lambda_{1D}t) + 0.4\exp(\lambda_{2D}t) \\ I_{Ac} &= 0.45 + 0.36\exp(-\lambda_{1Ac}t) + 0.16\exp(-\lambda_{2Ac}t) \\ I_{Ah} &= 0.33 + 0.5\exp(-\lambda_{1Ah}t) + 0.17\exp(-\lambda_{2Ah}t) \end{aligned} \quad (1)$$

Here I_D, I_{Ac}, I_{Ah} are the integrated intensities of EPR signals due to nitrogen, cubic boron, and hexagonal boron centers, respectively; t is the time (in min); and λ_i are the rates of exponential decay: λ_{1D} = 0.133 min^{-1}, λ_{2D} = 0.0025 min^{-1}, λ_{1Ac} = 0.48 min^{-1}, λ_{2Ac} = 0.002 min^{-1}, λ_{1Ah} = 1.1 min^{-1}, and λ_{2Ah} = 0.003 min^{-1}.

The dashed curves in Figs. 9 (a,b) plot the concentration decay of nonequilibrium charge carriers trapped in the nitrogen donors and boron acceptors, which were calculated from Eqs. (1). The experimental and theoretical curves demonstrate enough good agreement.
As seen from Figs. 9a and 9b, the decay curves of the nitrogen and boron EPR line intensities have slow and fast components. The slow components of the λ_{2D}, λ_{2Ac} and λ_{2Ah} decay curves have similar relaxation rates, which should be attributed to the effect of PR in the charge carrier, donor, and acceptor system.
The fast components of the decay curves for the boron acceptors (λ_{1Ac} and λ_{1Ah}) occupying positions c and h in the silicon carbide lattice have close relaxation rates, whereas the relaxation rate for the donors (λ_{1D}) is an order of magnitude smaller.
It should be noted that the higher relaxation rate components are seen at the smaller time scale indicating that the decay curves with time have super-exponential character.
The difference in the relaxation rates between the donors and acceptors may be due to the difference between their kinetic characteristics, as well as to the presence of traps for electrons and holes in the band gap, which brings about different residual EPR signal intensities from nitrogen and boron observed after termination of photo-excitation.
A comparison of the EPR spectra presented in Fig. 4 (curves b and c) suggests that after escaping from the nitrogen level the electrons become trapped into the level of the neighboring neutral trapping center, in this case, the P_1 and P_2 centers, where their lifetime is very long and reaches more than 1000 min. after termination of photo-excitation. Thus, the rate of recombination between the trapped electrons and the holes trapped into boron levels is negligible, which is corroborated by the course of the slow component of the boron EPR signal decay.
Unfortunately, it was not possible to study the evolution with time of the concentration of nonequilibrium charge carriers trapped by the P_1 and P_2 defects because of their low EPR signal intensities.

4.3. Decay kinetics of the PPC in HPSI 4H-SiC

Fig. 10 plots the decay with time of persistent photocurrent after termination of photo-excitation with interband light, which was obtained at room temperature (T = 300 K) on HPSI 4H-SiC samples cut from the plate used to prepare samples for photo-EPR measurements (a. Kalabukhova et al., 2006). This permits a comparison of the data derived from the evolution with time of the PPC and PR of nonequilibrium concentration of the paramagnetic centers and their treatment in terms of the same processes occurring in HPSI 4H-SiC samples under photo-excitation and after its termination.
The decay of photocurrent shown in Fig. 10 can be fitted by three exponentials

$$\mathbf{I_{curr} = 0{,}35exp(-\lambda_{1e}t) + 0{,}6exp(-\lambda_{2e}t) + 0{,}05exp(-\lambda_{3e}t),} \quad \text{(2)}$$

where λ_{1e} = 1.3 min^{-1}, λ_{2e} = 0.125 min^{-1}, and λ_{3e} = 0.002 min^{-1}.
Examining Eqs. (1) and (2), we see that the decay of the photocurrent is described through the same relaxation rates as the acceptor and donor concentration decays, so that $\lambda_{1e} \approx \lambda_{1Ah}$, $\lambda_{2e} \approx \lambda_{1D}$ and $\lambda_{3e} \approx \lambda_{2D}$.

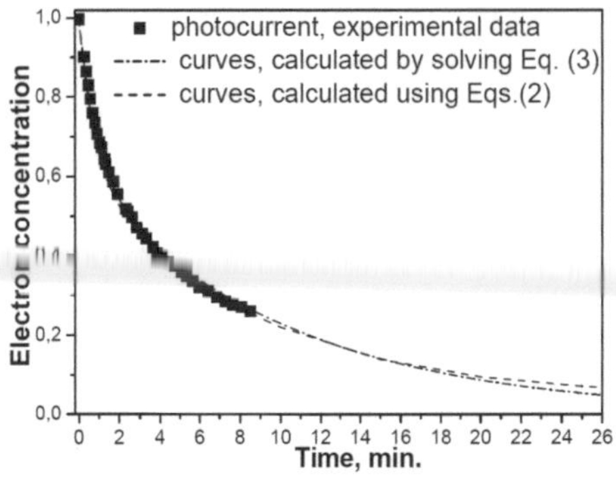

Fig. 10. Decay with time of persistent photocurrent in HPSI 4H-SiC samples at T = 300 K measured after termination of photo-excitation by interband light. After (a. Savchenko et al., 2009).

The empirical description of the PR and PPC decays cannot, however, be considered as satisfactory. It can give us only an idea about order-of-magnitude of the relaxation rates characterizing the processes in the donor and acceptor system while not providing any information about which of them are limiting.

A comprehensive theoretical analysis of the processes taking place in HPSI 4H-SiC samples under photo-excitation and after its termination is proposed in Section 5.

4.4 Electronic process describing the trapping and recombination process of the nonequilibrium charge carriers in HPSI 4H-SiC

Fig. 11 schematically shows possible electronic transitions between the impurities and defects in HPSI 4H-SiC samples under photo-excitation and after its termination.

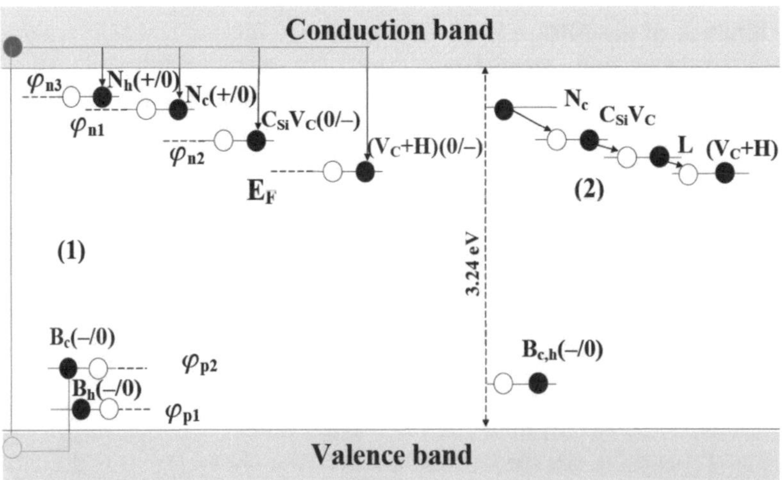

Fig. 11. Diagram of electronic transitions between impurities and defects in HPSI 4H-SiC samples stimulated by photo-excitation with interband light and after its termination. E_F is the Fermi level in the dark; φ_{n1}, φ_{n2}, φ_{n3}, φ_{p1}, and φ_{p2} are the Fermi quasi-levels under photo-excitation. V_{Si} energy level was not shown on diagram due to its small concentration.

Electronic transition 1 is involved in photo-excitation of HPSI 4H-SiC samples by interband light. In this case, electrons are excited from the valence band into the conduction band, with their subsequent trapping from the conduction band into the nitrogen levels N_c and N_h and intrinsic defects including $C_{Si}V_C$ pair, V_{Si}, (V_C+H) complex, and the holes, from the valence band into the boron levels B_c and B_h.
Estimates of the concentrations of the nitrogen and boron centers derived from the integrated intensities of the EPR spectra showed them to be comparable in magnitude.
Electronic transition 2 relates to the processes occurring in HPSI 4H-SiC samples after termination of their photo-excitation by interband light. Because the relaxation of EPR signals from boron acceptors and nitrogen donors is described by different exponentials, the probability of donor–acceptor recombination between nitrogen and boron is disregarded. The most probable appears to be cascade transfer of nonequilibrium charge carriers from nitrogen donors to deep donor trapping levels at the top of the band gap, where the charge carrier lifetime is very long.

4.5. Electronic process describing the trapping and recombination process of the nonequilibrium charge carriers in SI 6H-SiC

The electronic process describing the behavior of the photosensitive paramagnetic centers observed in SI 6H-SiC under excitation with the light of different photon energies is shown in Fig. 12. The transitions 1-5 describe the excitation process of the samples by below band gap light. Transition 1 corresponds to electrons excited from valence band into $C_{Si}V_C$ pair. At the same time the holes in the valence band are captured by boron and carbon vacancy (transition 2, 3), causing the corresponding appearance and disappearance of their EPR signals. Transitions 1 and 3 also describe charge transfer process between V_C and $C_{Si}V_C$ pair. Transition 4 is the direct excitation of electrons to the $C_{Si}V_C$ pair followed by the capture of the valence band holes by boron.

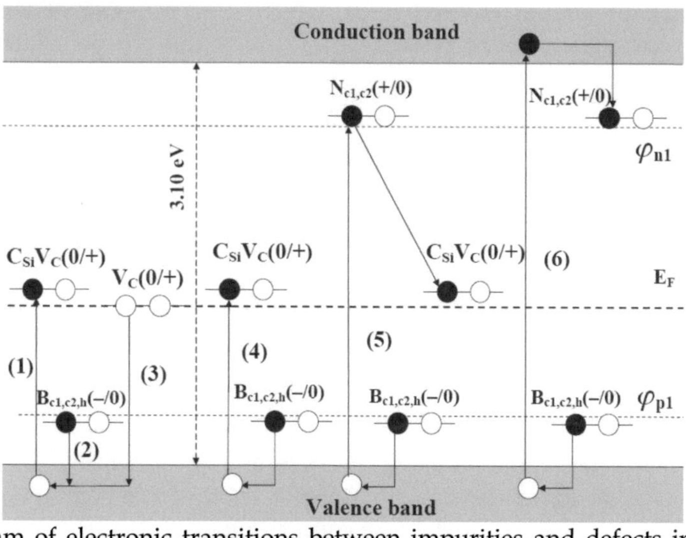

Fig. 12. Diagram of electronic transitions between impurities and defects in HPSI 6H-SiC samples stimulated by photo-excitation with interband light and after its termination. E_F is the Fermi level in the dark; φ_{n1}, and φ_{p1}, are the Fermi quasi-levels under photo-excitation.

Transition 5 describes the process of boron and nitrogen neutralization, which could be induced by excitation of an electron from the valence band directly to an ionized donor, followed by capture of the hole by boron. In addition, the charge transfer process between nitrogen and $C_{Si}V_C$ pair is shown.

The illumination of the sample by above band gap (UV) light (transition 6) induces the neutralization of the shallow donors and acceptors and as a result the EPR spectra of nitrogen and boron are observed simultaneously in the EPR spectrum.

5. Theory and comparison with experiment

5.1. Rate equations describing the electronic process occurring between impurities and defects in HPSI 4H-SiC after termination of the photo-excitation

In this chapter, the identification of the rate-limiting electronic processes shaping the behavior of PR and PPC has been presented for the HPSI 4H-SiC samples. To that end, we considered rate equations for the system of donors, acceptors, traps, and current charge carriers for the case where interband light is switched off and the transition to equilibrium state in the dark is studied.

By solving the rate equations and fitting them to the experimental curves displayed in Figs. 9, 10, we obtained kinetic characteristics of the photosensitive paramagnetic centers and the probabilities of the electronic and hole transitions illustrated in Fig. 11.

It should be noted that there is no influence of spin relaxation on the decay of the concentration of the paramagnetic centers observed in HP SI 4H-SiC after termination of the photo-excitation because the steady-state saturation method has been used for the registration of the EPR signals.

Assuming interaction among the main nitrogen donor centers, boron acceptors, traps, and charge carriers to be the most efficient processes, we write rate equations for neutral paramagnetic donors D^0, electrons n, neutral paramagnetic boron centers A^0, and holes p. The concentration of neutral electron traps, which after trapping of electrons become negatively charged paramagnetic centers, is given by the charge neutrality condition $n_t^0 = n + (D^0 - A^0) - p - (D - A) + n_t$:

$$\begin{aligned}
\frac{dD^0}{dt} &= -(w_{iD^0} + f_{Dt}n_t^0 + f_{eD}n)D^0 + (f_{eD}D + w_{nD})n; \\
\frac{dA^0}{dt} &= -(w_{iA} - f_{pA}(D-A) + f_{pA}(n + D^0 - A^0))A^0 + f_{pA}Ap \\
\frac{dn}{dt} &= N_{ph}S - (f_{eD}(D-D^0) + f_{et}n_t^0)n - f_{ep}np + w_{iD}D^0 + w_{it}(n_t - n_t^0); \\
\frac{dp}{dt} &= N_{ph}S + w_{iA}A^0 - p(f_{pA}(A-A^0) + f_{ep}n)
\end{aligned} \tag{3}$$

Here, n and p are the concentrations of free electrons and holes, respectively; D^0, A^0, and are the concentrations of neutral paramagnetic donors, acceptors, and neutral nonparamagnetic traps, respectively; D, A, and n_t are the total concentrations of donors, acceptors, and traps, respectively; $D^+(0) = D - D^0$ and $A^-(0) = A - A^0$ are the concentrations of ionized donors and acceptors, respectively, which are changed in the presence of the kinetic fluxes of electrons and holes; $f_{ij} = \sigma_{ij}v_i$ is the flux of charge carriers i reaching the corresponding trapping center j, where σ_{eD} and σ_{pA} are the cross sections of electron and hole trapping into the donor and acceptor levels, σ_{ep} is the cross section of the electron–hole recombination collision, and v_i is the charge carrier velocity. The product of f_{ij} by the concentration of particles gives the probability of leaving or arrival of particles from/to the level in the band gap; $N_{ph}S$ is the nonequilibrium charge carrier generation rate (N_{ph} is the photon concentration); w_{iD^0} and w_{iA^0} are the rates of ionization of nonequilibrium charge carriers from the levels of donors D^0 and acceptors A^0.

The transfer of charge carriers from a donor level to a lower-lying level of the trap closest to the donor is described by the probability. The probability w_{nD} in the first equation of coupled equations (3) describes the increase in the concentration of neutral donors through electrons becoming trapped by an ionized donor not directly from the conduction band but via cascade transitions from one level to another in the band gap.

In the Eqs. (3) we have taken into account the recombination process, which is proportional to the product n·p. The direct band-to-band recombination process is forbidden in our system because 4H-SiC is a semiconductor with indirect band structure. However, if the traps capture of both the electrons and holes they will act as a recombination center. The efficiency of this process will be estimated from the comparison of solution of Eqs. (3) with the experimental curves.

5.2. Rate parameters of the electronic process occurring in HPSI 4H-SiC after termination of photo-excitation derived from the solution of the rate equations

Experiments have shown that the evolution of the donor, acceptor, and charge carrier concentrations in the dark takes place with rates, which are equal in order of magnitude, and sometimes even in magnitude, for the different subsystems described by Eqs. (1) and (2). This suggests that the system may be considered closed.

The initial values for n, D^0, A^0, and p at time t = 0 after the termination of illumination were found as steady-state solutions of Eqs. (3) (i.e., for $t \to \infty$ all derivatives are zero) for the case of strong illumination, where the condition is satisfied $N_{ph}S/(f_{ep}D) >> D$. These values are given by

$$\mathbf{n(0) = p(0) = G_S},\ A^0(0) = \frac{G_S}{r_A + G_S/A},\ D^0(0) = \frac{G_S + r_t \cdot A^0(0) - r_t \cdot \Delta}{r_D + r_t + G_S/D}, \tag{4}$$

where $G_S = (N_{ph}S/f_{ep})^{1/2}$, $r_D = w_{iD}/(f_{eD}D)$, $r_A = w_{iA}/(f_{pA}A)$, $r_t = f_{et}/f_{eD}$, and $\Delta = n_t + A - D$.

The equilibrium values of the variables in the dark are obtained as steady-state solutions of Eqs. (3) for $N_{ph}S = 0$:

$$\mathbf{n_{eq} = 0};\ D^0_{eq} = 0,\ \mathbf{p_{eq} = A(A - D)/D};\ A^0_{eq} = r_A^{-1}A(A-D)/D,\ \textbf{for } \mathbf{A > D}$$
$$n_{eq} = n_t\sqrt{w_{it}/(n_t \cdot f_{Dt} + n_t \cdot f_{et})};\ \mathbf{D_{eq} = 0;\ A_{eq} = 0;\ p_{eq} = 0,}\ \textbf{for } \mathbf{A < D.} \tag{5}$$

The right-hand side of Eqs. (3) contains terms quadratic in the donor and acceptor concentrations. Because the samples are high resistive, all the variables n, D^0, A^0 << (D + A).

It would seem that one could now neglect the quadratic terms and solve the linear differential equations. In this case, we would obtain simple analytic solutions for system (3) in the form of a linear combination of exponentials.

While this approximate solution would not be valid for the total range of relaxation times, it would be acceptable for the interval within which the fastest-decaying quantities have completed their relaxation.

It can be shown that, in this case, a pair of coupled equations for acceptors and holes has one of the solutions of the secular equation for the decay rates, which is equal to zero. This relates to an extremely slow decay of the acceptor concentration. Indeed, in an experiment, one observes a considerable slowing down of the acceptor EPR signal 1000 min after the switching off the light. In this case, the residual acceptor concentration differs from the equilibrium value $A^0_{eq} = 0$ and the evolution of the EPR signal with time cannot be fitted with one exponential. Moreover, when we neglect quadratic terms, we actually disregard the electron–hole recombination processes.

All this suggests the importance of the quadratic terms on the right-hand side of Eqs. (3), which are actually responsible for the fast EPR signal intensity decay. For this reason, the coupled differential equations (3) were solved numerically with inclusion of all the quadratic terms by the finite difference method. In this approach, we increase the time by such a small increment Δt at which, in the Taylor expansion of the time dependent functions $n(t)$, $D^0(t)$, $A^0(t)$, and $p(t)$, one may restrict oneself to three first terms, including the term with $(\Delta t)^2$. The values of $n(t)$, $D^0(t)$, $A^0(t)$, and $p(t)$, as well as their first and second derivatives with respect to time, were assumed to be equal to the values reached at the preceding instant of time. The value of the first derivative was calculated as the right-hand side of rate equations (3), and that of the second derivative, as the derivative of the right-hand side of Eqs. (3). The solutions for the variables at the time $t + \Delta t$ thus obtained are substituted into the rate coefficients of Eqs. (3), and the procedure is repeated until the relaxation is complete.

Thus, the finite difference method chosen for solution of rate equations (3) has provided explanation for the existence of additional fast processes as due to the presence in the rate coefficients of time-varying terms. Graphic plots of the solutions obtained were fitted to the experimental curves shown in Figs. 9, 10.

The parameters of the rate equations were varied until the theoretical graphs matched fully the experimental EPR decay curves. The dash-dotted lines in Figs. 9, 10 plot the numerical solutions of Eqs. (3) for values of the rate coefficients which agree with experimental curves and are listed in the Table 7.

As can be seen from Table 7, numerical values of the rate parameters $f_{eD}D$, $f_{pA}A$ and $f_{Dt}n_t$ obtained from PPC data at 300 K and from the decay of EPR signal intensities of nitrogen and boron centers at 77 K are close to each other which indicates that the probability for charge carriers to be trapped by ionized nitrogen, boron centers and traps does not depend on the temperature.

On the contrary, the numerical values of the w_{iD}, w_{iA} describing the rates of ionization of charge carriers from the levels of donors D^0 and acceptors A^0 exponentially depend on the temperature: $\exp(E_i/kT)$, where E_i is the energy ionization of the trapping centers.

Parameter	PPC	Decay of EPR signal intensity, T = 77 K		
	T = 300 K	N_c	B_c	B_h
$f_{eD}D$, min^{-1}	0.0045	0.0033	0.0033	0.0033
w_{iD}, min^{-1}	$1.35 \cdot 10^{-3}$	$1.65 \cdot 10^{-4}$	$1.65 \cdot 10^{-4}$	$1.65 \cdot 10^{-4}$
$f_{pA}A$, min^{-1}	0.017	0.02	0.02	0.017
w_{iA}, min^{-1}	0.52	0.072	0.072	0.06
$f_{ep}D$, min^{-1}	0.044	0.014	0.014	0.014
$f_{Dt}n_t$, min^{-1}	0.13	0.16	0.16	0.16

Table 7. Rate parameters of the processes of recombination, trapping, and ionization of nonequilibrium charge carriers occurring in HPSI 4H-SiC samples after termination of photo-excitation.

5.3. Kinetic characteristics of the photosensitive impurities and defects in HPSI 4H-SiC

A comparison of the relaxation parameters of the donor, acceptor, trap, and charge carrier system listed in Table 7 with the relaxation times obtained through an empirical description of the experiment reveals that the rates of exponential decay I(T) derived for acceptors from relation (1) ($\lambda_{1Ac} = 0.48$ min^{-1} and $\lambda_{1Ah} = 1.1$ min^{-1}) are characteristic of none of the recombination, trapping, and ionization of nonequilibrium charge carrier processes occurring in HPSI 4H-SiC samples after termination of photo-excitation. The reason for this lies in that the law by which the integrated EPR signal intensities vary, rather than being described by a sum of exponentials, is actually superexponential because, for times $t < 10$ min, the rate coefficients for system (3) vary exponentially with time.

As seen from Table 7, the probability for an electron to be trapped from the conduction band by an ionized nitrogen donor, $f_{eD}D$, which enters the first and third equations of system (3) and can be determined with a fairly high confidence, turned out to be small. On the other hand, in the course of solving the coupled equations, it was established that the increase of the concentration of neutral donors D^0 should take place with an order-of-magnitude higher probability than $f_{eD}D$. Therefore, we added to the second term of the first equation of (3) the term w_{nD} which accounts for the probability of multi-step transitions of nonequilibrium charge carriers between levels in the band gap. This probability was found to be $w_{nD} = 0.0117$ min^{-1}.

As seen from Fig. 11, in addition to ionization of the neutral nitrogen donor, there is also a possibility of cascade transfer of nonequilibrium charge carriers from the nitrogen donor level to trapping centers, which can also affect the concentration of neutral donors. It is this process that is the fastest in the system under study, $f_{Dt}n_t = 0.16$ min^{-1}. The trap concentration n_t is 0.4 of the donor concentration D.

The data of the Table 7 suggest that the rate of electron–hole recombination $f_{ep}D$ plays an equally important role in recovery of equilibrium concentrations of all the participants of the process, and that it is comparable in magnitude with the probability of hole trapping by an ionized boron acceptor. Recombination favors fast relaxation of the holes, but after the holes have relaxed, a further recovery of the neutral acceptor concentration involves the quadratic terms in the rate coefficients associated with the concentration of donors and free charge carriers. Relaxation of the latter is restricted by the very low probability of ionization of the nitrogen neutral donor, $w_{iD} = 1.65 \cdot 10^{-4}$ min^{-1}. The slow hole relaxation, which sets in after the donors and electrons have recovered their equilibrium values, has a nearly constant value determined by the $w_{iA}/(f_{pA}A)$ ratio only. The larger this ratio, the smaller is the final level of EPR signal intensities due to the boron acceptors I_{Ac} and I_{Ah} observed after termination of photo-excitation. As seen from the Table 7, the probability for a hole to be trapped from the valence band into an ionized boron acceptor is higher by an order of magnitude than that for an electron to be trapped from the conduction band by an ionized nitrogen donor. Interestingly, the probability of direct hole trapping from the valence band by an ionized acceptor B_c was found to be higher than that by an ionized acceptor B_h. This corroborates the data derived from the temperature dependence of photo-EPR spectra, which suggest that the boron level in the position B_h is more shallow than that in B_c.

6. Conclusions

In this chapter the identification and kinetic properties of the photosensitive paramagnetic centers observed in HPSI 4H and 6H-SiC under photo-excitation and after its termination have been described. The HPSI 4H and 6H-SiC samples were shown to have four photosensitive paramagnetic centers, which can be photo-excited into a paramagnetic state and a nonparamagnetic state. Two of them are the well-known nitrogen and boron impurities, and the other two are thermally stable intrinsic defects with $S = 1/2$, labeled X and P_1, P_2 in 4H SiC and XX and PP in 6H SiC. The X and XX, P_1 and PP defects have similar g-tensors and symmetry features, respectively. The EPR spectrum of X and XX defect was observed in the dark while EPR signals due to P_1, P_2 and PP defects appeared in the EPR spectrum of HPSI 4H and 6H-SiC samples under photo-excitation.

The EPR parameters and ligand HF structure of the X defect residing at two inequivalent lattice sites (X_h and X_c) with ionization levels 1.36 and 1.26 eV below the conduction band were found to be in excellent agreement with those of carbon vacancy $V_C^{0/+}$, both at the c and h lattice sites known as ID1, D2 centers in HPSI 4H SiC and E15, E16 centers in electron irradiated 4H SiC. However, a significant discrepancy in the intensity ratio of lines with the smallest HF splitting was found for X_C-defect, ID1 and E15 center and explained by the presence of the hydrogen in the vicinity of the carbon vacancy. This conclusion is supported by the ionization energy of X-defect, which is in contrast to the $V_C^{0/+}$ close to that calculated for V_C with adjacent hydrogen (V_C+H). Thus, the X-defect which shows the donor-like behavior, was assigned to the hydrogenated carbon vacancy $(V_C+H)^{0/-}$ which occupies the c and h positions in the 4H-SiC lattice. In contrast to the X defect, the XX defect whose energy level is pinned in the lower half of the band gap (E_C – 1.84 eV) and shows acceptor-like behavior was attributed to the $V_C^{0/+}$ at three inequivalent positions.

The EPR parameters of the P_1 defect which were observed in EPR spectrum of HPSI 4H SiC under photo-excitation or due to the charge carrier transfer from the shallow nitrogen donor to a deep P_1 defect center after termination of the photo-excitation was found to be coincided with those of SI-5 center observed in HPSI 4H SiC and in electron irradiated n-type 4H SiC. Energy position of the P_1 defect coincides with that of SI-5 center which amounts to 1.1 eV below the conduction band and coincides with the ionization levels calculated from the first principles for the carbon antisite-vacancy pair in the negative charge state ($C_{Si}V_C^{0/-}$). Therefore, $C_{Si}V_C$ pair has been suggested as the most possible model for the P_1 defect.

By analogy with the P_1 defect, the PP defect was also attributed to the carbon AV pair but in the positive charge state ($C_{Si}\,V_C^{0/+}$). The P_2 EPR signal of small intensity with the isotropic g-factor similar to that observed for SI-11 center was tentatively attributed to the silicon vacancy in the negative charge state V_{Si}^{3-}.

The study of the kinetic properties of the photosensitive impurities and defects in HPSI 4H-SiC and 6H-SiC has shown that the lifetime of the nonequilibrium charge carriers trapped into the donor and acceptor levels of nitrogen and boron is very large (on the order of 30 h and longer) and the recombination rate of the photo-excited carriers is very small. Such a PR of the photo-response after termination of photo-excitation was found to be accompanied by the PPC phenomenon. To identify the rate-limiting electronic processes shaping the behavior of PR and PPC in HPSI 4H-SiC the rate equations describing the processes of recombination, trapping, and ionization of nonequilibrium charge carriers bound dynamically to shallow donors and acceptors (nitrogen and boron), as well as of charge carrier transfer from the shallow nitrogen donor to deep traps, have been solved. A comparison of calculations with experimental data has revealed the following efficient electron processes responsible for the PR and PPC in HPSI 4H-SiC samples.

(1) Multi-step transitions of nonequilibrium charge carriers from the nitrogen donor levels to trapping centers, among which the aggregated ($C_{Si}\,V_C^{0/-}$) centers play the main role. It turned out that the probability of this process is an order of magnitude higher than that of electron trapping from the conduction band by an ionized nitrogen donor. The probability of electron transfer from nitrogen donors to neighboring traps makes the latter the fastest relaxation process in the system under consideration.

(2) Electron–hole recombination, whose rate plays an equally essential role in the recovery of equilibrium concentrations of all centers involved in the fast PR and PPC processes.

(3) Ionization of boron acceptors (w_{iA}), as well as hole escape from, or arrival at the boron level ($f_{pA}A$). The ratio of probabilities of these processes, $w_{iA}/(f_{pA}A)$, mediates the rate of slow relaxation of the holes trapped into the boron acceptor levels. The higher the boron acceptor ionization probability, the faster is the PR process.

7. References

Aradi, B.; Gali, A.; Deak, P.; Lowther, J.E.; Son, N.T.; Janzen, E. & Choyke, W.J. (2001). Ab initio density-functional supercell calculations of hydrogen defects in cubic SiC. *Physical Review B,* Vol. 63, No. 24, June 2001, 245202-1-245202-18, ISSN 1098-0121

a. Bockstedte, M.; Heid, M. & Pankratov, O. (2003). Signature of intrinsic defects in SiC: Ab initio calculations of hyperfine tensors. *Physical Review B,* Vol. 67, No. 19, May 2003, 193102-1-193102-4, ISSN 1098-0121

b. Bockstedte, M.; Mattausch, A. & Pankratov, O. (2003). Ab initio study of the migration of intrinsic defects in 3C-SiC. *Physical Review B,* Vol. 68, No. 20, November 2003, 205201-1-205201-17, ISSN 1098-0121

Bockstedte, M.; Mattausch, A. & Pankratov, O. (2004). Ab initio study of the annealing of vacancies and interstitials in cubic SiC: Vacancy-interstitial recombination and aggregation of carbon interstitials. *Physical Review B,* Vol. 69, No. 23, June 2004, 235202-1-23520213, ISSN 1098-0121

Bockstedte, M.; Gali, A.; Umeda, T.; Son, N.T.; Isoya, J. & Janzen, E. (2006). Signature of the Negative Carbon Vacancy-Antisite Complex. *Materials Science Forum,* Vol. 527-529, October 2006, 539-542, ISSN 0255-5476

Bratus, V.Ya.; Petrenko, T.T.; Okulov, S.M. & Petrenko, T.L. (2005). Positively charged carbon vacancy in three inequivalent lattice sites of 6H-SiC: Combined EPR and density functional theory study. *Physical Review B,* Vol. 71, No. 12, March 2005, 125202-1-125202-22, ISSN 1098-0121

Carlsson, P.; Son, N.T.; Umeda, T.; Isoya, J. & Janzen, E. (2007). Deep Acceptor Levels of the Carbon Vacancy-Carbon Antisite Pairs in 4H-SiC. *Materials Science Forum,* Vol. 556-557, September 2007, 449-452, ISSN 0255-5476

Dissanayake, A.S.; Huang, S.X.; Jiang, H.X. & Lin, J.Y. (1991). Charge storage and persistent photoconductivity in a $CdS_{0.5}Se_{0.5}$ semiconductor alloy. *Physical Review B,* Vol. 44, No. 24, December 1991, 13343-13348, ISSN: 1098-0121

Dissanayake, A.S. & Jiang, H.X. (1992). Lattice relaxed impurity and persistent photoconductivity in nitrogen doped 6H-SiC. *Applied Physics Letters,* Vol. 61, No. 17, October 1992, 2048-2050, ISSN 0003-6951

Evwaraye, A.O.; Smith S.R.; Mitchel W.C. & Hobgood, H.McD. (1997). Boron acceptor levels in 6H-SiC bulk samples. *Applied Physics Letters,* Vol. 71, No. 9, September 1997, 1186-1188, ISSN 0003-6951

Evwaraye, A.O.; Smith, S.R. & Mitchel, W.C. (1995). Persistent photoconductance in n-type 6H-SiC. *Journal of Applied Physics,* Vol. 77, No. 9, May 1995, 4477-4481, ISSN 0021-8979

Evwaraye, A.O.; Smith, S.R. & Mitchel, W.C. (1996). Shallow and deep levels in n-type 4H-SiC. *Journal of Applied Physics,* Vol. 79, No. 10, May 1996, 7726-7730, ISSN 0021-8979

Gali, A.; Deak, P.; Son, N.T.; Janzen, E.; von Bardeleben, H.J., Monge, Jean-Louis (2003). Calculation of Hyperfine Constants of Defects in 4H SiC. *Materials Science Forum,* Vol. 433-436, September 2003, 511- 514, ISSN 0255-5476

Greulich-Weber, S. (1997). EPR and ENDOR Investigations of Shallow Impurities in SiC Polytypes. *Physica status solidi (a),* Vol. 162, No. 1, July 1997, 95-151, ISSN 1862-6300

Greulich-Weber, S.; Feege, F.; Kalabukhova, E.N.; Lukin, S.N.; Spaeth, J.-M. & Adrian, F.J. (1998). EPR and ENDOR investigations of B acceptors in 3C-, 4H- and 6H-silicon carbide. *Semiconductor Science and Technology,* Vol. 13, No. 1, January 1998, 59-70, ISSN 0268-1242

Kakalis, J. & Fritzsche, H. (1984). Persistent Photoconductivity in Doping-Modulated Amorphous Semiconductors. *Physical Review Letters,* Vol. 53, No. 16, October 1984, 1602-1605, ISSN 0031-9007

Kalabukhova, E.N.; Lukin, S.N.; Shanina, B.D. & Mokhov, E.N. (1990). Influence of Ge and excess Si on the ESR spectrum of nitrogen donor states in 6H-SiC. *Soviet Physics - Solid State,* Vol. 32, No. 3, Match 1990, 465-469, ISSN 0038-5654

Kalabukhova, E.N.; Lukin, S.N.; Saxler, A.; Mitchel, W.C.; Smith, S.R.; Solomon, J.S. & Evwaraye, A.O. (2001). Photosensitive electron paramagnetic resonance spectra in semi-insulating 4H SiC crystals. *Physical Review B,* Vol. 64, No. 23, December 2001, 235202-1-235202-4, ISSN: 1098-0121

Kalabukhova, E.N.; Lukin, S.N.; Savchenko, D.V.; Mitchel, W.C. & Mitchell, W.D. (2004). Photo-EPR and Hall measurements on Undoped High Purity Semi-Insulating 4H-SiC Substrates. *Materials Science Forum,* Vol. 457-460, June 2004, 501-504, ISSN 0255-5476

a. Kalabukhova, E.N.; Lukin, S.N.; Savchenko, D.V.; Sitnikov, A.A.; Mitchel, W.C.; Smith, S.R. & Greulich-Weber, S. (2006) Trapping Recombination Process and Persistent Photoconductivity in Semi-Insulating 4H-SiC. *Materials Science Forum,* Vol. 527-529, October 2006, 563-566, ISSN 0255-5476

b. Kalabukhova, E.N.; Lukin, S.N.; Savchenko, D.V.; Mitchel, W.C.; Greulich-Weber, S.; Rauls, E. & Gerstmann, U. (2006). Possible Role of Hydrogen within the So-Called X Center in Semi-Insulating 4H-SiC. *Materials Science Forum,* Vol. 527-529, October 2006, 559–562, ISSN: 0255-5476

Kalabukhova, E.N.; Lukin, S.N.; Savchenko, D.V.; Mitchel, W.C.; Greulich-Weber, S.; Gerstmann, U.; Pöppl, A.; Hoentsch, J.; Rauls, E.; Rozentzveig, Yu.; Mokhov, E.N.; Syväjärvi, M. & Yakimova, R. (2007). EPR, ESE and Pulsed ENDOR Study of Nitrogen Related Centers in 4H-SiC Wafers grown by Different Technologies. *Materials Science Forum,* Vol. 556-557, September 2007, 355-358, ISSN 0255-5476

Langhanki, B.; Greulich-Weber, S.; Spaeth, J.-M.; Markevich, V.P.; Clerjaud, B. & Naud, C. (2001). Magnetic resonance and FTIR studies of shallow donor centers in hydrogenated Cz-silicon. *Physica B,* Vol. 308–310, December 2001, 253–256, ISSN 0921-4526

Lang, D.V. & Logan, R.A. (1977). Large-Lattice-Relaxation Model for Persistent Photoconductivity in Compound Semiconductors. *Physical Review Letters,* Vol. 39, No. 10, September 1977, 635-639, ISSN 0031-9007

Litton, C.W. & Reynolds, D.C. (1964). Double-Carrier Injection and Negative Resistance in CdS. *Physical Review,* Vol. 133, No. 2A, January 1964, A536-A541, ISSN 1050-2947

Macfarlane, P.J. & Zvanut, M.E. (1999). Reduction and creation of paramagnetic centers on surfaces of three different polytypes of SiC. *Journal of Vacuum Science &Technology B,* Vol. 17, No. 4, July 1999, 1627-1631, ISSN 1071-1023

Müller, St.G.; Brady, M.F.; Brixius, W.H.; Glass, R.C.; Hobgood, H.McD.; Jenny, J.R.; Leonard, R.T.; Malta, D.P.; Powell, A.R.; Tsvetkov, V.F.; Allen, S.T.; Palmour, J.W. & Carter C.H.Jr. (2003). Sublimation-Grown Semi-Insulating SiC for High Frequency Devices. *Materials Science Forum,* Vol. 433-436, September 2003, 39-44, ISSN 0255-5476

Queisser, H.J. & Theodorou, D.E. (1986). Decay kinetics of persistent photoconductivity in semiconductors. *Physical Review B,* Vol. 33, No. 6, March 1986, 4027-4033, ISSN 1098-0121

Rauls, E.; Frauenheim, Th.; Gali, A. & Deak, P. (2003). Theoretical study of vacancy diffusion and vacancy-assisted clustering of antisites in SiC. *Physical Review B,* Vol. 68, No. 15, October 2003, 155208-1-155208-9, ISSN 1098-0121

Ryvkin, S.M. & Shlimak, I.S. (1973). A doped highly compensated crystal semiconductor as a model of amorphous semiconductors. *Physica status solidi (a),* Vol. 16, No. 2, April 1973, 515-526, ISSN 1862-6300

Savchenko, D.V.; Kalabukhova, E.N.; Lukin, S.N.; Sudarshan, T.S.; Khlebnikov, Y.I.; Mitchel, W.C. & Greulich-Weber, S. (2006). Intrinsic defects in high purity semi-insulating 6H SiC in Material Research Society Symposium Proceedings, Vol. 911, April, 2006, B05-07-1–B05-07-1-6, ISSN 02729172

Savchenko, D.V. & Kalabukhova, E.N. (2009). EPR diagnostics of Defect and Impurity Distribution Homogeneity in Semi-Insulating 6H-SiC. *Ukrainian Journal of Physics,* Vol. 54, No. 6, June 2009, 605-610, ISSN 2071-0186

a. Savchenko, D.V.; Shanina, B.D.; Lukin, S.N. & Kalabukhova, E.N. (2009). Kinetics of the Behavior of Photosensitive Impurities and Defects in High-Purity Semi-Insulating Silicon Carbide. *Physics of the Solid State,* Vol. 51, No. 4, April 2009, 733-740, ISSN 1063-7834

b. Savchenko, D.V.; Kalabukhova, E.N.; Kiselev, V.S.; Hoentsch, J. & Pöppl, A. (2009). Spin-coupling and hyperfine interaction of the nitrogen donors in 6H-SiC. *Physica Status Solidi B,* Vol. 246, No. 8, August 2009, 1908-1914, ISSN 0370-1972

Shik, A.Ya. (1975). Photo-conductivity of randomly-inhomogeneous semiconductors. *Soviet Physics - JETP,* Vol. 41, No. 5, May 1975, 932-940, ISSN 0038-5646

Sheinkman, M.K. & Shik, A.Ya. (1976). Long-time relaxations and residual conduction in semiconductors. *Soviet Physics - Semiconductors,* Vol. 10, No. 2, February 1976, 128-143, ISSN 0038-5700

Sridhara, S.G.; Clemen, L.L; Devaty, R.P.; Choyke, W.J.; Larkin, D.J.; Kong, H.S.; Troffer, T. & Pensl, G. (1998). Photoluminescence and transport studies of boron in 4H SiC. *Journal of Applied Physics,* Vol. 83, No. 12, June 1998, 7909-7919, ISSN 0021-8979

Son, N.T.; Magnusson, B.; Zolnai, Z.; Ellison, A. & Janzen, E. (2004). Defects in high-purity semi-insulating SiC, *Materials Science Forum,* Vol. 457-460, June 2004, 437–442, ISSN 0255-5476

Son, N.T.; Magnusson, B. & Janzen, E. (2002). Photoexcitation electron paramagnetic resonance studies of the carbon vacancy in 4H-SiC. *Applied Physics Letters,* Vol. 81, No. 21, November 2002, 3945-3947, ISSN 0003-6951

Suttrop, W.; Pensl, G.; Choyke, W.J.; Stein, R. & Liebenzeder S. (1992). Hall effect and infrared absorption measurements on nitrogen donors in 6H-silicon carbide. *Journal of Applied Physics,* Vol. 72, No. 8, October 1992, 3708-3713, ISSN 0021-8979

Torpo, L.; Marlo, M.; Staab T.E.M. & Nieminen, R.M. (2001). Comprehensive ab initio study of properties of monovacancies and antisites in 4H-SiC. *Journal of Physics: Condensed Matter,* Vol. 13, No. 28, July 2001, 6203-6231, ISSN 0953-8984

a. Umeda, T.; Isoya, J.; Morishita, N.; Ohshima, T. & Kamiya, T. (2004). EPR identification of two types of carbon vacancies in 4H-SiC. *Physical Review B,* Vol. 69, No. 12, March 2004, 121201-1-121201-4, ISSN 1098-0121

b. Umeda, T.; Isoya, J.; Morishita, N.; Ohshima, T.; Kamiya, T.; Gali, A.; Deak, P.; Son, N.T. & Janzen, E. (2004). EPR and theoretical studies of positively charged carbon vacancy in 4H-SiC. *Physical Review B,* Vol. 70, No. 23, December 2004, 235212-1-235212-6, ISSN 1098-0121

Umeda, T.; Ishitsuka, Y.; Isoya, J.; Son, N.T.; Janzen, E.; Morishita, N.; Ohshima, T. & Itoh, H. (2005). EPR and theoretical studies of negatively charged carbon vacancy in 4H-SiC. *Physical Review B,* Vol. 71, No. 19, May 2005, 193202-1-193202-4, ISSN 1098-0121

a. Umeda, T.; Son, N.T.; Isoya, J.; Morishita, N.; Ohshima, T.; Itoh, H. & Janzen, E. (2006). Electron paramagnetic resonance study of the HEI4/SI5 center in 4H-SiC. *Materials Science Forum,* Vol. 527-529, October 2006, 543-546, ISSN 0255-5476

b. Umeda, T.; Son, N.T.; Isoya, J.; Janzen, E.; Ohshima, T.; Morishita, N.; Itoh, H.; Gali. A. & Bockstedte, M. (2006). Identification of the Carbon Antisite-Vacancy Pair in 4H-SiC. *Physical Review Letters,* Vol. 96, No. 14, April 2006, 145501-1-145501-4, ISSN 0031-9007

Umeda, T.; Morishita, N.; Ohshima, T.; Itoh, H. & Isoya, J. (2007). Electron Paramagnetic Resonance Study of Carbon Antisite-vacancy pair in p-Type 4H-SiC. *Materials Science Forum,* Vols. 556-557, September 2007, 453-456, ISSN 0255-5476

2

One-dimensional Models for Diffusion and Segregation of Boron and for Ion Implantation of Aluminum in 4H-Silicon Carbide

Kazuhiro Mochizuki
Central Research Laboratory, Hitachi, Ltd.
Japan

1. Introduction

Silicon carbide with a poly-type 4H structure (4H-SiC) is an attractive material for power devices. While bipolar devices mainly utilize 4H-SiC p-n junctions, unipolar devices use p-n junctions both within the active region (to control the electric field distribution) and at the edges of the devices (to reduce electric-field crowding) (Baliga, 2005). In a p-type region, very high doping is necessary since common acceptors have deep energy levels (B: 0.3 eV; Al: 0.2 eV) (Heera et al., 2001). Boron is known to exhibit complex diffusion behaviour (Linnarsson et al., 2003), while aluminum has extremely low diffusivity (Heera et al., 2001). Precise modeling of boron diffusion and aluminum-ion implantation is therefore crucial for developing high-performance 4H-SiC power devices.

For carbon-doped silicon, a boron diffusion model has been proposed (Cho et al., 2007). Unfortunately, the results cannot be directly applied to boron diffusion in SiC because of the existence of silicon and carbon sublattices. In addition, knowledge of boron segregation in 4H-SiC is lacking, preventing improvement of such novel devices as boron-doped polycrystalline silicon (poly-Si)/nitrogen-doped 4H-SiC heterojunction diodes (Hoshi et al., 2007). Dopant segregation in elementary-semiconductor/compound-semiconductor heterostructures—in which point defects in an elementary semiconductor undergo a configuration change when they enter a compound semiconductor—has yet to be studied. A framework for such analysis needs to be provided.

With regards to aluminum distribution, to precisely design p-n junctions in 4H-SiC power devices, as-implanted profiles have to be accurately determined. For that purpose, Monte Carlo simulation using binary collision approximation (BCA) was investigated (Chakarov and Temkin, 2006). However, according to a multiday BCA simulation using a large number of ion trajectories, values of the simulated aluminum concentration do not monotonically decrease when the aluminum concentration becomes comparable to an n-type drift-layer-doping level (in the order of 10^{15} cm^{-3}). A continuous-function approximation, just like the dual-Pearson approach established for ion implantation into silicon (Tasch et al., 1989), is thus needed.

The historic development and basic concepts of boron diffusion in SiC are reviewed as follows. It took 16 years for the vacancy model (Mokhov et al., 1984) to be refuted by the interstitial model (Bracht et al., 2000). A "dual-sublattice" diffusion modeling, in which a different diffusivity is assigned for diffusion via each sublattice, was proposed next. At the same time, a "semi-atomistic" simulation, in which silicon interstitials (I_{Si}) and carbon interstitials (I_C) are approximated as the same interstitials (I) and silicon vacancies (V_{Si}) and carbon vacancies (V_C) are approximated as the same vacancies (V), was performed (Mochizuki et al., 2009). Although this approximation originally comes from the limitation of a commercial process simulator, it contributes to reducing the number of parameters needed in an atomistic simulation using a continuity equation of coupling reactions between I_{Si}, I_C, V_{Si}, V_C, and diffusing species.

After boron diffusion in 4H-SiC is discusssed, boron diffusion and segregation in a boron-doped poly-Si/nitrogen-doped 4H-SiC structure are investigated by combining the model described above and a reported model of poly-Si diffusion sources (Lau, 1990). The results of an experiment to analyze boron-concentration profiles near the heterointerface are presented. Care is taken in this experiment to avoid implantation damage by using in-situ doped poly-Si instead of boron-implanted poly-Si.

The latter half of this chapter is an analysis and modeling of aluminum-ion implantation into 4H-SiC. Owing to the extremely low diffusivity of aluminum, multiple-energy ion implantation is required to produce SiC layers with an almost constant aluminum concentration over a designed depth. First, the influence of the sequence of multiple-energy aluminum implantations into 6H-SiC (Ottaviani et al., 1999) is explained. Next, the dual-Pearson model, developed for ion implantation into silicon, is reviewed (Tasch et al., 1989). The experimental, as well as Monte-Carlo-simulated, as-implanted concentration profiles of aluminum are then presented. After that, aluminum implantation at a single energy is modelled by using the dual-Pearson approach.

To indicate the future direction of modeling and simulation studies on p-type dopants in 4H-SiC, state-of-the-art two-dimensional modeling of aluminum-ion implantation is discussed at the end of this chapter. The modeling and simulation described in this chapter will also provide a framework for analyzing n-type dopants (e.g., nitrogen and phosphorous) in SiC, group-IV impurities (e.g., carbon and silicon) in III-V compound semiconductors (e.g., GaAs and InP), and diffusion and segregation of any dopants in elementary-semiconductor/compound-semiconductor heterostructures (e.g., Ge/GaAs and C/BN).

2. Boron Diffusion and Segregation

2.1 Boron diffusion in 4H-SiC

(a) Historic background

The first analysis of boron diffusion in SiC was based on a boron-vacancy model of 6H-SiC (Mokhov et al, 1984). Detailed analysis of the boron-concentration profiles in nitrogen-doped 4H- and aluminum-doped 6H-SiC, however, indicated that I_{Si}, rather than V_{Si}, controls the diffusion of boron (Bracht et al., 2000). The I_{Si}-mediated boron diffusion was then reconsidered in light of evidence of participation of I_C (Rüschenschmidt et al., 2004). According to experiments on self-diffusion in isotopically enriched 4H-SiC, the diffusivities of I_{Si} and I_C are of the same order of magnitude, and it was proposed that under specific experimental conditions, either defect is strongly related to the preferred lattice site by boron. Theoretical calculations on 3C-SiC (Rurali et al., 2002; Bockstedte et al., 2003; Gao et al., 2004) also showed that I_{Si} and I_C are far more mobile than V_{Si} and V_C. Under the assumption that I_{Si} and I_C have the same mobility in 4H-SiC, boron diffusion, via I_{Si} and I_C, can be simulated from a certain initial distributuion of point defects.

Boron-related centers in SiC are known to have two key characteristics: a shallow acceptor with an ionization energy of about 0.30 eV and a deep level with an ionization energy of about 0.65 eV (Duijin-Arnold et al., 1998). While the nature of the shallow acceptor defect is accepted as an off-center substitutional boron atom at a silicon site (B_{Si}) (Duijin-Arnold et al., 1999), that of the deep boron-related level is not clear. The B_{Si}-V_C pair (Duijin-Arnold et al., 1998) was refuted by *ab initio* calculations that suggest a B_{Si}-Si_C complex as a candidate (Aradi et al., 2001). In addition, candidates such as a substitutional boron atom at a carbon site (B_C) and a B_C-C_{Si} complex were also put forward (Bockstedte et al., 2001). The boron-related deep center prevails in boron-doped 4H-SiC homoepitaxially experimentally grown under a silicon-rich condition (Sridhara et al., 1998), while similar experiments under a carbon-rich condition favor the shallow boron acceptor (Rüschenschmidt et al., 2004). Since the site-competition effect suggests that boron atoms can occupy both silicon- and carbon-related sites, it is assumed that the deep boron-related level originates from B_C (Rüschenschmidt et al., 2004).

According to the theoretical results on 3C-SiC (Rurali et al., 2002; Bockstedte et al., 2003), a mobile boron defect is a boron interstitial (I_B) rather than a boron-interstitial pair , which is considered to mediate boron diffusion in silicon (Sadigh et al., 1999; Windl et al., 1999). Although it is ideal to simulate diffusion of I_B in order to calculate boron-concentration profiles, the most relevant configuration of I_B in 4H-SiC is still not clear. To reproduce the experimentally obtained boron-diffusion profiles for designing 4H-SiC power devices, a boron-interstitial-pair diffusion model in a commercial process simulator, which is optimized mainly for the use with silicon, is applied. The reported boron-concentration profiles in 4H-SiC (Linnarsson et al., 2003; Linnarsson et al., 2004) are taken as an example because the annealing conditions for high-temperature (500°C)-implanted (200 keV/4×10^{14} cm^{-2}) boron ions were systematically varied.

(b) Dual-sublattice diffusion modeling

It is assumed that I_{Si} and I_C diffuse on their own sublattices in accord with the theoretical calculation on 3C-SiC (Bockstedte et al., 2004). The kick-out reactions forming an I_B from a boron atom at the silicon site (B_{Si}) and at the carbon site (B_C) are then expressed as

$$B_{Si} + I_{Si} \leftrightarrows I_B(\text{type-I}) \tag{1}$$

and

$$B_C + I_C \leftrightarrows I_B(\text{type-II}), \tag{2}$$

where the expression for the charge state is omitted. In the case of 3C-SiC, I_B(type-I) and I_B(type-II) can be a carbon-coordinated tetrahedral site, a hexagonal site, or a split-interstitial at the silicon site or the carbon site (Bockstedte et al., 2003). The reactions given by Eqs. (1) and (2) correspond to the following reactions in the boron-interstitial pair diffusion model (Bracht, 2007):

$$B_{Si}^{j} + I_{Si}^{m} \leftrightarrows (B_{Si}\, I_{Si})^{u} + (j + m - u)\, h^{+}, \tag{1a}$$

$$B_C^k + I_C^n \leftrightarrows (B_C\, I_C)^v + (k + n - v)\, h^+, \tag{2a}$$

with charge states $j, k, m, n, u, v \in \{0, \pm1, \pm2, \ldots\}$ and the holes h^+. According to the previous calculation (Bockstedte et al., 2003), I_{Si} in 3C-SiC can be charged from neutral to +4, and I_C from -2 to +2. If it is assumed that the variations in the charge states of I_{Si} and I_C in 4H-SiC are the same as those in 3C-SiC, the ranges of m and n in Eqs. (1a) and (2a) are limited to $m \in \{0, 1, 2, 3, \text{and } 4\}$ and $n \in \{0, \pm1, \text{and } \pm2\}$.

Boron diffusion in an epitaxially grown 4H-SiC structure with a buried boron-doped layer (Janson et al., 2003a) is modeled as shown in Fig. 1. In this case, the concentrations of point defects are considered to be in thermodynamic equilibrium. The Fermi model, in which all effects of point defects on dopant diffusion are built into pair diffusivities (Plummer et al., 2000), is thus applied. In the present case, the pair diffusivities are $(B_{Si}I_{Si})^u$ and $(B_CI_C)^v$ in eqs. (1a) and (2a). In general, when doping concentration exceeds intrinsic carrier concentration n_i (Baliga, 2005), where

$$n_i\,(T) = 1.70\times10^{16}\, T^{1.5} \exp[(-2.08\times10^4) \,/\, T]\ (\text{cm}^{-3}), \tag{3}$$

diffusion becomes concentration-dependent (Plummer et al., 2000). Diffusivity D of a pair (impurity A/interstitial I) is thus expressed as

$$D_{AI} = D_{AI}{}^0 + D_{AI}{}^+(p/n_i)^{+1} + D_{AI}{}^{++}(p/n_i)^{+2} + D_{AI}{}^-(p/n_i)^{-1} + D_{AI}{}^=(p/n_i)^{-2}, \tag{4}$$

where p is hole concentration, and superscripts "++" and "=" traditionally stand for +2 and

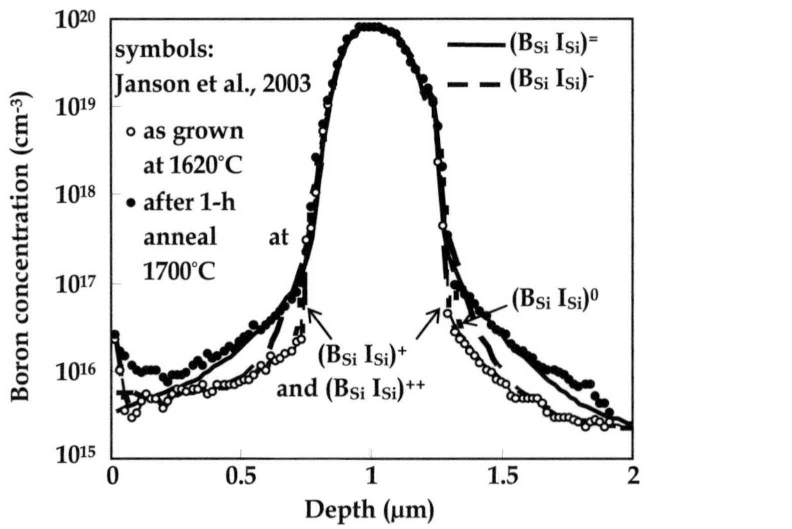

Fig. 1. Boron-concentration profiles in an epitaxially grown 4H-SiC structure with a buried boron-doped layer before (open circles) and after one-hour annealing at 1700°C [The profile simulated using diffusivity of a double-negatively charged $B_{Si}I_{Si}$ pair of 1×10^{-15} cm^2/s (solid curve) can precisely reproduce the experimental results (solid circles).]

–2. As described in section **2.1**(a), boron atoms are incorporated into silicon sites as shallow acceptors (B_{Si}^-) when a SiC structure is grown under a carbon-rich condition. Equations (1a) and (4) thus become

$$B_{Si}^- + I_{Si}^m \leftrightarrows (B_{Si}\, I_{Si})^u + (-1 + m - u)\, h^+, \tag{5}$$

$$D_{(BSi\ ISi)} = D_{(BSi\ ISi)^0} + D_{(BSi\ ISi)^+}(p/n_i)^{+1} + D_{(BSi\ ISi)^{++}}(p/n_i)^{+2} + D_{(BSi\ ISi)^-}(p/n_i)^{-1} + D_{(BSi\ ISi)^=}(p/n_i)^{-2}, \tag{6}$$

with $m \in$ {0, 1, 2, 3, and 4} and $u \in$ {0, ±1, and ±2}.

It is assumed that a single term in the right-hand side of Eq. (6) dominates the diffusion of ($B_{Si}I_{Si}$) pairs. The profile after one-hour annealing at 1700°C in Fig. 1 was fitted by using one of the diffusivities of the following five ($B_{Si}I_{Si}$) pairs: of neutral, single- and double-positively charged, and single- and double-negatively charged. As shown in Fig. 1, the profile simulated with the diffusivity of a double-negatively-charged ($B_{Si}I_{Si}$) pair of 1×10^{-15} cm²/s can precisely reproduce the experimentally obtained concentration profiles, while the profiles simulated using the other four diffusivities cannot. Therefore, $(B_{Si}I_{Si})^=$ is chosen to simulate the diffusion of B_{Si}^-.

The diffusion of B_C (Bockstedte et al., 2003) is modeled next. Since B_C can be regarded as an acceptor (Mochizuki et al., 2009), eq. (2a) becomes

$$B_C^- + I_C^n \leftrightarrows (B_C\, I_C)^v + (-1 + n - v)\, h^+, \tag{7}$$

with n and $v \in$ {0, ±1, and ±2}, and Eq. (4) becomes

$$D_{(BC\ IC)} = D_{(BC\ IC)^0} + D_{(BC\ IC)^+}(p/n_i)^{+1} + D_{(BC\ IC)^{++}}(p/n_i)^{+2} + D_{(BC\ IC)^-}(p/n_i)^{-1} + D_{(BC\ IC)^=}(p/n_i)^{-2}. \tag{8}$$

In p-type 6H-SiC, the diffusivity of in-diffused boron is proportional to p when the boron vapor pressure is low (Mokhov et al., 1984). Under the assumption that similar dependence is observable in 4H-SiC, the diffusivity of a single-positively charged pair $(B_CI_C)^+$ is chosen to simulate the diffusion of B_C^-.

(c) Semi-atomistic diffusion simulation

Diffusion of implanted boron ions is calculated from the initial point-defect distribution determined by Monte-Carlo simulation. In the calculation, the continuity equation

$$\partial/\partial t\, (C_I + C_{(BSi\ I)^=} + C_{(BC\ I)^+}) = -\nabla\cdot (J_I + J_{(BSi\ I)^=} + J_{(BC\ I)^+}) - K_r\, (C_I C_V - C_I^* C_V^*) \tag{9}$$

is solved with

$$J_{(BSi\ I)^=} = -D_{(BSi\ I)^=} \{-\nabla [C_{BSi^-} (C_I / C_I^*) + C_{BSi^-} (C_I / C_I^*)\, (q\, E / k_B\, T)]\} \tag{10}$$

and

$$J_{(BC\ I)^+} = -D_{(BC\ I)^+} \{-\nabla [C_{BC^-} (C_I / C_I^*) + C_{BC^-} (C_I / C_I^*)\, (q\, E / k_B\, T)]\}, \tag{11}$$

where C_I and C_V are interstitial and vacancy concentrations, C_I^* and C_V^* are equilibrium interstitial and vacancy concentrations, J_I is interstitial flux, K_r is interstitial-vacancy bulk recombination coefficient, q is electronic charge, E is electric field, k_B is Boltzmann's constant, and T is absolute temperature. As expressed in eqs. (10) and (11), both the fluxes of $(B_{Si}I)^{=}$ and $(B_{C}I)^{+}$ take the effect of electric field into account.

The first step of the simulation is to obtain as-implanted boron profiles along with the initial distribution of point defects. As explained in section **2.1**(a), once I_{Si} and I_C are created, they are treated as the same I (with an unidentified origin). Similarly, the created V_{Si} and V_C are treated as the same V. Equations (6) and (8) are therefore simplified as

$$D_{(BSi\ I)} = D_{(BSi\ I)^{=}}(p/n_i)^{-2} \tag{12}$$

and

$$D_{(BC\ I)} = D_{(BC\ I)^{+}}(p/n_i)^{+1}. \tag{13}$$

In the Monte-Carlo simulation, the surface of 4H-SiC was assumed to be misoriented by 8° from (0001) toward [11-20], and the boron-ion-beam divergence was set to 0.1°.The probabilities of the implanted boron ions occupying a silicon site (r_{Si}) or a carbon site (r_C) are specified as follows. For 200 keV/4×10^{14} cm^{-2} boron-ion implantation at 500°C (Linnarsson et al., 2003; Linnarsson et al., 2004), as-implanted concentration profiles of B_{Si}^{-}, B_{C}^{-}, I, and V are calculated under the tentative assumption that $r_{Si} = r_C = 0.5$ (Fig. 2).

The next step of the simulation is to solve Eq. (9). Both the time needed for increasing temperature before annealing and the time needed for decreasing temperature after annealing are neglected. Surface recombination of I and V, as well as surface evaporation of any species, are also neglected. The temperature dependences of n_i in Eq. (3) and the diffusivity of I (D_I) (Rüschenschmidt et al., 2004), where

$$D_I\,(T) = 4.8 \exp[-7.6\ (\text{eV}) / k_B T]\ (\text{cm}^2/\text{s}), \tag{14}$$

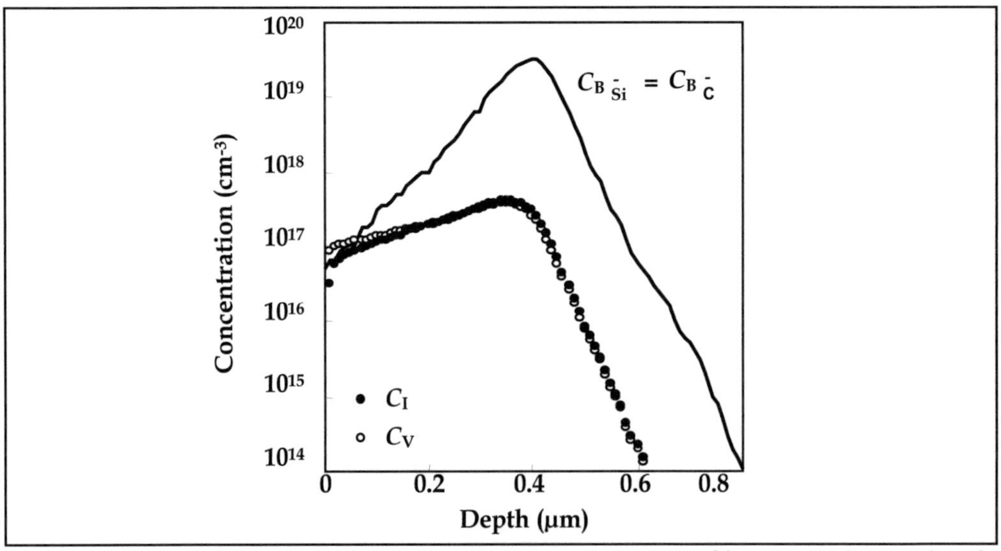

Fig. 2. Monte-Carlo simulated as-implanted concentration profiles in 4H-SiC under the assumption that the probability of implanted boron ions occupying a silicon site (r_{Si}) is 0.5 and that of occupying a carbon site (r_C) is 0.5

were used in the simulation. The diffusivity of V (D_V) was tentatively assumed to be the same as D_I, although the simulated profiles did not change with D_V.

Figure 3 shows simulated concentration profiles of B_{Si}^-, B_C^-, I, and V in the case of a background doping level N_b of 2×10^{15} cm^{-3} (n-type), T = 1900°C, and t = 15 min (Linnarsson et al., 2004). According to the assumption made at the beginning of section **2.1**(b), the concentration profiles of B_{Si}^- (dashed line) and B_C^- (solid line) were obtained separately. Total boron concentration was thus calculated as the sum of the B_{Si}^- and B_C^- concentrations.

In Fig. 3, C_I and C_V become equilibrium values (C_I^* and C_V^*, respectively) below a depth of 1.7 μm and are determined from the free energies of formation, F_I and F_V, as follows (Bockstedte et al., 2003; Bracht, 2007):

$$C_I^* = C_{sI} \exp(-F_I / k_B T), \tag{15}$$

$$C_V^* = C_{sV} \exp(-F_V / k_B T), \tag{16}$$

where C_{sI} and C_{sV} are the concentrations of the sites that are open to Is and Vs, respectively. In the case of silicon, C_{sI} = 5.0×10^{22} cm^{-3}, F_I = 2.36 eV, C_{sV} = 2.0×10^{23} cm^{-3}, and F_V = 2.0 eV have been conventionally used in a commercial process simulator. Even with these values, it is possible to fit the simulated profiles to the reported boron-concentration profiles in 4H-SiC, except for the reciprocal dependence of boron diffusion on p (Fig. 4). To explain the results in Fig. 4, the following values are employed: C_{sI} = 4×10^{30} cm^{-3}, F_I = 5.2 eV, C_{sV} = 2×10^{33} cm^{-3}, and F_V = 7.0 eV. The values of F_I and F_V, theoretically calculated in the case of 3C-SiC, are, respectively, in the ranges of 4 to 14 eV and 1 to 9 eV (Bockstedte et al., 2003). However, the values of C_{sI} and C_{sV} are 8 to 10 orders of magnitude larger than those in the case of silicon (as discussed later in this section).

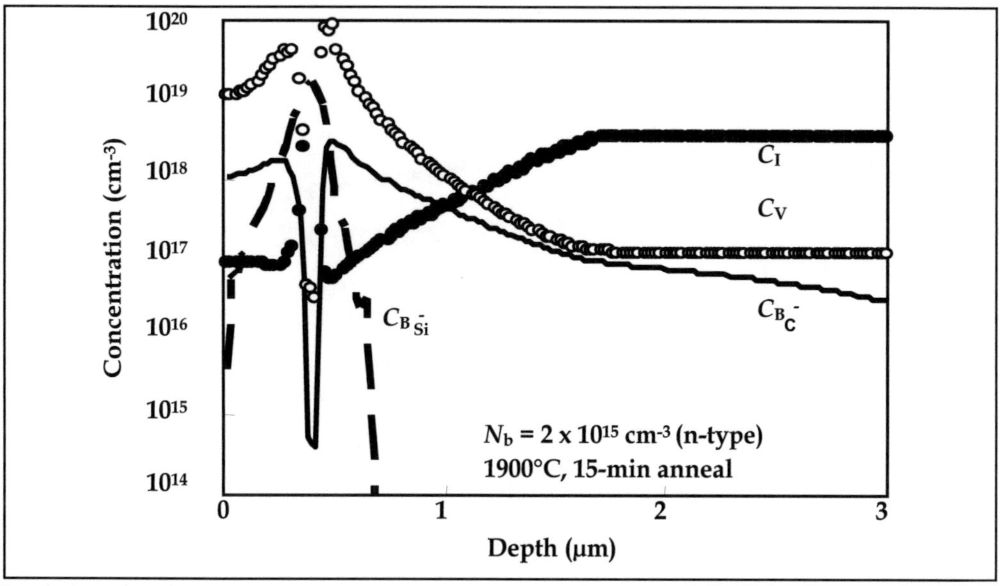

Fig. 3. Simulated concentration profiles of B_{Si}^-, B_C^-, I, and V in 2×10^{15}-cm^{-3}-doped n-type 4H-SiC after 15-min annealing at 1900°C simulated from the initial concentration profiles in Fig. 2

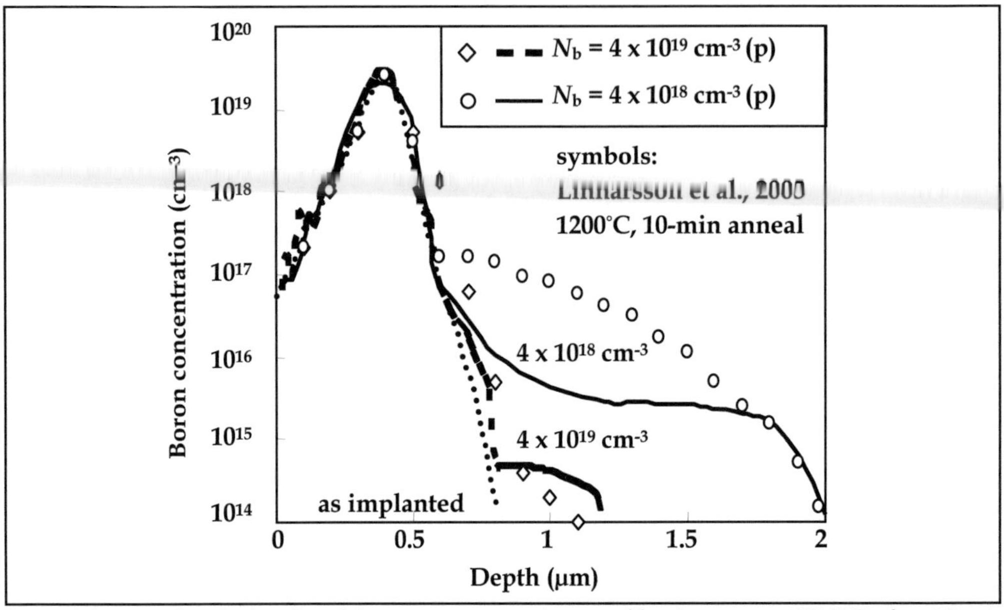

Fig. 4. Measured and simulated boron-concentration profiles in p-type 4H-SiC after 10-min annealing at 1200°C

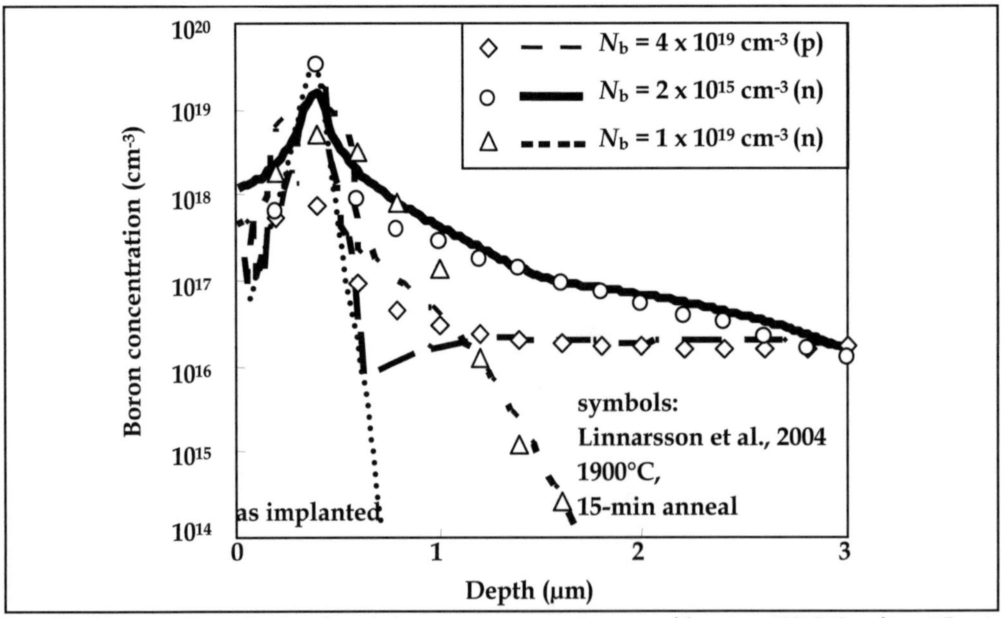

Fig. 5. Measured and simulated boron-concentration profiles in 4H-SiC after 15-min annealing at 1900°C

As shown in Fig. 5, owing to the introduction of the dual-sublattice modeling, the simulated boron-concentration profiles well describe the tail regions of the measured profiles (symbols) with background doping ranging from n- to p-type under conditions T = 1900°C and t = 15 min (Linnarsson et al., 2004). The fitting parameters used are the same as those for Fig. 3. The tail regions are represented mainly by B_{Si}^- [N_b = 1×10^{19} cm^{-3} (n-type)] and B_C^- [N_b = 2×10^{15} cm^{-3} (n-type) and 4×10^{19} cm^{-3} (p-type)].

To use the modeling (section **2.1**(b)) and simulation described here for optimizing the boron-diffusion process in regards to device fabrication, time-dependent diffusion has to be accurately simulated. Figure 6 shows annealing-time (5 ot 90 min) dependences of boron-concentration profiles for N_b = 4×10^{19} cm^{-3} (p-type) and T = 1400°C (Linnarsson et al., 2004). The measured time-dependent boron-diffusion profiles (symbols) are precisely reproduced with the parameters $D_{(B_{Si}I)^=}$ = 3×10^{-18} cm^2/s, $D_{(B_CI)^+}$ = 6×10^{-12} cm^2/s, and K_r = 3×10^{-16} cm^3/s and the parameters expressed by Eqs. (3) and (14).

One of the biggest challenges in understanding boron diffusion in 4H-SiC has been its reciprocal dependence on p, observed in the case of N_b = 4×10^{18} and 4×10^{19} cm^{-3} (p-type), T = 1200°C, and t = 10 min (symbols in Fig. 4). Even when the diffusivity of $(B_CI)^+$, which is proportional to p under thermodynamical equilibrium [Eq. (13)], is used, the dependence (Linnarsson et al., 2003) was successfully demonstrated, at least in the tail regions, with the parameters $D_{(B_{Si}I)^=}$ = 4×10^{-20} cm^2/s, $D_{(B_CI)^+}$ = 4×10^{-12} cm^2/s, and K_r = 6×10^{-16} cm^3/s and the parameters expressed in Eqs. (3) and (14). According to the theoretical calculation (Bockstedte et al., 2003), a variety of I_{Si}, I_C, V_{Si}, and V_C exists in 3C-SiC. If it is assumed that a similar variety of point defects also exists in 4H-SiC, it is possible that the values of C_I^* and C_V^* in 4H-SiC are much larger than those in silicon at the same temperature. Since Fig. 5 is the only experimental result showing the reciprocal dependence of diffusivity of boron on p, further experimental investigation is needed to revise the parameters C_{sI} and C_{sV}.

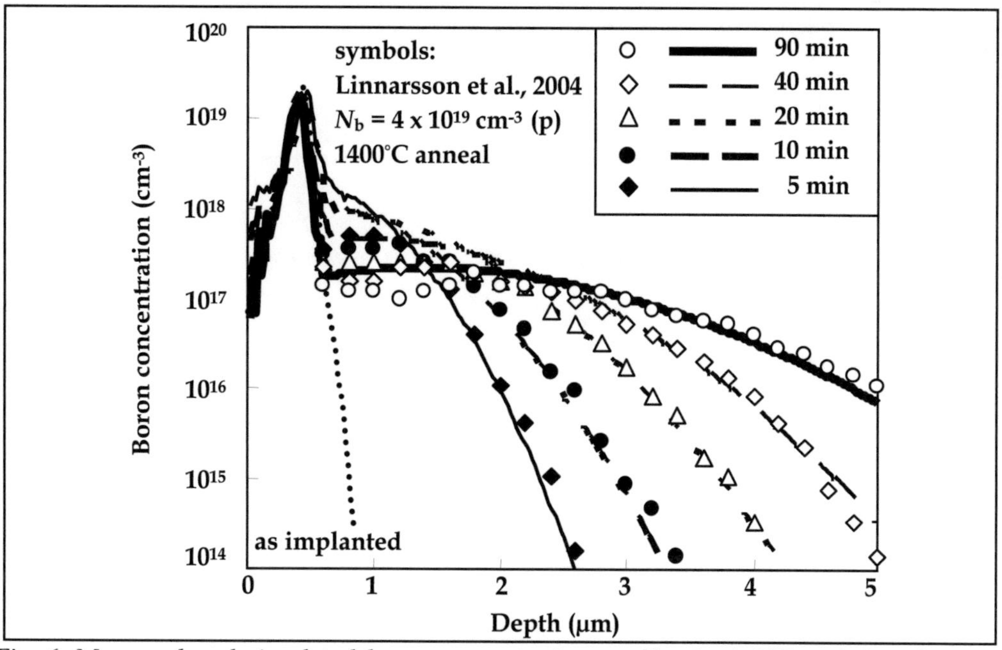

Fig. 6. Measured and simulated boron-concentration profiles in 4×10^{19}-cm^{-3}-doped p-type 4H-SiC after annealing at 1400°C

The following remaining issue should also be noted: optimization of r_{Si}/r_C for applying the developed semi-atomistic simulation to fit other experimentally obtained boron-concentration profiles. Since this optimization is strongly related to experimental conditions (Rüschenschmidt et al., 2004), r_{Si}/r_C needs to be optimized for each experimental condition.

2.2 Boron Diffusion and Segregation in Poly-Si/4H-SiC Structures

Diffusivities of a double-negatively charged B-I_{Si} pair and a single-positively charged B-I_C pair are extrapolated to less than 1000°C. Since the former diffusivity results in quite small values (Fig. 7), only the latter diffusivity is taken into account. Furthermore, only I_Cs coming from carbon atoms in native oxides that remained on 4H-SiC are treated since the diffusivity of I_C is also negligible (Rüschenschmidt et al., 2004).

(a) Model description

In regard to poly-silicon, three contributions to total boron concentration (C_B) were considered (Lau, 1990): active (C_{ga}) and inactive (C_{gi}) boron concentrations in grains, and boron concentration in grain boundaries (C_b). Since C_B was chosen to be 3×10^{20} cm^{-3}, which is larger than the maximum active concentration (C_{Si}^{sat}), $C_{ga} = C_{Si}^{sat}$; in the case of a poly-Si/Si structure, $k = C_{Si}^{sat}/C_{gi}$ [Fig. 8(a)].

In the case of a poly-Si/4H-SiC structure, C_{gi} profiles in poly-Si were calculated by using boron-interstitial pair diffusivities one hundred times larger than those in single-crystalline silicon (Plummer et al., 2000). In the case of 4H-SiC, C_{SiC}^{sat} is extrapolated, for example, to 4×10^{16} cm^{-3} at 850°C (Linnarsson et al., 2006). In accordance with the condition of k, as well as on the extent of diffusion of active (C_a) and inactive boron concentrations (C_i), C_B profiles change, as illustrated in Fig. 8(a) for silicon and in Figs. 8(b) to (d) for 4H-SiC.

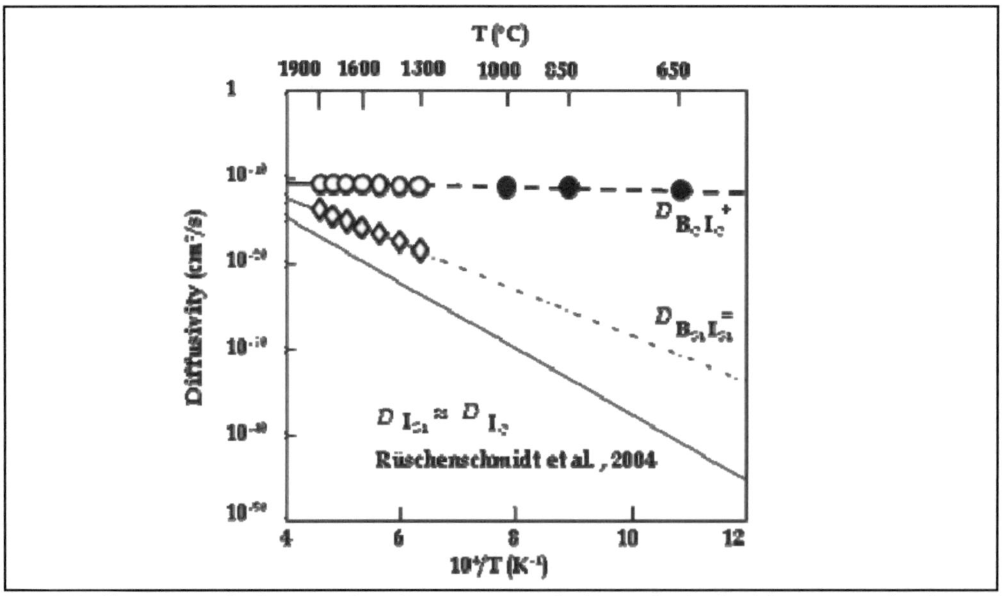

Fig. 7. Diffusivities of silicon and carbon interstitials (I_{Si} and I_C), double-negatively charged boron-I_{Si} pair, and single-positively charged boron-I_C pair (Open and closed symbols denote values used in section **2.1** and in this section, respectively.)

(b) Experiments and discussion

A 200-nm-thick boron-doped amorphous silicon film was formed on nitrogen-doped 4H-SiC (0001) substrates by chemical vapour deposition at 350°C. Annealing for post-crystallization in nitrogen ambient was performed, followed by in-depth concentration-profile analysis using an 8-keV O_2^+ beam in a secondary-ion mass spectrometer (SIMS).

Measured C_B profiles show a peak at the heterointerfaces but no tails corresponding to $C_{SiC}{}^{sat}$ (Fig. 9). This result indicates that inactive boron atoms with $k > 1$ (Fig. 8(c)) dominate boron diffusion and segregation. Under the assumption that the charge states of inactive boron atoms are neutral, C_B profiles of poly-Si/4H-SiC pn diodes annealed at 850°C for 30

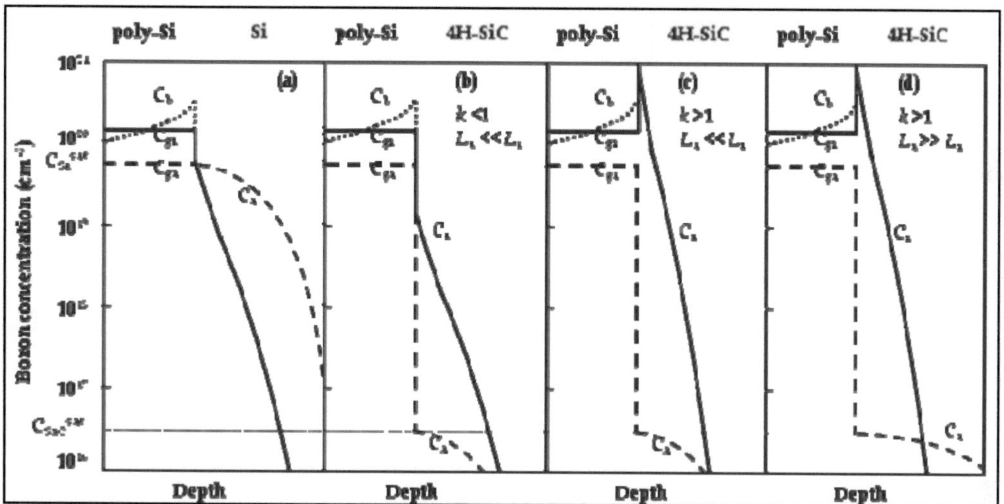

Fig. 8. Schematic illustrations of boron-concentration profiles in (a) poly-Si/Si and (b)–(d) poly-Si/4H-SiC pn diodes. In poly-Si/4H-SiC diodes, (b) corresponds to the case where segregation coefficient k is less than unity, and (c) and (d) correspond to the case that $k > 1$. When the diffusion length of active boron atoms (L_a) is much less than the diffusion length of inactive boron atoms (L_i), profiles of inactive boron concentration (C_i) dominate profiles of total boron concentration (C_B) in 4H-SiC ((b) and (c)); on the other hand, when L_a is much larger than L_i, profiles of active boron concentration (C_a) dominate the tail region of C_B profiles for 4H-SiC, as shown in (d).

C_{ga}: active boron concentration in grains; C_{gi}: inactive boron concentration in grains;
C_b: boron concentration at grain boundaries; $C_{Si}{}^{sat}$: maximum active boron concentration in silicon; $C_{SiC}{}^{sat}$: maximum active boron concentration in 4H-SiC.

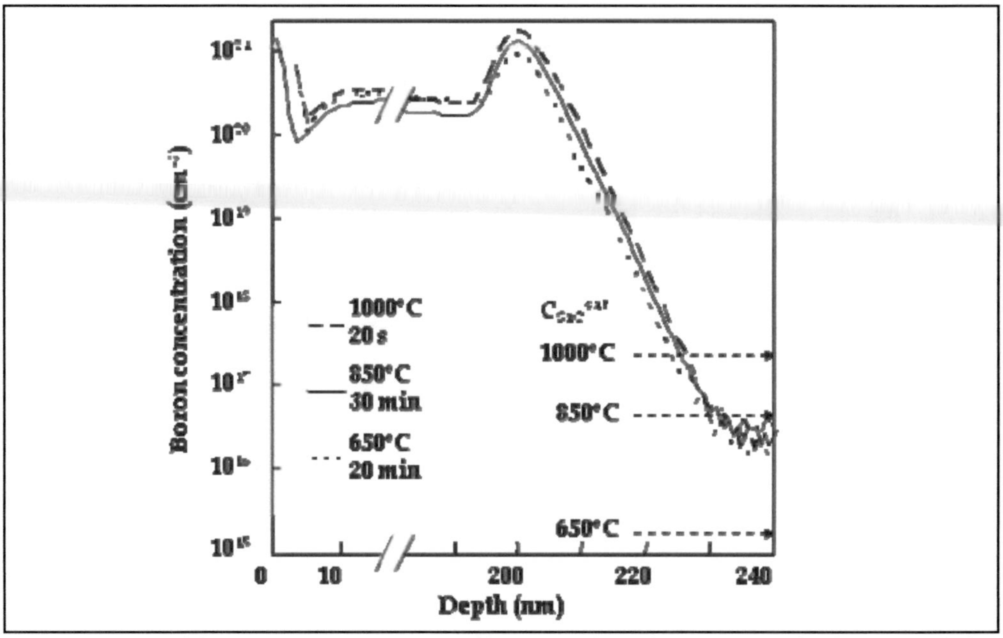

Fig. 9. Measured boron-concentration profiles in poly-Si/4H-SiC pn diodes annealed at 650–1000°C in nitrogen ambient

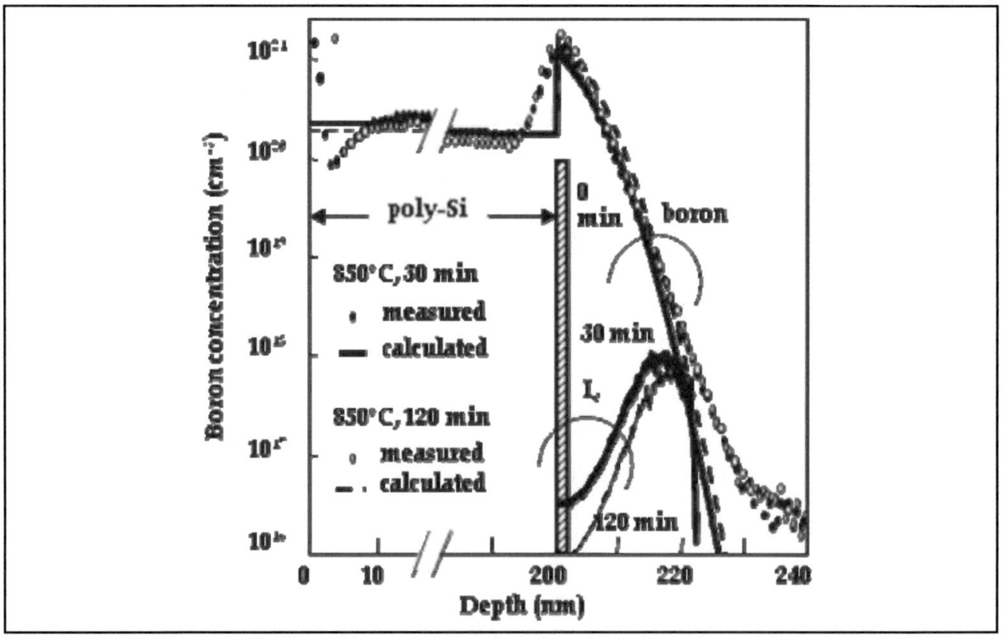

Fig. 10. Concentration profiles of boron and caluculated I_C assuming an Interstitial 1×10^{13} cm^{-2} I_C at the heterointerface [Interstitial-vacancy bulk recombination coefficient and equilibirium I_C concentration are extrapolated from the reported results (Mochizuki et al., 2009).]

and 120 min were calculated. Initial sheet concentration of I_C (N_s) of 1×10^{13} cm^{-2} at the heterointerface was found to reproduce the measured profiles, which show slight dependence on annealing time (Fig. 10). This N_s value was thus used to determine k in the temperature range of 650–1000°C (Mochizuki et al, 2010).

At 850°C, k of 6.7 is much larger than 0.7 for poly-Si/Si at 900°C (Rausch et al., 1983) and 1.7 for Si/3C-$Si_{0.996}C_{0.004}$ at 850°C (Stewart et al., 2005) (Fig. 11). The increased driving force for boron segregation with carbon mole fraction seems to support the previous model, in which boron is trapped at a carbon-related defect (Stewart et al., 2005). However, the positive activation energy of k (Fig. 12) indicates that a direct boron-carbon interaction (Liu et al., 2002) contributes to boron segregation. A recent report on suppression of boron diffusion by additional implantation of carbon (Tsirimpis et al, 2010) also supports the mechanism of direct boron-carbon interaction.

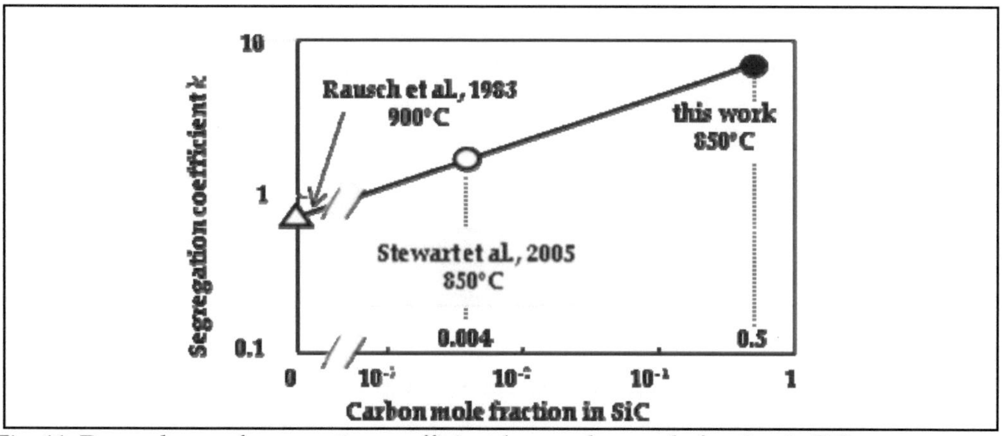

Fig. 11. Dependence of segregation coefficient k on carbon mole fraction in SiC

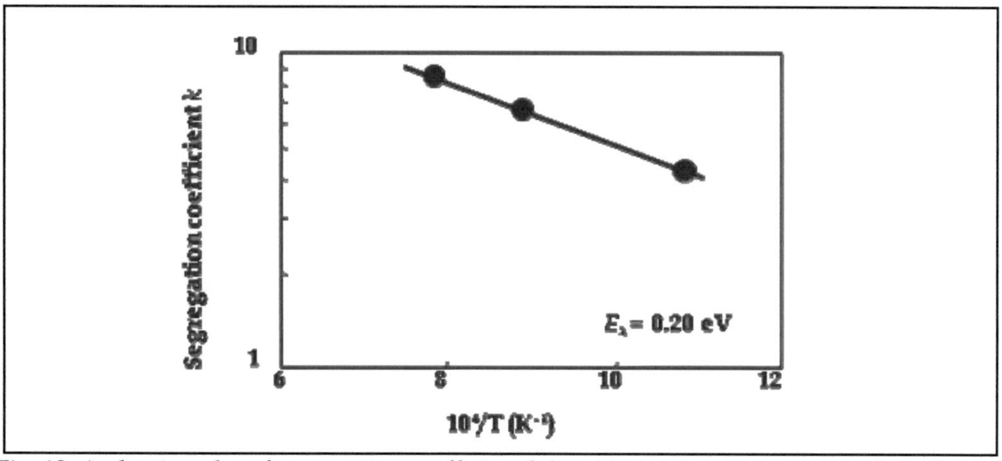

Fig. 12. Arrhenius plot of segregation coefficient k (positive activation energy is shown)

3. Aluminum-ion Implantation

(a) Historic background

The effect of a sequence of multiple-energy aluminum implantations into 6H-SiC on channeling was reported (Ottaviani et al., 1999). In that report, "scatter-in" channeling did not occur because of less channeling in the case of increasing order of implantation energy. Scatter-in channelling was first observed in boron implantation into silicon and was called "paradoxical profile broadening" (Park et al., 1991). That can occur during high-tilt-angle implantation (e.g., 7° for (100) Si) through a randomized surface layer. In the scatter-in process, some high-energy ions, which move in a random direction after crossing the surface layer, are scattered in a channeling direction and penetrate deeper into the undamaged underlying crystal. The reduced aluminum channeling in the case of increasing order of implantation energy was attributed to the amorphization caused by one implantation affecting the subsequent implantation (Ottaviani et al., 1999).

In the case of boron implantation into (100) Si, the influence of surface oxide layer is crucial (Morris et al., 1995). When the tilt angle is 0°, the depth of the as-implanted profile decreases with increasing oxide thickness because a well-collimated ion beam is scattered by the amorphous oxide layer. On the other hand, at higher tilt angles and at certain energies, the as-implanted profile becomes deeper with increasing oxide thickness because of the scatter-in channeling. In the case of aluminum implantation into 4H- and 6H-SiC, the substrate is usually misoriented from (0001) by 4°–8° to achieve step-flow epitaxial growth (Kuroda et al., 1987). Thus, there is such a high probability of the scatter-in channeling of aluminum that the effect of a surface oxide on channeling has to be calculated and demonstrated.

(b) Experiments

Experimental sample preparation was started by forming a 35-nm-thick SiO_2 layer by plasma-enhanced chemical-vapour deposition, on a 50.8-mm-thick 4H-SiC wafer misoriented by 8° from (0001) toward [11-20]. The SiO_2 layer was then removed from half of the wafer using a solution of buffered hydrofluoric acid. Subsequently, five-fold aluminum implantation was carried out at RT to achieve 0.3-μm-deep boxlike profiles with a mean plateau concentration of 1×10^{19} cm^{-3}. Implantation energies (keV) and corresponding doses ($\times10^{13}$ cm^{-2}) were 220/10, 160/5, 110/7, 70/6, and 35/3. A mechanical mask was used to form the following four implanted areas on the same wafer: without the SiO_2 layer in the case of decreasing order of implantation energy, without the SiO_2 layer in the case of increasing order of implantation energy, with the SiO_2 layer in the case of decreasing order of implantation energy, and with the SiO_2 layer in the case of increasing order of implantation energy.

To determine in-depth concentration profiles, SIMS analyses were carried out using an 8-keV O_2^+ beam. In addition to the experimentally obtained data, Monte-Carlo simulation using BCA was also utilized (Mochizuki and Onose, 2007).

The range parameters for Pearson frequency-distribution functions (Pearson, 1895) are sensitive to differences in SIMS background concentration levels (Janson et al., 2003b). Accordingly, the SIMS measured background levels (5×10^{14}–1×10^{15} cm^{-3}) were subtracted from the SIMS measured depth profiles of aluminum concentrations. The resultant depth profiles are compared to the BCA-simulated ones in Fig. 13. Very good agreements of the computationally obtained profiles with the experimentally determined ones confirm that the BCA simulation can be used to generate quasi-experimental data.

The BCA profiles in Figs. 13(a) to (d) are thus redrawn in Fig. 14. They demonstrate that the implantation without the SiO_2 layer in a decreasing energy order resulted in the least extended tail in the aluminum-concentration profiles. In the case of decreasing order of implantation energy, the tail is mainly determined by aluminum concentration profiles formed during the first energy step (220 keV). The tail of profiles for implantation with the SiO_2 layer extends deeper than that for implantation without the SiO_2 layer. This difference probably results from the scatter-in channeling. In the case of increasing order of implantation energy, on the other hand, little difference in the tail is observed between profiles for implantaions with and without the SiO_2 layer. Thus, the effect of the reported amorphization-suppressed channeling (Ottaviani et al., 1999) is considered to be less than that of the scatter-in channeling (as discussed later). The difference between the reported results (Ottaviani et al., 1999) and our results might be related to the differences in the tilt (3.5° vs. 8°) and rotation angles (-90° vs. 0°) during implantation (although that reasoning is yet to be confirmed).

(c) Dual-Pearson model

Pearson frequency distribution functions (Pearson, 1895) have been successfully applied to represent a wide selection of implanted ion profiles in 4H-SiC (Janson et al., 2003b). For such heavy ions as aluminum, however, large channeling tails of distributions deviate from the single-Pearson functions (Janson et al., 2003; Stief et al. 1998; Lee and Park, 2002). The

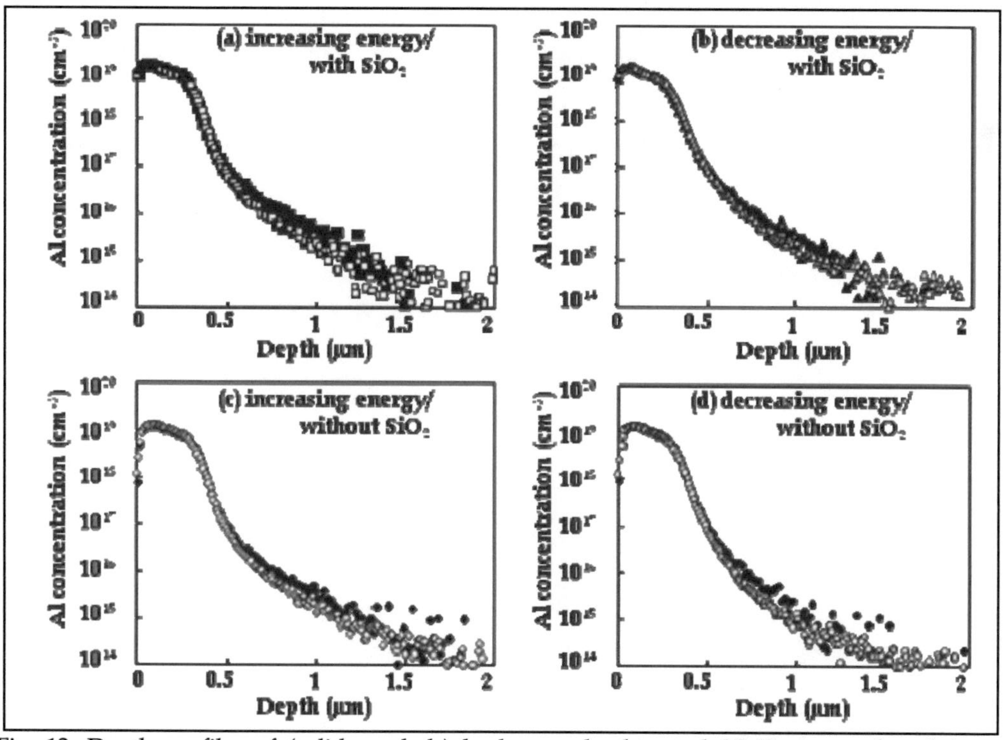

Fig. 13. Depth profiles of (solid symbols) background-subtracted SIMS-measured and (open symbols) BCA-simulated concentration profiles of five-fold aluminum implantation into 4H-SiC

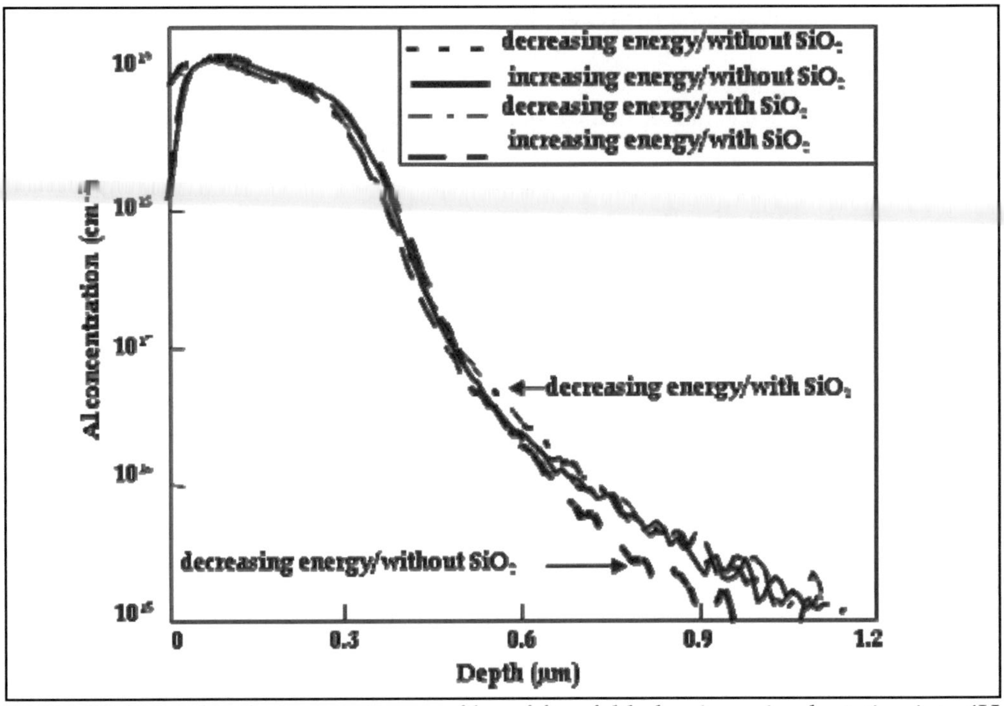

Fig. 14. BCA-simulated concentration profiles of five-fold aluminum implantation into 4H-SiC shown in Figs. 13(a) to (d)

dual-Pearson approach is thus extended to model the BCA-simulated profiles of aluminum implantations into 4H-SiC through a 35-nm-thick SiO_2 layer (Mochizuki and Onose, 2007) to implantations without the SiO_2 layer.

The dual-Pearson distribution is a weighted sum of two Pearson IV functions (Pearson, 1895) used to model the randomly scattered portion and the channeled portion of the profile (Morris et al., 1995). The depth profile of aluminum, $N(x)$, is represented by (Tasch et al., 1989)

$$N(x) = D_1 f_1(x) + D_2 f_2(x) \quad (17)$$

and

$$f_i(x) = K_i\,[1 + \{(x - R_{pi})/A_i - n_i/r_i\}^2]^{-m_i} \exp[-n_i \arctan\{(x - R_{pi})/A_i - n_i/r_i\}^2] \quad (i = 1, 2), \quad (18)$$

where f_1 and f_2 are, respectively, normalized Pearson IV distribution functions for the randomly scattered and channeled components of the profile, and D_1 and D_2 are the doses represented by each Pearson function. For Pearson IV functions, K_1 and K_2 are normalized constants. R_{p1} and R_{p2} are projected ranges, and n_1, n_2, r_1, r_2, A_1, A_2, m_1, and m_2 are parameters related to the range stragglings ΔR_{p1} and ΔR_{p2}, skewnesses γ_1 and γ_2, and kurtoses β_1 and β_2, as follows:

$$r_i = -(2 + 1/b_{2i}) \quad (19a)$$

$$n_i = -r_i\, b_{1i}/\sqrt{4\, b_{0i}\, b_{2i} - b_{1i}^2} \quad (19b)$$

$$m_i = -1/(2\, b_{2i}) \quad (19c)$$

$$A_i = m_i\, r_i\, b_{1i}/n_i \quad (19d)$$

$$b_{0i} = -\Delta R_{pi}^2\,(4\,\beta_i - 3\,\gamma_i^2)\, C \quad (19e)$$

$$b_{1i} = -\gamma_i\, \Delta R_{pi}\,(\beta_i + 3)\, C \quad (19f)$$

$$b_{2i} = -(2\,\beta_i - 3\,\gamma_i^2 - 6)\, C \quad (19g)$$

$$C = 1/[2\,(5\,\beta_i - 6\,\gamma_i^2 - 9)] \quad (i = 1, 2) \quad (19h)$$

Dose ratio, R, is defined as

$$R = D_1 / (D_1 + D_2). \quad (20)$$

To avoid arbitrariness of R_{p2} (Suzuki et al., 1998), R_{p2} was set equal to R_{p1}.

(d) Discussion

To understand the influence of the implantation energy sequence and the surface SiO_2 layer on channeling, BCA simulation of single-energy aluminum implantations into 4H-SiC were carried out, and the parameters of the dual Pearson model were fitted to the simulated data (Mochizuki et al., 2008). Concentration profiles of aluminum implanted with and without the SiO_2 layer are shown in Fig. 15. At an implantation energy of 35 keV, the profile of aluminum implanted with the SiO_2 layer becomes shallower than that without the SiO_2 layer when the aluminum concentration is more than 1×10^{15} cm^{-3}. On the other hand, at an implantation energy of 70 keV or more, the profile of aluminum implanted without the SiO_2 layer becomes shallower than that with the SiO_2 layer when the aluminum concentration is less than 1×10^{17} cm^{-3}. It is therefore concluded that in the case of the 35-nm-thick SiO_2 layer, the implantation energy at which the scatter-in channeling becomes more influential than the amorphization-suppressed channeling is between 35 and 70 keV.

The dual-Pearson parameters used to reproduce profiles in Fig. 15 are shown in Fig. 16, together with the reported parameters for single-Pearson models (Janson et al., 2003b; Stief et al. 1998; Lee and Park, 2002). Comparing the dual-Pearson parameters with the single-Pearson parameters shows that the dependences of R_p in the case of implantation without the SiO_2 layer, ΔR_{p1}, and γ_1 are almost the same as those stated in two reports (Janson et al., 2003b; Stief et al., 1998) but slightly differ from those stated in another report (Lee and Park, 2002). Although the β's of the reported single-Pearson model are not shown (to avoid complexity), the obtained relationship between β_1 and γ_1 in Fig. 16(e),

$$\beta_1 = 1.19\,\beta_{1o} \quad (21a)$$

$$\beta_{1o} = [39\gamma_1^2 + 48 + 6(\gamma_1^2 + 4)^{3/2}] / (32 - \gamma_1^2), \quad (21b)$$

is very similar to the following reported relationship (Lee and Park, 2002):

$$\beta = 1.30\,\beta_o. \quad (21c)$$

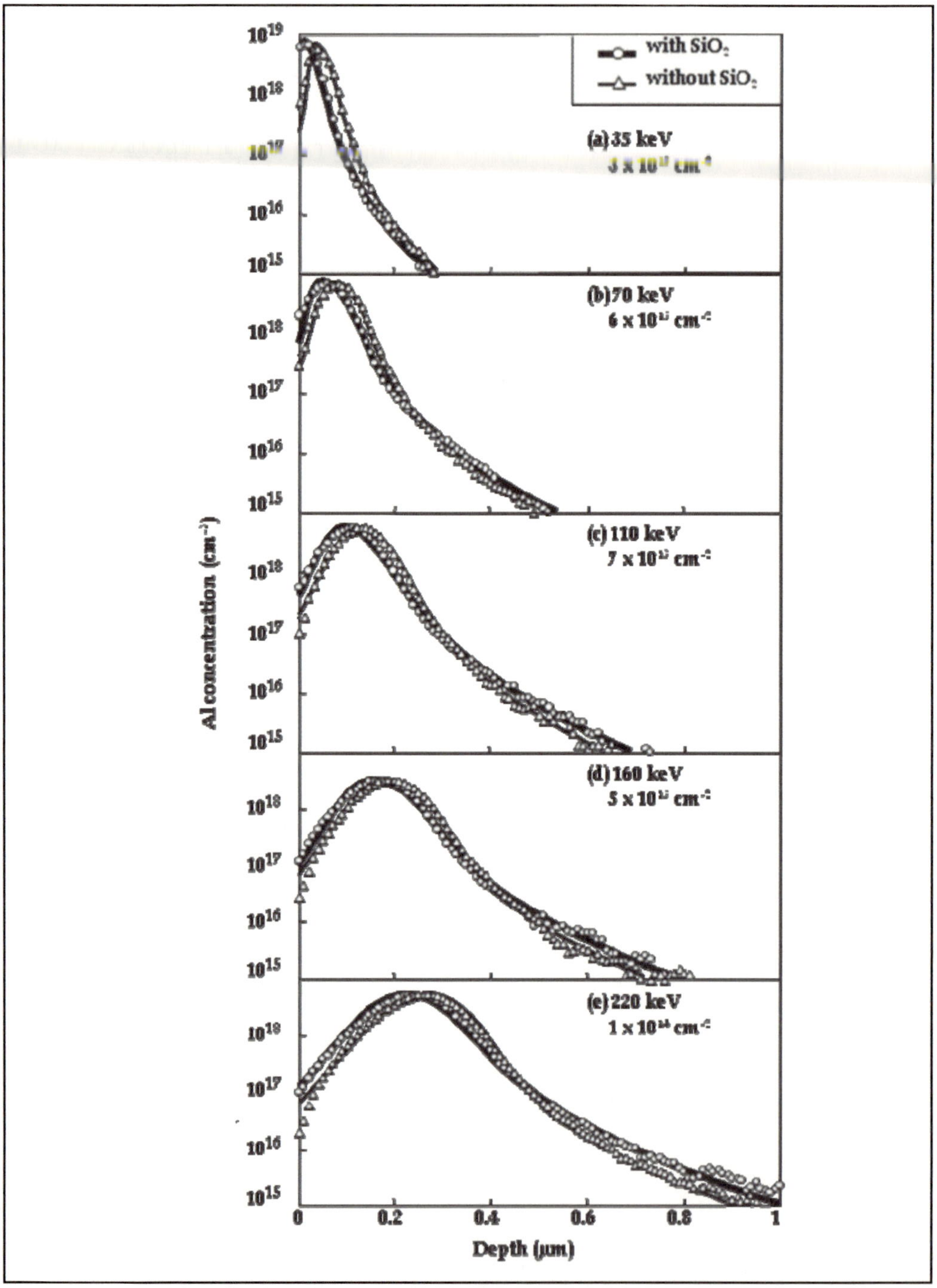

Fig. 15. BCA-simulated concentration profiles of single-energy aluminum implantations into 4H-SiC

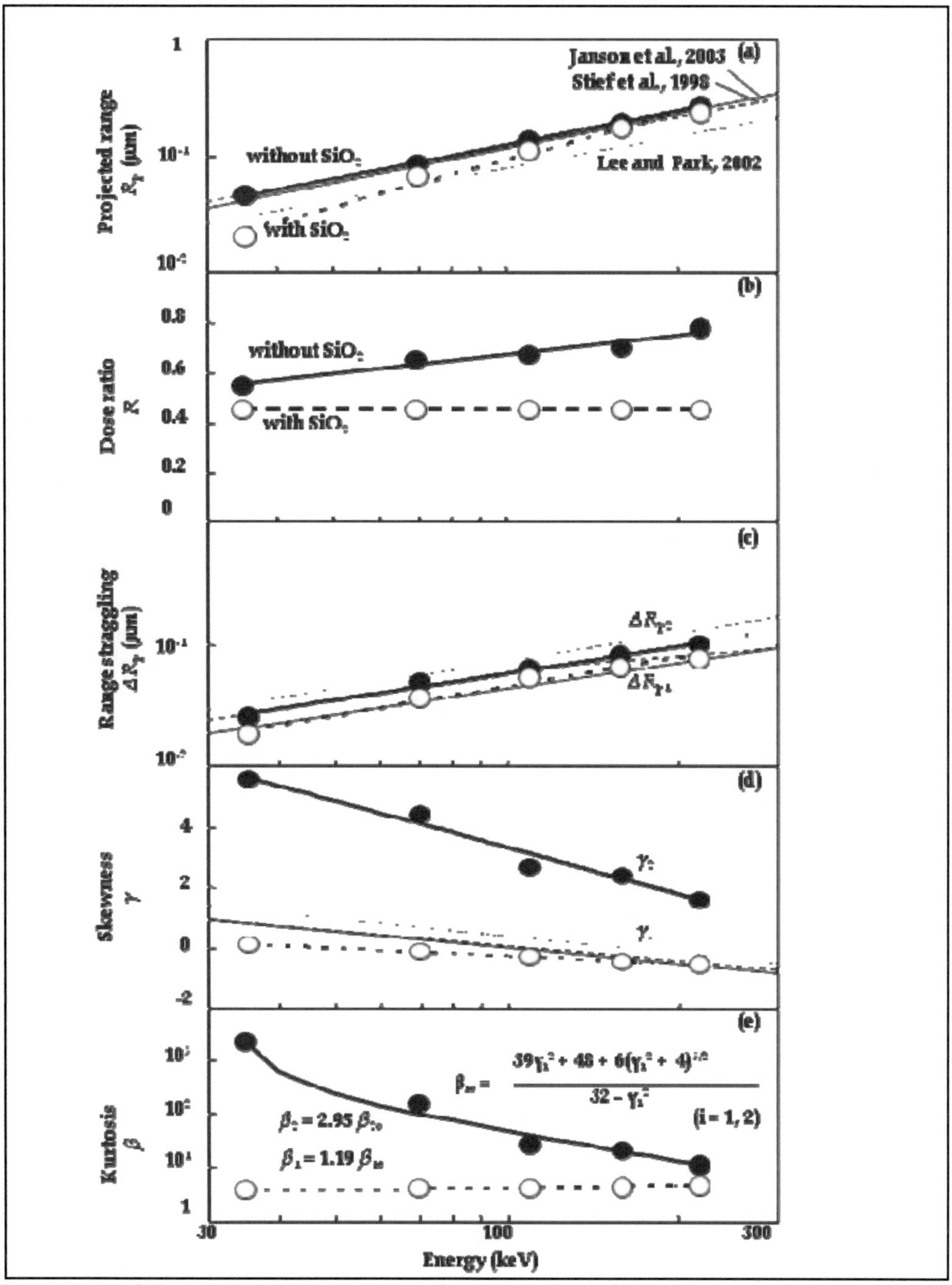

Fig. 16. Dual-Pearson parameters as a function of implantation energy (The projected ranges, range stragglings, and skewnesses from previous reports are also shown for comparison.)

In Fig. 16(a), R_p for implantations with the SiO_2 layer is smaller than that for implantations without the SiO_2 layer. This result is due to the existence of the SiO_2 layer during implantation. On the other hand, in Fig. 16(b), compared to R for implantations without the SiO_2 layer, R for implantations with the SiO_2 layer becomes smaller, i.e., more aluminum ions channel, owing to the scatter-in channeling. It should be noted that in the case of aluminum implantations with the 35-nm-thick SiO_2 layer, R monotonically increases under the following conditions (Mochizuki and Onose, 2007):

(1) dose of 1×10^{14} cm^{-2} and energy exceeding 300 keV, and
(2) dose of 1×10^{15} cm^{-2} or more and energy of 35 keV or more.

The influence of the amorphization-suppressed channeling (Ottaviani et al., 1999) might also increase under these conditions. To further investigate the effect of SiO_2, BCA simulation of 220-keV aluminum implantations through 0–40 nm SiO_2 layers was carried out, and the results were fitted with the dual-Pearson model using parameters shown in Figs. 16(c)–(e). With increasing thickness of the SiO_2 layer, R_p monotonically decreases, while the decrease in R tends to saturate (Fig. 17). The latter results may be helpful to understand the scatter-in channeling. Even in the case of implantation without the SiO_2 layer, R is not unity; 20% of the impinging aluminum ions channel because the ions basically encounter random atoms in the outermost part of the 8°-off 4H-SiC itself. This scatter-in effect is enhanced with increasing SiO_2 layer thickness until it saturates around 35 nm. This indicates that the thickness of 35 nm is sufficient to maximize the SiO_2-layer-induced scatter-in channeling at an implantation energy of 220 keV.

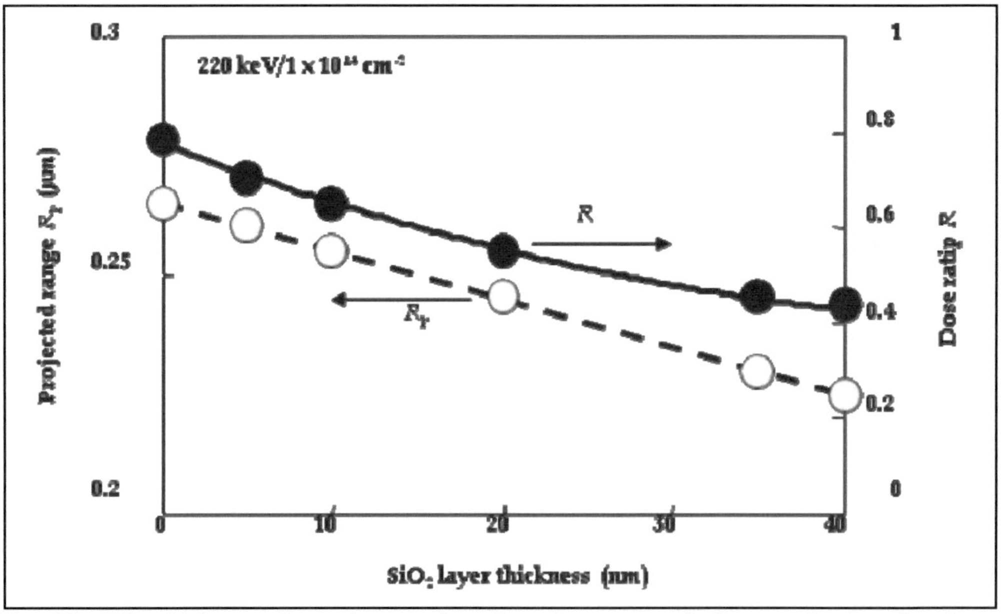

Fig. 17. SiO_2 thickness dependence of projected range and dose ratio for aluminum implantations at 220 keV with dose of 1×10^{14} cm^{-2}

On the basis of the above discussion of single-energy aluminum implantations into 4H-SiC, the tail in concentration profiles of multiple-energy aluminum implantations is discussed in the following. The symbols in Fig. 18 represent BCA-simulated profiles of aluminum implantations (a) without and (b) with the 35-nm-thick SiO_2 layer shown in Figs. 13 and 14. Concentration profiles of 220-keV implantations mainly determine the tail of the five-fold aluminum implantation, so the BCA-simulated results in Fig. 18 were fitted by varying R for 220-keV implantations only. In the case of the aluminum implantations without SiO_2 layers [Fig. 18(a)], R does not change from 0.78 in Fig. 16(b) for a decreasing energy order because the concentration profile in the first energy step (220 keV) determines the tail of the resultant five-fold implantation. On the other hand, R decreases to 0.60 for an increasing energy order, meaning that the amount of channeling during the final energy step (220 keV) increases from that during the single-energy implantation at 220 keV. This result indicates that partially amorphized SiC causes the scatter-in channeling.

In the case of aluminum implantations with the 35-nm-thick SiO_2 layer [Fig. 18(b)], BCA simulation exhibits a tendency opposite to that observed in Fig. 18(a). The concentration profile of aluminum implantations performed in increasing order of energy results in a slightly less extended tail compared to that of aluminum implantations performed in decreasing order of energy. In the case of increasing order of implantation energy, R increases to 0.57, and R for the decreasing energy order does not change from 0.45 in Fig. 16(b). However, the former value is still lower than R = 0.78 for the implantation without SiO_2 in a decreasing energy order [Fig. 18(b)]. This result indicates that although the amorphization-suppressed channeling (Ottaviani et al., 1999) occurred for implantations without SiO_2, the influence of the amorphization-induced scatter-in channeling is much larger.

When the two aluminum concentration profiles in Fig. 18(a) are compared vertically at a certain depth, the influence of channeling discussed above seems to be small. However, when they are compared horizontally, in the case of drift-layer doping in the order of 10^{15} cm^{-3}, the position of p-n junctions has about 10% error. The results in Fig. 18(a) also suggest that in the case of implantations without the SiO_2 layer in decreasing order of energy, concentration profiles of multiple-energy implantations can be obtained by simple summation of concentration profiles of single-energy implantations.

4. Conclusion and Future Studies

Diffusion and segregation of p-type dopants in 4H-SiC have been one-dimensionally modeled. Future studies should be directed toward two-dimensional models, which are challenging because of the great increase in the amount of data. The first two-dimensional model of aluminum-ion implantation into 4H-SiC (Mochizuki, K. & Yokoyama, N. 2011a and 2011b) is based on Monte-Carlo simulation, which revealed that iso-concentration contours of aluminum are independent of the orientation of the masking edge, as long as the aluminum dose is moderate (10^{11} to 10^{13} cm^{-2}). Lateral range straggling can be extracted by expressing the lateral-concentration profiles as a one-dimensional dual-Pearson-distribution function multiplied by a Gaussain distribution function. Such a two-dimensional model should contribute to efficiently simulating the current-voltage characteristics of 4H-SiC power devices.

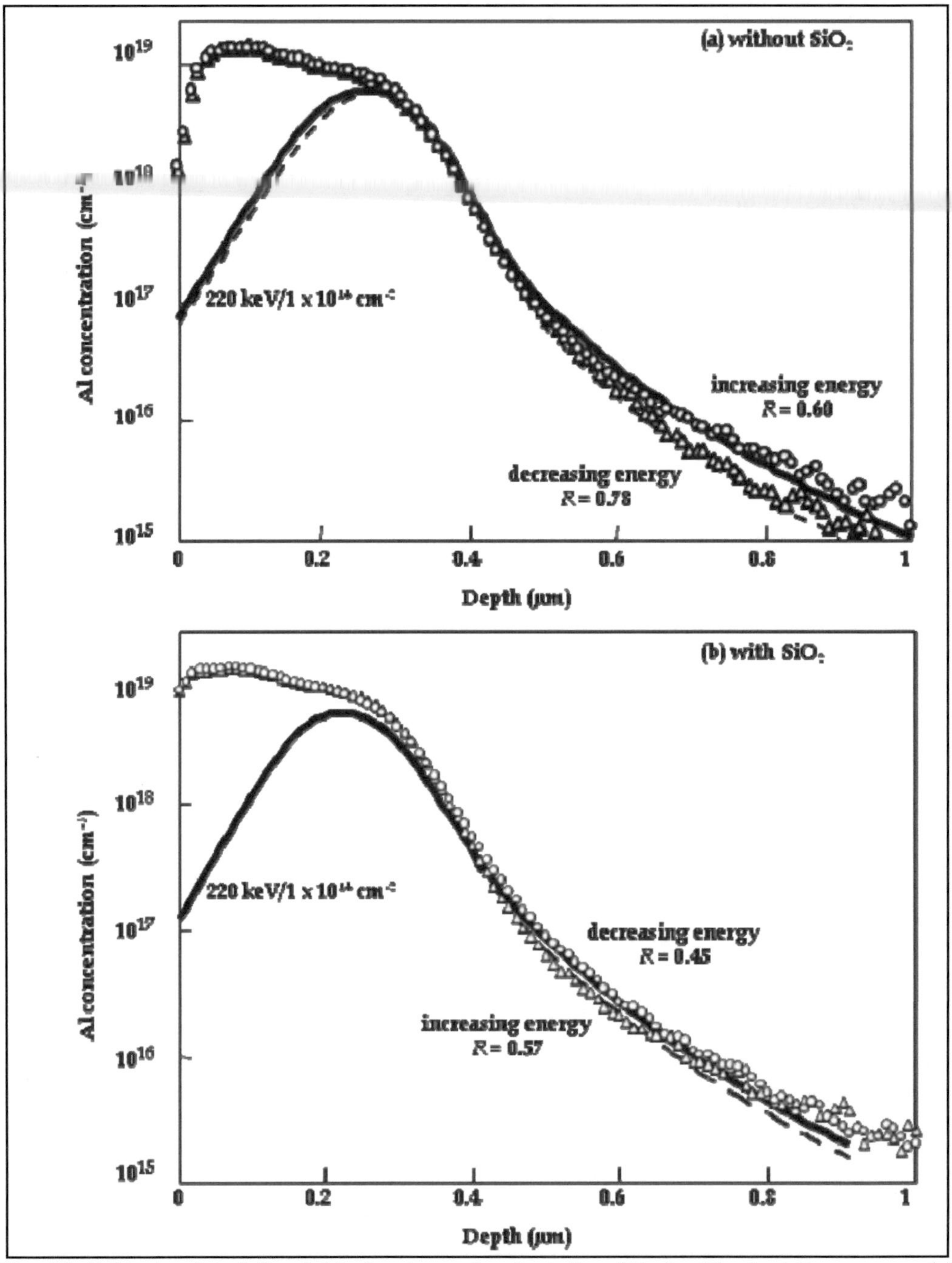

Fig. 18. BCA-simulated five-fold aluminum implantation fitted with dual-Pearson model with dose ratio of 220-keV implantation as parameter

5. References

Aradi, B., Gali, A., Deák, P., Rauls, E., Frauenheim, Th. & Son, N.T. (2001). Boron centers in 4H-SiC. *Materials Science Forum,* Vol. 353-356, (2001) 455-458, 0255-5476

Baliga, B. J. (2005). *Silicon Carbide Power Devices,* World Scientific, 978-981-256-605-8, Singapore

Bockstedte, M., Mattausch, A. & Pankratov, O. (2001). Boron in SiC: structure and kinetics. *Materials Science Forum,* Vol. 353-356, (2001) 447-450, 0255-5476

Bockstedte, M., Mattausch, A. & Pankratov, O. (2003). Ab initio study of the migration of intrinsic defects in 3C-SiC. *Physical Review B,* Vol. 68, No. 20 (Nov. 2003) 205201, 1098-0121

Bockstedte, M., Mattausch, A. & Pankratov, O. (2004). Different roles of carbon and silicon interstitials in the interstitial-mediated boron diffusion in SiC. *Physical Review B,* Vol. 70, No. 11 (Sept. 2004) 205201, 1098-0121

Bracht, H., Stolwijl, N. A. & Pensl, G. (2000). Diffusion of boron in silicon carbide: evidence for the kick-out mechanism. *Applied Physics Letters,* Vol. 77, No. 20, (Nov. 2000) 3188-3120, 0003-6951

Bracht, H. (2007). Self- and foreign-atom diffusion in semiconductor isotope heterostructures. I. Continuum theoretical calculations. *Physical Review B,* Vol. 75, No. 3 (Jan. 2007) 035210, 1098-0121

Chakarov, I. & Temkin, M. (1999). Modelling of ion implantation in SiC crystals, *Nuclear Instruments and Methods in Physics Research, Section B: Beam Interactions with Materials and Atoms,* Vol. 242, No. 1-2, (2006) 690-692, 0168-583X

Cho, Y., Zographos, N., Thirupapulivur, S. & Moroz, V. (2007). Experimental and theoretical analysis of dopant diffusion and C evolution in hjigh-C Si :C epi layers: optimazation of Si :C source and drain formed by post-epi implant and activation anneal, *Technical Digest of International Electron Devices Meeting,* pp. 959-962, 978-1-4244-1507-6, Washington, D.C., Dec. 2007, IEEE, Piscataway

Duijin-Arnold, A. v., Ikoma, T., Poluektov, O. G., Baranov, P. G., Mokhov, E. N. & Schmidt, J. (1998). Electronic structure of the deep boron acceptor in boron-doped 6H-SiC. *Physical Review B,* Vol. 57, No. 3 (Jan. 1998) 1607-1619, 1098-0121

Duijin-Arnold, A. v., Mol., J., Verberk, R., Schmidt, J., Mokhov, E.N. & Baranov, P. G. (1999). Spatial distribution of the electronic wave function of the shallow boron acceptor in 4H- and 6H-SiC. *Physical Review B,* Vol. 60, No. 23 (Dec. 1999) 15829-15847, 1098-0121

Gao, F., Weber, W. J., Posselt, M. & Belko, V. (2004). Atomistic study of intrinsic defect migration in 3C-SiC. *Physical Review B,* Vol. 69, No. 24 (June 2004) 245205, 1098-0121

Heera, V., Panknin, D. & Skorupa, W. (2001). P-type doping of SiC by high dose Al implantation–Problems and progress. *Applied Surface Science,* Vol. 184, No. 1–4, (Dec. 2001) 307–316, 0169-4332

Hoshi, M., Hayashi, Y., Tanaka, H. & Yamagami, S. (2007). Novel SiC power devices utilizing a Si/4H-SiC heterojunction, *Proceedings of Power Conversion Conference,* pp. 373-376, 1-4244-0844-X, Nagoya, April 2007, IEEE, Piscataway

Janson, M.S., Linnarsson, M.K., Hallén, A. & Svensson, B.G. (2003a). Transient enhanced diffusion of implanted boron in 4H-silicon carbide. *Applied Physics Letters,* Vol. 76, No. 11, (March 2003) 1434-1436, 0003-6951

Janson, M.S., Linnarsson, M.K., Hallén, A. & Svensson, B.G. (2003b). Ion implantation range distributions in silicon carbide. *Journal of Applied Physics,* Vol. 93, No. 11, (June 2003) 8903-8909, 0021-8979

Kinchin, G.H. & Pease, R.S. (1955). The displacement of atoms in solids by radiation. *Reports on Progress in Physics,* Vol. 18, (1955) 1-51, 0034-4885

Kuroda, N., Shibahara, K., Yoo, W.S., Nishino, S. & Matsunami, H. (1987). Step-controlled VPE growth of SiC single crystals at low temperatures, *Extended Abstracts of 19th Conference on Solid State Devices and Materials,* pp. 227-230, Tokyo, 1987, Japan Society of Applied Physics, Tokyo

Lau, F. (1990). Modeling of polysilicon diffusion sources, *Technical Digest of International Electron Devices Meeting,* pp. 67-70, 0163-1918, San Francisco, Dec. 1990, IEEE, Piscataway

Lee, S.-S. & Park, S.-G. (2002). Empirical depth profile model for ion implantation in 4H-SiC. *Journal of Korean Physical Society,* Vol. 41, No. 5, (Nov. 2002) L591-L593, 0374-4884

Linnarsson, M. K., Janson, M. S., Shoner, A. & Svensson, B.G. (2003). Aluminum and boron diffusion in 4H-SiC, *Materials Research Society Proceedings,* Vol. 742, paper K6.1, 1-55899-679-6, Warrendale, Dec. 2002, Materials Research Society, Boston

Linnarsson, M.K., Janson, M.S., Schnöer, A., Konstantinov, A. & Svensson, B.G. (2004). Boron diffusion in intrinsic, n-type amd p-type 4H-SiC. *Materials Science Forum,* Vol. 457-460, (2004) 917-920, 0255-5476

Linnarsson, M.K., Janson, M.S., Nordell, N., Wong-Leung, J. & Schöner, A. (2006). Formation of precipitates in heavily boron doped 4H-SiC. *Applied Surface Science,* Vol. 252, (2006) 5316-5320, 0169-4332

Liu, C.-I., Windl, W., Borucki, L. & Lu, S. (2002). Ab initio modeling and experimental study of C-B interactions in Si. *Applied Physics Letters,* Vol. 80, No. 1, (Jan. 2002) 52-54, 0003-6951

Mochizuki, K. & Onose, H. (2007). Dual-Pearson approach to model ion-implanted Al concentration profiles for high-precision design of high-voltage 4H-SiC power devices, *Technical Digest of International Conference on Silicon Carbide and Related Materials,* pp. Fr15-Fr16 (late news), Otsu, Oct. 2007

Mochizuki, K., Someya, T., Takahama, T., Onose, H. & Yokoyama, N. (2008). Detailed analysis and precise modelling of multiple-energy Al implantations through SiO_2 layers into 4H-SiC. *IEEE Transactions on Electron Devices,* Vol. 55, No. 8, (Aug. 2008) 1997-2003, 0018-9383

Mochizuki, K., Shimizu, H. & Yokoyama, N. (2009). Dual-sublattice modeling and semi-atomistic simulation of boron diffusion in 4H-slicon carbide. *Japanese Journal of Applied Physics,* Vol. 48, No. 3, (March 2009) 031205, 021-4922

Mochizuki, K., Shimizu, H. & Yokoyama, N. (2010). Modeling of boron diffusion and segregation in poly-Si/4H-SiC structures. *Materials Science Forum,* Vol. 645-648, (2010) 243-246, 0255-5476

Mochizuki, K. & Yokoyama, N. (2011a). Two-dimensional modelling of aluminum-ion implantation into 4H-SiC. To be published in *Materials Science Forum;* presented at *European Conference on Silicon Carbide and Related Materials,* paper WeP-47, Oslo, Aug. 2010

Mochizuki, K. & Yokoyama, N. (2011b). Two-dimensional analytical model for concentration profiles of aluminium implanted into 4H-SiC (0001). To be published in *IEEE Transactions on Electron Devices,* Vol. 58, (2011), 0018-9383

Mokhov, E.N., Goncharov, E.E. & Ryabova, G.G. (1984). Diffusion of boron in p-type silicon carbide. *Soviet Physics - Semiconductors,* Vol. 18, (1984) 27-30, 0038-5700

Morris, S.J., Yang, S.-H., Lim, D.H., Park, C., Klein, K.M., Manassian, M. & Tasch, A.F. (1995). An accurate and efficient model for boron implants through thin oxide layers into single-crystal silicon. *IEEE Transactions on Semiconductor Manufacturing,* Vol. 8, No. 4, (Nov. 1995) 408-413, 0894-6507

Ottaviani, L., Morvan, E., Locatelli, M.-L.., Planson, D., Godignon, P., Chante, J.-P. & Senes, A. (1999). Aluminum multiple implantations in 6H-SiC at 300K. *Solid-State Electronics,* Vol. 43, No. 12, (Dec. 1999) 2215-2223, 0038-1101

Park, C., Klein, K., Tasch, A., Simonton, R. & Lux, G. (1991). Paradoxical boron profile broadening caused by implantation through a screen oxide layer, *Technical Digest of International Electron Devices Meeting,* pp. 67-70, 0-7803-0243-5, Washington, D.C., Dec. 1991, IEEE, Piscataway

Pearson, K. (1895). Contributions to the mathematical theory of evolution, II: skew variation in homogeneous material. *Philosophical Transactions of the Royal Society of London, A,* Vol. 186, (1895) 343-414, 0080-4614

Plummer, G. H., Deal, M. D. & Griffin, P. B. (2000). *Silicon VLSI Technology,* 411, Prentice Hall, 9780130850379, Upper Saddle River

Rausch, W.A., Lever, R.F. & Kastl, R.H. (1983). Diffusion of boron into polycrystalline silicon from a single crystal source. *Journal of Applied Physics,* Vol. 54, No. 8, (Aug. 1983) 4405-4407, 0021-8979

Rurali, R., Godignon, P., Rebello, J., Ordejón, P. & Hernández, E. (2002). Theoretical evidence for the kick-out mechanism for B diffusion in SiC. *Applied Physics Letters,* Vol. 81, No. 16, (Oct. 2002) 2989-2991, 0003-6951

Rüschenschmidt, K., Bracht, H., Stolwijk, N. A., Laube, M., Pensl, G. & Brandes, G. R. (2004). Self-diffusion in isotopically enriched silicon carbide and its correlation with dopant diffusion. *Journal of Applied Physics,* Vol. 96, No. 3, (Aug. 2004) 1458-1463, 0021-8979

Sadigh, B., Lenosky, T. J., Theiss, S. K., Caturla, M.-J., de la Rubia, T. D. & Foad, M. A. (1999). Mechanism of boron diffusion in silicon: an ab initio and kinetic Monte Carlo study. *Physical Review Letters,* Vol. 83, No. 21 (Nov. 1999) 4341-4344, 0031-9007

Srindhara, S. G., Clemen, L. L., Devaty, R. P., Choyke, W. J., Larkin, D. J., Kong, H. S., Troffer, T. & Pensl, G. (1998). Photoluminescence and transport studies of boron in 4H-SiC. *Journal of Applied Physics,* Vol. 83, No. 12, (Jan. 1998) 7909-7920, 0021-8979

Stewart, E.J., Carroll, M.S. & Sturm, J.C. (2005). Boron segregation in single-crystal $Si_{1-x-y}Ge_xC_y$ and $Si_{1-y}C_y$ alloys. *Journal of Electrochemical Society,* Vol. 152, (2005) G500, 0013-4651

Stief, R., Lucassen, M., Schork, R., Ryssel, H., Holzlein, K.-H., Rupp, R. & Stephani, D. (1998). Range studies of aluminum, boron, and nitrogen implants in 4H-SiC, *Proceedings of International Conference on Ion Implantation Technology,* pp. 760-763, 0-7803-4538-X, Kyoto, June 1998, IEEE, Piscataway

Suzuki, K., Sudo, R., Tada, Y., Tomotani, M., Feudel, T. & Fichtner, W. (1998). Comprehensive analytical expression for dose dependent ion-implanted impurity concentration profiles. *Solid-State Electronics,* Vol. 42, No. 9, (Sept., 1998) 1671-1678, 0038-1101

Tasch, A.F., Shin, H., Park, C., Alvis, J. & Novak, S. (1989). An improved approach to accurately model shallow B and BF_2 implants in silicon. *Journal of Electrochemical Society,* Vol. 136, No. 3, (1989) 810-814, 0013-4651

Tsirimpis, T., Krieger, M., Weber, H.B. & Pensl, G. (2010). Electrical activation of B^+-ions implanted into 4H-SiC. *Materials Science Forum,* Vol. 645-648, (2010) 697-700, 0255-5476

Windle, W., Bunea, M.M., Stumpf, R., Dunham, S.T. & Masquelier, M.P. (1999). First-principles study of boron diffusion in silicon. *Physical Review Letters,* Vol. 83, No. 21 (Nov. 1999) 4345-4348, 0031-900

3

Low temperature deposition of polycrystalline silicon carbide film using monomethylsilane gas

Hitoshi Habuka
Yokohama National University
Yokohama, Japan

1. Introduction

Silicon carbide (Greenwood and Earnshaw, 1997) has been widely used for various purposes, such as dummy wafers and reactor parts, in silicon semiconductor device production processes, due to its high purity and significantly small gas emission. In many other industries, silicon carbide has been used for coating various materials, such as carbon, in order to protect them from corrosive environment. Recently, many researchers have reported the stability of silicon carbide micro-electromechanical systems (MEMS) under corrosive conditions consisting of various chemical reagents (Mehregany et al., 2000; Stoldt et al., 2002; Rajan et al., 1999; Ashurst et al., 2004).

For producing silicon carbide film, chemical vapour deposition (CVD) is performed at the temperatures higher than 1500 K (Kimoto and Matsunami, 1994; Myers et al., 2005). Because such a high temperature is necessary, various materials having low melting point cannot be coated with silicon carbide film. Thus, the development of the low temperature silicon carbide CVD technique (Nakazawa and Suemitsu, 2000; Madapura et al., 1999) will extend and create enormous kinds of applications. For this purpose, the CVD technique using a reactive gas, such as monomethylsilane, is expected.

Here, the silicon carbide CVD using monomethylsilane gas (Habuka et al., 2007a; Habuka et al., 2009b; Habuka et al., 2010) is reviewed. In this article, first, the thermal decomposition behaviour of monomethylsilane gas is clarified. Next, the chemical reactions are designed in order to adjust the composition of silicon carbide film. Finally, silicon carbide film is obtained at low temperatures, and its stability is evaluated.

2. Reactor and process

The horizontal cold-wall CVD reactor shown in Figure 1 is used for obtaining a polycrystalline 3C-silicon carbide film. This reactor consists of a gas supply system, a quartz chamber and infrared lamps. The height and width of quartz chamber are 10 mm and 40 mm, respectively. A (100) silicon substrate, 30 x 40 mm, is placed on the bottom wall of the quartz chamber. The silicon substrate is heated by halogen lamps through the quartz chamber walls.

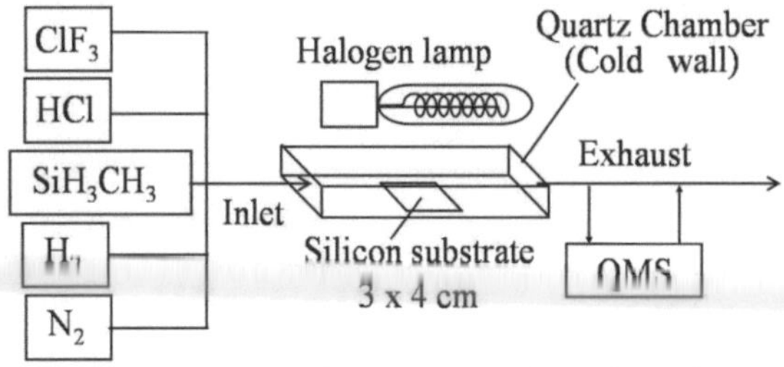

Fig. 1. Horizontal cold-wall CVD reactor for silicon carbide film deposition.

In this reactor, hydrogen gas, nitrogen gas, monomethylsilane gas, hydrogen chloride gas and chlorine trifluoride gas are used. Hydrogen is the carrier gas. It can remove the silicon oxide film and organic contamination presents at the silicon substrate surface. Hydrogen chloride gas is used for adjusting the ratio of silicon and carbon in the silicon carbide film. Throughout the deposition process, the hydrogen gas flow rate is 2 slm. Figures 2, 3 and 4 show the film deposition process, having Steps (A), (B), (C), (D) and (E).

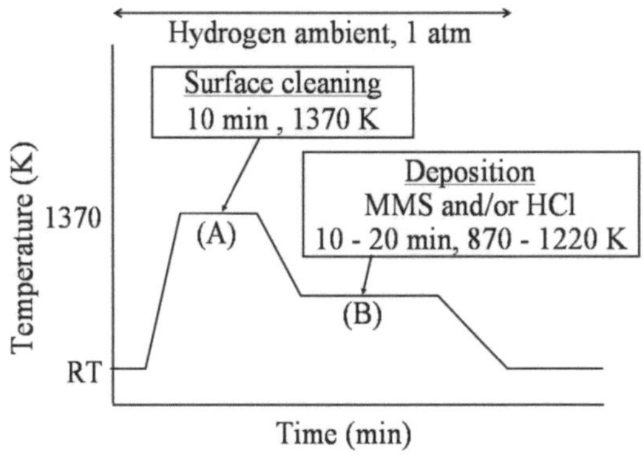

Fig. 2. Process of silicon carbide film deposition using gases of monomethylsilane, hydrogen chloride and hydrogen.

At Step (A), the silicon substrate surface is cleaned at 1370 K for 10 minutes in ambient hydrogen. Step (B) is the silicon carbide film deposition using monomethylsilane gas with or without hydrogen chloride gas at 870 - 1220 K. Step (C) is the annealing of the silicon carbide film in ambient hydrogen at 1270 K for 10 minutes.

In the process shown in Figure 2, Step (B) is performed after Step (A). In contrast to this, the process shown in Figure 3 involves first Step (A) and then the repetition of Steps (B) and (C). Figure 4 is the process for low temperature deposition and evaluation of the film, consisting of Steps (A), (D) and (E). Step (D) is the silicon carbide film deposition at low temperatures, room temperature - 1070 K, using a gas mixture of monomethylsilane and hydrogen chloride. At Step (E), the obtained film is exposed to hydrogen chloride gas at 1070 K for 10 minutes. Because hydrogen chloride gas can significantly etch silicon surface at 1070 K (Habuka et al., 2005) and does not etch silicon carbide surface, the stability of the obtained film is quickly evaluated by Step (E).

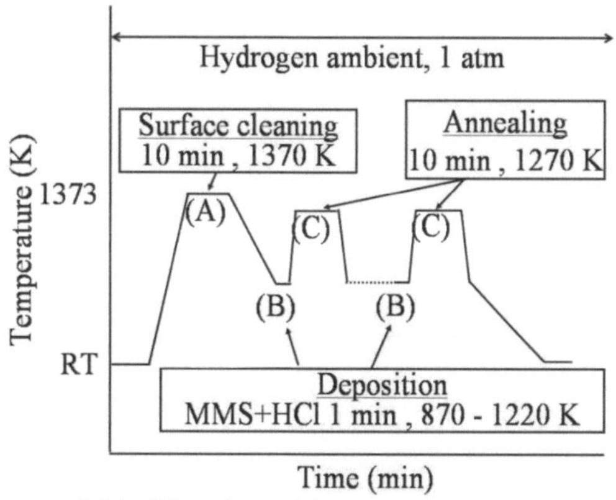

Fig. 3. Process of silicon carbide film deposition accompanying annealing step.

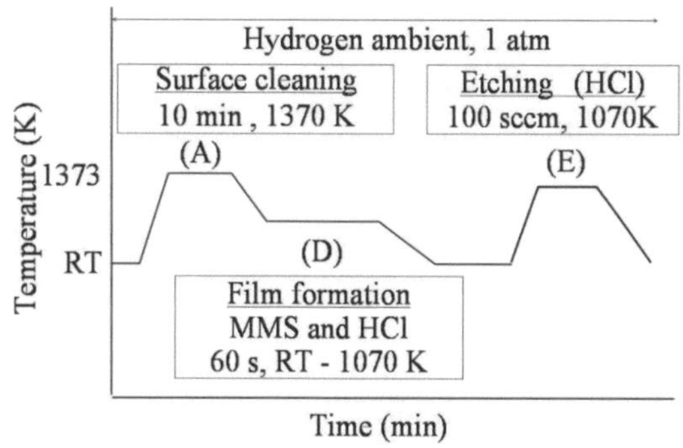

Fig. 4. Process of silicon carbide film deposition and etching.

The average thickness of the silicon carbide film is evaluated from the increase in the substrate weight. The surface morphology is observed using an optical microscope, a scanning electron microscope (SEM) and an atomic force microscope (AFM). Surface microroughness is evaluated by AFM. In order to observe the surface morphology and the film thickness, a transmission electron microscope (TEM) is used. The X-ray photoelectron spectra (XPS) reveal the chemical bonds of the silicon carbide film. Additionally, the infrared absorption spectra through the obtained film are measured.

In order to evaluate the gaseous species produced during the film deposition in the quartz chamber, a part of the exhaust gas from the reactor is fed to a quadrupole mass spectra (QMS) analyzer, as shown in Figure 1.

After finishing the film deposition, the quartz chamber is cleaned, using chlorine trifluoride gas (Kanto Denka Kogyo Co., Ltd., Tokyo, Japan) at the concentration of 10 % in ambient nitrogen at 670 - 770 K for 1 minute at atmospheric pressure.

3. Thermal decomposition of monomethylsilane

First, the thermal decomposition behavior of monomethylsilane gas is shown in order to choose and adjust the substrate temperature so that the silicon-carbon bond is maintained in the molecular structure during the silicon carbide film deposition.

Figure 5 shows the quadrupole mass spectra at the substrate temperatures of (a) 300 K, (b) 970 K, and (c) 1170 K. The concentration of monomethylsilane gas is 5% in ambient hydrogen at atmospheric pressure. The measured partial pressure is normalized using that of hydrogen molecule.

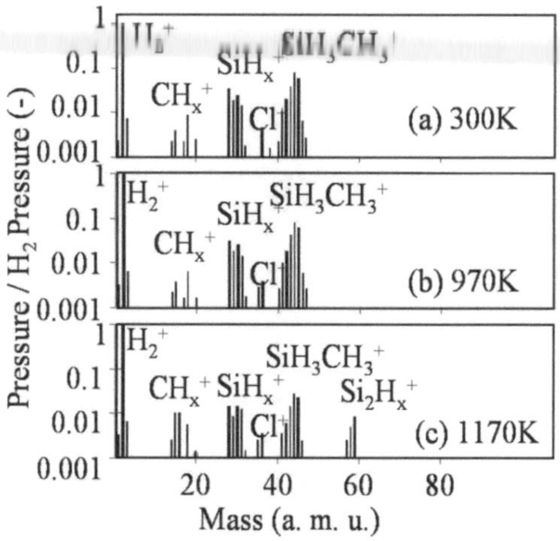

Fig. 5. Quadrupole mass spectra measured during silicon carbide film deposition at Step (B) in Figure 2. The substrate temperatures are (a) 300 K, (b) 970 K, and (c) 1170 K. The monomethylsilane concentration is 5%.

Figure 5 (a) shows the three major groups at masses greater than 12, 28 and 40 a. m. u., corresponding to CH_x^+, SiH_x^+ and $SiH_xCH_y^+$, respectively. Because no chemical reaction occurs at room temperature, CH_x^+ and SiH_x^+ are assigned to products due to the fragmentation in the mass analyzer. Cl^+ is detected, as shown in Figure 5 (a), because a very small amount of chlorine from the chlorine trifluoride, used for the *in situ* cleaning, remains in the reactor. Figure 5 (b) also shows that the three major groups of CH_x^+, SiH_x^+ and $SiH_xCH_y^+$ exist at 970 K without any significant change in their peak height compared with the spectrum in Figure 5 (a). Therefore, Figure 5 (b) indicates that the thermal decomposition of monomethylsilane gas is not significant at 970 K. However, at 1170 K, the partial pressure of the CH_x^+ group increases and that of the $SiH_xCH_y^+$ group significantly decreases, as shown in Figure 5 (c). Simultaneously, the $Si_2H_x^+$ group appears at a mass greater than 56. The appearance of $Si_2H_x^+$ is due to the formation of the silicon-silicon bond among SiH_x produced by the thermal decomposition of monomethylsilane.

4. Film deposition from monomethylsilane

From Figure 5, a substrate temperature lower than 970 K is expected to be suitable for suppressing the thermal decomposition of monomethylsilane gas. Therefore, the silicon carbide film deposition is performed at 950K following the process shown in Figure 2. Here, the monomethylsilane concentration is 5% in ambient hydrogen at the total flow rate of 2 slm. After the deposition, the chemical bond and the composition of the obtained film are evaluated using the XPS.

Figure 6 (a) and (b) show the XPS spectra of C 1s and Si 2p, respectively, of the film obtained from monomethylsilane gas. Because very large peaks due to the silicon-carbon bond exist near 282 eV and near 100 eV, most of the deposited film is shown to be silicon carbide. This coincides with the fact that the infrared absorption spectrum of this film showed a peak near 793 cm^{-1}, which corresponds to the silicon-carbon bond (Madapura et al., 1999).

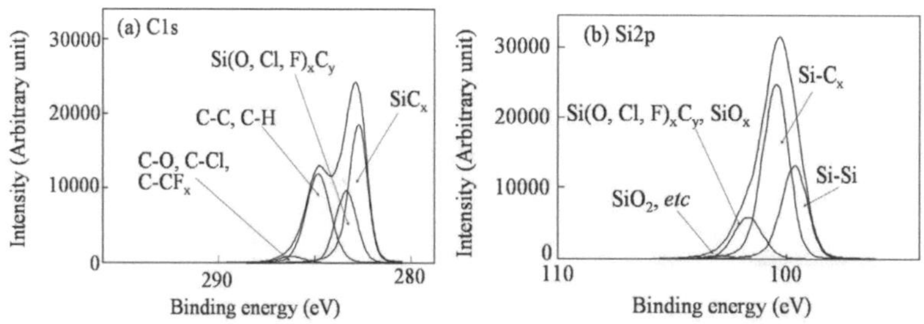

Fig. 6. XPS spectra of (a) C 1s and (b) Si 2p of silicon carbide film deposited at the monomethylsilane concentration of 5%, and at the substrate temperature of 950K.

In Figure 6, the peak corresponding to Si(O, Cl, F)$_x$C$_y$, SiO_x is detected. Because the gas mixture used for the film deposition do not include considerable amount of chlorine,and fluorine, and because the XPS measurements were performed *ex-situ*, the film surface oxidization may occur during its storage in air. This oxidation is attributed to monomethylsilane species remaining at the growth surface. The other peaks related to carbon are considered to be organic contamination on the film surface (Ishiwari et al., 2001). However, the existence of an XPS peak below 100 eV shows that this film includes a considerable amount of silicon-silicon bonds. The silicon-silicon bond can be formed due to the silicon deposition from the SiH_x produced in the gas phase. This indicates that the thermal decomposition of monomethylsilane gas in the gas phase at 950 K is not negligible, although it is significantly low at this temperature, as shown in Figure 5. Therefore, a method of reducing the excess silicon is necessary.

5. Film deposition from monomethylsilane and hydrogen chloride

Here, the method of reducing the excess silicon in the film is explained, adopting the process using hydrogen chloride gas shown in Figure 2.
Figure 7 shows the quadrupole mass spectrum measured during the silicon carbide film deposition using monomethylsilane gas and hydrogen chloride gas. The substrate temperature is 1090K, which is higher than 970 K used in the previous section. Because the higher temperature increases all the chemical reaction rates, any changes due to the addition of hydrogen chloride gas can be clearly recognized. At this temperature, a considerable number of silicon-carbon bonds can be maintained in monomethylsilane molecule, according to Figure 5 (c). Additionally, this temperature is near the optimum temperature for silicon carbide film growth using monomethylsilane gas, as reported by Liu and Sturm (Liu and Sturm, 1997). The gas concentrations of monomethylsilane and hydrogen chloride are 2.5% and 5%, respectively, in hydrogen gas at the flow rate of 2 slm. In Figure 7, the partial pressure of the various species is normalized using that of hydrogen molecule.

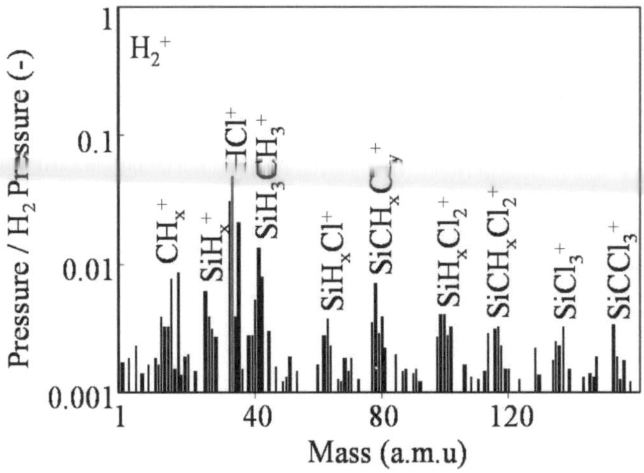

Fig. 7. Quadrupole mass spectra measured during silicon carbide film deposition by the process in Figure 2. The substrate temperature is 1090K. The monomethylsilane gas concentration is 2.3%. The hydrogen chloride gas concentration is 4.7%.

Figure 7 shows the $SiH_xCH_y^+$, CH_x^+, SiH_x^+ and HCl^+ groups, which are assigned to the monomethylsilane gas, its fragments and hydrogen chloride gas, respectively. In this figure, the $Si_2H_x^+$ group was not detected, unlike Figure 5. In addition to these, there are the chlorosilane groups (SiH_xCl_y) at masses over 63 (y=1), 98 (y=2) and 133 (y=3) and the chloromethylsilane group ($SiH_xCl_yCH_z$) at masses over 75 (y=1), 110 (y=2) and 145 (y=3). Therefore, the chlorination of monomethylsilane and silanes is concluded to occur in a monomethylsilane-hydrogen chloride system.

Figure 8 (a) shows the XPS spectra of C 1s of the obtained film. The carbon-silicon bond is clearly observed at 283 eV; its oxidized or chlorinated state, $Si(O, Cl, F)_xC_y$, also exists, as shown in this figure. The other peaks are related to the organic contamination on the film surface (Ishiwari et al., 2001). Figure 8 (b) shows the XPS spectra of Si 2p of the film obtained under the same conditions as those in the case of Figure 8 (a). Consistent with Figure 8 (a), Figure 8 (b) shows that the silicon-carbon bond and $Si(O, Cl, F)_xC_y$ bond exist on the film surface. Because the infrared absorption spectra through the obtained film showed a peak near 793 cm^{-1}, which corresponded to the silicon-carbon bond (Madapura et al., 1999), most of this film is determined to be silicon carbide. From a small number of silicon-oxygen bonds in Figure 8 (b), some of the silicon-carbon bonds in the remaining intermediate species show that it has oxidized during storage in air.

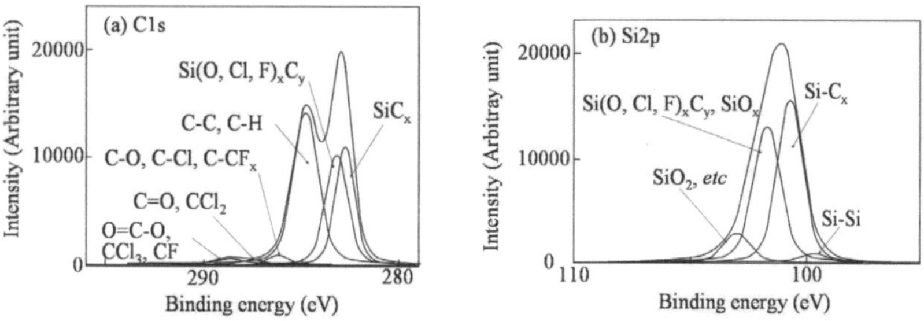

Fig. 8. XPS spectra of (a) C 1s and (b) Si 2p of silicon carbide film. The substrate temperature is 1090K. The monomethylsilane gas concentration is 2.3%. The hydrogen chloride gas concentration is 4.7%.

The most important information obtained from Figures 8 (a) and (b) is that the amount of silicon-silicon bonds are reduced at 1090 K, which is higher than that in Figure 6; many carbon-carbon bonds exist at the film surface. Therefore, this result shows that the hydrogen chloride plays a significant role in reducing the amount of excess silicon.

6. Chemical reaction in monomethylsilane and hydrogen chloride system

On the basis of the information obtained from Figures 5 - 8, the chemical reactions in the gas phase and at the substrate surface can be described as shown in Figure 9 and in Eqs. (1) - (9).

Thermal decomposition of SiH_3CH_3:

$$SiH_3CH_3 \rightarrow SiH_3+CH_3 \tag{1}$$

Si_2H_6 production:

$$2SiH_3 \rightarrow Si_2H_6 \tag{2}$$

Si production:

$$SiH_3 \rightarrow Si + (3/2)H_2 \tag{3}$$

Si production:

$$Si_2H_6 \rightarrow 2Si +3H_2 \tag{4}$$

Si etching (Habuka et al., 2005):

$$Si+3HCl \rightarrow SiHCl_3 +H_2 \tag{5}$$

Chlorination of SiH_3:

$$SiH_3+3HCl \rightarrow SiHCl_3+ (5/2)H_2 \tag{6}$$

Chlorination of SiH_3CH_3:

$$SiH_3CH_3 + 3HCl \rightarrow SiCl_3CH_3 + 3H_2 \quad (7)$$

Chlorination of Si_2H_6:

$$Si_2H_6 + 6HCl \rightarrow 2SiHCl_3 + 5H_2 \quad (8)$$

Silicon carbide production:

$$SiH_3CH_3 \rightarrow SiC + 3H_2 \quad (9)$$

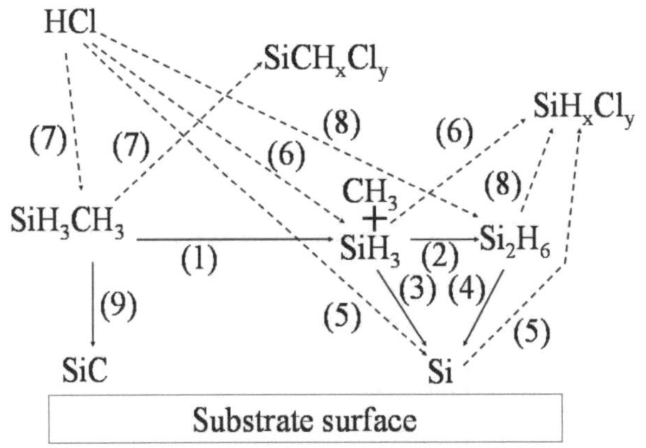

Fig. 9. Chemical process of silicon carbide film deposition using monomethylsilane gas and hydrogen chloride gas. (*i*) is the equation number.

In these chemical reactions, a small amount of monomethylsilane gas is thermally decomposed to form SiH_3, as shown by Eq. (1). SiH_3 forms silicon-silicon chemical bonds with each other to produce Si_2H_6 following Eq. (2). Both SiH_3 and Si_2H_6 can produce silicon in the gas phase and at the substrate surface, following Eqs. (3) and (4), respectively.

One of the possible origins of chlorosilanes, as shown in Figure 7, is the etching of silicon at the substrate surface, as described in Eq. (5), because the silicon etch rate using hydrogen chloride is considerably high (Habuka et al., 2005). Another reason for the production of chlorosilanes is the chemical reaction of hydrogen chloride gas with SiH_3 and Si_2H_6 in the gas phase, as described in Eqs. (6) and (8), respectively. Because chloromethylsilanes are simultaneously detected, monomethylsilane reacts with hydrogen chloride, as shown in Eq. (7). In addition to these reactions, silicon carbide is produced by the chemical reaction in Eq. (9).

The chemical reactions, Eqs. (1) - (8), can affect the film composition. Si_2H_x is very easily decomposed to produce silicon clusters in the gas phase and on the substrate surface, in Eq. (4). However, the formation of Si_2H_6 is suppressed by means of the production of $SiHCl_3$ from SiH_3, in Eq. (6), immediately after the SiH_3 formation. Therefore, the number of silicon clusters produced in the gas phase is reduced by adding the hydrogen chloride gas; this change can affect the composition of the film.

Here, the composition of the film measured by XPS shows that the film surface formed without using hydrogen chloride gas has greater silicon content than that of carbon, as shown in Figure 6. In contrast, the film surface obtained using hydrogen chloride gas has a smaller silicon content than that of carbon, as shown in Figure 8. This result shows that hydrogen chloride gas can reduce the excess silicon on the film surface; the film composition can be adjusted by changing the ratio of hydrogen chloride gas to monomethylsilane gas.

7. Film thickness

Figure 10 shows the relationship between the silicon carbide film thickness and the deposition time, using the process shown in Figure 2, at the substrate temperature of 1070 K. As shown in this figure, the film thickness is maintained at around 0.14 μm from 1 to 30 minutes. This shows that the film deposition stops within 1 minute. This coincides with those obtained by Ikoma et al. (Ikoma., 1999) and Boo et al. (Boo et al, 1999) using monomethylsilane gas.

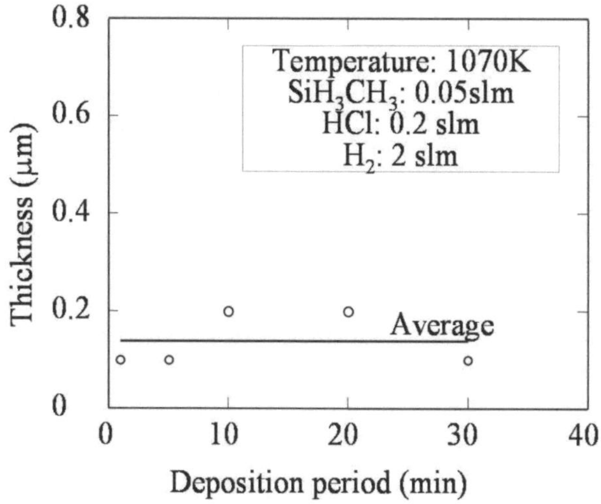

Fig. 10. Relationship between silicon carbide film thickness and deposition period, at the substrate temperature of 1070 K. The flow rate of monomethylsilane and hydrogen chloride is 0.05 slm and 0.2 slm, respectively, in hydrogen gas of 2 slm.

When the deposition stopped, the surface is assumed to have a major amount of carbon terminated with hydrogen. This assumption is consistent with the following results:

(1) The bonding energy between carbon and hydrogen is much higher than that of other chemical bonds among silicon, hydrogen and chlorine (Kagaku Binran, 1984).
(2) Hydrogen bonded with carbon remains at temperatures less than 1270 K (Nakazawa and Suemitsu, 2000).
(3) The silicon-hydrogen and silicon-chlorine chemical bonds cannot perfectly terminate the surface to stop the film deposition, because the silicon epitaxial film growth can continue in a chlorosilane-hydrogen system at 1070 K (Habuka et al., 1996).

In order to remove the hydrogen atoms bonded with carbon at the surface, high-temperature annealing is convenient. Using the process shown in Figure 3, the substrate is heated at 1270 K for 10 minutes, Step (C), before and after the film deposition at 1070 K. Here, the film deposition period in each step is 1 minute.

Figure 11 shows the thickness of silicon carbide film obtained by the process employing Step (C), between the film deposition steps, as shown in Figure 3. The flow rates of hydrogen gas and hydrogen chloride gas are fixed to 2 slm and 0.2 slm, respectively. The flow rate of monomethylsilane gas is 0.05 and 0.1 slm. The film deposition period at each step is 1 minute. The film thickness increases with the increasing flow rate of monomethylsilane gas. Simultaneously, the film thickness is increased by repeating the deposition and annealing. The thickness of the obtained film is greater than 2 μm with the total deposition period of 4 minutes.

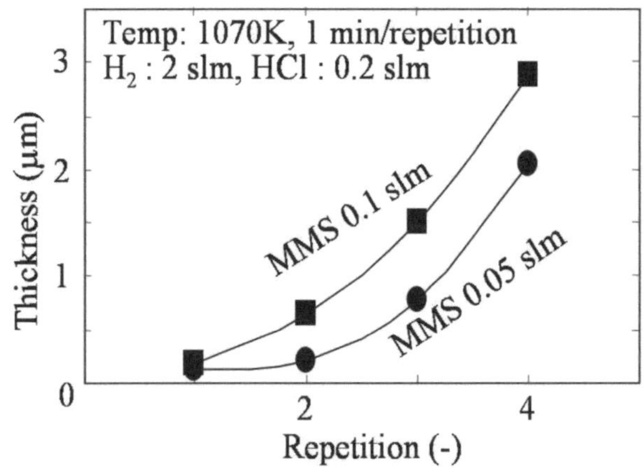

Fig. 11. Silicon carbide film thickness increasing with the repetition of the deposition using monomethylsilane gas with hydrogen chloride gas (Step (B)) at 1070 K and the annealing at 1270 K (Step (C)). The flow rate of monomethylsilane gas is 0.05 slm and 0.1 slm. The flow rate of hydrogen chloride and hydrogen is 0.2 slm and 2 slm, respectively.

Figure 12 shows the infrared spectra of the films corresponding to those at the monomethylsilane gas flow rate of 0.05 slm in Figure 11. The numbers in this figure indicate the number of repetitions of Steps (B) and (C) in Figure 3. Although these spectra are very noisy, a change in the transmittance clearly appears at the silicon carbide reststrahl band (700 - 900 cm^{-1}) (MacMillan et al., 1996). With the increasing number of repetitions of Steps (B) and (C), the transmittance near 793 cm^{-1} of 3C-silicon carbide (Madapura et al., 1999) significantly decreases while maintaining the wave-number having a very wide absorption bandwidth. Therefore, the thick film obtained by the process shown in Figure 3 is polycrystalline 3C-silicon carbide.

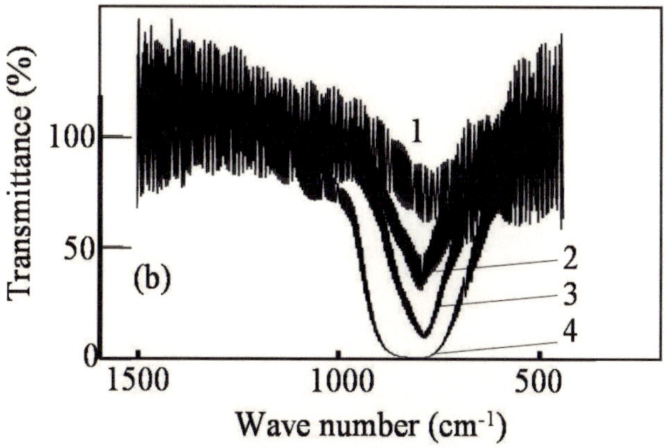

Fig. 12. Infrared absorption spectra of silicon carbide film after repeatedly supplying gas mixture of monomethylsilane and hydrogen chloride for 1 min at 1070 K (Step (B)) and annealing at 1270 K for 10 min (Step (C)). The flow rates of monomethylsilane and hydrogen chloride are 0.05 slm and 0.2 slm, respectively, in hydrogen gas of 2 slm.

Figures 13 and 14 show the surface of the film obtained at 1070K, corresponding to 4 repetitions of Steps (B) and (C) in Figures 11 and 12. The substrate surface is covered with the film having small grains, and it has neither porous nor needle-like appearance.

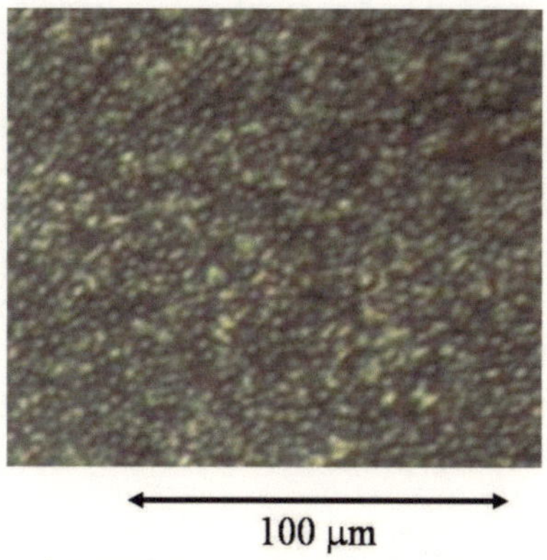

Fig. 13. Surface morphology of the silicon carbide film after four repetitions of Steps (B) and (C), observed using optical microscope. The condition of silicon carbide film is the same as that in Figure 12.

Figure 15 shows the morphology of the film surface which is obtained after (R1) one, (R2) two, (R3) three and (R4) four repetitions of Steps (B) and (C). At the deposition, substrate temperature is 1070 K; the flow rate of monomethylsilane gas is 0.05 slm. The flow rate of hydrogen chloride and hydrogen is 0.2 slm and 2 slm, respectively. With increasing the repetitions, the film surface tends to be slightly rough, and shows very small grains. However, no significant roughening is recognized to occur.

When the film deposition is governed by particles formed in the gas phase, the film deposition can continue as long as the monomethylsilane gas is supplied. However, the film deposition saturated. Therefore, the film having the small grain appearance is concluded to be formed dominantly by the surface process. Additionally, it is noted here that the roughening of silicon substrate surface due to etching by hydrogen chloride is not significant, because the film surface can be covered with silicon carbide, immediately after initiating the film deposition.

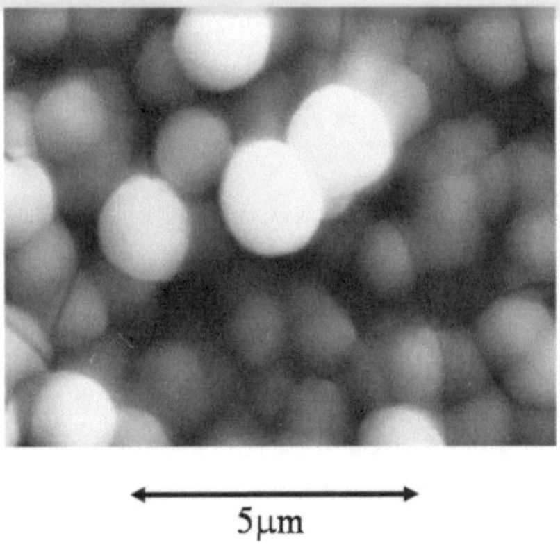

Fig. 14. SEM image of the film surface after four repetitions of Steps (B) and (C). The condition of silicon carbide film is the same as that in Figure 12.

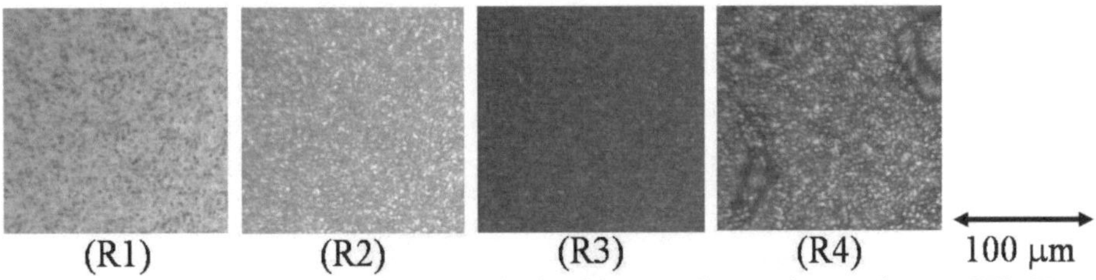

Fig. 15. Photograph of the silicon carbide film surface, obtained at 1070 K and at monomethylsilane gas flow rate of 0.05 slm. The flow rate of hydrogen chloride and hydrogen is 0.2 slm and 2 slm, respectively. (R1), (R2), (R3) and (R4) are obtained after one, two, three and four repetitions, respectively, of Steps (B) and (C) in Figure 3.

9. Surface chemical process: stop and restart deposition

The surface chemical process is discussed in relation to stopping and restarting the silicon carbide film growth.

The silicon carbide film deposition starts at the silicon substrate surface, as shown in Figure 16 (*i*). During Step (B) in Figure 3, silicon carbide film is formed, as shown in Figure 16 (*ii*). However, because the carbon-hydrogen bond tends to remain (Nakazawa and Suemitsu, 2000; Yoon et al., 2000), carbon at the surface can be terminated with hydrogen, as shown in Figure 16 (*iii*).

The hydrogen terminating the silicon carbide film surface is removed by means of high temperature annealing, as shown in Figure 16 (*iv*). Here, the bare silicon carbide surface can be formed; the process can return to the surface shown in Figure 16 (*ii*), at which Step (B) is possible.

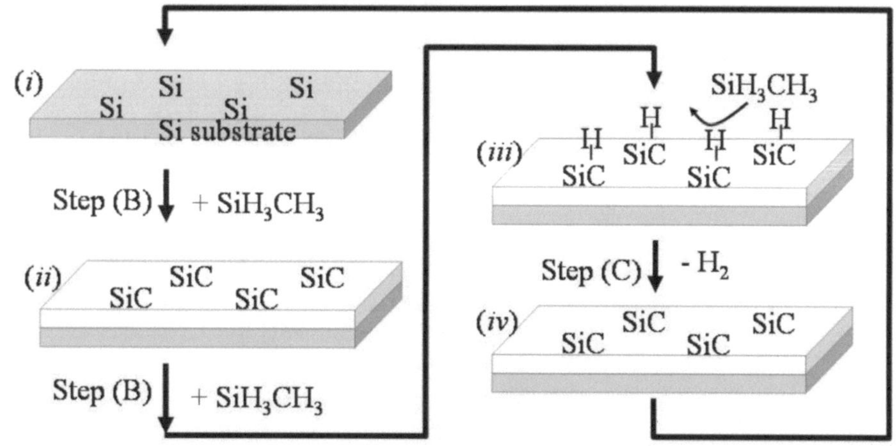

Fig. 16. Chemical process for silicon carbide film formation from monomethylsilane gas. (*i*): silicon substrate, (*ii*): silicon carbide deposition using monomethylsilane gas, (*iii*): surface termination by hydrogen, and (*iv*) desorption of hydrogen.

The effective method to increase the film thickness, other than the repetition of Steps (B) and (C), is to increase the growth rate at Step (B), while the hydrogen-terminated surface is built. Figure 11 shows that the obtained film thickness at the monomethylsilane gas flow rate of 0.1 slm is greater than that at 0.05 slm. Thus, the silicon carbide film growth rate increases with the monomethylsilane gas concentration.

10. Hydrogen chloride gas flow rate

The silicon carbide film thickness at various gas compositions of monomethylsilane and hydrogen chloride for 5 minutes at 1070 K is shown in Figure 17. The hydrogen gas flow rate is 2 slm; the hydrogen chloride gas flow rate is 0.1 slm (circle), 0.15 slm (square) and 0.2 slm (triangle).

In Figure 17, the film thickness entirely decreases with the increasing hydrogen chloride gas flow rate. The square and triangle show that the silicon carbide film thickness is very small but it gradually increases with the increasing monomethylsilane gas flow rate between 0.05 and 0.2 slm. In contrast to this, the silicon carbide film thickness obtained at the hydrogen chloride gas flow rate of 0.1 slm, indicated by the circle, shows a significant increase at the monomethylsilane gas flow rate greater than 0.1 slm. Simultaneously, the surface appearance of the film having such a significant thickness increase becomes dark and very rough.

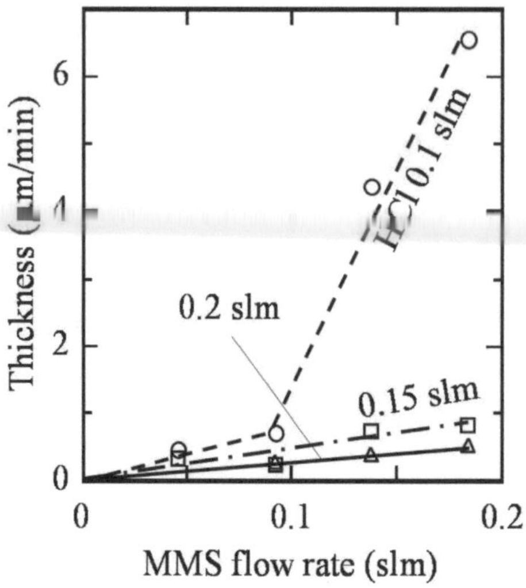

Fig. 17. Silicon carbide film thickness produced for 1 minute at 1070 K. Hydrogen chloride gas flow rate is 0.1 slm (circle), 0.15 slm (square) and 0.2 slm (triangle). Hydrogen gas flow rate is 2 slm.

Here, it should be noted that the silicon substrate surface was significantly etched by hydrogen chloride gas at its flow rate of 0.1 slm for 60 s, without monomethylsilane gas. This indicates that the silicon-silicon bond present at the film surface can be removed by hydrogen chloride gas. Thus, from these results, the amount of excess silicon in the silicon carbide film is decreased, however, the insufficient amount of hydrogen chloride gas can not sufficiently suppress the incorporation of excess silicon. From Figure 17, the amount of hydrogen chloride gas, comparable to or greater than that of the monomethylsilane gas, is necessary for the effective removal of the excess silicon. Because the hydrogen chloride flow rate larger than 0.15 slm is sufficient for the film formation at the monomethylsilane gas flow rate between 0.05 and 0.2 slm, the film thickness could linearly increase with the increasing monomethylsilane gas flow rate, as indicated using square and triangle in Figure 17.

11. Surface morphology

The surface morphology of the silicon carbide film is evaluated by the AFM, because some of the silicon carbide films obtained from monomethylsilane gas shows a mirror-like appearance by visual inspection. Figure 18 shows the AFM photograph of (a) silicon surface before the film formation, and (b) silicon carbide film surface with a thickness of 0.2 μm obtained at 1070 K for 5 min at the monomethylsilane gas flow rate of 0.092 slm and hydrogen chloride gas flow rate of 0.15 slm. The measured area was 0.2 x 0.2 μm.

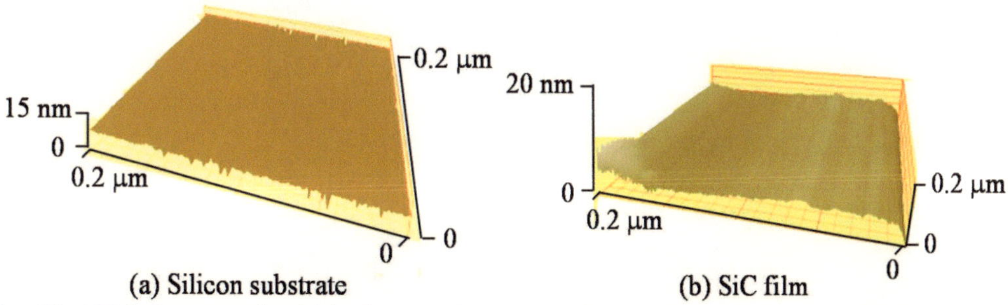

(a) Silicon substrate (b) SiC film

Fig. 18. AFM photograph of (a) silicon substrate surface and (b) silicon carbide film surface with the thickness of 0.2 μm obtained at 1070 K for 5 minutes at the monomethylsilane gas flow rate of 0.092 slm and hydrogen chloride gas flow rate of 0.15 slm. The Ra and RMS microroughness are 0.6 nm and 0.7 nm, respectively.

Figure 18 (a) shows that the silicon substrate surface before the film formation is very smooth with the average roughness (Ra) and the root-mean-square roughness (RMS) of 0.2 nm and 0.3 nm, respectively. After the silicon carbide film formation, the surface roughness slightly increases due to the formation of short hillocks, as shown in Figure 18 (b). However, its surface appearance is still specular by visual inspection. The Ra and RMS microroughness are 0.6 nm and 0.7 nm, respectively.

Some of the silicon carbide films obtained from monomethylsilane gas at 1070 K show a small grain-like surface, as shown in Figures 13, 14 and 15, but the other films often show a specular surface. Because the specular surface is expected to have a higher coating quality than that of a grain-like surface, the condition for obtaining the smooth surface with a high reproducibility should be studied in future.

12. Low temperature deposition

In this section, the low temperature silicon carbide film formation is described. For maintaining the gas condition in a series of film deposition, hydrogen chloride gas is introduced with monomethylsilane gas, even at room temperature, at which temperature hydrogen chloride gas hardly reacts with silicon. In the silicon carbide film formation for 60 seconds at various temperatures between 1070 K and room temperature following Steps (A) and (D) in Figure 4, the obtained film thickness was around 0.1 μm, and their surface often has a grain-like morphology, as shown in Figure 19 and a yellowish appearance indicating the existence of the silicon carbide film. Thus, the film formation at the lowest temperature, that is, at room temperature, is further explained.

The average film thickness obtained at room temperature, following Steps (A) and (D) in Figure 4, is 0.1 μm, which is comparable to the thickness obtained at 1070 K. In order to quickly evaluate the coating quality of the silicon carbide film, the film surface is further exposed to hydrogen chloride gas at 1070 K, following Step (E) in Figure 4. Because the film shows no decrease in weight and no change in its surface appearance, the film formed at room temperature is expected to be silicon carbide.

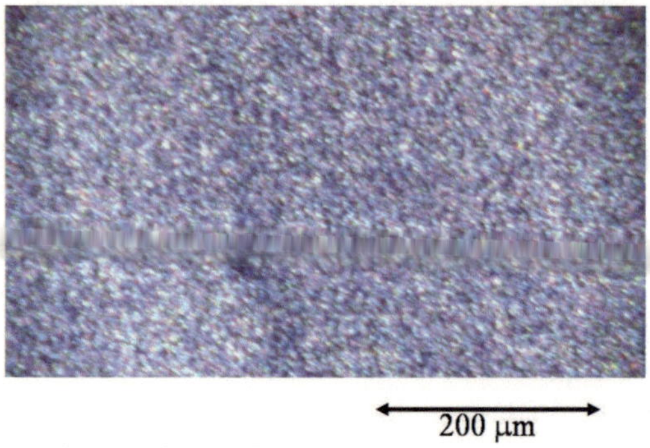

Fig. 19. Surface morphology of the film formed at room temperature for 60 s using monomethylsilane gas (0.092 slm) and hydrogen chloride gas (0.15 slm), immediately after the surface cleaning in ambient hydrogen at 1370 K for 10 min.

In order to show the necessary condition for the film formation at room temperature, monomethylsilane gas is supplied to silicon substrate skipping the silicon surface cleaning (Step (A)) of the process shown in Figure 4. This resulted in no weight increase to indicate no film formation; its surface was significantly etched by hydrogen chloride gas at 1070 K, by Step (E) in Figure 4. Thus, the surface cleaning in ambient hydrogen (Step (A)) takes an important role for the silicon carbide film formation from monomethylsilane gas at low temperatures.

In order to verify the silicon carbide film formation, Figure 20 shows the XPS spectra of C 1s of the 0.1 μm-thick deposited film which is obtained from monomethylsilane gas at room temperature, and further etched by hydrogen chloride gas at 1070 K for 10 minutes, following Steps (A), (D) and (E) shown in Figure 4. Figure 20 clearly shows the existence of the Si-C bond at 283 eV. Because silicon atom bonding with the carbon atom is consistently detected at 101 eV, the obtained film contains the Si-C bond.

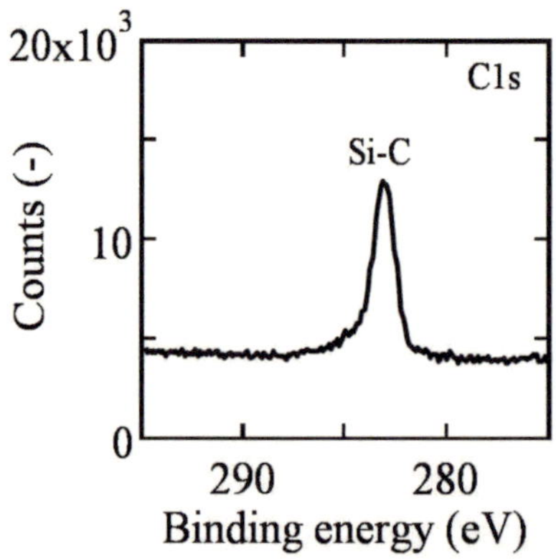

Fig. 20. XPS spectra of C1s of the silicon carbide film, obtained from monomethylsilane gas and hydrogen chloride gas on silicon surface at room temperature after annealing in hydrogen ambient. This film was further exposed to hydrogen chloride gas at 1070 K for 10 min before the XPS measurement.

13. Stability of film

In order to evaluate the stability of the silicon carbide film formed at room temperature, the obtained film is exposed to hydrogen chloride gas at 1070 K, following Step (C) in Figure 4; its surface is compared with that of a silicon substrate after exposed to hydrogen chloride gas.

Figure 21 shows SEM photograph of the silicon substrate surface and silicon carbide film. Figure 21 (a) is the silicon substrate surface after etching using hydrogen chloride gas at the flow rate of 0.1 slm diluted by hydrogen gas of 2 slm, at the substrate temperature of 1070 K for 10 min, without silicon carbide film formation. This figure shows the existence of many pits indicating the occurrence of etching by hydrogen chloride gas.

Figure 21 (b) shows the silicon carbide film surface formed using monomethylsilane gas of 0.069 slm at room temperature for 1 minute. This figure shows that there is no large pit at the film surface. Next, this surface is exposed to hydrogen chloride gas at the flow rate of 0.1 slm diluted in hydrogen gas of 2 slm, at 1070 K for 10 min. This condition is exactly the same as that performed for the silicon surface, shown in Fig, 21 (a). As shown in Figure 21 (c), a considerable morphology change is not observed at the deposited film surface, except of particles intentionally taken in order to clearly focus the surface for SEM observation.

Figure 22 is the TEM micrograph of the cross section of the silicon carbide film. The film, shown in this figure, was obtained from monomethylsilane gas and hydrogen chloride gas on silicon surface at room temperature after annealing at 1370 K in hydrogen ambient. This film was further exposed to hydrogen chloride gas at 1070 K for 10 min, before the TEM measurement.

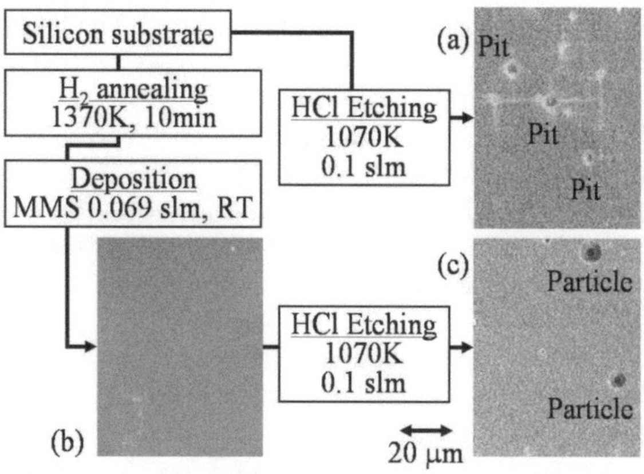

Fig. 21. Surface of (a) silicon substrate after etching using hydrogen chloride gas at 1070 K for 10 min, (b) deposited film using monomethylsilane gas of 0.069 slm at room temperature, and (c) the film of (b) further etched using hydrogen chloride gas at 1070 K for 10 min.

Figure 22 shows that the entire silicon substrate surface is sufficiently covered with the silicon carbide film consisted of arranged many grains, diameter of which is about 0.2 - 0.3 μm. The average film thickness in the observed area is about 0.3 μm. Additionally, there are no etch pit and pin-hole caused due to etching by hydrogen chloride gas at the silicon carbide-silicon interface. Thus, the silicon carbide film deposited at room temperature is stable in a hazardous ambient including hydrogen chloride gas.

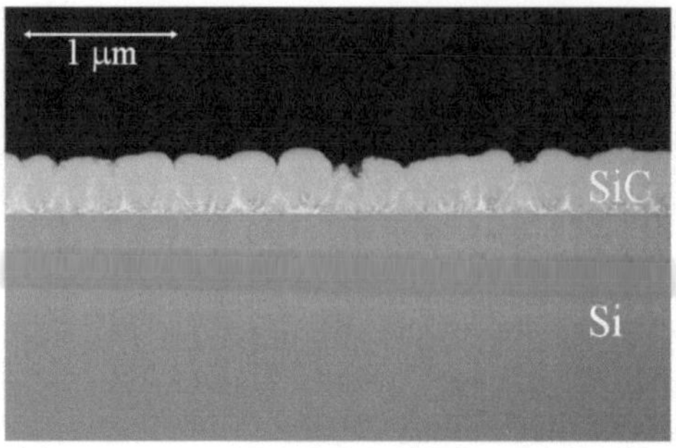

Fig. 22. TEM micrograph of the cross section of the silicon carbide film, shown in Figure 21 (c).

14. Film formation mechanism at room temperature

Based on the result that the surface cleaning in ambient hydrogen is necessary for producing the silicon carbide film, the surface chemical process for the low temperature silicon carbide formation using monomethylsilane gas is shown Figure 23.

The silicon carbide film formation is initiated by Step (*i*), as shown in Figure 23. At Step (*i*), monomethylsilane molecule approaches to silicon dimer present at hydrogen-terminated silicon surface. The silicon dimer is assumed to be broken in order to accept monomethylsilane molecule. Here, Step (*i*) is for an initiation of the surface chemical reaction; Steps (*ii*) and (*iii*) are for a repetition of the surface chemical reaction to produce multilayer film. After Step (*i*), Processes 1, 2 and 3, are expected to occur.

At Step (*ii*) in Process 1, silicon atom in monomethylsilane forms covalent bonds with silicon atom of the substrate. Here, hydrogen radicals are produced. The hydrogen radicals bond to the neighboring silicon atoms. At Step (*iii*) in Process 1, two hydrogen atoms are produced; dangling bonds remain at the neighboring silicon atoms.

At Step (*ii*) in Process 2, silicon atom in monomethylsilane forms covalent bonds with silicon atom of the substrate, similar to Process 1. Next, one of the hydrogen radicals produced can approach the hydrogen atom bonding with the carbon atoms in the chemisorbed monomethylsilane molecule. At Step (*iii*) in Process 2, two hydrogen atoms are produced; dangling bonds remain at the neighboring silicon atom and at the carbon atom in the monomethylsilane.

At Step (*ii*) in Process 3, silicon atom in monomethylsilane forms covalent bonds with silicon atom of the substrate, similar to Processes 1 and 2; one of the hydrogen radical produced can approach the hydrogen atom bonding with the silicon atom in the chemisorbed monomethylsilane molecule. At Step (*iii*) in Process 3, two hydrogen atoms are produced; dangling bonds remain at the neighboring silicon atom and at the silicon atom in the monomethylsilane. Because the dangling bonds formed after Step (iii) of Processes 1, 2 and 3, can accept more monomethylsilane molecules, chemisorption of monomethylsilane is expected to be spread and repeated over the substrate surface.

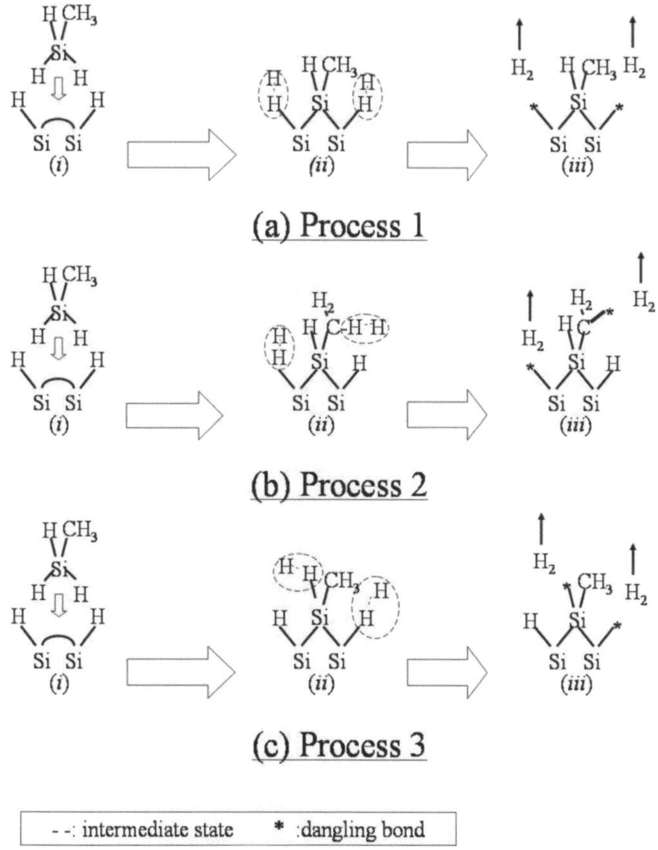

Fig. 23. Surface processes 1, 2 and 3 for low temperature silicon carbide film growth. (*i*) approach of monomethylsilane to silicon dimer at hydrogen-terminated silicon surface, (*ii*) chemisorption of monomethylsilane and production of hydrogen radicals, and (*iii*) production of hydrogen molecules, and dangling bonds.

When Process 2 is slower than Process 3, a larger amount of C-H bond remains at the film surface. Because this induces the C-H termination over the entire surface, the silicon carbide film formation finally stops.

Here, silicon dimer was reported to be very weak (Redondo and Goddard III, 1982); many research groups (Nakazawa and Suemitsu, 2000; Sutherland et al., 1997) reported the occurrence of the dissociative adsorption of organosilane on silicon dimer at room temperature. Additionally, Silvestrelli et al. (Silvestrelli et al., 2003) reported that SiH_2CH_3 can bond to silicon dimer, when monomethylsilane molecule vertically approached the surface. Taking into account these previous studies, the surface process, shown in Figure 23, is consistent with the results of the low temperature silicon carbide film formation and its saturation, using monomethylsilane gas.

15. Reactor cleaning using chlorine trifluoride gas

During the film deposition, the silicon carbide film is very often formed at various positions in the reactor other than the substrate. Particularly, the susceptor suffers significant deposition. When such a film becomes thick, small particles are produced from the film and they attach at the film surface. This behaviour causes the quality deterioration of film surface. Thus, the susceptor and the inner wall of the reactor should be cleaned after each deposition.

Because silicon carbide is very stable material, as shown in Figure 21 (c), the cleaning of the silicon carbide CVD reactor is quite difficult, except when using chlorine trifluoride gas (Habuka et al., 2009).

Figure 24 (a) shows the quartz chamber which has a thick dark-brown-coloured film formed at its inner surface. Because this thick film was formed from monomethylsilane gas at the substrate temperature of higher than 1000 K, it is a mixture of silicon carbide and silicon.

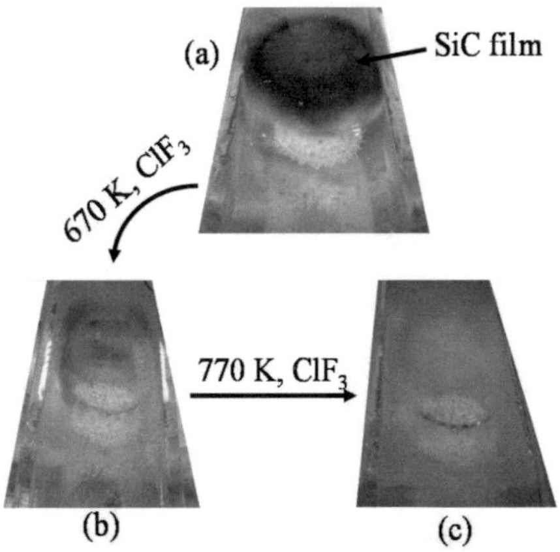

Fig. 24. Photograph of quartz chamber (a) after silicon carbide film deposition using monomethylsilane gas at high temperatures, (b) after cleaning using chlorine trifluoride gas at 10% in ambient nitrogen and at 670 K, and (c) after cleaning using chlorine trifluoride gas at 10% and 770 K.

Most of the deposited film is removed by chlorine trifluoride at its concentration of 10% gas at 670 K, as shown in Figure 24 (b), within 5 minutes, although very small amount of silicon carbide film remains. The remained film was removed again using chlorine trifluoride gas at 10 % and at 770 K, as shown in Figure 24 (c). Because very slight etching of quartz glass occurs, the cleaning condition has been discussed by Miura et al. (Miura et al., 2009).

16. Conclusions

The 3C-silicon carbide thin film is formed on silicon surface using monomethylsilane gas at the temperatures between room temperature and 1270 K. Although silicon, produced by thermal decomposition in gas phase and substrate surface, is incorporated into the silicon carbide film, it can be significantly reduced by means of the addition of hydrogen chloride gas. Although the silicon carbide film formation saturates within 1 minute due to the surface termination by C-H bonds, it can start again by means of annealing at 1270 K for removing hydrogen atoms. In order to develop the low-temperature silicon carbide film formation process, monomethylsilane gas is introduced to silicon substrate at room temperature. After the silicon surface is cleaned at 1370 K and cooled down in hydrogen ambient, monomethylsilane molecule can adsorb on the silicon surface to produce silicon carbide film, even at room temperature. Such the low temperature film formation is possible, because the hydrogen terminated silicon surface has silicon dimer. The silicon carbide film formed at room temperature is shown to be stable, because it can maintain after the etching using hydrogen chloride gas at 1070 K.

Acknowledgments

The studies written in this Chapter were performed with Dr. Yutaka Miura, Mr. Takashi Sekiguchi, Ms Satoko Kaneda, Mr. Mikiya Nishida, Ms. Mayuka Watanabe, Mr. Hiroshi Ohmori, Mr. Yusuke Ando of Yokohama National University.

17. References

Ashurst,W. R.; Wijesundara, M. B. J.; Carraro, C. and Maboudian, R. (2004) Tribological Impact of SiC Encapsulation of Released Polycrystalline Silicon Microstructures, *Tribology Lett*, 17, 195-198.

Boo, J. H.; Ustin, S. A. and Ho, W. (1999) Low-temperature epitaxial growth of cubic SiC thin films on Si(111) using supersonic molecular jet of single source precursors, *Thin Solid Films*, 343-344, 650-655.

Greenwood,N. N. and Earnshaw, A. (1997) Chemistry of the Elements, (Butterworth and Heinemann, Oxford).

Habuka, H.; Nagoya, T.; Mayusumi, M.; Katayama, M.; Shimada M. and Okuyama, K. (1996) Model on transport phenomena and epitaxial growth of silicon thin film in $SiHCl_3$---H_2 system under atmospheric pressure, *J. Cryst. Growth*, 169, 61-72.

Habuka, H.; Suzuki, T.; Yamamoto, S.; Nakamura, A.; Takeuchi, T. and Aihara, M. (2005) Dominant rate process of silicon surface etching by hydrogen chloride gas, *Thin Solid Films*, 489, 104-110.

Habuka, H.; Watanabe, M.; Miura, Y.; Nishida, M. and Sekiguchi, T. (2007a) Polycrystalline silicon carbide film deposition using monomethylsilane and hydrogen chloride gases, *J. Cryst. Growth*, 300 (2007) 374-381.

Habuka, H.; Watanabe, M.; Nishida, M. and Sekiguchi, T. (2007b) Polycrystalline Silicon Carbide Film Deposition Using Monomethylsilane and Hydrogen Chloride Gases, *Surf. Coat. Tech.*, 201, 8961-8965.

Habuka, H.; Tanaka, K.; Katsumi, Y.; Takechi, N.; Fukae, K. and Kato, T. (2009) Temperature-Dependent Behavior of 4H-Silicon Carbide Surface Morphology Etched Using Chlorine Trifluoride Gas, *J. Electrochem. Soc.*, 156, H971-H975.

Habuka, H.; Ohmori, H. and Ando, Y. (2010) Silicon Carbide Film Deposition at Low Temperatures Using Monomethylsilane Gas, *Surf. Coat. Tech.* 204, 1432-1437.

Ikoma, Y.; Endo, T.; Watanabe, F. and Motooka, T. (1999) Growth of Ultrathin Epitaxial 3C-SiC Films on Si(100) by Pulsed Supersonic Free Jets of CH_3SiH_3, *Jpn. J. Appl. Phys.*, 38, L301-303.

Ishiwari, S.; Kato, H. and Habuka, H. (2001) Development of Evaluation Method for Organic Contamination on Silicon Wafer Surfaces, *J. Electrochem. Soc.*, 148, G644-G648.

Kagaku Binran (Iwanami, Tokyo, 1984) 3rd ed. [in Japanese].

Kimoto, T. and Matsunami, H. (1994) Surface kinetics of adatoms in vapor phase epitaxial growth of SiC on 6H - SiC{0001} vicinal surfaces, *J. Appl. Phys.*, 75, 850-859.

Liu, C. W. and Sturm, J. C. (1997) Low temperature chemical vapor deposition growth of β-SiC on (100) Si using methylsilane and device characteristics, *J. Appl. Phys.*, 82, 4558-4567.

MacMillan, M. F.; Devaty, R. P.; Choyke, W. J. ; Goldstein, D. R.; Spanier, J. E. and Kurtz, A. D. (1996) Infrared reflectance of thick *p* - type porous SiC layers, J. Appl. Phys., 80 , 2412-2419.

Madapura, S.; Steckl, A. J. and Loboda, M. (1999) Heteroepitaxial Growth of SiC on Si(100) and (111) by Chemical Vapor Deposition Using Trimethylsilane, *J. Electrochem. Soc.*, 146, 1197-1202.

Mehregany, M.; Zorman, C. A.; Roy, S.; Fleischman, A. J.; Wu, C. H.; and Rajan, N. (2000) Silicon carbide for microelectromechanical systems, *Int. Mat. Rev.*,45,85-108.

Miura,Y. ; Kasahara, Y.; Habuka, H.; Takechi, N. and Fukae, K. (2009) Etching Rate of Silicon Dioxide Using Chlorine Trifluoride Gas, Jpn. J. Appl. Phys., 48, 026504.

Myers, R. L.; Shishkin, Y.; Kordina, O. and Saddow, S. E.; (2005) High growth rates (>30 μm/h) of 4H-SiC epitaxial layers using a horizontal hot-wall CVD reactor, *J. Cryst. Growth*, 285, 486-490

Nakazawa, H. and Suemitsu, M. (2000) Dissociative adsorption of monomethylsilane on Si(100) as revealed by comparative temperature-programmed desorption studies on H/, C_2H_2/, and MMS/Si(100), *Appl. Surf. Sci.* 162–163, 139-145.

Rajan, N.; Mehregany, M.; Zorman, C. A.; Stefanescu, S. and Kicher, T. P. (1999) Fabrication and testing of micromachined silicon carbide and nickel fuel atomizers for gas turbine engines, *J. Microelectromech. Sys.*, 8, 251-257

Redondo, A. and Goddard III, W. A. (1982) Electronic correlation and the Si(100) surface: Buckling versus nonbuckling, *J. Vac. Sci. Technol.*, 21, 344-350.

Sanchez, E. R. and Sibener, S. (2002) Low-Temperature Growth of Epitaxial β-SiC on Si(100) Using Supersonic Molecular Beams of Methylsilane, *J. Phys. Chem.* B, 106, 8019-8028.

Silvestrelli, P. L.; Sbraccia, C.; Romero, A. H. and Ancilotto, F. (2003) Dissociative chemisorption of methylsilane and methylchloride on the Si(1 0 0) surface from first principles, *Surf. Sci.*, 532, 957-962.

Stoldt, C. R.; Carraro, C.; Ashurst, W. R.; Gao, D.; Howe, R. T. and Maboudian, R. (2002) A low-temperature CVD process for silicon carbide MEMS, *Sensors and Actuators a-Physical*, 97-98, 410-415

Sutherland, D. G. J.; Terminello, L. J.; Carlisle, J. A.;Jimenez, I. ; Himpsel, F. J.; Baines, K. M.; Shuh, D. K. and Tong, W. M. (1997) The chemisorption of $H_2C[Si(CH_3)_3]_2$ and $Si_6(CH_3)_{12}$ on Si(100) surfaces, *J. Appl. Phys.*, 82, 3567-3571.

Yoon, H. G.; Boo, J. H.; Liu, W. L.; Lee, S. B.; Park, S. C. and Kang, H. (2000) *In situ* study of the formation of SiC thin films on Si(111) surfaces with 1,3-disilabutane: Adsorption properties and initial deposition characteristics, *J. Vac. Sci. Technol.* A18, 1464-1468.

4

Growth rate enhancement of silicon-carbide oxidation in thin oxide regime

Yasuto Hijikata and Hiroyuki Yaguchi
Saitama University
Japan

Sadafumi Yoshida
National Institute of Advanced Industrial Science and Technology
Japan

1. Introduction

Among wide band-gap semiconductor materials, only silicon carbide (SiC) can have SiO_2 layers, known as superior insulating films for metal-oxide-semiconductor (MOS) applications, on its surface by thermally oxidizing it, similarly to Si (2). In addition, its physical properties, such as high-breakdown electric field and high thermal conductivity, compared with Si, are suitable for high-speed switching and low-power-loss electronic devices. For these reasons, SiC metal-oxide-semiconductor field-effect transistors (MOSFETs) are expected to have superior specifications that cannot be obtained using Si. However, the electrical characteristics of SiC MOSFETs, such as on-resistance, are seriously poorer than those predicted from SiC bulk properties (1). It has been considered that these poor characteristics result from a high interface state density (3). Therefore, the clarification of the structure of SiC–oxide interfaces or the formation mechanism of the interface layer is one of the most important subjects to be studied to improve the electrical characteristics of SiC MOS devices.

In a previous work, we have performed real-time observation of SiC thermal oxidation using an *in-situ* ellipsometer (4). The results show that the oxidation-time dependence of oxide thickness can be represented using the Deal-Grove (D-G) model (5), which has been originally proposed for the explanation of Si oxidation. Song *et al.* (6) have modified the D-G model for application to SiC oxidation taking the presence of carbon into account. They have concluded that a linear-parabolic formula can also be applicable to SiC oxidation, although the parabolic term includes the contribution from the diffusion of CO or CO_2 molecules from the SiC–oxide interface to the surface as well as that of oxygen from the surface to the interface. On the other hand, it is well known that the oxidation behavior of Si cannot be explained using the D-G model, i.e., a simple linear-parabolic model, particularly at the initial oxidation stage. Accordingly, several models have been proposed for the explanation of Si oxidation (7–12).

In this work, we have studied 4*H*-SiC oxidation at the initial stage in more detail by performing *in-situ* spectroscopic ellipsometry and discussed the oxidation mechanism of SiC by comparing it with that of Si.

2. Experimental procedure

4*H*-SiC (000$\bar{1}$) C-face and (0001) Si-face epitaxial layers with 8° off-angles (*n*-type, $N_d - N_a =$ around 6×10^{15} cm^{-3}) were used in this study. After Radio Corporation of America (RCA) cleaning was executed, an epiwafer was placed into the *in-situ* spectroscopic ellipsometer and was heated at temperatures between 890 and 1220°C. The oxidation was performed also at oxygen partial pressures between 0.02 and 1.0 atm at 1150°C for C-face and between 0.1 and 1.0 atm at 1200°C for Si-face. Prior to the oxidation, ellipsometric parameters (Ψ, Δ) were measured in the argon atmosphere to determine the optical constants of SiC at the oxidation temperatures. Then, oxygen gas was introduced into the chamber and the oxidation time dependence of (Ψ, Δ) was monitored during dry oxidation. Spectroscopic ellipsometric measurements were carried out at wavelengths between 310 and 410 nm, and at an angle of incidence of 75.8°.

In the evaluation of oxide thickness from the ellipsometric measurements, we assumed three-layer structures, i.e., the sample is composed of a SiO_2 layer, an interface layer, and a SiC substrate. We have evaluated the thickness of SiO_2 layers under the assumption that the interface layer is 1 nm in thickness and has the optical constants obtained in the previous study (13). The details of the spectroscopic ellipsometric measurements and the evaluation of oxide thickness were described elsewhere (4; 14).

3. Results and Discussion

3.1 Oxidation rate at various oxidation temperatures

We have applied the D-G model to the relations between oxide thickness, X, and oxidation time, t, observed, and have obtained the values of the parameters B/A and B in the D-G equation (5) by fitting the calculated curve to the observed values in the entire thickness range. The fits are in general good at all of the oxidation temperatures, as reported previously (4). However, we found that, in the thickness range less than approximately 20 nm, there is a tendency for the observed values to be slightly larger than the calculated ones. To investigate these discrepancies in more detail, we have derived the oxidation rates dX/dt as a function of oxide thickness. Figure 1 shows the values of dX/dt of SiC(000$\bar{1}$) C-face as a function of oxide thickness at various oxidation temperatures. As shown in this figure, we have successfully obtained the values of the oxide growth rate even in the thin oxide thickness range of less than 10 nm by real-time *in-situ* observation. We have failed to fit the dX/dt against the X curves calculated using the D-G equation to the observed ones over the entire oxide thickness range measured at any oxidation temperature. However, as shown by the solid lines in Fig. 1, we can well fit the calculated curves to the observed ones when the thickness range of curve fitting is limited in the range thicker than approximately 20 nm. In this case, the figure clearly shows that the values of dX/dt in the thin thickness region are larger than the values given using the D-G equation at all the oxidation temperatures measured. These results suggest the existence of an oxidation process having a larger growth rate than that predicted by the D-G model.

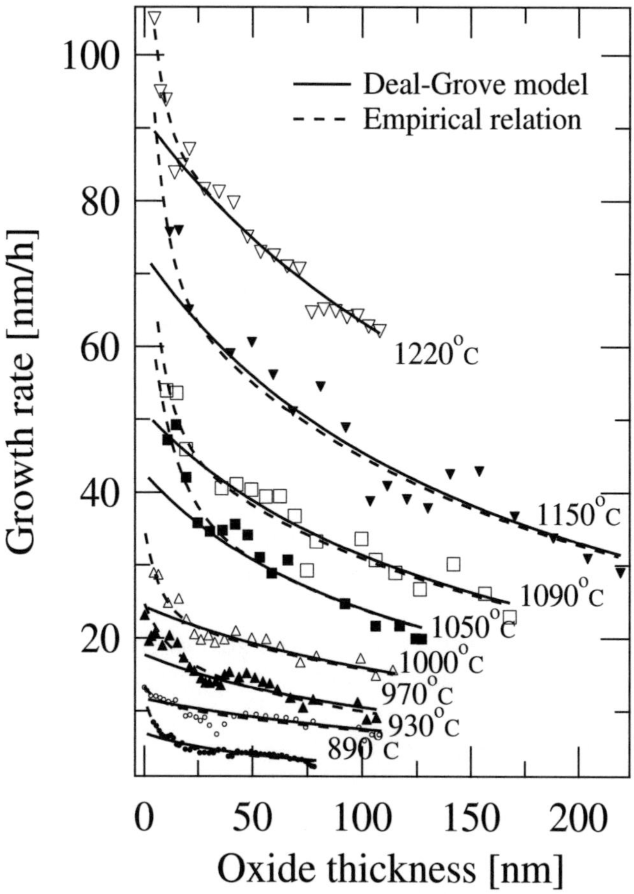

Fig. 1. Oxide thickness dependences of oxidation rate for SiC(000$\bar{1}$) C-face at various oxidation temperatures. The solid and dashed lines denote the values derived from the Deal-Grove model eq. (1) (5) and those from the empirical relation eq. (2) (8; 9), respectively.

Figure 2 shows the growth rate dX/dt of SiC(0001) Si-face as a function of the oxide thickness X at various oxidation temperatures (15). The figure shows that the values obtained are almost constant in the larger thickness range in this study at any oxidation temperature. The reason for this constant thickness dependence will be discussed later. While in the smaller thickness range, the values of dX/dt are not constant but increase with decreasing oxide thickness, i.e., the oxide growth rate enhancement occurs at any temperature in this study, as in the case of C-face.

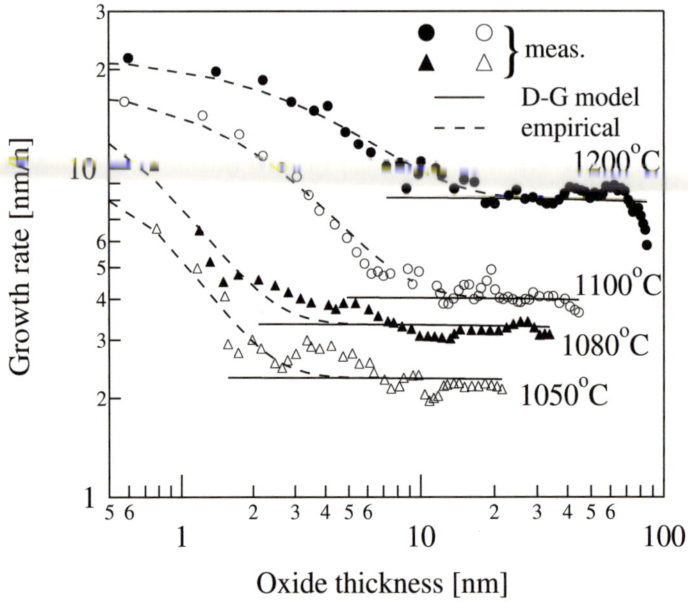

Fig. 2. Oxide thickness dependence of s growth rate for SiC (0001) Si-face at various oxidation temperatures (15). The solid and dashed lines denote the same values as those in Fig. 1.

3.2 Oxidation rate at various oxygen partial pressures

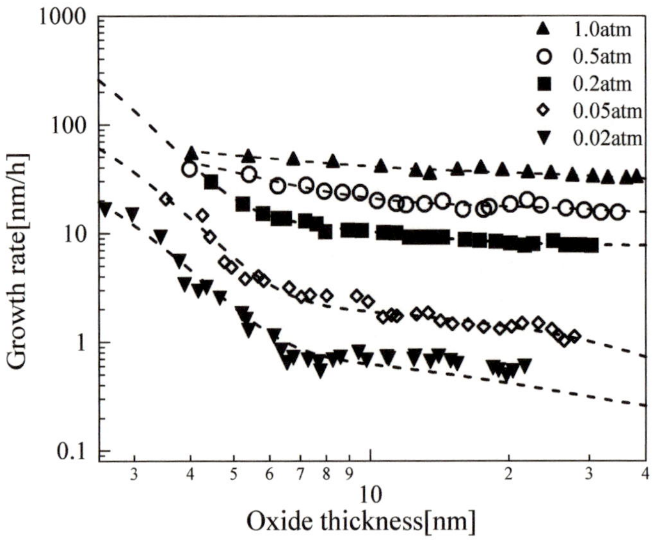

Fig. 3. Oxide growth rate of C-face as a function of oxide thickness at various oxygen partial pressures (17). The dashed curves denote the ones fitted to the experimental data using eq. (2).

Figure 3 shows the oxide growth rate dX/dt as a function of the oxide thickness for C-face at various oxygen partial pressures. Since the initial oxide growth rates at pressures higher than 0.2 atm was too fast, we could not accurately derive them, and thus the experimental data are not shown in this figure. As can be seen from this figure, the oxide growth rate slowly increases with decreasing oxide thickness in the thick oxide regime, and then, steeply increases with decreasing thickness. Figure 4 shows the oxide thickness dependence of dX/dt at various oxygen partial pressures for Si-face. This figure also shows that the oxide growth rate enhancement occurs at any partial pressure, as in the case of C-face.

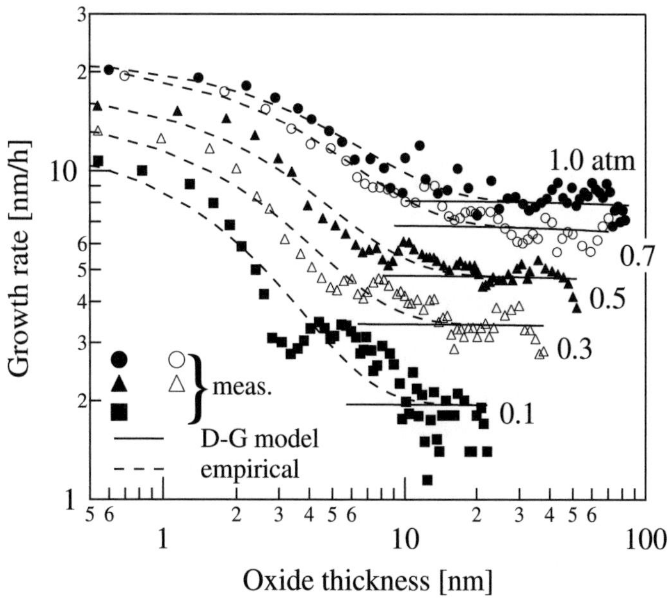

Fig. 4. Oxide thickness dependence of growth rate for Si-face at various oxygen partial pressures (15). The solid and dashed lines denote the same values as those in Fig. 1.

3.3 Massoud Empirical Relation

The kinetic model of Si oxidation that is most often taken as references is the Deal-Grove model proposed by Deal and Grove (5). According to this model, the beginning of oxidation is limited to interfacial oxidation reaction and, after oxidation proceeds, the rate-imiting process is transferred from the interfacial reaction to diffusion of oxidants in SiO_2. This process is expressed by the following equation given by Deal and Grove as (5)

$$\frac{dX}{dt} = \frac{B}{A + 2X}, \tag{1}$$

where B/A and B are denoted as the linear and parabolic rate constants of oxidation, respectively. It is noted that B/A and B are the rate coefficients for the interfacial reaction and the diffusion of oxidants rate-limiting step, respectively. However, it is well known that the oxidation rate of Si in the thin oxide thickness range cannot be reproduced by this equation and, hence, several models to describe the growth rate enhancement in thin oxide regime have

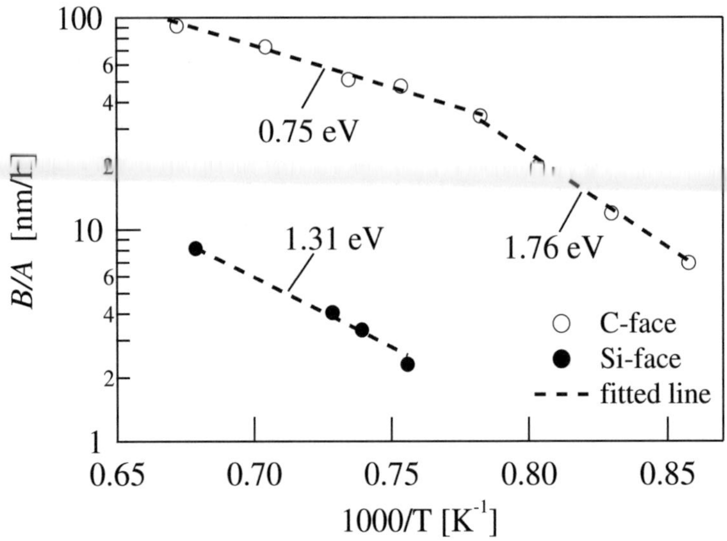

Fig. 5. Arrhenius plots of the linear rate constant B/A for C- and Si-faces.

been proposed (7–12). Among them, Massoud *et al.* (8; 9) have proposed an empirical relation for the oxide thickness dependence of oxidation rate, that is, the addition of an exponential term to the D-G equation,

$$\frac{dX}{dt} = \frac{B}{A+2X} + C\exp\left(-\frac{X}{L}\right), \tag{2}$$

where C and L are the pre-exponential constant and the characteristic length, respectively. We have found that it is possible to fit the calculated values to the observed ones using eq. (2) much better than using eq. (1) in any cases, as shown by the dashed and solid lines, respectively, in Figs. 1–4. We discuss the temperature and oxygen partial pressure dependencies of the four parameters B/A, B, C, and L below.

3.4 Arrhenius Plots of the Fitting Parameter

Figure 5 shows the Arrhenius plots of the linear rate constant B/A for C- and Si-faces. The values of B/A for Si-face are one order of magnitude smaller than those for C- face at any studied temperature, which is in agreement with the well-known experimental result indicating that the growth rate of Si-face is about 1/10 that of C-face. In the case of Si-face, the observed values of B/A are on a straight line with an activation energy of 1.31 eV. While for C-face, the values are on two straight lines, suggesting the existence of two activation energies, i.e., 0.75 and 1.76 eV, and the break point in the activation energy is around 1000°C (14). As we have measured the growth rates of SiC Si-face in the oxide thickness range less than 100 nm, the diffusion limiting-step regime, in which the growth rate is inversely proportional to X, does not appear regardless of the temperatures used in this study. Therefore, the precision in determining the values of B, related to the diffusion coefficient, is not sufficient, and thus, we do not discuss the value of B in this report.

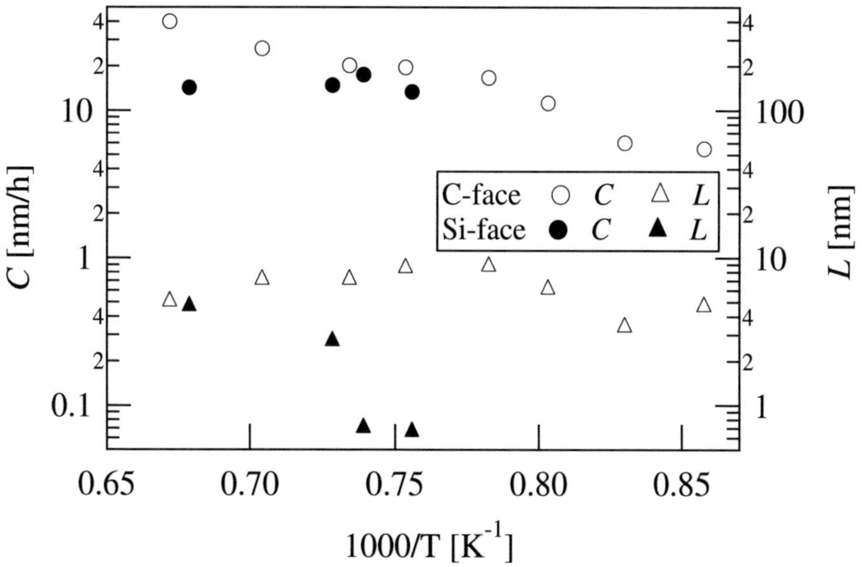

Fig. 6. Arrhenius plots of pre-exponential constant and characteristic length of the growth rate enhancement (C and L) for C- and Si-faces.

The values of $C/(B/A)$, which mean the magnitude of oxide growth enhancement, are around 2–6 for Si-face in the studied temperature range. On the other hand, those for C-face are less than 1. These results suggest that the growth rate enhancement phenomenon is more marked for Si-face than for C-face. The temperature dependences of the values of C and L for C- and Si-face shown in Fig. 6. Figure 6 shows that the values of C for Si-face are slightly smaller than those for C-face and almost independent of temperature, which is in contrast to the result for C-face. Figure 6 also shows that the values of L for the Si-face, around 3 nm at 1100°C, are smaller than those for C-face, around 6 nm at the same temperature, and increase with temperature, which is also in contrast to the result for C-face, i.e., almost independent of temperature. In the case of Si oxidation (8), the values of L are around 7 nm and almost independent of temperature, and the values of C increase with temperature. Therefore, it can be considered that the values of L and the temperature dependences of C and L for SiC C-face are almost the same as those for Si, but different from those for SiC Si-face. As seen in the oxide thickness dependence of the growth rate, the surface reaction-limiting-step regime, in which the growth rate is constant against the oxide thickness X, does not appear in the temperature range studied for SiC C-face (14; 16), as in the case for Si (8). This means that the oxidation mechanism of SiC C-face is in some sense similar to that of Si, but that of SiC Si-face is very different from that of Si. For SiC Si-face, the surface reaction rate is much smaller than the rate limited by oxygen diffusion, compared with the cases of SiC C-face and Si, which may cause the characteristics of the SiC Si-face oxidation to differ from those for SiC C-face and Si.

3.5 Oxygen Partial Pressure Dependencies of the Fitting Parameter

We examined the oxygen partial pressure dependence of oxide growth rate at $X = 0$, i.e., $C + B/A$, for C-face. As a result, the value of $C + B/A$ is propotional to oxygen partial pressure (17). When the oxide thickness X is nearly equal to 0, the oxide growth rate is essentially proportional to the quantity of oxidants that reach the interface between the oxide and SiC because this quantity is much lower than the number of Si atoms at the interface. Since the interfacial reaction rate when the oxide thickness X approaches 0 is considered to depend not on partial pressure but on oxidation temperature, the initial growth rate $C + B/A$ is represented by the following expression:

$$C + \frac{B}{A} \propto k_0 C^{\mathrm{I}}_{\mathrm{O2}} \tag{3}$$

where k_0 is the interfacial reaction rate when the oxide thickness X approaches 0, C_{O2} is the concentration of oxidants, and the superscript 'I' means the position at the SiC–SiO_2 interface. According to the Henry's law, the value of $C^{\mathrm{I}}_{\mathrm{O2}}$ is proportional to the oxygen partial pressure. Therefore, the initial growth rate $C + B/A$ should be proportional to the oxygen partial pressure, which is consistent with the experimental results obtained in this study.

While in the case of B/A, the oxygen partial pressure dependence showed a proportion to $p^{0.5-0.6}$ (16; 17). This non-linear dependence is also seen in the case of Si oxidation though the exponent is slightly higher. As will be described below, the value of B/A is considered as the quasi-state oxide growth rate and is determined by the balance between many factors, such as the quasi-steady concentration of C atoms at the interface, that of Si atoms emitted from the interface, interfacial reaction rate changing with oxide thickness. We believe that these are responsible for the non-linear dependence of B/A.

3.6 Discussion

Some Si oxidation models that describe the growth rate enhancement in the initial stage of oxidation have been proposed (10–12; 18). The common view of these models is that the stress near/at the oxide–Si interface is closely related to the growth enhancement. Among these models, 'interfacial Si emission model' is known as showing the greatest ability to fit the experimental oxide growth rate curves. According to this model, Si atoms are emitted as interstitials into the oxide layers accompanied by oxidation of Si, which is caused by the strain due to the expansion of Si lattices during oxidation. The oxidation rate at the interface is initially large and is suppressed by the accumulation of emitted Si atoms near the interface with increasing oxide thickness, i.e., the oxidation rate is not enhanced in the thin oxide regime but is quickly suppressed with increasing thickness. To describe this change in the interfacial reaction rate, Kageshima *et al.* introduce the following equation as the interfacial reaction rate constant, k (10; 18):

$$k = k_0 \left(1 - \frac{C^{\mathrm{I}}_{\mathrm{Si}}}{C^{0}_{\mathrm{Si}}}\right) \tag{4}$$

where $C^{\mathrm{I}}_{\mathrm{Si}}$ is the concentration of Si interstitials emitted at the interface and the C^{0}_{Si} is the solubility limit of Si interstitials in SiO_2. It is noted that, in the Deal-Grove model (5), the function k is assumed to be constant regardless of the oxidation thickness.

By the way, since the density of Si atoms in SiC (4.80×10^{22} cm^{-3}) (19) is almost the same as that in Si (5×10^{22} cm^{-3}) and the residual carbon is unlikely to exist at the oxide–SiC interface in the early stage of SiC oxidation, the stress near/at the interface is considered to be almost identical to the case of Si oxidation. Therefore, it is probable that atomic emission due to the interfacial stress also occurs for SiC oxidation and it also accounts for the growth enhancement

in SiC oxidation. In addition, in the case of SiC oxidation, we should take C emission as well as Si emission into account because SiC consists of Si and C atoms.
Recently, we have proposed a SiC oxidation model , termed "Si and C emission model", taking the Si and C emissions into the oxide into account, which lead to a reduction of interfacial reaction rate (20). Considering Si and C atoms emitted from the interface during the oxidation as well as the oxidation process of C, the reaction equation for SiC oxidation can be written as,

$$\begin{aligned} \mathrm{SiC} \quad &+ \left(2 - \nu_{\mathrm{Si}} - \nu_{\mathrm{C}} - \tfrac{\alpha}{2}\right) \mathrm{O_2} \rightarrow \left(1 - \nu_{\mathrm{Si}}\right) \mathrm{SiO_2} + \nu_{\mathrm{Si}}\mathrm{Si} \\ &+ \nu_{\mathrm{C}}\mathrm{C} + \alpha \mathrm{CO} + \left(1 - \nu_{\mathrm{C}} - \alpha\right) \mathrm{CO_2}, \end{aligned} \tag{5}$$

where ν and α denote the interfacial emission rate and the production rate of CO, respectively.
In the case of Si oxidation, the growth rate in thick oxide regime is determined by the parabolic rate constant B as is obvious if we consider the condition $A \ll 2X$ for eqs. (1,2). Song *et al.* proposed a modified Deal-Grove model that takes the out-diffusion of CO into account by modifying the parabolic rate constant B by a factor of 1.5 (called 'normalizing factor' (21)), and through this model, explained the oxidation process of SiC in the parabolic oxidation rate regime (6). For the Si and C emission model, the normalizing factor corresponds to the coefficient of the oxidant shown in eq. (5), i.e. $(2 - \nu_{\mathrm{Si}} - \nu_{\mathrm{C}} - \alpha/2)$. Since Song's model assumed that there is no interfacial atomic emission (i.e. $\nu_{\mathrm{Si}} = \nu_{\mathrm{C}} = 0$) and carbonaceous products consist of only CO (i.e. $\alpha = 1$), for this case, it is obvious that the coefficient of the oxidant in eq. (5) equals 1.5. Actually, it has been found in our study that this coefficient is 1.53 by fitting the calculated growth rates to the measured ones (20). Therefore, for C-face, the parameters $\nu_{\mathrm{Si}}, \nu_{\mathrm{C}}$, and α should be close to those assumed in the Song's model. While for Si-face, this coefficient results in a lower value. According to our recent work (20; 22), the most significant differences between C- and Si-face oxidation are those in k_0 and ν_{Si}. Therefore, it can be consider that the difference in ν_{Si} leads to that in the coefficient of oxidant. Anyway, it is believed that the different B from that of Si oxidation is necessary to reproduce the growth rate in the diffusion rate-limiting region (4; 6; 21; 23) because CO and CO_2 production is neglected.
In the case of Si oxidation, the interfacial reaction rate (i.e. eq. (4)) is introduced by assuming that the value of $C_{\mathrm{Si}}^{\mathrm{I}}$ does not exceed the C_{Si}^{0} though the reaction rate decreases with increase of $C_{\mathrm{Si}}^{\mathrm{I}}$. Based on this idea, the interfacial reaction rate for SiC is thought to be given by multiplying decreasing functions for Si and C (20):

$$k = k_0 \left(1 - \frac{C_{\mathrm{Si}}^{\mathrm{I}}}{C_{\mathrm{Si}}^{0}}\right) \left(1 - \frac{C_{\mathrm{C}}^{\mathrm{I}}}{C_{\mathrm{C}}^{0}}\right). \tag{6}$$

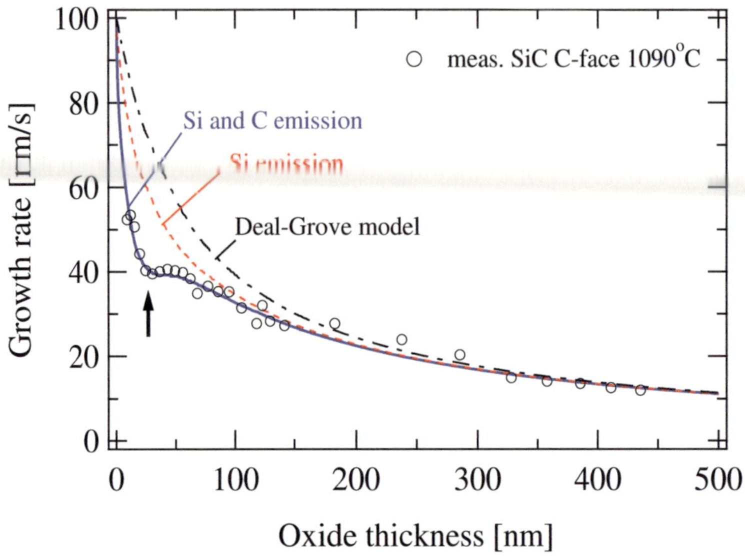

Fig. 7. Oxide thickness dependence of growth rates for C-faces.

This equation implies that the growth rate in the initial stage of oxidation should reduce by two steps because the accumulation rates for Si and C interstitials should be different from each other, and hence, the oxidation time when the concentration of interstitial saturates should be different between Si and C interstitial. This prediction will be evidenced in the next paragraph.

Figure 7 shows the oxide growth rates observed for C - face at 1090°C (circles). Also shown in the figure are the growth rates given by the Si and C emission model (solid lines), the Si emission model, and the model that does not take account of both Si and C emission, i.e., the Deal-Grove model (broken line and double broken line, respectively). We note that the same parameters were used for these three SiC oxidation models. Figure 7 shows that the Si and C emission model reproduces the experimental values better than the other two models. In particular, the dip in the thickness dependence of the growth rate seen around 20 nm (pointed by the arrow in the figure), which cannot be reproduced by the Si emission model or the Deal-Grove model no matter how well the calculation are tuned, can be well reproduced by the Si and C emission model. These results suggest that the C interstitials play an important role in the reduction of the oxidation rate, similarly to the role of the Si interstitials. Moreover, from the fact that the drop in growth rate in the initial stage of oxidation is larger for the Si and C emission model than in the case of taking only Si emission into account, we found that the accumulation of C interstitials is faster than that of Si interstitials and that the accumulation of C interstitials is more effective in the thin oxide regime.

4. Conclusion

By performing *in-situ* spectroscopic ellipsometry, we have, for the first time, observed the growth enhancement in oxide growth rate at the initial stage of SiC oxidation, which means that the D-G model is not suitable for SiC oxidation in the whole thickness regime, as in the case of Si oxidation. We have also observed the occurrence of the oxide growth rate enhancement at any oxidation temperature and oxygen partial pressure measured both in the cases of C- and Si-faces. We found that the growth rate of SiC for both polar faces can be well represented by the empirical equation proposed by Massoud *et al.* using the four adjusting parameters B/A, B, C, and L, and that the values of B/A, C, and L, and the temperature dependences of C and L for Si-face are different from those for C-face. Finally, we have discussed the mechanism of the growth rate enhancement in the initial stage of oxidation by comparing with the oxidation mechanism of Si.

5. References

[1] H. Matsunami: Jpn. J. Appl. Phys. Part 1 **43** (2004) 6835.
[2] S. Yoshida: *Electric Refractory Materials*, ed. Y. Kumashiro (Dekker, New York, 2000) 437.
[3] V. V. Afanas'ev and A. Stesmans: Appl. Phys. Lett. **71** (1997) 3844.
[4] K. Kakubari, R. Kuboki, Y. Hijikata, H. Yaguchi, and S. Yoshida: Mater. Sci. Forum **527-529** (2006) 1031.
[5] B. E. Deal and A. S. Grove: J. Appl. Phys. **36** (1965) 3770.
[6] Y. Song, S. Dhar, L. C. Feldman, G. Chung and J. R. Williams: J. Appl. Phys. **95** (2004) 4953.
[7] A. S. Grove: *Physics and Technology of Semiconductor Devices* (John Wiley & Sons, New York, 1967) 31.
[8] H. Z. Massoud, J. D. Plummer, and E. A. Irene: J. Electrochem. Soc. **132** (1985) 2685.
[9] H. Z. Massoud, J. D. Plummer, and E. A. Irene: J. Electrochem. Soc. **132** (1985) 2693.
[10] H. Kageshima, K. Shiraishi, and M. Uematsu: Jpn. J. Appl. Phys. Part 2 **38** (1999) L971.
[11] S. Ogawa and Y. Takakuwa: Jpn. J. Appl. Phys. **45** (2006) 7063.
[12] T. Watanabe, K. Tatsumura, and I. Ohdomari: Phys. Rev. Lett. **96** (2006) 196102.
[13] T. Iida, Y. Tomioka, M. Midorikawa, H. Tsukada, M. Orihara, Y. Hijikata, H. Yaguchi, M. Yoshikawa, H. Itoh, Y. Ishida, and S. Yoshida: Jpn. J. Appl. Phys. Part 1 **41** (2002) 800.
[14] T. Yamamoto, Y. Hijikata, H. Yaguchi, and S. Yoshida: Jpn. J. Appl. Phys **46** (2007) L770.
[15] T. Yamamoto, Y. Hijikata, H. Yaguchi, and S. Yoshida: Jpn. J. Appl. Phys. **47** (2008) 7803.
[16] T. Yamamoto, Y. Hijikata, H. Yaguchi, and S. Yoshida: Mater. Sci. Forum. **600-603** (2009) 667.
[17] K. Kouda, Y. Hijikata, H. Yaguchi, and S. Yoshida: Mater. Sci. Forum. **645-648** (2010) 813.
[18] M. Uematsu, H. Kageshima, and K. Shiraishi: J. Appl. Phys. **89** (2001) 1948.
[19] Y. Hijikata, H. Yaguchi, S. Yoshida, Y. Takata, K. Kobayashi, H. Nohira, and T. Hattori: J. Appl. Phys. **100** (2006) 053710.
[20] Y. Hijikata, H. Yaguchi, and S. Yoshida: Appl. Phys. Express **2** (2009) 021203.
[21] E. A. Ray, J. Rozen, S. Dhar, L. C. Feldman, and J. R. Williams: J. Appl. Phys. **103** (2008) 023522.
[22] Y. Hijikata, H. Yaguchi, and S. Yoshida: Mater. Sci. Forum **615-617** (2009) 489.
[23] Y. Hijikata, T. Yamamoto, H. Yaguchi, and S. Yoshida: Mater. Sci. Forum **600-603** (2009) 663.

Magnetic Properties of Transition-Metal-Doped Silicon Carbide Diluted Magnetic Semiconductors

Andrei Los[1, 2] and Victor Los[3]
[1]*ISS Ltd., Semiconductors and Circuits Lab*
[2]*Freescale Semiconductor Ukraine LLC*
[3]*Institute of Magnetism, National Academy of Sciences*
Kiev, Ukraine

1. Introduction

Possibility to employ the spin of electrons for controlling electronic device operation has long been envisaged as a foundation for future extremely low power amplifying and logic devices, polarized light emitting diodes, new generation magnetic field sensors, high density 3D magnetic memories, etc. (Gregg et al., 2002; Žutić et al., 2004; Bratkovsky, 2008). While metal-metal and metal-insulator spin-electronic (or spintronic) devices have already found their application as hard drive magnetic field sensors and niche nonvolatile memories, diluted magnetic semiconductors (DMSs), i.e. semiconductors with a fraction of the atoms substituted by magnetic atoms, are expected to become a link enabling integration of spin-electronic functionality into traditional electron-charge-based semiconductor technology. Following the discovery of carrier-mediated ferromagnetism due to transition metal doping in technologically important GaAs and InAs III-V compound semiconductors (Munekata et at., 1989; Ohno et al., 1996), a wealth of research efforts have been invested in the past two decades into investigations of magnetic properties of DMSs. Ferromagnetic semiconductors were, of course, not new at the time and carrier-mediated ferromagnetism, a lever allowing electrical control of the magnetic ordering, had also been demonstrated albeit only at liquid helium temperatures (Pashitskii & Ryabchenko, 1979; Story et al., 1986). The achievement of the ferromagnetic ordering temperature, the Curie temperature T_C, in excess of 100 K in (Ga, Mn) As compounds was a significant step towards practical semiconductor spintronic device implementation. A substantial progress has been achieved in increasing the ordering temperature in this material system and T_C as high as 180 K has been reported (Olejník et al., 2008). (Ga, Mn) As has effectively become a model magnetic semiconductor material with its electronic, magnetic, and optical properties understood most deeply among the DMSs. Still, however, one needs the Curie temperature to be at or above room temperature for most practical applications.

Mean-field theory of ferromagnetism (Dietl et al., 2000; Dietl et al., 2001), predicting that above room temperature carrier-mediated ferromagnetic ordering may be possible in certain wide bandgap diluted magnetic semiconductors, including a family of III-nitrides and ZnO, had spun a great deal of interest to magnetic properties of these materials. The resulting flurry of activities in this area led to apparent early successes in fabricating the DMS samples exhibiting ferromagnetism above room temperature (Pearton et al., 2003; Hebard et al., 2004). Ferromagnetic ordering in these samples was attributed to formation of homogeneous DMS alloys which, however, was in many cases later refuted and explained differently, by, for instance, impurity clustering, at the time overlooked by standard characterization techniques. Much theoretical understanding has been gained since then on the effects of exchange interaction, self-compensation, spinodal decomposition, etc. Given that various effects may mimic the "true DMS" behaviour, a careful investigation of the microscopic picture of magnetic moments formation and their interaction, as well as attraction of different complementary experimental techniques is required for a realistic understanding and prediction of the properties of this complex class of materials.

Silicon carbide is another wide bandgap semiconductor which has been considered a possible candidate for spin electronic applications. SiC has a long history of material research and device development and is already commercially successful in a number of applications. The mean field theory (Dietl et al., 2000; Dietl et al., 2001) predicted that semiconductors with light atoms and smaller lattice constants might possess stronger magnetic coupling and larger ordering temperatures. Although not applied directly to studying magnetic properties of SiC, these predictions make SiC DMS a promising candidate for spintronic applications.

Relatively little attention has been paid to investigation of magnetic properties of SiC doped with TM impurities, and the results obtained to date are rather modest compared to many other DMS systems and are far from being conclusive. Early experimental studies evidenced ferromagnetic response in Ni-, Mn-, and Fe-doped SiC with the values of the Curie temperature T_C varying from significantly below to close to room temperature (Theodoropoulou et al., 2002; Syväjärvi et al., 2004; Stromberg et al., 2006). The authors assigned the magnetic signal to either the true DMS behaviour or to secondary phase formation. Later experimental reports on Cr-doped SiC suggested this material to be ferromagnetic with the T_C ~70 K for Cr concentration of ~0.02 wt% (Huang & Chen, 2007), while above room temperature magnetism with varying values of the atomic magnetic moments was observed for Cr concentration of 7-10 at% in amorphous SiC (Jin et al., 2008). SiC doped with Mn has become the most actively studied SiC DMS material. Experimental studies of Mn-implanted 3C-SiC/Si heteroepitaxial structure (Bouziane et al., 2009), of C-incorporated Mn-Si films grown on 4H-SiC wafers (Wang et al., 2007), a detailed report by the same authors on structural, magnetic, and magneto-optical properties of Mn-doped SiC films prepared on 3C-SiC wafers (Wang et al., 2009) as well as studies of low-Mn-doped 6H-SiC (Song et al., 2009) and polycrystalline 3C-SiC (Ma et al., 2007) all suggested Mn to be a promising impurity choice for achieving high ferromagnetic ordering temperatures in SiC DMS. Researchers recently turned to studying magnetic properties of TM-doped silicon carbide nanowires (Seong et al., 2009).

Theoretical work done in parallel in an attempt to explain the available experimental data and to obtain guidance for experimentalists was concentrated on first principles calculations which are a powerful tool for modelling and predicting DMS material properties. Various *ab initio* computational techniques were used to study magnetic properties of SiC DMSs theoretically. Linearized muffin-tin orbital (LMTO) technique was utilized for calculating substitution energies of a number of transition metal impurities in 3C-SiC (Gubanov et al., 2001; Miao & Lambrecht, 2003). The researchers found that Si site is more favourable compared to C site for TM substitution. This result holds when lattice relaxation effects are taken into account in the full-potential LMTO calculation. Both research teams found that Fe, Ni and Co were nonmagnetic while Cr and Mn possessed nonzero magnetic moments in the 3C-SiC host. Calculation of the magnetic moments in a relaxed supercell containing two TM atoms showed that both Mn and Cr atoms ordered ferromagnetically. Ferromagnetic ordering was later confirmed for V, Mn, and Cr using ultrasoft pseudopotential plane wave method (Kim et al., 2004). In another *ab initio* study, nonzero magnetic moments were found for Cr and Mn in 3C-SiC using full potential linearized augmented plane wave (FLAPW) calculation technique and no relaxation procedure accounting for impurity-substitution-related lattice reconstruction (Shaposhnikov & Sobolev, 2004). The authors additionally studied magnetic properties of TM impurities in 6H-SiC substituting for 2% or 16% of host atoms. It was found that on Si site in 6H-SiC Cr and Mn possessed magnetic moments in both concentrations, while Fe was magnetic only in the concentration of 2%. Ultrasoft pseudopotentials were used for calculations of magnetic moments and ferromagnetic exchange energy estimations for the case of Cr doping of 3C-SiC (Kim & Chung, 2005). In a later reported study (Miao & Lambrecht, 2006) the authors used FP-LMTO technique with lattice relaxation to compare electronic and magnetic properties of 3C- and 4H-SiC doped with early first row transition metals. Spin polarization was found to be present in V, Cr, and Mn-doped SiC. The authors of (Bouziane et al., 2008) additionally studied the influence of implantation-induced defects on electronic structure of Mn-doped SiC. The results of the cited calculations were also somewhat sensitive to the particular calculation technique employed.

Here, we attempt to create a somewhat complete description of SiC-based diluted magnetic semiconductors in a systematic study of magnetic states of first row transition metal impurities in SiC host. Improving prior research, we do this in the framework of *ab initio* FLAPW calculation technique, perhaps one of the most if not the most accurate density functional theory technique at the date, combined with a complete lattice relaxation procedure at all stages of the calculation of magnetic moments and ordering temperatures. Accounting for the impurity-substitution-caused relaxation has been found crucial by many researchers for a correct description of a DMS system. We therefore are hopefully approaching the best accuracy of the calculations possible with the ground state density functional theory. We analyze the details of magnetic moments formation and of their change with the unit cell volume, as well as of the host lattice reconstruction due to impurity substitution. Such analysis leads to revealing multiple magnetic states in TM-doped SiC. We also study, for the first time, particulars of exchange interaction for different TM impurities and provide estimates of the magnetic ordering temperatures of SiC DMSs.

2. Methodology and computational setup

2.1 SiC-TM material system

Crystal lattice of any SiC polytype can be represented as a sequence of hexagonal close-packed silicon-carbon bilayers. Different bilayer stacking sequences correspond to different polytypes. For example, for the most technologically important hexagonal 4H polytype, the stacking sequence is ABAC (or, equivalently, ABCB), where A, B, and C denote hexagonal bilayers rotated by 120° with respect to each other (Bechstedt et al., 1997). The stacking sequence for another common polytype, the cubic 3C-SiC, is ABC. Although in all SiC polytypes the nearest neighbours of any Si or C atom are always four C or Si atoms, respectively, forming tetrahedra around the corresponding Si or C atom, there are two types of the sites (and layers) in SiC lattice, different in their next nearest neighbour arrangement or the medium range order. The stacking sequences for these different sites are ABC and ABA, where in the former the middle layer (layer B) has the cubic symmetry (sometimes also called quasi-cubic if one deals with such layer in a hexagonal polytype), while in the latter the symmetry of the middle layer is hexagonal (Bechstedt et al., 1997). There are only cubic layers in 3C-SiC, while other common polytypes such as 4H and 6H contain different numbers of both hexagonal and cubic layers. We will show below that site symmetry plays crucial role in TM d-orbital coupling and, therefore, ferromagnetic ordering temperatures of SiC DMSs.

Diluted magnetic semiconductor material systems usually have TM impurity concentrations of the order of several atomic per cent. Such concentrations, although very high for typical semiconductor device applications, are essential for securing efficient exchange interaction between TM impurities and thus achieving high ordering temperatures needed for practical spintronic device operation. In our calculations of the magnetic properties of SiC DMSs the effective TM impurity concentrations range from about 4% to 10%. Substitutional TM impurity in the host SiC lattice is assumed to reside at the Si sites in the SiC crystal lattice. The choice of the substitution site preference can be made according to the atomic radii which are much closer for Si and TM than for C and TM and, therefore, much smaller lattice distortion would be required in the case of Si site substitution. Results of prior studies of the TM substitution site preference (Gubanov et al., 2001; Miao & Lambrecht, 2003) support this intuitive approach.

It is important that the Fermi level position in a semiconductor system with diluted TM doping of several per cent is defined by the TM impurity itself, unless another impurity is present in the system in a comparable concentration. In other words, in such a DMS system, TM impurity pins the Fermi level and defines its own charge state and charge states of all other impurities. From the computational point of view this is, of course, automatically achieved by the self-consistent solution for the state occupation. If one were to vary the TM impurity charge state independently, this would require co-doping with a comparable amount of another donor or acceptor. On the other hand, reducing TM concentration to the typical heavy doping levels of, say, 10^{19} cm^{-3} would result in an "ultradiluted magnetic semiconductor", where an efficient exchange interaction would be hindered by large distances between the transition metal atoms, and this latter case is not considered here. Calculations of formation energies of different charge states of first row TM impurities in SiC (Miao & Lambrecht, 2003; Miao & Lambrecht, 2006) in their typical DMS concentrations indicate that they are expected to form deep donor and acceptor levels in the SiC bandgap, and be in their neutral charge states. This means that, contrary to the case of, for example, GaAs, TM impurities in SiC do not contribute free carriers which could mediate exchange interaction between the TM atoms.

2.2 Supercells

In this work, magnetic properties of TM-doped SiC are studied for 3C (Zincblende) and 4H (Wurtzite) SiC polytypes. We start with calculations of the lattice parameters and electronic structure of pure 3C- and 4H-SiC. The primitive cell of pure cubic 3C-SiC consists of single silicon-carbon pair. Another way of representing this type of lattice is with a sequence of hexagonal close-packed Si-C bilayers with the ABC stacking sequence. In that case the unit cell consists of 3 Si-C bilayers or 6 atoms. Both cells are, of course, equivalent from the computational point of view and must produce identical results. The unit cell of 4H-SiC consists of 4 Si-C bilayers or 8 atoms.

Then, the lattice parameters, electronic structure and magnetic properties are calculated for 3C- and 4H-SiC doped with TM impurity. In the calculations, to model the lattice of doped single crystal SiC, we employ the supercell approach. For a doped semiconductor, the minimum lattice fragment (the supercell), needed to model the material, includes one impurity atom, while the total number of atoms in the supercell is inversely proportional to the impurity concentration. For only one TM atom in the supercell, the solution sought will automatically be a ferromagnetically-ordered DMS (in case a nonzero magnetic moment is obtained on TM atoms), as the solution is obtained for an infinite lattice implicitly constructed from the supercells with identically oriented magnetic moments. This is convenient and sufficient for establishing the trends for achieving spin polarization in the SiC-TM system. Investigation of true (energetically more preferable) magnetic moment ordering type requires larger supercells with at least two TM atoms with generally different directions of their magnetic moments.

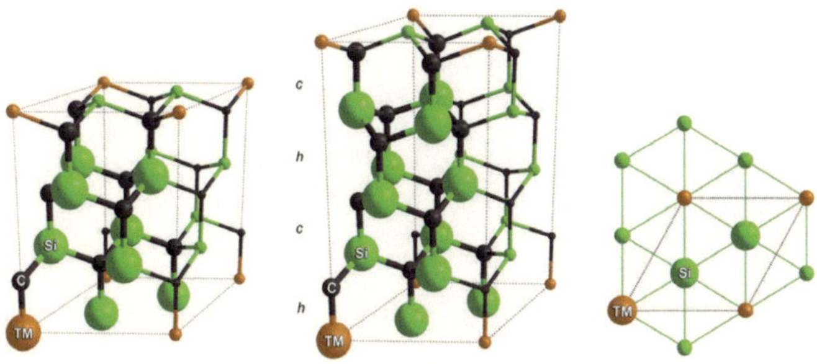

(a) (b) (c)

Fig. 1. Supercells of TM-doped (a) 3C-SiC and (b) 4H-SiC, and (c) in-plane TM atom placement used in the calculations of magnetic moments and properties of different magnetic states. TM, Si, and C atoms added by periodicity and not being part of the supercells are shown having smaller diameters compared to the similar atoms in the supercells. Layers with the hexagonal and quasicubic symmetries in 4H-SiC are marked by *h* anc *c*, respectively.

Investigation of the magnetic moment formation and related lattice reconstruction in SiC doped with TM impurities is done using Si_8C_9TM and $Si_{11}C_{12}TM$ supercells, containing a total of 18 and 24 atoms for 3C- and 4H-SiC, respectively. These supercells are shown in Fig. 1. Impurity atoms are placed in the centres of the adjacent close-packed hexagons, so that the distance between them equals $\sqrt{3}\,a$. In the *c*-axis direction, the distance between TM atoms is equal to one 3C or 4H-SiC lattice period (3 or 4 Si-C bilayers). The resultant impurity concentration calculated with respect to the total number of atoms is about 4 % in the case of 4H-SiC and about 5% in the case of 3C-SiC. As already mentioned, such concentrations are typical for experimental, including SiC, DMS systems.

Calculations of SiC DMS ordering temperatures require adding another TM impurity atom to the supercells. These supercells are shown in Fig. 2. In the case of 3C-SiC, we study magnetic ordering for two different spatial configurations of TM impurities in the SiC lattice. First, we simply double the 3C-SiC supercell shown in Fig. 1 in the *c*-direction so that the two TM atoms are at the distance of 14.27 a.u., while TM concentration is kept at 5 at % (Fig. 2 (a)). Next, we return to the original Si_8C_9TM supercell and introduce an additional TM atom as the nearest neighbor to the TM atom in the Si-TM plane (Fig. 2 (b)). The distance between the TM atoms in this case equals to 5.82 a.u. and their effective concentration is approximately 10%. Such TM configuration can also be thought of as a simplest TM nanocluster in SiC lattice. For Mn-doping, which we identify as the most promising for obtaining high temperature SiC DMS, we additionally study substitution of a pair of TM atoms at different, hexagonal and cubic, 4H-SiC lattice sites with varying distances between the impurities and impurity electronic orbital mutual orientations (Fig. 2 c-e). We show that the strength of exchange coupling and the Curie temperature depend not only on the distance between TM atoms but also significantly on the particular lattice sites the impurities substitute at.

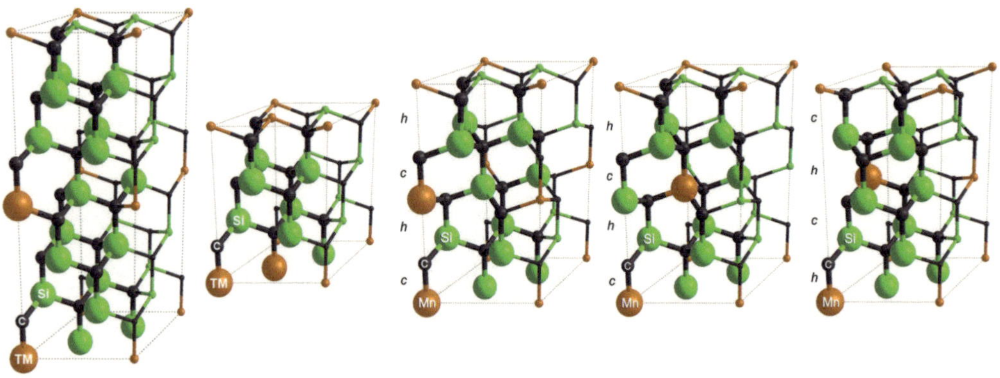

(a) (b) (c) (d) (e)

Fig. 2. SiC-TM supercells containing two TM atoms, which are used in the calculations of DMS ordering types and temperatures. Supercells (a) and (b) correspond to 3C-SiC, while supercells (c)-(e) correspond to 4H-SiC. TM, Si, and C atoms added by periodicity and not being part of the supercells are shown having smaller diameters compared to the similar atoms in the supercells. Layers with the hexagonal and quasicubic symmetries in 4H-SiC are marked by *h* anc *c*, respectively.

2.3 Computational procedure

In contrast to earlier studies of magnetic properties of TM-doped SiC, the calculations are done for the supercells with optimized volumes and atomic positions, i.e. with both the local and global lattice relaxations accounted for. As we will show below, not only the FM ordering temperature, but also the value and even the existence of the magnetic moments in TM-doped semiconductor sensitively depend on the semiconductor host lattice structure and its reconstruction due to impurity substitution. Furthermore, it will be shown that multiple states with different, including zero, magnetic moments can be characteristic for SiC DMSs (and, perhaps, the other DMS systems as well). The different states correspond to different equilibrium lattice configurations, transition between which involves reconstruction of the entire crystalline lattice. The reconstruction and the transition between the states may either be gradual or the states can be separated by an energy gap with the energy preference for either the magnetic or nonmagnetic state. The width of the energy gap between the different states varies across the range of impurities.

In the procedure, which is used for finding the optimized supercells, total energies are calculated for a number of volumes of isotropically expanded supercells. Additionally, at each value of the volume, the supercells are fully relaxed to minimize the intra-cell forces. Total energy-volume relationships for the relaxed supercells are then fitted to the universal equation of state (Vinet et al., 1989), and the minimum of the fitted curve corresponds to the supercell with the equilibrium volume and atomic positions in one of the magnetic states peculiar to that particular DMS. The supercells with such optimized equilibrium volumes and atomic positions are then used in the calculations of the DMS magnetic moments and ordering temperatures.

SiC DMS ordering type, either ferromagnetic (FM) or antiferromagnetic (AFM), and the corresponding values of the Curie or Néel critical temperatures are estimated from the total energy calculations for supercells containing a pair of TM impurities with their magnetic moments aligned in parallel (FM) or antiparallel (AFM). Employing the Heisenberg model, which should describe the orientational degrees of freedom accurately, for the description of the magnetic ordering, one can use the difference ΔE_{FM-AFM} between the total energies of the FM- and AFM-ordered supercells for estimating the value of the Curie or Néel temperature in the framework of the mean-field model using the following expression:

$$T_C = -\frac{1}{3k_B}\Delta E_{FM-AFM}, \quad (1)$$

where k_B is the Boltzmann constant. For the negative total energy difference ΔE_{FM-AFM} between the FM and AM states, the energy preference is for the FM state and (1) provides the Curie temperature value. When ΔE_{FM-AFM} is positive, the ground state of the DMS is AFM and the negative value of the critical temperature can be interpreted as the positive Néel temperature.

We note that the total energies of the FM and AFM states entering equation (1) are those of a simple ferromagnet and a two-sublattice Néel antiferromagnet models, describing only the nearest neighbour magnetic moment interactions. Taking into account the next neighbour interactions may result, particularly, in more complex, compared to the simple FM or AFM state, magnetic moment direction configurations. Using larger supercells with more TM atoms and taking into account more terms in the Heisenberg Hamiltonian would allow going beyond the model given by (1) and make the estimate of T_C more accurate. This, however, would lead to significantly larger computational resource requirements and is beyond the scope of the present work. The use of the simplest model is justified by the fact that usually the more long-range interactions of magnetic moments are noticeably weaker than the interactions between the nearest neighbours due to the rather localized nature of electronic shells responsible for the magnetic properties of TM atoms, as we will see below.

It is important that, since the calculations are done in the framework of the ground state density functional theory, only direct TM-TM exchange mechanisms can be accounted for in this case. Another important type of exchange interaction which, for example, is peculiar to

Mn-doped GaAs, namely free-carrier-mediated exchange, requires attraction of different computational techniques for its modelling. We also note that since the calculated FM and AFM total ground state energies in (1) are defined by the electronic density distribution, which also defines the corresponding magnetic moment values, one can in general expect magnetic moments to be different in the FM and AFM states (see Table 3 and discussion below).

According to Eq. (1), the difference in the total energy values between the FM and AFM states corresponding to a Curie or Néel temperature change of, say, 50 K is about 14 meV per pair of TM atoms (per supercell). The accuracy of the total energy calculations needs to be better than this and per atom accuracy requirement, which is usually specified in the total energy calculations, is more strict for larger supercells or, equivalently, smaller impurity concentrations; this also means that critical temperature estimations for DMSs with smaller TM concentrations are typically prone to larger errors. Additionally, as will be evident from the discussion below, we need to resolve magnetic and nonmagnetic solutions which are close on the energy scale, and thus the accuracy of the calculations needs to be higher than at least the differences in energy between these solutions. To satisfy these accuracy requirements we perform both stringent convergence tests with respect to all parameters influencing the calculations and choose the corresponding settings for the self-consistent calculations convergence criteria.

The calculations are performed using the FLAPW technique (Singh & Nordstrom, 2006) implemented in the Elk (formerly EXCITING) software package (Dewhurst et al., 2004). Exchange-correlation potential is calculated using the generalized gradient approximation according to the Perdew-Burke-Ernzerhof model (Perdew et al., 1996). The muffin-tin radii are set at 2.0 a.u. for Si and TM atoms and 1.5 a.u. for C atoms. The APW basis set includes 150 plane waves per atom. Within the atomic spheres, spherical harmonic expansions with angular momentum up to 8 are used for the wave function, charge density, and potential representation. Local orbitals are added to the APW basis set to improve convergence and accuracy of the calculations. The self-consistent calculations are performed for 12 (Fig. 1, Fig. 2 (b)), 6 (Fig. 2 (a)), 8 (Fig. 2 (c)) and 10 (Fig. 2 (d) and (e)) reciprocal lattice points in the irreducible wedge of the Brillouin zone and are considered converged when the RMS change in the effective potential is less than 10^{-6} and the total energy error is within 0.1 meV/atom.

To check the computational setup, lattice constants of undoped 3C-SiC were obtained using the optimization procedure described above. The calculated unit cell volume was found to be equal to 70.19 a.u.^{3}/atom. This is in an excellent agreement with the experimental value of 69.98 a.u.^{3}/atom (Bechstedt et al., 1997) confirming the effectiveness of the lattice optimization procedure used. Calculated indirect bandgap of 3C-SiC of about 1.62 eV is below the experimental value of approximately 2.3 eV. Such bandgap underestimation is known to be a problem of the ground state density functional theory, in particular when LDA or GGA exchange-correlation functionals are used. Other carefully constructed exchange-correlation functionals may rectify the problem somewhat (Sharma et al., 2008). We, however, warn the reader from taking the DFT bandgap values literally.

3. Multiple magnetic states of TM-doped SiC

3.1 Magnetic moment formation and lattice reconstruction

We begin the discussion of different magnetic states, which can be created by a TM impurity in SiC, with the case of Fe-doped SiC. Calculated total energies of (Si, Fe) C supercells as a function of the supercell volume for 3C- and 4H-SiC are presented in Fig. 3 (a). Two equilibrium solutions were found for the total energy - supercell volume dependencies, one of which is characterized by a nonzero supercell magnetic moment, while in the other the supercell is nonmagnetic. The magnetic state possesses a larger total energy, while the nonmagnetic state is the ground state in Fe-doped 3C- and 4H-SiC. The total energy difference between the nonmagnetic and magnetic states at equilibrium volume is approximately 33 meV/atom in 3C-SiC and 20 meV/atom in 4H-SiC. Magnetic moment of Fe in the magnetic state changes approximately linearly from about 1.9 to 2.3 μ_B per Fe in the range of the volumes from about 68 a.u.^{3}/atom to 78 a.u.^{3}/atom (see Fig. 5 below). It is important to note that the two solutions were not obtained in a constrained magnetic moment calculation. Rather, the calculations converge to the magnetic solution if starting from a sufficiently expanded lattice. The nonmagnetic solution is obtained in the lattice with the unit cell volume approximately corresponding to undoped SiC. Both total energy - supercell volume curves are obtained by using the solution at each volume value as the initial approximation for the next one. Thus, we let the calculations naturally follow the underlying physical processes.

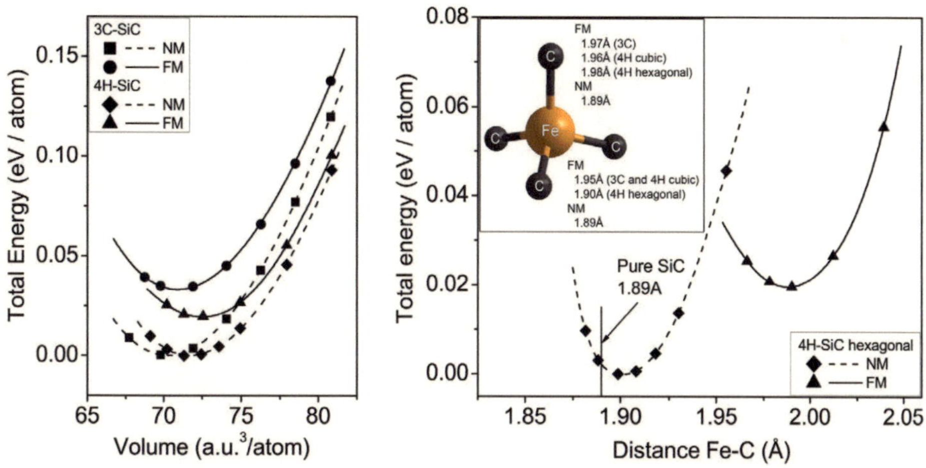

(a) (b)

Fig. 3. Total energy - supercell volume dependencies for Fe-doped 3C- and 4H-SiC in the nonmagnetic and magnetic states (a). Fe-C bond length relaxation in the different states for C atom above Fe atom at the hexagonal lattice site in 4H-SiC (b). The inset shows tetrahedron relaxation details in different SiC polytypes and at lattice sites with different symmetries.

Crystal lattice of (Si, Fe) C DMS in its magnetic state is significantly reconstructed compared both to (Si, Fe) C in its nonmagnetic state and to undoped SiC. Lattice relaxation of the (Si, Fe) C supercell in its nonmagnetic state is minimal compared to undoped SiC. It is important that the magnetic state cannot be reached by increasing the magnetic moment from zero while gradually expanding the lattice: different atomic and electronic structures are characteristic of the two states, which are separated by an energy gap. The average per atom volume in the magnetic state is approximately 1 % larger compared to the nonmagnetic state, while this difference is mostly due to the expansion of the tetrahedra formed by four C atoms around the Fe atoms; this expansion was found to be reaching 10 volume percent. Furthermore, TM bond length changes turn out to be different for different carbon atoms in the tetrahedra around Fe impurities. The relaxation additionally depends on SiC polytype and lattice site the impurity substitutes at. For example, for Fe at the hexagonal site in 4H-SiC, the relaxation in the magnetic state strongly affects C atom which is above the Fe atom in the *c*-axis direction, while the other three atoms in the tetrahedron shift noticeably less. This is illustrated by Fig. 3 (b). At the cubic sites in 4H-SiC and in 3C-SiC (in 3C-SiC all lattice sites possess cubic symmetry) the relaxation is nearly equilateral and C atoms in the elemental tetrahedra around Fe atoms shift more or less the same distance. The inset in Fig. 3 (b) presents the details of the elemental tetrahedron relaxation for Fe substituting at the lattice sites with different symmetries in 4H- and 3C-SiC. We thus conclude that TM-impurity-induced lattice reconstruction, which self-consistently defines the electronic and magnetic configuration of the DMS, depends not only on TM atom's nearest neighbors, which are the same for all SiC polytypes and inequivalent lattice sites, but also noticeably on the long-range stacking sequence.

Total energy – supercell volume dependencies for Mn-, Cr-, and V-doped 3C-SiC are presented in Fig. 4. Similarly to the case of (Si, Fe) C, SiC DMSs with these substitutional impurities can again be in either the nonmagnetic or magnetic state. However, the relationship between the two states is the opposite of that in (Si, Fe) C with the magnetic state lying lower on the energy scale and thus being energetically favorable. At the equilibrium volume the energy gap between the two states changes from a small value of 12 meV/atom in the case of V doping to 18 meV/atom in (Si, Mn) C and 45 meV/atom in (Si, Cr) C. The higher the energy gap between the states, the more "stable" the magnetic state is against thermal fluctuations. If, as we suggest below, magnetic-nonmagnetic state mixing takes place at nonzero temperatures, the average magnetic moment in (Si, Cr) C may be expected to be closest to its calculated zero temperature value.

Magnetic moments as a function of the supercell volume behave in a different manner for V, Cr and Mn (see Fig. 5). While the former two change only slightly in the range of the average volumes from 68 to 78 a.u.^{3}/atom, magnetic moment of Mn experiences a steep increase from 1.5 μ_B at 68 a.u.^{3}/atom and then almost saturates at approximately 2.5 μ_B for the larger values of the atomic volume. Such behavior of the magnetic moments can be related to the differences in the nature of the electronic orbitals of the TM atoms, the degree of their hybridization with the surrounding dangling bonds and, therefore, to the influence of the electronic structure of the host lattice. The same factors influence relaxation of the lattice as it approaches equilibrium after the impurity substitution. For instance, in (Si, V) C and (Si, Cr) C, the relaxation is mostly absorbed by the tetrahedra of four carbon atoms around the TM impurity, similarly to the case of (Si, Fe) C. For Mn-doped SiC there are noticeable changes in Si atom positions in the layers other than impurity substitution layer. For all three impurities, the equilibrium average lattice volume is larger for the magnetic state of the crystal, compared to the nonmagnetic state, while in the nonmagnetic state relaxation is, again, small compared to pure SiC.

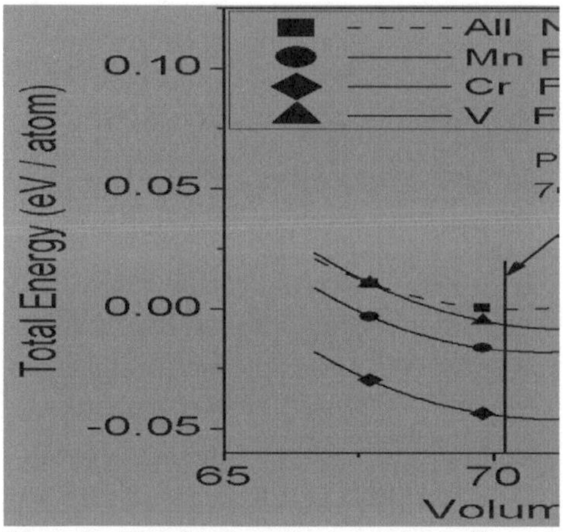

Fig. 4. Total energy - supercell volume dependencies for 3C-SiC doped with Mn, Cr, or V. The $E_{TOT}(V)$ curves for different impurities are brought to the same energy scale for easy comparison of the energies in the nonmagnetic and magnetic states.

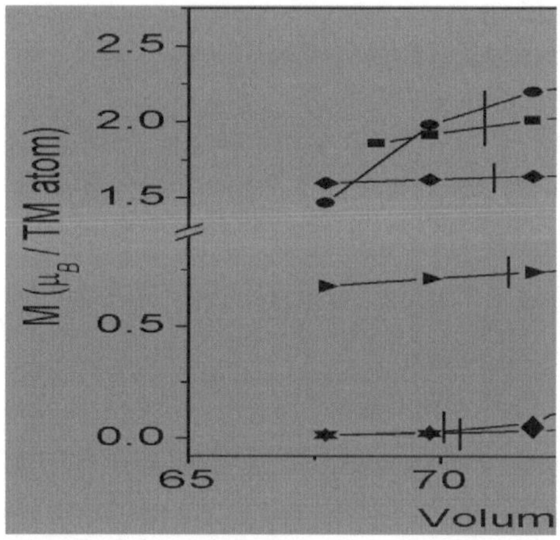

Fig. 5. Magnetic moment per impurity atom for different TM impurities substituting in 3C-SiC. Vertical lines mark equilibrium volume for each particular DMS.

Nickel and cobalt exhibit yet another behavior in the SiC matrix, different from that demonstrated by the other TM impurities. We found that for (Si, Co) C and (Si, Ni) C the magnetic and nonmagnetic solutions are very close on the energy scale. In the range of the supercell volumes from 68 to 78 a.u.^{3}/atom the difference between the solutions is not greater than just a few meV/atom, which is close to the accuracy of our calculations, so that we can state that the two solutions coincide. Both (Si, Co) C and (Si, Ni) C are practically nonmagnetic in their equilibrium lattice configurations, with the magnetic moments of about 0.02 μ_B /TM atom. The magnetic moments go up noticeably as (Si, Ni) C or (Si, Co) C lattice is expanded reaching almost 0.4 μ_B /Ni and more than 0.5 μ_B /Co at the average atomic volume of 78 a.u.^{3}/atom (see Fig. 5). Thus, in the case of Co and Ni substitution, transition from the practically nonmagnetic to a magnetic state occurs gradually, in contrast to Mn, Cr, V, and Fe-doped SiC, where the two states are separated by a significant potential step/barrier. Average equilibrium per atom volumes for both Ni- and Co-doped SiC are practically equal to those of pure SiC, however, host lattice reconstruction due to Ni and Co substitution turns out to be rather pronounced. The expansion of the tetrahedra around the impurity atoms reaches 3% for (Si, Co) C and is as large as 10% in (Si, Ni) C at the equilibrium volumes. This local relaxation is compensated by the corresponding changes in other, primarily Si, atom positions so that the average per atom lattice volume remains almost equal to that in pure SiC. As the lattice of (Si, Ni) C or (Si, Co) C is further expanded away from the equilibrium, which is accompanied by the magnetic moment growth, the tetrahedra containing Co or Ni atoms expand even further, with their relative (to pure SiC with the same unit cell volume) volumes reaching 13% for Ni and 4% for Co at the average volume value of 78 a.u.^{3}/atom. At the same time, relaxation of the rest of the supercell reduces with the Si atoms returning to the positions characteristic to pure material, indicating that these atoms provide the supercell rigidity preventing the TM-C tetrahedra from expanding further and the TM atoms from possibly reaching the high-spin state.

TM species	ΔE_{FM-NM} (meV/atom)	M_0 (μ_B/TM)	M_A (μ_B/TM)
V	-12	0.74	0.16
Cr	-45	1.63	1.39
Mn	-18	2.09	1.39
Fe	33	1.92	0.42
Co	0	0.03	0.03
Ni	0	0.02	0.02

Table 1. Properties of TM-doped 3C-SiC: total energy differences between ferromagnetic and nonmagnetic states, ΔE_{FM-NM}; magnetic moments at the equilibrium cell volume in the FM state, M_0; average over the states magnetic moment at room temperature and equilibrium cell volume, M_A, estimated assuming that the energy relationship between the FM and NM states remains the same as at zero temperature and using Boltzmann distribution function for calculation of the occupation probabilities.

3.2 Magnetic-nonmagnetic state relationship and mixing

Our calculations show that, similarly to what was known as a peculiarity of certain transition metals and alloys (Moruzzi & Markus, 1988; Moruzzi et al., 1989; Wassermann, 1991; Timoshevskii et al., 2004), TM-doped SiC can exist in one or another of the states differing in their structural and electronic properties, and in particular with the TM impurities possessing either zero or a nonzero magnetic moment. The states are characterized by significant differences in the lattice geometry, electronic density distribution, and thus chemical bond structure. As a result, coupling of the substitutional transition metal impurity electronic orbitals to the surrounding host atoms in the different states varies depending both on the substitution atom electronic structure and on the electronic configuration of the surrounding dangling bonds as well as more distant atoms in the lattice. In turn, substitution of a host atom by a TM atom results in SiC lattice reconstruction, which self-consistently defines the electronic and magnetic configuration of the impurity and SiC-TM DMS material. Transition between the different states can be gradual, such as in the case of (Si, Ni) C and (Si, Co) C, or the states can be separated by an energy gap – (Si, V) C, (Si, Fe) C, (Si, Mn) C, (Si, Cr) C, and the width of the gap varies for different impurities and SiC polytypes. It is not unreasonable to suggest that such multistate nature could also be peculiar to the other DMS material systems. Below in Fig. 6 we present partial densities of states for the nonmagnetic and magnetic configurations for the example of Mn-doped 3C-SiC. Clearly, the DOS spectra are completely different and while in the magnetic case the state filling is half-metallic with no occupancy in the minority spin channel, in the nonmagnetic case the DOS is not polarized at all and Mn d-electrons symmetrically fill the majority and minority states creating zero net magnetic moment. Below we provide a detailed analysis of DOS spectra of SiC doped with different TM impurities and deduce the exchange mechanisms responsible for the magnetic ordering in each particular case.

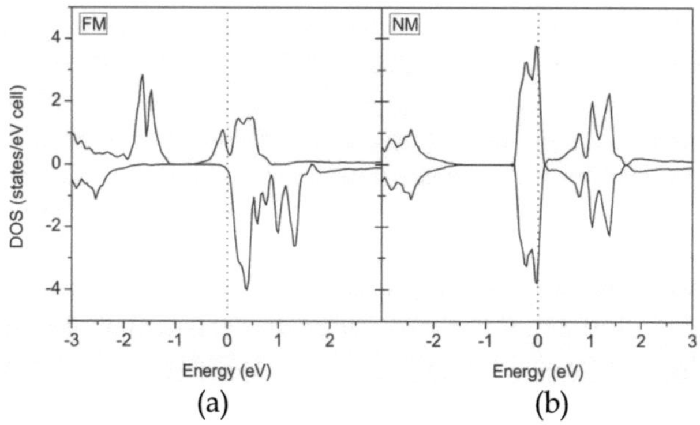

Fig. 6. Partial densities of states of Mn-doped 3C-SiC in its (a) ferromagnetic and (b) nonmagnetic states. Origin of the energy axis corresponds to the Fermi level position.

It is interesting to note that the magnetic state of SiC DMS is always realized in a lattice which is noticeably expanded compared to the nonmagnetic state (see discussion above), and one can therefore rather simplistically state that the TM atom "needs" more space for itself in the host lattice to acquire a magnetic moment. If the magnetic state is not the ground state of the DMS ((Si, Fe) C is an example), experimental realization of the magnetic state, which is a required prerequisite for spintronic device operation, may be achieved by either changing the lattice geometry in, for example, the lattice expanded due to ion implantation, or by making the magnetic state favourable by applying magnetic field under a high temperature equilibrium growth conditions. In the latter case, the magnetic state of the TM atoms leads to the corresponding lattice reconstruction, while nonequilibrium rapid lattice cooling to the point where the atom mobility is low could "freeze" the lattice in its magnetic state. If the magnetic state is the ground state it can of course be achieved under equilibrium growth conditions.

It is important to remember that calculations presented here were performed in the framework of the ground state density functional theory and describe SiC DMS material system at zero temperature conditions. At a nonzero temperature, the free energy of the crystal will additionally contain entropy terms with contributions due to both atomic position fluctuations and, in the magnetic case, due to the magnetic moment magnitude and direction fluctuations. Since configurational and magnetic fluctuations are due to similar thermal disordering processes, per atom values of the entropy terms quantifying fluctuations at these degrees of freedom of different nature should be of about the same order of magnitude. The experimental standard molar entropy values are equal to 16.5 J/mol K for hexagonal SiC and 16.6 J/mol K for cubic SiC (Lide, 2009). A simple estimate using these values yields the magnitude of the room temperature configuration disorder entropy of about 50 meV/atom. This exceeds the energy gap between the nonmagnetic and magnetic states for all impurities that we study and, consequently, at a high enough temperature the relationship between the two solutions can, in principle, change. This, for example, may lead to the magnetic state (ferromagnetic or antiferromagnetic) becoming energetically favourable, compared to the nonmagnetic state, in the case of Fe in SiC, for which the ground state solution is nonmagnetic. This possibility of changing the state preference may explain the magnetic response reported in several experimental studies of Fe-doped SiC (Theodoropoulou et al., 2002; Stromberg et al., 2006).

Furthermore, the fact that the magnetic and nonmagnetic solutions are separated by an energy gap comparable to room temperature thermal energy, allows us to suggest that a new equilibrium mixed state with a certain distribution of TM atoms between the magnetic and nonmagnetic states, and an average over this distribution magnetic moment, may be created at a nonzero temperature. Such state mixing may, in addition to other effects, explain the smaller, compared to calculated, values of the experimentally observed magnetic moments reported for GaN- and SiC-based DMS materials (Park et al., 2002; Lee et al., 2003, Singh et al., 2005; Bouziane et al., 2008). Somewhat similar mixed states can be realized in glasses where multiple unstable and easily thermally activated local atomic configurations exist. In these configurations atoms can occupy, with nearly equal probabilities, one of the several quasi-equilibrium positions and there exist a nonzero probability of tunneling transitions between these configurations (Anderson et al., 1972). If a mixed magnetic/nonmagnetic state is created, the average TM magnetic moments in this case will be lower compared to the purely magnetic case (Cf. Table 1). This leads to a reduction of the effective internal magnetic field stabilizing the magnetic order, which in turn means that the magnetic order will become weaker against thermal fluctuations or, equivalently, the mean-field Curie temperature will be reduced. Detailed analysis of the effects of state mixing and disorder is, however, rather involved and is beyond the scope of this study.

4. Magnetic ordering and critical temperatures

4.1 Magnetic coupling strength and range

We now turn to the calculations of the total energy of the ferromagnetic and antiferromagnetic states of the SiC-TM supercells containing a pair of TM atoms. We do not consider Co and Ni doping here, as magnetic moments of these impurities are practically zero at the equilibrium cell volume and are also rather small even in a significantly expanded lattice (Cf. Fig. 5). The strength of the exchange coupling and the values of the Curie or Néel temperature are estimated using the energy difference ΔE_{FM-AFM} between the FM and AFM states according to Eq. (1) above. We also note that the calculations of the total energy of the FM- and AFM-ordered supercells again, as in the case of magnetic moment calculations above, included a full inhomogeneous relaxation procedure. Since the FM and AFM states are in general characterized by different electronic density distribution and, self-consistently, atomic bond structure, we found accounting for the relaxation to be very important for an accurate calculation of the magnetic coupling strength and ordering temperatures. By separately relaxing the FM and AFM-ordered supercells we additionally find that magnetic moments in these two states are, in general, different.

SiC-TM FM-AFM energy differences and the corresponding values of the Curie or Néel temperature are summarized in Table 2 for TM substitution in 3C-SiC (supercells in Fig. 2 (a) and (b)). In the magnetic state, SiC DMS with substitutional Cr, Mn, and V orders ferromagnetically, while (Si, Fe) C orders antiferromagnetically. The strength and spatial extent of the exchange interaction vary significantly over the TM impurity range. For Cr the strength of magnetic coupling and, correspondingly, the Curie temperature drop from very high values in the nearest-neighbor arrangement to almost negligible values when the impurities are separated by two Si-C bilayers. There is even a change of sign and the interaction becomes antiferromagnetic in the latter case, however, such small FM-AFM energy difference values are comparable to the accuracy of our calculations and thus the values of the coupling energy and the Curie temperature can safely be considered to be zero. In contrast to Cr, Mn impurity exchange interaction is, although somewhat weaker, less dependent on the distance between impurity atoms. Similarly to the stronger, compared to Cr, volume dependence of Mn magnetic moments, this can be explained by the fact that in the case of Mn, an additional d-electron starts filling the antibonding t_2 orbitals, the symmetry of which is the reason for their delocalized nature and longer range exchange interaction (see Fig. 7 and a detailed discussion below). On the contrary, in (Si, Cr) C d-electrons fill the strongly localized nonbonding *e* orbitals, which explains the short ranged albeit strong exchange in (Si, Cr) C DMS. Thus, for a reasonable impurity concentration of 5% and a uniform impurity distribution, the Curie temperature of Mn-doped 3C-SiC DMS reaches 316 K, which makes this material a good candidate for achieving room temperature DMS behaviour. In the nearest-neighbour arrangement, with a much shorter Mn-Mn distance, the Curie temperature of (Si, Mn) C is comparable to the former case, which can be justified by a different mutual orientation of the electronic orbitals of Mn atoms and also by the fact that there is an about 15% reduction of the magnetic moments of Mn atoms in this configuration. In the case of V, the latter effect is even stronger and magnetic moments are completely quenched for V atoms as the nearest neighbors, so that the material becomes nonmagnetic. As mentioned above, we found magnetic coupling energy ΔE_{FM-AFM} and the corresponding critical temperature values to be strongly dependent on the supercell relaxation in both the FM- and the AFM-aligned magnetic moments. The dependence is particularly strong in the cases where the exchange coupling is long-ranged, such as in (Si, Mn) C, while lattice relaxation involves a large number of supercell atoms, not just the nearest neighbours as in, for example, (Si, Fe) C or (Si, Cr) C. This once again points out the importance of accounting for lattice relaxation in the calculations of magnetic properties of TM-doped semiconductors.

TM species	ΔE_{FM-AFM} (meV/cell)		Mean-field T_C (K)	
	d_{TM} = 14.27 a.u	d_{TM} = 5.82 a.u.	d_{TM} = 14.27 a.u	d_{TM} = 5.82 a.u.
Cr	5	-222	-19	859
Mn	-82	-92	316	357
V	-23	0	88	0
Fe	107	221	-412	-854

Table 2. Total energy differences between the ferromagnetic and antiferromagnetic states ΔE_{FM-AFM} and mean-field values of the critical temperature T_C for TM-doped 3C-SiC, calculated using Eq. (1) for two different distances between TM impurities d_{TM} (Fig. 2. (a) and (b)). Negative ΔE_{FM-AFM} and positive T_C is characteristic of the FM ordering, while for AFM ordering ΔE_{FM-AFM} is positive and T_C is negative, the latter can be interpreted as the positive Néel temperature.

We also note that the exchange coupling energy values ΔE_{FM-AFM}, although larger than the energy differences between the (ferro)magnetic and nonmagnetic states ΔE_{FM-NM} for the same impurities, cannot be compared to them directly. The exchange coupling energy, which is used in the corresponding mean-field Curie temperature expression, is the total energy difference between the FM- and AFM-ordered supercells containing a pair of TM atoms. This is the amount of energy needed to flip the spin of one of these atoms and thus change magnetic ordering of the supercell from FM to AFM or vice versa; therefore, this energy is counted *per supercell*. The second quantity ΔE_{FM-NM} is the amount of total energy needed to reconstruct the entire supercell (or the entire crystalline lattice) in such a way that the TM atoms in this lattice acquire or lose their magnetic moments. The reconstruction involves every atom in the supercell or lattice and therefore this quantity is counted *per atom*. If both the FM-AFM and FM-NM energy differences are brought to the same scale (per atom or per supercell) for comparison, the first difference becomes in all cases smaller than the second, indicating that the FM-AFM energy difference can be thought of as an additional "small splitting" of the corresponding magnetic state. Therefore, while for example the antiferromagnetic state of Fe atoms is more favourable compared to the ferromagnetic state, the total energy corresponding to either the FM or the AFM state is still significantly larger compared to the nonmagnetic state, which makes the nonmagnetic state the ground state in (Si, Fe) C.

4.2 Mechanisms of exchange interaction in SiC-TM

Understanding the nature of exchange coupling in DMS is an important problem in magnetic materials. As has been pointed out in (Belhadji et al., 2007), certain characteristic features of the DMS densities of states can be used not only for identification of the exchange mechanisms responsible for magnetic coupling but also for predicting the coupling strength and range. Due to the interaction with the crystal field in a tetrahedral lattice environment, such as in 3C-SiC lattice (the symmetry of 4H-SiC is lower than that of 3C-SiC), substitutional TM impurity d-orbitals form the so-called *e* and t_2 states creating localized energy levels in the semiconductor forbidden band or resonances in one of the allowed bands. The *e* and t_2 states, due to the different spatial distributions of their electronic clouds, experience different degrees of hybridization with the surrounding dangling bonds and thus the electrons filling these states experience different degrees of localization. The nature of the states TM d-electrons fill will determine the strength and range of exchange interaction between the d-electrons. In order to illustrate the details of TM state occupation, we calculated partial densities of states of V, Cr, Mn, and Fe in 3C-SiC. The nature of the particular impurity states is easiest to analyze in this polytype due its purely tetrahedral symmetry. In the other polytypes, such as in 4H-SiC, the T_d symmetry breaks, however, due to the short range tetrahedral environment the states will still approximately have the *e* or t_2 character. Thus, the analysis we present below is applicable to 4H-SiC as well. Partial densities of states for different TM impurities in 3C-SiC showing the impurity states of different nature are presented in Fig. 7.

When V substitutes for Si in SiC lattice, two of the three d-electrons completely fill the bonding t_2 states, while there is another d-electron half-filling the majority *e*-states. In such situation, when the Fermi level is in the middle of an impurity band, the energy is gained by aligning atomic spins in parallel. This leads to broadening of this band and shifting of the weight of the electron distribution to lower energies. This is Zener's double exchange mechanism which leads to ferromagnetism (Sato et al., 2004; Akai, 1998). The width of the impurity band and thus the energy gain due to double exchange scales as the square root of impurity concentration and is linear in the hopping matrix element between the neighbouring impurities. This square root dependence implies that the Curie temperature increases quickly already at small impurity concentrations and then almost saturates as a

function of impurity concentration. Another mechanism characteristic to the electronic structure of (Si, V) C DMS is hybridization between the *e* and t_2 states of the majority spin channel. This hybridization results in the filled *e* states shifting to lower energies. The resulting energy gain additionally stabilizes ferromagnetism by superexchange (Sato et al., 2004; Akai, 1998). This energy gain scales linearly with the impurity concentration and is proportional to the square of the hopping matrix element between the *e* and t_2 states, which is relatively small due to the localized nature of the *e* orbitals. An additional consideration following from the Heisenberg model is that rather small magnetic moment of V leads to a reduction of the exchange interaction between V atoms and, as a result, to a relatively small T_C. Localized nature of the nonbonding *e*-states is another reason for the reduction of T_C in (Si, V) C DMS.
In Cr-doped SiC there is an additional d-electron in the *e*-states and the Fermi level falls between the majority *e* and antibonding t_2 bands, the former being fully filled and the latter being empty. No energy can be gained by broadening the filled *e* impurity band and, therefore, double exchange is not important in this case. Hybridization between the *e* and t_2 states of the majority spin channel again, as in the case of (Si, V) C, results in the energy gain from filled *e* states shifting to lower energies and thus ferromagnetism is due to superexchange. The localized nature of the *e* orbitals and resulting quick fall-off of the exchange interaction with the TM-TM distance are the reasons for the strong concentration dependence of the ordering temperature in (Si, Cr) C (Miao & Lambrecht, 2006).

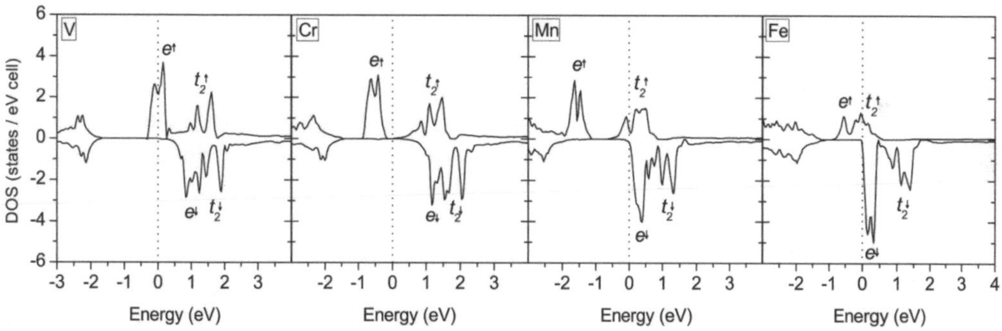

Fig. 7. Partial densities of states of TM impurities in 3C-SiC. Origin of the energy axis corresponds to the Fermi level position.

One more electron per TM in (Si, Mn) C is in the majority channel antibonding t_2 state and the Fermi level falls within the t_2 band. Thus, the double exchange kicks in again leading to ferromagnetism. Contrary to the case of (Si, V) C, however, the exchange interaction is between d-electrons in the less localized t_2 states, which results in a weaker TM-TM distance dependence and a more extended nature of the exchange. Ferromagnetic superexchange interaction due to the coupling between the majority *e* and almost empty t_2 states, although weaker than in (Si, Cr) C, because of the larger energy difference between these states, provides an additional stabilization of the ferromagnetic state in (Si, Mn) C. Another mechanism, which is characteristic to the electronic structure of (Si, Mn) C and which acts in competition with the double exchange, is the antiferromagnetic superexchange (Belhadji et al., 2007). The wave functions of the majority *e* and t_2 states hybridize with the antiferromagnetically aligned *e* and t_2 states of the opposite spin channel. As a result, the lower in energy states are shifted to even lower energies and the higher states are shifted to higher energies. The energy is gained by reduction of the energy of filled states. The mechanisms of ferromagnetic and antiferromagnetic superexchange interactions are similar; whether the parallel or the anti-parallel alignment will be more energetically favourable depends on the strength of hybridization of states in the same and in the opposite spin channels. The double-exchange-stabilized ferromagnetic state is more favourable if the Fermi level lies well within the impurity band, while the energy gain due to superexchange is larger when the Fermi level lies between the bands of the same or opposite spin channels or near the band edges (Belhadji et al., 2007). In (Si, Mn) C, the hybridization of the majority and minority *e* states is weak because of their localized nature and relatively large energy difference between the two. Hybridization between the t_2 states is stronger, but only a small fraction of these states is filled, so in (Si, Mn) C double exchange wins stabilizing ferromagnetism.

In Fe-doped SiC, there is again a competition between the ferromagnetic double exchange and antiferromagnetic superexchange mechanisms. However in this case, the Fermi level is close to the top of the majority bands and separates the majority and minority bands. The energy gain due to double exchange is reduced because of the nearly complete filling of the antibonding majority t_2 band. At the same time, the superexchange becomes more dominant due to the stronger hybridization between the *e* and t_2 bands in different spin channels. This explains why (Si, Fe) C orders antiferromagnetically.

Recalling that the calculations are performed for the ground state of the DMS, we note that only direct exchange mechanisms can be accounted for in our model. At a nonzero temperature and if a sufficient concentration of free carriers is created in the DMS, carrier-mediated exchange can also be important. Depending on the concentrations of TM and other impurities and the details of the material atomic and electronic structure, different exchange channels may play different roles in stabilizing an ordered magnetic state. In particular, different free carrier and TM atom concentrations may result in RKKY-type exchange interactions of different signs and having an oscillating character depending on the distance between magnetic TM impurities. Contribution of this carrier-mediated exchange may favour ferromagnetic, antiferromagnetic or spin-glass-like order (the latter is, however, specific for nonuniform TM impurity distributions). The resulting effective magnetic ordering will depend on the relative strengths of the direct and carrier-mediated exchange interactions.

5. Room temperature ferromagnetism in Mn-doped SiC

Having identified Mn as the most promising doping choice for achieving room temperature ferromagnetism in SiC DMS, we now turn to a detailed investigation of magnetic properties of various "flavours" of (Si, Mn) C DMS, i.e. with Mn substituting in different SiC polytypes and at lattice sites with different symmetries thus producing different mutual spatial configurations of Mn atoms (and, consequently, of their magnetic moments). In addition to the supercells in Fig. 2 (a) and (b), for which this was done in the previous section, we calculate the total energy differences ΔE_{FM-AFM} between the FM- and AFM-aligned states and the corresponding Curie temperature values for the supercells shown in Fig. 2 (c-e). We then compare the results for all the 3C- and 4H-SiC crystal structures.

The results of the calculations are summarized in Table 3. One can see that for most of the configurations the strength of the exchange interaction is sufficient to lead to above room temperature ferromagnetic ordering. As one could expect, there is a significant dependence of the exchange coupling strength on the substitution lattice site so that in some cases, such as for 4H-SiC with Mn atoms above each other in the *c*-axis direction (Fig. 2 (c)), the exchange is weaker although the impurity separation is relatively small. This lattice structure demonstrates the lowest Curie temperature, significantly below room temperature. Magnetic moments of substitutional Mn in this case are larger, closer to the atomic limit. For diagonally arranged impurities, either at the hexagonal or at the cubic sites in 4H-SiC (structures (d) and (e) in Fig. 2), exchange interaction becomes significantly stronger. In contrast to the 4H-SiC structure of Fig. 2 (c), in 3C-SiC the coupling between vertically arranged atoms (Fig. 2 (a)) is fairly strong with the corresponding value of the Curie temperature exceeding room temperature. According to the discussion above, Mn d-electrons in 3C-SiC occupy partially filled anti-bonding t_2 states. The symmetry of these states provides a significant coupling with the surrounding atoms and results in the extended range of exchange interaction between Mn atoms. This is in contrast to, for example, SiC-Cr, where d-electrons occupy strongly localized non-bonding *e*-states, while the anti-bonding t_2 states are empty, which leads to the very short-ranged albeit strong exchange and low Curie temperature. As already mentioned, although in 4H-SiC the T_d symmetry no longer applies, the d-states still have the *e*- or t_2-like character and the particulars of their filling and local bonding remains the same.

SiC-TM crystal structure (Cf. Fig. 2)	Minimum distance between Mn atoms (a.u.)	ΔE_{FM-AFM} (meV/cell)	Mean-field T_C (K) (Eq. (1))	Magnetic moments FM/AFM (μ_B/Mn)
(a) 3C-SiC, impurities above each other	14.27	-82	316	2.18/2.05
(b) 3C-SiC, impurities as nearest neighbours in-plane	5.82	-92	357	1.70/1.53
(c) 4H-SiC, c-site, impurities above each other	9.59	-53	205	2.36/2.24
(d) 4H-SiC, c-site, diagonal impurity arrangement	11.61	-142	550	2.28/2.04
(e) 4H-SiC, h-site, diagonal impurity arrangement	10.18	-104	403	2.29/2.04

Table 3. Energy differences between the FM- and AFM-ordered supercells, the corresponding values of the Curie temperature and magnetic moments of Mn atoms in different configurations in 3C- and 4H-SiC.

Although the general rules apply both to 3C- and 4H-SiC, there still is a difference between the polytypes as becomes evident from the significant Curie temperature differences. Due to the different medium- and long-range order, the wave functions of Mn t_2-like states, which are responsible for the exchange coupling between the TM atoms, are oriented differently with respect to each other in the different polytypes. Charge densities corresponding to Mn anti-bonding t_2 states in 3C-SiC, which were visualized in (Miao & Lambrecht, 2006), clearly illustrate that in this polytype d-electron wave function lobes are distributed approximately along the Mn bonds to the surrounding C atoms and extend significantly beyond the nearest neighbours. Charge density isosurfaces of filled anti-bonding Mn states, corresponding to an equal value of this quantity, are shown in Fig. 8 for two different 4H-SiC FM-ordered supercells of Fig. 2 (c) and (d), demonstrating the smallest and largest ordering temperatures, respectively. Calculations show that the spatial distribution of these orbitals in 4H-SiC is somewhat different compared to 3C-SiC, with the more localized nature for the vertical impurity configuration (Fig. 2 (c)) explaining the lower value of the Curie temperature for this configuration in 4H-SiC despite the shorter distance between Mn atoms compared to the diagonal configuration (Fig. 2 (d)). The total energies of the supercells with the impurities substituting at the hexagonal sites are about 13 meV/atom lower compared to the case of the impurities at the cubic sites, making the former configuration more energetically favourable. However, in real DMS materials at high enough temperatures impurity atoms will be distributed among the sites with the hexagonal and cubic symmetries creating a new state with the magnetic properties averaged over the different configurations. As mentioned above, Mn-doped SiC can also exist in the nonmagnetic state, which is characterized by different crystal lattice structure and is at zero temperature separated from the magnetic state by an energy gap (Cf. Fig. 4). The energy gap is comparable to the thermal energy at room temperature and, therefore, the nonmagnetic state may also mix in at a nonzero temperature.

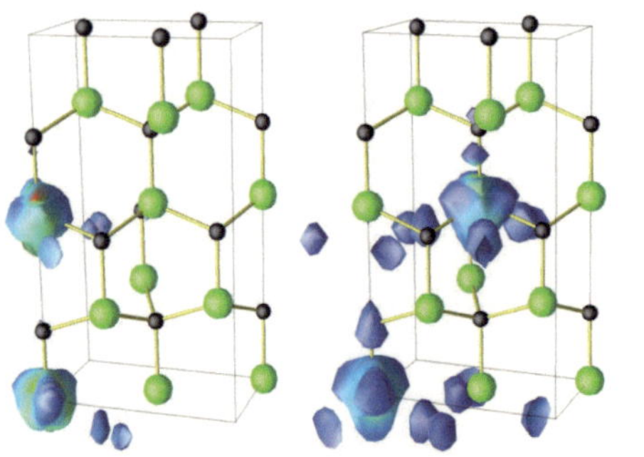

(a) (b)

Fig. 8. Charge density isosurfaces for impurity states of Mn atoms substituting at quasicubic sites in 4H-SiC in (a) vertical and (b) diagonal configurations corresponding to the supercells shown in Fig. 2 (c) and (d).

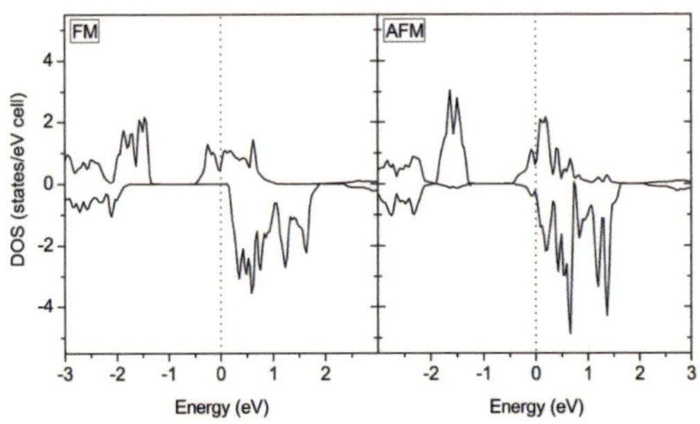

(a) (b)

Fig. 9. Partial densities of states of Mn in 4H-SiC, calculated for the crystal structure shown in Fig. 2 (d) in its (a) ferromagnetic and (b) antiferromagnetic states. Origin of the energy axis corresponds to the Fermi level position.

Another interesting observation following from Table 3 is that Mn magnetic moments are in all cases smaller in the AFM-ordered supercells compared to the FM-ordered supercells. Relating this fact to the partial densities of states of Mn in the FM and AFM states shown in Fig. 9, one can conclude that in the AFM-ordered supercell the half-metallic behaviour is lost and the minority states become partially occupied, which leads to the reduction of the DOS polarization. Recalling the Heisenberg expressions for the energies of the FM and AFM states, one can perhaps also argue that the reduction of the magnetic moments in the AFM state and their increase in the FM state can be one of the factors stabilizing the FM state in the SiC-Mn DMS.

6. Concluding remarks and outlook

We presented a study of structural, electronic, and magnetic properties of SiC doped with first row transition metal impurities. For this we used density functional theory FLAPW calculation technique which is probably one of the most accurate first principles techniques to date. One of the primary results that we have established is that SiC DMSs can exist in either a magnetic or nonmagnetic state. The different states are characterized by substantially different equilibrium crystal lattice structures. Lattice reconstruction compared to pure SiC is typically significant when SiC DMSs are in the magnetic states, while in the nonmagnetic state the reconstruction is relatively minor. For V, Cr, Mn, and Fe doping the states are separated by an energy gap which needs to be overcome in order to change the state from nonmagnetic to magnetic and vice versa. At zero temperature, the magnetic state is energetically favourable in V-, Cr-, and Mn-doped SiC, while for Fe doping the nonmagnetic state is the ground state. The energy differences between the magnetic and nonmagnetic states vary from impurity to impurity and are also different in different SiC polytypes, being in many cases comparable to thermal energy at room temperature. On the other hand, in Ni- and Co-doped SiC the two states practically coincide, while the magnetic moments are close to zero in equilibrium lattice and gradually increase as the lattice is expanded. We argue that at a nonzero temperature the relationship between the two states of SiC DMSs can change due to the free energy entropy term. Furthermore, we speculate that a mix of the magnetic and nonmagnetic states is possible, which would change the average magnetic moment of the system and self-consistently its magnetic order. Such existence of multiple magnetic states may be peculiar to the other DMS material systems as well.

The values of magnetic moments, exchange interaction strength and range vary significantly across the range of impurities studied. While magnetic moments of Ni and Co grow noticeably starting from very small values at equilibrium unit cell volume, they are still rather small (~0.5 μ_B) even in a significantly expanded lattice. Even though the calculated lattice parameters correspond to zero temperature conditions, given the values of thermal expansion coefficient of SiC (Taylor & Jones, 1960), such larger lattice volumes cannot be reached by thermal expansion at practical temperatures. Thus Ni- and Co-doped SiC can be considered practically nonmagnetic and will likely not be suitable for spin injection or other spin-electronic applications. V, Cr, Mn, and Fe can exist in high-spin states in SiC matrix. We found that magnetic Fe in SiC orders antiferromagnetically, while the rest of the impurities order ferromagnetically. Small magnetic moments of V atoms and localized nature of d-orbitals in (Si, V) C prevent this DMS from reaching high ordering temperatures. Magnetic interaction between Cr atoms was found to be strong but short-ranged, so that (Si, Cr) C is also not expected to exhibit high ordering temperatures at reasonable Cr concentrations, at least if only the direct exchange mechanism is responsible for the ordering. On the other hand, Mn-Mn interaction, although not as strong, was found to be relatively weakly dependent on the interatomic distance. The resulting mean-field Curie temperature was estimated to be above room temperature even for the case of the relatively modest Mn doping of 5%. Exchange interaction was found to be strongly dependent on the spatial distribution of Mn in SiC lattice with the value of the Curie temperature reaching 550 K in one of the atomic configurations.

Given the long-range character of Mn-Mn exchange interaction in SiC host, among the TM impurities Mn is perhaps the most promising candidate for achieving room temperature

DMS behaviour. The analysis of the band structure of (Si, Mn) C suggests that ferromagnetic coupling in Mn-doped SiC is mostly due to Zener's double exchange, which is related to the fact that the Fermi level in (Si, Mn) C lies within a Mn impurity band in the gap. This result is reliable in the sense that there is no spurious resonance of Mn levels in the allowed bands. The latter is often a problem due to the notorious underestimation of the bandgap by DFT+LDA or DFT+GGA and, as a result, the source of a mistakenly assumed long-range carrier-mediated exchange in, for example, diluted magnetic oxides such as Mn-doped ZnO (Zunger et al., 2010 and references therein). Such incorrect placement of TM levels within the allowed bands leads to the creation of a hostlike delocalized orbital that promises long-range exchange interaction (similarly to the classic case of (Ga, Mn) As (MacDonald et al., 2005; Sato et al., 2010), where the highest Curie temperatures to date were confirmed for the "true DMS" material and where Mn impurities serve as a source of both magnetic moments and valence band holes, the latter mediating magnetic moment interaction via an RKKY-type Zener's kinetic p-d exchange mechanism). Thus, the electronic structure of (Si, Mn) C is rather different from that of (Ga, Mn) As and more resembling that of wide bandgap (Ga, Mn) N DMS (Sato et al., 2007) with the Fermi level within an impurity band in the gap. Still, however, we find the remarkably long range interaction between Mn atoms in SiC leading to the Curie temperature values being safely above room temperature.

It is of course important to keep in mind the approximations imposed by the use of the Heisenberg and mean-field models for the description of the SiC DMS system. In particular, many-electron effects, which may be important in the case of narrow impurity bands, if taken into account, may lead to an improvement of the Curie temperature estimations. For disordered systems one may need to additionally take into account the problem of exchange interaction percolation, as in this case the mean-field model tends to overestimate Curie temperature values for low impurity concentrations. Such overestimation, however, may not be as pronounced in the case of the longer-range exchange, which is demonstrated by substitutional Mn in SiC. On the other hand, introduction of free carriers in order to achieve free-carrier-mediated exchange, which is usually sufficiently long ranged and could further stabilize the FM state in SiC DMS, may only turn out possible by co-doping with a shallow impurity in addition to a TM, since TM levels in SiC are expected to be deep donors and acceptors (Miao & Lambrecht, 2006). Shallow impurity concentration will need to be comparable to the TM concentration to overcome compensation and produce a sufficient concentration of free carriers. This entails further reconstruction of the host lattice and is subject to solubility difficulties. Given the very high, compared to those typical to semiconductor devices, TM concentrations which are usually required for achieving the needed exchange coupling efficiency, one will likely need to resort to nonequilibrium growth and doping. Co-doping may become another alternative for increasing TM solubility (Pan et al., 2010). The strong influence of the substitution site and lattice relaxation on magnetic properties of SiC DMS also suggests that magnetic properties of such materials are expected to be very sensitive to the growth and processing conditions and point out the high importance of a careful experimental approach to their synthesis.

Rather scarce experimental data available for SiC DMSs do not, unfortunately, allow to judge conclusively on how accurate our predictions are. In a recent communication (Jin et al., 2008) 10% Cr-doped amorphous SiC was reported to be ferromagnetic above room temperature with the average magnetic moment of 0.27 μ_B. Although Cr concentration of 10% approximately corresponds to the $Si_7C_9Cr_2$ supercell for which we obtained a large value of the ordering temperature, it should be kept in mind that effectively this result was obtained for the nearest neighbour dimers. As the magnetic interaction was found to drop off quickly beyond the nearest neighbours, for a random impurity distribution it would likely not percolate throughout the lattice. Although the authors reported their samples to exhibit hole conductivity, it may have been due to residual impurities or defects in the material since, according to the calculations (Miao & Lambrecht, 2006), both Cr donor and acceptor levels are deep in the gap and cannot be the source of free carriers. Nevertheless, the existence of free holes may have led to an additional, hole-mediated exchange interaction playing the role in stabilizing ferromagnetism at larger distances between the impurities. At least some of the experimental studies of Fe-doped SiC reported ferromagnetic DMS behaviour in Fe-implanted 6H-SiC (Theodoropoulou et al., 2002; Stromberg et al., 2006). Our calculations reveal nonmagnetic ground state and antiferromagnetic ordering in the magnetic state of (Si, Fe) C. Such discrepancy may be explained by several factors. First, as we already mentioned, state preference may change at nonzero temperatures. Second, our calculations were done for the TM impurities in the SiC lattice distributed either homogeneously or in the nearest neighbour pair configuration. Various other substitution TM configurations are possible and, as has been shown in (Cui et al., 2007) for (Ga, Cr) N DMS, different nanocluster complexes, including substitutional and interstitial TM atoms, as well as the other lattice defects, not only may be more favourable for formation compared to the homogeneous or random impurity distributions, but also can significantly alter magnetic properties of the DMS such as the ordering type and temperature. Various impurity complexes can coexist in a statistical distribution, further complicating the physical picture. Contribution of secondary phases to the magnetic signal can also not be ruled out completely (Stromberg et al., 2006).

SiC doped with Mn received more attention from the experimentalists compared to the other SiC DMSs and the reported promise of this impurity for obtaining high Curie temperatures is in-line with our conclusions. Most of the authors reported ferromagnetic ordering close to or above room temperature in Mn-doped SiC. These reports include ferromagnetism with the Curie temperatures of around 250 K in 5% Mn-implanted 6H-SiC (Theodoropoulou et al., 2002), low-Mn-doped polycrystalline 3C-SiC (Ma et al., 2007) and single crystalline 6H-SiC (Song et al., 2009). T_C around 300 K was reported for Mn-Si films synthesized on top of 4H-SiC and for Mn-diffused 3C-SiC (Wang et al., 2008). It should be stressed, however, that conclusions on the magnetic ordering and critical temperatures cannot be reliable if they are made using only magnetisation data, which is often the case. Lessons that have been learned in the last years suggest that the complex nature of DMS materials requires using a broad range of complementary techniques for an unambiguous all-around description of the material in-hand. Only such approach will allow avoiding possible pitfalls the experimentalist can encounter. Future theoretical work should also include studies of co-doping, impurity complexes and, perhaps, disorder to be able to provide an even more accurate guidance for the experimentalist. Recent detailed investigation of structural, magnetic and magneto-optical properties of thin Mn-doped SiC films (Wang et al., 2008) and investigation of Mn-defect complexes in 3C-SiC (Bouziane et al., 2008) are the first step towards understanding the nature of SiC DMSs, where impurities and defects constitute a significant portion of the crystal lattice atoms.

According to science historians, Wolfgang Pauli some 80 years ago said regarding impurities in semiconductors: "One shouldn't work on semiconductors, that's a filthy mess;

who knows whether any semiconductors exist." Modern semiconductor technology, which few these days can imagine the life without, managed to make an exquisite use of these once troublesome impurities. Will the researchers and technologists be able to continue the success story by integrating magnetism and harnessing the spin? We hope that the presented analysis of the magnetic states of SiC DMSs and the tendencies that were established may serve as a "road map" and motivation for experimentalists for implementing magnetism in silicon carbide, one of the oldest known semiconductors.

7. References

Akai, H. (1998). Ferromagnetism and Its Stability in the Diluted Magnetic Semiconductor (In, Mn)As. *Phys. Rev. Lett.*, 81, pp. 3002-3005.

Anderson, P.; Halperin, B. & Varma, C. (1972). Anomalous low-temperature thermal properties of glasses and spin glasses. *Philos. Mag.*, 25, pp. 1-9.

Bechstedt, F.; Käckell, P.; Zywietz, A.; Karch, K.; Adolph, B.; Tenelsen, K. & Furthmüller, J. (1997). Polytypism and Properties of Silicon Carbide. *Phys. Status Solid, B*, 202, pp. 35-62.

Belhadji, B.; Bergqvist, L.; Zeller, R.; Dederichs, P.; Sato, K. & Katayama-Yoshida, H. (2007). Trends of exchange interactions in dilute magnetic semiconductors. *J. Phys.: Condens. Matter*, 19, 436227.

Bouziane, K.; Mamor, M.; Elzain, M.; Djemia, Ph. & Chérif, S. (2008). Defects and magnetic properties in Mn-implanted 3C-SiC epilayer on Si(100): Experiments and first-principles calculations. *Phys. Rev. B*, 78, 195305.

Bratkovsky, A. (2008). Spintronic effects in metallic, semiconductor, metal–oxide and metal–semiconductor heterostructures. *Rep. Prog. Phys.*, 71, 026502.

Cui, X.; Medvedeva, J.; Delley, B.; Freeman, A. & Stampfl, C. (2007). Spatial distribution and magnetism in poly-Cr-doped GaN from first principles. *Phys. Rev. B*, 75, 155205.

Dewhurst, J.; Sharma, S. & Ambrosch-Draxl, C. (2004). *http://elk.sourceforge.net.*

Dietl, T.; Ohno, H.; Matsukura, F.; Cibert, J.; & Ferrand, D. (2000). Zener Model Description of Ferromagnetism in Zinc-Blende Magnetic Semiconductor. *Science*, 287, pp 1019-1022.

Dietl, T.; Ohno, H.; & Matsukura, F. (2001). Hole-mediated ferromagnetism in tetrahedrally coordinated semiconductors. *Phys. Rev. B* 63, 195205, pp 1-21.

Gregg, J.; Petej, I.; Jouguelet, E. & Dennis, C. (2002). Spin electronics – a review. *J. Phys. D: Appl. Phys.*, 35, pp. R121-R155.

Gubanov, V.; Boekema, C. & Fong, C. (2001). Electronic structure of cubic silicon–carbide doped by 3d magnetic ions. *Appl. Phys. Lett.*, 78, pp. 216-218.

Hebard, A.; Rairigh, R.; Kelly, J.; Pearton, S.; Abernathy, C.; Chu, S.; & Wilson, R. G. (2004). Mining for high *Tc* ferromagnetism in ion-implanted dilute magnetic semiconductors. *J. Phys. D: Appl. Phys.*, 37, pp 511-517.

Huang, Z. & Chen. Q. (2007). Magnetic properties of Cr-doped 6H-SiC single crystals. *J. Magn. Magn. Mater.*, 313, pp. 111-114.

Jin, C.; Wu, X.; Zhuge, L.; Sha, Z.; & Hong, B. (2008). Electric and magnetic properties of Cr-doped SiC films grown by dual ion beam sputtering deposition. *J. Phys. D: Appl. Phys.*, 41, 035005.

Kim, Y.; Kim, H.; Yu, B.; Choi, D. & Chung, Y. (2004). Ab Initio study of magnetic properties of SiC-based diluted magnetic semiconductors. *Key Engineering Materials*, 264-268, pp. 1237-1240.

Kim, Y. & Chung, Y. (2005). Magnetic and Half-Metallic Properties Of Cr-Doped SiC. *IEEE Trans. Magnetics*, 41, pp. 2733-2735.

Lee, J.; Lim, J.; Khim, Z.; Park, Y.; Pearton, S. & Chu, S. (2003). Magnetic and structural properties of Co, Cr, V ion-implanted GaN. *J. Appl. Phys*, 93, pp. 4512-4516.

Lide, D. (Editor). (2009). *CRC Handbook of Chemistry and Physics, 90th ed.* CRC, ISBN: 9781420090840, Boca Raton, FL.

Ma, S.; Sun, Y.; Zhao, B.; Tong, P.; Zhu, X. & Song, W. (2007). Magnetic properties of Mn-doped cubic silicon carbide. *Physica B: Condensed Matter*, 394, pp. 122-126.

MacDonald, A.; Schiffer, P. & Samarth, N. (2005). Ferromagnetic semiconductors: moving beyond (Ga,Mn)As. *Nature Materials*, 4, pp. 195–202.

Miao, M. & Lambrecht, W. (2003). Magnetic properties of substitutional 3d transition metal impurities in silicon carbide. *Phys. Rev. B*, 68, 125204.

Miao, M. & Lambrecht, W. (2006). Electronic structure and magnetic properties of transition-metal-doped 3C and 4H silicon carbide. *Phys. Rev. B*, 74, 235218.

Moruzzi V. & Marcus, P. (1988). Magnetism in bcc 3d transition metals. *J. Appl. Phys.*, 64, pp. 5598-5600.

Moruzzi, V.; Marcus, P. & Kubler, J. (1989). Magnetovolume instabilities and ferromagnetism versus antiferromagnetism in bulk fcc iron and manganese. *Phys. Rev. B*, 39, pp. 6957-6962.

Munekata, H.; Ohno, H.; von Molnar, S.; Segmüller, A.; Chang, L.; & Esaki, L. (1989). Diluted magnetic III-V semiconductors. *Phys. Rev. Lett.*, 63, pp. 1849–1852.

Olejník, K.; Owen, M.; Novák, V.; Mašek, J.; Irvine, A.; Wunderlich, J.; & Jungwirth, T. (2008). Enhanced annealing, high Curie temperature, and low-voltage gating in (Ga,Mn)As: A surface oxide control study. *Phys. Rev. B*, 78, 054403.

Ohno, H.; Shen, A.; Matsukura, F.; Oiwa, A.; Endo, A.; Katsumoto, S.; & Iye, Y. (1996). (Ga,Mn)As: A new diluted magnetic semiconductor based on GaAs. *Appl. Phys. Lett.*, 69, pp. 363-365.

Pan, H.; Zhang, Y-W.; Shenoy, V. & Gao, H. (2010). Controllable magnetic property of SiC by anion-cation codoping. *Appl. Phys. Lett.*, 96, 192510.

Park, S.; Lee, H.; Cho, Y.; Jeong, S.; Cho, C. & Cho, S. (2002). Room-temperature ferromagnetism in Cr-doped GaN single crystals. *Appl. Phys. Lett.*, 80, pp. 4187-4189.

Pashitskii E. & Ryabchenko S. (1979). Magnetic ordering in semiconductors with magnetic impurities. *Soviet. Phys. Solid State*, 21, pp. 322-323.

Pearton, S.; Abernathy, C.; Overberg, M.; Thaler, G.; Norton, D.; Theodoropoulou, N.; Hebard, A.; Park, Y.; Ren, F.; Kim, J.; & Boatner, L. A. (2003). Wide band gap ferromagnetic semiconductors and oxides. *J. Appl. Phys.*, 93, pp 1-13.

Perdew, J.; Burke, K. & Ernzerhof, M. (1996). Generalized Gradient Approximation Made Simple. *Phys. Rev. Lett.*, 77, pp. 3865-3868.

Theodoropoulou, N.; Hebard, A.; Chu, S.; Overberg, M.; Abernathy, C.; Pearton, S.; Wilson, R.; Zavada, J.; & Park, Y. (2002). Magnetic and structural properties of Fe, Ni, and Mn-implanted SiC. *J. Vac. Sci. Technol.*, A 20, pp. 579-582.

Sato, K.; Dederichs, P.; Katayama-Yoshida, H. & Kudrnovský, J. (2004). Exchange interactions in diluted magnetic semiconductors. *J. Phys.: Condens. Matter*, 16, pp. S5491-S5497.

Sato, K.; Fukushima, T. & Katayama-Yoshida, H. (2007). Ferromagnetism and spinodal decomposition in dilute magnetic nitride semiconductors. *J. Phys.: Condens. Matter,* 19, 365212.

Sato, K.; Bergqvist, L.; Kudrnovský, J.; Dederichs, P.; Eriksson, O.; Turek, I.; Sanyal, B.; Bouzerar, G.; Katayama-Yoshida, H.; Dinh, V.; Fukushima, T.; Kizaki, H. & Zeller, R. (2010). First-principles theory of dilute magnetic semiconductors. *Rev. Mod. Phys.*, 82, pp. 1633-1690.

Seong, H.; Park, T; Lee, S; Lee, K; Park, J.; & Choi, H. (2009). Magnetic Properties of Vanadium-Doped Silicon Carbide Nanowires. *Met. Mater. Int.*, 15, pp. 107-111.

Shaposhnikov, V. & Sobolev, N. (2004). The electronic structure and magnetic properties of transition metal-doped silicon carbide. *J. Phys.: Condens. Matter,* 16, pp. 1761-1768.

Sharma, S.; Dewhurst, J.; Lathiotakis, N. & Gross, E. (2008). Reduced density matrix functional for many-electron systems. *Phys. Rev. B*, 78, 201103.

Singh, D.; Nordstrom, L. (2006). *Planewaves, Pseudopotentials, and the LAPW Method, 2nd ed.,* Springer, ISBN: 978-0-387-28780-5, Berlin.

Singh, R.; Wu, S.; Liu, H.; Gu, L.; Smith, D. & Newman, N. (2005). The role of Cr substitution on the ferromagnetic properties of $Ga_{1-x}Cr_xN$. *Appl. Phys. Lett.*, 86, 012504.

Song, B.; Bao, H.; Li, H.; Lei, M.; Jian, J.; Han, J.; Zhang, X.; Meng, S.; Wang, W. & Chen, X. (2009). Magnetic properties of Mn-doped 6H-SiC. *Appl. Phys. Lett.*, 94, 102508.

Story, T.; Gałązka, R.; Frankel, R. & Wolff, P. (1986). Carrier-concentration–induced ferromagnetism in PbSnMnTe. *Phys. Rev. Lett.*, 56, pp. 777–779.

Stromberg, F.; Keune, W.; Chen, X.; Bedanta, S.; Reuther, H.; & Mücklich, A. (2006). The origin of ferromagnetism in ^{57}Fe ion-implanted semiconducting 6H-polytype silicon carbide. *J. Phys.: Condens. Matter,* 18, pp. 9881-9900.

Syväjärvi, M.; Stanciu, V.; Izadifard, M.; Chen, W.; Buyanova, I.; Svedlindh, P.; & Yakimova, R. (2004). As-Grown 4H-SiC Epilayers with Magnetic Properties. *Mater. Sci. Forum,* 457-460, pp. 747-750.

Taylor, A. & Jones, R. (1960). Silicon carbide (SiC), lattice parameters, thermal expansion, In: *Silicon Carbide - A High Temperature Semiconductor,* edited by J.R. O'Connor, J. Smiltens, pp.147-154, Pergamon Press, New York.

Timoshevskii, A.; Yanchitsky, B. & Bakai, A. (2004). Composition dependence of the low-temperature magnetic ordering and the hyperfine interactions in Fe–N austenite. *Low Temperature Physics,* 30, pp. 469-478.

Vinet, P.; Rose, J.; Ferrante, J. & Smith, J. (1989). Universal features of the equation of state of solids. *J. Phys.: Condens. Matter,* 1, pp. 1941-1963.

Wang, W.; Takano, F.; Akinaga, H. & Ofuchi, H. (2007). Structural, magnetic, and magnetotransport properties of Mn-Si films synthesized on a 4H-SiC(0001) wafer. *Phys. Rev. B*, 75, 165323.

Wang, W.; Takano, F.; Ofuchi, H. & Akinaga, H. (2008). Local structural, magnetic and magneto-optical properties of Mn-doped SiC films prepared on a 3C–SiC(001) wafer. *New Journal of Physics,* 10, 055006.

Wassermann E. The Invar problem. (1991). *J. Magn. Magn. Mater.,* 100, pp. 346-362.

Zunger, A.; Lany, S. & Raebiger, H. (2010). The quest for dilute ferromagnetism in semiconductors: Guides and misguides by theory. *Physics*, 3, 53.

Žutić, I.; Fabian, J.& Das Sarma, S. Spintronics: Fundamentals and applications. (2004). *Rev. Mod. Phys.*, 76, pp. 323-410.

6

Electrodynamical Modelling of Open Cylindrical and Rectangular Silicon Carbide Waveguides

L. Nickelson, S. Asmontas and T. Gric
Semiconductor Physics Institute of Center for Physical Sciences and Technology
Vilnius, Lithuania

1. Introduction

Silicon carbide (SiC) waveguides operating at the microwave range are presently being developed for advantageous use in high-temperature, high-voltage, high-power, high critical breakdown field and high-radiation conditions. SiC does not feel the impact of any acids or molten salts up to 800°C. Additionally SiC devices may be placed very close together, providing high device packing density for integrated circuits.

SiC has superior properties for high-power electronic devices, compared to silicon. A change of technology from silicon to SiC will revolutionize the power electronics. Wireless sensors for high temperature applications such as oil drilling and mining, automobiles, and jet engine performance monitoring require circuits built on the wide bandgap semiconductor SiC. The fabrication of single mode SiC waveguides and the measurement of their propagation loss is reported in (Pandraud et al., 2007).

There are not enough works proposing the investigations of SiC waveguides. We list here as an example some articles. The characteristics of microwave transmission lines on 4H-High Purity Semi-Insulating SiC and 6H, p-type SiC were presented as a function of temperature and frequency in (Ponchak et al, 2004). An investigation of the SiC pressure transducer characteristics of microelectromechanical systems on temperature is given in (Okojie et al., 2006). The high-temperature pressure transducers like this are required to measure pressure fluctuations in the combustor chamber of jet and gas turbine engines. SiC waveguides have also successfully been used as the microwave absorbers (Zhang, 2006).

The compelling system benefits of using SiC Schottky diodes, power MOSFETs, PiN diodes have resulted in rapid commercial adoption of this new technology by the power supply industry. The characteristics of SiC high temperature devices are reviewed in (Agarwal et al., 2006).

Numerical studies of SiC waveguides are described in an extremely limited number of articles (Gric et al., 2010; Nickelson et al., 2009; Nickelson et al., 2008). The main difficulty faced by researchers in theoretical calculations of the SiC waveguides is large values of material losses and their dependence on the frequency and the temperature. We would like to draw your attention to the fact that we take the constitutive parameters of the SiC material from the experimental data of article (Baeraky, 2002) at certain temperatures. Then for the frequency dependence, we take into account through the dependence of the imaginary part of the complex permittivity of semiconductor SiC material on the specific resistivity and the frequency by the conventional formula (Asmontas et al., 2009).

We would like to underline also that there are theoretical methods for calculation of strong lossy waveguides, but these methods were usually used for the electrodynamical analysis of metamaterial waveguides (Smith et al., 2005; Chen et al., 2006; Asmontas et al, 2009; Gric et al., 2010; Nickelson et al., 2008) or other lossy material waveguides (Bucinskas et al., 2010; Asmontas et al., 2010; Nickelson et al., 2009; Swillam et al., 2008; Nickelson et al., 2008; Asmontas et al., 2006).

In this chapter we present the electrodynamical analysis of open rectangular and circular waveguides. The waveguide is called the open when there is no metal screen. In sections 2 and 3 we give a short description of the Singular Integral Equations' (SIE) method and of the partial area method that we have used to solve the electrodynamical problems. Our method SIE for solving the Maxwell's equations is pretty universal and allowed us to analyze open waveguides with any arbitrary cross-sections in the electrodynamically rigorously way (by taking into account the edge condition and the condition at infinity). The false roots did not occur applying the SIE method. The waveguide media can be made of strongly lossy materials.

In order to determine the complex roots of the waveguide dispersion equations we have used the Müller's method. All the algorithms have been tested by comparing the obtained results with the results from some published sources. Some of the comparisons are presented in section 4.

Both of the methods allow solving Maxwell's equations rigorously and are suitable for making the full electrodynamical analysis. We are able to calculate the dispersion characteristics including the losses of all the modes propagating in the investigated waveguide and the distributions of the electromagnetic (EM) fields inside and outside of the waveguides. We used our computer algorithms based on two mentioned methods with 3D graphical visualization in the MATLAB language.

2. The SIE method

In this section, we describe the SIE method for solving Maxwell's equations in the rigorous problem formulation (Nickelson et al., 2009; Nickelson & Shugurov, 2005). Using the SIE method, it is possible to rigorously investigate to investigate the dispersion characteristics of main and higher modes in regular waveguides of arbitrary cross-section geometry containing piecewise homogeneous materials as well as the distribution of the EM field inside and outside of waveguides electrodynamically..

Our proposed method consists of finding the solution of differential equations with a point-source. Then the fundamental solution of the differential equations is used in the integral representation of the general solution for each particular boundary problem. The integral representation automatically satisfies Maxwell's differential equations and has the unknown density functions μ_e and μ_h, which are found using the proper boundary conditions. To present the fields in the integral form we use the solutions of Maxwell's equations with electric $\vec{j}_e$ and magnetic $\vec{j}_h$ point sources:

$$\mathrm{Curl}\,\vec{E} = -\mu_0\mu_r\frac{\partial\vec{H}}{\partial t} - \vec{j}_h, \qquad \mathrm{Curl}\,\vec{H} = \varepsilon_0\varepsilon_r\frac{\partial\vec{E}}{\partial t} - \vec{j}_e \qquad (1)$$

where $\vec{E}$ is the electric field strength vector and $\vec{H}$ is the magnetic field strength vector. Also ε_r is the relative permittivity and μ_r is the relative permeability of the medium. The electric and magnetic constants ε_0, μ_0 are called the permittivity and the permeability of a vacuum. The dependence on time t and on the longitudinal coordinate z are assumed in the form $\exp(i(\omega t - hz))$. Here h=h'-h''i is the complex longitudinal propagation constant, where h' is the real part (phase constant), $h'=2\pi/\lambda_w$, λ_w is the wavelength of investigated mode and h'' is the imaginary part (attenuation constant). The magnitude $\omega=2\pi f$ is the cyclic operating frequency and i is the imaginary unit ($i^2=-1$). Because of the equations linearity the general solution is a sum of solutions when $\vec{j}_h \neq 0$, $\vec{j}_e = 0$ and $\vec{j}_h = 0$, and $\vec{j}_e \neq 0$. The transversal components E_x, E_y, H_x, H_y of the EM field are being expressed through the longitudinal components E_z, H_z of EM field from Maxwell's equations as follows:

$$E_x = \frac{\mu_0\mu_r i\omega \frac{\partial H_z}{\partial y} + ih\frac{\partial E_z}{\partial x}}{\Delta}, \quad E_y = \frac{-\mu_0\mu_r i\omega \frac{\partial H_z}{\partial x} + ih\frac{\partial E_z}{\partial y}}{\Delta}, \tag{2}$$

$$H_x = \frac{-\varepsilon_0\varepsilon_r i\omega \frac{\partial E_z}{\partial y} + ih\frac{\partial H_z}{\partial x}}{\Delta}, \quad H_y = \frac{\varepsilon_0\varepsilon_r i\omega \frac{\partial E_z}{\partial x} + ih\frac{\partial H_z}{\partial y}}{\Delta}, \tag{3}$$

where $\Delta = h^2 - k^2\varepsilon_r\mu_r$.

The longitudinal components E_z, H_z, satisfy scalar wave equations, which are Helmholtz's equations:

$$\left(\Delta_\perp + k_\perp^2\right)H_z = 0, \qquad \left(\Delta_\perp + k_\perp^2\right)E_z = 0, \tag{4}$$

here $\Delta_\perp = \frac{\partial^2}{\partial x^2} + \frac{\partial^2}{\partial y^2}$ is the transversal Laplacian. Other magnitudes are $k_\perp^2 = -\Delta = k^2\varepsilon_r\mu_r - h^2$, $k=\omega/c$ and c is the light velocity in a vacuum. The fundamental solution of the second order differential equations (4) in the cylindrical coordinates (or in the polar coordinates, since the dependence on the longitudinal coordinate has already been determined) is the Hankel function of the zeroth order.

In Fig. 1 the points of the contour L where we satisfy the boundary conditions on the boundary line, dividing the media with the constitutive parameters of SiC: ε_r^{SiC}, μ_r^{SiC} and an environment area ε_r^a, μ_r^a are shown.

The problem is formulated in this way. We have in the complex plane a piecewise smooth contours L (Fig.1). The contour subdivides the plane into two areas; the inner S^+ and the outer S^- one. These areas according to the physical problem are characterized by different electrophysical parameters: the area S^+ has the constitutive parameters ε_r^{SiC}, μ_r^{SiC} and S^- has the constitutive parameters ε_r^a, μ_r^a of ambient air. Magnitudes ε_r^{SiC} = Re (ε_r^{SiC}) – Im (ε_r^{SiC}) and

μ_r^{SiC} = Re (μ_r^{SiC}) - Im (μ_r^{SiC}) are the complex permittivity and the complex permeability of the SiC medium. The positive direction of going round the contour is when the area S^+ is on the left side.

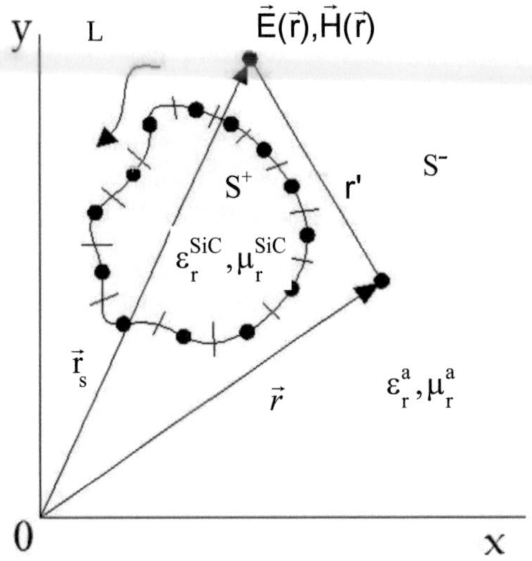

Fig. 1. Waveguide arbitrary cross section and designations for explaining the SIE method.

One has to determine in area S^+ solutions of Helmholtz's equation (4), which satisfy the boundary conditions for the tangent components of the electric and magnetic fields:

$$E^+_{tan}\big|_L = E^-_{tan}\big|_L , \tag{5}$$

$$H^+_{tan}\big|_L = H^-_{tan}\big|_L . \tag{6}$$

In the present work all boundary conditions are satisfied including the edge condition at the angular points of the waveguide cross-section counter and the condition at infinity.
The longitudinal components of the electric field and the magnetic field at the contour points that satisfied to the Helmholtz's equations (4) have the form:

$$E_z(\vec{r}) = \int_L \mu_e(\vec{r}_s) H_0^{(2)}(k_\perp r')ds, \tag{7}$$

$$H_z(\vec{r}) = \int_L \mu_h(\vec{r}_s) H_0^{(2)}(k_\perp r')ds, \tag{8}$$

where $E_z(\vec{r})$, $H_z(\vec{r})$ are the longitudinal components of the electric field and the magnetic field of the propagating microwave. Here $\vec{r} = \vec{i}x + \vec{j}y$ is the radius vector of the point, where the EM fields are determined, where $\vec{i}, \vec{j}$ are the unit vectors. The magnitudes $\mu_e(\vec{r}_s)$ and

$\mu_h(\vec{r}_s)$ are the unknown functions satisfying Hölder condition (Gakhov, 1977). Here $H_0^{(2)}(k_\perp r')$ is the Hankel function of the zeroth order and the second kind, where $r' = |\vec{r}_s - \vec{r}|$. The magnitude $\vec{r}_s = \vec{i}x_s + \vec{j}y_s$ is the radius vector (Fig. 1). Here ds is an element of the contour L and the magnitude s is the arc abscissa.
The expressions of all the electric field components which satisfy the boundary conditions are presented below. We apply the Krylov–Bogoliubov method whereby the contour L is divided into n segments and the integration along a contour L is replaced by a sum of integrals over the segments j=1…n. The expressions of all electric field components for the area S^+ and S^- are presented below:

$$E_z^+ = \sum_{j=1}^{n} \mu_e^+(s_j) \int_{\Delta L} H_0^{(2)}(k_\perp^+ r')ds\,, \tag{9}$$

$$E_z^- = \sum_{j=1}^{n} \mu_e^-(s_j) \int_{\Delta L} H_0^{(2)}(k_\perp^- r')ds\,. \tag{10}$$

We obtain that the transversal components of the electric field E_x, E_y after substituting formula (9) and (10) in the formulae (2) are:

$$\left(E_x\right)^+ = -\left(i\mu_0\mu_r^{SiC}\omega \Big/ \left(k_\perp^+\right)^2\right)\left[k_\perp^+ \sum_{j=1}^{n}\left(\mu_h^+(s_j)\right) \int_{\Delta L} H_1^{(2)}(k_\perp^+ r')\frac{y_s - y_0}{r'}ds\right] \tag{11}$$
$$+\left(ih \Big/ \left(k_\perp^+\right)^2\right)\left[k_\perp^+ \sum_{j=1}^{n}\left(\mu_e^+(s_j)\right) \int_{\Delta L} H_1^{(2)}(k_\perp^+ r')\frac{x_s - x_0}{r'}ds\right] - \frac{2\mu_0\mu_r^{SiC}\omega\cos\theta}{\left(k_\perp^+\right)^2}\mu_h^+(s_j),$$

$$\left(E_y\right)^+ = -\left(ih \Big/ \left(k_\perp^+\right)^2\right)\left[k_\perp^+ \sum_{j=1}^{n}\left(\mu_e^+(s_j)\right) \int_{\Delta L} H_1^{(2)}(k_\perp^+ r')\frac{y_s - y_0}{r'}ds\right] \tag{12}$$
$$-\left(i\mu_0\mu_r^{SiC}\omega \Big/ \left(k_\perp^+\right)^2\right)\left[k_\perp^+ \sum_{j=1}^{n}\left(\mu_h^+\left(s_j\right)\right)^- \int_{\Delta L} H_1^{(2)}(k_\perp^+ r')\frac{x_s - x_0}{r'}ds\right] - \frac{2\mu_0\mu_r^{SiC}\omega\cos\theta}{\left(k_\perp^+\right)^2}\mu_h^+(s_j),$$

$$\left(E_x\right)^- = -\left(i\mu_0\mu_r^a\omega \Big/ \left(k_\perp^-\right)^2\right)\left[k_\perp^- \sum_{j=1}^{n}\left(\mu_h^-\left(s_j\right)\right)^- \int_{\Delta L} H_1^{(2)}(k_\perp^- r')\frac{y_s - y_0}{r'}ds\right] \tag{13}$$
$$+\left(ih \Big/ \left(k_\perp^-\right)^2\right)\left[k_\perp^- \sum_{j=1}^{n}\left(\mu_e^-(s_j)\right) \int_{\Delta L} H_1^{(2)}(k_\perp^- r')\frac{x_s - x_0}{r'}ds\right] + \frac{2\mu_0\mu_r^a\omega\cos\theta}{\left(k_\perp^-\right)^2}\mu_h^-(s_j),$$

$$\left(E_y\right)^- = -\left(ih/\left(k_\perp^-\right)^2\right)\left[k_\perp^- \sum_{j=1}^{n}\left(\mu_e^-(s_j)\right) \int_{\Delta L} H_1^{(2)}(k_\perp^- r') \frac{y_s - y_0}{r'} ds\right] \tag{14}$$

$$-\left(i\mu_0\mu_r^a\omega\right)/\left(k_\perp^-\right)^2\left[k_\perp^- \sum_{j=1}^{n}\left(\mu_h^-(s_j)\right) \int_{\Delta L} H_1^{(2)}(k_\perp^- r') \frac{x_s - x_0}{r'} ds\right] + \frac{2\mu_0\mu_r^a\omega\cos\theta}{\left(k_\perp^-\right)^2}\mu_h^-(s_j).$$

The field components and the values of the functions $\mu_e(s_j)$ and $\mu_h(s_j)$ are noted in the upper-right corner with the sign corresponding to different waveguide area, for instance, the functions $\mu_e^+(s_j)$, $\mu_h^+(s_j)$ or $\mu_e^-(s_j)$, $\mu_h^-(s_j)$ (Fig.1). These functions at the same contour point are different for the field components in the areas S^+ and S^-, i.e. $\mu_h^+(s_j) \neq \mu_h^-(s_j)$. The magnitude $H_0^{(2)}$ is the Hankel function of the zeroth order and of the second kind, $H_1^{(2)}$ is the Hankel function of the first order and of the second kind. Here $k_\perp^+ = \sqrt{k^2\varepsilon_r^{SiC}\mu_r^{SiC} - h^2}$ and $k_\perp^- = \sqrt{h^2 - k^2\varepsilon_r^a\mu_r^a}$ are the transversal propagation constants of the SiC medium in the area S^+ and in the air area S^-, correspondingly (Fig.1). The segment of the contour L is ΔL=L/n, where the limits of integration in the formulae (9-14) are the ends of the segment ΔL. The angle θ is equal to g·90° with g from 1 to 4, if the contour of the waveguide cross-section is a rectangular one, then the result can be cos θ=±1 and sin θ=±1 in the formulae (11-14).
We obtain the transversal components of the magnetic field H_x and H_y using SIE method in the form analogical formulae (9) – (12) after substituting formula (8) in the formulae (3).
After we know all EM wave component representations in the integral form we substitute the component representations to the boundary conditions (5) and (6). We obtain the homogeneous system of algebraic equations with the unknowns $\mu_e^+(s_j)$, $\mu_h^+(s_j)$, $\mu_e^-(s_j)$ and $\mu_h^-(s_j)$. The condition of solvability is obtained by setting the determinant of the system equal to zero. The roots of the system allowed us to determine the complex propagation constants of the main and higher modes of the waveguide. After obtaining the propagation constant of some required mode, the determination of the electric and magnetic fields of the mode becomes possible. For the correct formulated problem (Gakhov, 1977) the solution is one-valued and stable with respect to small changes of the coefficients and the contour form (Nickelson & Shugurov, 2005).

3. The partial area method

The presentation of longitudinal components of the electric E_z^{SiC} and magnetic H_z^{SiC} fields that satisfies Maxwell's equations in the SiC medium (Nickelson et al., 2008; Nickelson et al., 2007) are as follows:

$$E_z^{SiC} = A_1 J_m\left(k_\perp^+ r\right)\exp(im\varphi), \quad H_z^{SiC} = B_1 J_m\left(k_\perp^+ r\right)\exp(im\varphi), \tag{15}$$

where J_m is the Bessel function of the m−th order, A_1 and B_1 are unknown arbitrary amplitudes. The longitudinal components of the electric field E_z^a and the magnetic field H_z^a that satisfy Maxwell's equations in the ambient waveguide medium (in air) are as follows:

$$E_z^a = A_2 H_m^{(2)}\left(k_\perp^- r\right)\exp(im\varphi)\ ,\ H_z^a = B_2 H_m^{(2)}\left(k_\perp^- r\right)\exp(im\varphi), \tag{16}$$

where A_2 and B_2 are unknown arbitrary amplitudes, $H_m^{(2)}$ is the Hankel function of the m−th order and the second kind, r is the radius of the circular SiC waveguide, m is the azimuthal index characterizing azimuthal variations of the field, φ is the azimuthal angle.
A further solution is carried out under the scheme of section 2 of present work. The resulting solution is the dispersion equation in the determinant form:

$$\begin{aligned}
&A_1 J_m\left(k_\perp^+ r\right) - A_2 H_m^{(2)}\left(k_\perp^- r\right) = 0,\\
&B_1 J_m\left(k_\perp^+ r\right) - B_2 H_m^{(2)}\left(k_\perp^- r\right) = 0,\\
&\frac{mh}{\left(k_\perp^+\right)^2 r} A_1 J_m\left(k_\perp^+ r\right) + \frac{i\omega\mu_0\mu_r^{SiC}}{k_\perp^+} B_1 J_m'\left(k_\perp^+ r\right)\\
&-\frac{mh}{\left(k_\perp^-\right)^2 r} A_2 H_m^{(2)}\left(k_\perp^- r\right) - \frac{i\omega\mu_0\mu_r^{a}}{k_\perp^-} B_2\left(H_m^{(2)}\left(k_\perp^- r\right)\right)' = 0,\\
&\frac{mh}{\left(k_\perp^+\right)^2 r} B_1 J_m\left(k_\perp^+ r\right) - \frac{i\omega\varepsilon_0\varepsilon_r^{SiC}}{k_\perp^+} A_1 J_m'\left(k_\perp^+ r\right)\\
&-\frac{mh}{\left(k_\perp^-\right)^2 r} B_2 H_m^{(2)}\left(k_\perp^- r\right) + \frac{i\omega\varepsilon_0\varepsilon_r^{a}}{k_\perp^-} A_2\left(H_m^{(2)}\left(k_\perp^- r\right)\right)' = 0.
\end{aligned} \tag{17}$$

We have used the Müller's method to find the complex roots. The roots of the dispersion equation give the propagation constants of waveguide modes. After obtaining the propagation constants of desired modes we can determine the EM field of these modes.

4. Validation of the computer softwares

We validated all our algorithms. Some of the validation results are presented in this section. We have created the computer software based on the method SIE (Section 2) in the MATLAB language. This software let us calculate the dispersion characteristics of waveguides with complicated cross-sectional shapes as well as the 3D EM field distributions. The computer software was validated by comparison with data from different

published sources, for example, the dispersion characteristics of the rectangular dielectric waveguide (Ikeuchi et al., 1981) and the modified microstrip line (Nickelson & Shugurov, 2005). In Figs. 2 and 3 we see that an agreement of the compared results is very good. In Fig. 2 the dispersion characteristics of the rectangular waveguide with sizes (15x5) $\cdot 10^{-3}$ m and the waveguide material permittivity equal to 2.06 are presented. Our calculations are the solid lines and the results from (Ikeuchi et al., 1981) are presented with points.
The dimensions of the microstrip line (Fig. 3) are given by: $d=3.17 \cdot 10^{-3}$ m, $w=3.043 \cdot 10^{-3}$ m, $l_1=l_2=5 \cdot 10^{-3}$ m, $t=3 \cdot 10^{-6}$ m. The permittivity of the microstrip line substrate is 11.8.

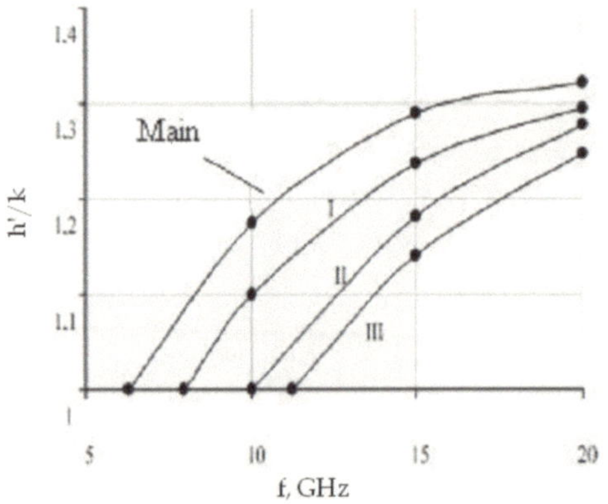

Fig. 2. Comparison of the dispersion characteristics of the rectangular dielectric waveguide calculated by SIE algorithm presented here and data from (Ikeuchi et. al., 1981)

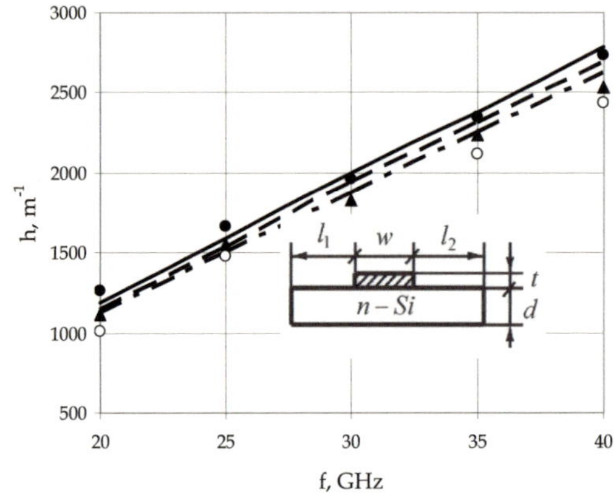

Fig. 3. Comparison of the microstrip line dispersion characteristics calculated by SIE algorithm and data from (Nickelson & Shugurov, 2005).

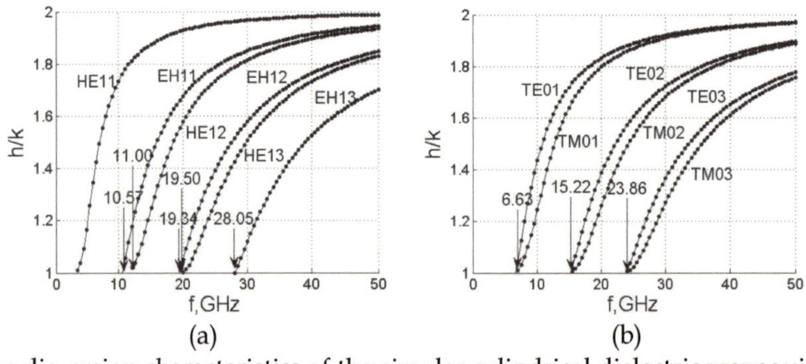

Fig. 4. The dispersion characteristics of the circular cylindrical dielectric waveguide: (a) – the hybrid modes and (b) – the axis-symmetric modes. The results taken from (Kim, 2004) are presented with solid lines and our calculations are presented with points.

In Fig. 3 our calculations are shown with dots (the main mode), with triangles (the first higher mode) and with circles (the second higher mode). In Fig. 3 the data from the book (Nickelson & Shugurov, 2005) is shown by the solid line (the main mode), dashed line (the first higher mode), dash-dotted line (the second higher mode).

We have also created the computer software on the basis of the partial area method (section 3) in the MATLAB language for calculations of the dispersion characteristics and the 2D &3D EM field distributions of circular waveguides. This software was validated by comparison with data from different published sources, for example, with (Kim, 2004). In Fig. 4 are shown dispersion characteristics of the circular cylindrical dielectric waveguide with a radius equal to 10^{-2} m and the permittivity of the dielectric equal to 4. In Fig. 4 (a) is shown six modes with the azimuth index m=1, are presented and in Fig 4 (b) six modes with the azimuth index m=0 are given. In Fig. 4. we see the good agreement between the simulations and the experimental results.

In Fig. 5 we demonstrate the validation of our computer program for calculations of the EM fields. We see that in work by (Kajfez & Kishk, 2002) the distribution of the electric field was presented by the arrows whose lengths are proportional to the intensity of the electric field at different points (Fig. 5(a)).

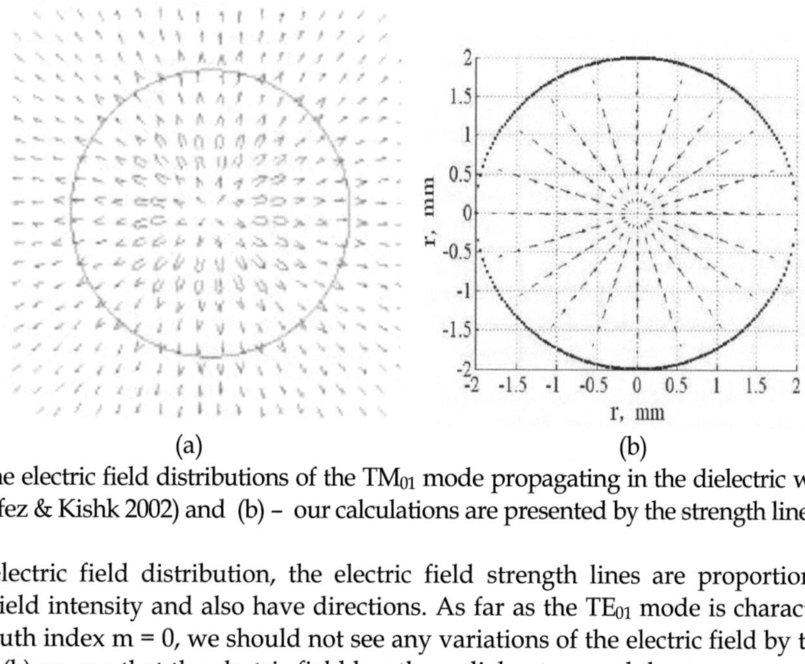

(a) (b)

Fig. 5. The electric field distributions of the TM_{01} mode propagating in the dielectric waveguide: (a) – (Kajfez & Kishk 2002) and (b) – our calculations are presented by the strength lines

In our electric field distribution, the electric field strength lines are proportional to the electric field intensity and also have directions. As far as the TE_{01} mode is characterized by the azimuth index m = 0, we should not see any variations of the electric field by the radius. In Fig. 5 (b) we see that the electric field has the radial nature and there are no variations of the electric field along the circular the radius.

The validation of our computer programs was made for different types of the waves having the different number of variations by the radius and the different azimuthal index. The distributions of the electric fields of other modes are also correct. It should be noticed that we have validation our computer programs for calculation of losses in the waveguide slowly increasing the losses of the material from Im (ε_r) = 0 and Im (μ_r) = 0 up to the required values.

5. The rectangular SiC waveguide

In this chapter we present the investigations of the electrodynamical characteristics of the open waveguides using the algorithm that is described in Section 2. Here we present our calculations of two SiC waveguides with different cross-sectional dimensions at different temperatures. We also present the distributions of the magnetic fields at the temperature of 500°C.

5.1. The investigation of the rectangular SiC waveguide with sizes $(2.5x2.5)\cdot10^{-3}$ m^2 at T=500°C

The dispersion characteristics of the rectangular SiC waveguide at the temperature 500 °C are presented in Fig. 6. The sizes of the rectangular SiC waveguide are 2.5x2.5 mm^2. The dispersion characteristics of the main mode are shown by solid lines. The dispersion characteristics of the first higher mode are shown by dashed lines. Here the complex longitudinal propagation constant is h = h'-h''i, the phase constant h' = Re (h) [rad/m] and the attenuation constant h'' = Im (h) determines the waveguide losses [rad/mm = 8.7dB/mm]. Here $h'=2\pi/\lambda_w$, where λ_w is the wavelength of the waveguide modes. In our calculations the azimuthal index is m = 1, because the main waveguide mode has the index equal to unity. The magnitude *k* is the wavenumber. The permittivity of the SiC material is 6.5 - 0.5i at the temperature 500°C and the frequency f = 1.41 GHz (Baeraky, 2002). The values of the permittivity depend upon frequency (Asmontas et al., 2009).
Concerning the fact that SiC is the material with large losses at certain temperatures and frequencies the complex roots of the dispersion equation were calculated by the Müller method. In Fig. 6(a) we see that the main and the first higher modes are slow waves (because $h'/k > 1$. The dependencies of losses of the both modes propagating in the rectangular SiC waveguide on the frequency range are pretty intricate (Fig. 6 (b)). The main mode has three loss maxima. We discovered that the minimum of the losses of the main mode is approximately at f=59 GHz. The losses of the first higher mode have the larger value at the frequency of 59 GHz. It means that the first higher mode is strongly absorbed in the waveguide at this frequency. Fig. 6 (b) shows the excellent properties of SiC waveguide at f = 59 GHz for creation of some devices on the base of the main mode. The SiC waveguide could be used for creation of single-mode devices.

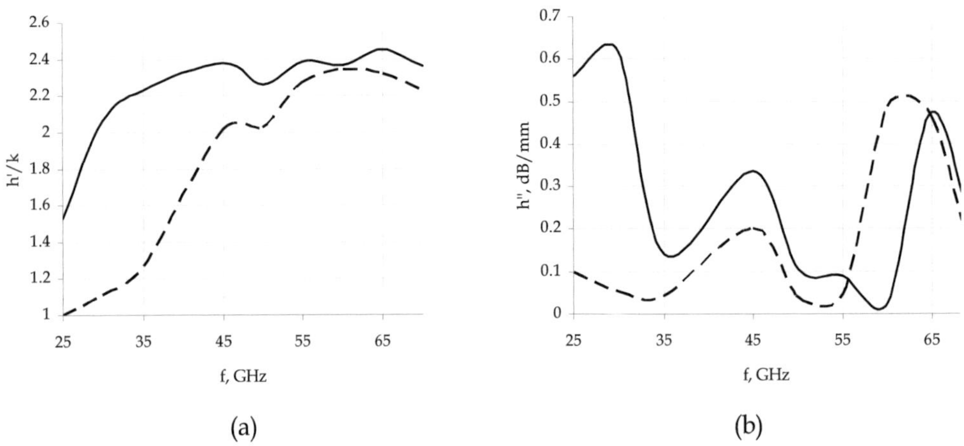

(a) (b)

Fig. 6. The dispersion characteristics of the rectangular SiC waveguide: (a) – the dependence of the normalized phase constant h'/k upon frequency and (b) - the dependence of the attenuation constant h'' upon frequency

The 3D magnetic field distribution of the main mode propagating in the rectangular SiC waveguide at T= 500°C and f=30 GHz is presented in Fig. 7.

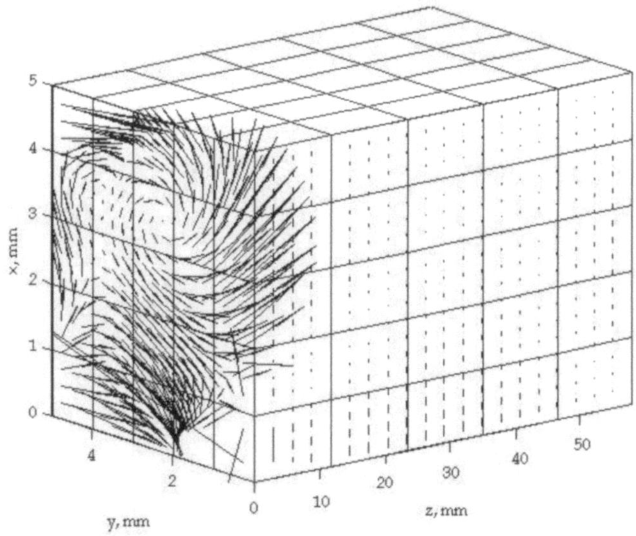

Fig. 7. The 3D vector magnetic field distribution of the main mode propagating in the open rectangular SiC waveguide at T= 500°C and f=30GHz

In Fig. 7 we see that the magnetic field of the main mode is distributed in the form of circles in the cross-section of the rectangular SiC waveguide. We can see the vertical magnetic field strength lines in the plane of a vertical waveguide wall along the z axis. The calculations were made with 100000 points in 3D space.

5.2 The investigation of the rectangular SiC waveguide with sizes (3x3)·10^{-3} m^2 at T=1000°C

The SiC waveguide with sizes (3x3) ·10^{-3} m^2 has been analyzed at the temperature T = 1000°C (Fig. 8). The values of permittivities depend upon temperature and were taken from (Baeraky, 2002). The dispersion characteristics of the rectangular SiC waveguide are presented in Fig. 8.

In Fig. 8(a) are shown the phase constants of two modes propagating in the circular waveguide with the azimuth index m=1 (Nickelson & Gric, 2009). We see that the cutoff frequency of the main mode is 21 GHz and the first higher mode is 27 GHz. In Fig. 8 (b) we see the dependences of losses of the main and the first higher modes on frequency. We see that the loss dependences have the waving character. When the frequency is lower than 30 GHz, the losses of the main mode are larger than losses of the first higher mode at the same frequency interval. When the frequency is higher than 30 GHz, the losses of these modes have approximately the same values. Comparing the modes depicted in Fig. 8 with the analogue modes propagating in the circular dielectric waveguide, we should notice that the main mode is the hybrid HE_{11} mode and the first higher mode is the hybrid EH_{11} mode.

Comparing Figs. 6 and 8 we see that the dispersion characteristics can be changed by changing temperatures and waveguide cross-section sizes. Especially, we would stress that the SiC waveguide operates in single-mode regime. And the waveguide broadband width is approximately 25%.

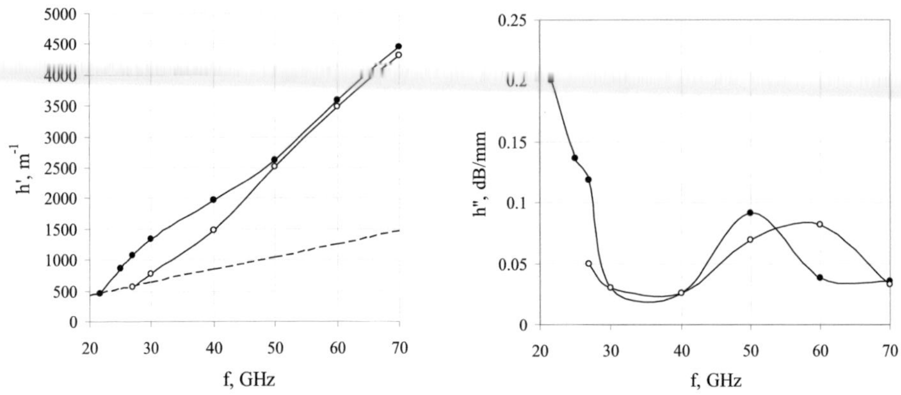

(a) (b)

Fig. 8. Dispersion characteristics of the main mode and the first higher mode of the rectangular SiC waveguide: (a) – dependence of the phase constants and (b) – dependence of the attenuation constants upon frequency. The main mode is denoted with points and the first higher mode is denoted with circles.

6. The circular SiC waveguide

Here we present the investigations of the electrodynamical characteristics of the open circular cylindrical waveguides using the algorithms presented in Chapter 3 (Gric et al., 2010; Asmontas et al., 2009; Nickelson et al., 2009; Nickelson et al. 2008). We propose our calculations of the dispersion characteristics of the SiC waveguide with different sizes of their cross-sections at several different temperatures. We also give here the 2D electric field distributions and the 3D magnetic field distributions.

6.1. The investigation of the circular cylindrical SiC waveguide with the radius 2.5 mm at three temperatures T=500°C, 1000°C and 1500°C

In Figs. 9 and 10 are presented the dispersion characteristics of the main and the first higher modes propagating in the open circular SiC waveguide at three different temperatures 500°C (solid line with crosses), 1000°C (solid line with black), 1500°C (solid line with circles). The permittivity of the SiC is 6.5 - 0.5i when T=500°C, 7-1i when T=1000°C and 8-2i when T=1500°C at f =11 GHZ. In Fig. 9(a) we see that the value of phase constant increases with increasing frequency. Comparing the dependence of the phase constant at 500°C, 1000°C and 1500°C we see that the higher the temperature is, the higher the values of the phase constant are.

The value of the attenuation constant increases with increasing the temperature and remains almost constant at frequencies above 20 GHz. This feature is important for operating of modulators and phase shifters which could be created on the basis of such waveguides. The EM signal propagating in the waveguide is not modulated by losses.

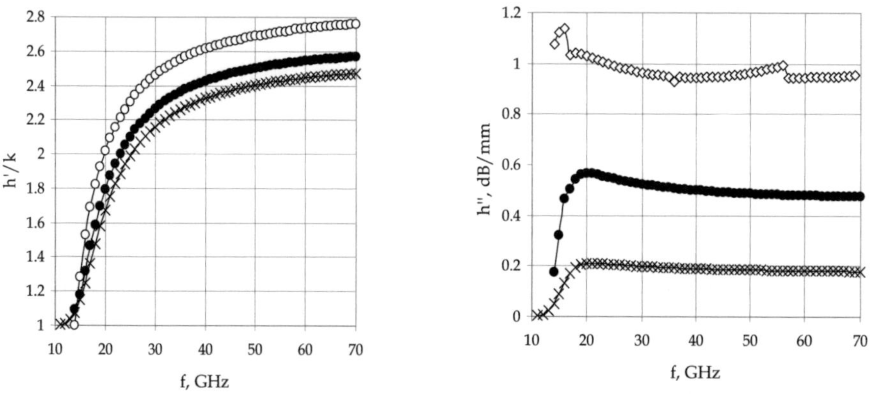

(a) (b)

Fig. 9. The dispersion characteristics of the main mode propagating in the circular SiC waveguide: (a) – the dependences of the normalized phase constant and (b)– the dependences of the attenuation constant upon the frequency at the temperatures 500°C, 1000°C and 1500°C.

In Fig. 10 we present the dispersion characteristics of the first higher modes propagating in the circular SiC waveguide at three different temperatures. We see from Fig. 10 (a) that at the low frequencies before 45 GHz the phase constant h′ can be higher at the temperature 1000°C in comparison with h′ at the temperatures 500°C and 1500°C. Thus the tendency of dependences of the complex longitudinal propagation constants upon frequency is destroyed for the first higher mode.

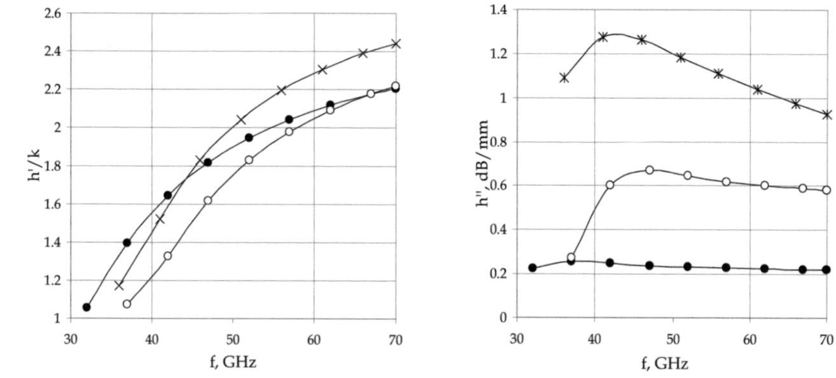

(a) (b)

Fig. 10. The dispersion characteristics of the first higher modes propagating in the circular SiC waveguide: (a) – the dependences of the normalized phase constant and (b) – the dependences of the attenuation constant upon the frequency at the temperatures 500°C, 1000°C, 1500°C

We see that the changes of attenuation constant curves (Fig. 10 (b) of the first higher mode are different compared to the same dependences of the main mode (Fig. 9(b)).

We have calculated all EM field components with 600 points inside and outside of the SiC waveguide. The 3D vector magnetic field distributions of the main mode propagating in the circular SiC waveguide at two temperatures 500°C and 1500°C and when f = 30 GHz are shown in Figs. 11 - 12. In Figs. 11(a) -12(a) the distribution of magnetic field lines in the waveguide cross-section as well as the magnetic field line projections on the vertical xOz and horizontal yOz planes in the longitudinal directions are shown. In Figs. 11(b) -12(b) the distributions of 3D vector magnetic field lines in the space of front quarter between the vertical xOz and horizontal yOz planes are presented. Calculations were fulfilled in the 300 points of every cross-section.

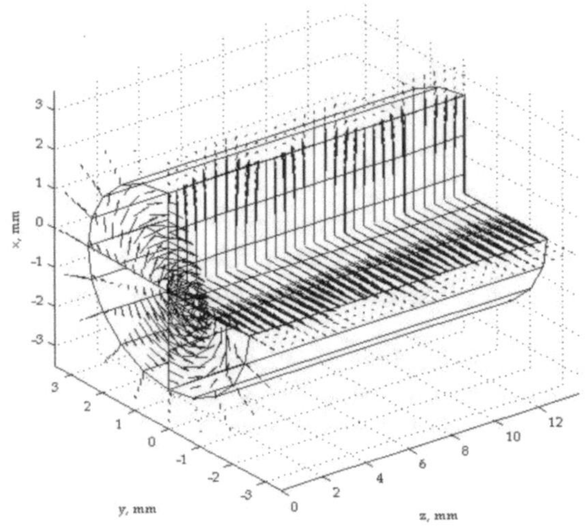

(a)

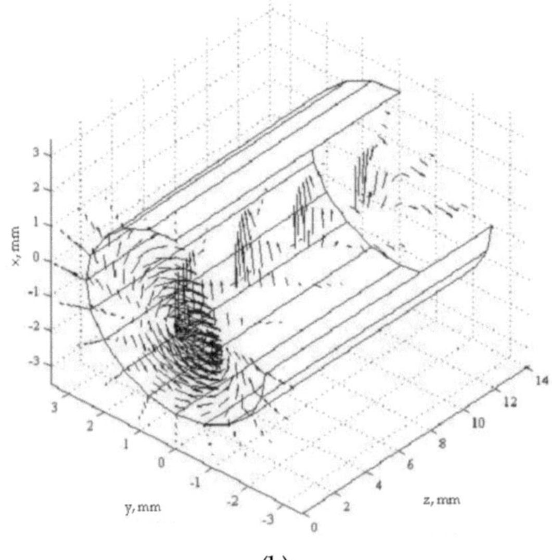

(b)

Fig. 11. The 3D vector magnetic field distributions of the main mode when T= 500 °C: (a) – the cross-section distribution of magnetic field lines as well as their projections at horizontal and vertical planes and (b) – the 3D vector magnetic field in the space of front quarter

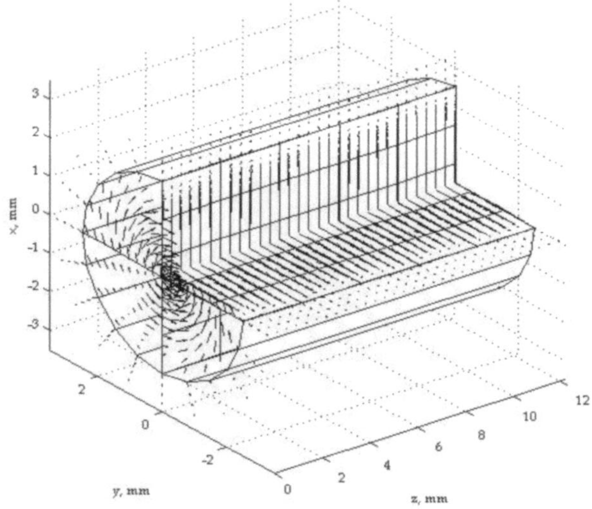

(a)

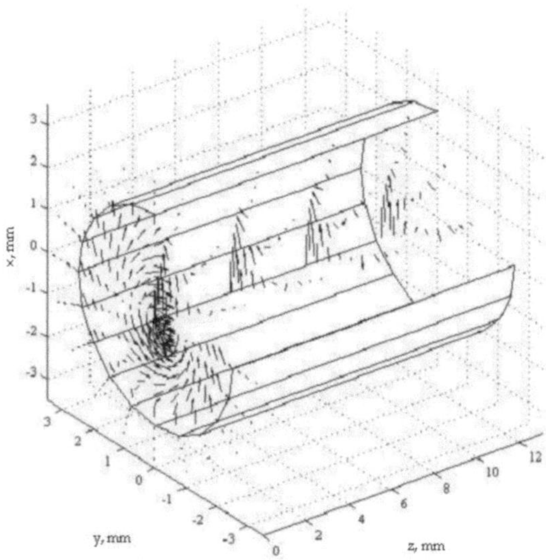

(b)

Fig. 12. The 3D vector magnetic field distributions of the main mode when T= 1500°C: (a) – the cross-section distribution of magnetic field lines as well as their projections at horizontal and vertical planes and (b) – the 3D vector magnetic field in the space of front quarter

The comparison of the magnetic fields at the temperatures 500°C, 1500°C (Fig. 11 and 12) shows that the magnetic field at 1500°C is weaker inside and outside of the waveguide. This happened due to the fact that the losses are larger at 1500°C.

6.2. The investigation of the circular SiC waveguide with radius 3 mm at T = 1000°C

Here we investigate the SiC rod waveguide (Nickelson et al., 2009). The radius of the SiC rod waveguide is 3 mm. The SiC waveguide has been analyzed at the temperature T = 1000°C. The permittivity of the SiC material at this temperature is 7-1i at f=11 GHz (Baeraky, 2002). The dependences of the phase constant and the attenuation constant of the SiC waveguide with the radius r=3 mm when T = 1000 °C on the operating frequency f are presented in Fig.13 (a) and (b).

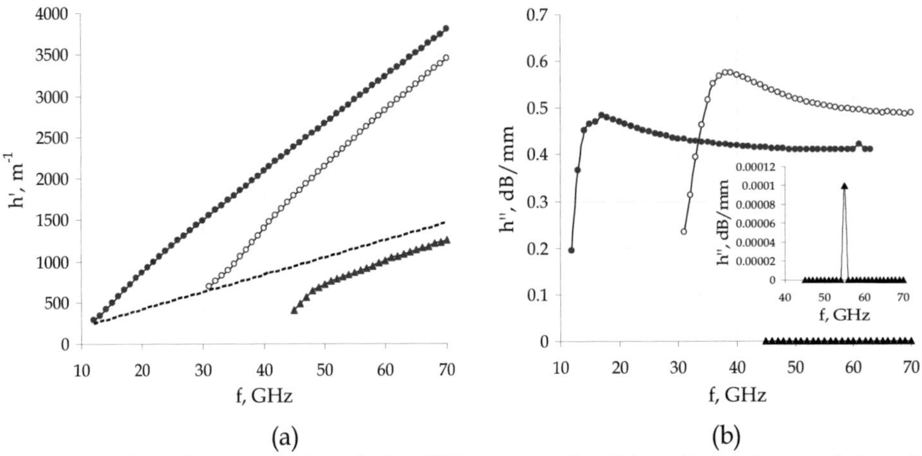

Fig. 13. Dispersion characteristics of the SiC waveguide: (a) - dependence of the phase constants and (b) - dependence of the attenuation constant upon frequency. The main mode is denoted with points, the first higher mode is denoted with circles, the fast mode is denoted with triangles

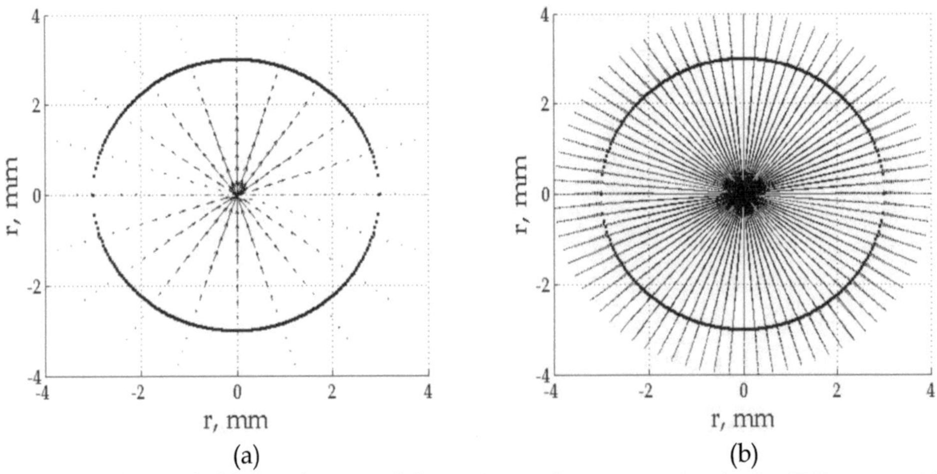

Fig. 14. The electric field distribution of the main mode propagating in the SiC waveguide at f = 55 GHz: (a) - the electric fields strength lines and (b) - the electric field intensities

The electric field distributions of all the propagated modes were calculated at the frequency f = 55 GHz. The obtained results are presented in Figs. 14 - 16. Here we present the electric field strength lines and the electric field intensities. The intensity of the electric field is expressed through a module of the transversal electric field components.

In Fig. 14 we see that the electric field distribution of the main mode has one variation by radius. The strongest electric field of this mode concentrates in the centre of the waveguide at a small enough radius.

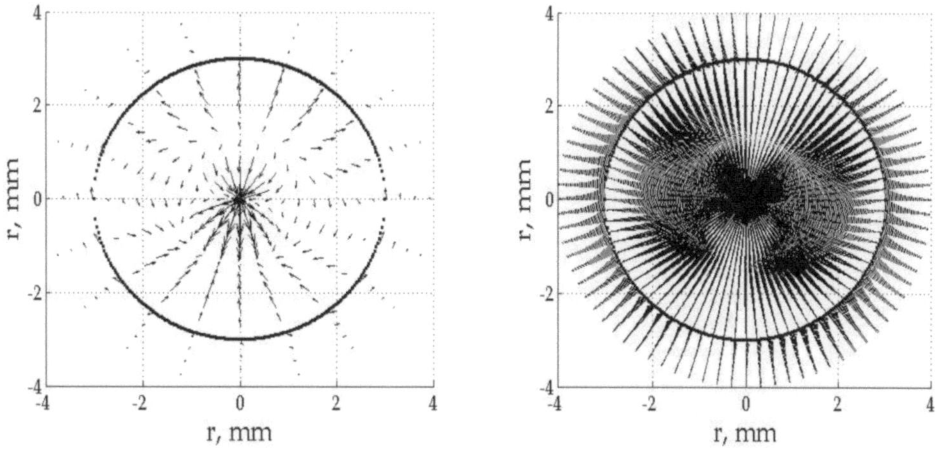

(a) (b)

Fig. 15. The electric field distribution of the first higher slow mode propagating in the SiC waveguide at f = 55 GHz: (a) - the electric fields strength lines and (b) - the electric field intensities

In Fig. 15 we see that the strongest electric field of this mode concentrates in the large part of the waveguide in the form of two twisted lobes.

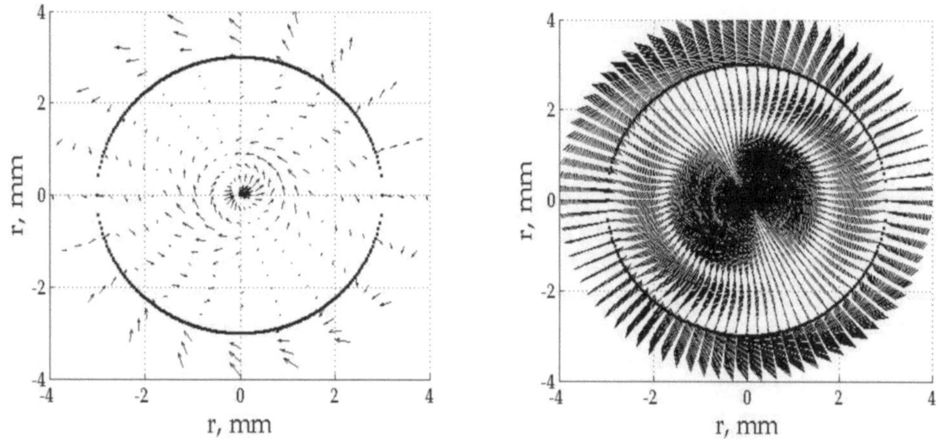

(a) (b)

Fig. 16. The electric field distribution of the second higher fast mode propagating in the SiC waveguide at f = 55 GHz: (a) - the electric fields strength lines and (b) - the electric field intensities

In Fig. 16 we see that the electric field distribution of the fast mode has two variations by radius. The strongest electric field of this mode concentrates in the centre of the waveguide in the form of two small lobes and outside it. We see that the electric field outside the waveguide is stronger in the places where the inner waveguide electric field is weaker.

6.3. The investigation of the circular SiC waveguide with radius 2.5 mm when T = 1800°C

The dependences of the phase constant and the attenuation constant of the SiC waveguide with the radius r=2.5 mm when T = 1800⁰C on the operating frequency f are presented in Figs. 17 (a) and (b).

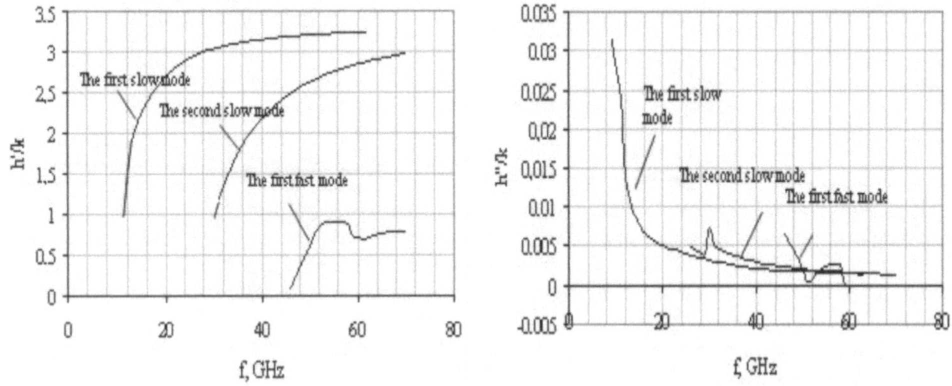

(a) (b)

Fig. 17. Dispersion characteristics of the SiC waveguide: (a) – dependence of the normalized phase constant and (b) – dependence of the normalized attenuation constant upon frequency

There are dispersion curves of three waveguide modes in Fig. 17. The main and first higher waveguide modes are slow modes, because their h'/k >1. The third depicted mode is a first fast mode because the value h'/k <1 for this mode.

The cutoff frequencies of the two slow modes are f_{cut}=12.5 GHz and 30 GHz respectively. The cutoff frequency of the first fast mode is f_{cut} = 46 GHz.

The propagation losses of all analyzed modes (see Fig. 17 (b)) were calculated in the assumption that the imaginary part of the complex permittivity Im (ε_r^{SiC}) is equal to 7 at the operating frequency 12.5 GHz. The value Im (ε_r^{SiC}) decreases when the operating frequency f increases, because this magnitude is inversely proportional to the value f (Asmontas et al., 2009). Analyzing the propagation losses of the slow and fast modes we see that the first slow mode has the largest propagation losses in the area of its cutoff frequency. We see peaks on the loss curves of the second and third modes. Our research has shown that the position of these peaks depends on the waveguide radius also. At smooth reduction of waveguide radius the peak of propagation losses will be smoothly displaced to the right side as a function of increasing frequencies.

The electric field distributions of the slow and fast modes are presented in Figs. 18 – 20. The distributions of the electric field were calculated in 10000 points. The electric field strength

lines are presented in Figs 18(a) – 20(a). Visualizations of the electric field intensity are shown in Figs 18(b) – 20(b).
The values of the electric E_r, E_φ, E_z and magnetic H_r, H_φ, H_z field components of these modes are summarized in Table 1.

	Slow modes		
The main	m = 1, f = 15 GHz		
	E_r, V/m	E_φ, V/m	E_z, V/m
	$2.2 \cdot 10^{-2} - 8 \cdot 10^{-3}i$	$3.9 \cdot 10^{-2} + 4 \cdot 10^{-2}i$	$8 \cdot 10^{-2} + 1.46 \cdot 10^{-1}i$
	H_r, A/m	H_φ, A/m	H_z, A/m
	$-5.943 \cdot 10^{-4} - 6.49 \cdot 10^{-4}i$	$-1.6 \cdot 10^{-5} + 4.267 \cdot 10^{-5}i$	$4.148 \cdot 10^{-4} - 4.424 \cdot 10^{-4}i$
The first higher	m = 1, f = 50 GHz		
	E_r, V/m	E_φ, V/m	E_z, V/m
	$1.349 \cdot 10^{-4} - 1.582 \cdot 10^{-4}i$	$1.554 \cdot 10^{-4} + 1.204 \cdot 10^{-4}i$	$4.205 \cdot 10^{-4} + 3.623 \cdot 10^{-4}i$
	H_r, A/m	H_φ, A/m	H_z, A/m
	$-1.604 \cdot 10^{-6} - 1.127 \cdot 10^{-6}i$	$1.32 \cdot 10^{-6} - 1.643 \cdot 10^{-6}i$	$8.011 \cdot 10^{-7} - 1.115 \cdot 10^{-6}i$
	Fast mode		
The second higher	m = 1, f = 51 GHz		
	E_r, V/m	E_φ, V/m	E_z, V/m
	$2.5 \cdot 10^{-2} - 1.4 \cdot 10^{-2}i$	$2 \cdot 10^{-2} - 1.6 \cdot 10^{-2}i$	$2.95 \cdot 10^{-1} - 3 \cdot 10^{-2}i$
	H_r, A/m	H_φ, A/m	H_z, A/m
	$-3.434 \cdot 10^{-4} + 6.650 \cdot 10^{-5}i$	$10^{-3} - 10^{-3}i$	$-2.887 \cdot 10^{-4} + 1.958 \cdot 10^{-4}i$

Table 1. The EM field components in the fixed point (r = 2.4 mm, φ=0, z=0) of the SiC waveguide cross-section when T = 1800^{0}C.

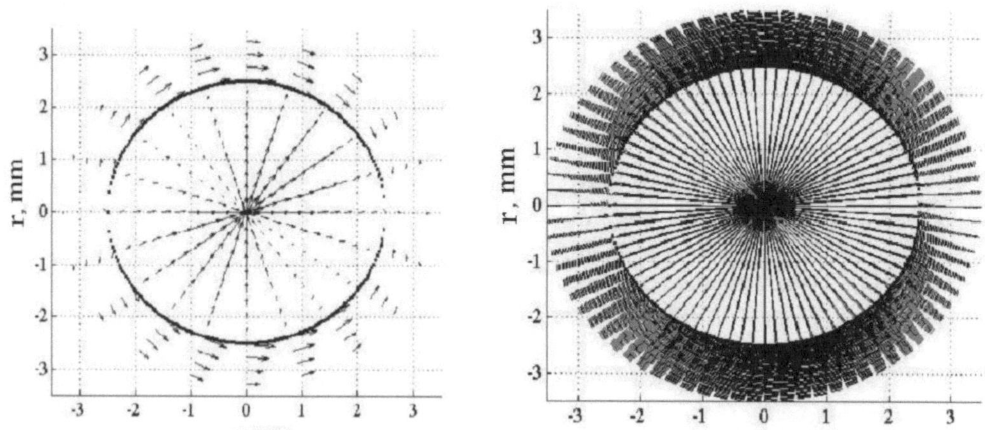

(a) (b)

Fig. 18. The electric field distribution of the first slow mode propagating in the SiC waveguide at f = 15 GHz: (a) - the electric fields strength lines and (b) - the electric field intensities

The electric field distributions of the first slow mode (the main mode) are presented in Fig. 18. We should notice that the electric field distribution of the first slow mode depicted in Fig. 18 (a) is rotated clockwise by 90 degrees respectively to the electric field distribution of the same mode propagated in the analogical waveguide made of lossless material SiC with Im (ε_r^{SiC})=0. In Fig. 18 (a) we can see that electric field strength lines are directed clockwise in the I and II quarters and counterclockwise in the III and IV quarters. The electric field strength lines are directed radially inside the SiC waveguide. We see that there is only one variation of the electric field on the waveguide radius. In Fig. 18 (b) we can see that the strongest electric field concentrates in the two areas. These ones are in the waveguide center and on the waveguide boundary. We see that the field is inhomogeneously distributed along the perimeter of the waveguide boundaries.

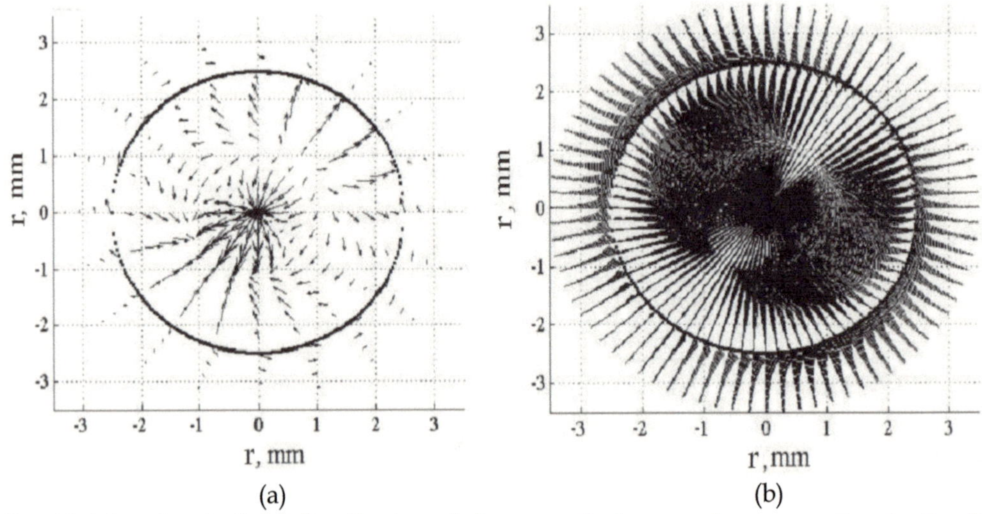

Fig. 19. The electric field distribution of the second slow mode propagating in the SiC waveguide at f = 50 GHz: (a) – the electric fields strength lines and (b) – the electric field intensities

In Fig. 19 we can observe an interesting behavior of the electric field distribution of the second slow mode (the first higher waveguide mode) inside the SiC waveguide as well as outside it close to its boundary. In Fig. 19 (a) we see that there are two variations of the electric field on the waveguide radius. The electric field intensity distribution inside the waveguide has an intricate picture in the shape of two lobes. We see that when the distance from the waveguide becomes larger the electric field becomes smaller outside the waveguide. We should notice that there is a third slow mode (the third higher waveguide mode) in the frequency range of 1 - 100 GHz. The cutoff frequency of this mode is f_{cut} = 51 GHz. The third slow mode has two variations by the radius. The analysis of the third slow mode is beyond the present work.

In Fig. 20 we can see the electric field distribution of the first fast mode propagating in the SiC waveguide. The electric field strength lines of the first fast mode (the second higher waveguide mode that we study here) have three variations along the radius Fig. 20(a). In Fig. 20(b) we can observe that the electric field distribution of the first fast mode inside the

SiC waveguide is in the form of two lobes. The strongest electric field concentrates outside the waveguide. The number of variations of a field along the waveguide radius for all the modes corresponds to the current understanding about the main and higher modes of dielectric waveguides.

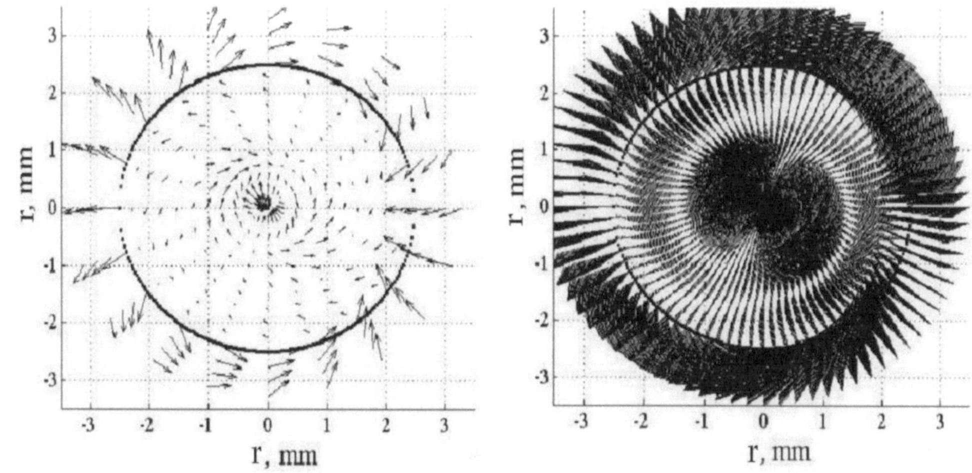

Fig. 20. The electric field distribution of the first fast mode propagating in the SiC waveguide at f = 51 GHz: (a) – the electric fields strength lines and (b) – the electric field intensities

The comparison of Figs 18-20 shows us that the electric field distributions of different waveguide modes are strongly different.

7. The hollow-core SiC waveguide

The electrodynamical solution of the Helmholtz equation of analogical waveguides has already been presented in section 3 (Nickelson et al., 2008). We should mention that the argument of the Hankel function changes its sign if there is lossy material outside the waveguide.

The dispersion characteristics of the main and the first higher modes propagating in the hollow-core SiC waveguide with r = 1 mm at the temperatures T equal to 20°C, 1250°C, 1500°C, 1800°C are presented in Figs. 21 and 22.

In Fig. 21 we see the dependencies of the phase constant and the attenuation constant of the main mode on the frequency f. In Fig. 21(a) we see that the phase constants at temperatures T = 1250°C and T = 1500°C practically coincide. We can see that when the temperature decreases the cutoff frequencies of the main mode moves to the range of higher frequencies. The behavior of the attenuation constant (waveguide losses h'') are different at temperatures T = 1250°C and T = 1500°C (Fig. 21 (b)). The waveguide losses of the main modes became lower with increasing frequency until a certain value and after that losses started to increase again. We can see that the minima of waveguide losses at temperatures T = 20°C, 1250°C, 1500°C and 1800°C correspond to frequencies f = 118, 109, 79 and 56 GHz. The waveguide losses of the main modes (at different T) decrease in the beginning part of the dispersion curves until their minima because the waveguide material has features of a polar dielectric in this frequency range. The waveguide losses of main modes increase after points of their minima because the waveguide material behaves as conductive media in this frequency range.

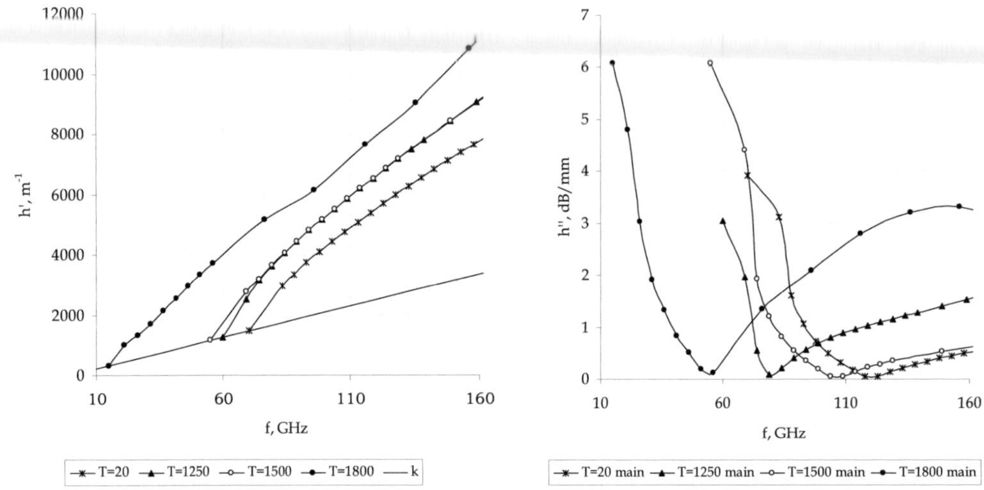

(a) (b)

Fig. 21. The complex dispersion characteristics of the main mode at the different temperatures

The dispersion characteristics of the first higher modes propagating in the hollow-core SiC waveguide with r = 1 mm at the different temperatures are presented in Fig. 22. We see that the behavior of waveguide losses at T = 1800°C is like for waveguide of conductive media.

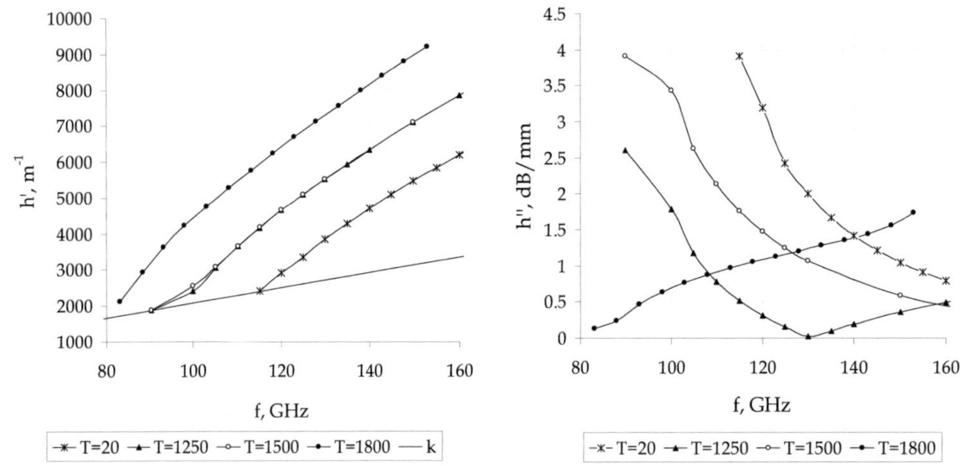

(a) (b)

Fig. 22. The complex dispersion characteristics of the first higher mode at the different temperatures

The features of the first higher mode losses at T = 20°C, 1250°C, 1500°C (Fig. 22b) can be explained in the same way as for the main mode (Fig. 21b).
The electric field distributions of the main hybrid HE_{11} mode and the first higher hybrid EH_{11} mode propagating in the hollow-core SiC waveguide at T = 20°C and 1800°C are presented, correspondingly, in Figs. 23, 24 and 25, 26. In all of these figures, the left side shows the electric field strength lines and the right (b) shows the intensity of the electric field. All electric field distributions were calculated at frequencies which are close to the cutoff frequencies of each investigated mode. The calculation of electric field in the waveguide cross-section was executed with 10000 points. We see from Figures 23 (a)-26 (a) that HE_{11} and EH_{11} modes have only one variation on the waveguide radius.

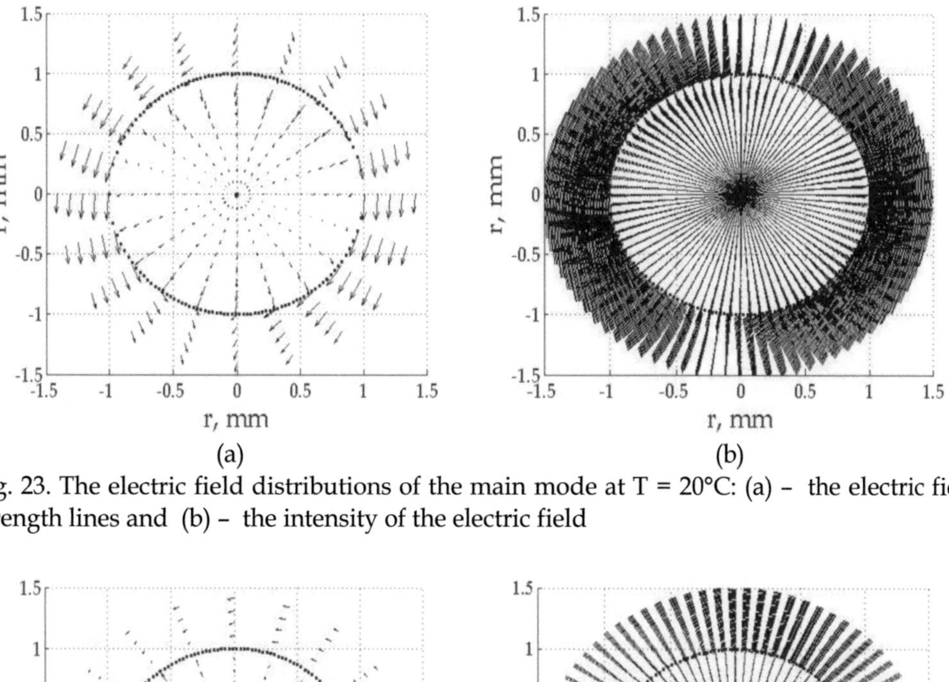

(a) (b)

Fig. 23. The electric field distributions of the main mode at T = 20°C: (a) - the electric field strength lines and (b) - the intensity of the electric field

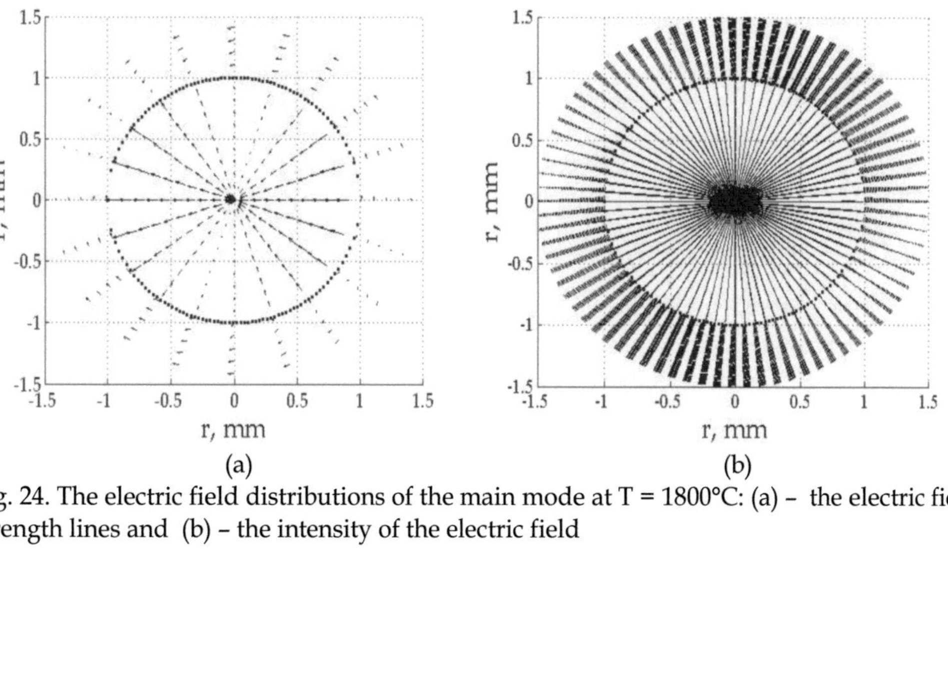

(a) (b)

Fig. 24. The electric field distributions of the main mode at T = 1800°C: (a) - the electric field strength lines and (b) - the intensity of the electric field

The electric field distribution of the main mode (at T = 20°C) presented in Fig. 23 was calculated at f = 103 GHz. The electric field distribution of the main mode (at T = 1800°C) presented in Fig. 24 was calculated at f = 31 GHz. Comparing Fig. 23, 24 we can make a conclusion that the electric fields of the main modes propagating in the hollow-core cylindrical waveguides at T = 20°C and T = 1800°C are different. The strongest electric field of the main mode propagating at T = 20 °C concentrates in the center of hollow-core and closely to the boundary separating SiC and air. However the strongest electric field of the main mode propagating at T = 1800°C concentrates in the air area of the hollow-core waveguide.
The electric field distributions of the first higher modes propagating in the hollow- core SiC waveguide at T = 20°C, shown in Fig. 25, was calculated at f = 135 GHz.

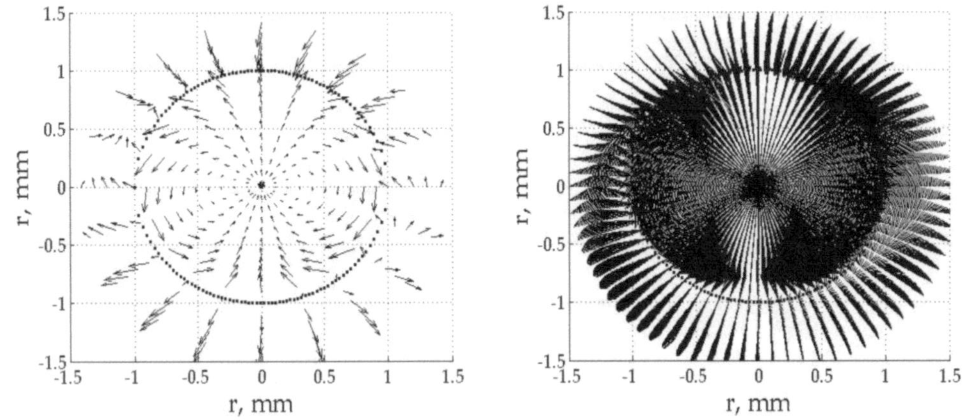

(a) (b)

Fig. 25. The electric field distributions of the first higher mode at T = 20 °C: (a) – the electric field strength lines and (b) – the intensity of the electric field

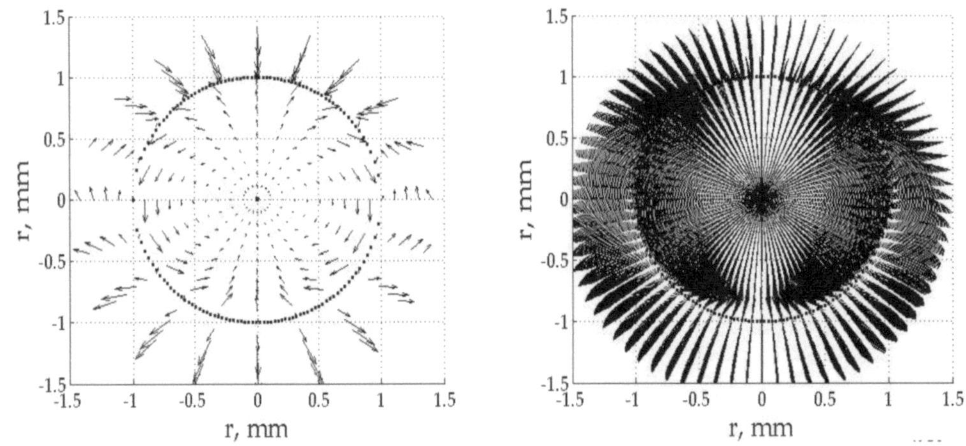

(a) (b)

Fig. 26. The electric field distributions of the first higher mode at T = 1800 °C: (a) – the electric field strength lines and (b) the intensity of the electric field

The electric field distribution of the first higher mode presented at Fig. 26 was calculated at f = 98 GHz. Comparing Figure 25, 26 we can make a conclusion that the electric fields of the first higher modes propagating in the hollow-core cylindrical waveguides at T = 20 °C and T = 1800 °C have similar response.

8. Conclusions

We have fulfilled the electrodynamical analysis of rectangular and circular SiC waveguides. The rectangular waveguides have been investigated by the SIE method, which is very useful and allowed us to analyze waveguides with cross-sections of any intricate shapes. The circular waveguides have been investigated by the partial area method.

Our electrodynamical analysis consists of two steps:

- The calculation of the dispersion characteristics,
- The calculation of the EM field distributions.

The eigenmodes of all investigated SiC waveguides are hybrid modes (Table 1).

We have noted that losses of the main modes propagating in the rectangular waveguides are smaller than losses of the main mode propagating in the circular waveguides at the certain frequencies (compare Figs. 6(b) with 9(b) and Figs. 8 with 13(b)).

We would like to point out that it is possible to find some conditions when the SiC waveguide operates in the single-mode regime and the waveguide broadband width is approximately 25% (Fig. 8).

The losses of the main mode propagating in the circular SiC waveguide have practically invariable values for frequencies ranging from 20-70 GHz. (Fig. 9(b)). This feature is important for operating of modulators and phase shifters which could be created on the basis of such waveguides. The EM signal propagating in the waveguide is not modulated by losses.

Extremely low-loss has a fast wave SiC waveguide. This feature could be used in practice for creation of feeder lines and specific devices that require low losses and low distortions of transmitted signals (Fig. 13(b)).

Thus we can conclude that we have carried out the full electrodynamical analysis of some SiC waveguides. Our investigations can be useful for creation of SiC waveguide devices.

9. References

Agarwal, A. & Ryu, S.-H. (2006). Status of SiC power devices and manufacturing issues, *Proceedings of CS MANTECH Conference*, pp. 215-218, Vancouver, Canada, April 2006

Asmontas S.; Nickelson L.; Bubnelis A.; Martavicius R. & Skudutis J. (2010). Hybrid Mode Dispersion Characteristic Dependencies of Cylindrical Dipolar Glass Waveguides on Temperatures, *Electronics and Electrical Engineering*, Vol. 106, No 10, 83-86, ISSN 1392-1215

Asmontas, S.; Nickelson, L. & Gric, T. (2009). Electric field distributions in the open cylindrical silicon carbide waveguides. *Acta Physica Polonica A*, Vol. 115, No. 6, 1160-1161, ISSN 0587-4254

Asmontas, S.; Nickelson, L. & Plonis D. (2009). Dependences of propagation constants of cylindrical *n*-Si rod on the material specific resistivity, *Electronics and Electrical Engineering*, Vol. 94, No. 6, 57-60, ISSN 1392-1215

Asmontas, S.; Nickelson, L.; Gric, T. & Galwas, B. A. (2009). Solution of Maxwell's equations by the partial area method for the electrodynamical analyses of open lossy metamaterial waveguides, *Proceedings of International Conference Differential Equations and their Applications*, pp. 17-20, ISBN 978-9955-25-747-9, Panevezys, Lithuania, September 2009, KTU, Kaunas, Lithuania

Asmontas, S.; Nickelson, L. & Malisauskas, V. (2006). Investigation of magnetized semiconductor and ferrite waveguides. *Electronics and Electrical Engineering*, Vol. 66, No. 2, 56-61, ISSN 1392-1215

Baeraky, T. A. (2002). Microwave measurements of the dielectric properties of silicon carbide at high temperature. *Egyptian Journal of Solids*, Vol. 25, No. 5, 263–273, ISSN 1012-5566

Bucinskas J.; Nickelson L. & Sugurovas V. (2010). Microwave diffraction characteristic analysis of 2d multilayered uniaxial anisotropic cylinder. *Progress In Electromagnetics Research*, Vol. 109, 175-190, ISSN: 1070-4698

Bucinskas, J.; Nickelson, L. & Sugurovas, V. (2010). Microwave scattering and absorption by a multilayered lossy metamaterial-glass cylinder. *Progress In Electromagnetics Research*, Vol. 105, 103-118, ISSN: 1070-4698

Chen, H.; Wu, B.-I. & Kong A. J. (2006). Review of electromagnetic theory in left-handed materials. *J. of Electromagn. Waves and Appl.*, Vol. 20, No. 15, 2137–2151, ISSN 0920-5071

Gakhov, F.D. (1977). *The boundary problems*, 640 p., Science, Moscow (In Russian)

Gric, T.; Nickelson, L. & Asmontas, S. (2010). Three dimensional magnetic field distributions and dispersion characteristics of rectangular and circular SiC waveguides. *Proceedings of 18th International Conference on Microwaves, Radar and Wireless Communication MIKON-2010*, Vol. 2, pp. 722-725, ISBN 978-9955-690-20-7, Vilnius, June 2010, JUSIDA, Vilnius

Gric T., Nickelson L. & Asmontas S. (2010). 3D vector electric field distributions and dispersion characteristics of open rectangular and circular metamaterial waveguides. *Proceedings of 18th International Conference on Microwaves, Radar and Wireless Communication MIKON-2010*, Vol. 2, pp. 578-581, ISBN 978-9955-690-20-7, Vilnius, Lithuania, June 2010, JUSIDA, Vilnius

Gric T.; Nickelson L., & Asmontas S. (2010). Waveguide Modulator. Patent 5710. Application of Invention 2010-040 is given in *Official bulletin of the state patent bureau of the republic of Lithuania* 2010/11, 2010-11-25, ISSN 1648-9985, Vilnius

Ikeuchi, M.; Sawami, M. & Niki, H. (1981). Analysis of open–type dielectric waveguides by the finite–element iterative method. *IEEE Trans.*, MTT–29, No. 3, 234–239, ISSN 0018-9480

Kajfez, D & Kishk, A. A. (2002). Dielectric resonator antenna – possible candidate for adaptive antenna arrays. *Proceedings of Telecommunications, Next Generation Networks and Beyond International Symposium*, Portoroz, Slovenia, May 2002

Kim, K. Y. (2004). Guided and Leaky Modes of Circular Open Electromagnetic Waveguides: Dielectric, Plasma, and SiC Columns: *Thesis for the Degree of Ph.D.*, 201 p., Kyungpook National University. Daegu, Korea

Kong, J.A. (2008). *Electromagnetic wave theory*, 1016 p., EMW Publishing, ISBN 0-9668143-9-8, Cambridge, Massachusetts, USA

Nickelson L., Galwas B.A., Gric T. & Ašmontas S. (2008). Electric field distributions in the cross sections of the metamaterial hollow-core and rod waveguides, *Proceedings of 17th International Conference on Microwaves, Radar and Wireless Communication MIKON-2008,* Vol. 2, pp. 497-500, ISBN 9781424431229, Wrocław, Poland, May 2008, MDruk, Warszawa, Poland

Nickelson L., Gric T., Asmontas S. & Martavicius R. (2008). Electrodynamical analyses of dielectric and metamaterial hollow-core cylindrical waveguides, *Electronics and Electrical Engineering*, Vol. 82, No 2, 3-8, ISSN 1392-1215

Nickelson L; Gric T; Asmontas S & Martavicius R. (2009) Analyses of the Gyroelectric Plasma Rod Waveguide, *Proceedings of 17th IEEE International Pulsed Power Conference*, (PPC2009) pp. 724-727, Washington

Nickelson L. & Gric, T. (2008). Dispersion characteristics and electric field distributions of modes propagating in the open electrically gyrotropic semiconductor rod waveguide. *Proceedings of 17th International Conference on Microwaves, Radar and Wireless Communication MIKON-2008,* pp. 501-504, ISBN 83-906662-7-8, IEEE Catalog Number: CFPO8784-PRT, Wroclaw, Poland, May 2008, MDruk, Warszawa

Nickelson, L. & Shugurov, V. (2005). *Singular integral equations' method for the analysis of microwave structures*, 348 p., VSP Brill Academic Publishers, ISBN 90-6764-410-2, Leiden-Boston, The Netherlands

Nickelson, L.; Asmontas, S.; Gric, T. & Martavicius, R. (2009). Analysis of Slow and Fast Modes of Lossy Ceramic SiC Waveguides, *Proceedings of Progress In Electromagnetics Research Symposium*, pp. 573-576, ISSN 1159-9450, Moscow, Russia, August 2009, PIERS, Cambridge

Nickelson, L.; Asmontas, S.; Malisauskas, V. & Martavicius, R. (2009). The dependence of open cylindrical magnetoactive p-Ge and p- Si plasma waveguide mode cutoff frequencies on hole concentrations. Journal of Plasma Physics. Vol. 75, 35-51, ISSN 0022-3778

Nickelson, L.; Gric, T.; Asmontas, S. & Martavicius, R. (2009). Electric Field Distributions of the Fast and Slow Modes Propagated in the Open Rod SiC Waveguide. *Electronics and Electrical Engineering,* Vol. 93, No. 5, 87-90, ISSN 1392-1215

Okojie, R. S.; Page, S. M. & Wolff, M. (2006). Performance of MEMS-DCA SiC Pressure Transducers under Various Dynamic Conditions. *Proceedings of IMAPS International High Temperature Electronics Conference*, pp. 70-75, Santa Fe, NM, May 2006

Pandraud, G.; Pham, H. T. M.; French, P. J. & Sarro, P. M. (2007). PECVD SiC optical waveguide loss and mode characteristics. *Optics and Laser Technology*, Vol. 39, No. 3, 532-536

Ponchak, G. E.; Schwartz, Z. D.; Alterovitz, S. A. & Downey, A. N. (2004). Measured Attenuation of Coplanar Waveguide on 6H, p-type SiC and High Purity Semi-Insulating 4H SiC through 800 K. *Proceedings of 12th GAAS Symposium*, pp. 439-442, Amsterdam, the Netherlands, 11-15 October, 2004

7

Silicon Carbide Based Transit Time Devices: The New Frontier in High-power THz Electronics

Moumita Mukherjee
Centre of Millimeter-wave Semiconductor Devices and Systems,
Institute of Radio Physics and Electronics,
University of Calcutta, India
e-mail: mm_drdo@yahoo.com

1. Introduction

In recent years, the field of Terahertz (THz) science and technology has entered a completely new phase of unprecedented expansion that is generating ever-growing levels of broad-based international attention. Indeed, the plethora of activities that have arisen recently in both the technology and scientific arenas associated with the THz frequency domain - i.e., between 1 millimeter (300 GHz) and 100 micrometers (3 THz), suggest that the field might be attempting to undergo a dramatic transition that could lead to long-awaited payoffs in a number of application areas. The inherent advantages and potential payoffs of the THz regime for military & security as well as industry relevant applications have long stood as an important driver of interest in this science and technology area. This extremely expansive and spectrally unique portion of the EM spectrum had initial application in space-based communications, upper atmospheric sensing and potentially for short-range terrestrial communications and non-intrusive package screening. However, the very rapid growth in more recent years is arguably most closely linked to the potential payoffs of THz sensing and imaging for an array of military, security and industrial applications. These applications include the spectroscopic-based detection identification and characterization of chemical and biological agents and materials, remote and standoff early-warning for chemical-biological warfare threats, and imaging of concealed weapons and explosives, just to name a few. In addition, THz-regime finds its application possibilities in industry and private-sector areas as food-industry process control, pharmaceutical industry, biological science, medical diagnostics and security screening.

Systems for rapidly emerging applications at THz frequencies thus require reliable high-power sources. In the last few years, the development of suitable sources for this frequency regime is being extensively explored worldwide. There are broadly two technology roadmaps for THz semiconductor devices. Approaching from the lower frequency range in the THz regime, electronic devices such as, Gunn diode, Resonant Tunneling diode (RTD) and nanometer Field Effect Transistors (FET) based on plasma wave have been widely investigated for THz frequency generation. From higher portion of the THz frequency spectrum, the photonics-based device Quantum Cascade Laser (QCL) extends the emission wavelength to Terahertz spectral range. The other approach to THz generation is through femtosecond lasers incident on materials with non-linear optical properties or on photoconductors such as InP. Parametric amplifiers are also being used for the purpose.

All the above efforts are to pursue the effective generation of THz signals. Most of the available THz sources are complex and bulky. QCL, on the other hand, has the advantage of small size, though they require low temperature operation to directly generate THz. Thus it seems that there is lack of availability of small-sized suitable THz source to serve a useful purpose. So, the development of high-power, low-cost and compact semiconductor sources in THz regime has attracted the recent attention of researchers working in this field.

Nowadays two-terminal solid-state Avalanche Transit-Time (ATT) devices are finding increasing applications in advanced RADAR, missile seekers and MM-wave communication systems. The performance of conventional Si (Silicon) and GaAs (Gallium Arsenide)-based IMPact ionization Avalanche Transit Time (IMPATT) devices are limited by power, operating temperature and especially by operating frequency. Recently, there is a global demand for THz-frequency applications and this warrants a new class of IMPATT oscillators which would outclass conventional Si and GaAs IMPATTs. Investigations on the prospects of Wide Bandgap (WBG) semiconductor materials, particularly IV-IV SiC semiconductors, for developing devices for high-frequency, high-power and high-temperature applications have been started recently.

Considering all the above facts, the author has made an attempt to study the THz-frequency characteristics of SiC-based IMPATT oscillators. In this Chapter, light will be thrown on the reliability and experimental feasibility of this new class of devices in the THz-region. Also, photo-sensitivity of these new classes of devices will be included in the chapter.

2. Importance of SiC as a base semiconductor material for IMPATT fabrication

2.1 Hexagonal SiC

SiC is recognized as a semiconductor of great importance in electronic applications because of its distinct properties, the possibility of easy growth on a native oxide, and the presence of numerous polytypes [1-4]. Silicon carbide is made up of equal parts silicon and carbon. Both are period IV elements, so they will prefer a covalent bonding such as in the left figure. Also, each carbon atom is surrounded by four silicon atoms, and vice versa. This will lead to a highly ordered configuration, a single polarized crystal. However, whereas Si or GaAs has only one crystal structure, SiC has several. The SiC family of semiconductor contains the same semiconductor material grown in many polytypes. The most commonly grown SiC materials are 4H-SiC, 6H-SiC, 3C-SiC. SiC, although of varied polytypes, generally have high carrier saturation velocity and high thermal conductivity, which make them suitable for high-temperature (above 900K), high-frequency (Terahertz region) applications [5]. Table 1 compares four semiconductors: silicon, gallium arsenide, silicon carbide and gallium nitride. Gallium Nitride is included here since in some respects it is perhaps a better material than SiC. It is also of interest to combine GaN with SiC, though the big difference is the energy bandgap. Standard semiconductors have almost three times smaller bandgaps than the wide bandgap materials SiC and GaN. However, it is probably the ten times larger critical field for breakdown which makes the biggest difference. There are no large differences in the other parameters, except the high mobility of GaAs. Cree Research Inc. [6]

was the first commercial vendor of SiC wafers which are commercially available as 4–inch wafers of 4H-SiC. It is well known that SiC wafer quality deficiencies are delaying the realization of outstandingly superior 4H-SiC high-power semiconductor devices. While efforts to date have centered on eradicating micropipes, 4H-SiC wafers and epilayers also contain elementary screw dislocations in densities on the order of thousands per cm^2, nearly 100 fold micropipe densities. While not nearly as detrimental to SiC device performances as micropipes, it was shown earlier that diodes containing elementary screw dislocations exhibit a 5% to 35% reduction in breakdown voltage, higher pre-breakdown reverse leakage current, softer reverse breakdown I-V knee and concentrated microplasmic breakdown current filaments when measured under DC testing conditions. At present, the micropipe densities have decreased to less than 1 cm^{-2} in 4-inch wafers. 4H- and 6H- are the easiest to grow and are usually epitaxially grown on a Si substrate. Nowadays the commonly used method to grow SiC epitaxial layer is the Chemical Vapor deposition (CVD) technique. It provides good structural quality and excellent doping control. Recent advances in crystal growth and thin film epitaxy of SiC, allow the development of high-quality layers. The cubic phase, 3C-SiC, however, is difficult to grow because of lack of a suitable substrate, thus it receives less attention. However, in recent years, there has been some little interest in 3C-SiC, resulting in both experimental and theoretical works. The most difficult to grow is 2H-SiC, because of its high formation energy. The most common donors in SiC are nitrogen (N) and phosphorous (P). N substitutes on C sites in the lattice, while P on Si sites. The most common acceptors are aluminum (Al) and boron (B) which substitutes on Si sites.

Semiconductor	Si	GaAs	4H-SiC	3C-SiC	WZ-GaN	Diamond
Bandgap (E_g) (eV)	1.12	1.42	3.26	2.3	3.45	5.45
Electric Breakdown field (E_C)(10^7 V.m^{-1})	3.0	4.0	30.0	22.0	50.0	100.0
Relative dielectric constant (ε_r)	11.9	13.1	9.7	9.72	8.9	5.5
Electron mobility (μ_n) (m^2 $V^{-1}s^{-1}$)	0.15	0.85	0.10	0.08	0.125	0.22
Hole mobility (μ_p) (m^2 $V^{-1}s^{-1}$)	0.04	0.03	0.011	0.004	0.085	0.085
Saturated drift velocity of electrons (v_{sn}) (10^5 ms^{-1})	1.0	1.2	2.0	2.0	2.5	2.7
Thermal Conductivity (K) (Wm^{-1} K^{-1})	150.0	46.0	490.0	450.0	225.0	1200.0

Table 1. Comparison of material parameters of different semiconductors

SiC was considered to be a promising material for fabrication of IMPATT diodes for the first time in 1973 by Keys [7]. In 1998, Konstantinov et al. fabricated epitaxial p-n diodes in 4H-SiC with uniform avalanche multiplications and breakdown [8]. They have performed photo-multiplication measurements to determine electron and hole ionization rates. P-n junction diodes were fabricated from p+ -n^0 -n+ epitaxial structures grown by vapor phase epitaxy (VPE); n^0 and n^+ layers were deposited on the p^+ substrates. The substrates were oriented in (0001) crystal plane with a small off-orientation angle, 3.5^0 or lower. The photo-multiplication measurement revealed that impact ionization in 4H-SiC appears to be dominated by holes, a hole to electron ionization co-efficient ratio up to 40-50 was observed. This ionization rate asymmetry was related to band-structure effects, to the discontinuity of the conduction band or the electron momentum along the c-direction. The results had a qualitative agreement with earlier studies of impact ionization in 6H-SiC. In 6H-SiC also, electron impact ionization was strongly suppressed and that was contributed to the discontinuity of the electron energy spectrum in the conduction band. Earlier problems in SiC device development due to poor material quality and immature device processing techniques was greatly overcome with the availability of production-quality substrates and the progress made in the processing technology. Though excellent microwave performances were demonstrated in SiC MESFETs and Static Induction transistors (SIT) [9], no experimental work was reported for SiC IMPATT devices before 2000. First experimental success of 4H-SiC based pulsed mode IMPATT was achieved by Yuan et al. in the year 2001 [10]. The DC characteristics of the high-low diodes exhibited hard, sustainable avalanche breakdown, as required for IMPATT operation. The fabricated 75 μm diameter SiC diodes were found to oscillate at 7.75 GHz at a power level of 1 mW. However, the output power level was significantly lower than the expected simulated value. They pointed out that the low-power problem is related to the measurement systems, particularly the design of the bias line. Optimization of the microwave circuit, in which the diode is embedded, is very important to properly evaluate the device performance. Any dispute in circuit optimization causes severe reduction in output power level. Thus, Yuan et al. made a comment that the measured low power, as obtained by their group, does not reflect the true power capability of SiC IMPATT [10]. Vassilevski et al. also fabricated 4H-SiC based IMPATT [11]. Microwave pulsed power of 300 mW was measured at 10 GHz. Though a comparatively higher power level was achieved, the power conversion efficiency was found to be very low ~0.3%. To increase the output power level, Ono et al. later [12] introduced a highly resistive guard ring that surrounds the diode periphery. The advantage of this guard ring is to reduce the electric field at the p-n junction edge of the junction periphery. A high current can thus be supplied through the diode without any destruction. Output power of 1.8W at 11.93 GHz was obtained from their fabricated diode and which is to date the highest reported output power from 4H-SiC IMPATT diodes. Nevertheless this power level is much lower than that expected. To increase the output power level, the residual series resistance should be minimized. No theoretical or experimental works on lo-hi-lo type 4H-SiC-based diodes have been published by other workers. The author has first time investigated the prospects of such devices in THz-frequency region for the first time.

2.2 Cubic SiC

As discussed earlier, the high breakdown field and high thermal conductivity of all the polytypes of SiC coupled with high operational junction temperatures, theoretically permit extremely high power densities and efficiencies to be realized in SiC devices. Among all the polytypes of SiC, from the technological point of view, cubic (β)-SiC has certain advantages over hexagonal (α)-SiC. Although small area SiC wafers are commercially available for 4H-SiC and 6H-SiC hexagonal polytypes, their cost is 1000 times higher than that of 6″ Si substrates. Moreover, device quality 4H-SiC and 6H-SiC wafers are produced mainly by bulk-crystallization, including a process involving high substrate temperature (> 2000^0 C). This high temperature growth forms high density channeled defects, known as micropipes in α-SiC. Presence of such high density defects micropipes in α-SiC is a major problem, as it greatly degrades the device quality.

On the other hand, β-SiC appears as a potential candidate, since it can be grown at a lower temperature. As there is no suitable substrate for growth of β-SiC crystals, the alternative is to use Si wafers which exist with good surface crystalline quality and with large surface area free of defects. Hetero-epitaxial growth of β-SiC on Si is a possible solution to overcome the problem of micropipes present in α-SiC polytypes. Moreover, the growth of good-quality 3C-SiC epilayers on Si would make it a cheaper alternative to costly 6H-SiC and 4H-SiC commercial epilayers and also makes it compatible with present Si technology. Additionally the β-SiC/Si heterostructures hold the promise for developing novel SiC/Si heterojunction devices and monolithic circuits combining SiC and Si devices. Also, the temperature coefficient of breakdown voltage of a p-n junction formed in 3C-SiC shows a positive value. A positive temperature coefficient is highly desirable to prevent runaway if devices reach the breakdown point. This indicates that IMPATT diodes can possibly be made with β-SiC, because the positive temperature coefficient is the direct result of an impact ionization process, required for the IMPATT diodes. Despite all of its advantages, the prospect of 3C-SiC as a base material for IMPATT fabrication has still not been explored. For the first time, the authors have simulated 3C-SiC based Single Drift flat profile ($p^+ n n^+$) IMPATT diode and the corresponding DC and terahertz characteristics of the device are also reported here. The authors have deposited p and n type 3C-SiC epilayers on Si substrate by Rapid Thermal Processing Chemical Vapour Deposition (RTPCVD) technique at a growth temperature as low as 800^0C. A p-n junction has been grown successfully and the characterization of the grown 3C-SiC film has been completed. The corresponding results are reported here.

3. SiC (both Hexagonal and Cubic) based THz IMPATT

3.1 Design Approach

SiC IMPATT diodes are designed and optimized through a generalized double iterative simulation technique used for analysis of IMPATT action [13]. The fundamental device equations, i.e. the one-dimensional Poisson's equation and the combined current continuity equations under steady-state conditions, have been numerically solved subject to appropriate boundary conditions, through an accurate and generalized double iterative computer algorithm. Iteration over the value and location of field maximum are carried out until the boundary conditions of electric field E(x) and normalized current density P(x) = $[J_P(x) - J_n(x)]/J_0$ profiles are satisfied at both the edges of diode active layer. The DC solution gives the electric field E(x) profile, normalized current density P(x) profile, the maximum electric field (E_m), drift voltage drop (V_D), breakdown voltage (V_B) and avalanche zone width (x_a). The breakdown voltage (V_B) is calculated by integrating the spatial field profile over the total depletion layer width. The boundary conditions for current density profiles are fixed by assuming a high multiplication factor ($M_{n,\,p}$) ~ 10^6, since it is well known that, avalanche breakdown occurs in the diode junction when the electric field is large enough such that the charge multiplication factors (M_n, M_p) become infinite. The edges of the depletion layer are also determined accurately from the DC analysis.

The high-frequency analysis of the IMPATT diodes provide insight into the dynamic performance of the diodes. The range of frequencies exhibiting negative conductance of the diode can easily be computed by Gummel-Blue method [14]. From the DC field and current profiles, the spatially dependent ionization rates that appear in the Gummel-Blue equations are evaluated and fed as input data for the small-signal analysis. The edges of the depletion layer of the diode, which are fixed by the DC analysis, are taken as the starting and end points for the small-signal analysis. The spatial variation of high-frequency negative resistivity and reactivity in the depletion layer of the diode are obtained under small-signal conditions by solving two second order differential equations in R(x, ω) and X(x, ω). R(x, ω) and X(x, ω) are the real and imaginary part of diode impedance Z (x,ω), such that, Z (x,ω) = R(x, ω) + j X(x, ω). A modified Runge-Kutta method is used for numerical analysis. The total integrated diode negative resistance (Z_R) and reactance (Z_x) at a particular frequency (ω) and current density J_0, are computed from numerical integration of the R(x) and X(x) profiles over the active space-charge layer. The diode total negative conductance (G) and susceptance (B) are calculated from the equations.

The basic mechanism of optical control of IMPATT diode is that, the leakage current entering the depletion region of the reversed biased p-n junction of an un-illuminated IMPATT diode is only due to thermally generated electron-hole pairs and it is so small that the multiplication factors ($M_{n,p}$) become very high. When optical radiation of suitable wavelength (photon energy $hc/\lambda > E_g$) is incident on the active layer of the device, the leakage current increases significantly due to photo-generation of charge carriers. The leakage current densities due to optically generated electrons and holes, $J_{ns,\ ps\ |opt}$ depend on the incident optical power according to the following equation :

$$(J_{ns} \text{ or } J_{ps})_{opt} = q\,\eta\,P_{opt}/\,Ah\nu, \tag{1}$$

where, η is quantum efficiency and A is the surface area over which absorption of incident optical power P_{opt} takes place corresponding to photon energy hν (ν is the frequency of incident radiation). If recombination is neglected, a linear response of the avalanche breakdown can be assumed, and $(J_{ns})_{opt}$ or $(J_{ps})_{opt}$ would increase linearly with P_{opt} over a particular range of wavelengths in which appreciable absorption takes place. The enhancement of the leakage current under optical illumination of the devices is manifested as the lowering of $M_{n,p}$.

In order to assess the role of leakage currents in controlling the dynamic properties of IMPATT oscillators at THz frequency, simulation experiments were carried out on the effect of electron current multiplication factor, M_n , (keeping hole current multiplication factor M_p very high ~ 10^6) and M_p (keeping M_n very high ~ 10^6) on (i) the small-signal admittance characteristics, (ii) the negative resistivity profiles, (iii) quality factor at peak frequencies (Q_p), (iv) device negative resistance at peak frequencies ($-Z_{RP}$) and (v) maximum power output of DDR SiC (both 4H- and 6H-) IMPATTs.

3.2 Results (Hexagonal SiC: un-illuminated diodes)

The DC and small-signal properties of the designed diodes are shown in Table 2. The E(x) profiles of the THz DDR diodes are compared in Figure 1. It is found that the peak electric field (E_m) increases from 4.25×10^8 Vm^{-1} to $5.9\times10^8 Vm^{-1}$, as the design frequency increases from 0.3 THz to 1.85 THz. The breakdown voltage and normalized voltage drop decrease from 135.0 V to 40.0 V and from 44% to 30%, with this increase of optimum operating frequency. The decrease in normalized voltage drop from 44% to 30% results in the decrease of efficiency, as expected. It is observed that the device is 14.0% efficient at 0.3 THz, whereas, with the increase of operating frequency to 1.85 THz, efficiency reduces to 9.5%. The values of negative conductance ($-G_P$) at the corresponding peak frequencies are shown in Table 2. It is found that $|-G_P|$ increase while $|-Z_{RP}|$ decrease significantly with the increase of operating frequency. The study reveals that as the optimum frequency changes from 0.3 THz to 1.85 THz, $-G_P$ increases by almost 20 times whereas, $|-Z_{RP}|$ decreases from $23.8 \times 10^{-10}\ \Omega\ m^2$ to $0.58 \times 10^{-10} \Omega\ m^2$. The admittance characteristics of the THz IMPATTs are shown in Figure 2(a-d). It is moreover interesting to observe that the designed diodes are capable of delivering high output power density ($3.69 \times 10^{11}\ Wm^{-2}$ at 0.3 THz and $6.4 \times 10^{11}\ Wm^{-2}$ at 1.85 THz), even at higher THz region. The high-power capability of the 4H-SiC based devices at high-frequencies is thus established.

4H-SiC based DDR diode	Background doping concentration (n region) ($10^{23}\ m^{-3}$)	Background doping concentration (p region) ($10^{23}\ m^{-3}$)	Width of the n-region (W_n) (nm)	Width of the p-region (W_P) (nm)	Bias current density (J0) ($10^9\ Am^{-2}$)
Designed at 0.3 THz	6.5	6.5	250.0	250.0	3.4
Designed at 0.5 THz	9.5	9.5	160.0	160.0	6.0
Designed at 0.7 THz	30.0	30.0	80.0	80.0	13.7
Designed at 1.80 THz	63.0	61.0	50.0	50.0	75.0

Table 2. Design parameters of the THz 4H-SiC based IMPATT diodes

Initially the author has estimated the values of series resistances (R_S, barring the contribution of ohmic contact resistance) at different THz frequencies. The effects of parasitic resistance on the maximum exploitable power level of the devices are also simulated. The results are shown in Table 3. In order to realistically estimate the values of R_S, the author has incorporated the contribution of ohmic contact resistances. A very recent study showed that ohmic contacts to n-SiC are formed by using pure Ni based layer with a thin underlying Si layer. By this technique a stable and low n-SiC contact resistivity of $\sim 10^{-11}\ \Omega\ m^2$ can be realized in practice. Very low specific contact resistance for p-SiC has not been achieved until now. Using alloy composition such as Ni/Al to p-SiC, a contact resistivity $\sim 10^{-10}\ \Omega m^2$ can be realized in practice. It is worthwhile to mention that, in order to get appreciable power from a THz source, low specific contact resistance ($\sim 10^{-11}\ \Omega\ m^2$) must be achieved, since at the THz region the intrinsic diode negative resistance is usually very small. It may be predicted that, further increasing the doping concentration of p-SiC semiconductor material, a desired contact resistivity $\sim 10^{-11}\ \Omega\ m^2$ may be achieved in reality. Hence, more realistic values of effective parasitic series resistance ($R_{S,total}$ including the contribution from contact resistance) of the designed devices are presented in Table 3. It is interesting to notice that even in the presence of aforementioned $R_{S,\ total}$, the THz devices are still capable of generating appreciable output power, as shown in Table 2. The effects of $R_{S,\ total}$ on the maximum power density of the devices are shown in Figure 3(a-d). It is very important to note that device negative resistances are much higher (at least 3.0 times) than the positive series resistance , which is an essential criteria for oscillations to take place. This observation also opens the prospect of 4H-SiC-based devices in the THz region.

4H-SiC based DDR diode	Designed at 0.3 THz	Designed at 0.5 THz	Designed at 0.7 THz	Designed at 1.8 THz
Electric field maximum (E_m) (10^8 V m^{-1})	4.25	4.30	5.25	5.9
Breakdown Voltage (V_B) (V)	135.0	96.0	55.3	40.0
Normalized voltage drop (V_D/V_B) (%)	44.0	37.7	32.97	30.0
Efficiency (η) (%)	14.0	12.0	10.5	9.5
Avalanche frequency (fa) (THz)	0.170	0.425	0.6	1.2
Peak frequency (f_P) (THz)	0.325	0.515	0.7	1.85
Negative conductance ($-G_p$) (10^8 Sm^{-2})	1.62	2.90	7.4	32.0
P_{max} (R_S =0.0Ω) (10^{11} Wm^{-2})	3.69	3.34	2.82	6.4
P_{max} ($R_S = R_{S,\ total}$) (10^{11} Wm^{-2})	3.37	2.9	2.5	5.75
Negative resistance ($-Z_{RP}$) (10^{-10}Ω m^2)	23.8	7.3	3.8	0.58
$-Q_P$	1.26	1.95	1.6	2.15

Table 3. DC and Small-signal results of 4H-SiC based THz IMPATTs

4H-SiC DDR Diode	Negative conductance (-G) (10^8 Sm^{-2})	Susceptance (B)(10^8 Sm^{-2})	Estimated load conductance (G_L)(10^8Sm^{-2})	Negative resistance ($-Z_R$) (10^{-10}Ωm^2)	Series Resistance (R_S)(10^{-10} Ωm^2)	Total Series Resistance ($R_{S,total}$) (10^{-10} Ωm^2)
Designed at 0.3 THz	1.53	1.50	1.30	33.4	10.2 (estimated at 0.3 THz)	10.40
Designed at 0.5 THz	2.75	3.00	2.42	16.60	3.66 (estimated at 0.5 THz)	3.86
Designed at 0.7 THz	7.4	12.0	6.7	3.8	0.49 (estimated at 0.7 THz)	0.69
Designed at 1.80 THz	27.0	30.0	26.8	1.66	0.022 (estimated at 1.5 THz)	0.22

Table 4. Estimation of Series resistance of the THz IMPATTs

The plots of variation of impedance of the diodes with frequencies are shown in Figure 4 (a-d). The graphs show that all the devices in the THz regime, posses negative resistance for all frequencies above the avalanche frequency (f_a), where its reactance is capacitive. This is due to the fact that, in the oscillating frequency range, the magnitude of Z_R is found to be small compared to Z_X. Figures indicate that the values of $|-Z_R|$ and $|-Z_X|$ decrease as the operating frequency increases. The diode total negative resistance is thus an important parameter for designing a suitable THz device, as it determines the output power level and oscillation condition of the device.

The simulation program has been used to obtain the negative resistivity profiles (R(x)) of the DDR diodes shown in Figure 5 (a-d). It is observed that the R(x) profile in each case is characterized by two negative resistivity peaks (R_{max}) in the middle of the each drift layer along with a central negative resistivity minimum (R_{min}) located near the metallurgical junction. The magnitudes of the two peaks in the electron and hole drift layers are found to decrease when the operating frequency increases. Thus the diode negative resistances,$-Z_{RP}$, i.e. the area under the R(x) profiles, for the DDR diodes decrease very sharply with the increase of the frequency of operation. This observation is also reflected in Table 2, where the values of $-Z_{RP}$ are shown at different THz frequencies. Furthermore, the negative resistivity peaks produced by holes in the hole drift layers are appreciably higher than the peaks produced by electrons in the electron drift layers. Relative magnitudes of hole and electron ionization co-efficient in 4H-SiC at the operating electric field range are found to be correlated with the above effect.

For practical realization of THz IMPATT, self-heating of the device is a major problem that will be discussed here. IMPATT devices are designed to be operated at a large current density in order to generate appreciable power in the THz region. It is found from the present simulation experiment, that the diodes are capable of generating high power density in the THz region. But, as the conversion efficiencies of the devices are within 10-14%, a small fraction of DC input power is converted into output power and the rest is dissipated as heat raising the junction temperature of the device. The optimized design however requires that the device will dissipate the power without increasing the junction temperature much over the ambient (300K), since the enhancement of junction temperature degrades the over-all performances of the THz devices. The safe operating temperature for Si (K=150 W $m^{-1}K^{-1}$) based IMPATT device is 573K. The value of K for SiC (490.0 W $m^{-1}K^{-1}$) is much higher (~ 3.3 times) than that of Si (Table 1). Hence, the SiC based IMPATT devices are expected to withstand much higher junction temperature, before they burn out. Thus the higher thermal conductivity allows 4H-SiC based devices to handle higher power. Moreover, if the SiC based THz diodes are mounted on a semi-infinite diamond (K Diamond = 1200 $Wm^{-1}K^{-1}$) heat-sink of much larger diameter as compared to the diameter of the device, the effective thermal conductivity will be very high and as a result it may be expected that the diode will be capable of dissipating the large amount of heat quickly from the junction, without increasing the junction temperature substantially. So the rise in junction temperature can be limited by proper heat-sink design and by using multiple mesas or ring geometry so that the thermal resistances of the diode and the heat-sink are reduced. The best way to resolve the self-heating problem is to operate the device under pulsed mode conditions with a small duty cycle so that the device does not heat up significantly and degrade in performance in the THz region.

All the above observations are studied by the author for the first time and these definitely establish the prospects of 4H-SiC based devices in the THz regime, where conventional Si and GaAs diodes cannot perform.

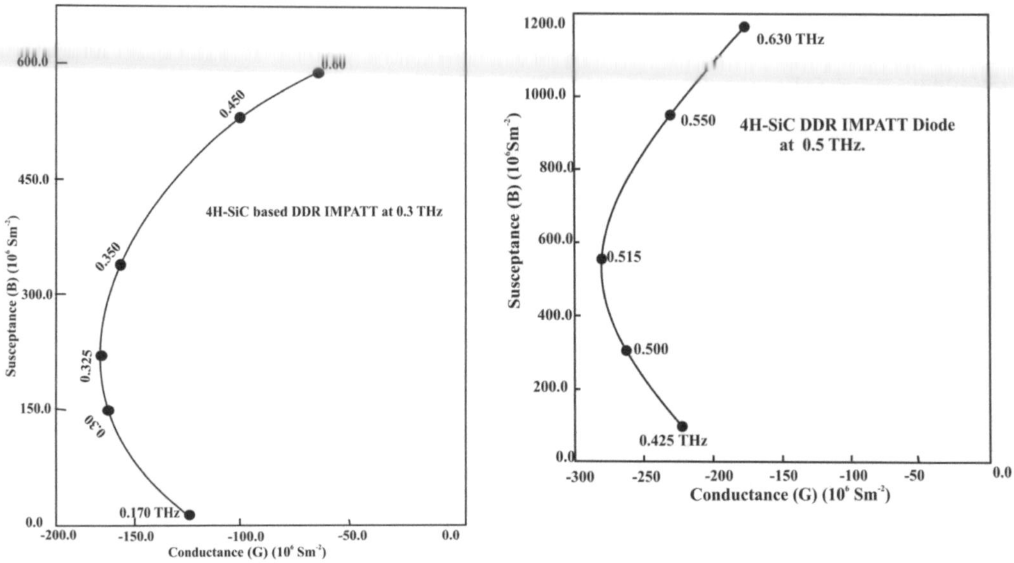

Fig. 1. Electric field profiles of SiC THz IMPATTs

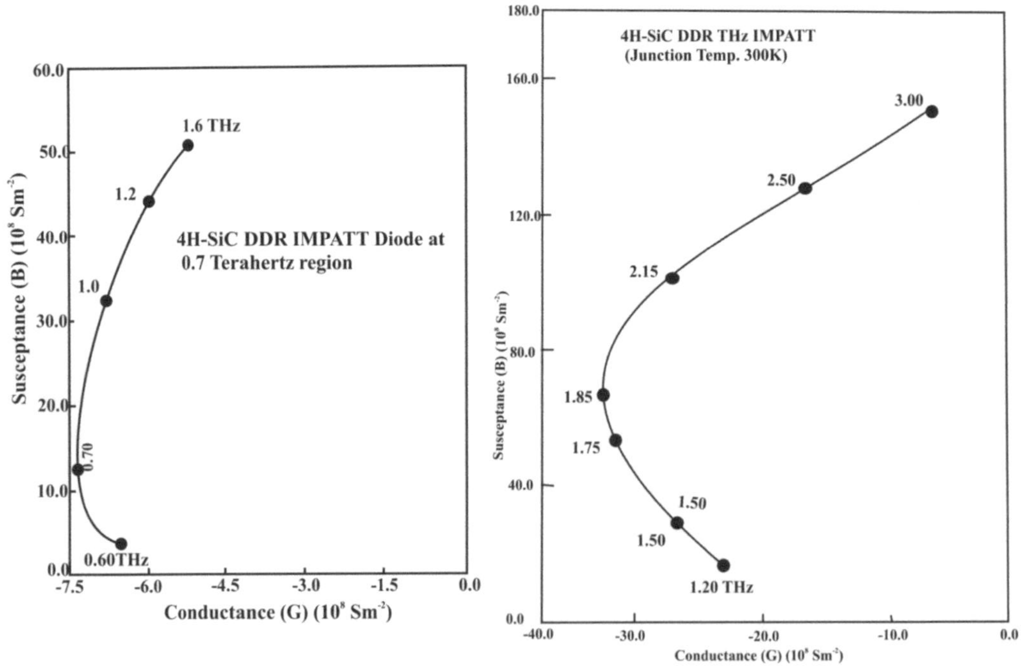

Fig. 2(a-d). Admittance plots of SiC THz IMPATTs within 0.3 – 1.8 THz

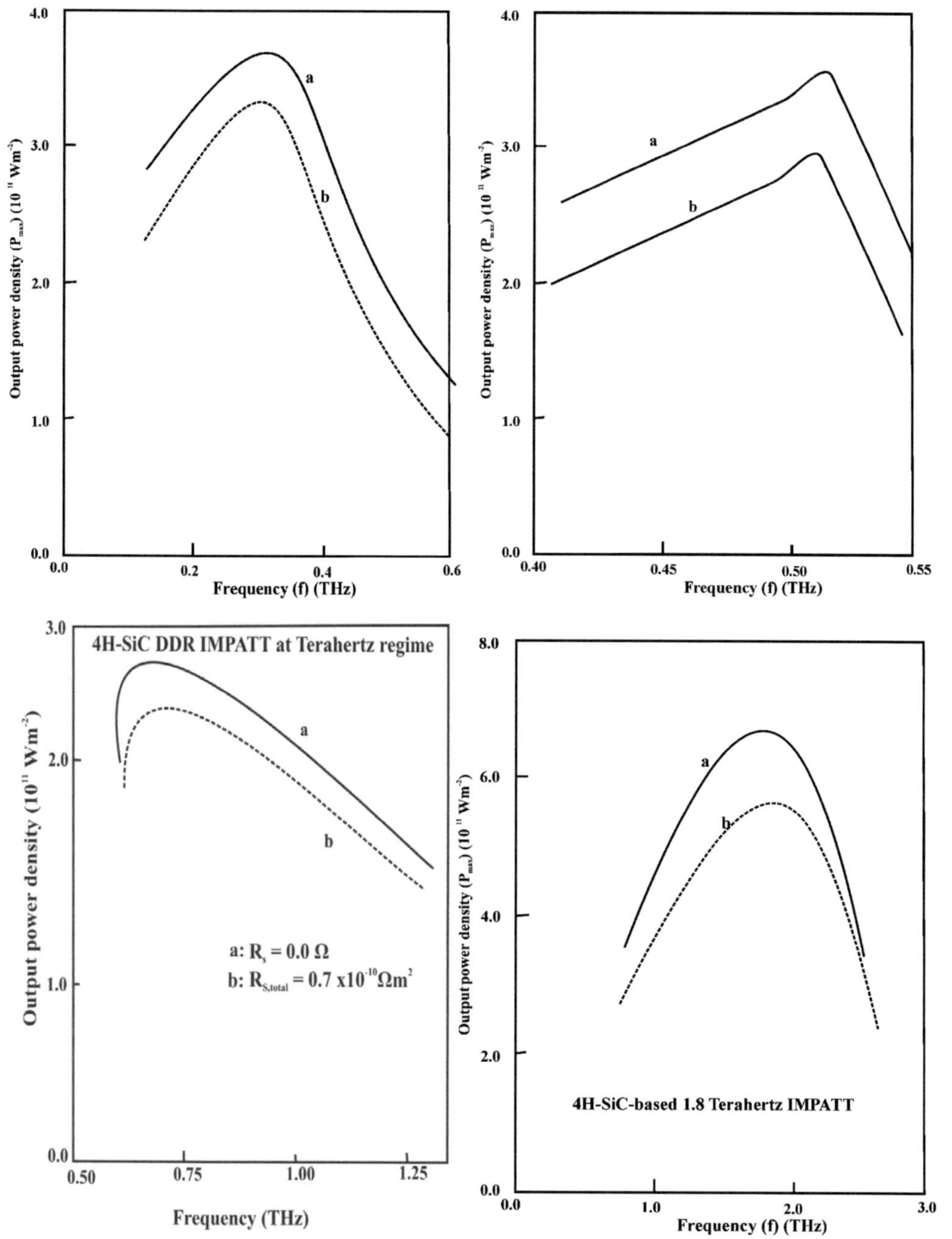

Fig. 3(a-d). Effects of series resistance on output power density of SiC THz IMPATTs within 0.3 – 1.8 THz (a: without R_S, b: with R_S)

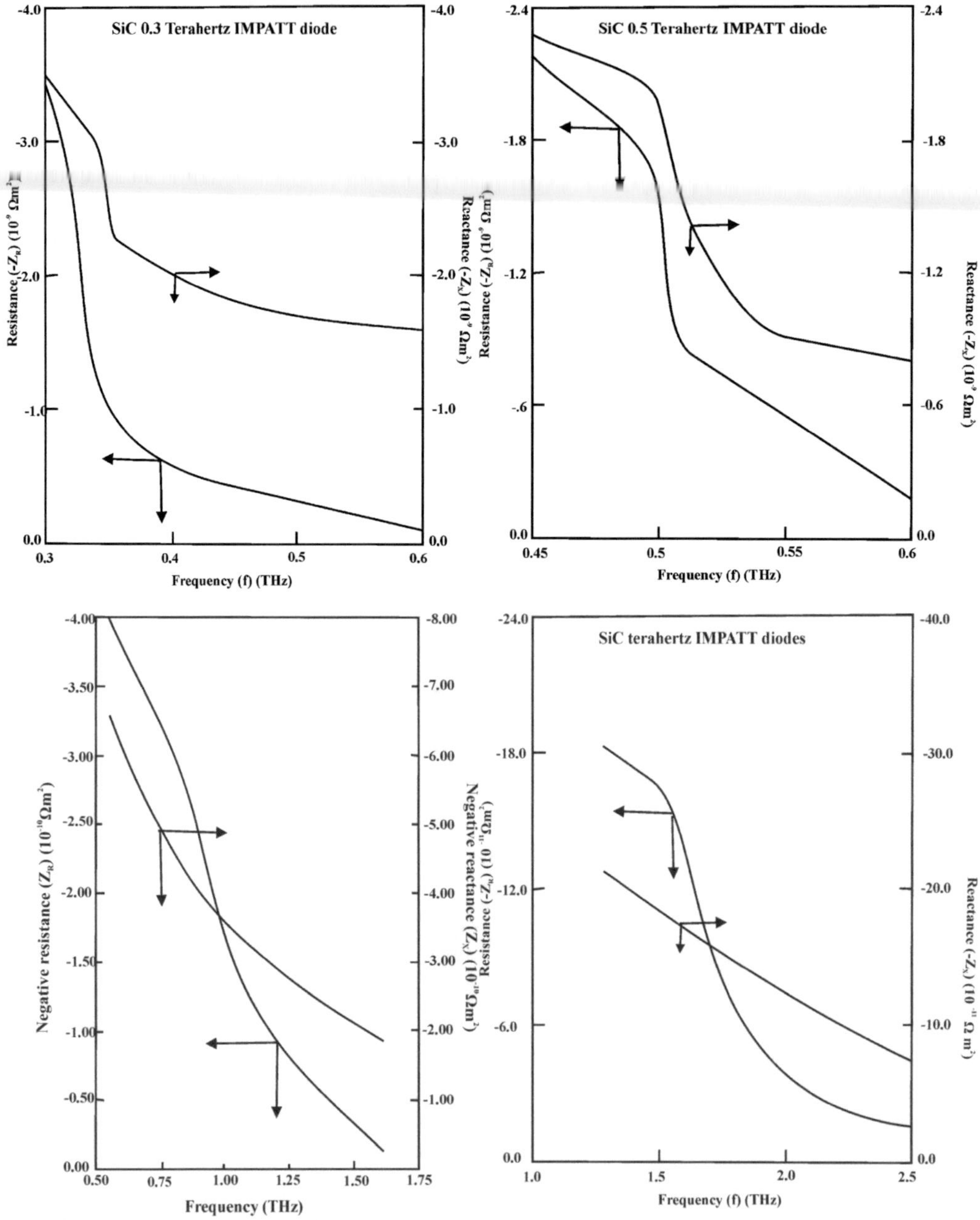

Fig. 4(a-d). Impedance plots of SiC THz IMPATTs within 0.3 – 1.8 THz.

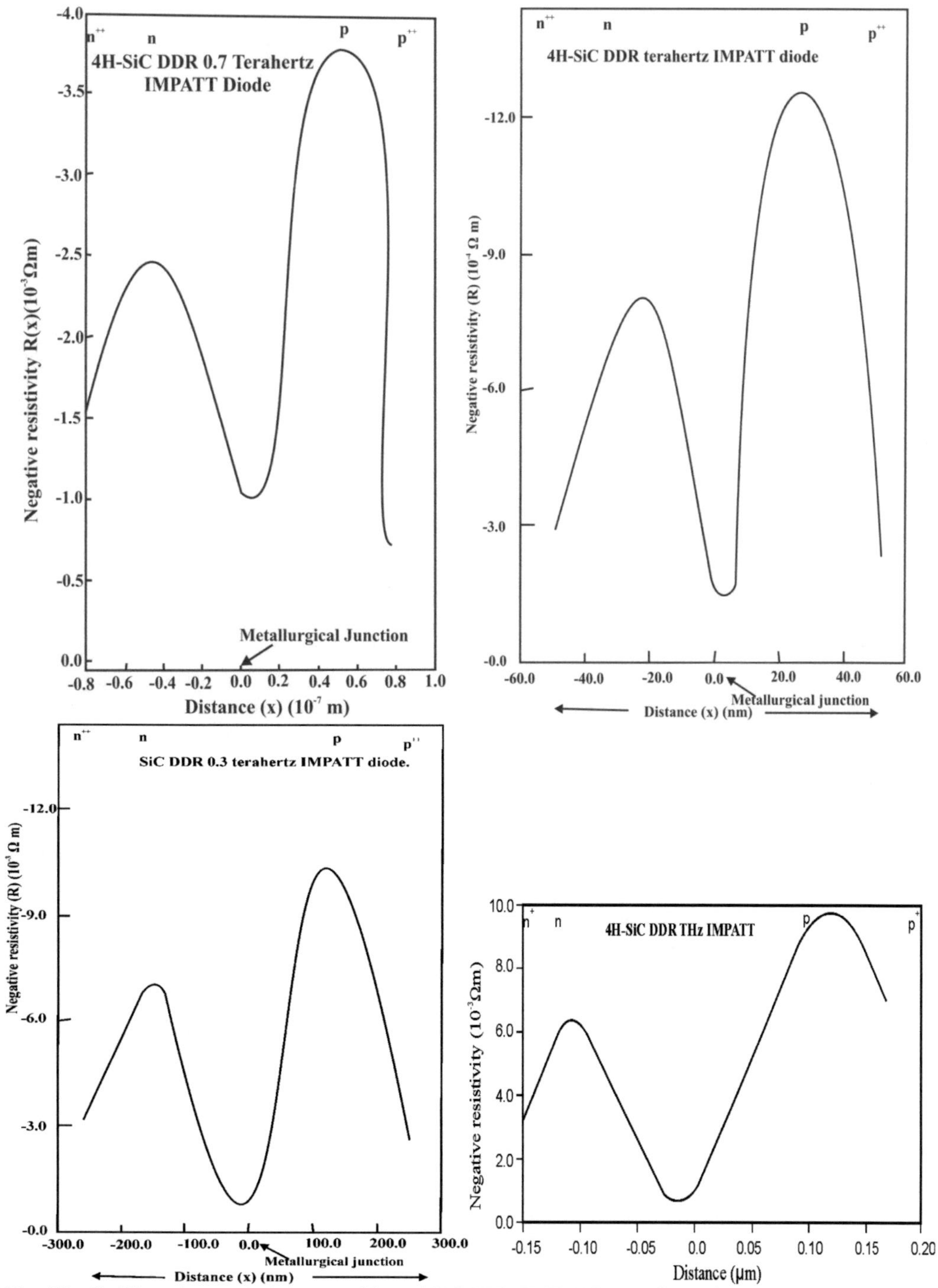

Fig. 5(a-d). Negative resistivity profiles of SiC THz IMPATTs within 0.3 – 1.8 THz.

3.3 Results (Cubic SiC: Un-illuminated diode)

The cubic (3C)-SiC based IMPATT diode has been optimized for low punch-through condition. Figure 6 shows the electric field profiles within the depletion layer of the $p^+ n n^+$ Terahertz IMPATT diode for the same bias current density (2 $x10^9$ Am^{-2}) and different active layer background doping concentration ranging from $0.7x10^{24}$ m^{-3} to $4x10^{24}$ m^{-3}. The figure clearly indicates that the decrease in n-layer doping concentration increases the punch-through and decreases the maximum breakdown field. The corresponding efficiencies of the diodes at different doping densities are also shown in Figure 6. The efficiency (13%) is found to be higher for the diode structure showing lower punch through (background doping density = 4 $x10^{24}$ m^{-3}). The low punch through diode is found to be most efficient in the THz region. Thus, the DC and high frequency properties of the optimized low punch through diode are further simulated and the corresponding results are summarized in Table 2. The diode admittance plots, with and without R_S , are shown in Figure 7 and they indicate that the diode negative conductance at 0.33 THz (peak operating frequency) will be 353.0 $x10^6$ Sm^{-2}. It is also observed from Figure 7 that in the presence of R_S, the device negative conductance decreases significantly. The RF power output for the optimized device is found to be 63.0W at 0.33 THz.

The parasitic series resistance for the 3C-SiC SDR diode was also determined and its effect on RF power is shown in Figure 8. It is evident from Figure 8 that due to the presence of R_S, the CW power reduces by ~9%. The value of total negative resistance ($-Z_R$) is found to be much higher than R_s at 0.3 THz, which is an essential condition for diode oscillation.

Figure 9 shows the negative resistivity profile at the peak frequency for the optimized 3C-SiC SDR IMPATT device, with and without R_S. Negative resistivity profiles give a physical insight into the region of the depletion layer that contributes to RF power. These figures show that for both the diode structures, the profiles exhibit negative resistivity peaks in the middle of the drift layer with dips in the avalanche layer close to the junction. The peak of the profile indicates that drift region contributes major role to higher negative resistance of the diode.

Ellipsometry measurement reveals that Ge layers of thickness 300- 1000 A^0 are formed on the Si substrate. The thickness of the SiC layer, as measured by ellipsometry is found to vary from 0.72 to 0.80 µm. The crystallinity of the doped films, studied by X-ray diffraction (XRD) technique, reveals that the SiC layers are perfectly crystalline (Figure 10). The peaks represent reflections from (left to right) Si (200), SiC (200), Si (400) and SiC (400) crystal planes. A Ge peak close to the SiC (400) peak is also observed. The angular positions of SiC (200), Si (400) and SiC (400) planes, as shown in the XRD spectrum match close with the corresponding angular position found by other researchers. The AFM pictures of the samples are shown in Figures 11 (a-b). Figure 11(a) reveals that there are large spots which are not visible, but these regions are visible for SiC layer grown on a Ge modified Si substrate (Figure 11(b)). Such large spots are indicative of holes or deeper lying parts on the SiC surface. Such features are connected with voids beneath the SiC layers. Therefore in the case of Ge pre-deposition the void formation is suppressed. Consequently the surface roughness of doped SiC film is found to be 16.0 nm, which is lower than the surface roughness of the film grown on Si surface without Ge pre-deposition. Transmission Electron Microscopy (TEM) measurement of sample without Ge incorporation is shown in Figure 12 (a). To improve the contrast of the TEM picture, the scale of the same sample in TEM mode is changed and shown in Figure 12 (b). In Figure 12(b), a comparison is made between TEM measurements of samples with and without Ge incorporation. It is found that there is a noticeable improvement on the quality of SiC layer (Figure 12(b)). Scanning TEM of samples with Ge and without Ge are shown in Figure 13. It can be observed that Ge layer displaces a better contrast under STEM mode, as shown in Figure 13. Hence, it is expected that Ge layer acts as a barrier between Si/SiC interfaces, and prevents out-diffusion of substrate Si into the SiC films. These improvements are likely to reduce the leakage current flowing through the p-n junction. From Figure 14 it is observed that Ge-layer displays a better contrast under STEM mode (b). Figure 15 shows the SIMS profiling of doped 3C-SiC film on Si. It is evident that a clear p-n junction with doping concentration of 4.5 x 10^{24} atoms m^{-3} is formed, which is very close to the simulation data. The p-and n-layer thickness are (in µm) 1.38 and 0.472, respectively. The location of p-n junction is found to be 0.99µm. The location of Ge layer is 1.28 µm.

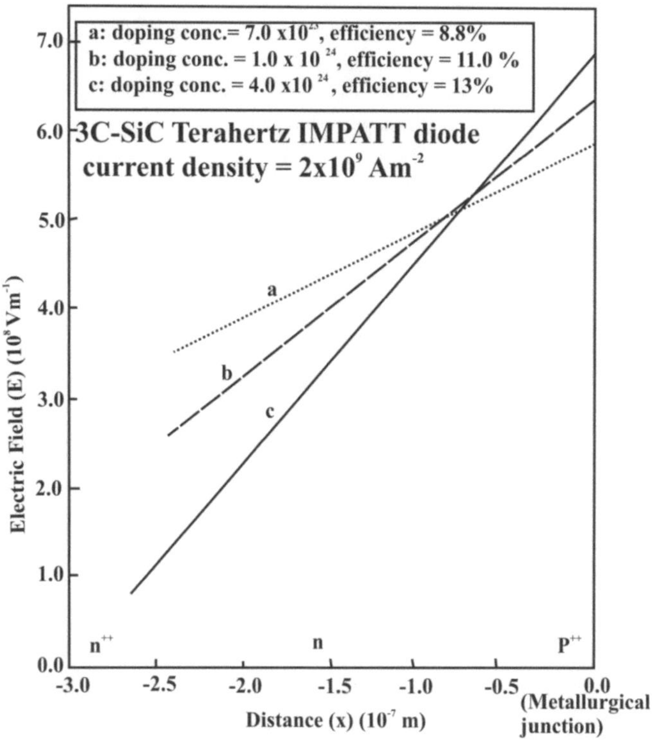

Fig. 6. Electric field profiles of 3C-SiC THz IMPATTs.

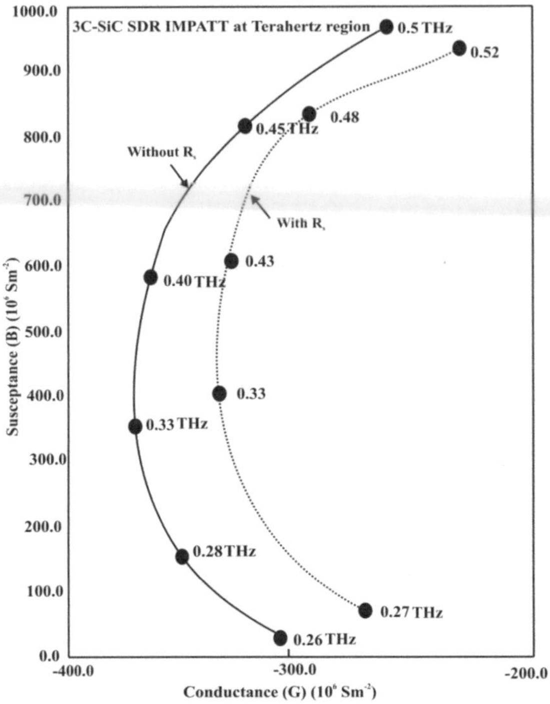

Fig. 7. Admittance plots of 3C-SiC THz IMPATT.

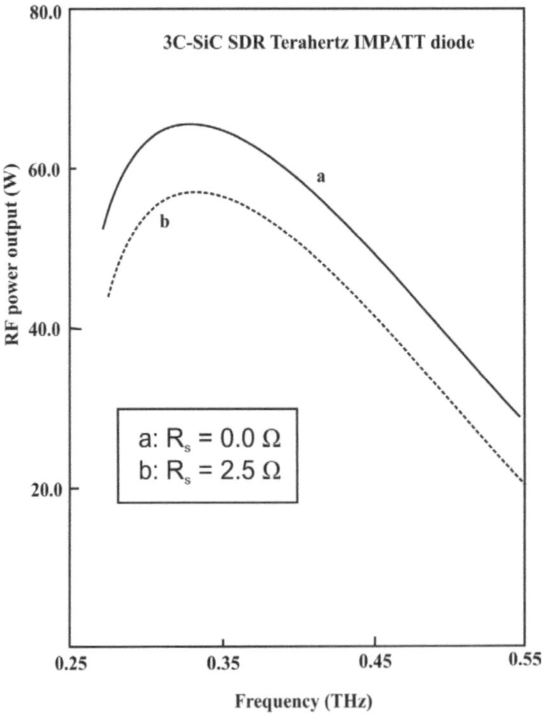

Fig. 8. Series resistance effects on RF power of 3C-SiC THz IMPATT.

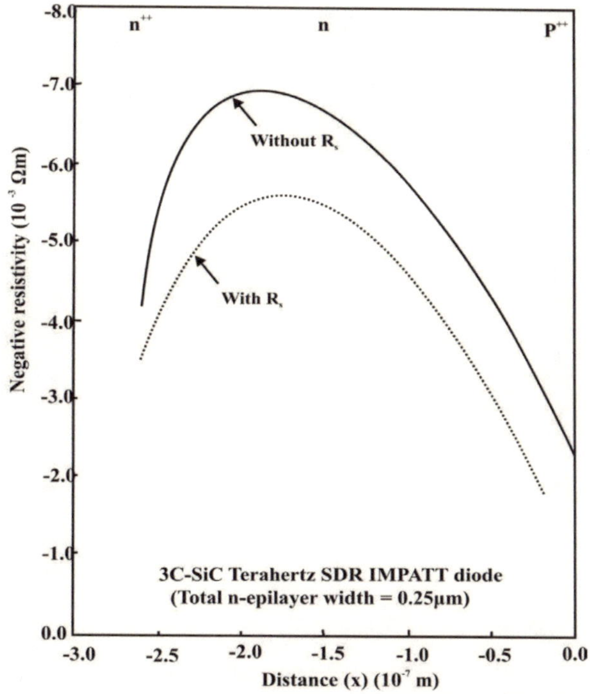

Fig. 9. Series resistance effects on negative resistivity of 3C-SiC THz IMPATT.

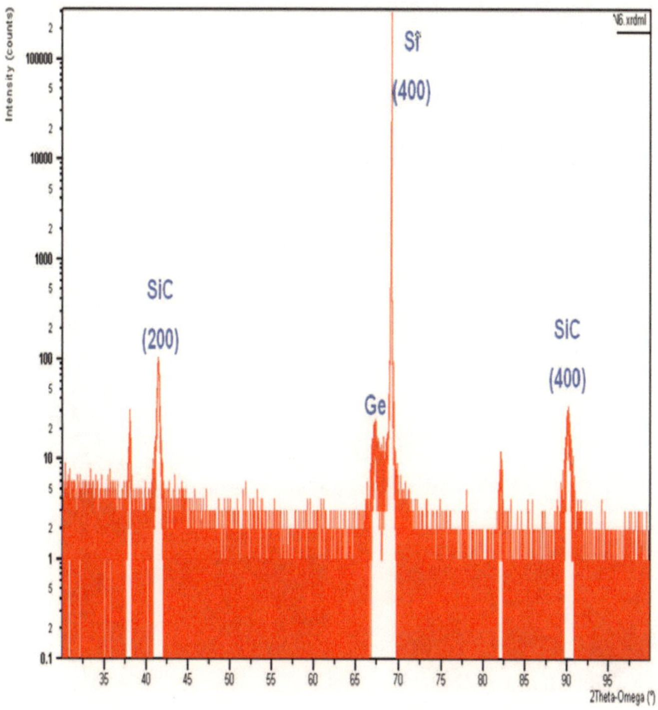

Fig. 10. XRD spectrum of β-SiC grown on Ge-modified Si <100> substrate.

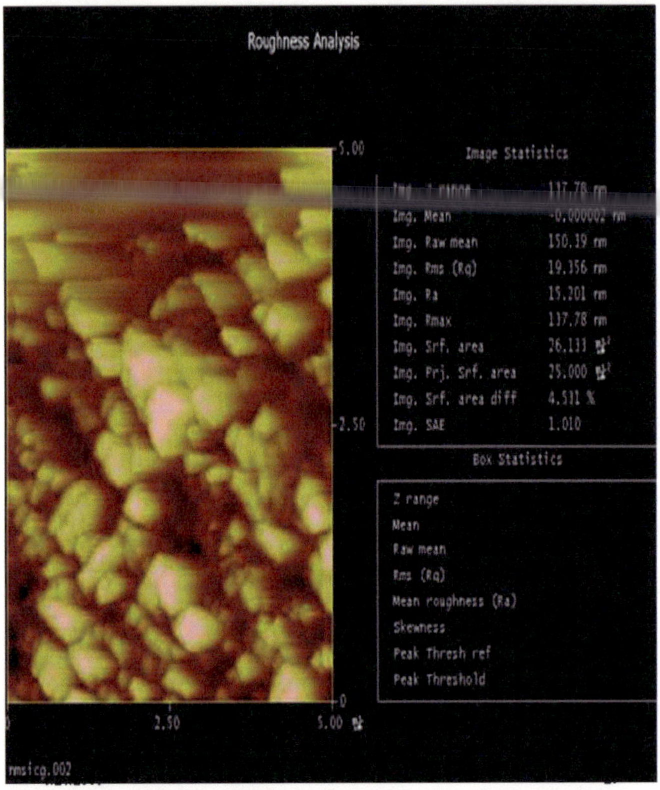

Fig. 11.(a) AFM picture taken from sample without Ge pre-deposition.

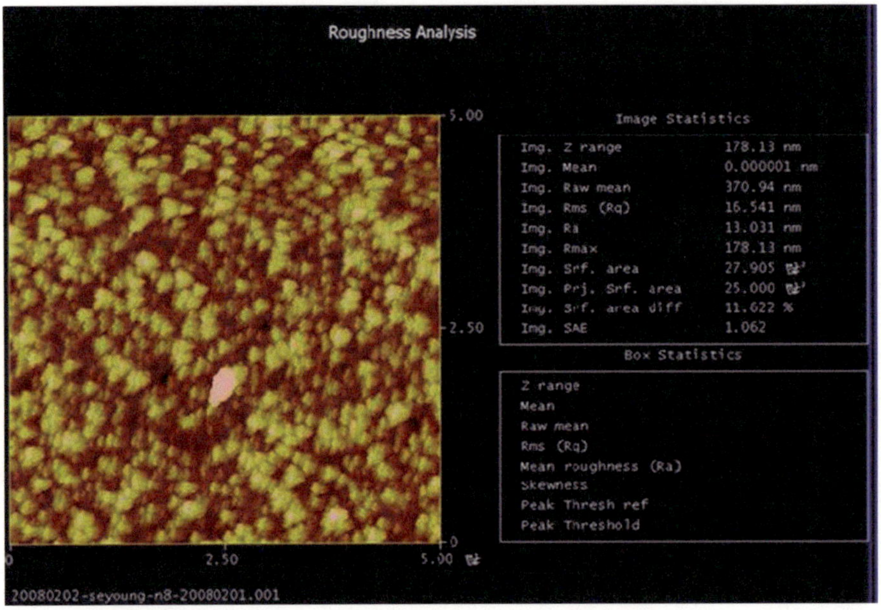

Fig. 11.(b) AFM picture taken from sample with Ge pre-deposition.

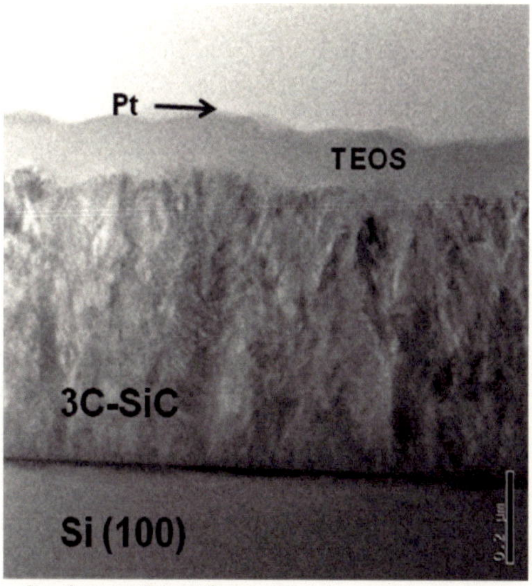

Fig. 12.(a) TEM micrograph of sample without Ge.

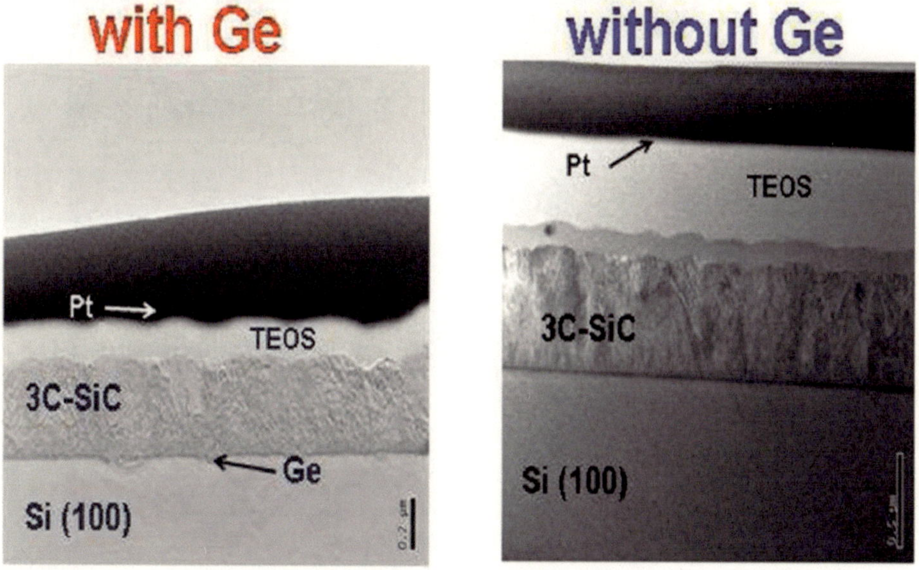

Fig. 12.(b) TEM micrographs of samples with Ge and without Ge.

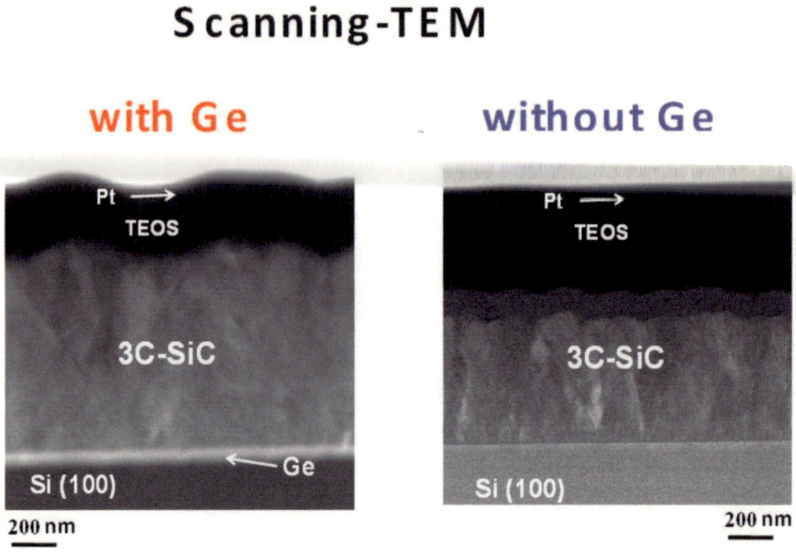

Fig. 13. Scanning-TEM of samples (a) with Ge & (b) without Ge. The arrow in Fig. 13(a) indicates a Ge layer formed at the interface of Si and 3C-SiC.

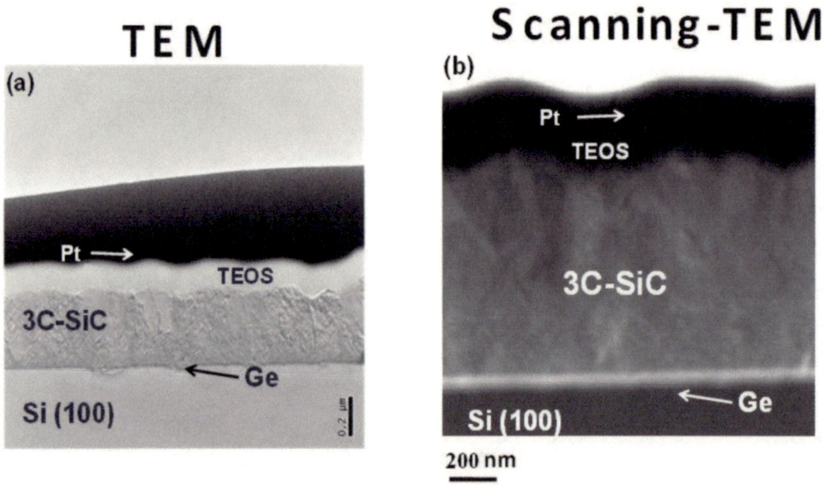

Fig. 14. 3C-SiC on Ge-modified Si: TEM and Scanning TEM analysis. Ge layer displays better contrast under STEM mode (figure 14(b))

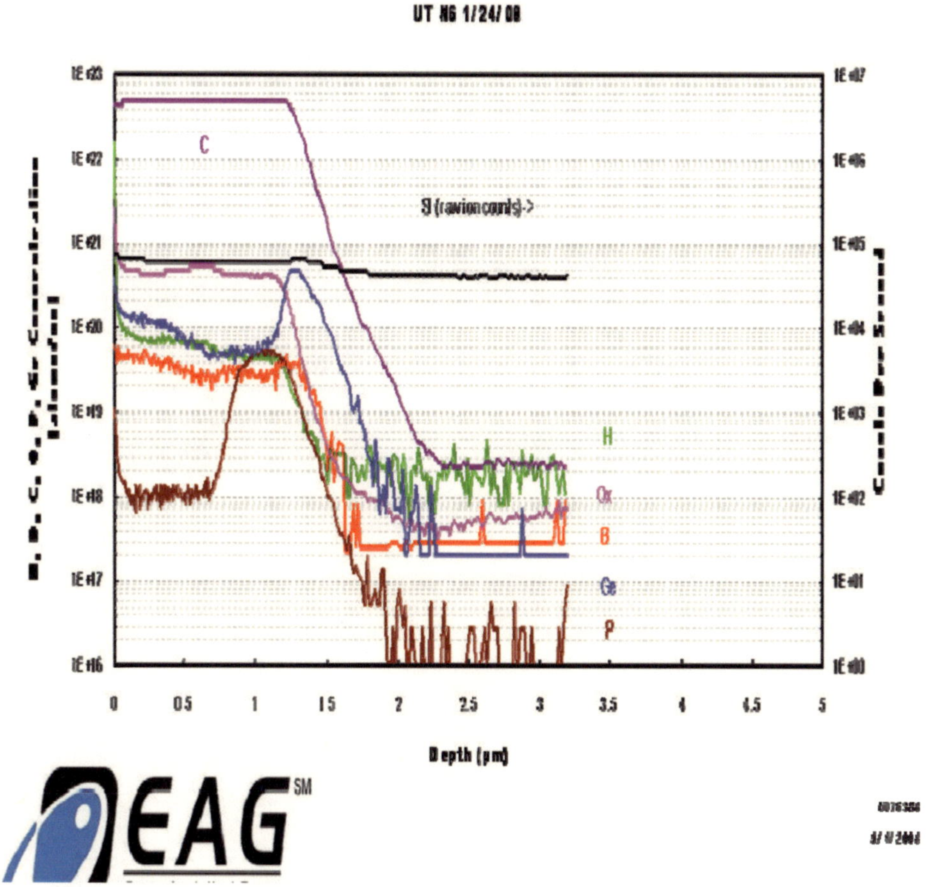

Fig. 15. SIMS profile of β-SiC grown on Ge-modified Si <100> substrate.

3.3 Results (illuminated diodes)

The effects of optical illumination on the THz behavior of the designed diode at 0.3 THz are shown in Figures 16 and 17. The computed values of $|-G_P|$, $|-Z_{RP}|$, P_{max}, f_P and $|-Q_P|$ for different electron and hole current multiplication factors are shown there. Admittance plots of the SiC DDR IMPATT under the three different illumination conditions, as mentioned before, are shown in *Figure 6*. It is evident from the figures that the values of $|-G_P|$ at the optimum frequencies decrease with the lowering of M_n or M_p. At the same time, the frequency ranges over which the devices exhibit negative conductance, shift towards higher frequencies with the lowering of M_n and M_p. In the present calculation M_n and M_P changes from 10^6 to 25, i.e. by a factor of 4×10^4 due to photo-illumination, thus indicating an increase of leakage current by the same factor. The leakage current can be increased in reality by varying the incident optical power. Previous experiments on illuminated Si IMPATTs revealed that the leakage current increase from 1 nA to 500 μA, that is by a factor of 5×10^5 by varying the incident optical power.

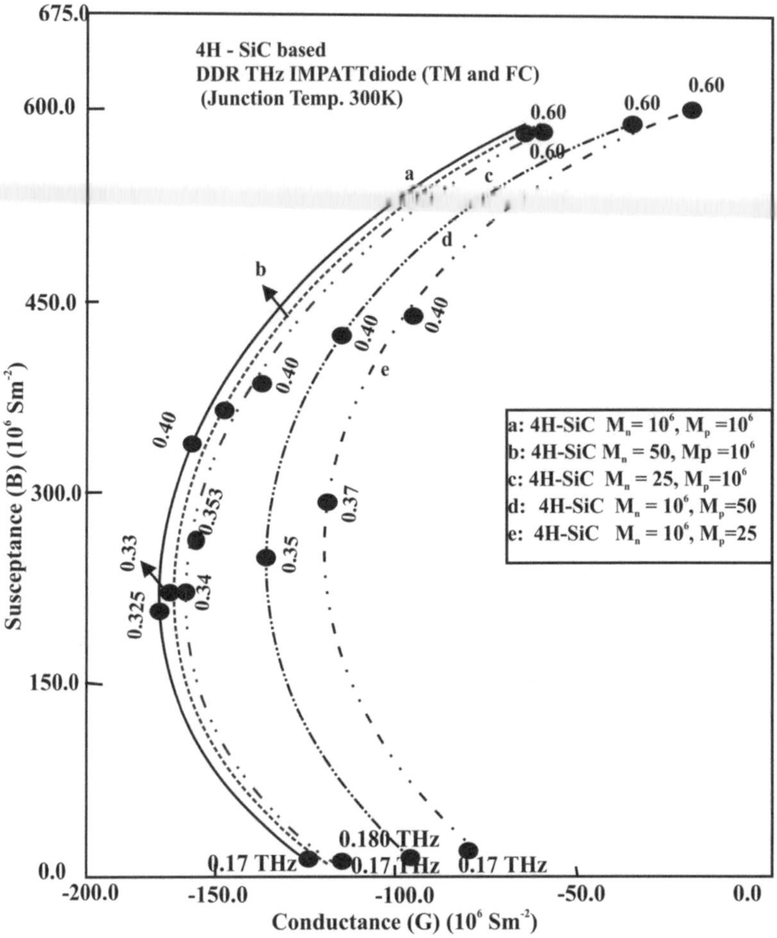

Fig. 16. Admittance plots of illuminated SiC THz IMPATTs at 0.3 THz.

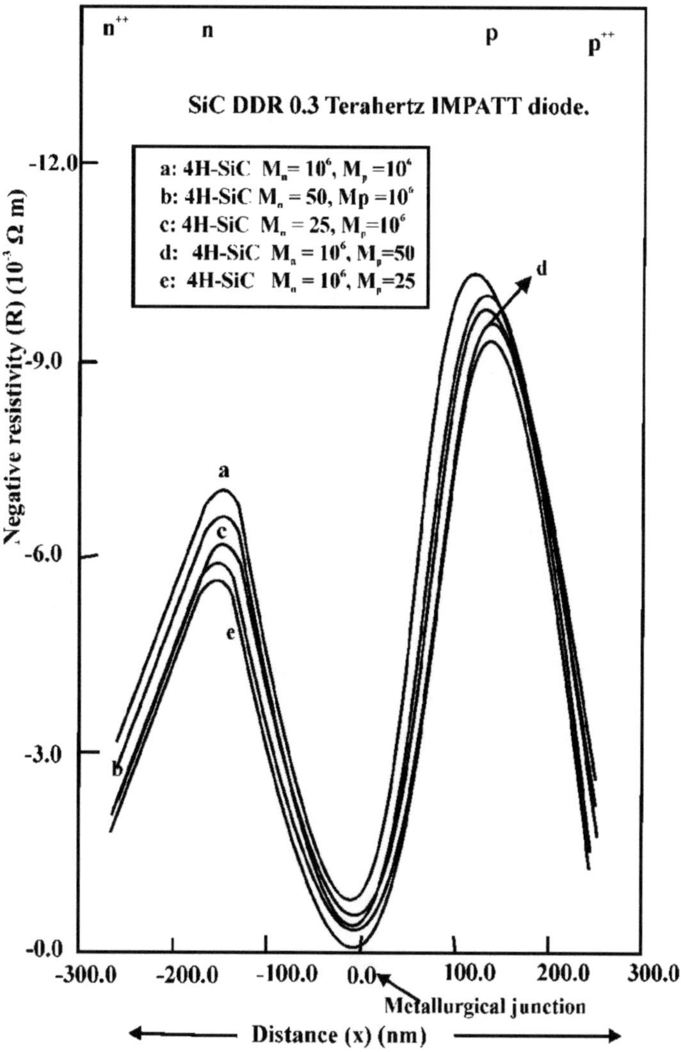

Fig. 17. Negative resistivity profiles of illuminated SiC THz IMPATTs at 0.3 THz.

Similar trends are observed in case of illuminated SiC THz IMPATTs at 0.5 THz, 0.7 THz and 1.85 THz.

4. Experimental feasibility

SiC epiwafer (n^{++} substrate and n-type epilayer) can be procured from Cree Inc., Durham, NC, USA. The n-type doping is usually realized at Cree using nitrogen gas as the precursor. A SiC IMPATT device can be fabricated on the epiwafer following the process steps described below and published elsewhere [M. Mukherjee, N. Mazumder and S. K. Roy 2008].

(i) Growth of p^{++} 4H-SiC layer: In order to assist p-type ohmic contact formation, the p^{++} 4H-SiC layer can be grown on top of the n-type film by Al^{2+} ion implantation; the doping concentration should be $N_A \geq 2 \times 10^{19}$ cm^{-3}. The post-implantation annealing may be performed at ≈ 1600 ° C for 45 minutes in argon atmosphere.
(ii) ***Formation of low resistive p^{++} and n^{++} contacts:*** The power dissipation of IMPATT devices strongly depend on the contact resistance. Following step (i), the samples may be cleaned by a "piranha" solution. After rinsing in DI water, the samples may be dipped in dilute hydrofluoric (HF) acid solution for 30 seconds and dried. Immediately after the cleaning, a SiO_2 layer on the p^{++} side can be grown by Plasma Enhanced Chemical Vapor Deposition (PECVD) at 285⁰C. Through photolithographic process, windows can be opened inside the oxide layer. Using photolithography and lift-off techniques, contact metals (Al/Ti/Al) can then be deposited in the oxide windows by an electron beam evaporator. In order to obtain ohmic contacts the samples may be annealed for 3 minutes in a Rapid Thermal Anneal (RTA) furnace in nitrogen atmosphere at 950 ⁰C. The post-deposition annealing at high temperature is generally preferred to reduce the specific contact resistance. As mentioned in the literature, Ni is the preferred material for ohmic contact on n-type SiC (Roccaforte et al., 2006). Hence, for n++-type contact, a Ni layer of 200 nm thickness may be evaporated on the back-side (n++-side) of the wafer, followed by RTA treatment for 3 minutes at 950 0C. The choice of the metallic composition is based on the formation of Ni2Si alloy. As mentioned in several publications (Konishi et al, 2003), the higher the concentration of the Ni_2Si in the contacts, the lower is the specific contact resistance. Finally, a composition of Ti/Au contact overlay is deposited on the p^{++}-side by an electron beam evaporator. The specific contact resistance can be determined from transmission line measurement (TLM) data.
(iii) The oxide layer from the p^{++}-side can then be removed by Buffered Oxide Etch (BOE).
(iv)Formation of diode mesa using Reactive Ion Etching (RIE): Si-C bonds show high chemical inertness, hence, wet etching is not efficient for reaching deep trenches. The more appropriate dry plasma etching, Reactive Ion Etching (RIE), may be used in separating the diode mesa. Patterning of mesas may be carried out with photoresists: AZ 5214 (standard) and TI 35 ES (special photoresist for deep RIE Si etching) (Lazar et al., 2006). A titanium/nickel (Ti/Ni) bilayer metal can be evaporated onto the sample. The metallic bilayer may be used as RIE mask that covers areas that will form mesa diodes. The dry etching can be performed in the Plasma Therm 790 reactor (say), an RIE reactor with a plasma (composed of SF_6 and O_2 gases) source generated at 13.56 MHz operating at a maximum power of 300W. The depth and etch rate of the mask may be determined by profilometry measurements with a Tencor Alpha Step.
(v) On-Wafer DC Testing: After finishing the fabrication process, on-wafer DC testing are performed before the diodes are packaged. DC testing may serve as the initial screening step of the devices and the test results can be used for process evaluation.
(vi)SiC IMPATT Device Packaging: The packaging should provide a low thermal resistance between the SiC diode chip and wave guide mount and should be mechanically rugged and hermetically sealed. The device can be bonded to a pill-type package. In pill-type configuration, the diode is bonded to a heat sink, which is usually gold plated. A ceramic or quartz ring encloses the diode and separates the heat sink from the package cap.

4.1 Terahertz transmission and measurement

It is highly desirable that a THz transmission system should have low signal attenuation, low radiation loss, and high mode-confining property. Sommerfeld wires exhibit very low attenuation and dispersion, but suffers from large field extension into the surrounding medium. Moreover, it suffers from significant radiation loss at bendings. Metallic parallel plate waveguides, on the other hand, exhibit undistorted wave propagation of broadband sub-picosecond pulses even at bendings, but the propagating wave is confined in one transverse dimension giving rise to divergence loss. Transmission of THz signals from an IMPATT device may be possible with a dielectric coplanar wave guide, the device being an integral part of the waveguide. For THz (0.1-1 THz) signal transmission, a metallic slit waveguide, fabricated by sawing a 270 μm wide slit through a 140 mm wide and 300 μm thick silicon slabs may be used (Wachter et al., 2007). The metallic slit waveguide shows dispersion less transmission of THz signals with very low attenuation. However, the residual edge roughness from wafer sawing can cause a small degree of radiation loss. For higher frequency (3 THz) power transfer, a ribbon-like structure fabricated from ceramic alumina may be utilized (Yeh et al., 2005).
Measurements of THz power and frequency may be done with a THz VNA (Vector Network Analyzer) or, by employing Photoconductive Method. In the Terahertz regime, vector network measurements are challenging because of the reduced wavelength. Also, large phase errors resulting from temperature fluctuation can occur. In addition, flexing of the cable linking the scanning waveguide probe to the measurement equipment adds to the phase uncertainty. There are other systems for measurement and detection of THz signals for example Hot Electron Bolometer mixer- receiver and THz Time Domain Spectroscopy (THz-TDS). For detection and measurement of THz beam by photoconductive method, a THz-TDS set-up may be used (Wachter et al, 2007). The photoconductive setup can avoid the input and output coupling of free space Terahertz beams, giving rise to a compact and versatile setup. THz Time Domain Spectroscopy measurement require multiple time delay scans, which is time consuming and can result in a systemic error caused by the intensity fluctuations of the THz beam. THz-TDS has smaller spectral range than Fourier Transform Spectroscopy (FTS) system and provide lower resolution than narrowband THz spectroscopy (Ferguson and Zhang, 2002).

5. Conclusion

It may be concluded that, the results reported here, reveal the potential of hexagonal and cubic -SiC IMPATTs as high-power THz source and the design may be used for experimental realization of SiC THz IMPATT. Based on the design, the authors have also successfully grown β-SiC p-n junction on Ge modified Si <100> substrate. Efforts have been made to reduce the lattice mismatch at Si/SiC interface by incorporating Ge. The preliminary results are encouraging and need further improvement to obtain 3C-SiC based IMPATTs with minimum reverse leakage current, additional results will be the subject of another paper.

6. Acknowledgment

The author wishes to acknowledge (late) Prof. S. K. Roy, *founder Director*, CMSDS, University of Calcutta and (late) Prof. N. Mazumder, Professor, West Bengal University of Technology, for their support and keen interest in the work. The author is also grateful to Prof. Sanjay K. Banerjee, Director MRC, UT Austin, USA for providing experimental support. Thanks are also due to Seyoung Kim and Shagandeep Kour for their assistance in 3C-SiC layer growth and characterization. The assistance of Domingo Ferrer for TEM measurements is also gratefully acknowledged. The author also gratefully acknowledges DRDO (*Defence Research and Development Organisation)*, Ministry of Defence, Govt. of India, and University Grants Commission (Govt. of India) for their financial support to carry out this work.

7. References

[1] Bhatnagar M ; and Baligla B, (1993) "Comparison of 6H-SiC, 3C-SiC, and Si for power devices," *IEEE Transactions on Electron Devices*, vol. 40, no. 3, pp. 645–55.

[2] Weitzel, C.E ; Palmour, J. W ; Carter, J. C. H ; Moore, K. ; Nordquist, S ; Allen, C. Thero, and Bhatnagar, M ; (1996), "Silicon carbide high-power devices,"*IEEE Transactions on Electron Devices*, vol. 43, no. 10, pp. 1732–41.

[3] Cooper, J.A ; (1997) "Advances in SiC MOS technology," *Physica Status Solidi A*, vol. 162, no. 1, pp. 305–320.

[4] Elasser A ; and Chow, T.P, (2002) "Silicon carbide benefits and advantages for power electronics circuits and systems," *Proceedings of the IEEE*, vol. 90, no. 6, pp. 969–86.

[5] Ruff, M ; Mitlehner, H ; and Helbig, R ; (1994) "SiC devices: physics and numerical simulation," *IEEE Transactions on Electron Devices*, vol. 41, no. 6, pp. 1040–54.

[6] www. Cree. Com

[7] Keyes, R.W ; (1973) "Silicon Carbide from the prospective of physical limits on semiconductor devices", *Proc. Silicon Carbide Conference*: 1973, pp. 534 -542.

[8] Konstantinov, A.O ; Waheb, Q ; Nordell N ; and Lindefelt, U ; (1998) "Ionization rates and critical fields in 4H-SiC junction devices", *Mat. Sci. Forum*, vol. 264 -268, pp. 513-515.

[9] Brandt, C.D ; Clarke, R.C ; Siergieg, R.C ; Casady, J.B ; Morse, A.W ; Sriram S ; and Agarwal, A. K ; "SiC for applications in high-power electronics", *in Semiconductors and Semimetals*, Editor - Y. S. Park, New York, Academic, vol. 52, pp. 195 - 236, 1998.

[10] Yuan, L ; Copper, J.A ; Melloch M.R ; and Webb, K. J. (2001) "Experimental demonstration of a Silicon Carbide IMPATT oscillator", *IEEE Electron Device Letters*, vol. 22, pp. 266 -268.

[11] Vassilevski, k ; Zorenko A ; et al., (2001) "4H-SiC IMPATT diode fabrication and testing", *Tech. Digest of Int. Conf. on SiC and Related Materials- ICSCRM, Tsukuba,* Japan, p. 713, 2001.

[12] Ono, S ; Arai M ; and Kimura, C ; (2005) "Demonstration of high power X-band oscillation in p+ n - n+ 4H- SiC IMPATT diodes with guard-ring termination", *Materials Science Forum*, Vol. 483-485, pp.981-984.

[13] Mukherjee M ; Mazumder N and Roy S. K ; (2008) "Photo-sensitivity analysis of GaN and SiC Terahertz IMPATT oscillators: comparason of theoretical reliability and study on experimental feasibility", *IEEE Trans. Device and Materials Reliability*, vol. 8, p. 608.

[14] Gummel H.K ; and Blue, J. L. ; (1967) "A small signal theory of avalanche noise on IMPATT diodes", *IEEE Trans. Electron Devices*, vol. ED -14, p. 569, 1967.

[15] F. Roccaforte, F. La Via, and V. Raineri, "SiC Materials and Devices; vol. 1, eds: M. Shur, S. Rumyantsev, and M. Levinshtein, World Scientific Publishing, p. 40, 2006.

[16] R. Konishi, R. Yasokuchi, O. Nakatsuka, Y. Koide, M. Moriyama, M Murakami "Development of Ni/Al and Ni/Ti/Al ohmic contact materials for p-type 4H-SiC" Mater. Sci. & Eng. B, vol. 98, pp. 286-293, 2003

[17] M. Lazar, H. Vang, P.Brosselard, C. Raynaud, P. Cremillieu, J.-L Leclercq, A.Descamps, S. Scharnholz, D. Planson,. "Deep SiC etching with RIE", Supperlattices and Microstructures, vol. 40, pp. 388-392, 2006

[18] M. Wachter, M. Nagel and H. Kurz, "Metallic slit waveguide for dispersion-free low-loss terahertz signal transmission", Applied Physics Letters, vol. 90, pp. 061111-061111-3, February 2007

[19] C. Yeh, F. Shimabukuro and P. H. Siegel, " Low-loss terahertz ribbon waveguides", Applied Optics, vol.44, No.28, pp. 5937-5946, October 2005

[20] M. Wachter, M. Nagel and H. Kurz, "Metallic slit waveguide for dispersion-free low-loss terahertz signal transmission", Applied Physics Letters, vol. 90, pp. 061111-061111-3, February 2007

[21] B. Ferguson and X. Zhang, "Materials for THz science and technology", Nat. Matr, vol. 1, pp.26-33, 2002.

8

Contact Formation on Silicon Carbide by Use of Nickel and Tantalum from a Materials Science Point of View

Yu Cao and Lars Nyborg
Department of Materials and Manufacturing Technology,
Chalmers University of Technology
SE-41296 Gothenburg, Sweden

1. Introduction

The advantageous electrical, thermal and mechanical properties make silicon carbide (SiC) a promising semiconductor for high temperature, high power and high frequency applications. Nickel (Ni) and tantalum (Ta) can be used to form both ohmic and Schottky contact. Since metallization represents one of the most important steps in the fabrication of electronic devices, the knowledge of the interaction between Ni, Ta and SiC are of primary importance for understanding and optimising the device performance.

In this chapter, an introduction of thermodynamics in Ni (or Ta)-Si-C system will be given. The reaction process and mechanisms of Ni-SiC during annealing will be reviewed. The phases existing in the film or at the interface and the distribution of elements in-depth will be clarified. The impact of pre-treatment on SiC substrate and Ni layer thickness on phase distribution will be summarized. The nature of the thermally induced solid-state reactions between Ta or Ni/Ta bilayer and SiC substrate over a wide temperature range will also be discussed.

2. Ni-SiC

2.1 Ni silicides

Nickel silicides play a significant role in the modern electronic device technology (Colgan et al., 1996; Gambino & Colgan, 1998; M. d'Heurle & Zhang, 2007), the reasons being their chemical and thermal stability as well as low electric resistivity. Among all the silicides, NiSi has the lowest specific resistivity (10.5-18 μΩ cm) which is close to that of Ni metal (7-10 μΩ cm) (Lavoie et al., 2004). The next lowest ones are those of Ni_2Si (24-30 μΩ cm) and $NiSi_2$ (34-50 μΩ cm), respectively. The Ni rich silicides have the highest specific resistivity which is 80-90 μΩ cm for Ni_3Si and 90-150 μΩ cm for $Ni_{31}Si_{12}$, respectively. Nickel silicides are also attractive candidates as coating materials for protection of Ni-base superalloys and stainless steel (Lou et al., 2006; Lu & Wang, 2004) because of their excellent combination of wear, corrosion and oxidation resistance.

The Ni-Si equilibrium phase diagram (Nash & Nash, 1992) predicts six stable intermetallic compounds: Ni_3Si, $Ni_{31}Si_{12}$, Ni_2Si, Ni_3Si_2, NiSi and $NiSi_2$. Only three of compounds melt congruently namely $Ni_{31}Si_{12}$, Ni_2Si and NiSi. The others form via the peritectic reaction. The synthesis method of the nickel silicides in Ni-Si system includes conventional melting and casting and solid state reaction between Ni and Si, the latter of which has been realised in two different ways, thin films and bulk diffusion couples. Other techniques such as ion beam mixing (Hsu & Liang, 2005) and mechanical alloying (Lee et al., 2001) can also be used. In the case of thin film reactions, (Ottaviani, 1979; Zheng et al, 1983; Chen et al, 1985; Lee et al, 2000; Yoon et al, 2003), the formation of the compounds depends on the relative amounts of the Ni and Si available for the reactions, the annealing temperature, the atmosphere, and impurities contained in the layers. Important characteristics include sequential appearance of phases, i.e. one compound is formed first and the second starts to form later on, and the absence of certain phases. Ni_2Si is always the first phase to form and Ni_3Si_2 is always absent in thin film experiments. After one of the elements is consumed, the next compound is richer in the remaining element.

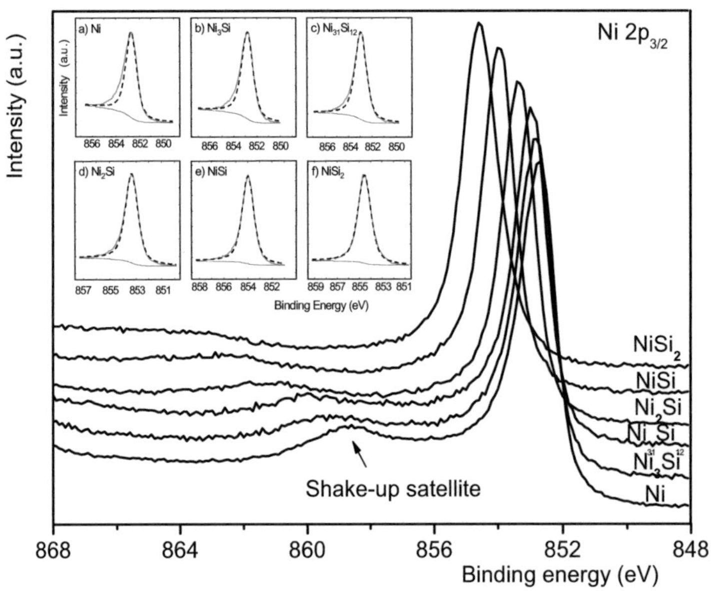

Fig. 1. Normalized XPS core level spectra from different silicides. Surface silicides were prepared by means of thin film solid-state reactions controlling the heating procedure in vacuum and the right sample preparation. (Cao et al., 2009)

XPS (X-ray photoelectron spectroscopy) can be used as a fingerprint for correct phase identification at the surface. The XPS core level spectra of Ni $2p_{3/2}$ in different silicides are shown in Fig. 1. In comparison to the Ni $2p_{3/2}$ peak (852.7 eV) representing the metal, the core level shift ΔEc are 0.1 eV for Ni_3Si, 0.3 eV for $Ni_{31}Si_{12}$, 0.7 eV for Ni_2Si, 1.2 eV for NiSi and 1.9 eV for $NiSi_2$, respectively. With higher amount of Si in the silicides, higher binding energy position and more symmetrical line shape (see insert in Fig. 1) are obtained. The shakeup satellite is shifted to higher binding energy upon increasing Si content as well.

Meanwhile, this structure is weaker and is actually smeared out over a larger binding energy region in the spectrum in the case of $NiSi_2$.

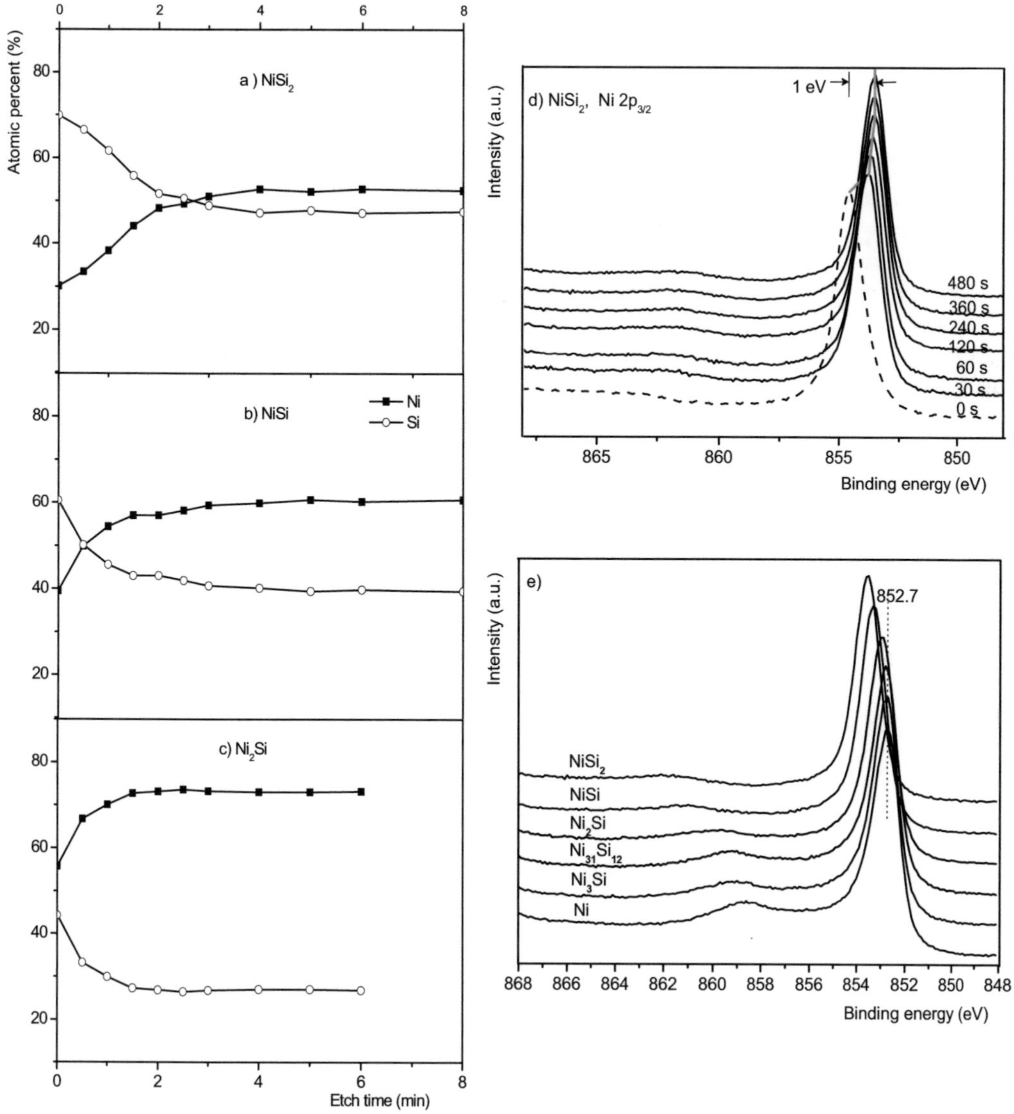

Fig. 2. Depth profiles of a) $NiSi_2$; b) NiSi and c) Ni_2Si derived from successive ion etchings and analysis of the Si 2p and Ni $2p_{3/2}$ levels in XPS. The content of C, O and Si from the surface contamination is not shown here. d) Evolution of XPS Ni $2p_{3/2}$ peaks in $NiSi_2$ during the process of argon ion etching. The spectra are normalized. e) Comparison of normalized Ni $2p_{3/2}$ peaks after 6 min argon ion etching in different silicides. The etch rate calibrated on Ta_2O_5 under these conditions is 4.7 nm / min. Ar ion beam energies of 4 keV are used.

Depth profiling by argon ion etching is a widespread method in studies of film structure and composition. Argon ion etching is a collisional process involving particle-solid interactions. It induces structural and chemical rearrangement for all the silicides at the surface. Figure 2 a)-c) shows the apparent atomic concentrations of Ni and Si in the silicides vs. etch time (Cao et al., 2009) derived from successive ion etchings and analysis of the Si 2p and Ni $2p_{3/2}$ levels in XPS. During the initial time period of argon ion etching, the surface composition for all the Ni silicides changes with increasing etching time; preferential sputtering of Si occurs, resulting in enrichment of the heavier element Ni. The effect of preferential sputtering decreases with increasing ion beam energy (Cao & Nyborg, 2006). After the prolonged ion etching, the Ni level becomes constant and reaches saturation level (Cao et al., 2009). The smallest preferential sputtering of Si occurs for Ni_3Si, whereas it is most evident for $NiSi_2$. Clearly, the preferential sputtering effect increases with increasing Si content. Moreover, during the process of argon ion etching, all the Ni $2p_{3/2}$ XPS peaks from silicides are moved to a lower binding energy positions until the steady state is reached. For $NiSi_2$, the Ni $2p_{3/2}$ peak is moved downwards in binding energy as much as 1 eV compared to that of the peak without argon ion etching, as shown in Fig. 2d). The corresponding values for NiSi, Ni_2Si and $Ni_{31}Si_{12}$ are 0.6, 0.4 and 0.2 eV, respectively. The steady state position of Ni $2p_{3/2}$ peak for ion etched Ni_3Si is also shifted downwards slightly. Therefore, not only the surface composition is changed with the ion etching, but also the surface chemical states are apparently modified. The comparison of peaks recorded after 6 min argon ion etching of the different silicides is illustrated in Fig. 2e). Clearly, the modified Ni $2p_{3/2}$ line position for ion etched $NiSi_2$, NiSi and Ni_2Si in the steady state can still be used as a fingerprint for correct phase identification. However, the Ni $2p_{3/2}$ peak shifts with respect to that of metallic Ni are different in these two cases, i.e. with and without argon ion etching.

2.2 Thermodynamics of Ni-Si-C system

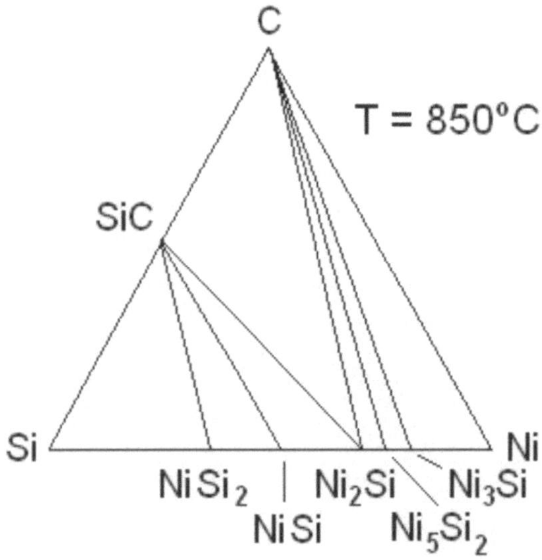

Fig. 3. Isothermal section of the Ni-Si-C at 850°C (La Via et al.; 2002).

Figure 3 shows the equilibrium isothermal section of the ternary Ni-Si-C phase diagram at 850ºC, which is characterised by the absence of both Ni-C compounds and ternary phase. Furthermore, only Ni_2Si can be in equilibrium with both C and SiC.

The elements Si and Ni have a strong affinity to one another. The thermodynamic driving force for the Ni/SiC reactions originates from the negative Gibb's free energy of nickel silicide formation. However, the strong Si-C bond provides an activation barrier for silicide formation. It is necessary to break the Si-C bonds before the reaction. Moreover, the interfacial energy of C/Ni-silicide is also positive and need to be overcome. Silicide formation can therefore only be expected at higher temperatures when enough thermal energy is available, and the activation barrier can be overcome completely.

The expressions for the Gibb's energies ΔG (Lim et al., 1997) for the various reactions within the Ni-Si-C system are illustrated in Table 1. Considering the reaction between SiC and Ni from room temperature to ~ 1600K, the formation of Ni_2Si shows the most negative ΔG value, and can thus occur by solid state reaction relatively more easily. Free C is liberated at the same time.

Possible reactions	Gibb's energy as a function of temperature T (kJ/mol Ni)
$Ni+\frac{1}{3}SiC \rightarrow \frac{1}{3}Ni_3C+\frac{1}{3}Si$	$30.793 + 0.0018 \cdot T \cdot \log T - 0.0103 \cdot T$
$Ni+2SiC \rightarrow NiSi_2+2C$	$22.990 + 0.0108 \cdot T \cdot \log T - 0.0454 \cdot T$
$Ni+SiC \rightarrow NiSi+C$	$-30.932 + 0.0054 \cdot T \cdot \log T - 0.0195 \cdot T$
$Ni+\frac{2}{3}SiC \rightarrow \frac{1}{3}Ni_3Si+\frac{2}{3}C$	$-38.317 + 0.0036 \cdot T \cdot \log T - 0.0158 \cdot T$
$Ni+\frac{1}{2}SiC \rightarrow \frac{1}{2}Ni_2Si+\frac{1}{2}C$	$-41.8 + 0.0027 \cdot T \cdot \log T - 0.0119 \cdot T$

Table 1. Possible reactions and their Gibb's free energies (ΔG_T) for the reaction between SiC and Ni. (Lim et al., 1997)

2.3. Bulk Ni-SiC diffusion couple

The interface reactions between bulk SiC and bulk Ni metal diffusion couples have been studied by several authors (see e.g. refs. Backhaus-Ricoult, 1992; Bhanumurthy & Schmid-Fetzer, 2001; Park, 1999). In the reaction zone, it has been observed that the diffusion couple shows alternating layers of C and Ni-silicides (900°C, 24 h or 40 h) (Bhanumurthy & Schmid-Fetzer, 2001; Park et al., 1999), or alternating silicide bands and silicide bands with embedded C (950°C, 1.5 h) (Backhaus-Ricoult, 1992). From the back-scattered electron imaging (BSE) (Park et al., 1999) of a Ni/SiC reaction couple annealed at 900°C for 40 h, the sequence of phases in bulk diffusion couples was observed to be $Ni/Ni_3Si/Ni_5Si_2+C/Ni_2Si+C/SiC$. The approximate width of the bands was about 5-10 µm. A schematic BSE image of SiC/Ni reaction couple annealed at 900°C is shown in Fig. 4. $NiSi_2$ is not observed because of the positive Gibb's free energies for its formation at the temperature studied, see Table 1. The absence of NiSi phase, however, is probably due to the insufficient annealing (kinetic reason) used by the author since the thermodynamic conditions are met. NiSi has been observed in the thin film Ni-SiC system.

The formation of Ni_2Si follows the parabolic rate law $d = kt^{1/2}$ (d: thickness of silicide, k: parabolic rate constant, t: time) with $k = 6.27 \times 10^{-8}$ cm^2/s at 950°C (Backhaus-Ricoult, 1992). This means that the global reaction is diffusion controlled. Nickel is the mobile species in Ni_2Si and its diffusion via its own sub-lattice by the vacancy mechanism is supposed to control the Ni_2Si growth (Ciccariello et al., 1990). The activation energies for Ni lattice and grain boundary diffusion have been found to be 2.48 eV and 1.71 eV, respectively. The diffusion of Ni along grain boundary is thus more important in the formation of Ni_2Si. The formation of NiSi is also diffusion controlled, while that of $NiSi_2$ is nucleation controlled (Lee et al., 2000).

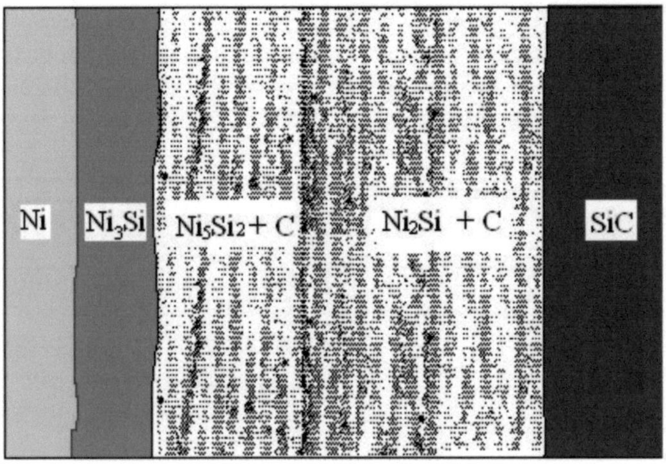

Fig. 4. Schematic BSE image of SiC/Ni reaction couple annealed at 900°C for 40 h (Park et al., 1999)

The formation mechanism of periodic bands is not very clear, but it is generally accepted that it depends on the diffusivities of the reacting elements. Metal is the most dominant diffusing species and C atoms are practically immobile (Bhanumurthy & Schmid-Fetzer, 2001; Park et al., 1999). After the formation of silicide, the Ni concentration at the SiC reaction interface decreases [Chou et al., 1990]. In order to further decompose SiC, the critical concentration level of Ni has to be satisfied. At the same time, the C, in front of the SiC reaction interface, forms small clusters and aggregates as a layer to minimize the interfacial energy. The continuation of this process will give rise to the formation of alternating Ni-silicide and C layers. The systems which show the tendency of the formation of periodic bands have relatively large parabolic rate constant k and k_0 values (intercept of the linear ln k versus 1/T plot) (Bhanumurthy & Schmid-Fetzer, 2001).

2.4. Ni film on SiC

2.4.1. Reaction products

A number of studies of the interfacial reactions between a Ni film and SiC have been reported (see e.g. Ohi et al., 2002; Gasser et al., 1997; Roccaforte et al., 2001; Madsen et al.,

1998; Litvinov et al., 2002; Marinova et al., 1996 & Cao et al., 2006). The dominant phase formed is almost independent of the polytype, the polarity of the SiC and the details of the annealing cycle.

In the Ni/SiC system, Ni reacts with SiC to form Ni silicides and C. Dissociation of SiC occurs at around 500°C (Kurimoto & Harima, 2002). Generally, Ni_2Si is the dominant species in a large temperature range between 600 and 950°C (Ohi et al., 2002; Gasser et al., 1997; La Via et al. 2002; Abe et al., 2002; Roccaforte et al., 2001; Cao et al., 2006 & Kestle et al., 2000), as shown in the X-ray diffraction (XRD) spectra in Fig. 5. Similar as thin film Ni-Si system, silicides is formed sequentially, i.e. one compound is formed first and the second starts to form later on during the annealing. The phase sequence is $Ni_{23}Si_2$+$Ni_{31}Si_{12}$ → $Ni_{31}Si_{12}$ → $Ni_{31}Si_{12}$+Ni_2Si → Ni_2Si (Madsen et al., 1998 & Bächli et al., 1998). This is the reason why $Ni_{31}Si_{12}$ has been found at the surface in some cases, see eg. Refs. (Han & Lee, 2002; Han et al., 2002). Silicon rich silicides can be observed at the interface of Ni_2Si and SiC (Cao et al., 2005). Increasing temperature to above 1000°C results in the formation of a NiSi thin film (Litvinov et al., 2002; Kestle et al., 2000 & Marinova et al., 1996).

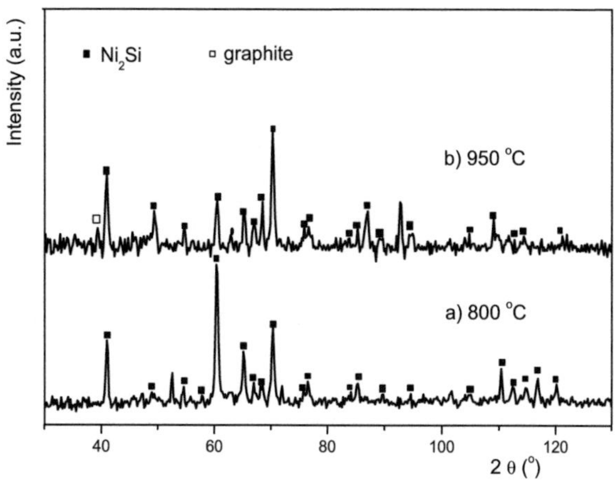

Fig. 5. XRD spectra of samples with ~ 100 nm Ni thickness on 4H-SiC after annealing. Glancing angle 3° with Cr k_α radiation (λ = 2.29Å)

2.4.2. Formation of Ni_2Si and its mechanisms

In the Ni/SiC system, the formation of Ni_2Si through the reaction 2Ni+SiC = Ni_2Si+C may consist of two stages (Cao et al., 2006) which are controlled by reaction and diffusion rate respectively.

The thermodynamic driving force for the Ni/SiC reaction originates from the negative Gibb's energy of Ni-silicide formation (Table 1). Before the formation of Ni_2Si by solid state reaction, however, it is necessary to break SiC bonds. The existence of Ni may help the dissociation of SiC at the temperatures lower than its dissociation value. It is known that the thermal expansion coefficient of SiC is 3-4 times higher than that of Ni (Adachi, 2004). This expansion difference results in thermal strain at higher temperatures for SiC sample coated with Ni, which corresponds to compression at the Ni side and tensile at the SiC side. It is thus possible that some Ni atoms slightly penetrate into the SiC side at the interface with the help of the thermal energy. The theoretical calculation on the chemical bonding in cubic SiC (Yuryeva & Ivanovskii, 2002) has shown that Ni impurities weaken the covalent character of the SiC crystal, resulting in a decrease in the stability of the SiC adjacent to the Ni layer. The decomposition of SiC, which starts at the interface, is therefore possible at a temperature lower than its dissociation value. However, the stability of SiC must be lowered to certain degree before the decomposition of SiC. In other words, an incubation period exists. Following the decomposition of the SiC, Si and C released will diffuse into the Ni due to the expected low diffusion coefficient of the Ni in SiC. This has been proved by the expansion of metal Ni lattice prior to the appearance of Ni silicides in ultra thin Ni/SiC system (Su et al., 2002; Iwaya et al., 2006). The opposite Ni flux into the SiC may not be dominant in this stage. The mixture of Si and Ni occurs very rapidly, provided Si atoms are available. In fact, an amorphous interlayer (~ 3.5 nm) which is a mixture of Ni and Si has been observed in the Ni/Si system even at room temperature by solid-state diffusion (Sarkar, 2000). Therefore, the formation of new phase Ni_2Si in the first stage is determined by the speed of bond breakage, i.e., by the supply of Si from the decomposition of SiC. This is a reaction-rate controlled process.

With the progress of the reaction, heat is released by the formation of Ni_2Si. More SiC is then decomposed and more Si atoms become available. The supply of Si atoms is then no longer the dominant factor in the formation of Ni_2Si, because Ni is the dominant diffusing species through Ni_2Si (Ciccariello et al., 1990). The growth of thin Ni_2Si films is controlled mainly by the diffusion of Ni along the silicide grain boundaries. Nickel is then provided at the Ni_2Si/SiC interface where the silicide formation takes place. This interface advances by the arrival of new Ni atoms. The formation obeys the parabolic rate law. In this case, the Ni flux increases relative to fluxes of Si and C from SiC and the mechanism of reaction changes to a diffusion controlled one, corresponding to the second stage of the reaction.

In addition, the Ni_2Si formed by annealing possesses textured structure to some degree, which was confirmed by XRD [Cao et al., 2006].

2.4.3. Formation of C and its chemical states

After the reaction between Ni and SiC, C present in the consumed SiC layer should precipitate. A number of studies of the chemical state of C after annealing have been reported (Gasser et al., 1997; La Via et al, 2003; Han & Lee, 2002; Marinova et al, 1996; Marinova et al, 1997). Figure 6a) shows the C1s XPS region spectra at the surface after heat treatment at 800°C and 950°C in vacuum. It is seen that C is mainly in the chemical state analogous to that of graphite in the surface region for both temperatures (Cao et al. 2006). To investigate further the chemical states of the C species inside the contact, C1s XPS peaks have been recorded after successive Ar ion etchings, as shown in Fig. 6b. It is revealed that the C1s binding energy value recorded from the sample heated at 950°C was slightly higher than that from lower temperature, implying the possible difference of the chemical state. Considering binding energy of C1s XPS peak decreases with decreasing structure order in C species (Rodriguez et al, 2001), a less ordered structure below the surface could be possible in the case of 800°C heat treatment. Further evidence can be obtained by means of Raman spectroscopy, as shown in Fig. 7. Compared with graphite standard, the broadened and shifted G and 2D peaks as well as the appearance of an additional D peak indicate the formation of nanocrystalline graphite cluster in annealed Ni-SiC samples (Cao et al, 2006). This is consistent with the result of Kurimoto and Harima (Kurimoto & Harima, 2002). Close examination of line position and shape of G and 2D Raman peaks together with the intensity ratio I_D/I_G obtained at different temperatures indicate that more highly graphitised and less disordered carbon is promoted by a higher annealing temperature at 950°C. Similar results have been reported in ref. (Ohi et al, 2002; Kurimoto & Harikawa, 2002). For temperatures of 600 and 800°C, Ohi et al. found the formation of C with modified π bonds when compared to graphite. The π sub-band has different density of states from that of graphite.

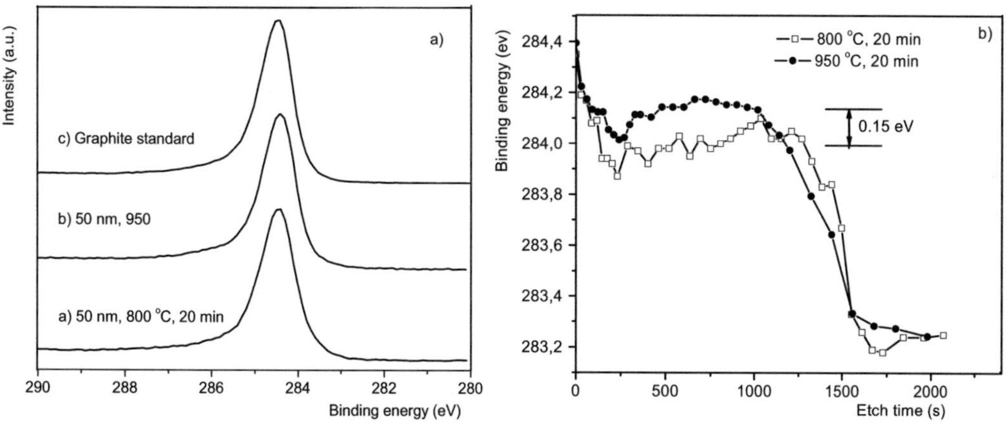

Fig. 6. a). C1s XPS spectra at the surface; b) C1s XPS peak position recorded by successive Ar ion etchings. Ni/4H-SiC samples annealed in vacuum. t_{Ni} = 50 nm. The etch rate calibrated on Ta_2O_5 under the experimental condition is 5.6 nm / min.

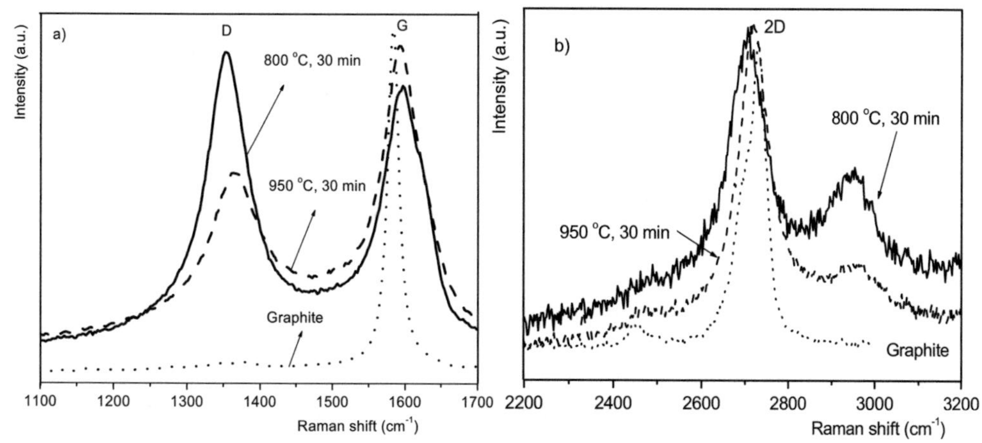

Fig. 7. Raman first-order a) and second-order b) spectra of graphite and vacuum annealed Ni-4H SiC samples. t_{Ni}= 200 nm.

In the process of formation of C, Ni acts as an effective catalyst for graphitisation (Lu et al, 2003). In fact, once silicide has formed, not only can Ni act as mediating agent but also the reaction product, the silicides (Hähne & Woltersdorf, 2004), can do so. The driving force for the graphitisation process is the decrease of free energy by the conversion of amorphous C to graphite. The graphitisation process is a gradual disorder-order transformation. It includes the rearrangement of disordered C atoms, released from the formation of silicide, to hexagonal planar structures and the formation of ordered stacking structures along c axis. The structure of C is less complete at lower temperature.

2.4.4. Distribution of phases in the reaction products and the effect of pre-treatment and Ni layer thickness

Carbon is released from the SiC during the silicide formation. The redistribution of C after annealing is one of the most controversial aspects in studying the Ni/SiC reactions. The main opinions are: a) Carbon atoms are distributed through the contact layer and accumulated at the top surface (Kurimoto & Harima, 2002; Han & Lee, 2002; Bächli et al., 1998; Han et al., 2002). b) Carbon in graphite state is present in the whole contact layer with a maximum concentration at the contact/SiC interface (Marinova et al., 1997). c). Carbon agglomerates into a thin layer far from the silicide/SiC interface after annealing (La Via et al., 2003). d). Carbon is almost uniformly distributed inside the silicide layer (Roccaforte et al., 2001).

To authors' opinion, the C distribution is dependent on several factors, such as annealing environment, pre-treatment on SiC substrate and Ni layer thickness. The in-situ depth profiles by XPS study for vacuum annealed Ni/SiC sample without exposure to the air reveal that there is a C layer at the external surface in all cases, as shown in Fig. 8 and 9 (Cao et al., 2005; Cao et al, 2006, Cao & Nyborg, 2006). The carbon diffuses mainly through the non-reacted Ni film towards the external surface at the beginning of reaction. The external surface acted as an effective sink for C accumulation. According to the Ellingham diagram, the equilibrium partial pressure of oxygen for reaction $2C + O_2 = 2CO$ at 800°C is ~ 10^{-20} atm (Shifler, 2003), which is much lower than the partial pressure of oxygen in the normal vacuum annealing furnace (~10^{-9}-10^{-10}atm). The driving force for the C moving to the free surface is thus provided. In the equilibrium state, the C at the free surface will disappear by reacting with oxygen to form CO. However, some C still exists and is thus in a metastable state. Besides the experimental error, one possible reason for the discrepancies in the literature regarding C distribution could be the annealing atmosphere having different reactivity with C. The use of unsuitable analysis methods, such as EDX, could also be a cause.

The surface pre-treatment of the SiC substrate has certain influence on the C distribution (Cao et al., 2005; Cao et al., 2006). In the case of SiC substrate without pre-treatment or with chemical cleaning, the in-situ depth profile obtained is illustrated in Fig. 8. For very thin Ni layers (less than ~ 10 nm), a C-depleted zone separates a thin C surface layer from the SiC substrate (Fig. 8a). For thicker Ni layers, a further accumulation of C is also observed below the surface region (Fig. 8b). The maximum C concentration is away from the silicide/SiC interface at a certain distance. The reason is as follows. After a continuous layer of silicide with certain thickness has formed, the rate of accumulation of C to the free surface decreases due to the expected low diffusivity of C in silicide. It is known that the diffusion coefficient of C in Ni at 800°C is 1.6×10^{-8} cm^2s^{-1} (Smithells, 1967). However, the diffusivity of C in N–doped n–type hexagonal SiC at 800°C extrapolated from the data at 1850-2180°C is as low as 1.1×10^{-31} cm^2s^{-1} (Matzke & Rondinella, 1999). Carbon is therefore much more mobile in metal Ni than in 4H–SiC. As the Ni_2Si/SiC interface advances, C phase is also buried within the silicide. To minimize the total interfacial energy between C and Ni-silicide, the C phase would tend to form clusters in the direction opposite to the external surface as well (Fig. 8b).

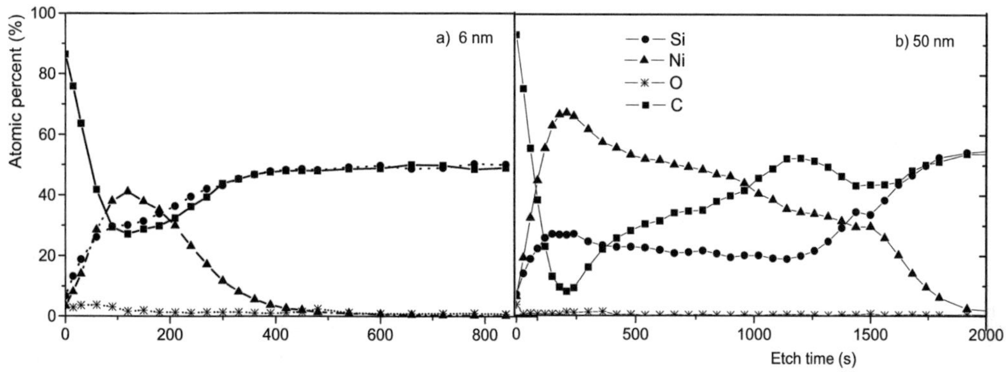

Fig. 8. In-situ depth profiles of samples with Ni layer thickness a) 6 nm and b) 50 nm (Cao et al., 2006). The samples were heated at 800°C for 20 min in vacuum. The SiC substrate is in the as-delivered state from manufacturer. The etch rate calibrated on Ta_2O_5 under the experimental condition is 5.6 nm / min.

However, for the sample experiencing Ar ion etching before the Ni deposition there is a different phase distribution in the reaction product (Fig. 9). The argon ion bombardment deposited a large amount of energy on the surface and created many excitations, including ionization of secondary ions and neutral particles and ejection of electrons. All these energetic particles could in principle transfer energy into SiC and facilitate its dissociation. The energetic particles mentioned above might also provide energy to enhance the diffusion of the Ni atoms into the bulk. It is known that nickel is the dominant diffusion species in nickel silicides and controls the rate of Ni_2Si formation in the second reaction stage. As a result of fast dissociation of SiC and enhanced diffusion of Ni, Ni_2Si is formed quicker under the action of argon ion pre-treatment. Consequently, there is less C agglomerated at the surface because C is much less mobile in Ni_2Si than in metal Ni.

For the thinnest Ni layer (d_{Ni} = 3 nm), heat treatment lead to the formation of surface graphitic carbon layer and silicide below with low carbon content (Fig. 9a). With the Ni thickness doubled to 6 nm (Fig.9b), there is a carbon rich layer below the surface region, which is clearly different from Fig. 8a. In Fig. 9c (d_{Ni} = 17 nm), a silicide layer with carbon deficiency develops adjacent to the interface. The maximum C content is ~ 4 nm away from the silicide/SiC interface. Increasing Ni thickness even more results in a repeated maximum of carbon intensity corresponding to the minimum of the nickel intensity, i.e., a multi-layer structure, consisting of silicide rich layer/ carbon rich layer / silicide rich layer / ···· (Fig. 9d). The silicide layer adjacent to the interface is deficient of C. The depth profiles indicate that there is a minimum Ni thickness (~ 15 nm) for the formation of such multi-layer structure. The development of such a structure can be explained by the quicker formation of Ni_2Si under such a condition. It is then difficult for free C released from the SiC to move long distance due to the low diffusivity and low solid solubility of C in silicide. In order to minimize the interfacial energy between C and Ni-silicide, as a compromise, the dissociated C atoms might form small clusters and aggregated as a layer.

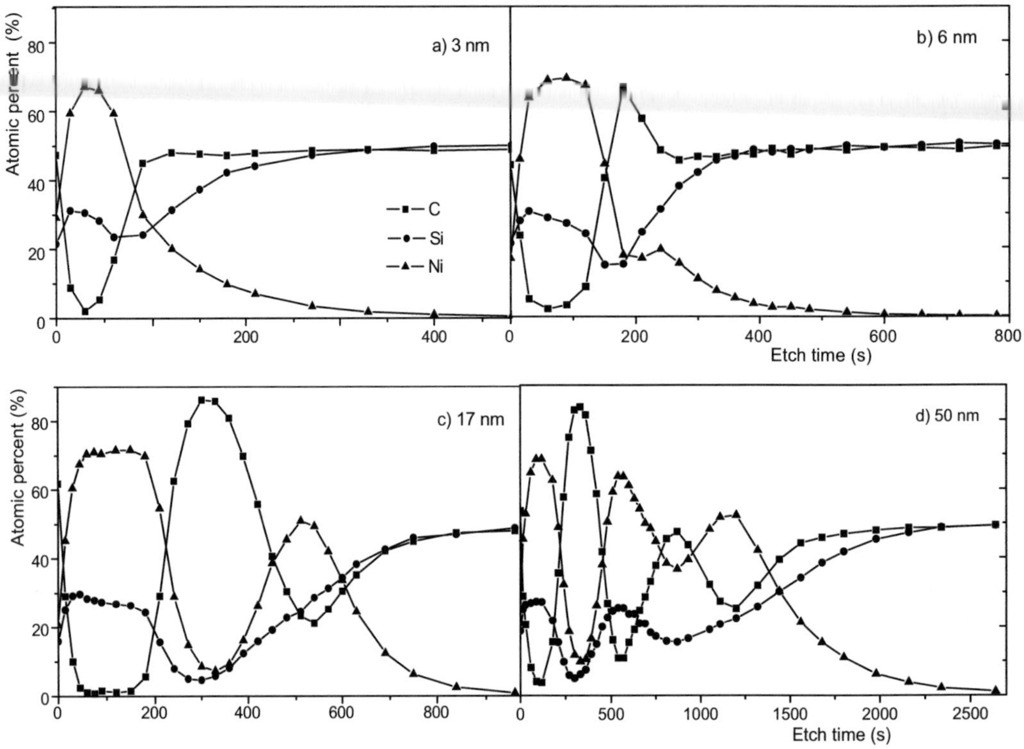

Fig. 9. In-situ depth profiles of samples with different Ni layer thickness (Cao et al., 2005). The SiC substrate was cleaned by Ar ion etching with 4 keV energy before Ni deposition. The samples were then heated at 800°C for 20 min in vacuum. The etching rate calibrated on Ta_2O_5 under the experimental condition is 5.6 nm / min.

It is also interesting to identify the silicide (Ni_2Si) morphology for thin Ni film samples. Figure 10 presents the Si2p peak from Ni/SiC samples with different Ni layer thickness (Cao et al., 2006). After annealing thin Ni layer sample (t_{Ni} = 3 nm) at 800°C for 20 min in vacuum, it is known from XPS curve fitting results that the Si2p peaks are composed of three chemical states (Fig. 10 a): the main part being Si in SiC, and the other two small parts being Si in SiO_2 and Ni_2Si, respectively. The existence of SiO_2 is due to the slight oxidation in the furnace. Considering that the deposited Ni film is continuous and uniform, the appearance of strong carbide signal (from Si in SiC) suggests that Ni_2Si tended to form islands during the annealing. With the Ni thickness doubled (Fig. 10b), the amount of Ni_2Si increases obviously and the detected amount of SiC decreases. The Ni silicide island can grow both laterally and vertically. Increasing Ni thickness even more (Fig. 10c) results in the disappearance of SiC signal and Ni_2Si is dominant. The above results indicate that the silicide becomes continuous with increasing Ni film thickness.

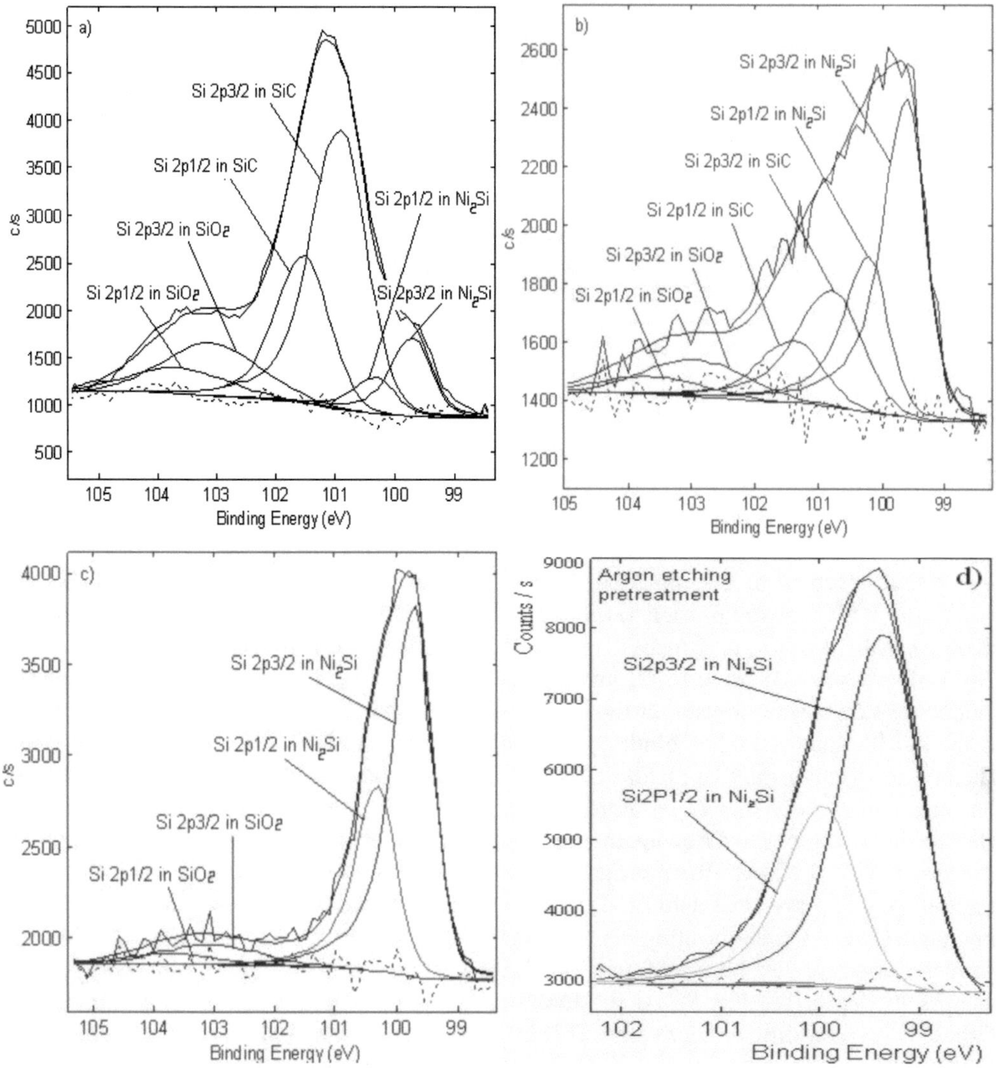

Fig. 10. In-situ Si2p XPS spectra of Ni/SiC samples after annealing at 800°C for 20 min in vacuum. a) t_{Ni} = 3 nm b) t_{Ni} = 6 nm c) t_{Ni} = 17 nm d) t_{Ni} = 6 nm. In Fig. a-c), the Ni thin films were deposited on as-delivered SiC substrate. In Fig. d), the SiC substrate was cleaned by Ar ion etching with 4 keV energy before Ni deposition.

Fig. 10b) and d) give the XPS Si 2p peak recorded from the samples with same initial Ni layer thickness (d_{Ni} = ~ 6 nm) but different pre-treatment on SiC substrate. From the comparison it has been found that the shoulder at higher binding energy representing Si in SiC disappears when the Ni thin film is deposited on an argon ion etched SiC substrate. This is again related to the fast dissociation of SiC and enhanced diffusion of Ni under the action of argon ion pre-treatment. The nucleation and growth of Ni_2Si are promoted. Therefore, the silicides formation kinetics is affected and a continuous silicide layer develops quicker.

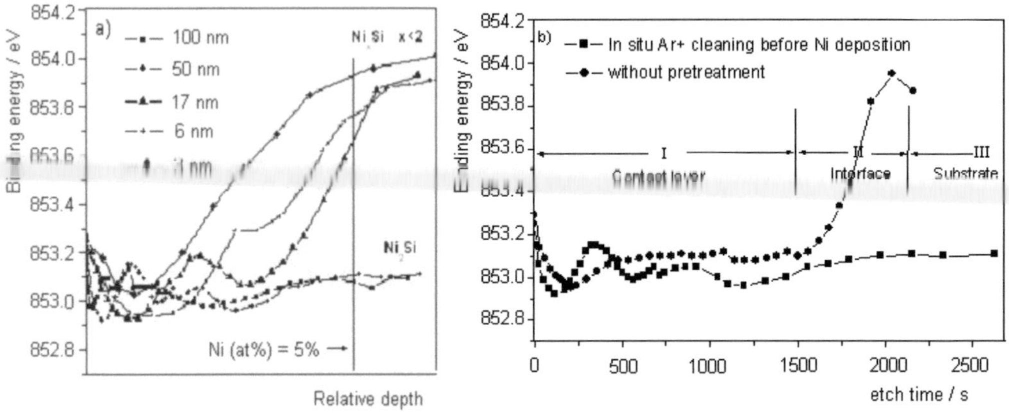

Fig. 11. Binding energy of Ni $2p_{3/2}$ peaks as function of (a) Ni layer thickness (the SiC substrate was cleaned by Ar ion etching with 4 kev energy before Ni deposition), and (b) pre-treatment (d_{Ni} = 50 nm)

The silicides formed at the interface depend also on the Ni layer thickness and the pre-treatment on SiC substrate prior to the Ni deposition. Figure 11 shows the development of Ni $2p_{3/2}$ peak position as function of initial Ni layer thickness and pre-treatment. From Fig.11 a), we see that for thin Ni layers (d_{Ni}= 3, 6, 17 nm), NiSi, $NiSi_2$ or even higher Si containing silicides are formed at the interface. It has been known that higher amount of Si in the silicides gives higher binding energy position (Fig. 2e). The reason why Si-richer silicides are formed may be attributed to the considerable consumption of nickel. Because the metal supply is likely to be more limited, one could expect the formation of Si-richer silicides following Ni_2Si. Anyway, the total amount of Si rich silicide is small because of the low availability of Ni near the interface. For fixed Ni film thickness (50 nm), the influence of argon ion etching pre-treatment on the type of interfacial silicide is shown in Figure 11b). In the contact layer (I), the binding energy fluctuation (as also observed in Fig. 11 a) results from the effect of ion bombardment (at the beginning) and the alternating composition changes in depth (see Fig. 9). At the interface (II), the sample without pre-treatment has higher Ni $2p_{3/2}$ binding energy value. This implies that a compositional gradient existed and that Si-richer silicides are formed at the interface. The reason may be also attributed to the limited availability of nickel. On the other hand, argon ion etching pre-treatment enhances the Ni diffusion and accelerate the supply of Ni and almost keeps the same kind of silicide all the time.

3. Ta (or Ni/Ta)-SiC

Tantalum (Ta) is a refractory metal with high melting point (around 3000°C) and it exhibits two crystalline phases, bcc α-phase and tetragonal β-phase. The α-phase has high toughness and ductility as well as low electrical resistivity and corrosion resistance, while the β-phase is hard and brittle and less desirable. Tantalum can form both stable carbides and silicides with attractive properties with respect to oxidation resistance and general physical behaviour. There exist two stable carbides in the Ta-C system, Ta_2C and TaC, with the melting points of 3330°C and 3985°C, respectively. Both these carbides are interstitial compounds and thermally very stable. For example, TaC has been used for reinforcing Ni superalloys (Berthod et al., 2004). The research on contacts involving Ta on SiC is not as extensive as that on Ni contacts. Attempts have been made to create ohmic contacts on SiC by using elemental Ta, and its silicide or carbide (Olowolafe et al, 2005;, Guziewicz, 2006, Jang et al., 1999; Cao et al, 2007[a,b]).

3.1 Thermodynamics of Ta-Si-C system

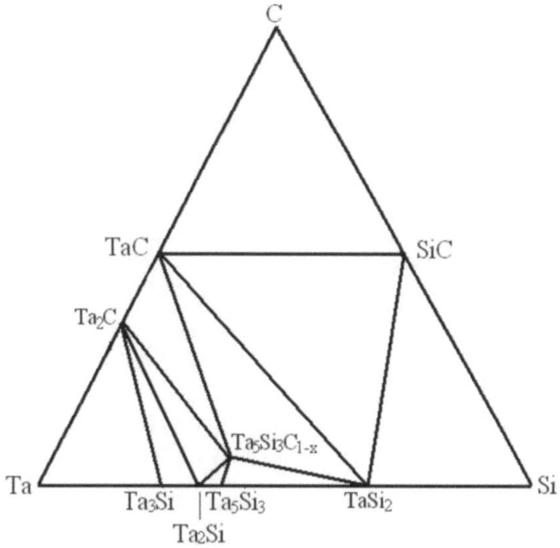

Fig. 12. Simplified isotherm ternary phase diagram of Ta-Si-C at 1000°C (Schuster, 1993-1994).

An isothermal section of Ta-Si-C at 1000°C is shown in Fig. 12 (Schuster, 1993-1994). It might apply at temperatures up to 1827°C (Brewer and Krikorian, 1956). It can be seen from the figure that SiC can be in equilibrium with both TaC and $TaSi_2$. The author proposed the existence of a ternary compound $Ta_5Si_3C_{1-x}$ ($x \approx 0.5$) which can coexist with TaC, Ta_2C, Ta_2Si, Ta_5Si_3 and $TaSi_2$. However, the status of this compound is in doubt (Laurila et al., 2002), since it is not clear if it is a real ternary compound or simply the metastable Ta_5Si_3 with carbon solubility.

Compound	ΔH (kcal/g atom)*	Reaction	ΔH_R (kcal/g atom)*
SiC	-26.7	4Ta+SiC = Ta_2C+Ta_2Si	- 4.9
Ta_2Si	-10.1	3Ta+SiC = TaC+Ta_2Si	-4.3
Ta_5Si3	-9.0	5Ta+2SiC = 2Ta_2C+$TaSi_2$	-5.2
$TaSi_2$	-8.0	3Ta+2SiC = 2TaC+$TaSi_2$	-4.4
Ta_2C	-46	11Ta+3SiC = 3Ta_2C+Ta_5Si_3	-3.9
TaC	-38		

*ΔH: Standard heats of formation

ΔH_R: Enthalpy change for the reaction of Ta and SiC at 800°C.

Table 2. Thermodynamic data in Ta-Si-Ta system (Geib et al., 1990)

The thermodynamic driving force for the Ta/SiC reactions also originates from the negative Gibb's free energy of Ta silicide or carbide formation. The standard heats of formation and calculated enthalpy changes, ΔH_R, for the various reactions within the Ta-Si-C system are illustrated in Table 2 (Geib et al., 1990). It can be seen that the standard heats of formation of Ta carbides Ta_2C and TaC are larger than that of SiC. Assuming the entropy contribution to be small, the change of Gibb's free energy can be approximated by ΔH_R. The negative values at this temperature imply that the reaction between Ta and SiC is energetically favorable.

3.2 Reaction between Ta film and SiC

Table 3 summarizes the evolution of the thermal reaction between Ta film and SiC substrate up to 1200°C (Chen et al., 1994, Cao et al., 2007[a]). Similar phase sequence has been reported (Feng et al., 1997) for SiC/Ta/SiC couples at 1500°C, while α-Ta_5Si_3 was not observed. Ta_5Si_3:C here is a carbon-stabilized phase with a hexagonal Mn_5Si_3 structure according to the XRD powder diffraction database. It is in fact the same compound as the ternary compound $Ta_5Si_3C_x$ reported by some researchers.

Temperature / time	Reaction products in Ta/SiC system	Reaction products in Ni/Ta/SiC system	Initial film thickness on SiC
650 °C /0.5 h	Ta	Amorphous Ni-Ta, Ta, Ta_2C	100 nm
800 °C /0.5 h	Ta + Ta_2C	Ta_2C, Ta_5Si_3:C, Ta(?)	100 nm
900 °C / 1 h	Ta + Ta_2C + Ta_5Si_3:C		320 nm
950°C/0.5 h	Ta + Ta_2C + Ta_5Si_3:C	Ta_2C, TaC, Ta_5Si_3:C, Ta_5Si_3, TaNiSi, Ni_2Si	100 nm
1000 °C/1 h	TaC + Ta_5Si_3:C + α-Ta_5Si_3 + Ta_2C		320 nm
1100 °C/0.5 h	TaC + Ta_5Si_3:C		320 nm
1200 °C/1 h	TaC + $TaSi_2$		320 nm

Table 3. Evolution of the thermal reaction between Ta (or Ni/Ta) and SiC (Chen et al., 1994, Cao et al., 2007[a,b]). In Ni/Ta/SiC system, the thickness ratio of Ni:Ta is ~3:5.

One important feature in the thin film Ta/SiC system is the development of layered structure at high temperatures. Figure 13 shows the Auger depth profiles obtained after annealing at different temperatures in vacuum. Compared to the as-deposited film (Fig. 13a), less sharp interface in Fig. 13b suggests that inter-diffusion has occurred at 650°C. Carbon, which corresponds to the formation of Ta_2C carbide, is observed in the contact layer after annealing at 800°C (see Fig. 13c). Clearly, more Ta_2C is formed at 950°C (Fig. 13d). No Si signal is detected in the near surface region, confirming the lack of silicide on the top surface. This can be further confirmed by the XRD analyses at 950°C (Fig.14), in which Ta_5Si_3:C is not able be observed by using grazing angle 1°. However, it can be detected by using grazing angle 3°. Ta_5Si_3:C must tend to form at a certain distance below the surface. A mixture layer of Ta_5Si_3:C and Ta_2C is present there. The reaction zone is shown to have a layered structure of Ta_2C/Ta_2C+Ta_5Si_3:C/SiC after heating at 950°C (Cao et al., 2007[a]). With the temperature increasing to 1000°C, it changes to a well defined four-layered structure Ta_2C/α-Ta_5Si_3/Ta_5Si_3:C/TaC/SiC (Chen et al., 1994). In fact, the layered structure is inclined to form when the vanadium group metals (V, Nb and Ta) react thermally with SiC at high temperatures. The layered structure can be attributed to the requirement of the minimization of the interfacial energy.

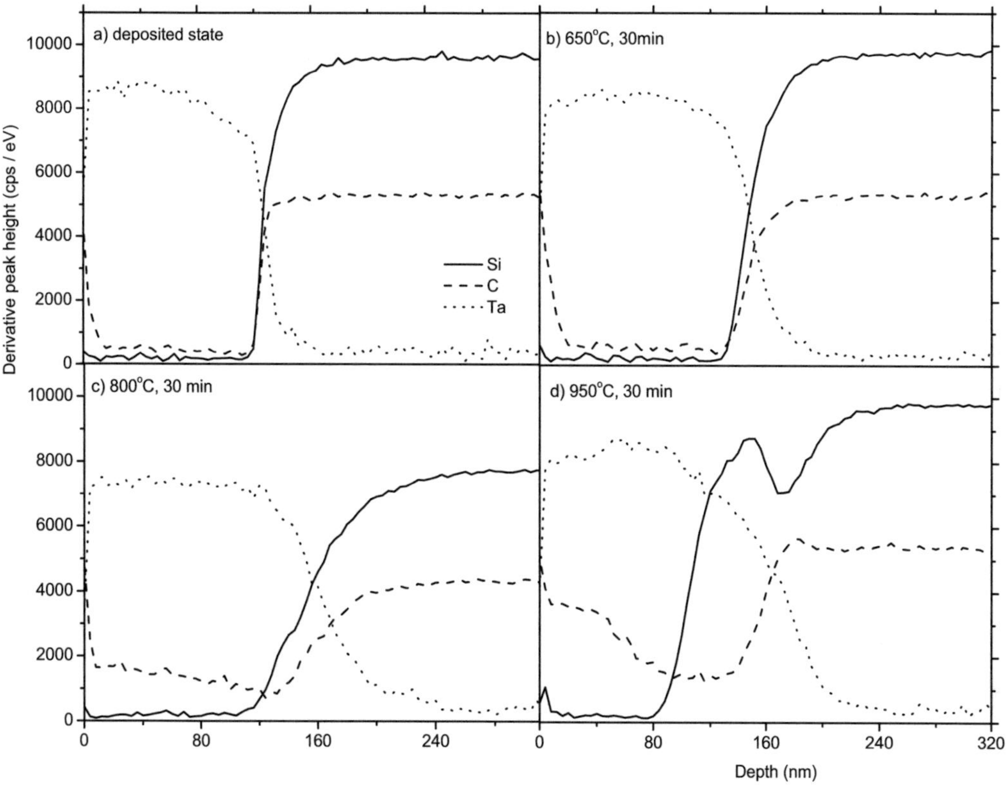

Fig. 13. Depth profiles of Ta/SiC before and after annealing in vacuum at different temperatures with Ta thickness of 100 nm. The depth scale is given by using the etch rate of Ta_2O_5 with known thickness under the same condition (8 nm/min).

Another important feature in the thin film Ta/SiC system is the formation of the C deficiency region in the near interface side of the SiC. The apparent atomic concentrations of C and Si in the thermal reaction product are given in Fig. 15 by considering the area percentages for the corresponding chemical states obtained from the curve fitting of XPS C1s and Si2p peak. The distribution of phases obtained at 800 and 950°C is further confirmed. Besides un-reacted metal Ta, as mentioned before, the reaction zone is shown to have a layered structure of $Ta_2C/Ta_2C+Ta_5Si_3{:}C/SiC$. Importantly, it can be seen that the total amount of C is higher than that of Si in the reaction layers formed at both 800 °C and 950°C. This means more C diffuses towards the Ta layer from the substrate. Carbon has a high diffusivity in Ta ($D = 6.7\times10^3 \exp(-161.6/RT)$ cm^2/s, for T = 463-2953K and R = 8.314 J mol^{-1} K^{-1}) (Le Claire, 1999). It is the dominant moving species and reacts more rapidly with Ta than Si. In other words, C deficiency region is preferentially produced in the near interface side of the SiC.

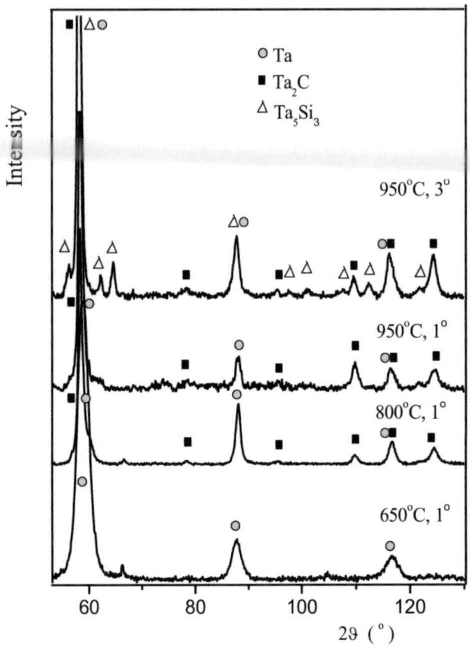

Fig. 14. Grazing angle XRD spectra with Cr k_α radiation of Ta/SiC after annealing in vacuum. The thickness ratio of Ni:Ta is ~3:5 and the total film thickness is ~100 nm.

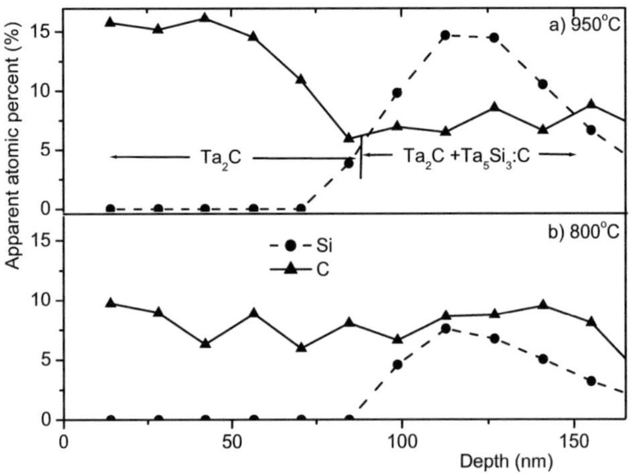

Fig. 15. Apparent atomic concentration of C and Si in the reaction layer with the depth. Ta/SiC samples with 100 nm Ta thickness were annealed at different temperatures in vacuum. Si and C from the SiC substrate are not included in the Fig. The depth is given by using the etch rate of Ta_2O_5. (Cao et al., 2007a).

3.3. Effect of Ni incorporation in thin film Ta/SiC system

Table 3 and Fig. 16 summarize the evolution of the thermal reaction between a dual metal Ni/Ta films on SiC substrate up to 950°C by means of grazing angle XRD. The deposited Ta and Ni films usually have crystalline structure. However, no reflection from metallic Ni can be observed for the samples treated at 650°. A broad diffraction peak in the region of 56-65° superimposed on the α-Ta and Ta_2C diffraction peaks (Fig. 16a) probably come from the amorphous phase produced by solid-state reaction. The driving force for amorphous phase formation is generally expected from the large negative enthalpy of mixing for the Ni/Ta system (Zhang et al., 2000). It has been reported that annealing polycrystalline Ni/Ta multilayers at between 673 and 773 K led to the formation of amorphous phase (Liu & Zhang, 1994; Hollanders et al., 1991). Amorphous Ni-Ta can also be achieved by mechanical milling (Lee et al., 1997), rapid quenching (Fedorov et al., 1986) and ion irradiation (Liu & Zhang, 1994). With increasing temperature, crystallization of Ni-Ta amorphous is expected. The crystallization temperature is related to the composition of the amorphous alloy and activation energy exhibits a maximum near the eutectic composition.

For the film annealed at 800°C (Fig. 16b) diffractions of Ta_2C and metastable carbon-stabilized Ta_5Si_3:C are observed. However, as certain Ta peaks overlap with those of Ta_2C, a minor amount of un-reacted Ta could also be present. It is difficult to ascertain any crystallised Ni-Ta compound or Ni silicide by the XRD pattern. Annealing at 950°C results in the presence of both the stable carbides Ta_2C and TaC (Fig. 16c), the latter of which is dominant. Ta_5Si_3:C and stable α-Ta_5Si_3 are also confirmed. In spite of the XRD peak overlapping, the formation of binary Ni_2Si and ternary TaNiSi silicides could be possible.

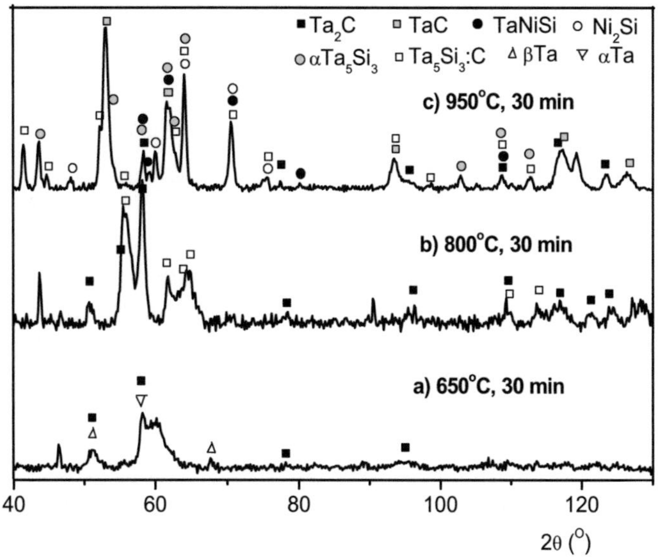

Fig. 16. Grazing angle XRD spectra of Ni/Ta films on SiC after annealing in vacuum. The thickness ratio of Ni:Ta is ~3:5 and the total film thickness is ~100 nm. Glancing angle 3° with Cr k_{α} radiation (λ = 2.29Å)

By comparison of the phases formed between Ta/SiC and Ni/Ta/SiC system at the same temperatures (Table 3), it is found that the dual layer Ni/Ta lowers the temperature at which Ta_2C, Ta_5Si_3:C, TaC or α-Ta_5Si_3 can be detected by grazing angle XRD. It is thus concluded that the existence of Ni promotes the reaction between Ta and SiC, lowering the formation temperature of Ta carbide and silicide.

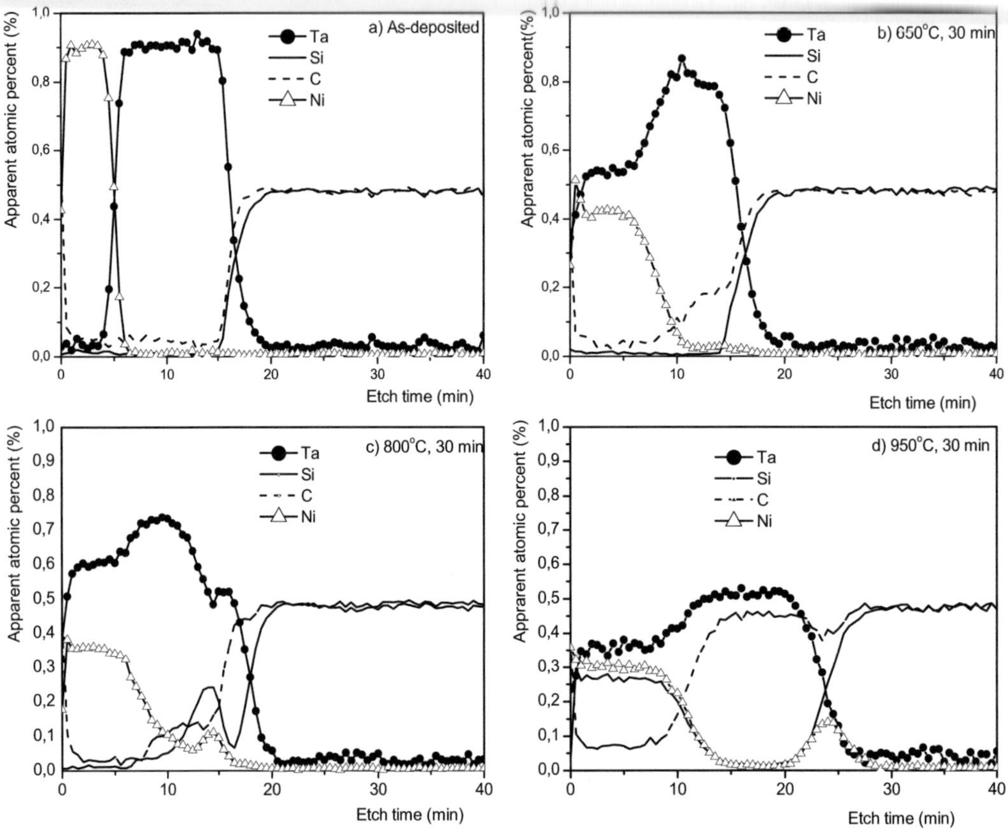

Fig. 17. Auger depth profiles of annealed Ni/Ta/SiC samples. The thickness ratio of Ni:Ta is ~3:5 and the total film thickness is ~100 nm. The etch rate calibrated on Ta_2O_5 with known thickness is 8 nm/min. The atomic concentration is obtained by using the measured Auger peak-to-peak intensities and relative sensitive factors deduced from the elementary intensities of the as-deposited sample.

Figure 17 gives the apparent atomic concentration vs. etch time. The as-deposited sample (Fig. 17a) possessed a rather sharp interface. Annealing at 650°C results in evident change of element distribution (Fig. 17b). The reaction zone is composed of two parts: an upper region of mainly Ni-Ta amorphous mixture and an inner region of metallic Ta and Ta_2C. The Ta_2C layer is supposed to be related to the fast diffusion of C being released from SiC. Apparently, as no Si is evident from the reaction layers within the depth profile, it is suggested that the C moves faster outwards through the coated layer and that it also reacts more readily with the deposited Ta. There is minor amount of Ni dissolved in the inner Ta phase as well. The Ni can act as mediating agent to trigger readily the reaction between Ta and SiC. Compared to the Ta atom, the Ni atom has smaller size (atomic radii of Ni and Ta are 1.62Å and 2.09Å, respectively). Some dissolved Ni atoms might partially penetrate into the SiC substrate at the interface with the help of the thermal activation. As discussed for Ni-SiC system, such kind of defect decreases the stability of the SiC, leading to the more weakened Si-C covalent bonds adjacent to the metal layer compared to Ta/SiC system. The decomposition of SiC is therefore easier at 650°C, accelerating the reactions between the Ta and SiC.

For the sample annealed at 800°C (Fig. 17c), the reaction between Ta and SiC proceeds further to form Ta_2C and Ta_5Si_3:C in the inner part. The increase in C concentration next to the interface implies the possible formation of TaC, although this is not the dominant carbide. The adjacent increase in Ni and Si concentrations suggests the presence of Ni silicide.

A well-defined layer structure is formed at 950°C as shown in Fig. 17d. This high temperature is required for outwards Si diffusion and subsequent initiation of the reaction with both Ni and Ta to form the top TaNiSi layer. This layer probably also has embedded Ta_2C, Ta_5S_3:C and α-Ta_5Si_3. A thin Ni-rich layer is observed close to the coating/substrate interface, probably corresponding to the Ni_2Si phase. The intermediate region between these two layers is composed of TaC. The layered structure can be attributed to the requirement of minimization of the interfacial energy. It has been found that annealing a Ta-Ni alloy on a Si substrate results in $TaSi_2$ phase in the outer region and Ni silicide in the inner region next to the Si (Hung et al., 1986). Annealing amorphous Ni-Ta alloy film on GaAs (Lahav et al., 1987) leads to similar phase separation with an upper region of Ta(Ni)As and an inner region of NiGa. The common factor in these two systems is the layer structure with near-noble metal Ni compounds in contact with the substrate and refractory metal Ta compounds in the outer region. It appears that the sample heated at 950°C has this trend of phase separation. Whether the top TaNiSi disappears or not and how the binary Ni or Ta compounds evolve progressively upon further annealing is of interest.

In the Ta/SiC system, the C atoms move faster than Si outwards through the coated layer and also react more readily with the deposited Ta. The total C content was higher than that of Si in the reaction product when the temperature is at or above 650°C, indicating the formation of C deficiency region in the near interface side of the SiC. It has been proposed that the transition of Schottky to ohmic contacts during high temperature annealing is due to the creation of sufficient C vacancies in the near interface region of the SiC (Han et al., 2002; Nikitina et al., 2005). The C vacancies then act as donors to increase the net concentration of electrons and thus change the electrical properties of the SiC in the near surface region, resulting in the formation of ohmic contact. However, rectifying behaviour at 650°C (Cao et al., 2007[b]) suggests that this mechanism is doubtful since plenty of C vacancies have formed at this temperature. Furthermore, rather deep ionization energy level of 0.51 eV (Aboelfotoh & Doyle, 1999) makes the carbon vacancy mechanism questionable as well. Instead of simple isolated C vacancies, more complicated defect configuration might be responsible for the formation of ohmic contacts.

4. Conclusion remarks

In the Ni/SiC system, 1) The formation of textured Ni_2Si via the reaction Ni + SiC = Ni_2Si + C consists of initial reaction-rate and subsequent diffusion controlled stages. The Ni_2Si islands are present for ultra thin initial Ni layer. The silicides formed at the interface depend on the Ni layer thickness and substrate surface condition. 2) The C released owing to the Ni_2Si formation reaction forms a thin graphite layer on the top of the surface and also tends to form clusters inside the reaction layer. The overall degree of graphitisation is higher at higher temperatures. 3) For the annealed Ni/SiC samples, argon ion etching before Ni deposition helps the formation of the multi-layer structure with less C agglomerated at the surface due to the quicker formation of the silicide Ni_2Si.

In Ta (or Ni/Ta)/SiC system, both silicides and carbides are formed after annealing. The reaction zone consists of a layered structure. A carbon deficiency region is preferentially produced in the near interface region of the underlying SiC. Amorphous Ni-Ta can be formed in Ni/Ta/SiC by the solid-state interfacial reaction. The existence of Ni lowers the formation temperature of carbide and silicide containing Ta, promoting the reaction between Ta and SiC. Although the sample heated at 950°C showed the trend of phase separation. Whether the top TaNiSi disappears or not and how the binary Ni or Ta compounds evolve progressively upon further annealing in Ni/Ta/SiC is of further interest.

The carbon vacancy mechanism is likely to be questionable for the transition of Schottky to ohmic contacts during high temperature annealing. However, the sufficient alteration of the SiC subsurface with more complicated defect configuration by the reaction is important to the formation of the ohmic contact. Detailed investigation related to this issue is interesting.

5. References

Abe, K.; Sumitomo, M.; Sumi, T.; Eryu, O. & Nakashima, K.(2002). *Mater. Sci. Forum,* 389-393, 909-912.

Aboelfotoh, M. O. & Doyle, J. P. (1999). *Phys. Rev.,* B59, 10823–10829.

Adachi, S. (2004): in: *Handbook on Physical Properties of Semiconductors,* vol. 1, 269, Norwell, Kluwer Academic Publishers.

Backhaus-Ricoult, M. (1992). Acta Metal. Mater., 40, s95-103.

Berthod, P.; Aranda, L.; Vébert, C. & Michon, S. (2004). *Computer Coupling of Phase Diagrams and Thermochemistry,* 28, 159-166.

Bhanumurthy, K. & Schmid-Fetzer, R. (2001). *Composites,* A32, 569-574.

Brewer, L. & Krikorian, O. (1956). J. Electrochem. Soc., 103, 38-51.

Bächli, A.; Nicolet, M.-A.; Baud, L.; Jaussaud, C.& Madar, R. (1998). Mater. Sci. Eng., B56, 11-23.

Cao, Y.; Nyborg, L; Jelvestam, U. & Yi, D. (2005). *Appl. Surf. Sci.,* 241, 392–402.

Cao, Y.; Nyborg, L.; Yi, D. & Jelvestam, U. (2006). Mater. Sci.& Techno., 22, 1227-1234.

Cao, Y.; Nyborg, L. (2006). *Surf. Interface Anal.,* 38, 748–751.

Cao, Y., Pérez-García, S.A. & Nyborg, L. (2007)[a]. *Mater. Sci. Forum,* 556-55, 713-716.

Cao, Y., Pérez-García, S.A. & Nyborg, L. (2007)[b]. *Appl. Surf. Sci.,* 254, 139–142.

Cao, Y.; Nyborg. L. & Jelvestam U. (2009). *Interface Anal.,* 41, 471–483.

Chen, J.S.; Kolawa, E.; Nicolet, M.-A. Ruiz, R.P.; Baud, L.; Jaussaud, C. & Madar, R. (1994). *J. Appl. Phys.*, 76, 2169-2175.

Chen, S.H.; Zheng, L.R.; Carter, C.B. & Mayer, J.W. (1985). *J. Appl. Phys.*, 57, 258-263.

Chou, T.C.; Joshi, A. & Wadsworth, J. (1990). *J. Vac. Sci. Technol*, A9, 1525-1534.

Ciccariello, J.C.; Poize, S. & Gas, P. (1990). *J. Appl. Phys.*, 67, 3315-3322.

Colgan, E.G.; Gambino, J.P.; Hong & Q.Z. (1996). *Mater. Sci. Eng.*, R16, 43-96.

d'Heurle, C.L.F.M. & Zhang, S.-L (2007). Silicides, In: *Handbook of Semiconductor Manufacturing Technology*, Doering, R; Nishi, Y (Ed.), p10-1 - 10-52, Taylor & Francis Group, LLC, ISBN 978-1-5744-4675-3, Boca Raton.

Fedorov, V.V.; Yedneral, A.F., Kachalov, V.M. & Borisov, V.T. (1986). *Phys. Met. Metall.*, 62, 96-99.

Feng, J.C.; Naka, M. & Schuster, J.C. (1997). *J. Mater. Sci. Lett.*, 16, 1116-1117.

Gambino, J.P. & Colgan, E.G. (1998). Silicides and ohmic contacts. In *Material Chemistry and Physics*, 52, 99-146.

Gasser, S.M.; Bächli, A.; Garland, C.M.; Kolawa, E. & Nicolet, M.-A.(1997). *Microelectronic Engineering*, 37/38, 529-534.

Geib, K.M.; Wilson, C.; Long, R.G. & Wilmsen, C.W. (1990). *J. Appl. Phys.*, 68, 2796-2800.

Guziewicz, M.; Piotroowska, A. Kaminska, Grasza, E. K.; Diduszko, R. Stonert, A. Turos, A.; Sochacki, M. & Szmidt, J. (2006). *Mater. Sci. Eng.*, B135, 289-293.

Han, S.Y. & Lee, J.-L. (2002). J. Electrochem. Soc., 149, G189-193.

Han, S.Y.; Kim, N.-K.; Kim, E.-D. & Lee, J.-L. (2002). *Mater. Sci. Forum*, 389-393, 897-900.

Han, S.Y.; Shin, J.-Y.; Lee, B.-T. & Lee. J.-L. (2002). *J. Vac. Sci. Technol.*, B20, 1496-1500.

Hollanders, M.A.; Duterloo, C.G.; Thijsse, B.J. & Mittemeijer, E.J. (1991). *J. Mater. Res.*, 6, 1862-1873.

Hsu, J.Y. & Liang, J.H. (2005). *Nucl. Instrum. Methods Phys. Res.*, B 241, 543–547.

Hung, L. S.; Saris, F. W.; Wang, S. Q. & Mayer, J. W. (1986). *J. Appl. Phys.*, 59, 2416.

Hähnel, A. & Woltersdorf, J. (2004). *Mater. Chem. Phys.*, 83, 380-388.

Kestle, A.; Wilks, S.P.; Dunstan, P.R.; Pritchard, M.; Pope, G.; Koh, A. & Mawby, P.A. (2000). *Mater. Sci. Forum*, 338-342, 1025-1028.

Kurimoto, E. & Harima, H. (2002). *J. Appl. Phys.*, 91, 10215-10217.

Iwaya, M.; Kasugai, H.; Kawashima, T.; Iida, K.; Honshio, A.; Miyake, Y.; Kamiyama, S.; Amano, H. & Akasaki, I. (2006). *Thin Solid Films*, 515, 768-770.

Jang, T.; Porter, L.M.; Rutsch, G.W.M. & Odekirk, B. (1999). *Appl. Phys. Lett.*, 75, 3956-3958.

Lahav, A.; Eizenberg, M. & Komen, Y. (1987). *J. Appl. Phys.*, 62, 1768.

Laurila, T.; Zeng, K. & Kivilahti, J.K. (2002). *J. Appl. Phys.*, 91, 5391-5399.

La Via, F.; Roccaforte, F.; Makhtari, A.; Raineri, V.; Musumeci, P. & Calcagno, L. (2002). *Microelectron. Eng.*, 60, 269-282.

La Via, F.; Roccaforte, F.; Raineri, V.; Mauceri, M.; Ruggiero, A.; Musumeci, P.; Calcagno, L; Castaldini, A. & Cavallini, A. (2003). *Microelectron. Eng.*, 2003, 70, 519-523.

Lavoie, C; Detavernier, C. & Besser, P. (2004). Nickel silicide Technology, In *Silicide Technology for Integrated Circuits*, Chen, Lih J. (Ed.), IEEE, London.

Le Claire, A.D. (1999). Diffusion in solid metals and Alloys, in *Numerical Data and Functional Relationships in Science and Technology*, Mehrer, H. (Ed.), III/26, Springer, 478.

Lee, P.S.; Mangelinck, D.; Pey, K.L.; Ding, J.; Dai, J.Y.; Ho, C.S. & See, A. (2000). *Microelectr. Eng.*, 51-52, 583-594.

Lee, P.-Y.; Yang, J.-L.; Lin, C.-K.& Lin, H.-M. (1997). *Metall. Mater. Trans.*, 28A, 1429-1435.

Lee, W.; Lee, J.; Bae, J.D.; Byun, C.S. & Kim, D.K. (2001). *Scripta Mater.*, 44, 97-103.

Lim, C.S.; Nickel, H; Naoumidis, A. & Gyarmati, E. (1997). *J. Mater.Sci.*, 32, 6567-6572.

Litvinov, V.L.; Demakov, K.D.; Agueev, O.A.; Svetlichnyi, A.M.; Konakova, R.V.; Lytvyn, P.M. & Milenin, V.V.(2002). *Mater. Sci. Forum,* 389-393, 905-908.

Liu, B.X. & Zhang, Z.J. (1994). Phys. Rev. B, 49, 12519-12527.

Lou, D.C., Akselsen, O.M., Solberg, J.K., Onsoien, M.I.; Berget, J. & Dahl, N. (2006). *Surf. Coatings Technol.*, 200, 3582-3589.

Lu, W.; Mitchel, W.C.; Landis, G.R.; Crenshaw, T.R. & Eugene Collins, W. (2003). *Solid-State Electron.*, 47, 2001-2010.

Lu, X.D. & Wang, H.M. (2004). *Acta Mater.*, 52, 5419-5426.

Madsen, L.D.; Svedberg,E.B.; Radamson, H.H.; Hallin, C.; Hjörvarsson, B.; Cabral, Jr., C.; Jordan-Sweet, J.L. & Lavoie, C.(1998). *Mater. Sci. Forum*, 264-268, 799-804.

Marinova, Ts.; Krastev, V.; Hallin, C.; Yakimova, R. & Janzén, E.(1996). *Appl. Surf. Sci.*, 99, 119-125.

Marinova, Ts.; Kakanakova-Georgieva, A.; Krastev, V.; Kakanakov, R.; Neshev, M.; Kassamakova, L.; Noblanc, O.; Arnodo, C.; Cassette, S.; Brylinski, C.; Pecz, B.; Radnoczi, G. & Vincze, Gy. (1997). *Mater. Sci. Eng.*, B46, 223-226.

Matzke, Hj. & Rondinella. V.V.(1999). in *Landolt Börnnstein. Added t.p.:Numerical Data and Functional Relationships in Science and Technology,* Beke, D.L. (Ed.), vol. III/33B1, 5-5, 6, Berlin, Springer.

Nash, P. & Nash, A. (1992). Alloy Phase Diagrams, In *ASM Handbooks*, Baker, H. (Ed.), Vol. 3, p2859-2860, ASM International, Materials Park, Ohio.

Nikitina, I.P.; Vassilevski, K.V.; Wright, N.G.; Horsfall, A.B. & O'Neill, A.G. (2005). *J. Appl. Phys.*, 97, 083709.

Ohi,A; Labis, J.; Morikawa, Y.; Fujiki, T.; Hirai, M.; Kusaka, M.& Iwami M.(2002). *Appl. Surf. Sci.*, 190, 366-370.

Olowolafe, J.O.; Solomon, J.S.; Mitchel, W. & Lampert, W.V. (2005). *Thin Solid Films*, 479, 59-63.

Ottaviani, G. (1979). *J. Vac. Sci. Technol.* 16, 1112-1119.

Park, J.S.; Landry, K. & Perepezko, J.H. (1999). *Mater. Sci. Eng.*, A259, 279-286.

Roccaforte, F.; La Via, F.; Raineri, V.; Calcagno, L. & Musumeci. P. *Appl. Surf. Sci.*, 2001, 184, 295-298.

Rodriguez, N.M.; Anderson, P.E.; Wootsch, A.; Wild, U.; Schlögl, R. & Paál, Z. (2001). *Journal of Catalysis*, 197, 365-377.

Sarkar, D.K.; Falke, M.; Giesler, H.; Teichert, S.; Beddies G. & Hinneberg, H.-J. (2000). *Appl. Phys.*, A70, 681-684.

Schuster, Julius C. (1993-1994). *Int. J. of Refractory Metals & Hard Materials,* 173-177.

Shifler, D.A. High-Temperature Gaseous Corrosion Testing, in *Corrosion: Fundamentals, Testing, and Protection, ASM Handbook,* Vol 13A, ASM International, 2003, p 650–681.

Smithells, C.J. (1967). Metals Reference book, vol. 2, 664, London, Butterworth & Co. Ltd.

Su, Y-K.; Chiou, Y-Z.; Chang, C-S.; Chang, S-J.; Lin, Y-C. & Chen, J. F. (2002). *Solid-State Electron.*, 46, 2237-2240.

Yoon, J-K; Byun, J-Y; Kim, G-H; Lee, J-K; Yoon, H-S & Hong, K-T. (2003). *Surf. Coating Technol.*, 168, 241-248.

Yuryeva, E.I. & Ivanovskii, A.L. (2002). *Russ. J. Coord. Chem.*, 28, 881-888.

Zhang, Q.; Lai,W. S. & Liu, B. X. (2000). Phys. Rev., B 61, 9345–9355.

Zheng, L.R.; Hung, L.S. & Mayer, J.W. (1983). *J. Vac. Sci. Technol.*, A1, 758-761.

Part 2

Other applications: Electrical, Structural and Biomedical

9

Properties and Applications of Ceramic Composites Containing Silicon Carbide Whiskers

Brian D. Bertram and Rosario A. Gerhardt
Georgia Institute of Technology
United States of America

1. Introduction

In the early 1980s, researchers at Oak Ridge National Laboratory developed a new type of ceramic composite reinforced with dispersed silicon carbide (SiC) whiskers. The majority-phase ("matrix") material of these composites was aluminum oxide (Al_2O_3) and the minority-phase ("filler") SiC was grown in whisker-form by a pyrolysis technique using organic waste such as rice hulls. Inclusion of a small volume fraction of silicon carbide whiskers (SiC_w) in Al_2O_3 had a large impact on the properties and allowed for interesting applications which otherwise would not be possible with Al_2O_3 alone. The whisker form of SiC made the composites exceptionally non-brittle compared to conventional monolithic ceramics due to a variety of toughening mechanisms within the composites. Many years later, it was discovered that the whiskers also imparted useful electrical properties.

The invention of the composites roughly coincided with the U.S. Department of Energy's development of the 1983 "Ceramic Technology Program for Advanced Heat Engines" (Sinott, 1987). A few years earlier, President Carter pushed for the development of more fuel-efficient automobiles to mitigate the energy problems of the United States(Carter, 1977) which were earlier noted by President Ford(Ford, 1975). Ceramic engines operating at high temperatures have the potential for high energy efficiency due to the thermodynamics of engine operation(DeHoff, 1993) and the whisker-reinforced composites were candidate materials for this purpose. However, subsequent research found that significant degradation of the whisker-laden composites occurred during extended exposure to stress and air at high temperatures due to reaction of the SiC whiskers, leading to extensive deformation by creep phenomena. Such structural degradation was correlated to worsening of mechanical properties and the development of Al_2O_3-SiC_w composites for engines stalled. However, Al_2O_3-SiC_w composites have been found to be useful and commercially viable as cutting tools and microwave-heating inserts. In working towards these applications, a large amount of research has been conducted on this composite system: the literature contains many different perspectives on the material. In considering the implications of the different perspectives, one can develop new insights. The goal of this chapter is to distill the most significant results of the literature and create a big-picture description of these materials for readers having some background in materials science.

2. SiC Whiskers

2.1. SiC Whisker Fabrication

The term 'whisker' is commonly used to describe a small single-crystalline rod. SiC whiskers may be synthesized by various methods, such as sol-gel, plasma deposition, chemical vapor deposition, and pyrolysis of agricultural waste. The latter method seems to be preferred for industrial-scale production of SiC whiskers. Generally, agricultural waste containing both silica (SiO_2) and carbon is used. Waste sources include rice hulls, coconut shells, bean-curd refuse, sugarcane leaf, and rice straw. Silicon carbide whiskers can also be fabricated from mixtures of silica and carbon powders, carbon and silicon halides, polymer precursors, or other materials like silicon nitride. In addition to SiC whiskers, the synthesis process also usually produces some amount of (usually unwanted) non-whisker SiC particles which varies depending on processing details. After synthesis, whiskers are often cleaned by ultrasonication (Raju & Verma, 1997) and/or acid washing (Raju & Verma, 1997; Ye et al., 2000) to remove contamination of particulates (Lee & Sheargold, 1986) and trace amounts of amorphous silica which often remain on the whiskers (Raju & Verma, 1997).

Synthesis from rice hulls may be conducted in the 1000-1500°C range in an inert gas ambient, e.g. N_2 (Raju & Verma, 1997). In another work, a 700-800°C carbonization pretreatment of the organic waste was performed to increase carbon surface area before conducting the 1400°C pyrolysis in argon (Singh et al., 1998). Reaction occurs when silica (SiO_2) undergoes a carbothermic reduction to SiC. The overall reaction is given by Reaction 1 and the related sub-reactions (2-4) are:

$$(1):\ SiO_2 + 3C \rightarrow SiC + 2CO_{(g)} \qquad (3):\ SiO_{(g)} + C \rightarrow Si + CO_{(g)}$$

$$(2):\ SiO_{(g)} + 2C \rightarrow SiC + CO_{(g)} \qquad (4):\ Si + C \rightarrow SiC$$

It is believed that silica is first reduced to gaseous SiO (i.e. $SiO_2 + C \rightarrow SiO_{(g)} + CO_{(g)}$), and then forms SiC either by a replacement reaction (#2) or through the continued reduction to Si (Reaction #3) and subsequent addition reaction (#4) (Raju & Verma, 1997). The initial reaction between carbon and the silica (or silica derivative) depends upon the physical contact between the two. Ideally, the silica should be intimately mixed with the cellulose. As the reduction reaction proceeds, the intimate contact between silica and carbon is lost and further reduction of silica depends mainly on the gas-phase carbon source. Certain waste, such as coconut shells, are believed to produce more carbonaceous vapours (e.g. CO) for more efficient carbothermal reduction and higher yield. It is believed that the porous cellular structure of rice hulls facilitates vapor-phase transport. To achieve high-yield, it seems important to achieve a certain balance of silica:carbon in the starting materials (Raju & Verma, 1997; Singh et al., 1998).

A small amount of metal catalyst may be mixed in with the source materials for SiC whisker synthesis. This can be done by soaking the raw ingredients in an appropriate metal-salt solution. Examples of metals include iron, Pd, Ni, or Co. Catalysts generally promote whisker growth by the vapor-liquid-solid (VLS) mechanism, which is fairly well-established in the growth of "1-dimensional" materials such as nanowires and nanotubes (Wagner & Ellis, 1964). For SiC whiskers, the theory implies that Si- and C-containing vapors dissolve into the molten metal catalyst balls and SiC precipitates when the solution becomes sufficiently supersaturated. This is then followed by unidirectional growth in the energetically-favored crystallographic direction. However, it should be noted that it is possible to grow SiC whiskers without such catalysts and these are believed to grow by a vapor-solid mechanism (Singh et al., 1998). Such whiskers tend to exhibit pointed tips at the ends and are not capped with catalyst balls, unlike VLS-grown whiskers.

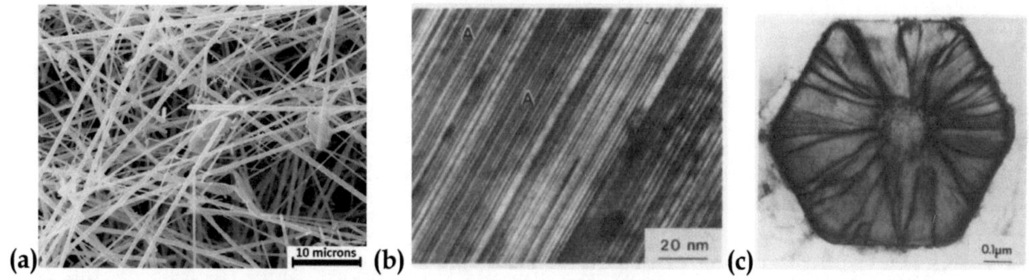

Fig. 1. Electron micrographs of (a) loose as-grown SiC whiskers, (b) planar defects perpendicular to the length of a single SiC whisker, and (c) a whisker cross-section showing partial dislocations and core cavities. *Sources*: (a) the Greenleaf Corporation and (b & c) S. R. Nutt, 1988; John Wiley & Sons, with permission.

2.2. SiC Whisker Structure

Figure 1a shows an image of as-grown SiC whiskers. Typically, SiC whiskers have lengths and diameters on the order 5-50 μm and 0.2-1.0 μm, respectively. The crystal structure of SiC may be described by a polytype. Broadly, the SiC polytypes may be categorized as β (cubic, i.e. zinc-blende) or α (non-cubic, e.g. hexagonal). However, there are many other SiC_w structural characteristics to consider. Generally, the whisker morphology and structure of the SiC particles results from the growth process (Nutt, 1984, 1988; Porter & Davis, 1995).

The structure and defects of silicon carbide whiskers made from rice hulls have been studied by transmission electron microscopy and this provided insight into their atomic- and nano-structure (Nutt, 1984). The whiskers were found to have grown in the <111> direction and exhibited a high density of planar defects such as stacking faults and twins on the {111} planes. As shown in Figure 1b, the spacings between planar faults in SiC_w may be as small as a few nm.The presence of these planar defects mean that the whisker structure may be described as a complex mixture of α and β polytypes arranged in thin lamellae perpendicular to the growth axes. However, it is common for whiskers to be reported as being of a single polytype (ACM website; Bertram & Gerhardt, 2010; Mebane & Gerhardt, 2004) based on x-ray diffraction phase identifications and reports from the manufacturer. Currently, the leading SiC_w manufacturer in the United States is Advanced Composite Materials LLC in Greer, SC.

Other common SiC_w defects include core cavities and partial dislocations,(Nutt, 1984) as shown in Figure 1c. The size range for core-cavities and voids was 2-20 nm. Whiskers were found to have hexagonal facetting, likely traceable to anisotropy in the surface energy during SiC crystal formation. Other studies indicate that as-fabricated whiskers may contain small pockets of silicon oxycarbide glass (Karunanithy, 1989). Generally speaking, whisker structure and arrangement can have significant effects on composite properties. Rice provides a more general review of ceramic (and metal) whisker growth and points out that for general growth of inorganic crystals, the morphology and growth habits are affected by impurities in the growth environment (Rice, 2003).

2.3. Health Risk of Loose SiC Whiskers

Finally, it should be noted that the loose whisker morphology (i.e. when the whiskers are not incorporated in solid composites) gives rise to a serious respiratory health risk and that unsafe handling of loose SiC whiskers can lead to deadly lung diseases (Rodelsperger & Bruckel, 2006; Vaughan & Trently, 1996). Generally, the elongated shapes (high aspect ratio), small sizes (micron level), and the facts that they are chemically inert and cannot be readily removed by natural human-body processes make SiC whiskers behave similarly to asbestos when in the lungs. Therefore it is of utmost importance to take care when working with loose whiskers or whisker-laden powders and to avoid stirring them up into the air. Also, one should always employ engineering controls (e.g. filtered respirators, specialized hoods) to minimize human inhalation and exposure during handling. However, certain whisker-containing composites used for cooking food have been NSF-51 approved for food contact. Currently, this technology is being marketed as Silar© by Advanced Composite Materials LLC.

3. Ceramic Composites Reinforced with Silicon Carbide Whiskers

The incorporation of SiC_w into ceramic composites generally results in an improvement in properties compared to the monolithic matrix. However, the behavior of SiC_w-filled ceramic composites depends on many factors and this requires one to consider them from numerous different perspectives in order to understand the behavior. To this end, we begin by considering their ultimate purposes / applications. Once, Al_2O_3-SiC_w composites had prospective application (Sinott, 1987) in high-temperature ceramic engines, and for this one would be interested in responses to creep, thermal shock, and fatigue. Currently, the commercial applications of Al_2O_3-SiC_w are in cutting tools and microwave-heating inserts.

3.1. Cutting-Tool Application

Images of Al_2O_3-SiC_w cutting-tool inserts are shown in Figure 2a. Cutting tools usually must be hot-pressed to achieve full density and the necessary thermal-mechanical properties. Such tools are cost-effective for shaping difficult-to-machine metals such as steel (Figure 2b) and Ni-based superalloys for turbines in aerospace applications. For the cutting tool application, one is generally interested in the following properties of the composite: fracture toughness, thermal conductivity, abrasive wear resistance, chemical inertness, and thermal shock resistance (which depends on fracture toughness and thermal conductivity). Compared to conventional ceramics, Al_2O_3-SiC_w composites provide improvements (increases) in all the above properties except chemical inertness. During cutting with tools of Al_2O_3-SiC_w, damage is accompanied by whisker toughening mechanisms and reaction may occur between SiC and Fe. This generally limits cutting of ferrous materials to lower speeds (Collin & Rowcliffe, 2001; Thangaraj & Weinmann, 1992).

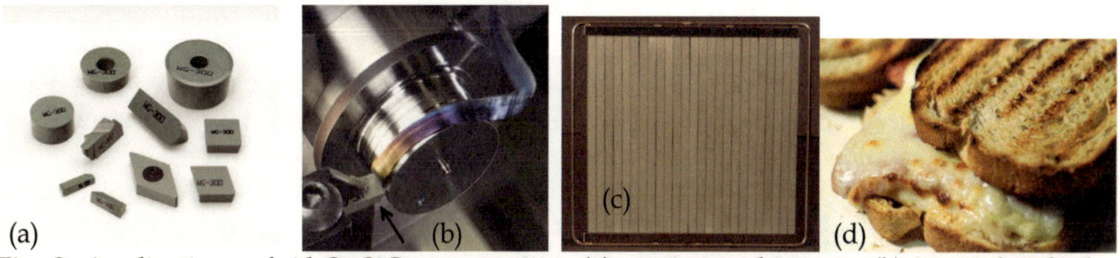

Fig. 2. Applications of Al_2O_3-SiC_w composites: (a) cutting-tool inserts, (b) insert (marked with arrow) machining hardened tool steel, (c) microwave flatstone, (d) panini sandwich speed-grilled with Silar© technology. Images provided by Greenleaf Corporation in Saegertown, PA (a,b) and Advanced Composite Materials LLC in Greer, SC (c,d).

Notably, the general cutting process is complicated and depends on many factors (Li, 1994; Mehrotra, 1998), including the defect distribution in the tool, tool shape, and the properties of the piece being cut. Also, there are different types of tool degradation which depend on the location on the tool. For SiC_w-reinforced tools, literature suggests that the residual stresses in the microstructure are rather important in determining mechanical and wear response, e.g. at the sample surface and at whisker-matrix interfaces(Billman et al., 1988; Thangaraj & Weinmann, 1992). It has also been found that wear performance depends on sample orientation relative to the processing direction and fracture toughness at the microstructural level, not the fracture toughness of the bulk sample (Dogan & Hawk, 2000). Moreover, one may use a lamination process to increase control of the Al_2O_3-SiC_w microstructure and thus improve cutting tool performance (Amateau et al., 1995). Also, tools employing specially-coated whiskers can provide increased toughness, tool life, and cutting speed. These may be obtained from the Greenleaf Corporation.

3.2. Application in Microwave Heating and Cooking

Currently, the microwave application is led by Advanced Composite Materials LLC in Greer, SC (Quantrille, 2007, 2008). For this application, density and mechanical properties are not as important and composite rods or bars ~1 foot long and ~½ inch wide are made by extrusion and pressureless sintering. Several rods are then arranged with a metal clip to form a grill or flatstone like that shown in Figure 2c which is then placed in a microwave oven having a clip-compatible design. Food products such as fresh dough pizza and panini sandwiches can then be cooked faster to yield a high quality result (e.g. crispy crust, grill marks, etc.) with application of microwave radiation (e.g. at 0.915, 2.45 GHz) in the cavity. An example is shown in Figure 2d. The microwaves preferentially heat the ceramic composite compared to the food. This is because, generally, the solid food on the grill has a much-reduced dielectric loss compared to the ceramic composite. The dielectric loss of the solid food can be expected to follow mainly from moisture and oil content (Basak & Priya, 2005; Metaxas & Meredith, 1983; Quantrille, 2008). For the composites, the dielectric loss primarily depends on the properties of the dispersed SiC_w population and is proportional to the microwave heating rate of the composite. The literature suggests that composite thermal conductivity and thermal shock response may influence food heating during cooking and the lifetime of the parts, respectively (Basak & Priya, 2005; Parris & Kenkre, 1997) (McCluskey et al., 1990) (W. J. Lee & Case, 1989; Quantrille, 2007, 2008).

Quantrille has analyzed the heat transfer into the food during cooking and reported results of various microwave heating tests (Quantrille, 2007, 2008). The ceramics heat quickly, e.g. at ~2.3-4.6 °C/s from 900 W incident on powders blends of 7.5-15 wt% SiC_w in Al_2O_3. Food heating results from three processes: (1) dielectric loss in the food itself (2) air convection from the ceramic and (3) thermal conduction across the thermal gradient at the ceramic-food interface. Due to (3) and the fact that food moisture content declines as cooking proceeds, it is possible to sear the food with grill marks at the end of cooking. It was found that pizzas and paninis could be "grilled, toasted, and cooked to perfection" in ~80 and ~90 seconds, respectively (Quantrille, 2008). This method can also reduce the energy cost of cooking compared to conventional methods.

3.3. Introduction to Structure-Property Relations of Composites

To understand different types of models for SiC_w composites, it is useful to know the broader context of composite-material modeling (Jones, 1984; Mallick, 2008; Runyan et al, 2001a, 2001b; Taya, 2005). Most models depend in some way on the spatial distribution of the phases. Common distributions are shown in Figure 2.

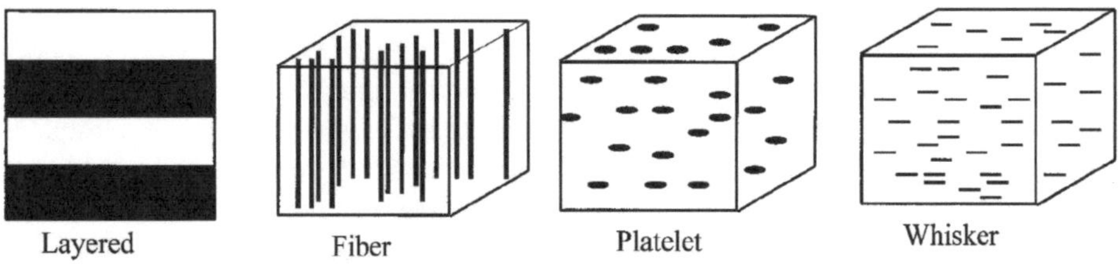

Fig. 3. Common types of composite structures. Here, "Fiber" implies continuous unidirectional fibers. After Runyan et al., 2001a, with permission (John Wiley & Sons).

Model complexity can vary a great deal based on the extent of assumptions made in the model development. The simplest models are the mixing rules which are applicable when the second phase has a unidirectional and continuous morphology, e.g. layered and fiber composites, as shown in Figure 3. These models result in the composite material response properties being predicted as volume-fraction (V) weighted averages of the properties of the constituent phases. Specifically, many properties may be modeled via

$$G_M = V_0G_0 + V_1G_1 \tag{5}$$

$$1/H_M = V_0/H_0 + V_1/H_1 \tag{6}$$

where G relates to the response along the fiber/phase alignment direction and H relates to the response in the transverse direction. Here, subscripts '0' and '1' denote phase 0 (the matrix) and phase 1 (the filler)and 'M' denotes the composite mixture. Equation 5 can be applied to model how mechanical stress is distributed among the two phases of a fiber composite loaded in the fiber direction when the isostrain condition is applicable. Or, it may be used to model the electrical resistivity of two phases in series (e.g. layered composites). Equation 6 can be used to model the effective elastic modulus of fibrous composites in the direction perpendicular to the fibers. Or, it may be used to model the effective resistivity of two phases in parallel (e.g. layered composites). In principle, these mixing laws can predict many properties if the materials and structure are consistent with the assumptions of the mixing law. The interested reader should consult the following references, especially for mechanical properties (Jones, 1999; Mallick, 2008). Taya provides a nice treatment of electrical modeling and the physics, continuum mechanics, and mathematics principles which underlie modeling efforts in general (Taya, 2005).

For composites having a discontinuous dispersed second phase (e.g. platelet- or whisker-like filler, as shown in Figure 3) effective medium theory is more applicable. Compared to mixing rules, effective medium theories employ a different perspective: they use descriptions of the effects of inclusions on the relevant stress and/or strain fields in the bulk material to deduce related macroscopic materials properties. For example, they may relate the local electric and magnetic fields around conductive filler particles to the electromagnetic response of the composite, or alternatively, the mechanical stress-strain fields around reinforcement particles to the composite mechanical response. In other words, these theories attempt to generalize outward from descriptions of the small-scale situations to predict the effective macroscopic response of the composite mixtures. State-of-the-art theories (Lagarkov & Sarychev, 1996) have been found to provide fair to very-good agreement with experimental data from complicated dispersed-rod composite structures.(Lagarkov et al, 1997, 1998; Lagarkov & Sarychev, 1996) For fracture-toughness modeling, one reference stands out (Becher et al., 1989). Newcomers to electrical modeling may find that other references provide a better introduction to these topics (Gerhardt, 2005; Jonscher, 1983; Metaxas & Meredith, 1983; Runyan et al., 2001a, 2001b; Gerhardt et al., 2001; Streetman & Banerjee, 2000; Taya, 2005; von Hippel, 1954).

Many additional perspectives and models for electrical response are available in the literature and cannot be reviewed thoroughly here (Balberg et al., 2004; Bertram & Gerhardt, 2009, 2010; Connor et al., 1998; Gerhardt & Ruh, 2001; Lagarkov & Sarychev, 1996; Mebane & Gerhardt, 2006; Mebane et al., 2006; Panteny et al., 2005; Runyan et al., 2001a, 2001b; Tsangaris et al., 1996; C. A. Wang et al., 1998; Zhang et al., 1992). Most of these models adopt a single perspective for considering the structure. In one type, the composite itself is considered as an electrical circuit consisting of a large number of passive elements, such as resistors and capacitors, which are themselves models of individual microstructural features (e.g. SiC_w/matrix/SiC_w structures) and the associated electrical processes. Models of this type are sometimes called equivalent circuits or random-resistor networks (Panteny et al., 2005) Analysis of a random network of passive elements typically starts with basic principles of circuit analysis. Such analyses have provided insight into the electrical response of the systems in question (Bertram & Gerhardt, 2010; Mebane & Gerhardt, 2006).

The filler material may be accounted for in other ways as well. One model took into account both the percolation of the filler particles and the fractal nature of filler distribution in non-whisker particulate composites and related it to the ac and dc electrical response (Connor et al., 1998). The Maxwell-Wagner model (Bertram & Gerhardt, 2010; Gerhardt & Ruh, 2001; Metaxas & Meredith, 1983; Runyan et al., 2001b; Sillars, 1937; Tsangaris et al., 1996; von Hippel, 1954) originally considered the frequency-dependent ac electrical response of a simple layered composite structure (von Hippel, 1954) based on polarization at the interface of the two phases. Generally, it was found that this model gave similar but not identical results to the Debye model (von Hippel, 1954) for a general dipole polarization. The Maxwell-Wagner model has been extended to consider more complicated geometries for the filler distribution (Sillars, 1937). Recent studies (Bertram & Gerhardt, 2010; Runyan et al., 2001b) have revealed that the Cole-Cole modification (Cole & Cole, 1941) of the Debye model can be applied to describe non-idealities observed experimentally for the Maxwell-Wagner polarizations in SiC-loaded ceramic composites. Unfortunately, there do not seem to be many composite models which account for the semiconductive (Streetman & Banerjee, 2000) character of SiC whiskers. Our description of Schottky-barrier blocking between metal (electrode) and semiconductor (SiC) junctions at whiskers on Al_2O_3-SiC_w composite surfaces is an exception (Bertram & Gerhardt, 2009). The interested reader may also consult other works concerning modeling transport in systems that may be relevant to Al_2O_3-SiC_w but which will not be described here (Calame et al., 2001; Goncharenko, 2003).

3.4. Percolation of General Stick-filled Composites

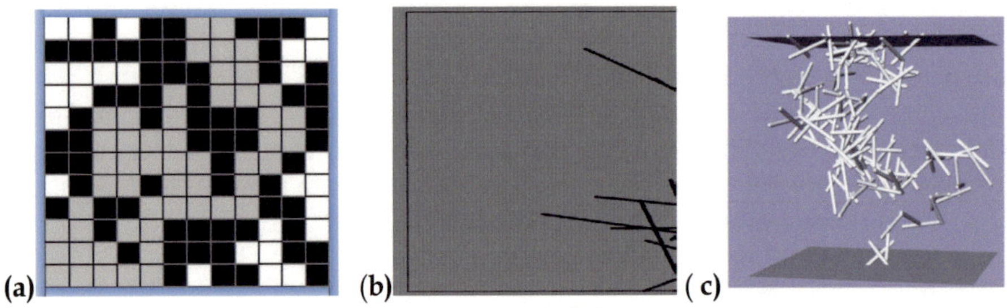

Fig. 4. Top-to-bottom percolation pathways in models of increasing complexity: (a) Binary black-and-white composite on a square grid, where white-percolation is darkened to gray. (b) Two-dimensional stick percolation, (c) Three-dimensional model based on stereological measurements of the length-radii-orientation distribution of SiC whiskers.

Sources: (b) Lagarkov & Sarychev's Fig 1b, *Phys. Rev. B*. 53 (10) 6318, 1996. Copyright 1996 by the American Physical Society. (c) Mebane & Gerhardt, 2006. John Wiley & Sons, with permission.

The fundamentals of the old statistical physics problem of percolation are discussed elsewhere (Stauffer & Aharony, 1994). In Figure 4, it is shown that the percolation transition in composites may be understood on various levels of conceptual complexity. As complexity increases to more accurately describe the stick morphology of the filler, the model changes from (a) a two-dimensional binary pixel array to (b) one-dimensional (1D) rods in 2D space, to (c) 2D rods in 3D space. Electrically, percolation amounts to an insulator-conductor transition (Gerhardt et al., 2001). Percolation also causes a significant change in creep response (de Arellano-Lopez et al., 1998) and hinders densification during composite sintering (Holm & Cima, 1989).

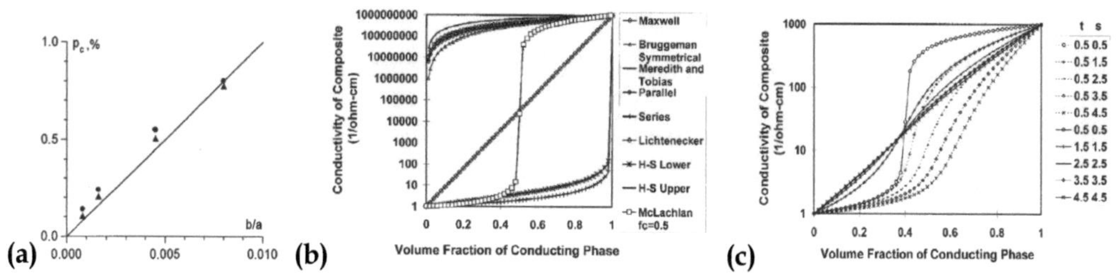

Fig. 5. (a) Linear dependence of percolation threshold (ϕ_c=p_c) on inverse aspect ratio (b/a). (b) Comparison between the McLachlan model of percolation vs. various other models for electric composites. (c) Effect of McLachlan parameters on the shape of the percolation curve when p_c=0.4. *Sources:* (a) Lagarkov et al., 1998. American Institute of Physics. (b-c) Runyan et al, (2001a). John Wiley & Sons.

The actual value of the percolation threshold depends on the shape of the percolating particles, the dimensionality of the structure, the definition of connectivity, and for real composites, the details of processing. In Figure 4a, it is easy to imagine that percolation of white pixels along a particular direction could be achieved with the lowest possible ratio of white-to-black in the overall grid if the white pixels are arranged in a straight line along the direction of interest. This fact relates to the percolation of sticks in 3D space. For sticks having lengths 'a' and diameters 'b', the stick aspect ratio is a/b and is related to the percolation threshold ϕ_c=p_c via

$$p_c \propto b/a \tag{7}$$

Figure 5a demonstrates this with experimental data from chopped-fiber composites (Lagarkov et al., 1998; Lagarkov & Sarychev, 1996). This relation has been concluded by several investigators, can be proven with an excluded volume concept (Balberg et al., 1984; Mebane & Gerhardt, 2006), and has important implications for real composite materials. Generally speaking, simulated and experimental results for percolation thresholds indicate that the percolation process is strongly dependent on the geometric features of the problem and that analytical and computer models are often useful for understanding of specific situations.

The Generalized Effective Medium (GEM) equation first proposed by McLachlan provides a useful model of the percolation transition for general insulator-conductor composites and uses semi-empirical exponents to account for variation in the shapes observed for experimental percolation curves (McLachlan, 1998; Runyan et al., 2001a; Wu & McLachlan, 1998). It may be written as

$$(v_i)\frac{\sigma_i^{1/s}-\sigma_M^{1/s}}{\sigma_i^{1/s}+\left(\frac{1}{\phi_c}-1\right)\sigma_M^{1/s}}+(v_c)\frac{\sigma_c^{1/t}-\sigma_M^{1/t}}{\sigma_c^{1/t}+\left(\frac{1}{\phi_c}-1\right)\sigma_M^{1/t}}=0 \tag{8}$$

where ϕ_c =p_c is the (critical) percolation threshold, σ is dc electrical conductivity, σ_c refers to the conductive phase, v indicates volume fraction, *s* and *t* are semi-empirical exponents,'M' denotes the composite mixture, and 'i' denotes the insulator. The McLachlan equation predicts a drastically different response compared to many other composite models, as shown in Figure 5b. Figure 5c shows some effects of the semi-empirical parameters on the shape of the percolation curve. The percolation of rods having specific size and orientation distributions has been simulated (Mebane & Gerhardt, 2006) and this model required certain assumptions about the nature of interrod connectivity, e.g. a "shorting distance" concept.

3.5. Structural Characteristics of Al_2O_3-SiC_w Composites

What should one focus on when considering the complicated microstructure of a Al_2O_3-SiC_w composite? There are many options, including density, whisker size and orientation, whisker percolation, interwhisker distance, whisker-matrix interface properties, and whisker defects. Many of these will be considered in this section, and a discussion of interface effects and SiC_w defects is given in Section 3.8. Properties of the Al_2O_3 and sintering additives seem to have less impact on the final properties (assuming high density is achieved) and will not be considered here.

3.5.1. Percolation of SiC Whiskers

The formation of a continuous percolated network of SiC whiskers across a sample has important implications for electrical/thermal transport and mechanical properties (de Arellano-Lopez et al., 1998, 2000; Gerhardt et al., 2001; Holm & Cima, 1989; Mebane & Gerhardt, 2004, 2006; Quan et al., 2005). For hot-pressed ceramic composites, it seems that percolation tends to occur in the 7 to 10 vol% range. One study, which considered creep response, associated these lower and upper bounds with point-contact percolation and facet-contact percolation respectively (de Arellano-Lopez et al., 1998). This range is also consistent with experimental data for electrical percolation (Bertram & Gerhardt, 2010; Mebane & Gerhardt, 2006). Most work investigating percolation is based on electrical response because it is (arguably) much easier to perform the needed experiments compared to those for mechanical response. Electrically, the SiC_w are at least ~9 orders of magnitude more conductive than Al_2O_3. Thus, they are likely to carry much more current than the matrix and the majority of the current through the sample. Therefore, the percolation of the SiC_w is of principal importance in the determination of the composite electrical properties. Composites having such contrast in conductivity between the filler and the matrix undergo a drastic change in electrical response when the conductive filler becomes interconnected within the sample such that a continuous pathway of filler spans the sample. However, percolation is also known to affect mechanical (creep) response (see Section 3.6.2.) and reduce sinterability of composites due to the formation of a rigid interlocking network (Holm & Cima, 1989).

Consideration of this fact raises a question: for a dispersed binary composite (e.g. Al_2O_3-SiC_w), does the percolation threshold depend on the property or process of interest? This question can be reframed in terms of (1) mechanical percolation vs. electrical percolation, or (2) general percolation theory in regards to how one defines a "connection" between two squares on the black-and-white grid of Figure 4a. For electrical percolation in composites of dispersed particles that are much more conductive than the matrix, the prevailing theories(Balberg et al., 2004; Connor et al., 1998; Sheng et al., 1978) generally propose that direct physical contact between the particles is *not* required, and that charge transport takes place by tunneling or hopping across interfiller gaps. Thus, for electrical considerations, one may consider two whiskers to be connected even if they are separated by physical space. For mechanical percolation, the underlying concept of a rigid percolated network implies intimate physical contact between SiC_w spanning the entire sample and that electrical percolation could exist without mechanical percolation. If true, electrical and mechanical percolation thresholds for Al_2O_3-SiC_w and similar composites need not coincide. In order to verify such a difference, a study investigating electrical and mechanical percolation on the same set of samples is required.

3.5.2. Properties of the Spatially Dispersed SiC-Whisker Population

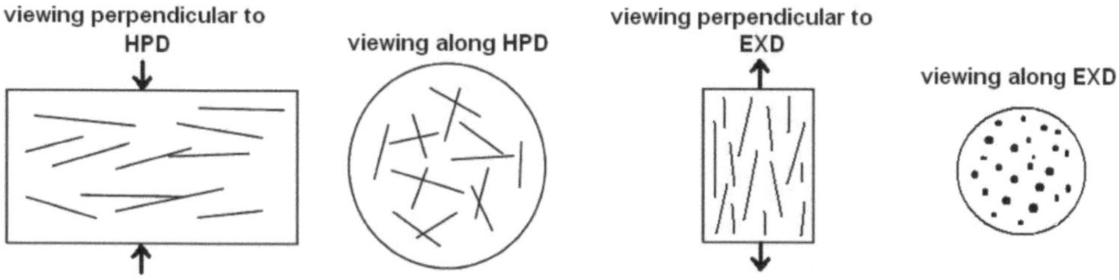

Fig. 6. Schematics showing the preferred orientations of SiC whiskers which result from the hot-pressing and extrusion-based processing methods. The processing directions (HPD and EXD, respectively) are marked by arrows. HP figures after Mebane and Gerhardt, 2006 (John Wiley & Sons).

The whisker sizes and orientations generally affect the properties of interest for the composite applications and so the Al_2O_3-SiC_w structure has often been discussed as such. The ball-milling process often used for mixing the component powders seems to result in a lognormal distribution of whisker lengths peaking around ~10 μm (Farkash & Brandon, 1994; Mebane & Gerhardt, 2006). The preferred whisker orientation depends on the fabrication method. In hot-pressed composites, whiskers tend to be aligned perpendicular to the hot-pressing direction (HPD) and have random orientation in planes perpendicular to the HPD (Park et al., 1994; Sandlin et al., 1992). In extruded samples, the whiskers are expected to be approximately aligned with the extrusion direction (EXD). These preferred orientations are shown in Figure 6. Such material texture generally results in anisotropy in both electrical and mechanical properties and has been shown for hot-pressed samples (Becher & Wei, 1984; Gerhardt & Ruh, 2001).

In consideration of the SiC_w dispersion as the most important aspect of the microstructure, one can characterize the associated trivariate length-radii-orientation distribution with a comprehensive stereological method (Mebane et al., 2006). Other methods also exist for determining the orientation distributions of SiC_w or estimating the overall degree of preferred alignment (texture) and generally result in a unitless orientation factor between 0 and 1. One measurement based on x-ray-diffraction-based texture analysis was effectively correlated to the composite resistivity (C. A. Wang et al., 1998). Another orientation factor based on a different stereological method increased linearly with the length/diameter ratio of the extrusion needle, and thus seemed effective (Farkash & Brandon, 1994). However, for both of these methods, information about the coupling of whisker size and orientation distributions which is known to exist (Mebane et al., 2006) is lost.

3.5.3. Microstructural Axisymmetry

The preferred orientation of SiC_w in hot-pressed and extruded samples has been studied by multiple investigators (Park et al., 1994; Sandlin et al., 1992) and means that the composites tend to be symmetrical around the processing direction (e.g. the HPD or EXD). In other words, both hot-pressed and extruded samples possess a single symmetry axis in regards to the SiC_w distribution and therefore have axisymmetric microstructures. Such composite materials can be considered to have only two principal directions in terms of property anisotropy: (1) the processing direction, and (2) the set of all directions which are perpendicular to the processing direction and are therefore equivalent.

Fig 7. Scanning electron micrographs of the microstructure for an Al_2O_3-SiC_w sample containing 14.5 vol% SiC_w. In (a), the white arrow points along the hot-pressing direction. In (b), the microstructure is viewed along this direction. Part (c) shows the average distance between SiC inclusions along the HPD and perpendicular direction. The inset shows a schematic of the microstructure. *Source:* Bertram & Gerhardt, 2010.

Recently, a simple but useful stereological characterization method was developed and applied to Al_2O_3-SiC_w composite microstructures like those shown in Figures 7a and 7b (Bertram & Gerhardt, 2010). In this method, the distributions of distances between the SiC phase are characterized with stereological test lines as a function of principal direction in the microstructure. We propose that the results implicitly contain information about whisker sizes, orientations, dispersion uniformity and agglomeration and should be generally relevant for transport properties dominated by the SiC phase. For example, anisotropy in average interparticle distance (Fig. 7c) was strongly correlated to electrical-resistivity anisotropy (not shown).

3.5.4. Performance-based Perspective on Composite Structure

For complicated materials such as dispersed-rod composites, it is especially important to remember that structure determines properties and properties determine performance. To meet performance specifications for an application, certain properties must be optimized. After mixing component powders, Al_2O_3-SiC_w composites are usually consolidated and solidified by dry-pressing followed by hot-pressing or extrusion followed by pressureless-sintering. For these materials, the cutting and microwave-heating applications imply that the process engineer often desires the following structural characteristics for the final ceramic:

- a density as close to theoretical as possible (porosity has deleterious effects on properties)
- uniform whisker dispersion (for better toughness, strength, conductivity)
- minimal whisker content to minimize cost while achieving the desired properties
- percolation, for conductive electrical response and improved mechanical response
- "medium" whisker aspect ratios (e.g. 10<length/width<20) to balance the needs for percolation and high sintered density
- no particulate SiC, only whiskers (SiC particulates do not improve properties as much)
- silica- and glass-free (clean) interfaces between the whiskers and the matrix (for toughness)
- a ceramic matrix material that is both inexpensive and environmentally friendly

3.6 Selected Mechanical, Thermal, and Chemical Behavior

3.6.1 Effects of SiC Whiskers on Mechanical Properties and Deformation Processes

Inclusion of SiC whiskers in a ceramic matrix generally increases the material strength but is mainly done to reduce the brittle character of the ceramic. Failure of brittle materials usually results from crack growth, a process driven by the release of elastic stress-strain energy of the atomic network (e.g. the crystal or glass) and retarded by the need to produce additional surface energy (Richerson, 1992). The whiskers tend to increase the fracture toughness and work of fracture by redirecting crack paths and diverting the strain energy which enables crack growth. Toughening mechanisms include modulus transfer, crack deflection, crack bridging, and whisker pull-out. Some examples of these are shown in Figure 8.

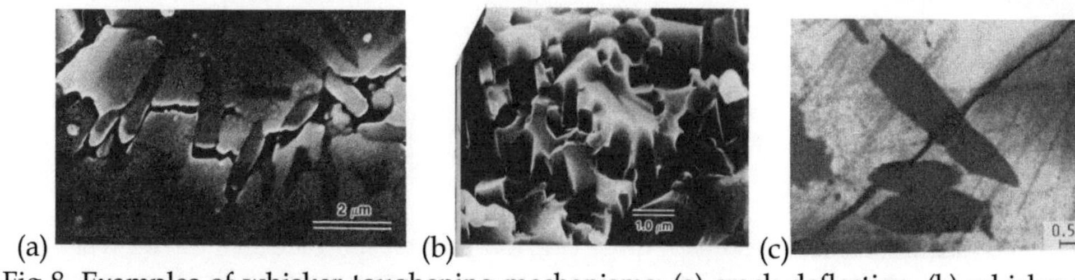

Fig 8. Examples of whisker toughening mechanisms: (a) crack deflection, (b) whisker pull-out, (c) crack bridging. *Sources:* (a) J. Homeny et al., 1990. John Wiley & Sons (b) F. Ye et al., 2000. Elsevier. (c) R.H. Dauskardt et al., 1992. John Wiley & Sons.

In modulus transfer, the stress on the matrix is transferred to the stiffer and stronger whiskers. In crack deflection, cracks are forced to propagate around whiskers due to their high strength, effectively debonding them from the matrix and creating new free surfaces. In crack bridging, whiskers which span the wakes of cracks impart a closing force, absorbing some of the stress-strain energy (concentrated at the crack tip) and thereby reducing the impetus for crack advancement. Some bridging whiskers debond from the matrix or rupture, resulting in pull-out. The associated breakage of interatomic bonds and frictional sliding also increase the work of fracture.

One study (Iio et al., 1989) showed that the inclusion of SiC whiskers up to 40 vol% increased the fracture toughness of pure alumina from ~3.5 up to ~8 MPa·m$^{1/2}$ for samples hot-pressed at 1850°C. However, such large gains only apply when the crack plane is parallel to the hot-pressing direction. This orientation provides for more interactions between the whiskers and the crack in comparison to when the crack plane is perpendicular to the hot pressing direction. In the latter situation, the cracks can avoid the whiskers more easily and only modest improvements in toughness are achieved. It was also found that the fracture strength increases from ~400 MPa to ~720 MPa in going from 0% to 30 vol% SiC_w and decreased as whisker content increased to 40 vol%. The improvement is the result of the increased likelihood of whisker-crack interactions when whisker content is increased. They found that changing the hot-pressing temperature to 1900°C, despite improving densification, significantly reduced the toughness for 40 vol% samples and correlated this to a reduction in crack deflections and load-displacement curves being characteristic of brittle failure with no post-yield plasticity (unlike samples pressed at 1850°C). Scanning and transmission electron microscopy revealed different whisker-matrix interfacial structure for the different temperatures and suggested that matrix grain growth at 1900°C led to whisker agglomerations (Iio et al., 1989). The authors concluded that the distribution of SiC whiskers is of principal importance in determining the strength and toughness of the material and that whisker agglomerations may act as stress concentrators adversely affecting toughening mechanisms and initiating failure. From the 1850°C pressing temperature, analysis revealed failure initiating at clusters of micropores from incomplete densification.

Modeling work indicates that whisker bridging in the wake of the crack tip is the most important toughening contribution(Becher et al., 1988, 1989). Two different modeling approaches (stress intensity and energy-change) were used to determine the following relationship between the bridging-based toughness improvement (dK_{Ic}) imparted to the composite from the whiskers and several parameters:

$$dK_{I_c} = \sigma_{w,f}\left(\frac{V_w r_w}{6(1-\nu^2)}\frac{E_c}{E_w}\frac{G_m}{G_i}\right)^{\frac{1}{2}} \quad (9)$$

Here, r_w V_w and E_w are the radius, volume fraction and elastic modulus of the whiskers, respectively. The whisker strength at fracture is $\sigma_{w,f}$. The Poisson ratio and elastic modulus of the composite are ν and E_c, respectively. Strain-energy release rates are given by G, where subscripts indicate the matrix (m) and the whisker-matrix interfaces (i).

3.6.2. Creep Response and High-Temperature Chemical Instability

Prolonged high-temperature exposure to mechanical stress leads to creep, degradation of Al_2O_3-SiC_w composite microstructures, and worsening of mechanical properties. Creep in these composites has been studied at temperatures from 1000°C to 1600°C and resistance to creep is generally much better than that of monolithic alumina (Tai & Mocellin, 1999). Figure 9a shows that, for a given stress level, the creep strain rate at 1300°C is much reduced if whisker content is increased. However, stressing for long times at elevated temperature has resulted in composite failure well-below the normal failure stress at a given temperature. For example, one study found failure occurring at 38% of the normal value after stressing in flexure for 250 hours at 1200°C (Becher et al., 1990).

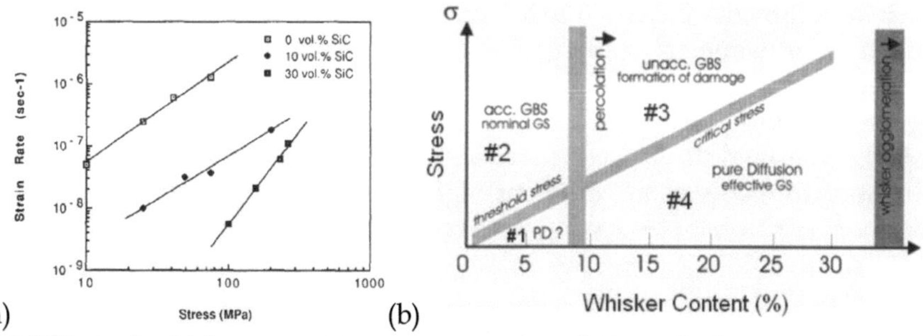

(a) (b)

Fig. 9. (a) Effect of whisker content on stress-strain relations during compressive creep at 1300°C. (b) Creep deformation mechanism map showing the effects of stress and whisker content on the dominant deformation mechanism. *Sources:* (a) Nutt & Lipetzky, 1993. (b) De Arellano-Lopez et al., 1998 (region numbers added). From Elsevier, with permission.

Creep in polycrystalline ceramics often proceeds by grain boundary sliding (Richerson, 1992). Due to rigid particles (whiskers) acting as hard pinning objects against grain boundary surfaces, such sliding is impeded in alumina-SiC_w composites and creep resistance is improved (Tai & Mocellin, 1999). For such systems, one model (Lin et al., 1996; Raj & Ashby, 1971) predicts the steady state creep strain rate ($\dot{\varepsilon}$) to be

$$\dot{\varepsilon} = C\frac{\sigma_a{}^n}{d^u r_w^q V_w}\exp\left(-\frac{Q}{RT}\right) \tag{10}$$

where C is a constant, σ_a is the applied stress, d is the grain size, r_w is the whisker radius, V_w is the whisker volume fraction, Q is the apparent activation energy, R is the gas constant, and T is the absolute temperature. The exponents n, u, and q are phenomenological constants. The stress exponent n is of particular interest because it has been correlated to the dominant deformation mechanism (de Arellano-Lopez et al., 2001; Lin et al., 1996; Lin & Becher, 1991).

The qualitative dependence of deformation mechanism on stress, temperature, and whisker content is best understood by considering the deformation map of Figure 9b, which was developed after many years of research on the creep response (de Arellano-Lopez et al., 1998). The applicability of this map does not seem to depend on whether or not creep is conducted in flexure or compression. In region #1, the dominant mechanism is Liftshitz grain boundary sliding, which is also called pure diffusional creep (PD) in the literature. In this process, grains elongate along the tensile axis and retain their original neighbors. Above the threshold stress, which relates to whisker pinning, Rachinger grain boundary sliding (GBS) becomes the dominant mechanism (region #2). In this process, grains retain their basic shapes and reposition such that the number of grains along the tensile axis increases. Grain rotation has been observed in some cases (Lin et al., 1996; Lin & Becher, 1990). Both the Liftshitz and Rachinger processes are accommodated by diffusion that is believed to be rate-limited by that of Al^{3+} ions through grain boundaries (Tai & Mocellin, 1999).

When whisker content exceeds the percolation threshold, the creep rate does not depend on the nominal alumina grain size but rather on an effective grain size which is defined by the volumes between whiskers (de Arellano-Lopez et al., 2001; Lin et al., 1996). The percolated whisker network provides increased resistance to GBS and this relates to the critical stress, below which deformation proceeds by PD (region #4). Increasing the stress above this critical value marks a transition to grain boundary sliding that can no longer be accommodated by diffusional flow and requires the formation of damage (region #3). Both the critical and threshold stress decrease with increasing temperature. Also, one should note that the boundaries between the different regions are not strict. For example, Rachinger sliding has occurred in samples having whisker fractions as high as 10 vol% (de Arellano-Lopez et al., 2001).

In region #3, whiskers seem to promote damage such as cavitation, cracking, and increased amounts of silica-rich glassy phases at internal boundaries and cavities (de Arellano-Lopez et al, 1990; Lin & Becher, 1991). Examples of such damage are shown in Figures 10a-b. The cavitation and cracking is believed to be result of whiskers acting as stress concentrators and the presence of glass pockets is from the thermal oxidation of whiskers (de Arellano-López et al., 1993). High stress results in less grain-boundary sliding and promotes cracking, separation of matrix grains from whiskers, and the formation of cavities within glass pockets at whisker/whisker and whisker/matrix interfaces (Lipetzky et al., 1991). Whisker clusters containing cavitation and crack growth have also been reported for higher (≥25 vol%) loadings of SiC_w (O'Meara et al., 1996).

Fig. 10. (a) TEM of glass-filled cavities which formed near SiC_w in a 20 vol% SiC_w composite during creep (b) TEM of cavitation and cracking near whiskers (w) during creep (c) Whisker hollowing (red arrows) on a fracture surface from the bulk after annealing in air at 1000°C for 1 hr. *Sources*: (a-b) de Arellano-Lopez et al., 1993. John Wiley & Sons (c) S. Karunanithy, 1989 (arrows added). From Elsevier, with permission.

The ion diffusion that assists grain boundary sliding is believed to be accelerated by the presence of intergranular glassy regions and further complemented by the viscous flow of this glass. The amount of glassy phase in as-fabricated composites is generally small but increases in size after creep deformation in air ambient due to oxidation of SiC whiskers, resulting in a SiO_2-rich glassy phase surrounding the whiskers. Such whisker oxidation has been observed to occur in the bulk of samples and is attributed to residual oxygen on whisker surfaces, enhanced oxygen diffusion through already-formed glass, and short-circuit transport through microcracks. Also, much of the glass found in the bulk is believed to originate from oxidation scales on the composite surfaces flowing into internal interfaces. During creep, the glassy phase apparently seeps into grain boundaries, interfaces, and triple grain junctions (de Arellano-Lopez et al., 1990; Lin et al., 1996; Lin & Becher, 1991; Nutt, 1990; Tai & Mocellin, 1999).

It is believed that glass formation begins with the oxidation of SiC. This yields silica and/or volatile silicon monoxide products which subsequently react with alumina to produce aluminosilicate glass. This glass has been observed to contain small precipitates of crystalline mullite and is believed to accelerate oxygen diffusion compared to pure silica glass. Such an acceleration might explain the overall fast oxidation rate of the composite, which exceeds that of pure SiC by more than an order of magnitude (Jakus & Nair, 1990; Karunanithy, 1989; Luthra & Park, 1990; Nutt et al., 1990). The following reactions can account for these transformations:

$$SiC(s) + 3/2\, O_2(g) \rightarrow SiO_2(l) + CO(g) \quad (11)$$
$$SiC(s) + O_2(g) \rightarrow SiO(g) + CO(g) \quad (12)$$
$$2SiC(s) + 3Al_2O_3(s) + 3O_2(g) \rightarrow 3Al_2O_3 \cdot 2SiO_2(s) + 2CO(g) \quad (13)$$
$$SiC(s) + O_2(g) \rightarrow SiO_2(l) + C(s) \quad (14)$$

As indicated above, the presence of oxygen in the atmosphere affects creep and results in higher creep rates compared to inert gas. Another process that may make use of oxygen was observed after heat treatment in air at 1000°C: whisker-core hollowing. Core hole diameters ranged from 200 to 500 nm and were attributed to metallic impurities and the decomposition of oxycarbide in whisker-core cavities that are known to exist in as-fabricated whiskers. The core-hollowing phenomenon was seen on fracture surfaces after breaking heat-treated samples with a hammer, suggesting its occurrence throughout the bulk of the sample. Such a fracture surface is shown in Figure 10c (Karunanithy, 1989).

Generally, the oxidation occurring on the composite surface as a result of the ambient has been found to depend on the partial pressure of the oxidizing agent. In most studies, molecular oxygen is the oxidizing agent of interest. When the oxygen partial pressure is low, active oxidation occurs and Reaction 12 is operable: SiC is lost into the gas phase. Otherwise, passive oxidation occurs in the form of Reaction 14, which results in free carbon and liquid SiO_2, which then mixes with alumina to form aluminosilicate glass, from which mullite often precipitates. When passive oxidation occurs, this can result in surface scales and crack blunting/healing which can mitigate the degradation of mechanical properties (Shimoo et al., 2002; Takahashi et al., 2003). Another study (Kim & Moorhead, 1994) showed that the oxidation of SiC at 1400°C may significantly increase or decrease the strength depending on the partial pressure of oxygen.

In air at 1300-1500°C, (Wang & Lopez, 1994) oxidation results in the formation of layered scales, consisting of a porous external layer on top of a thin ($0 < 6$ μm) layer of partially oxidized SiC_w and glassy phases. Intergranular cracking occurred in this thin layer and was attributed to volume changes associated with the oxidation reaction. The rates of scale-thickening and weight gain were parabolic and rate constants and activation energies were calculated and ascribed to diffusion of oxidant across the porous region. At 600-800°C, oxidation obeys a linear rate law for the first 10 nm of oxide growth (P. Wang, et al., 1991). From 1100-1450°C, oxidation has also been found to degrade the elastic modulus of samples. It has also been found that the presence of nitrogen in the ambient at elevated temperatures may similarly result in the chemical degradation of the SiC_w and composite properties (Peng et al., 2000).

3.6.3 Thermal Conductivity

Thermal conductivities of hot-pressed Al_2O_3-SiC_w composites along the hot-pressing direction typically range from ~35-50 W/m·K for 20-30 vol% composites at 300 K and the highest ever reported was 49.5 W/m·K (Collin & Rowcliffe, 2001; Fabbri et al., 1994; McCluskey et al., 1990). Such volume fractions of whiskers likely translate to SiC_w percolation, but the associated thermal conductivity values are not much higher than those of the Al_2O_3 (~30 W/m·K). This can be understood in terms of modeling results for conductive nanowire composites which show that percolation does not lead to a dramatic increase in thermal conductivity because of phonon-interface scattering (Tian & Yang, 2007). In an experimental study of Al_2O_3-SiC_w composites, it was estimated that the SiC_w phase has a thermal conductivity that is 90-95% lower than that of ideal cubic SiC based on a model for overall composite thermal conductivity and planar defects spaced 20 nm apart in the SiC_w could account for this discrepancy (McCluskey et al., 1990). Such a spacing agrees well with stacking fault densities observed in the whiskers (Nutt, 1984). It has also been found that the preferred orientation of the SiC_w results in the composites having significant anisotropy in thermal diffusivity and conductivity (Russell et al., 1987).

3.6.4 Cycling Fatigue from Thermal Shock

In ceramic matrix composites, fatigue during thermal shock cycling is believed to be the result of the thermal-gradient imposed stresses and thermal expansion mismatch between the two phases. The addition of silicon carbide whiskers results in significant improvements in thermal shock resistance compared to monolithic alumina. One study found that a thermal shock quench with a temperature difference of 700°C caused a 61% decrease in flexural strength for pure alumina (Tiegs & Becher, 1987). For samples having 20 vol% SiC_w, the strength loss after ten similar quenches was only 13% of the original value. The improvement is generally attributed to whiskers toughening effects on shock-induced microcracks. A reduction in thermal gradient due to an increase in composite thermal conductivity provided by the whiskers is also believed to contribute to the improved thermal shock resistance but is expected to be of less importance (Collin & Rowcliffe, 2001).

Another study (Lee & Case, 1989) found that the amount of thermal shock damage saturates as a function of the number of increasing thermal shock cycles and that the saturation value depends strongly on the shock temperature difference, ΔT. This is shown in Figure 11, which displays the effects of these factors on elastic modulus. Since crack density is expected to vary inversely with the elastic modulus (Budiansky & Oconnell, 1976; Salganik, 1973), Figure 11 suggests that ΔT is more important than the number of cycles in determining the extent of damage.

The modulus values were obtained by the sonic resonance method (Richerson, 1992) which is sensitive to the distribution of cracks in each sample. This method allowed for a reduction in sample size in comparison to that needed for measurements based on the stochastic process of brittle fracture. It was possible to reuse samples by annealing out damage at 950°C and near-full recovery of elastic properties was achieved (Lee & Case, 1989). However, only samples with 20 vol% whisker loading were investigated in this work.

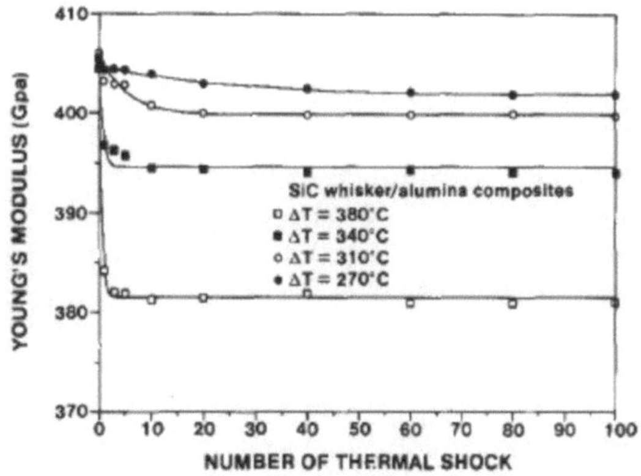

Fig. 11. Dependence of composite elastic modulus on thermal-shock quenching temperature difference, ΔT, and the number of quenches performed. *Source:* Lee & Case, 1989. Elsevier.

In contrast to the thermal-shock work on hot-pressed composites discussed above, composites made by extrusion and sintering were reported to withstand shocks of almost 500°C and lost only 12% of strength after 400 shocks of 230°C. Strength slowly decreased with cycling and did not plateau (Quantrille, 2007).

3.7. Electrical Response

3.7.1. Relevant Formalisms

In general, complex relative dielectric constant ε_r^* of a material is composed of a frequency-dependent real part ε_r' (the dielectric constant) and imaginary part $j\varepsilon_r''$ and may be written as

$$\varepsilon_r^* = \varepsilon_r' - j\varepsilon_r'' \tag{15}$$

where $-\varepsilon_r''$ is the dielectric loss, $j = \sqrt{-1}$ and the subscript "r" indicates relation to the vacuum permittivity ε_0, i.e. the complex permittivity of a general medium is $\varepsilon^* = \varepsilon_0\varepsilon_r^*$. The complex permittivity is related to the complex impedance ($Z^*=Z'-jZ''$) via

$$Z^* = 1/j\omega C_0\varepsilon_r^* \tag{16}$$

where ω is the radial frequency and C_0 is the geometric capacitance. The real impedance (Z') may be converted to resistivity or conductivity after geometric normalization. The volumetric microwave-heating rate of a material at a particular frequency depends mainly on the dielectric loss and is given by

$$\left(\frac{dT}{dt}\right)_V = \frac{2\pi f E^2\left(-\varepsilon_r''\right)\varepsilon_0}{DC_p} \tag{17}$$

where f is the frequency, E is the magnitude of electric field, C_p is the heat capacity at constant pressure, and D is the material density (Bertram & Gerhardt, 2010; Metaxas & Meredith, 1983). Thus, $-\varepsilon_r''$ is the material property that determines the power delivered per unit volume (i.e. the numerator in Equation 17) and generally includes loss contributions from all forms of frequency-dependent polarization or resonance and the dc conductivity σ_{dc} of the material, i.e.

$$(-\varepsilon_r'') = (-\varepsilon_r'')^{di} + (-\varepsilon_r'')^{ion} + (-\varepsilon_r'')^{el} + (-\varepsilon_r'')^{MW} + \frac{\sigma_{dc}}{2\pi f \varepsilon_0} \qquad 18)$$

Here, the superscripts refer to dipolar (di), ionic (ion), electronic (el) and Maxwell-Wagner (MW) polarizations and the last term in the sum follows from the Maxwell-Wagner theory of a two-layer condenser (Metaxas & Meredith, 1983; von Hippel, 1954). The complex dielectric constant resulting from a Maxwell-Wagner interfacial space-charge polarization $(\varepsilon_r^*)^{MW}$ can be written as

$$(\varepsilon_r^*)^{MW} = \varepsilon_\infty' + \frac{\varepsilon_s' - \varepsilon_\infty'}{1+(j\omega\tau)^{1-\alpha}} - j\frac{\sigma_{dc}}{\omega\varepsilon_0} \qquad (19)$$

where ε_∞' and ε_s' are the real parts of the complex relative dielectric constant at the high- and low-frequency extremes, $\omega=2\pi f$, τ is the characteristic relaxation time, and $(1-\alpha)$ is an empirical fitting exponent which accounts for a distribution of relaxation times (Bertram & Gerhardt, 2010; Runyan et al., 2001b). For an ideal Maxwell-Wagner polarization with a single relaxation time, $\alpha=0$. The explicit effects of α on the real and imaginary parts of ε_r^* were determined more than 60 years ago (Cole & Cole, 1941). It is also common to apply an empirical exponent (k) to the frequency (ω) of the $j\sigma_{dc}/\omega\varepsilon_0$ term of Equation 19 to better fit the low-frequency part of the loss spectrum.

The standard frequencies for microwave oven operation are 2.45 and 0.915 GHz. At these frequencies, loss-based heating in the composites will occur preferentially in the SiC_w phase compared to Al_2O_3 due to much greater values of $-\varepsilon_r''$. One study found that loss values are ~180 times greater for SiC compared to Al_2O_3 (Basak & Priya, 2005). However, the dielectric properties of SiC are known to vary widely depending on the crystalline structure and impurity content. Notably, the SiC_w made by Advanced Composite Materials LLC are known to be exceptionally lossy (Quantrille, 2007).

3.7.2. Preliminary Work on SiC_w Composites

In 1992, it was found that the resistance of the composite decreased linearly with temperature. This was attributed to the semiconductive nature of SiC (Zhang et al., 1992). On a similar system, for which the matrix was mullite instead of alumina, the dc conductivity and frequency-dispersions of the dielectric constant depended on measurement orientation relative to preferred orientation of whiskers (Gerhardt & Ruh, 2001). This paper also showed the effect of volume fraction on dielectric-property spectra of the composites in the 10^2-10^7 Hz frequency range. Later, the orientation dependence was confirmed for the Al_2O_3-SiC_w system (Mebane & Gerhardt, 2004). Both studies noted that representative impedance spectra contain two semicircles in the complex impedance plane and that dc conductivity increased with whisker volume fraction.

In another work, SiC_w composite properties were measured in the 2-18 GHz range with a coaxial airline system (Ruh & Chizever, 1998). Adding ~30 vol% SiC_w to mullite greatly increased the dielectric loss from ~0.0-0.2 up to 6-10 and the dielectric constant increased by 2-3 times. Adding 32.4 vol% SiC_w to a spinel matrix increased the loss from ~0 to 13-20 and increased the dielectric constant by 5-6 times. These results suggest that the matrix can play a significant role in macroscopic electrical response.

3.7.3. Recent Investigations Combining Measurements and Modeling

For composites with SiC_w, multiple interpretations have been proposed for the two complex-impedance semicircles which are generally observed during parallel-plate measurements on percolated samples. Presently, the accepted interpretation of this result is that the higher-frequency semicircle is associated with the bulk conductivity through the whisker network and the lower frequency semicircle relates to blocking processes at the electrodes (Mebane & Gerhardt, 2006). The establishment of these semicircles as being related to specific processes has allowed for modeling of the associated contributions to the electrical response.

3.7.4. Bulk Electrical Response

Recently, the higher-frequency impedance semicircle was considered to be a function of the trivariate length-radius-orientation distribution of the whiskers measured by stereology (Mebane & Gerhardt, 2006). A computer simulation using such distribution information from stereological measurements was used to convert the experimental whisker network in simulation space to a resistor network based on an estimate of $SiC_w/Al_2O_3/SiC_w$ interfacial conductance assuming a tunneling-mechanism-based "shorting distance" for defining interwhisker connectivity. Kirchoff's Laws were then applied to the equivalent circuit generated for different volume fractions of SiC_w. The results were then converted to effective conductivities for the simulated whisker networks and these were compared to the experimental dc-conductivities. It was found that the simulation overpredicted the rate of conductivity-increase with whisker content, which suggests that a conductivity-limiting factor was not accounted for in the simulation. It was also found that the size of the low-frequency impedance semicircle depended on electrode area but not sample thickness, thereby implying that this semicircle was related to the electrodes. In this work, the percolation threshold was experimentally determined to be ~9 vol%.

In another work, the real and imaginary parts of the dielectric response of hot-pressed Al_2O_3-SiC_w were concurrently modeled using a similar formalism (Bertram & Gerhardt, 2010). In this work, ac electrical measurements were conducted over a wide frequency range and indicate that interfiller interactions are probed between ~0.1-1 MHz and ~1 GHz. The response changed from dielectric to conductive across the percolation threshold (5.8-7.7 vol%), as shown in Figure 12a. Non-percolated samples were characterized by a Maxwell-Wagner interfacial polarization with a broad loss peak indicating a distribution of relaxation times and a dc-conductivity tail which was more prominent for samples nearer to the percolation threshold. These features are shown in Figure 12b, along with the real part of the relaxation. Modeling the response based on Equation 19 revealed that the Cole-Cole non-ideality exponent α was larger for the hot-pressing direction. Compared to earlier investigations, the microstructure of the whisker network was interpreted with a different perspective, i.e. as two distributions of interfiller distances along the two principal directions of the axisymmetric microstructure. The orientation and composition dependencies of the relaxation time were explained in terms of the different average distances for high-frequency capacitive coupling between fillers, i.e. SiC-Al_2O_3-SiC interactions. This distance was found to be shorter along the microstructural symmetry axis (i.e. the HPD) compared to the perpendicular direction, as seen in Figure 7c. But, Figure 12a shows that the resistivity was actually higher along the HPD. The inset of Figure 12a shows the relation between the orientation-based differences in average distance and dc resistivity. This trend could be modeled as cubic or exponential and suggests that the whisker network becomes increasingly isotropic as volume fraction increases.

The relevance of percolation to the composition dependence and the relevance of the Maxwell-Wagner interfacial polarization model for the dielectric response of SiC_W composites has been known for several years (Gerhardt & Ruh, 2001). The frequency dependence of the dielectric constant for a similar system with SiC platelets has also been modeled with a Maxwell-Wagner formalism modified with a Cole-Cole exponent to account for non-ideality in the experimental response (Runyan et al., 2001b).

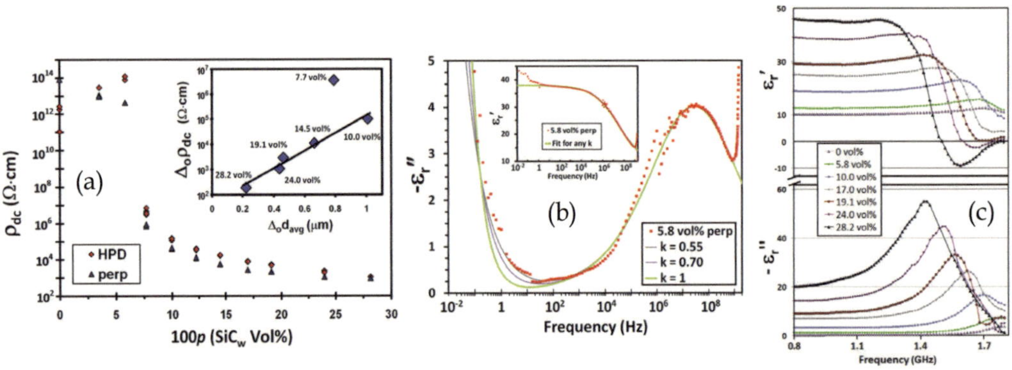

Fig 12. (a) Percolation curves for dc resistivity in the hot-pressing direction (HPD) and perpendicular direction. Inset shows the relation between the direction-based differences in resistivity and inter-SiC distance from Figure 7c. (b) Dielectric loss spectrum of 5.8 vol% SiC_W composite fitted with Equation 19 and various *k* values. (c) Real and imaginary parts of the observed dielectric resonance. All parts a-c from Bertram & Gerhardt, 2010.

The dependence of dc conductivity on SiC_W content for percolated samples fit well to models for fluctuation-induced tunneling theory. The percolation transition from insulating to conducting caused the Maxwell-Wagner dc-conductivity tail to greatly increase in prominence such that it concealed the related dielectric loss peak via superposition (not shown). The frequency-dependent dielectric constant and loss were generally smaller in the HPD compared to the perpendicular direction (not shown).

Also, a damped resonance was observed between 1.4 - 1.7 GHz and was found to shift to lower frequency and increase in magnitude as whisker content increased, as shown in Figure 12c. However, the source of the resonance has not been conclusively established. It may relate to dopants in the SiC_w, sintering-additive reaction products at grain boundaries, long-range interactions characteristic of conductive-stick composites, or an anomalous instrumental effect. Generally speaking, the commercial application of these composite materials in microwave heating is probably influenced by the resonance, the relaxation, and the dc-conductivity tail. All of these are related to the SiC content.

3.7.5. Electrode/Contact Electrical Response

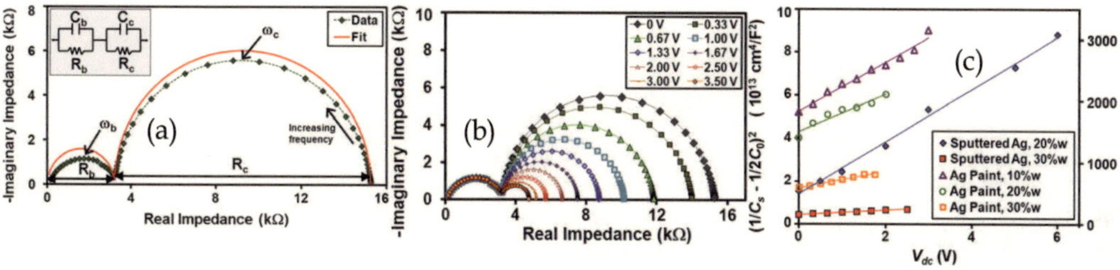

Fig 13. (a) Typical complex impedance data for percolated Al_2O_3-SiC_w material, with a fit based on the equivalent-circuit model in the inset. (b) Effect of dc bias on the complex impedance data. (c) Effect of dc bias on the Mukae capacitance function, revealing the applicability of symmetrical Schottky blocking at the electrodes. *Source:* Bertram & Gerhardt, 2009. Copyright American Institute of Physics.

The electrical response at the electrodes has also been studied in detail (Bertram & Gerhardt, 2009). Electrical contacts to Al_2O_3-SiC_w composites are manifested in the complex impedance plane by a semicircle that is well-modeled by a parallel resistor-capacitor circuit having a characteristic relaxation time. This appears in Figure 13a on the right side, next to the higher-frequency bulk-response semicircle on the left. As dc bias is increased, the bulk response (related to the whisker network) is unaffected and the electrode semicircle shrinks, as shown in Figure 13b. The dc capacitance-voltage data associated with the electrode semicircle evidenced that the underlying physical process was blocking by symmetrical Schottky barriers (at the opposing electrodes). Figure 13c shows this with straight-line fits to the bias-dependent capacitance function derived for symmetrical Schottky blocking (Mukae et al., 1979).

The effect of the bulk electrode material was also studied. Compared to sputtered metal electrodes, samples having silver-paint electrodes had inferior properties, i.e. higher values of area-normalized resistance and lower values of area-normalized capacitance. These differences were attributed to more limited interfacing between metal particles and whiskers on the composite surface for painted samples. Sputtered contacts of Pt were superior to those of Ag, suggesting a dependence on metal work function. Charge carrier concentrations were estimated to be 10^{17}–10^{19} cm^{-3} in the whiskers and related to *p*-type dopants in the SiC. For all types of electrodes, the electrode semicircle was unstable at room temperature and grew with time. Reducing the whisker content in the composite increased the resistance and decreased the capacitance associated with the electrodes by factors that were agreeable with the percolation-network sensitivity proposed above. Also, the dc-bias dependencies of both the resistance and relaxation time associated with the electrode semicircle were exponential and were related through the equivalent circuit model. However, for sputtered electrodes these exponential dependences are often so weak that they appear linear. This is because sputtered contacts contribute little to the total resistance and thus (by Kirchoff's Voltage Law) only a small fraction of the applied bias is applicable to the electrode response. Correction factors were needed to obtain reasonable estimates of the Schottky barrier height as 0.2-1.6 eV for sputtered Ag. Figure 14c shows a proposed band diagram for Schottky barriers at the opposing electrical contacts, assuming p-type conductivity dominates the bulk whisker network.

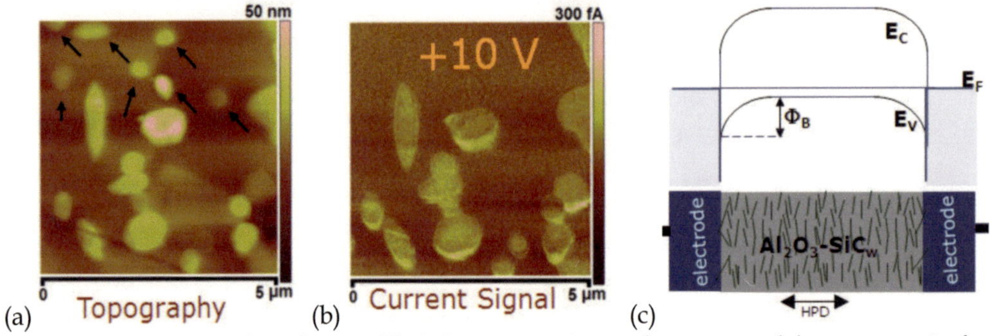

(a) (b) (c)

Fig. 14. (a) Topography of 20 vol% SiC_w composite surface scanned by an atomic force microscope. (b) Current signal from a scan on the same area using a 10 V dc bias, showing the absence of the SiC_w marked with arrows in 'a'. (c) Simple equilibrium band diagram model for symmetrical Schottky barriers at the electrodes of bulk samples. *Source for (a-b):* Bertram & Gerhardt, 2009. Copyright American Institute of Physics.

3.8. Effects of Whisker Defect Structures and Interfaces on Various Properties

In Al_2O_3-SiC_w composites, the whiskers are the most important phase because they have the greatest influence on the composite properties. These properties can therefore be expected to significantly depend on defects within the SiC whiskers and whisker interfaces with the matrix and each other. However, when incorporated in composites, it may be difficult to separate these effects from those related to characteristics of the microstructure previously described in Section 3.5. Nevertheless, such issues are worthy of discussion.

To this end, one must first know that the hot-pressing process and the sintering process which follows extrusion are conducted at high temperatures (~1600-1800°C). There is reason to believe that the chemical compositions of the SiC_w are altered during such high temperature processes and this is of interest because it might affect intra- and inter-whisker transport and microwave dielectric loss. Energy dispersive x-ray spectroscopy of whisker surfaces in composites hot-pressed at 1600°C suggest aluminum diffusion from Al_2O_3 into the whiskers to a concentration of 1-2 atomic percent (Zhang et al., 1996). This suggests very high p-type (semiconductive) doping and is expected to affect both electrical and thermal conductivity. Since this doping level persisted throughout the full measurement depth of ~350 nm into the whisker, a value on the order of the typical whisker diameter, one might infer that this doping is achieved throughout entire whisker cross sections. Evidence of aluminum doping has also been found in similar composites of liquid-phase sintered silicon carbide which used alumina as a sintering aid. However, in these studies the Al doping was only found around the rims of the SiC grains and did not persist into the grain cores (Sanchez-Gonzalez et al., 2007).

In another study, it was found that high-frequency dielectric loss of certain SiC composites decreases with increasing concentrations of aluminum and nitrogen dopants in the SiC (Zhang et al., 2002). These results can be understood in terms of the proposed underlying mechanism of dielectric loss in SiC: dipole relaxation and/or reorientation of vacancy (V_C-V_{Si}) and antisite (Si_C-C_{Si}) defect pairs. Smaller concentrations of such defects are expected as temperature decreases or impurity concentration increases in the SiC_w.

Another possible electrical issue is scattering of charge carriers at stacking faults within the SiC_w. One might expect that the short-spacing between faults would severely limit the mean free path of carriers and therefore the mobility and conductivity of the SiC_w. Also, band-structure calculations reveal that stacking faults in SiC introduce states inside the bandgap which are localized at about 0.2 eV below the conduction band (Iwata et al., 2002). Due to this position relative to the band edges, it was concluded that stacking faults hinder electron transport in n-type SiC but not hole transport in p-type SiC. As they noted, this assertion is the only way to rationalize other experimental results which show that SiC electrical resistivity strongly depends on transport direction relative to stacking-fault orientation for n-type SiC but not for p-type SiC (Takahashi et al., 1997). This result may be applicable to Al_2O_3-SiC_w composites since the SiC_w might be doped during high-temperature composite fabrication. Since SiC is a wide-bandgap semiconductor, one would naturally expect the doping of the SiC whiskers to determine their electrical conductivity and influence the electrical response of composites which contain them.

In a study of the internal potential fluctuations associated with the stacking faults in SiC nanowires, the wires were considered as alternating stacks of perfect crystal and defective regions based on stacking-fault observation during TEM investigation (Yoshida et al., 2007). The position-dependence of the inner potential of the SiC rods was measured by electron holography and correlated to the perfect-crystal and defective regions. The perfect crystal regions had significantly lower inner potential. They speculated that Al-P impurity segregation occurred and led to the self-organized formation of p-n junctions along the lengths of the SiC nanowires. If true, this would certainly affect the electrical response of the SiC wires themselves and likely also that of ensembles of such wires in composites.

Additionally, at high temperatures, β-SiC is known to undergo a phase transformation to form the α-SiC and this is relevant to ceramic-processing aspects of making the composites because the two polytypes tend to exhibit different morphologies. One study investigated this process in the 1925-2000°C range (Nader et al., 1999). The transformation rate decreased as the reaction proceeded such that the kinetics are adequately described by a nucleation-and-growth equation for α-SiC grain growth based on the general equation developed by Johnson and Mehl for grain growth processes. Also, the reaction was found to be "slow" (only 94% completion after 14 hours) and was affected by the chemical composition (impurities and non-stoichiometry) of the SiC crystallite.

It has also been found that coating SiC_w prior to inclusion in composites can provide improved properties and cutting-tool performance. Such high-performance tools may be obtained from Greenleaf Corporation. In one study, precoating the SiC_w with alumina by an aqueous precipitation technique allowed for better dispersion during ultrasonic mixing in ethanol (Kato et al., 1995). The more-homogeneous mixing resulted in improved sintered density and improved mechanical strength. In another study, carbon coatings were applied to SiC_w and the resulting composites exhibited improved fracture resistance, possibly due to reduced residual thermal stresses from post-sintering cool down (Thompson & Krstic, 1993).

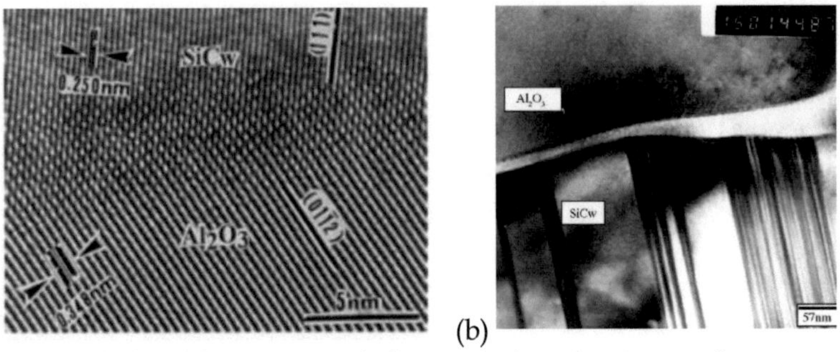

(a) (b)

Fig 15. (a) High-resolution TEM image of clean whisker-matrix interface in a composite made from whiskers etched in HF prior to hot-pressing. (b) TEM image showing a glassy coating formed after hot-pressing due to residual oxygen on the whisker surface. *Sources:* (a) F. Ye et al., 2000. (b) V. Garnier et al., 2005. Copyright Elsevier.

The whisker-matrix interface seems to determine to what extent the thermal-expansion difference between Al_2O_3 and SiC_w dictates the stress distributions in the composites. Because the alumina matrix has a higher thermal expansion coefficient compared to the SiC whiskers, the matrix is in tension and the SiC_w are in compression upon cooling from hot-pressing temperatures (Becher et al., 1995). However, it has also been proposed that matrix$\rightarrow SiC_w$ stress transfer can explain observations of (tensile) whisker fracture instead of pullout: specifically, that the higher-modulus SiC whiskers unload the matrix and reduce stress around defects in the composite microstructure and result in an overall improvement in composite mechanical properties (Björk & Hermansson, 1989). The mechanical stress distribution within the SiC whiskers might also affect electrical response of composites because elastic strain was found to result in changes in electrical response of β-SiC films and used to explain the observed back-to-back Schottky-barrier response (Rahimi et al., 2009).

It thus seems that the interfaces of whiskers with neighboring materials are of considerable importance. Even without coatings, it has been found that treating whisker surfaces to control surface chemistry prior to composite fabrication can increase final-composite toughness from 4.0 up to 9.0 MPa·$m^{1/2}$. This study used different gas mixtures during heat treatments of whiskers and it was found that reducing the residual oxygen and silica on whisker surfaces resulted in the highest fracture toughness (Homeny et al., 1990). Free carbon and silicon nitride on surfaces resulted in lower toughness, and silicon dioxide resulted in the lowest. In another work, pretreating whiskers with an HF acid etch caused surface smoothening and a reduction in residual oxygen and other impurities on whisker surfaces (Ye et al., 2000). This led to a relatively clean SiC-alumina interface in the fabricated composites, as shown in Figure 15a. Without such cleaning, it is common for amorphous interfacial phases composed of Al, Si, and O to form around the whiskers during hot pressing and this generally results in composites exhibiting inhibited whisker pull-out, crack bridging, and deflection due to excessively-high interfacial bonding strength. An example of such a glassy phase coating a whisker is shown in Figure 15b and was correlated to a fracture toughness reduced by 22% (Garnier et al., 2005).

4. Conclusions and Future Work

A great deal of research has been conducted on Al_2O_3-SiC_w composites to understand the basic physical processes on which their successful applications in cutting-tool inserts and microwave cooking are based. The initial reason for development of these materials, an application in ceramic engines, did not appear to pan out due to insufficient stability under conditions of high-temperature and stress. Future investigators might attempt to improve properties of the composites by applying principles covered in this chapter (and elsewhere) to more carefully control the microstructure of the composites. It has already been seen that use of whisker coatings can significantly improve mechanical performance. As the electrical application has motivated the most recent work, it may be interesting to investigate the effects of sintering additives and controlled whisker doping on electrical response. Alternatively, one might be able to manipulate the SiC_w aspect-ratio distribution to reduce the percolation threshold. Finally, improvements in microwave sintering of larger ceramic-composite parts having tuned dielectric properties might lead to interesting applications.

Acknowledgements

This work was funded by the National Science Foundation under DMR-0604211. BDB received some additional support from a Presidential Fellowship from Georgia Tech and a Shackelford Fellowship from Georgia Tech Research Institute. Advanced Composite Materials LLC and Greenleaf Corporation are thanked for images and sharing expertise about the applications.

5. References

ACM website. http://www.acm-usa.com/whiskers-properties.html. 10/13/2010.

Amateau, M.; Stutzman, B; Conway, J. & Halloran, J. "Performance of laminated ceramic composite cutting tools." *Ceramics International,* Vol. 21, No. 5, (1995) 317-323.

Balberg, I.; Anderson, C.; Alexander, S. & Wagner, N. "Excluded volume and its relation to the onset of percolation." *Physical Review B,* Vol. 30, No. 7, (1984) 3933-3943.

Balberg, I.; D. Azulay, D.; Toker, D. & Millo, O. "Percolation and tunneling in composite materials," *International Journal of Modern Physics B,* Vol. 18, No. 15, (2004) 2091-2121.

Basak, T. & Priya, A. "Role of ceramic supports on microwave heating of materials." *Journal of Applied Physics,* Vol. 97, No. 8, (2005) 083537.

Becher, P.; Angelini, P.; Warwick, W. & Tiegs, T. "Elevated-temperature delayed failure of alumina reinforced with 20 vol% silicon-carbide whiskers." *Journal of the American Ceramic Society,* Vol. 73, No. 1, (1990) 91-96.

Becher, P.; Hsueh, C.; Angelini, P. & Tiegs, T. "Theoretical and experimental analysis of the toughening behavior of whisker reinforcement in ceramic matrix composites." *Materials Science and Engineering A-Structural Materials Properties Microstructure and Processing,* Vol. 107, (1989) 257-259.

Becher, P.; Hsueh, C.; Angelini, P. & Tiegs, T. "Toughening behavior in whisker-reinforced ceramic matrix composites." *Journal of the American Ceramic Society,* Vol. 71, No. 12, 1050-61 (1988).

Becher, P.; Hsueh, C.; & Waters, S. "Thermal-expansion anisotropy in hot-pressed SiC-whisker-reinforced alumina composites." *Materials Science and Engineering a-Structural Materials Properties Microstructure and Processing,* Vol. 196, No. 1-2, (1995) 249-251.

Becher, P. & Wei, G. "Toughening behavior in SiC-whisker-reinforced alumina." *Journal of the American Ceramic Society,* Vol. 67, No. 12, (1984) C267-C269.

Bertram, B. & Gerhardt, R. "Room temperature properties of electrical contacts to alumina composites containing silicon carbide whiskers." *Journal of Applied Physics,* Vol. 105, No. 7, (2009) 074902.

Bertram, B. & Gerhardt, R. "Effects of frequency, percolation, and axisymmetric microstructure on the electrical response of hot-pressed alumina-silicon carbide whisker composites." *J. Am. Ceram. Soc.,* (2010) accepted.

Billman, E.; Mehrotra, P.; Shuster, A. & Beeghly, C. "Machining with Al_2O_3-SiC-whisker cutting tools." *American Ceramic Society Bulletin,* Vol. 67, No. 6, (1988) 1016-1019.

Björk, L. & Hermansson, L. "Hot Isostatically Pressed Alumina-Silicon Carbide-Whisker Composites." *Journal of the American Ceramic Society,* Vol. 72, No. 8, (1989) 1436-1438.

Budiansky, B. & O'Connell, R. "Elastic moduli of a cracked solid." *International Journal of Solids and Structures,* Vol. 12, No. 2, (1976) 81-97.

Calame, J.; Abe, D.; Levush, B. & Danly, B.; "Variable temperature measurements of the complex dielectric permittivity of lossy AlN-SiC composites from 26.5-40 GHz." *Journal of Applied Physics,* Vol. 89, No. 10, (2001) 5618-5621.

Carter, J. "Proposed Energy Policy for United States." (televised speech) 1977.

Cole, K. & Cole, R. "Dispersion and absorption in dielectrics I. Alternating current characteristics." *Journal of Chemical Physics,* Vol. 9, No. 4, (1941) 341-351.

Collin, M. & Rowcliffe, D. "Influence of thermal conductivity and fracture toughness on the thermal shock resistance of alumina-silicon-carbide-whisker composites." *Journal of the American Ceramic Society,* Vol. 84, No. 6, (2001) 1334-1340.

Connor, M.; Roy, S.; Ezquerra, T. & Baltá Calleja, F. "Broadband ac conductivity of conductor-polymer composites," *Physical Review B,* Vol. 57, No. 4, (1998) 2286.

de Arellano-Lopez, A.; Cumbrera, F.; Dominguez-Rodriguez, A.; Goretta, K. & Routbort, J. "Compressive creep of SiC-whisker-reinforced Al_2O_3." *Journal of the American Ceramic Society,* Vol. 73, No. 5, (1990) 1297-1300.

de Arellano-López, A.; Domínguez-Rodríguez, A.; Goretta, K. & Routbort, J. "Plastic deformation mechanisms in SiC-whisker-reinforced alumina." *Journal of the American Ceramic Society,* Vol. 76, No. 6, (1993) 1425-1432.

de Arellano-Lopez, A.; Dominguez-Rodriguez, A. & Routbort, J. "Microstructural constraints for creep in SiC-whisker-reinforced Al_2O_3." *Acta Materialia,* Vol. 46, No. 18, (1998) 6361-6373.

de Arellano-Lopez, A.; Melendez-Martinez, J.; Dominguez-Rodriguez, A. & Routbort, J. "Creep of Al_2O_3 containing a small volume fraction of SiC-whiskers." *Scripta Materialia,* Vol. 42, No. 10, (2000) 987-991.

de Arellano-Lopez, A.; Melendez-Martinez, J.; Dominguez-Rodriguez, A.; Routbort, J.; Lin, H. & Becher, P. "Grain-size effect on compressive creep of silicon-carbide-whisker-reinforced aluminum oxide." *Journal of the American Ceramic Society,* Vol. 84, No. 7, (2001) 1645-1647.

DeHoff, R. (1993). *Thermodynamics in materials science,* McGraw-Hill, ISBN 0070163138,New York.

Dogan, C. & Hawk, J. "Influence of whisker toughening and microstructure on the wear behavior of Si_3N_4- and Al_2O_3-matrix composites reinforced with SiC." *Journal of Materials Science,* Vol. 35, No. 23, (2000) 5793-5807.

Fabbri, L.; Scafe, E. & Dinelli, G. "Thermal and elastic properties of alumina-silicon carbide whisker composites." *Journal of the European Ceramic Society,* Vol. 14, No. 5, (1994) 441-446.

Farkash, M. & Brandon, D. "Whisker alignment by slip extrusion." *Materials Science and Engineering A-Structural Materials Properties Microstructure and Processing,* Vol. 177, No. 1-2, (1994) 269-275.

Ford, G. "Address on the State of the Union before a Joint Session of the 94th Congress." (1975).

Garnier, V.; Fantozzi, G.; Nguyen, D.; Dubois, J. & Thollet, G. "Influence of SiC whisker morphology and nature of SiC/Al_2O_3 interface on thermomechanical properties of SiC reinforced Al_2O_3 composites." *Journal of the European Ceramic Society,* Vol. 25, No. 15, (2005) 3485-3493.

Gerhardt, R. "Impedance Spectroscopy and Mobility Spectra," *Encyclopedia of Condensed Matter Physics.* Elsevier, (2005) 350-363.

Gerhardt, R. & Ruh, R. "Volume fraction and whisker orientation dependence of the electrical properties of SiC-whisker-reinforced mullite composites." *Journal of the American Ceramic Society,* Vol. 84, No. 10, (2001) 2328-2334.

Gerhardt, R.; Runyan, J.; Sana, C.; McLachlan, D. & Ruh, R. "Electrical Properties of Boron Nitride Matrix Composites: III, Observations near the Percolation Threshold in BN-B_4C Composites" *Journal of the American Ceramic Society,* Vol. 84, No. 10, (2001) 2335-42.

Goncharenko, A. "Generalizations of the Bruggeman equation and a concept of shape-distributed particle composites." *Physical Review E,* Vol. 68, No. 4. (2003) 041108.

Holm, E. & Cima, M. " Two-Dimensional Whisker Percolation in Ceramic-Matrix Ceramic-Whisker Composites." *Journal of the American Ceramic Society,* Vol. 72, No. 2, (1989) 303-305.

Homeny, J. & Vaughn, W. & Ferber, M. "Silicon-carbide whisker alumina matrix composites - Effect of whisker surface treatment on fracture toughness." *Journal of the American Ceramic Society,* Vol. 73, No. 2, (1990) 394-402.

Iio, S.; Watanabe, M.; Matsubara, M. & Matsuo, Y. "Mechanical properties of alumina silicon-carbide whisker composites." *Journal of the American Ceramic Society,* Vol. 72, No. 10, (1989) 1880-1884.

Iwata, H.; Lindefelt, U.; Oberg, S. & Briddon, P. "Localized electronic states around stacking faults in silicon carbide." *Physical Review B,* Vol. 65, No. 3, (2002) 033203.

Jakus, K. & Nair, S. "Nucleation and growth of cracks in SiC/Al_2O_3 composites." *Composites Science and Technology,* Vol. 37, No. 1-3, (1990) 279-297.

Jones, R. (1999) *Mechanics of Composite Materials,* Taylor & Francis, ISBN 156032712X, Philadelphia, PA.

Jonscher, A. (1983). *Dielectric Relaxation in Solids,* Chelsea Dielectrics Press Ltd., ISBN 0950871109, London.

Karunanithy, S. "Chemical processes that degrade composites of alumina with SiC whiskers." *Materials Science and Engineering A-Structural Materials Properties Microstructure and Processing,* Vol. 112, (1989) 225-231.

Kato, A.; Nakamura, H.; Tamari, N.; Tanaka, T. & Kondo, I. "Usefulness of alumina-coated SiC whiskers in the preparation of whisker-reinforced alumina ceramics." *Ceramics International,* Vol. 21, No. 1, (1995) 1-4.

Kim, H. & Moorhead, A. "Oxidation behavior and effects of oxidation on the strength of SiC-whisker reinforced alumina."*Journal of Materials Science,* Vol. 29, No. 6, (1994) 1656-1661.

Lagarkov, A.; Matytsin, S.; Rozanov, K. & Sarychev, A. "Dielectric permittivity of fiber-filled composites: comparison of theory and experiment." *Physica A,* Vol. 241, No. 1-2, (1997) 58-63.

Lagarkov, A.; Matytsin, S.; Rozanov, K. & Sarychev, A. "Dielectric properties of fiber-filled composites." *Journal of Applied Physics,* Vol. 84, No. 7, (1998) 3806-3814.

Lagarkov, A. & Sarychev, A. "Electromagnetic properties of composites containing elongated conducting inclusions." *Physical Review B,* Vol. 53, No. 10, (1996) 6318-6336.

Lee, K. & Sheargold, S. "Particulate matters in silicon-carbide whiskers." *American Ceramic Society Bulletin,* Vol. 65, No. 11, (1986) 1477-1477.

Lee, W. & Case, E. "Cyclic thermal shock in SiC-whisker-reinforced alumina composite." *Materials Science and Engineering A-Structural Materials Properties Microstructure and Processing,* Vol. 119 (1989) 113-126.

Li, X. "Ceramic Cutting Tools - An Introduction." *Key Engineering Materials,*Vol. 96, (1994) 1-18.

Lin, H.; Alexander, K. & Becher, P. "Grain size effect on creep deformation of alumina-silicon carbide composites." *Journal of the American Ceramic Society,* Vol. 79, No. 6, (1996) 1530-1536.

Lin, H. & Becher, P. "Creep behavior of a SiC-whisker-reinforced alumina." *Journal of the American Ceramic Society,* Vol. 73, No. 5, (1990) 1378-1381.

Lin, H. & Becher, P. "High-temperature creep deformation of alumina SiC-whisker composites." *Journal of the American Ceramic Society,* Vol. 74, No. 8, (1991) 1886-1893.

Lipetzky, P.; Nutt, S.; Koester, D. & Davis, R. "Atmospheric effects on compressive creep of SiC-whisker-reinforced alumina." *Journal of the American Ceramic Society,* Vol. 74, No. 6, (1991) 1240-1247.

Luthra, K. & Park, H. "Oxidation of silicon carbide-reinforced oxide-matrix composites at 1375°C to 1575°C." *Journal of the American Ceramic Society,* Vol. 73, No. 4, (1990) 1014-1023.

Mallick, P. (2008). *Fiber-Reinforced Composites,* CRC Press, ISBN 0849342058, Dearborn, MI.

McCluskey, P.; Williams, R.; Graves, R. & Tiegs, T. "Thermal diffusivity/conductivity of alumina-silicon carbide composites." *Journal of the American Ceramic Society,* Vol. 73, No. 2, (1990) 461-464.

McLachlan, D. "Analytic scaling functions for percolative metal-insulator phase transitions fitted to Al_xGe_{1-x} data." *Physica B,* Vol. 254, No. 3-4, (1998) 249-255.

Mebane, D. & Gerhardt, R. "Orientation Dependence of Resistivity in Anisotropic Ceramic Composites." *Ceramic Transactions,* Vol. 150, (2004) 265-272.

Mebane, D. & Gerhardt, R. "Interpreting impedance response of silicon carbide whisker/alumina composites through microstructural simulation." *Journal of the American Ceramic Society,* Vol. 89, No. 2, (2006) 538-543.

Mebane, D.; Gokhale, A. & Gerhardt, R., "Trivariate, stereological length-radius-orientation unfolding derived and applied to alumina-silicon carbide whisker composites." *Journal of the American Ceramic Society,* Vol. 89, No. 2, (2006) 620-626.

Mehrotra, P. "Applications of ceramic cutting tools." *Advanced Ceramic Tools for Machining Application - Iii,"* Vol. 138, No. 1, (1998) 1-24.

Metaxas, A. & Meredith, R. (1983). *Industrial Microwave Heating.* Peter Peregrinus Ltd. on behalf of the Institution of Electrical Engineers, ISBN 0906048893, London.

Mukae, K.; Tsuda, K. & Nagasawa, I. "Capacitance-vs-voltage characteristics of ZnO varistors." *Journal of Applied Physics,* Vol. 50, No. 6, (1979) 4475-4476.

Nader, M.; Aldinger, F. & Hoffmann, M. "Influence of the alpha/beta-SiC phase transformation on microstructural development and mechanical properties of liquid phase sintered silicon carbide." *Journal of Materials Science,* Vol. 34, No. 6, (1999) 1197-1204.

Nutt, S. "Defects in silicon carbide whiskers," *Journal of the American Ceramic Society,* Vol.67, No. 6, (1984) 428-431.

Nutt, S. " Microstructure and growth model for rice-hull-derived SiC whiskers." *Journal of the American Ceramic Society,* Vol. 71, No. 3, (1988) 149-156.

Nutt, S.; Lipetzky, P. & Becher, P. "Creep deformation of alumina-SiC composites." *Materials Science and Engineering A-Structural Materials Properties Microstructure and Processing,* Vol. 126, (1990) 165-172.

Nutt, S. & Lipetzky, P. "Creep deformation of whisker-reinforced alumina." Materials Science and Engineering A, Vol. 166, (1993) 199-209.

O'Meara, C.; Suihkonen, T.; Hansson, T. & Warren, R. "A microstructural investigation of the mechanisms of tensile creep deformation in an Al_2O_3/SiC_w composite." *Materials Science and Engineering a-Structural Materials Properties Microstructure and Processing,* Vol. 209, No. 1-2, (1996) 251-259.

Panteny, S.; Stevens, R. & Bowen, C. "The frequency dependent permittivity and ac conductivity of random electrical networks." *Ferroelectrics,* Vol. 319, (2005) 199-208.

Park, H.; Kim, H. & Kim, D. "Evaluation of whisker alignment in axisymmetrical SiC_w-reinforced Al_2O_3 composite materials." *Journal of the American Ceramic Society,* Vol. 77, No. 11, (1994) 2828-2832.

Parris, P. & Kenkre, V."Thermal runaway in ceramics arising from the temperature dependence of the thermal conductivity." *Physica Status Solidi B-Basic Research,* Vol. 200, No. 1, (1997) 39-47.

Peng, X.; Zhang, Y.; Huang, X.; Song, L. & Hu, X. "Nitrogen corrosion in silicon carbide whisker-reinforced alumina composites." *Advanced Engineering Materials,* Vol. 2, No. 7, (2000) 448-450.

Porter, L. & Davis, R. "A critical review of ohmic and rectifying contacts for silicon carbide." *Mater. Sci. Eng. B,* Vol. 34, No. 2-3, (1995) 83-105.

Quan, G.; Conlon, K. & Wilkinson, D. "Investigation of anelastic creep recovery in SiC whisker-reinforced alumina composites." *Journal of the American Ceramic Society,* Vol. 88, No. 11, (2005) 3104-3109.

Quantrille, T. (2007) "Novel Composite Structures for Microwave Heating and Cooking." Proceedings of 41st Annual Microwave Symposium. Vancouver, BC, 2007, International Microwave Power Institute.

Quantrille, T. (2008) "Ceramic Composites for Microwave Grilling and Speed Cooking." Proceedings of 42nd Annual Microwave Symposium, New Orleans, LA, 2008, International Microwave Power Institute.

Rahimi, R.; Miller, C.; Raghavan, S.; Stinespring, C. & Korakakis, D. "Electrical properties of strained nano-thin 3C-SiC/Si heterostructures." *Journal of Physics D-Applied Physics,* Vol. 42, No. 5, (2009) 055108.

Raj, R. & Ashby, M. "Grain boundary sliding and diffusional creep." *Metallurgical Transactions,* Vol. 2, No. 4, (1971) 1113.

Raju, C. & Verma, S. "SiC whiskers from rice hulls: formation, purification, and characterisation." *British Ceramic Transactions,* Vol. 96, No. 3, (1997) 112-115.

Rice, R. (2003). *Ceramic Fabrication Technology,* Marcel Dekker, ISBN 0824708539, New York, NY.

Richerson, D. (1992) *Modern Ceramic Engineering,* Marcel Dekker, ISBN 1574446932, New York, NY.

Rodelsperger, K. & Bruckel, B. "The Carcinogenicity of WHO fibers of silicon carbide: SiC whiskers compared to cleavage fragments of granular SiC." *Inhalation Toxicology,* Vol. 18, No. 9, (2006) 623-631.

Ruh, R. & Chizever, H. "Permittivity and permeability of mullite-SiC-whisker and spinel-SiC-whisker composites." *Journal of the American Ceramic Society,* Vol. 81, No. 4, (1998) 1069-1070.

Runyan, J.; Gerhardt, R. & Ruh, R. "Electrical properties of boron nitride matrix composites: I, analysis of McLachlan equation and modeling of the conductivity of boron nitride-boron carbide and boron nitride-silicon carbide composites." *Journal of the American Ceramic Society,* Vol. 84, No. 7, (2001a) 1490-1496.

Runyan, J.; Gerhardt, R. & & Ruh, R. "Electrical properties of boron nitride matrix composites: II, dielectric relaxations in boron nitride-silicon carbide composites." *Journal of the American Ceramic Society,* Vol. 84, No. 7, (2001b) 1497-1503.

Russell, L.M.; Johnson, L.F.; Hasselman, D.P.H. & Ruh, R. "Thermal Conductivity/Diffusivity of Silicon Carbide Whisker Reinforced Mullite." *Journal of the American Ceramic Society,* Vol. 70, No. 10, (1987) C226-C229.

Salganik, R. "Mechanics of bodies with many cracks." *Mechanics of Solids,* Vol. 8, No. 4, (1973) 135-143.

Sanchez-Gonzalez, J.; Ortiz, A.; Guiberteau, F. & Pascual, C. "Complex impedance spectroscopy study of a liquid-phase-sintered alpha-SiC ceramic." *Journal of the European Ceramic Society,* Vol. 27 (2007) 3935-3939.

Sandlin, M.; Lee, E. & Bowman, K. "Simple geometric model for assessing whisker orientation in axisymmetrical SiC-whisker-reinforced composites." *Journal of the American Ceramic Society,* Vol. 75, No. 6, (1992) 1522-1528.

Sheng, P.; Sichel, E. & Gittleman, J. "Fluctuation-induced tunneling conduction in carbon-polyvinychloride composites." *Physical Review Letters,* Vol. 40, No. 18, (1978) 1197-1200.

Shimoo, T.; Okamura, K. & Morisada, Y. "Active-to-passive oxidation transition for polycarbosilane-derived silicon carbide fibers heated in Ar-O_2 gas mixtures." *Journal of Materials Science,* Vol. 37, No. 9, (2002) 1793-1800.

Sillars, R. "The Properties of a Dielectric Containing Semiconducting Particles of Various Shapes." *J. Inst. Elect. Engrs.,* Vol. 80, (1937) 378-394.

Singh, P.; Selvam, A. & Nair, N. "Synthesis of SiC whiskers from a mixture of rice husks and coconut shells." *Advances in Powder Metallurgy & Particulate Materials,* Vol. 3-53 (1998) 53-64.

Sinott, M. (1987) "Ceramic Technology for Advanced Heat Engines." Edited by N. R. C. National Materials Advisory Board. Washington D.C.

Stauffer, D. & Aharony, A. (1994). Introduction to Percolation Theory: Revised Second Edition, CRC Press, ISBN 0748402535, London and Philadelphia.

Streetman, B. & Banerjee, S. (2000) *Solid State Electronic Devices,* Prentice Hall, Inc. ISBN 013149726X, Upper Saddle River, NJ.

Tai, Q. & Mocellin, A. "Review: High temperature deformation of Al_2O_3-based ceramic particle or whisker composites." *Ceramics International,* Vol. 25, No. 5, (1999) 395-408.

Takahashi, J.; Ohtani, N.; Katsuno, M. & Shinoyama, S. "Sublimation growth of 6H- and 4H-SiC single crystals in the $[1\bar{1}00]$ and $[11\bar{2}0]$ directions," *Journal of Crystal Growth,* Vol. 181, No. 3, (1997) 229-240.

Takahashi, K.; Yokouchi, M.; Lee, S. & Ando, K. "Crack-healing behavior of Al_2O_3 toughened by SiC whiskers." *Journal of the American Ceramic Society,* Vol. 86, No. 12, (2003) 2143-2147.

Taya, M. (2005). Electronic Composites: Modeling, Characterization, Processing, and Applications, Cambridge University Press, ISBN 0521057318, Cambridge.

Thangaraj, A. & Weinmann, K. "On the wear mechanisms and cutting performance of silicon-carbide whisker-reinforced alumina." *Journal of Engineering for Industry-Transactions of the ASME,* Vol. 114, No. 3, (1992) 301-308.

Thompson, I. & Krstic, V. "Effect of carbon coating on mechanical strength of SiC whisker-reinforced alumina composites." *Theoretical and Applied Fracture Mechanics,* Vol. 19, No, 1, (1993) 61-67.

Tian, W. & Yang, R. "Effect of interface scattering on phonon thermal conductivity percolation in random nanowire composites." *Applied Physics Letters,* Vol. 90, No. 26, (2007) 263105.

Tiegs, T. & Becher, P. "Thermal shock behavior of an alumina-SiC whisker composite." *Journal of the American Ceramic Society,* Vol. 70, No. 5, (1987) C109-C111.

Tsangaris, G.; Kouloumbi, N. & Kyvelidis, S. "Interfacial relaxation phenomena in particulate composites of epoxy resin with copper or iron particles." *Materials Chemistry and Physics,* Vol. 44, No. 3, (1996) 245-250.

Vaughan, G. & Trently, S. "The toxicity of silicon carbide whiskers: a review." *Journal of Environmental Science and Health Part a-Environmental Science and Engineering & Toxic and Hazardous Substance Control,* Vol. 31, No. 8, (1996) 2033-2054.

von Hippel, A. (1954). *Dielectrics and Waves,* Wiley, ISBN 1580531229, New York, NY.

Wagner, R. & Ellis, W. "Vapor-liquid-solid mechanism of single crystal growth." *Applied Physics Letters,* Vol. 4, No. 5, (1964) 89-90.

Wang, C.; Huang, Y.; Li, Y.; Zhang, Z.; Xie, Z. & Li, J. "Resistivity controlled by SiC whisker orientation in Si_3N_4 matrix composites." *Journal of Materials Science Letters,* Vol. 17, No. 10, (1998) 829-831.

Wang, D. & Lopez, H. "Morphological and kinetic aspects of thermal-oxidation of SiC whisker-reinforced Al_2O_3." *Materials Science and Technology,* Vol. 10, No. 10, (1994) 879-885.

Wang, P.; Hsu, S. & Wittberg, T. "Oxidation kinetics of silicon carbide whiskers studied by x-ray photoelectron spectroscopy." *Journal of Materials Science,* Vol. 26, No. 6, (1991) 1655-1658.

Wu, J. & McLachlan, D. "Scaling behavior of the complex conductivity of graphite-boron nitride percolation systems." *Physical Review B,* Vol. 58, No. 22, (1998) 14880.

Ye, F.; Lei, T. & Zhou, Y. "Interface structure and mechanical properties of Al_2O_3-20vol%SiC_w ceramic matrix composite." *Materials Science and Engineering A-Structural Materials Properties Microstructure and Processing,* Vol. 281, No. 1-2, (2000) 305-309.

Yoshida, H.; Kohno, H.; Ichikawa, S.; Akita,T. & Takeda, S. "Inner potential fluctuation in SiC nanowires with modulated interior structure." *Materials Letters,* Vol. 61, No. 14-15, (2007) 3134-3137.

Zhang, B.; Li, J.; Sun, J.; Zhang, S.; Zhai, H. & Du, Z. "Nanometer silicon carbide powder synthesis and its dielectric behavior in the GHz range." *Journal of the European Ceramic Society,* Vol. 22, No. 1, (2002) 93-99.

Zhang, J.; Huang, H.; Cao, L.; Xia, F. & Li, G. "Semiconductive property and impedance spectra of alumina silicon-carbide whisker composite." *Journal of the American Ceramic Society,* Vol. 75, No. 8, (1992) 2286-2288.

Zhang, Z.; Shan, H.; Huang, Y. & Jiang, Z. "Characterisation of interfacial bonding in Al_2O_3 coated SiC whisker reinforced TZP composites." *British Ceramic Transactions,* Vol. 95, No. 3, (1996) 125-128.

10

Spectroscopic properties of carbon fibre reinforced silicon carbide composites for aerospace applications

Davide Alfano
Italian Aerospace Research Centre
Italy

1. Introduction

The use of Ceramic Matrix Composites (CMCs) such as carbon fibre reinforced silicon carbide composites (C/SiC), carbon fibre reinforced carbon composites (C/C) but also silicon carbide fibre reinforced silicon carbide composites (SiC/SiC) is mandatory within the aerospace sector whenever the transfer of mechanical loads at high temperatures (up to 1900 K in air) is required and any metallic material (e.g. refractory metals) or intermetallic materials cannot be employed.

CMCs are constituted by the coupling of long fibres reinforcement (usually carbon fibres) and a refractory ceramic matrix (i.e. carbon, or silicon carbide) and represent a class of ceramic materials characterized by good mechanical properties, thermo-mechanical stability and fracture toughness: their fracture behaviour sets them apart from conventional monolithic ceramics, allowing for a variety of uses in which damage tolerance is the main requirement.

The extremely good high temperature fracture toughness of CMCs is provided by the crack bridging effect of the carbon fibres: stress concentrations, e.g. notches or holes, are reduced by stress redistribution and inelastic deformation. In case of overloading, monolithic ceramics break immediately, while CMC materials are still able to carry load even if the elastic mechanical load range is exceeded. Such a damage tolerant behaviour constitutes an important point for the safety issues in particular for space re-entry vehicles.

Typical applications for CMCs within aerospace sector include thermo stable satellite structures for optical components, ultra lightweight scanning mirrors (Harnisch et al., 1998), hot structures for atmospheric re-entry vehicles (Laux et al., 2001), and very high temperature parts for hypersonic and rocket engines (Mühlratzer & Leuchs, 2001 Naslain & Schneider, 2002).

One of the most relevant application fields of carbon fibre reinforced silicon carbide composites (C/SiCs) concerns the manufacturing of structural thermal protection systems (TPSs) for planetary re-entry vehicles, where good thermo-mechanical properties at high temperature associated with oxidation resistance are required. During the re-entry phase in Earth's atmosphere nose tips and leading edges are the structural parts of spacecrafts subject

to high heat fluxes and then high temperatures which, in any case, depend on re-entry trajectory as well as the shape and dimension of the structural part.
The design of shape and dimensions of that element must consider a lot of parameters such as the temperatures, pressures and thermal fluxes reached at the stagnation point, but also surface properties. In particular, if the stagnation pressure depends on the re-entry trajectory, the surface temperature, and the heat flux are also greatly influenced by the emissivity and recombination efficiency (catalycity) values of the employed material. Consequently the choice of the material is not only the starting point for the design of TPSs but it is also a strategic point since it can strongly influence the value of a series of parameters that define the re-entry conditions.
This chapter aims to present the results of emissivity and catalytic measurements carried out reproducing the experimental conditions, in terms of pressure ad temperature values, typical of a re-entry phase in atmosphere for a carbon fibre reinforced silicon carbide composite produced by polymer vapour infiltration and then coated with SiC by chemical vapour deposition. In particular, the tested samples are produced by MT Aerospace and marketed under the brand name Keraman®. Moreover microstructural investigations performed on post-test samples are illustrated in order to discuss about the resistance to oxidation processes of C/SiC in terms of active and passive oxidation of silicon carbide.

2. Aerospace applications for silicon carbide

Silicon carbide is defined by the Engineered Materials Handbook (Reinhart, 1987) as "reinforcement, in whiskers, particulate, and fine or large fibre, that has application as metal matrix reinforcement because of its high strength and modulus, density equal to that of aluminium, and comparatively low cost". For aerospace applications SiC can be employed as matrix in reinforced composites (CMCs) or as particulate filler in massive ceramic composites also indicated as Ultra High Temperature Ceramics (UHTCs).

2.1 SiC-based Ceramic Matrix Composites

Silicon carbide is often combined with carbon fibres in order to obtain Ceramic Matrix Composites (CMCs) with non-oxide matrix materials for high temperature applications. Carbon fibres show no degradation up to temperatures over 3000 K in non oxidizing atmosphere. So that if carbon fibres are protected from oxidation reactions, they become a useful material in space vehicle applications where temperatures up to 2000 K occur (for instance during the re-entry phase in atmosphere) . The main role of SiC matrix in CMCs is to protect the carbon fibres from oxidation processes, which already become active starting from about 800 K, by the formation of a protective silica-based glassy layer. SiC can be also employed as protective coating in the case of carbon fibre reinforced carbon composites (C/C). Reinforced Carbon-Carbon (RCC) have been used as Thermal Protection System (TPS) for wing leading edges and nose cap of the Space Shuttle where the temperature values can increase up to 1800-1900 K during the re-entry phase in atmosphere. In order to provide oxidation resistance for reuse capability, the outer layers of the RCC are converted to silicon carbide.
The Institute of Structures and Design of DLR (German Aerospace Centre) has been developing fibre reinforced ceramic matrix composites via the liquid silicon infiltration process for more than a decade. The materials manufactured using this processing technique

are suitable for a broad range of applications. In general, this material exhibits excellent thermal shock resistance, high mass specific values and possess a dense matrix. In the framework of the project EXPRESS (realised within the scope of a German-Japanese cooperation), DLR developed and produced the tile called CETEX (Ceramic Tile EXperiment) which is a fibre reinforced ceramic component made of C/C-SiC, a special ceramic material, applying the liquid silicon infiltration process. CETEX was attached to the stagnation point of a capsule's ablative heat shield. Qualification tests with samples and components took place in several plasma wind tunnels: in the PWK 2 wind tunnel of the Institute for Space Systems of the University of Stuttgart the maximum applied temperature was reportedly 3000 K. Although the Japanese launcher did not work as intended, the (uncontrolled) re-entry of the capsule was successful, leading to about 2500 K in CETEX without apparently revealing any problems (Hald & Winkelmann, 1995; Hald & Winkelmann, 1997; Hald, 2003).

MT Aerospace (Augsburg, Germany) produces a carbon reinforced silicon carbide for aerospace applications and markets it under the name Keraman®. This materials was qualified during NASA X-38 project, in the form of CMC body flaps and leading edges for the X-38 vehicle (Coperet et al., 2002; Dogigli et al., 2002a,b; Pfeiffer & Peetz, 2002; Steinacher et al., 2007). On the same space vehicle the nose cap was manufactured in C/SiC provided by DLR. Nose cap and body flaps were produced by a Chemical Vapour Infiltration process (CVI). Also Snecma Propulsion Solide (Bordeaux, France) has developed a CMC shingle TPS (Pichon et al., 2006); a flat panel was tested in arc-jet reaching the maximum temperature of 1500 K.

In the frame of the European eXPErimental Re-entry Testbed (EXPERT) project conducted by the European Space Agency (ESA), SiC-coated C/C composite manufactured by DLR has been chosen as material to produce the nose cap of the vehicle whose goal will be to collect data of different physical phenomena during the re-entry phase in atmosphere (Reimer & Laux, 2005).

A load bearing aeroshell in C/SiC for hypersonic flight was developed in the project Sustained Hypersonic Flight Experiment (SHyFE) financed by UK Ministry of Defence (Dadd et al., 2006; Goodman & Ireland, 2006). The aeroshell was fabricated by MT Aerospace utilizing CVI.

In the framework of the Sharp Hot Structures project, a technology project within the Italian Unmanned Space Vehicle (USV) program, the Italian Aerospace Research Centre (Capua, Italy) has studied and developed, during last ten years, a nose cone ceramic demonstrator for re-entering Low Earth Orbit (LEO) space vehicles, whose structure is divided into a conical/hemispherical part (nose tip) manufactured in massive UHTC and a layered conical part (dome) in C/SiC (Russo & Marino, 2003; Scatteia et al., 2005; Del Vecchio et al., 2006). Moreover a ceramic composite containing SiC particles dispersed in a ZrB_2 matrix was deposited on the dome surface by plasma spraying in order to improve its oxidation resistance at high temperature (Bartuli et al., 2002; Tului et al., 2006).

2.2 SiC as additive for Ultra High Temperature Ceramics

Ceramic compounds based on metal borides, such as zirconium diboride (ZrB_2) and hafnium diboride (HfB_2) have been commonly referred to as Ultra High Temperature Ceramics (UHTCs). UHTCs represent a class of promising materials for use in extreme applications because of their high melting point and relatively good oxidation resistance in re-entry conditions. UHTCs are characterized by high melting temperatures (ZrB_2 3518 K, HfB_2 3653 K), solid state stability, good thermo-chemical, and thermo-mechanical properties (Schneider, 1991). These extremely promising high performance materials are also characterized by hardness above 20 GPa, high wear resistance, high emissivity, high electrical conductivity, excellent corrosion resistance, and good thermal shock resistance (Mroz, 1994; Fahrenholtz et al., 2007a). Leading applications are currently found in aerospace, more specifically in the possibility to employ them to realize sharp-shaped hot structures like wing leading edges and nose caps able to withstand the severe thermal requirements of next generation of hypersonic re-entry vehicles. The highly thermal demanding trajectories foreseen for future spaceplane-like winged re-entry vehicles dictate the need for base materials able to sustain operating temperatures approaching 2500 K, to resist evaporation, erosion and oxidation in the harsh re-entry environment.

The research on this class of materials began in the 60's in the frame of Air Force contracts (Kuriakose, & Magrave, 1964; Tripp & Graham, 1971). The early works were devoted to the production of dense materials by mean of pressure assisted sintering, and to investigate the influence of a variety of additives, including carbon and silicon carbide, on the processing and oxidation resistance of Hf and Zr diborides. These works showed that the addition of SiC as secondary reinforcing phase provides significant enhancements to the oxidation resistance of UHTCs (Tripp, et al., 1973). Moreover the SiC addition was also found to improve the processing by lowering sintering temperatures (Monteverde et al., 2003; Chamberlain et al., 2004; Monteverde, 2006). Then, when combined with SiC, ZrB_2 and HfB_2–based composites exhibit indeed excellent refractoriness, high oxidation resistance, and are as such good potential candidates for the above-mentioned application. An important parameter such as the upper limit of the service temperature is strongly related to the characteristics of secondary phases. For example, above 1500 K the oxide scale formed in air on the surface of pure MB_2, with M=Zr or Hf, is unstable and non-protective due to intensive volatility of B_2O_3 (Kuriakose, & Magrave, 1964; Tripp & Graham, 1971; Opeka et al., 1999; Ban'kovskaya & Zhabrev, 2005; Chamberlain et al., 2005; Fahrenholtz, 2005; Fahrenholtz et al., 2007) while the SiC-containing MB_2 showed enhanced resistance to oxidation up to 1900 K. In fact for temperature higher than 1500 K, the addition of SiC promotes, on the exposed surface, the formation of borosilicate glass which gives much more oxidation protection than B_2O_3 alone. Several studies have dealt with the thermal stability and physical properties of ultra-refractory MB_2-based ceramics in oxidizing environments, and highlighted the role of composition and microstructure on the mechanisms governing the materials response to hostile environments (Opeka et al., 1999; Levine et al., 2002; Fahrenholtz et al., 2004; Gasch et al., 2004; Opeka et al., 2004; Opila et al., 2004; Ban'kovskaya & Zhabrev, 2005; Chamberlain et al., 2005; Monteverde & Bollosi, 2005; Rezaire et al., 2006; Fahrenholtz, 2007b; Rezaire et al., 2007; Han et al., 2008; Zhang et al., 2008; Carney et al., 2009; Hu, et al., 2009; Karlsdottir & Halloran, 2009).

NASA started in 1990 a research program on UHTCs and ended up in 1997 and 2000 demonstrating the use of ZrB_2 and HfB_2 for sharp leading edge in the Sharp Hypersonic Aero-thermodynamic Research Probe Ballistic experiments (SHARP-B1 and B2) (Rasky et al., 1998). UHTCs were also tested by the flights of Delta Clipper (DC- X and DC-XA) in order to evaluate their potential application on new entry vehicles (Smith et al., 1997) .

During the 90s a wide range research activity on UHTC materials was conducted in Italy, mainly by the Italian National Research Council Institute of Ceramic Materials (CNR-

ISTEC). The CNR-ISTEC investigated new processing routes based on pressure assisted sintering, on the adoption of sintering aids and secondary reinforcing phases in order to obtain dense bodies characterized by superior oxidation resistance and mechanical properties.

Since 2000, the Italian Aerospace Research Centre (CIRA) has studied, developed, and tested massive UHTCs in the frame of the Unmanned Space Vehicle (USV) National Program (Russo & Marino, 2003; Savino, et al., 2005; Scatteia et al., 2005; Del Vecchio et al., 2006; Monteverde & Scatteia, 2007; Monteverde et al., 2008; Scatteia et al., 2010).

The poor fracture toughness of UHTCs can be still considered the main limitation of this class of materials for aerospace applications. In these last years the activities of several research groups on UHTCs have been focused on the improvement of the fracture toughness by using SiC whiskers or SiC chopped fibers as reinforcing adds (Chen et al., 2009; Zhang et al., 2009; Guicciardi et al., 2010; Silvestroni et al., 2010).

3. C/SiC employed in emissivity and catalycity tests

Emissivity and catalycity tests were carried out on a SiC-coated two-dimensional C/SiC ceramic matrix composite produced by MT Aerospace and marketed under the name Keraman® (Fig.1a).

The C/SiC composite is produced by Polymer Vapour Infiltration process (PVI) while the SiC-coating, applied by Chemical Vapour Deposition method (CVD), is characterised by a thickness of about 25 μm. In Fig. 1b one example of cross-section SEM micrograph used to estimate the SiC-coating thickness is reported. In the same image the two orthogonal directions of carbon fibres are also evident.

Fig. 1. CVDed SiC coated C/SiC: a) picture of specimen employed to perform emissivity and catalycity tests, b) cross-section SEM micrograph of a specimen wherein SiC-coating and 2D orthogonal carbon fibres are indicated.

4. Emissivity

The heat transfer by radiation from the surface of such space vehicle becomes a significant part of the total heat transferred when the surface temperature is high, and when the convective heat transfer is low as at high altitudes. Since radiative heat transfer is an important method of cooling under such conditions, a knowledge of the emissivity values of a surface is required whenever theoretical simulatons involving radiant heat are to be made. At each given temperature and wavelength, there is a maximum amount of radiation that a surface can emit which is known as a blackbody radiation, and can be theoretically predicted by Planck's law. However, most surfaces are not blackbodies, and emit some fraction of the amount of thermal radiation that a blackbody would. This fraction is known as *emissivity*. Then emissivity of a body is the rate between the energy emitted and an ideal emitter or blackbody at the same temperature. Hence emissivity (ε) may be expressed as follows:

$$\varepsilon\,(\lambda, \theta, \varphi, T) = i_b(\lambda, \theta, \varphi, T)/i_b(\lambda, T) \quad (1)$$

where $i_s(\lambda, \theta, \varphi, T)$ is the energy emitted by the sample per unit time, per unit area, per solid angle per wavelength interval, at temperature T, and $i_b(\lambda, T)$ (given by the Planck's law) is the energy emitted by a blackbody per unit time, per unit area, per solid angle, per wavelength interval at the same temperature T of the sample. $\varepsilon\,(\lambda, \theta, \varphi, T)$ is known as *spectral directional emissivity.* By integrating over the angle variables θ and φ one can define the *spectral hemispherical emissivity* while integrating only on wavelength the *total directional emissivity* is obtained. *Total hemispherical emissivity* ($\varepsilon^{\cap}$) is obtained by integrating both on angle variables and on wavelength range.

4.1 Experimental set-up for emissivity measurements on C/SiC samples

Total hemispherical emissivity, as a function of temperature, have been measured in the range 1000-1800 K with the *Moyen d'Essai et de DIagnostic en Ambiance Spatiale Extrême* (MEDIASE) set-up developed at PROMES-CNRS laboratory in France (Paulmier et al., 2001; Balat-Pichelin et al., 2002; Paulmier et al., 2005).

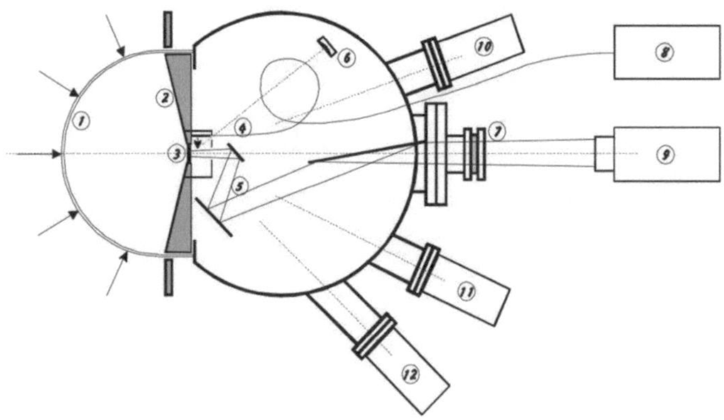

Fig. 2. MEDIASE test facility: 1) hemispherical silica window, 2) water-cooled front shield and sample-holder, 3) sample, 4) optical fibre, 5) 3-mirrors goniometer, 6) quartz crystal microbalance, 7) viewport, 8) pyro-reflectometer, 9) radiometer, 10) UV source position, 11) and 12) ion gun positions.

The device in Fig. 2 is composed of a vacuum chamber equipped with a hemispherical silica glass window (35 cm in diameter) for solar irradiation, a water-cooled front shield surrounding a sample holder. The specimen, in our experiments 40 mm diameter and 2 mm thick, is heated by concentrated solar energy at the focus of the 1 MW solar furnace. On the back face of the sample, total (wavelength range 0.6-40 μm) directional (0 to 80° by 10° step) radiance is measured by mean of a moving three-mirrors goniometer that collects the radiation emitted from the sample at different angles. The total directional radiance $i_s(\theta, \varphi, T)$ is measured with a radiometer calibrated against a reference blackbody.
The total directional emissivity is then given by:

$$\varepsilon(\theta, \varphi, T) = i_s(\theta, \varphi, T)/i_b(T) \quad (2)$$

where $i_b(T)$ is the theoretical blackbody radiation at temperature T.
The surface temperature is measured with a pyro-reflectometer, developed at PROMES-CNRS laboratory (Hernandez, 2005), collecting radiation from the centre of the sample. The total hemispherical emissivity ($\varepsilon^{\cap}$) is readily obtained by angular integration of the directional ones. Emissivity experiments have been performed in air at the pressure of 4 and 200 Pa.

4.2 Emissivity values of C/SiC samples

Emissivity values of C/SiC composites obtained in the temperature range 1000-1800 K and at 4 and 200 Pa are listed in Table 1 and plotted in Fig. 3 (Alfano et al., 2009).

Total Pressure [Pa]	Temperature [K]	Total Hemispherical Emissivity ($\varepsilon^{\cap}$) (0.6-40 μm)
200	1019±53	0.57±0.03
	1220±37	0.71±0.04
	1391±40	0.73±0.04
	1627±59	0.73±0.04
	1695±66	0.72±0.04
	1897±92	0.74±0.04
4	1114±41	0.59±0.03
	1220±32	0.69±0.03
	1417±38	0.71±0.04
	1620±68	0.72±0.04
	1714±68	0.70±0.03

Table 1. Values of total hemispherical emissivity measured at 4 and 200 Pa on CVDed SiC coated C/SiC samples produced by MT Aerospace (from Alfano et al., 2009).

In the examined temperature range, the observed differences between the emissivity values measured at 4 Pa and at 200 Pa lie in the experimental uncertainty. In both cases the emissivity values show a similar trend increasing from ~0.6 to ~0.7 in the temperature range 1000-1300 K then staying almost constant in the temperature range 1300-1800 K with an average value of about 0.7.

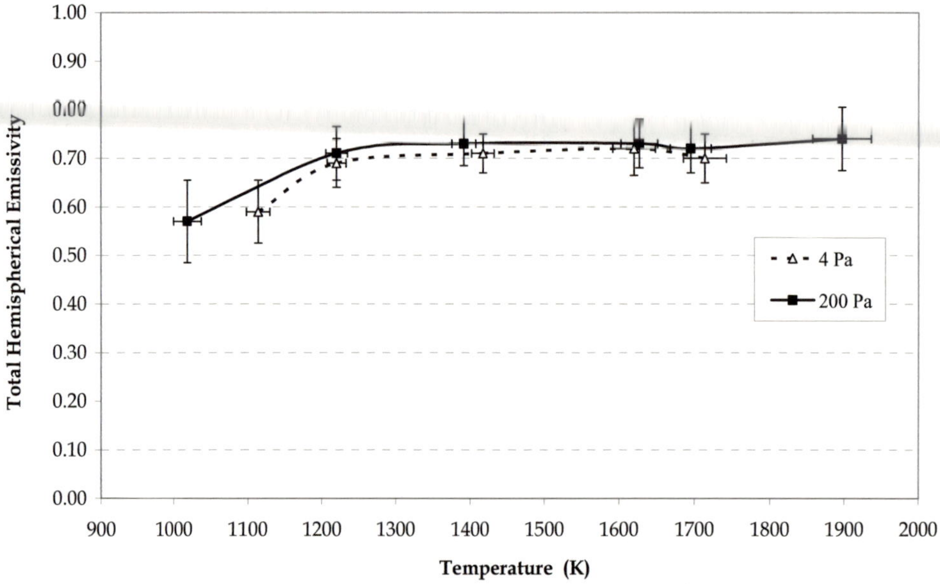

Fig. 3. Total hemispherical emissivity values of CVDed SiC coated C/SiC samples measured at 4 Pa (open triangles) and at 200 Pa (filled squares) from Alfano et al., 2009.

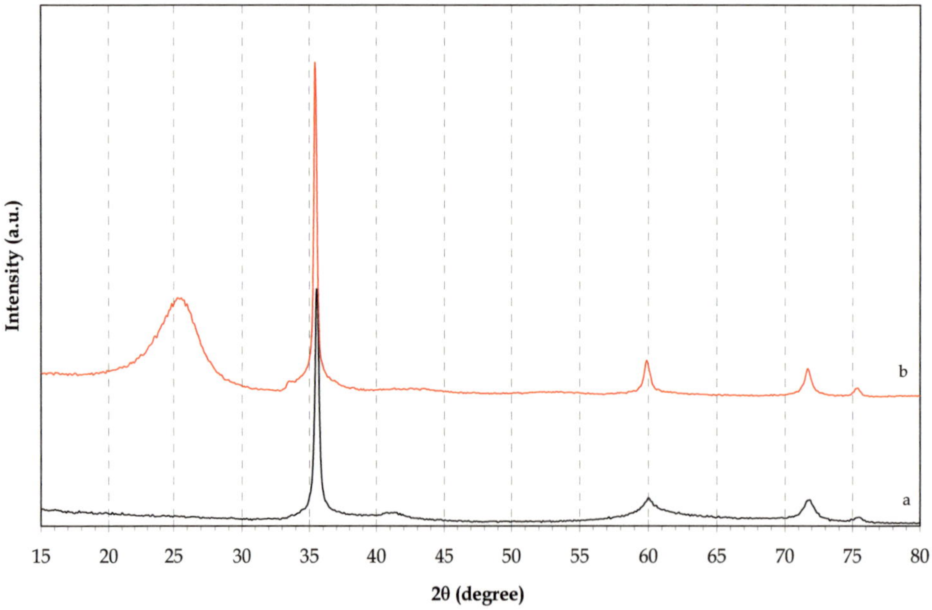

Fig. 4. X-ray diffraction patterns of the CVDed SiC coated C/SiC sample a) before and b) after the emissivity measurements performed at 200 Pa.

In Fig. 4 X-ray diffraction (XRD) patterns of the pristine specimen and tested one in the MEDIASE set-up are compared. In the XRD pattern of the tested sample the presence of the peak at $2\theta=26.2°$ related to the reflection [101] of quartz (SiO_2) is evident as well as the characteristic signals of β-SiC at $2\theta=35.5°$, 60.0°, 71.9°, and 79.9° related to [111], [220], [311], and [222] reflections, respectively. SiO_2 is the result of oxidation processes occurring during the emissivity measurements.
At the testing conditions the oxide layer is not thermodynamically stable: actually, the partial pressures of both gaseous SiO_2 and SiO are not negligible. In fact, as Heuer and Lou theoretically demonstrated by means of the volatility diagrams of β-SiC (Heuer & Lou, 1990), at 1800 K and at total air pressure of 200 Pa silicon carbide exhibits active oxidation with formation of oxides SiO and CO. This result is confirmed by several experimental data obtained by tests performed under standard air on massive and CVDed β-SiC: at 200 Pa the threshold temperature for the transition passive/active oxidation of CVDed samples is at about 1600-1700 K (Balat et al., 1992; Balat, 1996; Morino et al., 2002; Charpentier et al., 2010).
By Fig. 5, wherein the optical picture of the post test sample is illustrated, the surface corrosion is evident in particular if compared to the picture of the virgin specimen shown in Fig. 1a. These observations are also confirmed by SEM analysis: by comparing between the SEM cross-section micrograph carried out on the tested sample and shown in Fig. 6a to the analogous SEM image recorded on the pristine specimen and reported in Fig. 1b, the absence of the SiC-coating after emissivity tests becomes evident. Moreover SEM analysis also confirms that the fraction of SiO_2 remaining on the sample surface and detected by XRD measurements is not thermodynamically stable since a uniform protective silica-based glassy layer has not been observed.

Fig. 5. Picture of the emitting surface of the sample CVDed SiC coated C/SiC after the emissivity experiment carried out at air pressure of 200 Pa increasing the temperature from 1000 K up to about 1900 K.

In Fig. 6b fibre cracks can be observed probably due to high temperatures reached during the test and expansion mismatch between the SiC matrix and the carbon fibers. Therefore during emissivity tests, the sample surface undergoes ablation process due to formation of SiO and CO by active oxidation of SiC and contemporary partial evaporation of SiO_2.

Fig. 6. SEM micrographs of a) cross-section and b) surface of the sample CVDed SiC coated C/SiC after the emissivity experiment carried out at air pressure of 200 Pa increasing the temperature from 1000 K up to about 1900 K.

5. Catalytic efficiency (catalycity)

Catalycity could be defined as the catalytic efficiency shown by a material with respect to the recombination on the surface of atomic species due to the chemical reactivity of unsaturated valences of surface atoms.

The vehicle entering in a planet atmosphere by hypersonic trajectory creates a shock wave leading to very high temperatures. The created excited species (ions, atoms, molecules, electrons) diffuse in the boundary layer and react with the materials of the vehicle. Atomic oxygen (major species) and nitrogen, present for terrestrial entries, can strike the surface of the vehicle and recombine to form molecular species (O_2, N_2, NO) by exothermic reactions which have the effect to produce an increase of the surface temperature and to damage the integrity of the material. Then during the phase of development of a material as TPS, the contribution due to the recombination of atomic species on surfaces to estimation of thermal fluxes must be taken into account to predict heat rates on the hot parts of a re-entry spacecraft (Scott, 1985; Carleton & Marinelli, 1992; Barbato et al., 2000). Furthermore considering that atomic recombination reactions are typical surface reactive processes usually described by means of models for heterogeneous catalysis (Kovalev & Kolesnikov, 2005), the surface molecular structure and the surface morphology of structural part of aerospace vehicle involved in atomic recombination processes play a major part on the catalytic efficiency.

5.1 Experimental set-up to evaluate the atomic oxygen recombination coefficient of C/SiC samples

The catalytic efficiency of the material for the recombination of atomic oxygen was studied by means of a direct method for the measurement of the recombination coefficient γ_O, defined as the ratio of the flux of atomic oxygen that recombines on the surface to that of the total atomic oxygen impinging the surface of the sample. The *Moyen d'Essai Solaire d'Oxydation* (MESOX) experimental set-up, developed at French PROMES-CNRS laboratory, was used for this kind of experiments. In this apparatus, the atomic oxygen recombination coefficient γ_O is determined by measuring the concentration gradient of atomic oxygen in proximity of the sample surface by means of actinometry and Optical Emission Spectroscopy (OES) techniques.

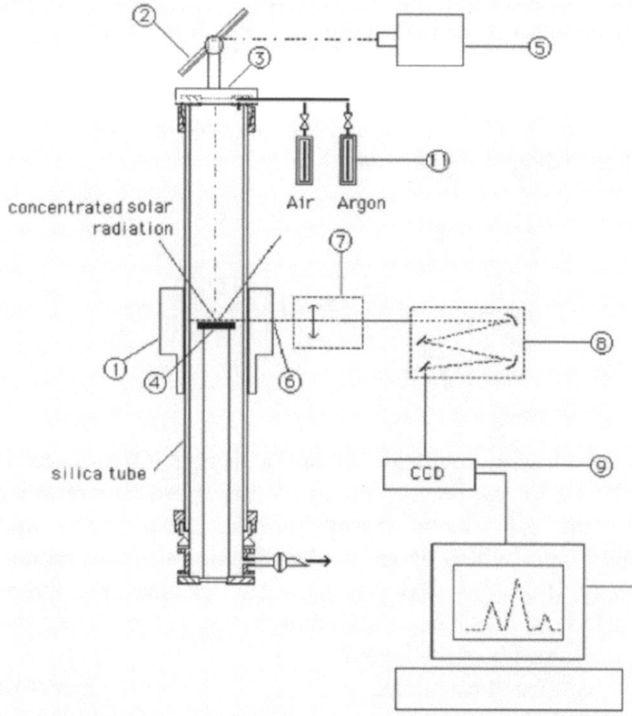

Fig. 7. MESOX set-up for optical emission spectroscopy measurements :1) waveguide, 2) mirror, 3) viewport, 4) sample, 5) pyrometer, 6) aiming slit, 7) lens, 8) spectrometer, 9) CCD-3000, 10) computer, 11) mass flowmeters.

The MESOX apparatus, shown in Fig. 7, is described in detail elsewhere (Balat et al., 2003): here we will only outline its main features. The ceramic composite sample (25 mm diameter and 2 mm thickness) is put in a plasma reactor consisting of a silica tube (quartz), 50 cm length and 5 cm diameter with CaF_2 viewports. The air plasma is generated by a 300 W microwave discharge surrounding the sample. A regulator, a gauge and a vacuum pump allow the precise control of the total pressure during experiments. The airflow, coming from the top of the reactor and then pumped downward, is fixed at 4 L/h. The sample is placed on a sample-holder at the centre of the plasma discharge. The reactor is positioned in such a way to let the sample be at 25 mm above the theoretical focus of the 6 kW solar furnace equipped with a variable opening shutter. Due to this shutter, the available incident concentrated solar flux can reach 5 MW m^{-2}.
The relative atomic oxygen concentration in the reactor is determined by means of OES combined with actinometry. The actinometry technique allows to follow the relative atomic oxygen concentration profile along the discharge. A small amount of argon (5%) is introduced in the flow and the evolution of the ratio I_O/I_{Ar}, where I_O and I_{Ar} are the intensities of oxygen and argon emission spectral lines respectively, is measured along the

discharge zone. The transitions at 844.6 nm and at 842.4 nm are chosen for atomic oxygen and argon respectively. Under well-defined conditions, the ratio I_O/I_{Ar} is proportional to the oxygen concentration in a wide range of temperature and in a broad region of the reactor. The spectroscopic bench is composed of an optical sampling system including a lens and a mirror, and a monochromator (spectrometer Triax 550 Jobin-Yvon) equipped with an Optical Multichannel Analyzer (OMA). The spectral resolution is 0.2 nm and the spatial resolution is around 270 μm.
Once the concentration profile of the atomic oxygen has been measured, the recombination coefficient is obtained according to the relation:

$$\gamma_O = \left(\frac{\left.\frac{I_O}{I_{Ar}}\right|_{x=L}}{\left.\frac{I_O}{I_{Ar}}\right|_{x=0}} \cdot \frac{T_S}{T_L} - 1 \right) \cdot \frac{4D_{O,air}}{VL} \qquad (3)$$

where $D_{O,air}$ is the binary diffusion coefficient of atomic oxygen (O) in air calculated for each gas temperature by the Chapman-Enskog theory, V the mean square velocity of atomic oxygen calculated by the gas kinetic theory (rarefied gas), L the thickness of the recombination boundary layer which enters in the diffusion-reaction model and which is measured experimentally, $(I_O/I_{Ar})_{x=L}$ and $(I_O/I_{Ar})_{x=0}$ are the measured intensities ratios at distance L from the surface and on the surface respectively, T_s the surface temperature and T_L the gas temperature at the edge of the layer L.
The uncertainties $\Delta\gamma/\gamma$ have been calculated taking into account the errors on I_O/I_{Ar} (10%) and L (5%) but also on the flow parameters: the diffusion coefficient $D_{O,air}$ and the mean square atomic velocity V. The accuracy on these two last values is due essentially to that of the gas temperature (5%), measured by OES (N_2 rotational temperature) giving a total accuracy on the recombination coefficient γ of ± 30%.

5.2 Recombination coefficient of atomic oxygen on surface of C/SiC samples

Two identical C/SiC specimens (labelled as A and B) were employed to perform catalycity tests in MESOX facilities. In Fig. 8 the logarithm of γ_O for the two samples with respect to the reciprocal of the absolute temperature is plotted in accordance with the well-known Arrhenius' equation (Arrhenius, 1889):

$$\gamma_O = A \exp(-E_a/RT) \qquad (4)$$

where A is the pre-exponential coefficient, E_a is the activation energy of the atomic oxygen recombination process, R the universal gas constant, and T is the absolute temperature.

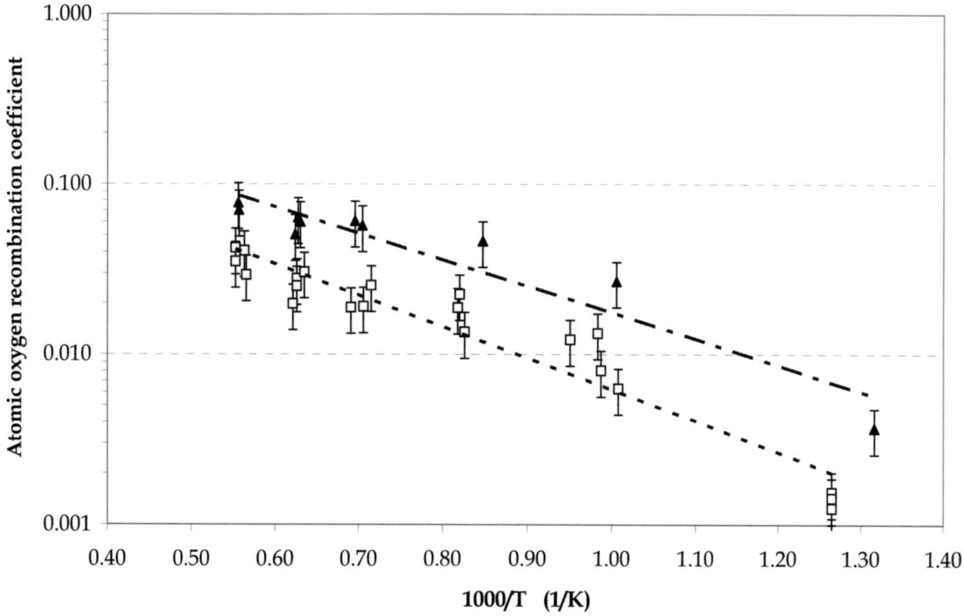

Fig. 8. Recombination coefficient for atomic oxygen versus reciprocal absolute temperature for the two CVDed SiC coated C/SiC samples A (open squares) and B (filled triangles). Both the measurements were performed at 200 Pa of total air pressure (from Alfano et al., 2009).

Both samples conform approximately to the Arrhenius law that predicts a linear behaviour in the examined temperature range. The Arrhenius parameters (A and E_a) used to fit by the equation 4 and the values of γ_o measured at the maximum temperature reached during the tests are summarized in Table 2.

Sample	Pre-exponential coefficient (A)	E_a/R [K]	Correlation coefficient	Activation Energy [kJ/mol]	Mean γ_o measured at 1800 K
A	0.436	4.251	0.913	35.3	0.0367
B	0.629	3.574	0.899	29.7	0.0744

Table 2. Parameters of the Arrhenius type expression (Eq. 4) used to fit the measured recombination coefficients (γ_o) plotted on Fig. 8, and mean measured values of γ_o at 1800 K related to the two tested samples of CVDed SiC coated C/SiC shown in Fig. 9 (Alfano et al., 2009).

The values of γ_O coefficients exponentially increase with the temperature starting from about 0.001 at 800 K to about 0.07 at 1800 K. Although the two series of γ_O coefficients differ more for low temperatures, the fitting curves are characterised by comparable slope values confirming the reproducibility of the catalycity experiments. This behaviour could be explained considering that each tested sample has a different surface roughness which strongly depends on the manufacturing process. In Fig. 9 the pictures of the two samples manufactured by PVI and employed to perform emissivity tests are shown: the different roughness structure of the two samples is evident by the simple macroscopic and optical examination. The activation energy for the atomic oxygen recombination process on C/SiC surface has been estimated equal to about 30 kJ/mol.

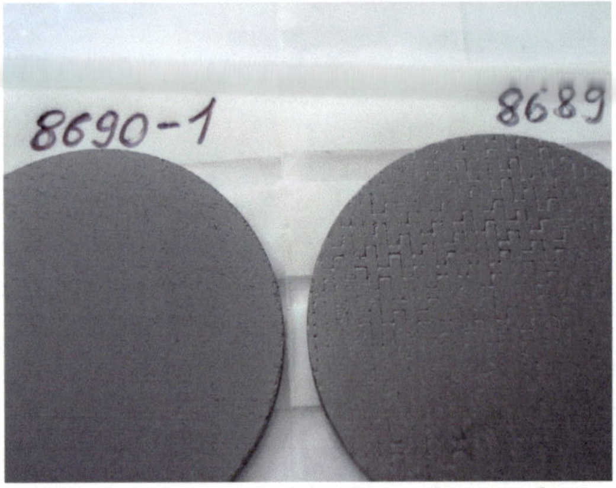

Fig. 9. Pictures of two CVDed SiC coated C/SiC specimens obtained by physical vapour infiltration.

The experimental conditions applied during the catalycity tests lie on the boundary line between the active and passive oxidation process (Balat, 1996; Morino et al., 2002). Then the plasma flux activates the passive oxidation process of the C/SiC samples associated with the formation of a silica glassy oxide scale (Fig. 10), and, at the same time, the partial ablation of the samples due to the loss of silicon as SiO_2 and SiO. The final results are the net weight loss that for the two tested samples has never been higher than 0.6 wt% and the formation of a glass coating with the thickness of 25-35 μm.

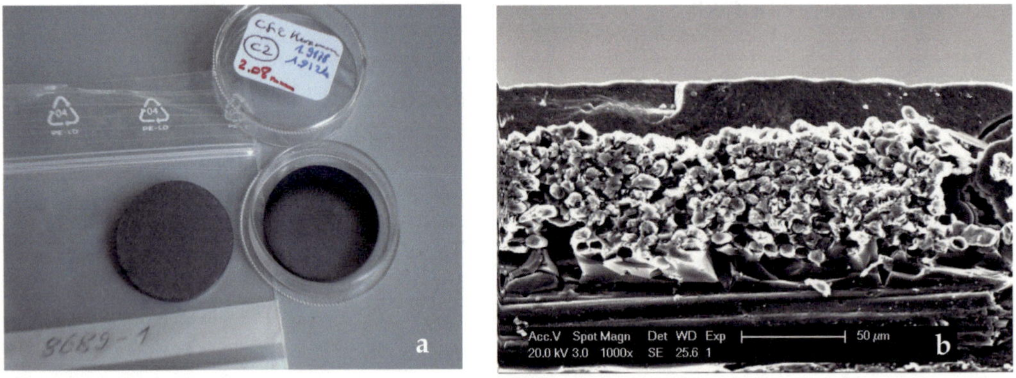

Fig. 10. a) Picture of post test C/SiC sample labelled as A and a virgin one, b) cross-section SEM micrograph after the catalycity test at 1800 K and 200 Pa.

6. Conclusions

The evaluation of emissivity and catalycity efficiency of materials employed in aerospace applications is necessary to estimate heat fluxes and surface temperature values reached by spacecraft vehicles during the re-entry phase in a planet atmosphere. The target is to design and then employ materials characterised by high emissivity values and low catalycity efficiency. In this chapter the measurement of hemispherical emissivity and atomic recombination coefficients of carbon fibre reinforced silicon carbide composites, used as thermal protection system for space vehicles, have been illustrated.

The C/SiC samples tested in MEDIASE facility in the temperature range 900-1900 K and both at 4 than 200 Pa are characterised by relatively high average emissivity values of about 0.7. This value is in line with data retrieved on silica-covered surfaces on different ceramic materials, confirming that the radiative behaviour of the SiC-coated C/SiCs is mainly dictated by the surface glassy oxide layer.

The catalycity measurements carried out using the MESOX facility have on the other hand shown a low oxygen recombination coefficient at high temperature (about 0.07 at 1800 K). The tests performed on two samples have also shown the strong dependence of the recombination coefficient on the surface morphology, which may slightly vary from sample to sample due to manufacturing issues. While samples of the same production batch have indeed shown different values of the recombination coefficient, the overall catalycity trend is the same allowing, moreover, to evaluate the activation energy of atomic oxygen recombination reaction that is about 30 kJ/mol.

The low catalycity and high emissivity exhibited by the investigated C/SiC further confirm its suitability for the intended application in the manufacturing of hot structures for re-entry vehicles. In any case the control of the manufacturing process becomes really mandatory in order to obtain surfaces characterised by a specific morphology which is able to guarantee defined emissivity and catalycity values.

Degradation effects due to the emissivity and catalycity tests on surface of the samples have been also evaluated by electron microscopy analysis and XRD analysis. The characterisation of post-test surface modifications becomes particularly important whenever the reuse capability of a material must be verified. To this end a lot of research groups are making many efforts to improve the resistance to oxidation of C/SiC composites in particular by deposition of ceramic coatings.

7. References

Alfano, D.; Scatteia, L.; Cantoni, S. & Balat-Pichelin, M. (2009). Emissivity and catalycity measurements on SiC-coated carbon fibre reinforced silicon carbide composite. *Journal of the European Ceramic Society*, Vol. 29, (July 2009), 2045-2051.

Arrhenius, S. (1889). Uber die reaktionsgeschwindigkeit bei der inversion von Rohrzucker durch Säuren. *Z. Phys. Chem.*, Vol.4, 226-248.

Barbato, M.; Reggiani, S.; Bruno, C. & Muylaert, J. (2000). Model for heterogeneous catalysis on metal surfaces with applications to hypersonic flows. *Journal of thermophysics and Heat Transfer*, Vol. 14, No. 3, (July-September 2000) 412-420.

Balat, M.; Flamant, G.; Male, G. & Pichelin, G. (1992). Active to passive transiation in the oxidation of silicon carbide at high temperature and low pressure in molecular and atomic oxygen. *Journal of Materials Science*, Vol. 27, (January 1992) 687-703

Balat, M. J. H. (1996). Determination of the active-to-passive transition in the oxidation of silicon carbide in standard and microwave-excited air. *Journal of the European Ceramic Society*, Vol. 16, No. 1, (May 1996) 55-62.

Balat-Pichelin, M.; Hernandez, D.; Olalde, G.; Rivoire, B. & Robert, J. F. (2002). Concentrated solar energy as a diagnostic tool to study materials under extreme conditions. *Journal of Solar Energy Engineering*, Vol. 124, No. 3, (August 2002) 215–222.

Balat-Pichelin, M.; Badie, J. M.; Berjoan, R. & Boubert, P. (2003). Recombination coefficient of atomic oxygen on ceramic materials under earth re-entry conditions by optical emission spectroscopy. *Chemical Physics*, Vol. 291, No. 2, (June 2003) 181–194.

Ban'kovskaya, I. B. & Zhabrev, V. A. (2005). Kinetic analysis of the heat resistance of ZrB_2–SiC composites. *Glass Physics and Chemistry*, Vol. 31, No. 4, (July 2005) 482-488.

Bartuli, C.; Valente, T. & Tului, M. (2002). Plasma spray deposition and high temperature characterization of ZrB_2-SiC protective coatings. *Surface and Coatings Technology*, Vol. 155, No. 2-3, (June 2002) 260–273.

Carleton, K. L. & Marinelli, W. J. (1992). Spacecraft thermal energy accommodation from atomic recombination. *Journal of Thermophysics and Heat Transfer*, Vol. 6, No. 4, (October-December 1992) 650- 655.

Carney, C. M.; Mogilvesky, P. & Parthasarathy, T. A. (2009). Oxidation behaviour of zirconium diboride silicon carbide produced by the spark plasma sintering method. *Journal of the American Ceramic Society*, Vol. 92, No. 9, (September 2009) 2046-2052.

Chamberlain, A.; Fahrenholtz, W. & Hilmas, G. (2004). High-strength zirconium diboride-based ceramics. *Journal of the American Ceramic Society*, Vol. 87, No. 6, (June 2004) 1170-1172.

Chamberlain, A.; Fahrenholtz, W.; Hilmas, G. & Ellerby, D. (2005). Oxidation of ZrB_2-SiC ceramics under atmospheric and reentry conditions. *Refractories Applications Transactions*, Vol. 1, No. 2, (July-August 2005) 1-8.

Charpentier, L., Balat-Pichelin, M.; Glénat, H.; Bêche, E.; Laborde, E. & Audubert, F. (2010). High temperature oxidation of SiC under helium with low-pressure oxygen. Part 2: CVD β-SiC. *Journal of the European Ceramic Society*, Vol. 30, No. 12, (September 2010) 2661-2670.

Chen, D.; Xu, L.; Zhang, X.; Ma, B. & Hu, P. (2009). Preparation of ZrB_2 based hybrid composites reinforced with SiC whiskers and SiC particles by hot-pressing. *Journal of Refractory Metals & Hard Materials*, Vol. 27, No. 4, (July 2009) 792-795.

Coperet, H.; Soyris, P.; Lacoste, M.; Garnett, J. & Tidwell, D. (2002). MMOD Testing of C-SiC Based Rigid External Insulation of the X-38/CRV Thermal Protection System, *Proceedings of the 53rd International Astronautical Congress*, Houston, Texas, 10-19 October 2002, IAC-02-I.3.06, Curran Associates Inc, Ohio.

Dadd, G.; Owen, R.; Hodges, J. & Atkinson, K. (2006). Sustained Hypersonic Flight Experiment (SHyFE), *Proceedings of the 14th AIAA/AHI space planes and hypersonic systems and technologies conference*, Canberra, Australia, 6-9 November 2006, AIAA-2006-7926.

Del Vecchio, A.; Di Clemente, M.; Ferraiuolo, M.; Gardi, R.; Marino, G.; Rufolo, G. & Scatteia, L. (2006). Sharp hot structures project current status. *Proceedings of the 57th*

53rd International Astronautical Congress, Valencia, Spain, 2-6 October 2006, IAC-06-C2.4.05.

Dogigli, M.; Pfeiffer, H.; Eckert, A. & Fröhlich, A. (2002a). Qualification of CMC body flaps for X-38, *Proceedings of the 52nd International Astronautical Congress*, pp. 1-11, Toulose, France, 1-5 October 2001, IAF-01-I.3.02.

Dogigli, M.; Pradier, A. & Tumino, G. (2002b). Advanced key technologies for hot control surfaces in space re-entry vehicles, *Proceedings of the 53rd International Astronautical Congress*, pp. 1-13, Houston, Texas, 10-19 October 2002, IAC-02-I.3.02, Curran Associates Inc, Ohio.

Fahrenholtz, W. G.; Hilmas, G. E.; Chamberlain, A. L. & Zimmermann, J. W. (2004). Processing and characterization of ZrB_2-based ultra-high temperature monolithic and fibrous monolithic ceramics. *Journal of Materials Science*, Vol. 39, No. 19, (March 2004) 5951-5957.

Fahrenholtz, W. G. (2005). The ZrB_2 Volatility Diagram. *Journal of the American Ceramic Society*, Vol. 88, No. 12, (December 2005) 3509-3512.

Fahrenholtz, G.; Hilmas, G. E.; Talmy, I. G. & Zaykoski, J. A. (2007a). Refractory diborides of zirconium and hafnium. *Journal American Ceramic Society*, Vol. 90, No. 5, (May 2007) 1347-1364.

Fahrenholtz, W. G. (2007b). Thermodynamic analysis of ZrB_2-SiC oxidation: formation of a SiC-depleted region. *Journal of the American Ceramic Society*, Vol. 90, No. 1, (January 2007) 143-148.

Gasch, M.; Ellerby, D.; Irby, E.; Beckman, S.; Gusman, M. & Johnson, S. (2004). Processing, properties and arc jet oxidation of hafnium diboride/silicon carbide ultra high temperature ceramics. *Journal of Materials Science*, Vol. 39, No. 19, (October 2004) 5925-5937.

Goodman, J. & Ireland, P. (2006). Thermal Modelling for the Sustained Hypersonic Flight Experiment, *Proceedings of the 14th AIAA/AHI space planes and hypersonic systems and technologies conference*, Canberra, Australia, 6-9 November 2006, AIAA-2006-8071.

Guicciardi, S.; Silvestroni, L.; Nygren, M. & Sciti, D. (2010). Microstructure and toughening mechanisms in spark plasma-sintered ZrB_2 ceramic reinforced by SiC whiskers or SiC chopped fibers. *Journal of the American Ceramic Society*, Vol. 93, No. 8, (August 2010) 2384-2391.

Hald, H. & Winkelmann, P. (1995). TPS development by ground and reentry flight testing of CMC materials and structures, *Proceeding of the 2nd European workshop on thermal protection systems*, Stuttgart, Germany, 23-27 October 1995, ESA.

Hald, H. & Winkelmann, P. (1997). Post mission analysis of the heat shield experiment CETEX for the EXPRESS capsule, *Proceeding of the 48th IAF-Congress*, Turin, Italy, 6-10 October 1997, IAF-97-1.4.01.

Hald, H. (2003). Operational limits for reusable space transportation systems due to physical boundaries of C/SiC materials. *Aerospace Science and Technology*, Vol. 7, No. 7, (October 2003) 551-559, 10.1016/S1270-9638(03)00054-3.

Han, J.; Hu, P.; Zhang, X; Meng, S. & Han, W. (2008). Oxidation-resistant ZrB_2-SiCcomposites at 2200°C. *Composites Science and Technology*, Vol. 68, No. 3-4, (March 2008) 799-806.

Harnisch, B.; Kunkel, B.; Deyerler, M.; Bauereisen, S. & Papenburg, U. (1998). Ultra-lightweight C/SiC mirrors and structures. *ESA bulletin*, 95, 4-8.

Hernandez, D. (2005). A concept to determine the true temperature of opaque materials using a tricolor pyroreflectometer. *Review of Scientific Instruments*, Vol. 76, No. 2, (January 2005) 024904-024911.

Heuer, A. H. & Lou, V. L. K. (1990). Volatility diagrams for silica, silicon nitride, and silicon carbide and their application to high temperature decomposition and oxidation. *Journal of the American Ceramic Society*, Vol. 73, No. 10, (October 1990) 2785–3128.

Hu, P.; Goulin, W. & Wang, Z. (2009). Oxidation mechanism and resistance of ZrB_2-SiC composites. *Corrosion Science*, Vol. 51, (November 2009) 2724-2732.

Karlsdottir, S. N. & Halloran, J. W. (2009). Oxidation of ZrB_2-SiC: influence of SiC content on solid and liquid oxide phase formation. *Journal of the American Ceramic Society*, Vol. 92, No. 2, (February 2009) 481-486.

Kovalev, V. L. & Kolesnikov, A. F. (2005). Experimental and theoretical simulation of heterogeneous catalysis in aerothermochemistry. *Fluid Dynamics*, Vol. 40, No. 5, (September 2005) 669-693.

Kuriakose, A. K. & Magrave, J. L. (1964).The oxidation kinetics of zirconium diboride and zirconium carbide at high temperatures. *Journal Electrochemical Society*, Vol. 111, No. 7, (July 1964) 827-831.

Laux, T.; Ullmann, T.; Auweter-Kurtz, M.; Hald, H. & Kurz, A. (2001). Investigation of thermal protection materials along an X-38 re-entry trajectory by plasma wind tunnel simulations, *Proceedings of the 2nd International Symposium on Atmospheric Reentry Vehicles and Systems*, pp. 1-9, Arcachon, France, 26-29 March 2001, AAAF.

Levine, S. R.; Opila, E. J.; Halbig, M. C.; Kiser, J. D.; Singh, M. & Salem, J. A. (2002). Evaluation of ultra-high temperature ceramics for aeropropulsion use. *Journal of the European Ceramic Society*, Vol. 22, No. 14-15 , (June 2002) 2757-2767.

Monteverde, F.; Guicciardi, S. & Bellosi, A. (2003). Advances in microstructure and mechanical properties of zirconium diboride based ceramics. *Material Science and Engineering*, Vol. 346, No. 1-2, (April 2003) 310-319.

Monteverde, F. & Bollosi, A. (2005). The resistance to oxidation of an HfB_2–SiC composite. *Journal of the European Ceramic Society*, Vol. 25, No. 7, (May 2005) 1025-1031.

Monteverde, F. (2006). Beneficial effects of an ultra-fine α-SiC incorporation on the sinterability and mechanical properties of ZrB_2. *Applied Physics A: Materials Science & Processing*, Vol. 82, No. 2, (February 2006) 329-337.

Monteverde, F. & Scatteia, L. (2007). Resistance to Thermal Shock and to Oxidation of Metal Diborides–SiC Ceramics for Aerospace Application. *Journal of the American Ceramic Society*, Vol. 90, No. 4, (April 2007) 1130-1138.

Monteverde, F.; Bellosi, A. & Scatteia L. (2008). Processing and properties of ultra-high temperature ceramics for space applications. *Materials Science and Engineering: A*, Vol. 485, No. 1-2, (June 2008) 415–421.

Morino, Y.; Yoshinaka, T.; Auweter-Kurtz, M.; Hilfer, G.; Speckmann, H.-D. & Sakai, A. (2002). Erosion characteristics of SiC coated C/C materials in arc heated high enthalpy air flow. *Acta Astronautica*, Vol. 50, No. 3, (February 2002) 149-158.

Mroz, C. (1994). Zirconium diboride. *American Ceramic Society Bulletin*, Vol. 73, No. 6, (1994) 141-142.

Mühlratzer, A. & Leuchs, M. (2001). Applications of Non-Oxide CMCs, *Proceedings of the 4th International conference on high temperature ceramic matrix composites*, pp. 288-298, Munich, Germany, 1-3 October 2001, Wiley-VCH, Weinheim.

Naslain, R. & Schneider, H. (2002). *High temperature ceramic matrix composites*, Wiley-VCH, 3527303200, Weinheim.

Opeka, M. M.; Talmy, I. G.; Wuchina, E. J.; Zaykoski, J. A. & Causey, S. J. (1999). Mechanical, thermal, and oxidation properties of refractory hafnium and zirconium compounds. *Journal of the European Ceramic Society*, Vol. 19, No. 13-14, (October 1999) 2405-2414.

Opeka, M. M.; Talmy, I. G. & Zaykoski, J. A. (2004). Oxidation-based materials selection for 2000°C + hypersonic aerosurfaces: Theoretical considerations and historical experience. *Journal of the Materials Science*, Vol. 39, No. 19, (October 2004) 5887-5904.

Opila, E.; Levine, S. & Lorincz, J. (2004). Oxidation of ZrB_2- and HfB_2-based ultra-high temperature ceramics: effect of Ta additions. *Journal of Materials Science*, Vol. 39, No. 19, (October 2004) 5969-5977.

Pfeiffer, H. & Peetz, K. (2002). All-ceramic body flap qualified for space flight on the X-38, *Proceedings of the 53rd International Astronautical Congress The World Space Congress*, Houston, TX, 10-19 October 2002, IAF-02-I.6.b.01, Curran Associates Inc, Ohio.

Pichon, T.; Soyris, P.; Faucault, A.; Parenteau, J.M.; Prel, Y. & Guedron, S. (2006). C/SiC based rigid external thermal protection system for future reusable launch vehicles: generic shingle, pre-X/FLPP anticipated development test studies, *Proceedings of the 5th European Workshop on Thermal Protection Systems and Hot Structures*, Noordwijk, The Netherlands, 17-19 May 2006, ESA Publications Division, Noordwijk.

Paulmier, T.; Balat-Pichelin, M.; Le Quéau, D.; Berjoan, R. & Robert, J. F. (2001). Physico-chemical behaviour of carbon materials under high temperature and ion irradiation. *Applied Surface Science*, Vol. 180, No. 3-4, (August 2001) 227–245.

Paulmier, T.; Balat-Pichelin, M. & Le Quéau, D. (2005). Structural modifications of carbon-carbon composites under high temperature and ion irradiation. *Applied Surface Science*, Vol. 243, No. 1-4, (April 2005) 376–393.

Rasky, D. J.; Salute, J.; Kolodziej, P. & Bull, J. (1998). The NASA Sharp Flight Experiment, *Proceedings of the 3rd European Workshop on Thermal Protection Systems*, Noordwijk, The Netherlands, 25-27 March 1998, ESA Publications Division, Noordwijk.

Reimer, T. & Laux, T. (2006). Thermal and Mechanical Design of the EXPERT C/C-SiC Nose. *5th European Workshop on Thermal Protection Systems and Hot Structures*, Noordwijk, The Netherlands, 17-19 May 2006, ESA Publications Division, Noordwijk

Reinhart, T. J. (1987). *Engineered Materials Handbook*, ASM International, 0-87170-279-7, Ohio.

Rezaire, A.; Fahrenholtz, W. G. & Hilmas, G. E. (2006). Oxidation of zirconium diboride silicon carbide at 1500°C at a low partial pressure of oxygen. *Journal of the American Ceramic Society*, Vol. 89, No. 10, (October 2006) 3240-3245.

Rezaire, A.; Fahrenholtz, W. G. & Hilmas, G. E. (2007). Evolution of structure during the oxidation of zirconium diboride–silicon carbide in air up to 1500°C. *Journal of the European Ceramic Society*, Vol. 27, No. 6, (November 2007) 2495-2501

Russo, G. & Marino, G. (2003). The USV Program & UHTC development, *Proceedings of the 4th European Workshop on Thermal Protection Systems for Space Vehicles*, pp. 157-163, Palermo, Italy, 26-29 November 2002, European Space Agency, Paris.

Savino, R.; De Stefano Fumo, M.; Paterna, D. & Serpico, M. (2005). Aerothermodynamic study of UHTC-based thermal protection systems. *Aerospace Science and Technology*, Vol. 9, (January 2005) 151–160.

Scatteia, L.; Del Vecchio, A.; De Filippis, F.; Marino, G. & Savino, R. (2005). PRORA-USV SHS: Development of sharp hot structures based on ultra high temperature metal diborides current status, *Proceedings of the 56th International Astronautical Congress*, Fukuoka, Japan, 17-21 November 2005, IAC-05-C2.3.05, Curran Associates Inc, Ohio.

Scatteia, L.; Alfano, D.; Cantoni, S.; Monteverde, F.; De Stefano Fumo, M. & Di Maso, A. (2010). Plasma Torch Test of an Ultra-High-Temperature Ceramics Nose Cone Demonstrator. *Journal of Spacecraft and Rockets*, Vol. 47, No. 2, (March-April 2010) 271-279.

Scott, C. D. (1985). Effects of nonequilibrium and wall catalysis on Shuttle heat transfer. *Journal of Spacecraft and Rockets*, Vol. 22, No. 5, (September-October 1985) 489-499.

Schneider, S. J. (1991). *Engineered Materials Handbook: Ceramics and Glasses, Volume 4*, ASM International, 0-87170-282-7, Ohio.

Silvestroni, L.; Sciti, D.; Melandri, C. & Guicciardi, S. (2010). Toughened ZrB_2-based ceramics through SiC whisker or SiC chopped fiber additions. *Journal of the European Ceramic Society*, Vol. 30, No. 11, (August 2010) 2155-2164.

Smith, D.; Caroll, C.; Marschall, J. & Pallix, J. (1997). Materials testing on the DC-X and DC-XA. *NASA Technical Memorandum 110430*, (January 1997).

Steinacher, A.; Lange, H.; Weiland, S. & Hudrisier, S. (2007). Development of CMC body flaps for future re-entry vehicles, *Proceedings of the 58th International Astronautical Congress*, Hyderabad, India, 24-28 September 2007, IAC-07-C2.4.01, Curran Associates Inc, Ohio.

Tripp, W. C. & Graham, H. C. (1971). Thermogravimetric study of the oxidation of ZrB_2 in the temperature range of 800°C to 1500°C. *Journal Electrochemical Society*, Vol. 118, No. 7, (July 1971) 1195-1199.

Tripp, W. C.; Davis, H. H. & Graham, H. C. (1973). Effect of SiC addition on the oxidation of ZrB_2. *American Ceramic Society Bulletin*, Vol. 52, No. 8, (1973) 612-616.

Tului, M.; Marino, G. & Valente, T. (2006). Plasma spray deposition of ultra high temperature ceramics. *Surface and Coatings Technology*, Vol. 201, No. 5, (October 2006) 2103-2108.

Zhang, X.; Hu, P.; Han, J. & Meng, S. (2008). Ablation behaviour of ZrB_2–SiC ultra high temperature ceramics under simulated atmospheric re-entry conditions. *Composites Science and Technology*, Vol. 68, No. 7-8, (June 2008) 1718–1726.

Zhang, X.; Xu, L.; Han, W.; Weng, L.; Han, J. & Du, S. (2009). Microstructure and properties of silicon carbide whisker reinforced zirconium diboride ultra-high temperature ceramics. *Solid State Sciences*, Vol. 11, No. 1, (January 2009) 156-161.

11

Effect of Self-Healing on Fatigue Behaviour of Structural Ceramics and Influence Factors on Fatigue Strength of Healed Ceramics

Wataru Nakao
Yokohama National University
Japan

1. Introduction

Silicon carbide particles embedded in ceramic matrices give rise to self-healing function in the structural ceramics operated at high temperature in air. This feature is taken advantage of to enhance life time of the ceramic components with high mechanical reliability.

Ceramics are well-known to tend to have brittle fracture that usually occurs in a rapid and catastrophic manner. Brittle fracture is usually caused by the stress concentration at the tip of the flaws. For brittle fracture under pure mode I loading, under which crack is subjected to opening, the fracture criterion is that the stress intensity factor, K_I, is equal to the fracture toughness, K_{IC}. Since the value of K_I is determined from the flaw size and the geometry between flaw and loading, one can understand that the fracture strength of ceramic components is not an intrinsic strength but is determined from fracture toughness and flaw size. Especially surface cracks are the most severe flaws because surface cracks lead to the highest stress concentration. If surface cracks are introduced during service, e.g., crash or thermal shock, the strength of ceramic components decreases significantly. The behaviour leads to low mechanical reliability of ceramic components.

The ceramic composites containing silicon carbide particles can heal surface cracks by themselves, as shown in Fig. 1 (Nakao et al., 2010, Ando et al., 2004). Surface cracking allows the silicon carbide particles on the crack walls to contact the oxygen in the surrounding atmosphere. If the components operated at high temperature, the contact would cause the oxidation of silicon carbide. The oxidation includes almost two times volume expansion of the condensed phases and the huge exothermic heat. Due to the volume expansion, the space between the crack walls can be completely filled with the formed oxide. Furthermore, the reaction heat leads to strong bonding at the interface between the matrix and the formed oxide. As a result, the self-healing induced by the oxidation of the embedded silicon carbide particles can recover the degraded strength completely and can enhance the life time of the ceramic components with high mechanical reliability.

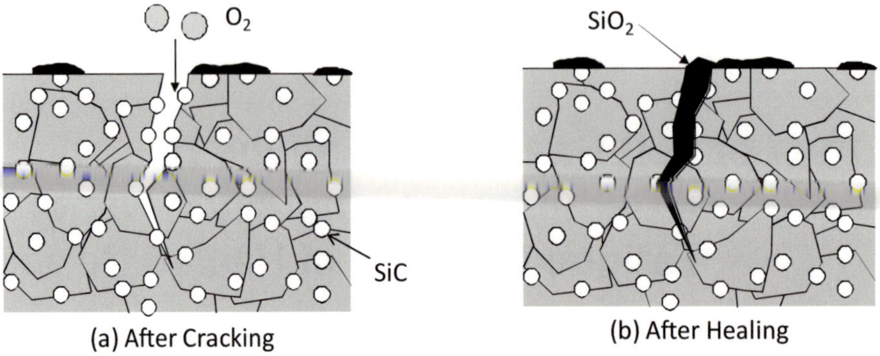

Fig. 1. Schematic diagram of the self crack-healing mechanism

To discuss the life time of the ceramic components, it is also important to know the fatigue behaviour that caused the crack growth when the stress intensity factor is lower value of K_{IC}. The mechanism has been analyzed to describe the slow crack growth behaviour including chemical reaction kinetics, and it is called as stress corrosion cracking (SCC). Figure 2 shows the typical example of SCC, in which Si-O bonds in silica glass are de-bonded by the SCC with the moisture in the surrounding atmosphere. At the stressed crack tip, the accumulated elastic energy activates the Si-O bond, thereby enhancing the hydrolysis of the bond. As a result, surface cracks propagate with the progression of the hydrolysis. This suggests that the fatigue degradation of the structural ceramics has also been generated by the presence of surface cracks, which not only leads to the highest stress concentration but it is also able to react with the reactant in the surrounding atmosphere. Thus, one can understand fatigue strength and life time are also determined by the size of surface cracks.

This chapter will introduce the effect of self crack-healing on fatigue strength in structural ceramics. As mentioned above, fatigue strength is also significantly influenced by the presence of surface cracks. Therefore, self-healing of surface cracks gives large advantage to the fatigue strength in structural ceramics, and the knowledge of the effects contributes to the realization of a long life time with high strength integrity of ceramic components.

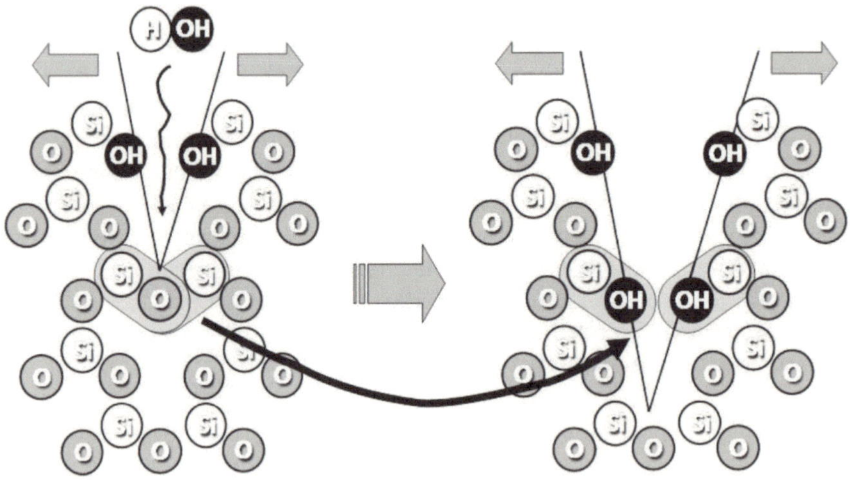

Fig. 2. Schematic illustration of the mechanism of stress corrosion cracking

2. Fatigue Behaviour of Crack-Healed Surface

2.1 Effect of Self Crack-Healing on Dynamic Fatigue Behaviour

The fatigue strength enhancement by the self crack-healing has been clearly found in dynamic fatigue behaviour, as shown in Fig. 3 (Nakao et al., 2006).

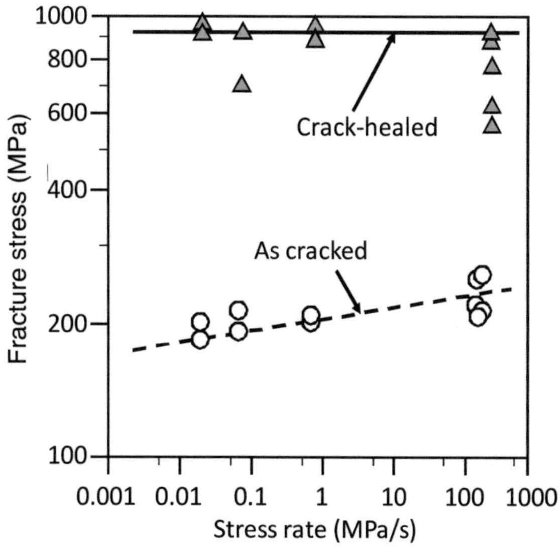

Fig. 3. Dynamic fatigue results of the crack-healed mullite containing 15 vol% SiC whiskers and 10 vol% SiC particles composite with that of the composite having a semi-elliptical crack of 100 μm in surface length

Dynamic fatigue behaviour can be obtained from the fracture strength as a function of the applied stress rate. If the specimen exhibits the slow crack growth due to SCC, lower stress rate allows the surface cracks to progress larger by the applied stress until the fracture, thereby giving lower fracture strength. Thus, demonstrating the logarithmic plot of the fracture strength versus the stress rate, one can find the positive slope in the materials exhibiting the SCC crack propagation. The gradient of the slope implies the indicator of the fatigue sensitivity.

Figure 3 shows the dynamic fatigue results of the crack-healed mullite containing 15 vol% SiC whiskers and 10 vol% SiC particles composite (MS15W10P), which possesses the excellent self crack-healing ability (Nakao et al., 2006) and high crack growth resistance by SiC whiskers reinforcement. In order to test the self crack-healing effect, the specimen contained a semi-elliptical surface crack having surface length of 0.1 mm, which comes from the prolongation of the diagonal line of the indentation introduced by the Vickers indentation, and the indenter test, pre-crack was completely healed by the high temperature heat treatment (Nakao et al., 2006) at 1300 °C for 2 h in air. In comparison, Figure 3 also shows the dynamic fatigue behaviour of the as-cracked mullite based composite, i.e., the specimens were subjected to no healing treatment, thereby exhibiting the SCC crack growth.

The crack-healed MS15W10P sample shows a constant fracture strength over whole the stress rate, while the as-cracked MS15W10P exhibits the positive slope in the dynamic fatigue curve.

Furthermore, the fracture initiation of the crack-healed MS15W10P is not the healed pre-crack but the embedded flaws, e.g., the aggregation of SiC particles. The embedded flaws cannot be reacted with the moisture in the surrounding atmosphere. Then the flaws cannot propagate by the SCC crack growth. A similar behaviour was reported in the fatigue behaviour of the sintered alumina in toluene (Evans, 1972). Therefore, the result demonstrates clearly that the self crack-healing makes the fatigue sensitivity decrease significantly.

2.2 Effect of Surface Morphology on the Fatigue Strength of Self Crack-Healed Specimens

Surface morphology of the healed specimen was found to affect the fatigue strength in the situation when the continuous stress is applied for a long period. Here, the effect demonstrates the fatigue behaviour of three alumina- 30 vol% SiC composite having different SiC whiskers content (20% in this case). These composites possess excellent self crack-healing ability (Nakao et al., 2005).

Static fatigue testing, in which the constant stress is continuously applied, is well-known to be the most severe fatigue situation for alumina based ceramics, because SCC crack growth is mechanically enhanced by only the stress intensity factor at the crack tip, and not enhanced by the fluctuation of the applied stress. According to Japan Industrial Standard (JIS) R1632, the optimal test finish time is 100 h, and the maximum stress under which the specimen survived until the test finish time is determined as the static fatigue limit.

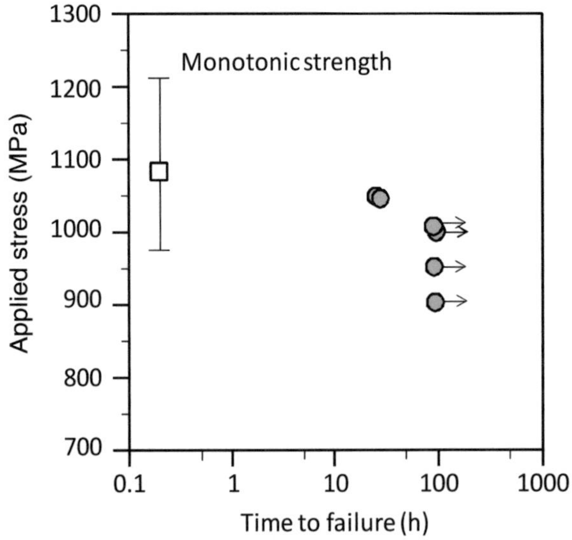

Fig. 4. Stress- time to failure diagram of the crack-healed alumina- 20 vol% SiC whiskers- 10 vol% SiC particles composite, with its monotonic strength

The stress-time to failure diagram of the crack-healed alumina- 20 vol% SiC whiskers and 10 vol% SiC particles composite (AS20W10P) is shown in Fig. 4 (Sugiyama et al., 2008). In order to test for the self crack-healing effect, the specimen contained the indentation pre-crack having surface length of 0.1 mm. The indentation pre-crack was completely healed by the high temperature heat treatment (Nakao et al., 2005) at 1300 °C for 5 h in air. For comparison, the monotonic strength is also shown in Fig. 4 using open squares. Under the applied stress below 1000 MPa, all the specimens survived up to test finish time, while two specimens fractured after 30 h under 1050 MPa. Therefore, the static fatigue limit of the crack-healed AS20W10P has been determined to be 1000 MPa. The fatigue limit present in the distribution range of the monotonic strength. Furthermore, the fracture initiation was found not to be the healed pre-crack.

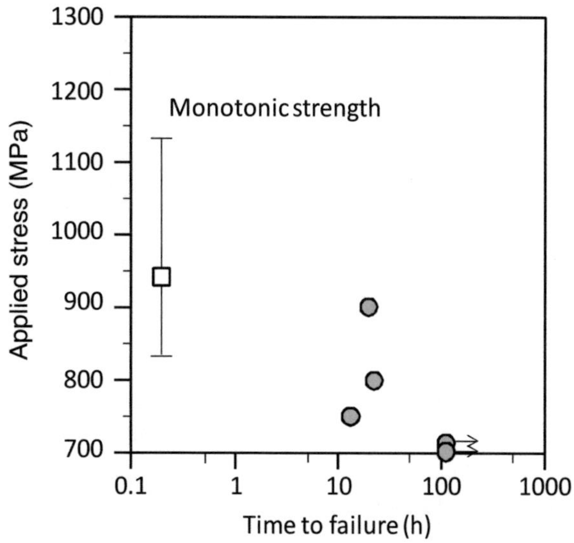

Fig. 5. Stress- time to failure diagram of the crack-healed alumina- 30 vol% SiC whiskers composite, with its monotonic strength

Alternatively, the static fatigue limit of the healed alumina-30 vol% SiC whiskers composite (AS30W) containing the healed pre-crack is less than the monotonic strength as shown in Fig.5. Also the healed pre-crack was found not to act as the fracture (fatigue) initiation in the healed AS30W, but large strength degradation due to static fatigue occurs from the tensile surface. The strength degradation results from that the surface morphology of the healed surface, which affects significantly the static fatigue behaviour.

The fatigue strength degradation has a relation to the surface morphology, such as surface roughness, as shown in Fig. 6. The progression of the SiC oxidation, inducing the self crack-healing, generated the island like formed oxide on the healed surface, as shown in Fig. 1. If coarse SiC particle, for example SiC whisker, exist on the surface, the surface roughness significantly increases with the progression of the oxidation. Thus, Sugiyama et al. (Sugiyama et al. 2009) concluded that the fatigue crack of the healed ceramics which has large surface roughness is initiated from the "valley" in the rough surface, which can generate too high a stress concentration to induce SCC reaction. Although the healed surface of alumina-30 vol% SiC particles composite (AS30P) has low surface roughness, the composite exhibits large fatigue strength degradation. From the SEM investigation, AS30P had the grain-pullout traces, which were about 3 μm in diameter, on the healed surface as shown in Fig. 7. The pullouts were introduced during polishing and cannot be eliminated by crack healing. Therefore, the fatigue crack initiated from these traces.

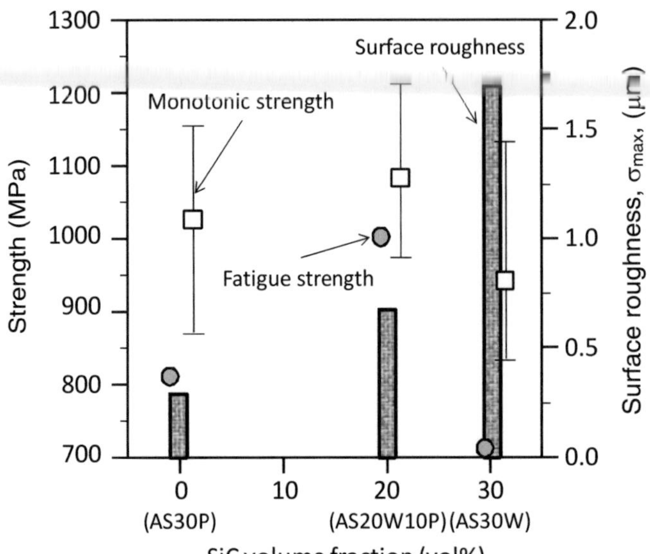

Fig. 6. The influence of SiC whiskers content on the surface roughness of the healed surface, monotonic strength and static fatigue limit

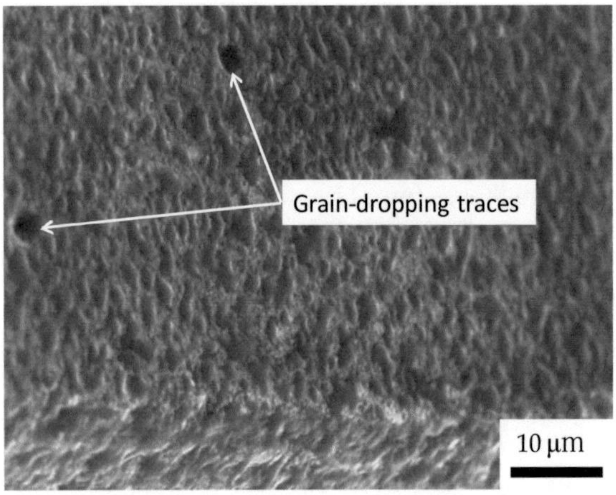

Fig. 7. The grain-pullout traces on the surface of the crack-healed alumina- 30 vol% SiC particles composite

From the above results, it is noted that if the design requires extreme low fatigue sensitivity, it is necessary not only to heal the whole surface cracks but also to manage the surface roughness of the healed surface.

3. Fatigue Behaviour with Self Crack-Healing

In high temperature fatigue, there is another interesting phenomenon, in order that self crack-healing occurs at the same time as fatigue damage. For example, figure 8 (Nakao et al., 2006) shows a logarithmic plot of life time in terms of the applied stress for the crack-healed mullite containing 15 vol.% SiC whiskers and 10 vol.% SiC particles composite at 1000 °C. In general, i.e. the slow crack growth is included, the life time increases as the applied stress decreases.

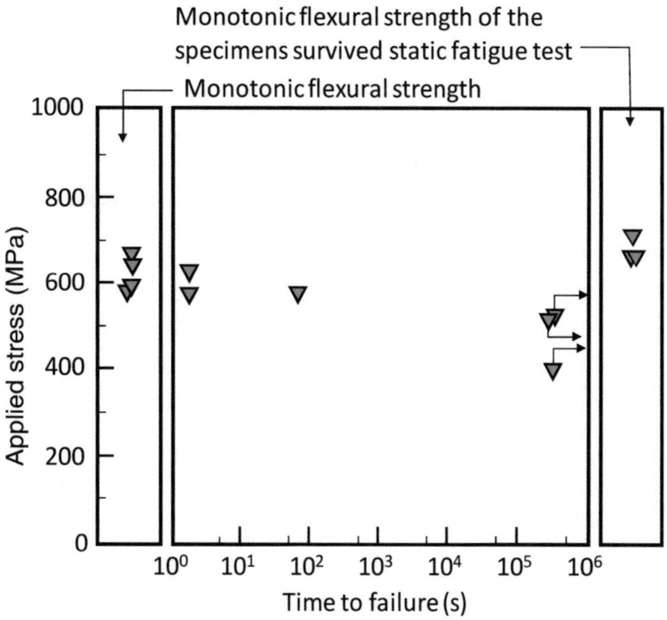

Fig. 8 . Logarithmic plot of life time in terms of the applied stress for the the crack-healed mullite containing 15 vol% SiC whiskers and 10 vol% SiC particles composite at 1000 °C

However, all crack-healed test specimens survived up to finish time of 100 h under static stresses of 50 MPa less than the lower bound of the monotonic strength at the same temperature. Alternatively, the specimens fractured at less than 100 s under stresses corresponding to the lower bound of the flexural strength. This failure is not fatigue but rather rapid fracture. Therefore, it is confirmed that the crack-healed composite is not degraded by the static fatigue at 1000 °C. The behaviour would result from the self crack-healing that occurs rapidly compared with the fatigue damage.

4. Summary

Self-healing induced by the high temperature oxidation of the dispersed silicon carbide particle can eliminate the surface cracks, leading to a great benefit to ensure the high monotonic strength of ceramics. Since most mechanical and structural components usually work under applied continuous stresses, it is important to know the self-healing effect on fatigue behaviour.

The fatigue sensitivity, which corresponds to the strength degradation is caused by the continuous loading, has been decreased significantly in the healed ceramics, because the self-healing of surface cracks can prevent stress corrosion cracking from occurring in the original surface cracks. Furthermore, the improvement on the fatigue sensitivity has been affected by the surface roughness of the healed ceramics. Another attractive phenomenon has occurred in high temperature fatigue behaviour. This results from the fact that self-healing occurs at the same time as fatigue damage. As a result, the fatigue limit has been equal to the minimum monotonic strength at the same temperatures.

5. References

Nakao, W., Takahashi, K. & Ando, K. (2010). Chapter 6 Self-healing of surface Cracks in Structural Ceramics, In: *Self-healing Materials: Fundamentals, Design Strategies, and Applications*, Ghosh, S.K. (Ed.), 183-217, WILEY-VCH Verlag GmbH & Co. KGaA, ISBN : 978-3-527-31829-2, Weinheim, German

Ando, K., Kim, B.S., Chu, M.C., Saito, S. & Takahashi, K. (2004). Crack-healing and Mechanical Behaviour of Al_2O_3/SiC Composites at Elevated Temperature. *Fatigue and Fracture of Engineering Materials and Structures*, Vol. 27, 533-541

Nakao, W., Nakamura, J., Yokouchi, M., Takahashi, K. & Ando, K. (2006). Static Fatigue Limit of Crack-Healed Mullite/ SiC Multi-Composite. *Transaction of JSSE*, Vol. 51, 20-26.

Nakao, W., Mori, S., Nakamura, J., Yokouchi, M., Takahashi, K. & Ando, K. (2006). Self-Crack-Healing Behavior of Mullite/SiC Particle/SiC Whisker Multi-Composites And Potential Use for Ceramic Springs. *Journal of the American Ceramic Society*, Vol. 89, No. 4, 1352-1357

Evans, A.G. (1972). A Method for Evaluating the Time-Dependent Failure Characteristics of Brittle Materials- and Its Application to Polycrystalline Alumina. *Journal of the Materials Science*, Vol. 7, No. 10, 1137-1146

Nakao, W., Osada, T., Yamane, K., Takahashi K. & Ando, K. (2005). Crack-healing Mechanism by Alumina/ SiC particles/ SiC Whiskers Multi-Composite. *Journal of the Japan Institute of Metals*, Vol. 69, No. 8, 663-666

Sugiyama, R., Yamane, K., Nakao, W., Takahashi, K. & Ando, K. (2008). Effect of Difference in Crack-healing Ability on Fatigue Behavior of Alumina/ Silicon Carbide Composites. *Journal of Intelligent Material Systems and Structures*, Vol. 19, 411-415

12

Contribution to the Evaluation of Silicon Carbide Surge Arresters

Arnaldo Gakiya Kanashiro and Milton Zanotti Jr.
University of São Paulo
Brazil

1. Introduction

Basically, there are two types of surge arresters installed in the power system: the silicon carbide (SiC) surge arresters and the zinc oxide (ZnO) ones. The SiC surge arresters are being replaced by the latter but a large number are still operating in the electrical system. Due to high costs and operational difficulties it is not possible to replace all the SiC surge arresters in a short term and the utilities have to manage both technologies (Carneiro, 2000). Therefore, in order to avoid failure, it is very important to know the level of degradation of the SiC surge arresters, most of them with 20 or more years of service. The diagnostic techniques applied to the ZnO surge arresters have been developed for many years and they are presented in (Heinrich & Hinrichsen, 2001) and (Kannus & Lahti, 2005). The main techniques are: thermovision and leakage current measurement. The leakage current measurement can give valuable information about the degradation of the ZnO surge arresters. The level of degradation is proportional to the magnitude of the resistive component of the total leakage current. Regarding the SiC surge arresters, there is no recommendation according to the (IEC-60099-5, 2000) standard.
This chapter is focused on the application of the leakage current aiming the diagnostic of the SiC surge arresters. In the electrical utilities, the thermovision technique is also applied to the SiC surge arresters but it is very difficult to conclude, among several arresters, which of them are more degraded in order to replace them and to avoid failure (Almeida et al., 2009). This chapter describes the methodology and the results of the investigation that was performed aiming the application of the leakage current technique to the SiC surge arresters. The basic idea was to correlate the level of degradation of the SiC surge arresters with the magnitude of the third harmonic (3^a H) of the total leakage current. Laboratory tests were performed in arresters that had been replaced in the substations. The characteristics of the leakage current showed a good correlation with the degradation of the arresters. Some of them were disassembled in order to verify their internal components. Afterwards, measurements were carried out in substation aiming to verify the feasibility of the technique in the field. Another substation was also evaluated, and based on the leakage current measurements obtained some SiC arresters were replaced. The results obtained in the investigation might help the utilities to develop more adequate maintenance programs.

2. Methodology

The methodology of the investigation was based on laboratory tests and visual inspection of the internal components of the surge arresters (Kanashiro et al., 2009). The first step of the research was to analyze the characteristics of the SiC surge arresters in service, considering their age, manufacturer, rated voltage, etc. Then, some arresters were selected to be tested at the laboratory, considering the nominal voltages of 88 kV, 138 kV, 230 kV, 345 kV and 440 kV, from five manufacturers. These SiC surge arresters were no longer being used by the utility in its electrical system because they were either replaced by ZnO surge arresters or presented abnormal heating during thermovision periodical measurements.

The surge arresters operated during about 20 up to 25 years in the electrical system; however, there wasn't any information about their behaviour during this time period. Therefore, the values of the power frequency spark-over voltage and lightning spark-over voltage tests, obtained at the laboratory, were firstly used as reference to indicate degradation, taking into account the requirements of the manufacturers and of the (IEC 60099-1, 1999) standard.

Afterwards, measurements of the leakage current were carried out, with the amplitude (I_{peak}) and the 3^a H component being obtained. The phase difference between the leakage current and the voltage applied to the surge arrester was also determined. Then, radio influence voltage (RIV) tests, with frequency of 500 kHz and impedance of 300 Ω, and thermovision measurements were also performed for some selected arresters.

In some of the 88 kV arresters, surface erosion was observed which is a typical sign of electric arc damage on the porcelain, as shown in Fig. 1. Also, most of them had some degree of corrosion in their metallic parts, as shown in Fig. 2. Finally, in order to check the viability of measuring the leakage current in the field, some SiC surge arresters installed at Paraibuna substation were chosen. After this, due to the explosion of one 88 kV SiC surge arrester at Mairiporã substation, leakage current measurements were also performed in this substation.

Fig. 1. Typical sign of electric arc damage on the porcelain of the surge arrester.

Fig. 2. Corrosion in the metallic parts of the surge arrester.

In the next section, the main results of the investigation, related to the diagnostic of the SiC surge arresters, are presented. The results refer to the 88 and 138 kV surge arresters.

3. Results and Discussion

3.1 Laboratory tests

Because some of the SiC surge arresters presented signs of electric arc damage on the porcelain and, also, corrosion in their metallic parts, prior to the laboratory tests, measurements of insulation resistance and of Watts losses, were carried out on several samples, aiming at a preliminary check of their general condition. The results of the power frequency spark-over voltage and lightning spark-over voltage tests are shown in Table 1 and Table 2. According to the manufacturer's requirements, all 88 kV samples failed the power frequency spark-over voltage test. In this test, samples A5 and A6 presented unstable behaviour.

The 138 kV samples A7, A8, A9, C5, D3 and D5 failed the power frequency spark-over voltage test. Regarding the lightning spark-over voltage test, all samples (88 kV and 138 kV) fulfilled the manufacturer's requirements.

Manufacturer A	Power frequency spark-over voltage (kV)	Lightning spark-over voltage (kV)	
		Positive	Negative
A1	134	182	181
A2	105	171	168
A3	85	178	178
A4	102	172	167
A5	-----	173	172
A6	-----	173	188

Table 1. 88 kV surge arresters.

Manufacturers A/B/C/D	Power frequency spark-over voltage (kV)	Lightning spark-over voltage (kV)	
		Positive	Negative
A7	193	227	227
A8	170	222	228
A9	178	225	224
B1	244	284	272
B2	246	279	272
B3	242	287	294
B4	233	234	225
B5	237	234	229
B6	241	271	269
B7	232	272	272
C1	226	382	354
C2	219	374	363
C3	224	364	359
C4	218	340	322
C5	188	349	344
C6	233	355	344
D1	274	374	367
D2	273	376	372
D3	268	376	366
D4	271	372	369
D5	262	378	369

Table 2. 138 kV surge arresters.

Afterwards, measurements of the total leakage current were carried out, with the I_{peak} values and the 3ª H component being obtained. The phase difference between the total leakage current and the voltage applied to the sample was also determined. The results, with the exclusion of samples A5 and A6, are shown in Table 3.

Manufacturers A/B/C/D	Power frequency spark-over voltage (kV)	Leakage current		Phase difference (degree)
		I_{peak} (mA)	3ª H (%)	
A1	134 (F)	0.172	6.7	89
A2	105 (F)	0.192	10.1	65
A3	85 (F)	0.412	24.9	54
A4	102 (F)	0.696	32.9	47
A7	193 (F)	0.278	2.6	85
A8	170 (F)	0.268	5.6	70
A9	178 (F)	0.246	6.8	71
B1	244	0.226	4.8	72
B2	246	0.252	5.7	70
B3	242	0.370	6.0	77
B4	233	0.234	6.4	68
B5	237	0.251	6.8	68
B6	241	0.230	8.5	63
B7	232	0.261	9.4	53
C1	226	0.363	5.6	73
C2	219	0.456	5.8	75
C3	224	0.346	6.8	79
C4	218	0.332	6.9	68
C5	188 (F)	0.430	7.5	83
C6	233	0.726	18	51
D1	274	0.364	1.9	89
D2	273	0.357	2.1	89
D3	268 (F)	0.357	2.1	82
D4	271	0.330	2.5	84
D5	262 (F)	0.331	3.8	78

Table 3. Leakage current measurement.

In Table 3, (F) means that the sample failed the power frequency spark-over voltage test. After the measurements above, some arresters were selected to be submitted to the radio influence voltage (RIV) and thermovision tests.

In the three tests, leakage current, RIV and thermovision, the phase-to-ground voltages 51 kV and 80 kV were applied to the 88 kV and 138 kV samples, respectively.

The thermovision were carried out after the samples had been energised for a time period of 5 to 7.5 hours, depending on the manufacturer. One measurement was carried out for each of four different sides of the sample. Each measurement corresponds to the thermal imaging obtained along the sample, from top to bottom. Each of the four sides of the sample had its maximum and minimum temperatures determined, and the difference (Δt) between these temperatures was calculated. The greatest difference value found was named "Δt_{max}".

The highest temperature value obtained in the sample was named "t_{max}". The results are shown in Table 4, where (F) means that the sample failed the power frequency spark-over voltage test, (*) means that significant results were not observed in the RIV test and (**) that the sample was not tested. Fig. 3 shows an example of a thermal image measurement.

Surge arresters	Leakage current		RIV (μV)	Thermovision (°C)	
	I_{peak} (mA)	3ª H (%)		t_{max}	Δt_{max}
A1	0.172	6.7	< 25	20.8	2.0
A2	0.213	10.1	*	21.6	2.0
B2	0.252	5.7	< 25	28.0	4.6
B3	0.370	6.0	*	28.3	4.3
B6	0.230	8.5	< 25	27.9	4.4
C3	0.346	6.8	< 25	19.9	2.6
C5 (F)	0.430	7.5	4518	19.3	2.8
C6	0.726	18	6381	32.6	17.6
D1	0.364	1.9	< 25	18.1	1.7
D3 (F)	0.357	2.1	64	18.2	1.9

Table 4. Results of the leakage current measurement, RIV and thermovision.

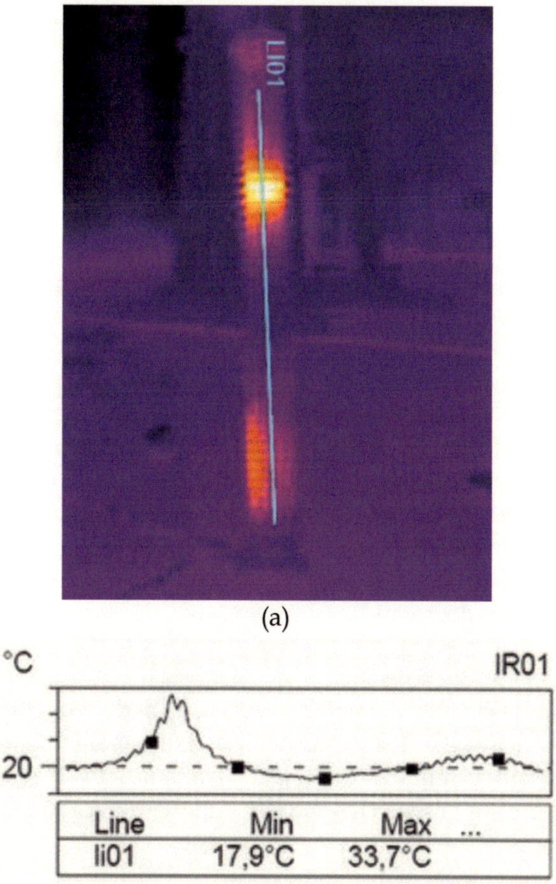

(a)

(b)

Fig. 3. Example of a thermal image measurement, (a) thermal image of the surge arrester and (b) temperature along the surge arrester.

The following aspects can be pointed out, concerning the results shown in Table 3 and Table 4:

Manufacturer A - 88 kV surge arresters:

- all surge arresters failed the power frequency spark-over voltage test;
- surge arrester A1 presented the highest power frequency spark-over voltage value (134 kV), the lowest amplitude value of the leakage current (0.172 mA), the lowest 3ª H component (6.7 %) and the greatest phase difference (89^0);
- on the other hand, surge arrester A4, which showed the greatest amplitude of the leakage current (0.696 mA), had the greatest 3ª H component (32.9 %) and the lowest phase difference (47^0).

Manufacturer A – 138 kV surge arresters:

- all surge arresters failed the power frequency spark-over voltage test;
- surge arrester A7, which presented the highest power frequency spark-over voltage value (193 kV), also had the lowest harmonic distortion (2.6 %) and the greatest phase difference (85^0);

- significant results were not observed in the RIV and thermovision measurements.

Manufacturer B – 138 kV surge arresters:
- all surge arresters were successful in the power frequency spark-over voltage tests;
- surge arresters B6 and B7 presented harmonic distortion values (8.5 % and 9.4 %, respectively) greater than the values obtained with other samples of the same manufacturer. Smaller phase difference values were also obtained (63^0 and 55^0, respectively);
- significant results were not obtained in the RIV and thermo vision measurements.

Manufacturer C – 138 kV surge arresters:
- surge arrester C5 failed the power frequency spark-over voltage test and presented 3^a H component of 7.5 % and phase difference of 83^0;
- although surge arrester C6 was succesful in the power frequency spark-over voltage test, it presented leakage current amplitude of 0.726 mA, distortion of 18 % and phase difference of 51^0, which may indicate some degradation of its internal components;
- surge arresters C5 and C6 had high RIV values, suggesting the presence of internal electrical discharges. In spite of this, the thermovision measurement showed higher temperature only in surge arrester C6.

Manufacturer D – 138 kV surge arresters:
- surge arresters D3 and D5 failed the power frequency spark-over voltage test;
- surge arrester D5, which presented the lowest power frequency spark-over voltage value, had the greatest leakage current distortion (3.8 %) and the smallest phase difference (78^0);
- significant results were not observed in the RIV and thermovision measurements.

3.2 Internal components of the surge arresters

Some of the surge arresters were disassembled in order to verify the correlation between the presence of deterioration in their internal parts and the results obtained in the laboratory tests. The following surge arresters were selected: A2, A4, A6 (manufacturer A) and C1, C3 and C5 (manufacturer C).

In the surge arresters A2, A4 and A6 there are permanent magnets in parallel with gap electrodes. Nonlinear resistors of SiC are placed between the gap electrodes. The dismantled surge arrester of manufacturer A can be seen in the Fig. 4.

In the SiC surge arresters of manufacturer C, the gap electrodes are divided in groups. In each group a tape is applied to fix the gap electrodes. A nonlinear resistor is placed in parallel with each group to equalize the voltage potential of the gap electrodes.

The internal components of the surge arrester C can be seen in Fig. 5. At the edges are placed coils in order to facilitate arc extinguishing. Fig. 6 shows one group of gap electrodes.

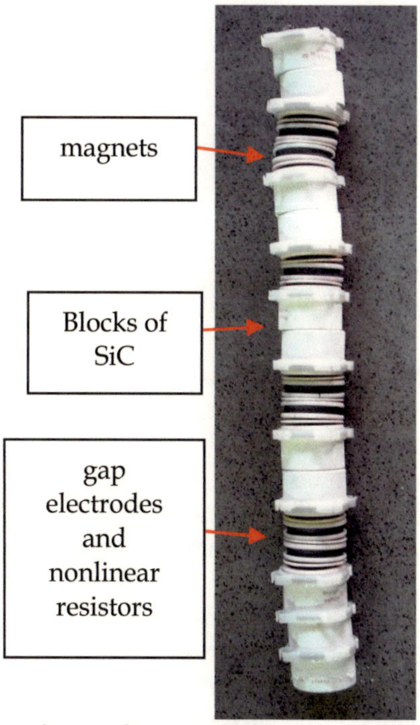

Fig. 4. Surge arrester of manufacturer A.

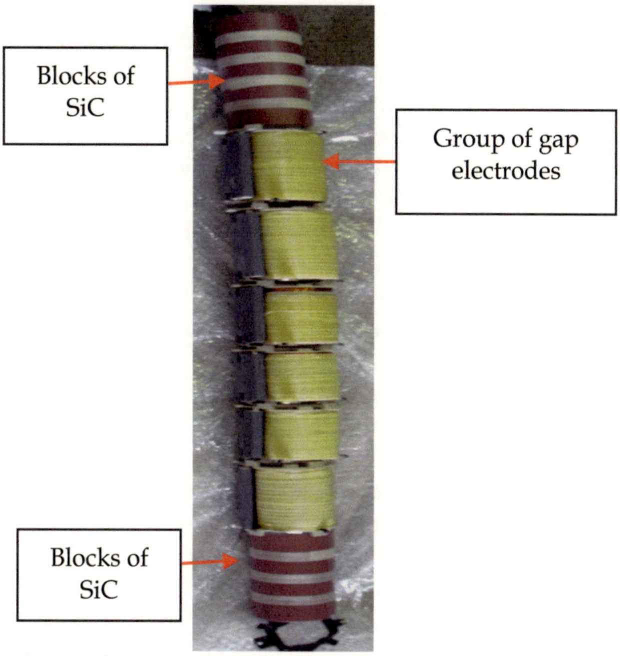

Fig. 5. Surge arrester of manufacturer C.

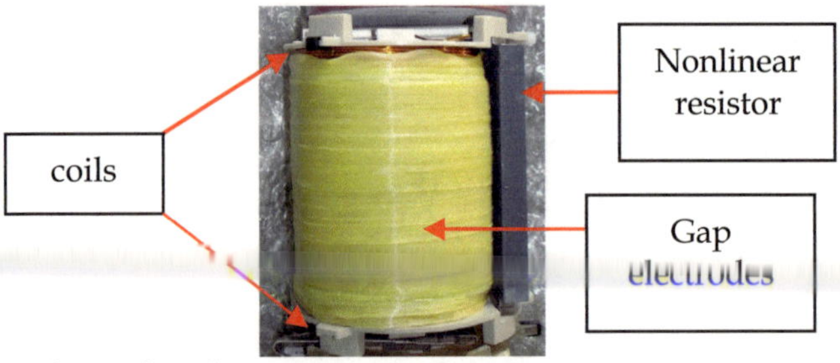

Fig. 6. Group of gap electrodes of surge arrester C.

In general, it was noticed that moisture was presented in the internal components of the arresters. Some traces of discharges on the surface of the blocks were also observed. Some of the surge arresters presented signs of discharges in the gap electrodes. During the visual inspection, it was also observed that some nonlinear resistors were damaged.

The surge arrester A6 (manufacturer A) was more degraded in comparison with A2 and A4. The arrester C5 (manufacturer C) was the worst in comparison to the surge arresters C1 and C3.

The surge arrester C5 presented some damaged nonlinear resistors and, probably, this was the reason for the high level of RIV (4,518 μV), shown in Table 4. This surge arrester also failed the power frequency spark-over voltage test. In Fig. 7 and Fig. 8 it is possible to visualize the condition of the components of the surge arresters, considering manufacturers A and C, respectively.

As a general conclusion, it was observed that the surge arresters of manufacturers A and C presented evidence of ingress of moisture and signs of discharges. Moisture ingress may have deteriorated the SiC material (McDermid, 2002) and (Grzybowski, 1999).

Afterwards, surge arresters of manufacturer B were also dismantled and it was observed that the internal components were in good condition. These results mean that they could have remained in service until they needed to be replaced by the ZnO surge arresters.

After disassembling the surge arresters, the following aspects can be pointed out, concerning the results shown in Table 3 and Table 4:

- the highest values of the leakage current, in terms of amplitude and harmonic distortion, corresponded to the degradation of the surge arresters;
- the thermovision technique, RIV tests and also the leakage current, considering the C6 sample, showed that this surge arrester was degraded. The visual inspection of its internal components confirmed this assumption;
- the surge arresters C5 presented high RIV values, suggesting the presence of internal electrical discharges. In spite of this, the thermovision measurement showed higher temperature only in surge arrester C6;
- the B1 to B7 surge arresters were successful in all tests but samples B6 and B7 presented greater harmonic distortion values and should be removed first from the electrical system;
- the leakage current values, in terms of the amplitude and the third harmonic component, could be used to select the SiC surge arresters to be replaced by the ZnO ones.

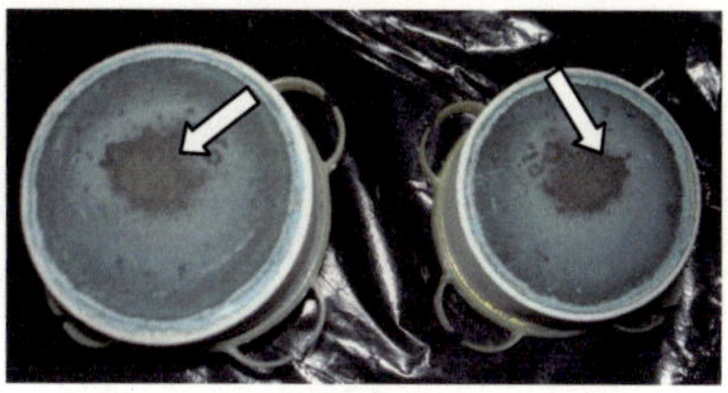

(a)

(b)

(c)

(d)

Fig. 7. Surge arresters of manufacturer A, (a) blocks: signs of discharge, (b) gap electrodes: signs of discharge, (c) block: presence of moisture and (d) gap electrode: signs of discharge.

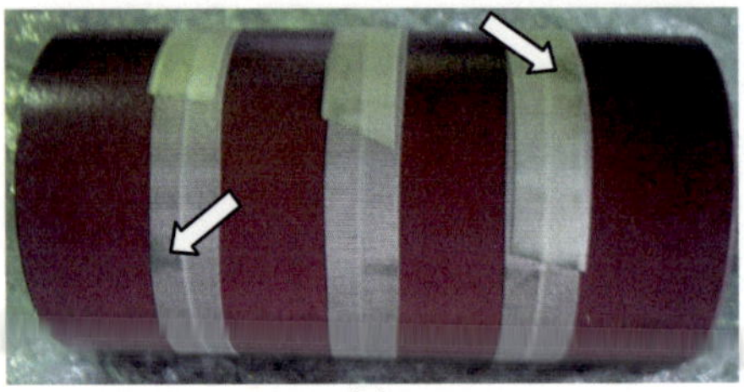

(a)

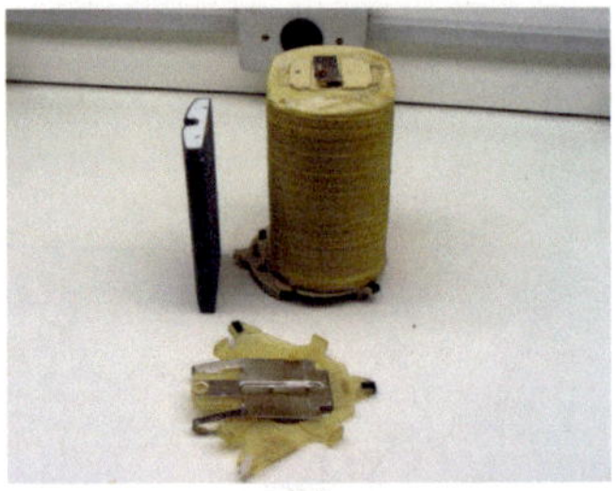

(b)

(c)

(d)

Fig. 8. Surge arresters of manufacturer C, (a) block surface: presence of moisture, (b) group of gap electrodes: damaged, (c) nonlinear resistor: broken and (d) nonlinear resistor: broken.

4. Measurements at Substation

Leakage current measurements in 88 kV SiC surge arresters, in service, were performed in the Paraibuna substation first, aiming to check the viability of this technique. Details of the SiC surge arresters installation were considered, such as presence of counter discharges, grounding cable of the surge arresters, the presence of insulators in the assembled surge arresters, etc. These aspects have important influence on the results. A device, consisting of a current transformer (CT) and a digital instrument, was used in the field. The CT was placed in the grounding cable, between the discharges counter and the bottom part of the surge arrester (position 1) or after the discharge counter (position 2), as shown in Fig. 9. The aim was to investigate the interference of the installation in the results. The leakage current was measured using 60 Hz and 180 Hz frequencies. When the CT was placed in the position 2, there was interference, as shown in the oscillograms of Fig. 10.

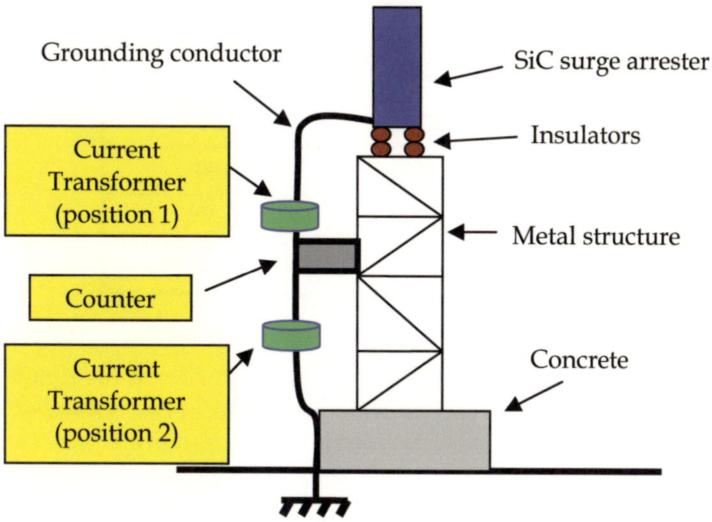

Fig. 9. Leakage current measurement at the substation.

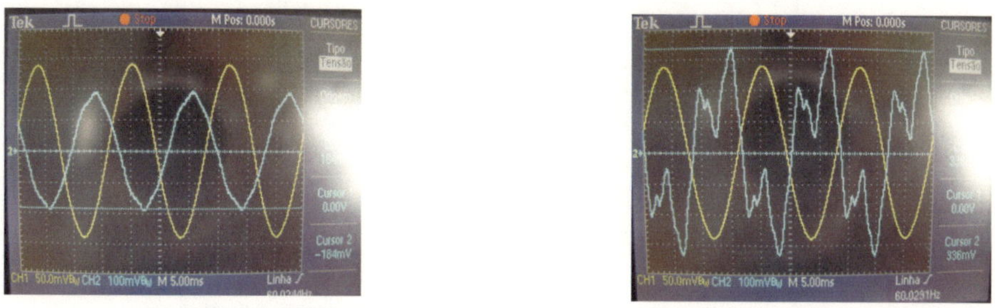

(a) (b)

Fig. 10. Waveforms of the leakage current (blue) and of the applied voltage (yellow), (a) CT in the position 1 and (b) CT in the position 2.

The SiC surge arresters were installed in the 88 kV, circuits TAU-01, JAG-01 and JAG-02. The three phases of each circuit were named as a, b and c. Table 5 shows the results. The comparison between the results from the field and from the laboratory is not so easy because the manufacturers of the surge arresters are not the same, therefore, it is possible to observe that the values are relatively low.

Surge arresters	Leakage current 60 Hz rms (mA)	Leakage current 180 Hz rms (mA)
TAU-01a	0.132	0.005
TAU-01b	0.128	0.004
TAU-01c	0.112	0.004
JAG-01a	0.170	0.004
JAG-01b	0.078	0.002
JAG-01c	0.163	0.005
JAG-02a	0.122	0.003
JAG-02b	0.088	0.002
JAG-02c	0.070	0.002

Table 5. 88 kV arresters - Paraibuna substation.

Afterwards, due to the explosion of one 88 kV SiC surge arrester at Mairiporã substation, several measurements of the leakage current were performed in that substation. 88 kV and 138 kV SiC surge arresters in service, were measured and the results are presented in Table 6 and Table 7, respectively.

Surge arresters	Leakage current 60 Hz rms (mA)	Leakage current 180 Hz rms (mA)
JAG-2a	0.182	0.013
JAG-2b	0.126	0.011
JAG-2c	0.124	0.013
JAG-1a	0.078	0.010
JAG-1b	0.092	0.014
JAG-1c	0.281	0.055

Table 6. 88 kV arresters - Mairiporã substation.

Surge arresters	Leakage current 60 Hz rms (mA)	Leakage current 180 Hz rms (mA)
BRP-1a	0.122	0.011
BRP-1b	0.096	0.012
BRP-1c	0.114	0.013
BRP-2a	0.089	0.009
BRP-2b	0.092	0.009
BRP-2c	0.102	0.009
SAA-1a	0.110	0.011
SAA-1b	0.079	0.010
SAA-1c	0.088	0.011
SAA-2a	0.083	0.008
SAA-2b	0.070	0.009
SAA-2c	0.055	0.009
CAV-1a	0.095	0.005
CAV-1b	0.061	0.012
CAV-1c	0.100	0.007
CAV-2a	0.082	0.006
CAV-2b	0.075	0.006
CAV-2c	0.076	0.007
SAI-1a	0.085	0.002
SAI-1b	*	*
SAI-1c	0.145	0.005
SAI-2a	0.143	0.004
SAI-2b	*	*
SAI-2c	0.161	0.005

Table 7. 138 kV arresters – Mairiporã substation.

It can be observed in Table 6 that the surge arrester, installed in the circuit JAG-1c, presented high values of leakage current and, probably, the degradation of its internal components is higher than the other arresters of the same circuit. Then, the arresters were removed from the substation. The thermovision measurements, uncluding the surge arresters of the circuit

(JAG-1a, JAG-1b, JAG-1c) indicated heating in the surge arresters JAG-1b and JAG-1c. In Table 7, the leakage current measurement was not performed in the surge arresters of the circuits SAI-1b and SAI-2b. All the surge arresters were made by the same manufacturer, except the arresters of the circuit SAI-1 and SAI-2. The leakage current values are low and the thermovision measurements did not indicate heating in the surge arresters. The RIV test is very difficult to apply in the field, then, in Mairiporã substation, measurements of the conducted electromagnetic field, generated by partial discharges, were performed. The aim was to identify the SiC surge arresters with internal electrical discharges.

5. Conclusion

This chapter shows results of laboratory tests and substations measurements concerning the diagnostic of the 88 kV and 138 kV SiC surge arresters. The results showed that the leakage current measurement, one of the techniques used to evaluate the ZnO surge arresters, can also be used to assess the SiC surge arresters, having obtained important information about their condition. This conclusion might help the electrical utilities to develop more adequate maintenance programs and to more accurately select the SiC surge arresters that need replacement in the substations.

6. References

Almeida, C. A. L., Braga, A. P., Nascimento, S., Paiva, V., Martins, H. J. A., Torres, R. & Caminhas, W. M. (2009). Intelligent thermographic diagnostic applied to surge arresters: a new approach. *IEEE Transactions on Power Delivery*, Vol. 24, No. 2, (April 2009) 751-757, ISSN 0885-8977.

Carneiro, J. C. (2007). Policy for renewal of power system substations silicon carbide (SiC) surge arresters: a new technical economical vision, *Proceedings of the IX International Symposium on Lightning Protection (IX SIPDA)*, pp. 294–299, ISSN 2176-2759, Foz do Iguaçu, September 2007, IEE/USP, São Paulo.

Grzybowski, S. & Gao, G. (1999). Evaluation of 15-420 kV substation lightning arresters after 25 years of service, *Proceedings of the IEEE Southeastcon'99*, pp. 333–336, ISBN 0-7803-5237-8, Lexington, March 1999.

Heinrich, C. & Hinrichsen, V. (2001). Diagnostics and monitoring of metal-oxide surge arresters in high-voltage - comparison of existing and newly developed procedures. *IEEE Transactions on Power Delivery*, Vol. 16, No. 1, (January 2001) 138-143, ISSN 0885-8977.

Kanashiro, A. G., Zanotti Junior, M., Obase, P. F. & Bacega, W. R. (2009). Diagnostic of silicon carbide surge arresters of substation. WSEAS Transactions on Systems. Vol. 8, No. 12, (December 2009) 1284-1293, ISSN 1109-2777.

Kannus, K. & Lahti, K. (2005). Evaluation of the operational condition and reliability of surge arresters used on medium voltage networks. *IEEE Transactions on Power Delivery*, Vol. 20, No. 2, (April 2005) 745-750, ISSN 0885-8977.

McDermid, W. (2002). Reliability of station class surge arresters, *Proceedings of the 2002 IEEE International Symposium on Electrical Insulation*, pp. 320-322, ISBN 0-7803-7337-5, Boston, April 2002.

13

Silicon Carbide Neutron Detectors

Fausto Franceschini* and Frank H. Ruddy**
*Westinghouse Electric Company LLC, Research and Technology Unit, Cranberry Township, Pennsylvania 16066 USA
**Ruddy Consulting, 2162 Country Manor Dr., Mt. Pleasant, South Carolina 29466 USA

1. Introduction

The potential of Silicon Carbide (SiC) for use in semiconductor nuclear radiation detectors has been long recognized. In fact, the first SiC neutron detector was demonstrated more than fifty years ago (Babcock, *et al.*, 1957; Babcock & Chang, 1963). This detector was shown to be operational in limited testing at temperatures up to 700 °C. Unfortunately, further development was limited by the poor material properties of SiC available at the time.

During the 1990's, much effort was concentrated on improving the properties of SiC by reducing defects produced during the crystal growing process such as dislocations, micropipes, *etc.* These efforts resulted in the availability of much higher quality SiC semiconductor materials. A parallel effort resulted in improved SiC electronics fabrication techniques.

In response to these development efforts, interest in SiC nuclear radiation detectors was rekindled in the mid 1990's. Keys to this interest are the capability of SiC detectors to operate at elevated temperatures and withstand radiation-induced damage better than conventional semiconductor detectors such as those based on Silicon or Germanium. These properties of SiC are particularly important in nuclear reactor applications, where high-temperature, high-radiation measurement environments are typical.

SiC detectors have now been demonstrated for high-resolution alpha particle and X-ray energy spectrometry, beta ray detection, gamma-ray detection, thermal- and fast-neutron detection, and fast-neutron energy spectrometry.

In the present chapter, emphasis will be placed on SiC neutron detectors and applications of these detectors. The history of SiC detector development will be reviewed, design characteristics of SiC neutron detectors will be outlined, SiC neutron detector applications achieved to date will be referenced and the present status and future prospects for SiC neutron detectors will be discussed.

2. Background

The initial efforts to develop SiC radiation detectors were directed towards neutron monitoring in nuclear reactors (Babcock, *et al.*, 1957; Babcock & Chang, 1963). Reactor neutron monitoring must often be carried out in high-temperature environments and intense radiation fields which lead to detector radiation damage concerns. Using crude detectors constructed by applying resistive contacts to SiC crystals, the authors were able to demonstrate detection of alpha particles. In anticipation of the high-temperature monitoring locations that would be encountered in nuclear reactors, these measurements were extended to temperatures up to 700 °C with only minimal changes in the detector response.

In follow-on work (Ferber & Hamilton, 1965), a SiC p-n diode coated with ^{235}U was exposed to thermal neutrons in a low-power research reactor. Good agreement was observed between the axial neutron flux profile measurements made with conventional gold-foil activation methods and the SiC detector measurements. The SiC neutron detector was also shown to have a linear response to reactor power in the 0.1 W to 1 kW range. Detector alpha response was observed to be acceptable after a thermal-neutron fluence of 6×10^{13} cm^{-2}.

Further development of SiC detectors was hindered by the poor quality of the available SiC materials available at the time.

Efforts at developing SiC detectors were renewed by Tikhomirova and co-workers in 1972 (Tikhomirova, *et al.*, 1972; Tikhomirova, *et al.*, 1973a; Tikhomirova, *et al.*, 1973b). Beryllium diffused 6H-SiC detectors with low, 1 nanoampere leakage currents were shown to be capable of 8% energy resolution for 4.8 MeV alpha particles (Tikhomirova, *et al.*, 1972).

The effects of neutron damage on a ^{235}U-coated, beryllium-diffused 6H-SiC diode were examined (Tikhomirova, *et al.*, 1973b). The detector response did not change significantly up to a thermal-neutron fluence of 10^{13} n cm^{-2}. At higher neutron fluences, the detector count rate decreased dramatically. The observed response changes were likely a result of fission-fragment induced radiation damage in the detector. The fission-fragment dose corresponding to a thermal-neutron fluence of 10^{13} cm^{-2} is approximately 10^{8} cm^{-2}.

Increases in SiC detector leakage currents as a result of neutron irradiation were reported by Evstropov, *et al.*, 1993.

In the 1990's, long-term development work resulted in the demonstration of technologies for producing high-quality SiC both in chemical vapour deposited (CVD) and large-wafer form. As a result of this development, some of the last major obstacles to commercial fabrication of high-performance SiC semiconductor devices were overcome.

The first use of these developments in high-quality CVD epitaxial SiC detectors was by Ruddy, *et al.*, 1998. Si substrate layers doped with n^- donor atoms (nitrogen) were overlayered with a lightly doped epitaxial layer containing a nitrogen concentration of 10^{15} cm^{-3}. The epitaxial layer thicknesses ranged from 3 μm to 8 μm. Detectors with 200 μm and 400 μm diameters were tested. Although detectors with diameters up to 1 mm were fabricated, the presence of defects in the form of micropipes limited the performance of detectors with diameters greater than 400 μm. Nickel Schottky metal contacts covered by gold were applied to the epitaxial layers to form Schottky diodes, and thin (1 μm) p+ layers were applied to the n- epitaxial layers to form p-n junction detectors. Both the Schottky diodes and p-n junctions were demonstrated as alpha detectors with ^{238}Pu sources. No drift in the pulse-height response was observed in the temperature range from 18 °C to 89 °C.

Similar results were reported by Nava, *et al.*, 1999. Alpha-particle response measurements were carried out for ^{241}Am using Schottky diodes fabricated on 4H-SiC epitaxial layers. Charge-carrier collection efficiency was shown to increase linearly with the square root of the detector reverse bias.

Rapid development of epitaxial SiC ensued leading to the development of high-resolution SiC alpha detectors (Ivanov, *et al.*, 2004; Ruddy, *et al.*, 2009b), high-resolution and temperature insensitive X-ray detectors (Bertuccio, *et al*, 2001; Bertuccio, *et al*, 2003; Bertuccio, *et al*, 2004a; Bertuccio, *et al*, 2004b; Bertuccio, *et al*, 2005; Bertuccio, *et al*, 2010, Phlips, *et al.*, 2006; Lees, *et al.*, 2007) and detectors for minimum ionizing particles (Bruzzi, *et al.*, 2003: Moscatelli, *et al.*, 2006) as well as neutron detectors, which will be emphasized in this chapter.

High-quality SiC diodes are now readily available with diameters up to 6 mm and depletion layer thicknesses of 100 μm (Ruddy, *et al.*, 2009a)

Epitaxial SiC detectors have also been shown to operate reliably in ambient temperatures up to 375 °C (Ivanov, *et al.*, 2009).

Comprehensive reviews of SiC detector design and development can be found in Nava, *et al.*, 1998 and Strokan, *et al.*, 2009.

3. Silicon Carbide Nuclear Radiation Detectors

3.1 Silicon Carbide Neutron Detector Design

SiC neutron detectors are usually based on Schottky or p-n diodes. (Ruddy, *et al*, 1998; Nava, *et al.*, 1999; Manfredotti, *et al.*, 2005) A schematic drawing of a SiC Schottky diode detector is shown in Figure 1. The SiC substrate layer consists of high-purity material containing a residual n+ doping concentration that is typically about 10^{18} cm^{-3} of nitrogen. The epitaxial layer is applied to the substrate layer and contains a much lower nitrogen concentration, typically 10^{14} - 10^{15} cm^{-3}. Lower n- concentrations are necessary if the thickness of the epitaxial layer is greater than 10 μm in order to limit the voltage required to fully deplete the layer and collect the radiation-induced charge from this layer. An ohmic back contact and a Schottky front contact are applied. The front contact typically consists of a thin layer of titanium or nickel (~800 Å) covered by thicker layers of platinum (~1000 Å) and gold (~9000 Å). (see, for example, Ruddy, *et al.* [2006]) The thicker layers are needed to protect and ruggedize the Schottky metal layer. The optional convertor layer is used to obtain increased neutron sensitivity.

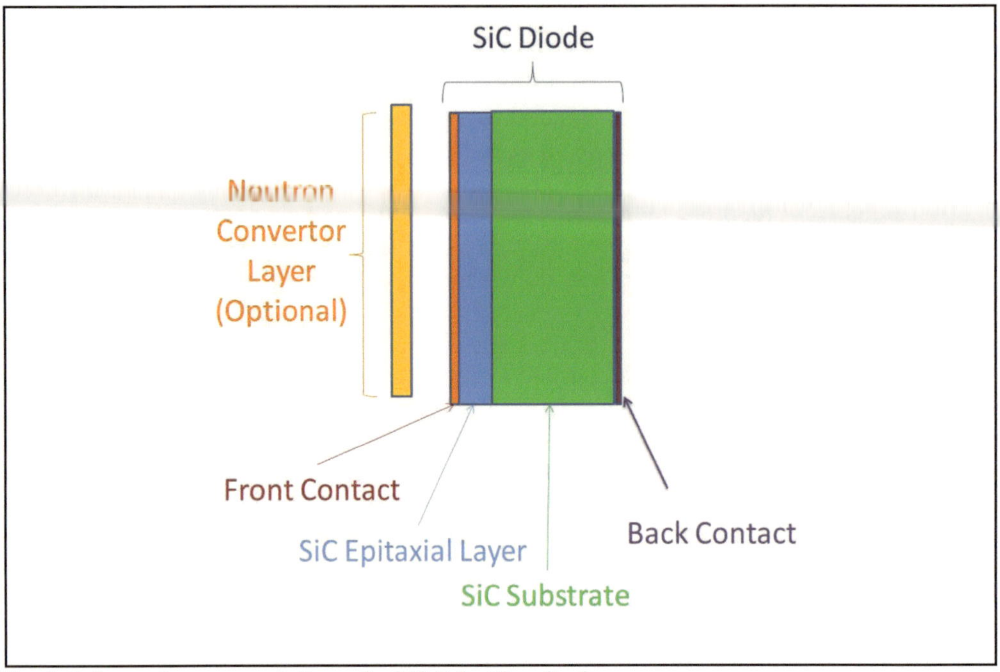

Fig. 1. Schematic representation of a SiC Schottky diode.

3.2 Silicon Carbide Thermal and Epithermal Neutron Detectors

A convertor layer with high thermal-neutron and epithermal-neutron cross sections is juxtaposed in front of the detector. In this way the likelihood of neutron-induced nuclear reactions leading to detectable ionization within the detector active volume is enhanced. For example, ^{6}Li has a thermal neutron cross section of 941 barns and can be used as a thin juxtaposed ^{6}Li layer as depicted in Figure 2.

Thermal neutrons interact with ^{6}Li to produce the following reaction:

$$^{6}Li + n \rightarrow {}^{4}He + {}^{3}H$$

The energetic alphas (^{4}He) and tritons (^{3}H) produced in the reaction can enter the detector active volume (epitaxial layer) and produce ionization in the form of electron-hole pairs. When a reverse bias voltage is applied to the detector as shown in Figure 2, the ionization is collected in the form of a charge pulse, which comprises the detector response signal. The tritons and alpha particles both contribute to the detector response as shown by the pulse-height spectrum in Figure 3 (Ruddy, *et al.*, 1996).

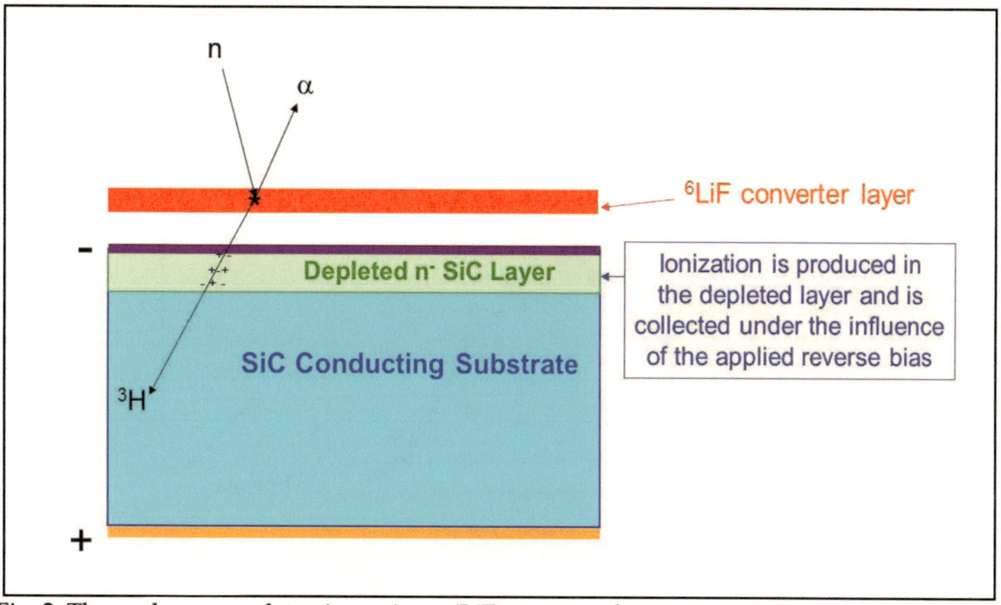

Fig. 2. Thermal neutron detection using a ^{6}LiF convertor layer.

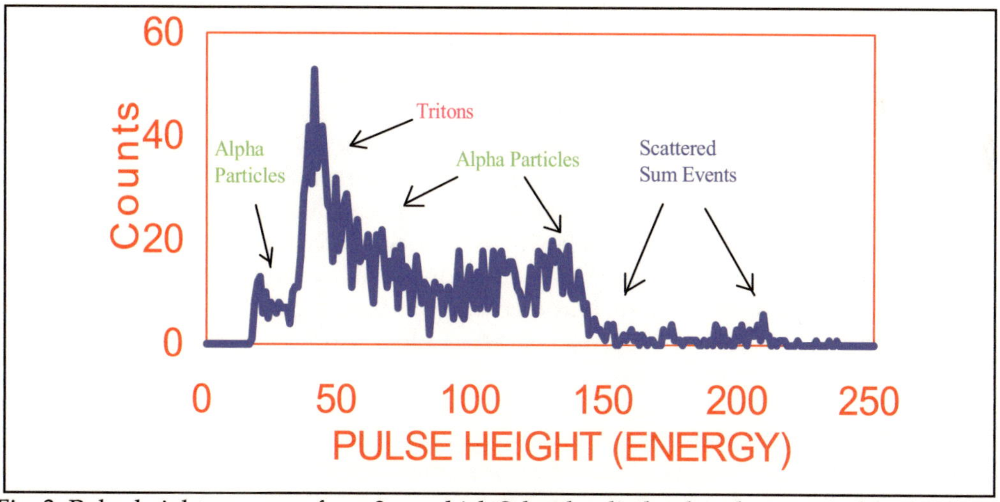

Fig. 3. Pulse height response for a 3-μm thick Schottky diode placed next to a thin ^{6}LiF layer and exposed to thermal neutrons (data from Ruddy, *et al.*, 1996).

Other nuclides with high neutron cross sections, such as ^{10}B and ^{235}U, can also be used in converter layers. The pulse-height response for a $Zr^{10}B_2$ layer positioned adjacent to a Schottky diode with a 3-μm active layer is shown in Figure 4 (Ruddy, *et al.*, 1996). The response is to charged particles from the following reaction:

$$^{10}B + n \rightarrow {}^{7}Li + {}^{4}He$$

Both ^{7}Li and ^{4}He ions are present in the spectrum. Two reaction branches are observed corresponding to production of ^{7}Li in the ground state (E_α = 1.78 MeV) and production of a 0.48-MeV excited state in ^{7}Li (E_α = 1.47 MeV). The former branch occurs in 6% of the reactions, whereas 94% populate the excited state.

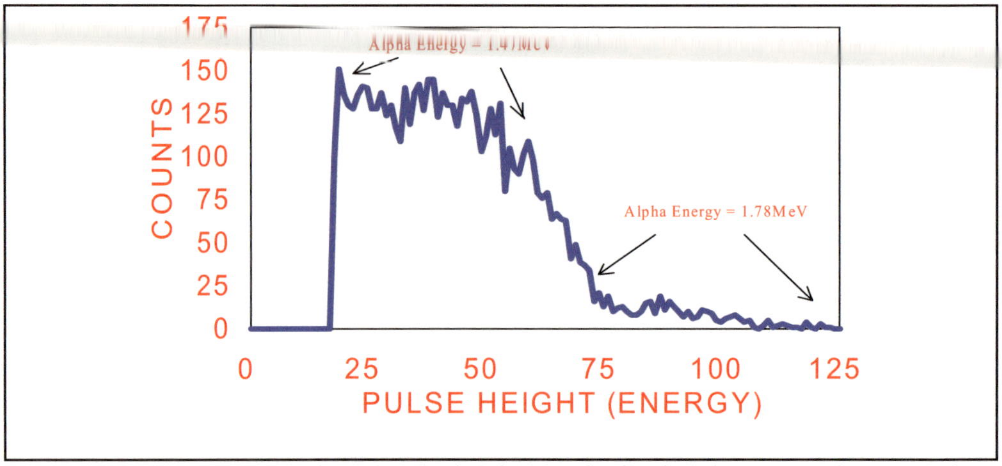

Fig. 4. Pulse height response for a 3-μ thick Schottky diode placed next to a thin $Zr^{10}B_2$ layer and exposed to thermal neutrons (data from Ruddy, *et al.*, 1996).

The pulse-height response for a thin ^{235}U layer placed adjacent to a Schottky diode with a 3-μm active layer is shown in Figure 5 (Ruddy, *et al.*, 1996). In this case, the pulse-height response is primarily to energetic fission fragments from thermal-neutron induced fission of ^{235}U. The fission process is asymmetric resulting predominantly in two fission fragments with different mass and kinetic energy: a heavy-mass peak with an average mass of 139 amu and average energy of 56.6 MeV and a low-mass peak with an average mass of 95 amu and an average energy of 93.0 MeV. Both peaks are clearly visible in the pulse-height spectrum. An additional low pulse-height peak is also visible. This peak is produced by alpha decay of the U^{235} enriched uranium used as the converter. Alpha particles from the decay of both ^{234}U and ^{235}U contribute to this peak.

^{235}U provides by far the most robust pulse-height response. However, the highly charged and energetic fission fragments produce a large amount of radiation damage in the detector active volume: the charge trapping sites produced by dislocation of the Si and C atoms from their original lattice positions degrade the pulse-height spectrum thereby limiting the service lifetime of the detector.

Although one may anticipate that ^{10}B with a thermal-neutron cross section of 3838 barns would produce a higher sensitivity than ^{6}Li with a thermal-neutron cross section of 941 barns, ^{6}Li produces a higher response as demonstrated by the data in Figure 6. (Ruddy, *et al.* 1996) The count rate for $Zr^{10}B_2$ levels off at about 1 μm, while the count rate for ^{6}LiF increases over the entire range of the measurements. The increasing ^{6}LiF sensitivity compared to $Zr^{10}B_2$ is a result of the greater range of the ^{6}Li reaction products (2.73-MeV ^{3}H plus 2.05-MeV ^{4}He) compared to ^{10}B (0.84-MeV ^{7}Li plus 1.47-MeV ^{4}He).

A calculation of the relative neutron sensitivity as a function of ^{6}LiF thickness using the SRIM code (Ziegler & Biersack, 1996) is shown in Figure 7 (Ruddy, *et al.* 1996). The neutron sensitivity levels off at thicknesses greater than 20 µm as a result of the fact that the 2.73 MeV tritons from the $^6Li(n,\alpha)^3H$ reaction have a range of 25 µm in LiF. Use of LiF converter layers thicker than 25 µm will not increase the neutron sensitivity and will, in fact, decrease it as a result of thermal neutron absorption by the ^{6}Li in the LiF layer. Thermal neutron attenuation is about 10% at 20 µm and increases rapidly with LiF thickness (Ruddy, *et al.*, 1996).

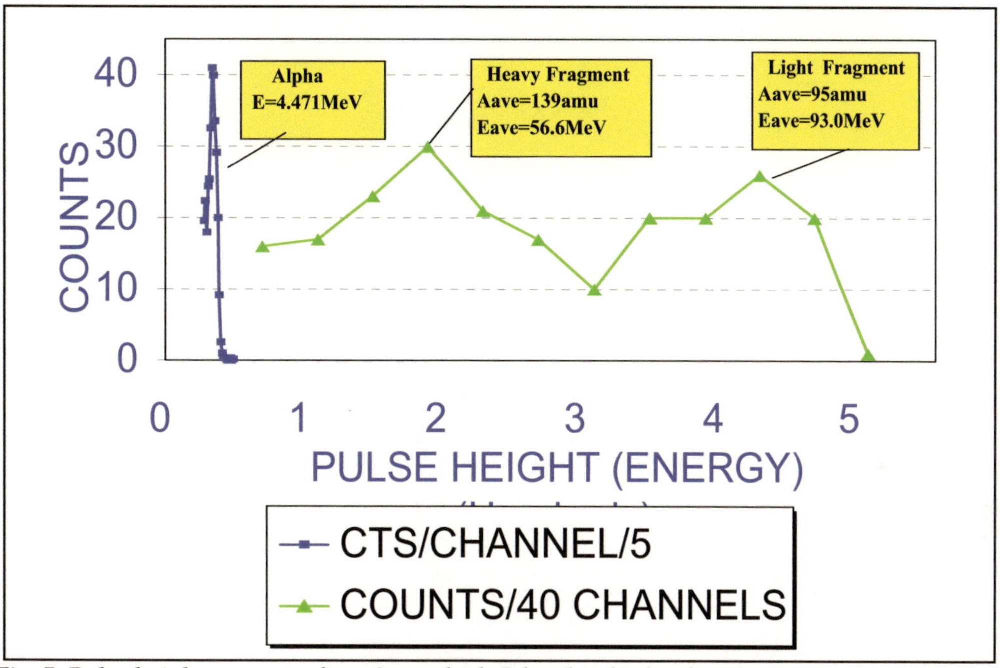

Fig. 5. Pulse height response for a 3-µm thick Schottky diode placed next to a thin ^{235}U layer and exposed to thermal neutrons (data from Ruddy, *et al.*, 1996).

Other materials containing ^{6}Li can provide greater neutron sensitivity than LiF if the range of the neutron-induced tritons in the material is greater than in LiF. A listing of materials with greater triton ranges is contained in Table 1 (Ruddy, *et al.*, 2000). A calculation of neutron sensitivity as a function of layer thickness for each of these materials is shown in Figure 8. The relative sensitivity increases proportionally with the number of tritons escaping the material layer. It can be seen that the relative sensitivity can be increased by factors of two and 4 for LiH and Li, respectively, if used instead of LiF. However, these materials may be less suitable for use in a neutron detector because of their chemical properties. For example, Li is a highly reactive alkali metal and would need to be passivized by encapsulation within a layer of a less reactive metal. LiH is chemically unstable and likely not suitable for use in a neutron detector (Ruddy, *et al.*, 2000).

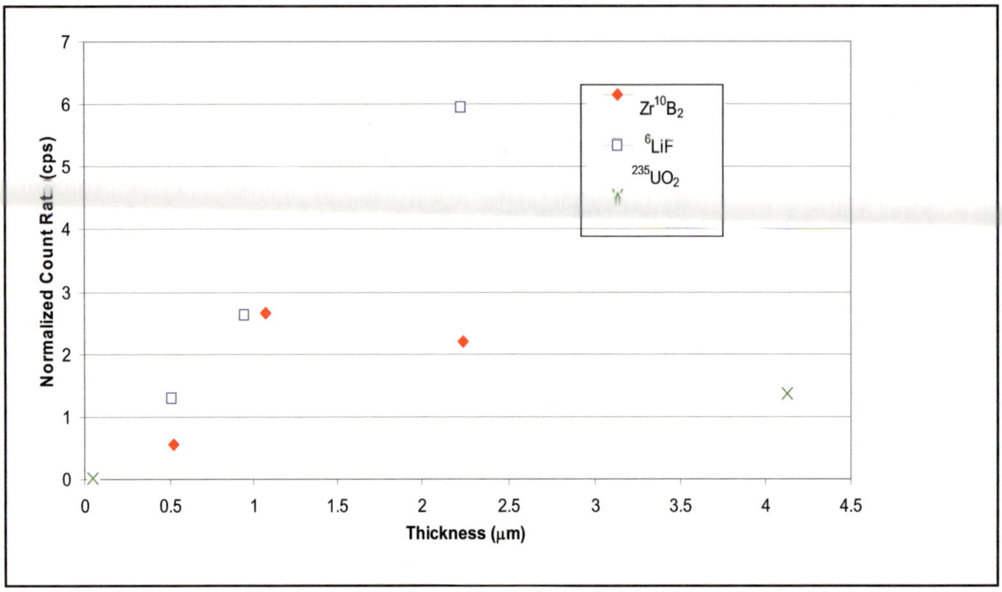

Fig. 6. Count rate as a function of thickness for selected thermal-neutron converter materials (data from Ruddy, *et al.*, 1996).

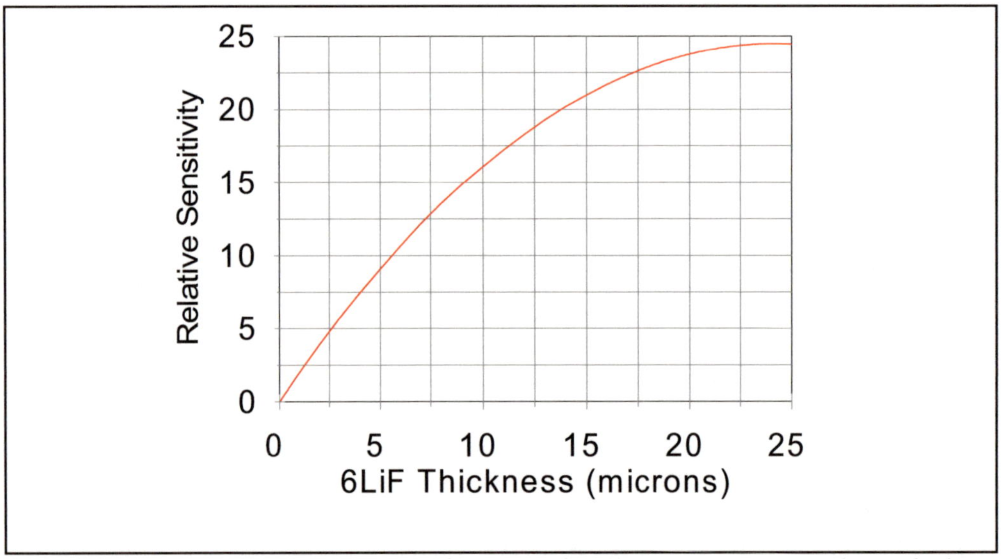

Fig. 7. Relative neutron sensitivity as a function of ^{6}LiF neutron converter layer thickness (data from Ruddy, *et al.*, 1996).

Material	**Range (μm)**
Li	117.88
LiH	60.4
Li_3N	51.95
Li_2C_2	41.58
Li_2O	35.87
LiF	30.77

Table 1. Triton Ranges in Different Materials Containing ^{6}Li (calculations from Ruddy, *et al.*, 2000).

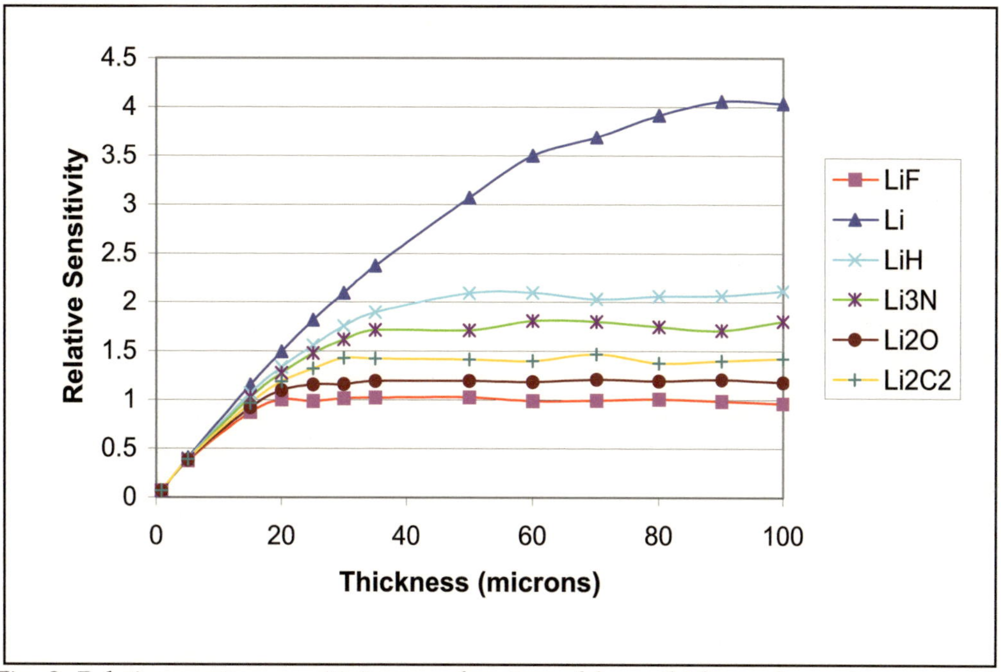

Fig. 8. Relative neutron sensitivity as a function of layer thickness for various materials containing ^{6}Li (calculations from Ruddy, *et al.*, 2000).

3.3 Silicon Carbide Fast Neutron Detectors

At the high energy range pertaining to fast neutrons, several neutron-induced threshold reactions directly with the Si and C atoms of the detector become viable. These reactions lead to the creation of ionizing particles within or close to the detector active volume which carry part of the kinetic information of the incoming neutron thereby enabling neutron detection and, to some extent, neutron spectroscopy. These fast-neutron induced reactions include:

$$^{28}Si + n \rightarrow {}^{28}Si + n'$$
$$^{12}C + n \rightarrow {}^{12}C + n'$$
$$^{28}Si + n \rightarrow {}^{28}Al + p$$
$$^{12}C + n \rightarrow {}^{12}B + p$$
$$^{28}Si + n \rightarrow {}^{25}Mg + {}^{4}He$$
$$^{12}C + n \rightarrow {}^{9}Be + {}^{4}He$$

The list includes only the most prevalent fast-neutron reactions in SiC. Other more complex reactions resulting in the emission of two or more particles will also occur. Also, reactions are listed only for the most abundant Si and C isotopes in the natural elements. Silicon consists of 92.23% ^{28}Si, 4.87% ^{29}Si and 3.10% ^{30}Si. Carbon consists of 98.90% ^{12}C, 1.10% ^{13}C and a negligible amount of ^{14}C. Fast-neutron reactions similar to those listed above can occur with the less abundant isotopes.

The first two reactions listed include elastic and inelastic neutron scattering. In elastic scattering, the neutron interacts with the target nucleus and transfers a variable fraction of its momentum while preserving the overall kinetic energy of the two particles. In inelastic scattering, the neutron elevates the target nucleus to an excited state and transfers momentum without preserving the kinetic energy of the system. The ^{28}Si or ^{12}C recoil atoms are energetic charged particles, which can produce ionization in the active layer of the SiC detector. The secondary neutrons resulting from these reactions however generally escape from the system before inducing any further reactions due to the combined effects of low cross sections and small detector volume. In both elastic and inelastic scattering, the amount of kinetic energy transferred to the ionizing particle is not fixed and a continuum of recoil ion energies will result in the response. While this continuum makes fast-neutron detection still possible, it will not convey an adequate amount of information to infer the energy of the incoming neutron. This is enabled by the other reactions listed above, as discussed in Section 5.

The last four reactions listed result in charged particles, which will all produce ionization in the detector active volume. If the incident fast-neutron energy is monoenergetic, these reactions will produce a fixed response, and a peak will be observed in the pulse-height response spectrum. Such reaction peaks have been observed for SiC and will be discussed in Section 4.2.

The sensitivity for any detector that responds directly to fast neutrons, such as SiC, can be enhanced by juxtaposing a neutron converter layer. Generally, the most effective converter is a layer containing a hydrogenous material, such as polyethylene, because of the high fast -neutron cross section for ^{1}H and the large recoil ranges of the protons produced via the following neutron scattering reaction:

$$^{1}H + n \rightarrow {}^{1}H + n'$$

The recoil protons can produce ionization in the detector active volume and add to the detector response.

SiC fast-neutron response measurements using hydrogen converter layers were carried out by Flammang, *et al.*, 2007.

4. Neutron Response Measurements

4.1 Thermal and Epithermal Neutron Response Measurements

SiC thermal-neutron response measurements have been performed (Dulloo, *et al.*, 1999a; Dulloo, *et al.*, 2003). SiC Schottky diodes with 200μm and 400μm diameters and 3μm thick active layers were used. Converter layers with 6LiF thicknesses of 8.28 μm and 0.502 μm were used. These measurements demonstrated that when compared to United States National Institute for Standards & Technology (NIST) measurements in NIST standard neutron fields, thirty SiC thermal-neutron responses were linear over neutron fluence-rates ranging from 1.76 x 10^4 cm^{-2} s^{-1} to 3.59 x 10^{10} cm^{-2} s^{-1}. The relative precision of the measurements over this range was $\pm$0.6%. The measurements also demonstrated that pulse-mode operation with discrimination of gamma-ray pulses was possible in a gamma-ray field of approximately 433 Gy Si h^{-1} at a thermal-neutron fluence rate of 3.59 x 10^{10} cm^{-2} s^{-1}. In addition, the thermal-neutron response of a SiC neutron detector previously irradiated with a fast-neutron (E > 1 MeV) fluence of 1.3 x 10^{16} cm^{-2} was indistinguishable from that of an unirradiated SiC detector. The NIST measurements and additional low fluence rate measurements using a ^{252}Cf source are shown in Figure 9. With the ^{252}Cf source results, the linear response spans nine orders of magnitude in fluence rate.

The thermal-neutron response of a prototype SiC ex-core neutron detector was shown to be linear over eight orders of magnitude in neutron fluence rate at the Cornell University Reactor by Ruddy, *et al.*, 2002.

The epithermal response of SiC detectors was measured using cadmium covers by Dulloo, *et al.* 1999b. The epithermal-neutron response was linear as a function of reactor power over the range from 50 watts to 293 watts at the Penn State Brazeale reactor. The relative response of SiC detectors compared to the reactor power instrumentation over the range of the measurements was $\pm$1.7%.

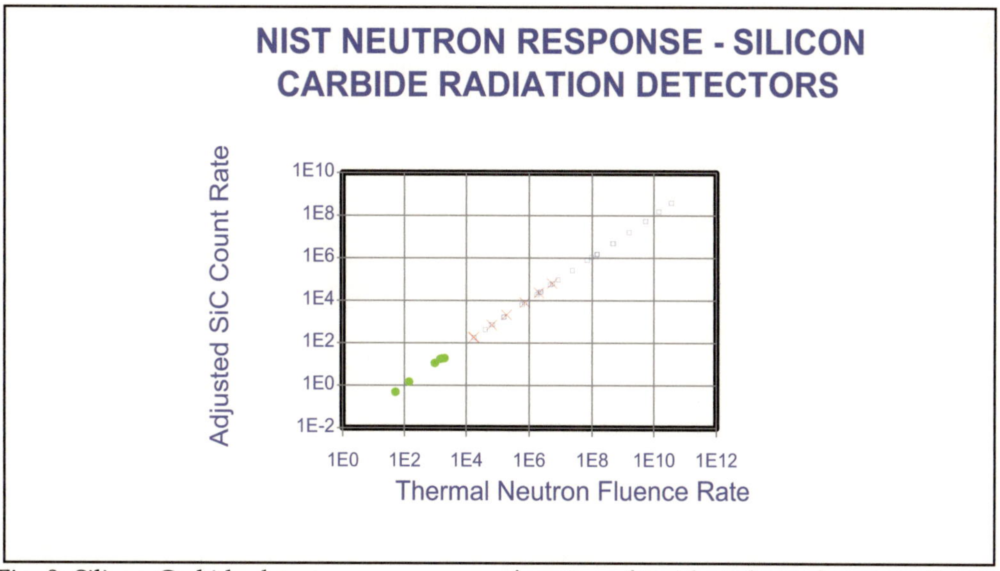

Fig. 9. Silicon Carbide detector response as a function of incident thermal-neutron fluence rate. The NIST response results for an unirradiated SiC detector are shown in blue. The NIST response results for a detector previously irradiated with a fast-neutron (E>1 MeV) fluence of 1.3 x 10^{16} cm^{-2} are shown in red. The response results for thermalized neutrons from a ^{252}Cf source are shown in green.

4.2 Fast-Neutron Response Measurements

Fast-neutron response measurements to ^{252}Cf fission neutrons (E_{AVE} = 2.15 MeV), ^{241}Am-Be (α, n) neutrons (E_{AVE} = 4.5 MeV), 14 MeV neutrons from an electronic deuterium-tritium neutron generator and cosmic-ray induced neutrons were carried out by Ruddy, *et al.*, 2003. A Schottky diode SiC detector with a 28 mm^2 area and a 10 μm active-volume thickness was used without a proton-recoil converter layer. The results are shown in Figure 10.

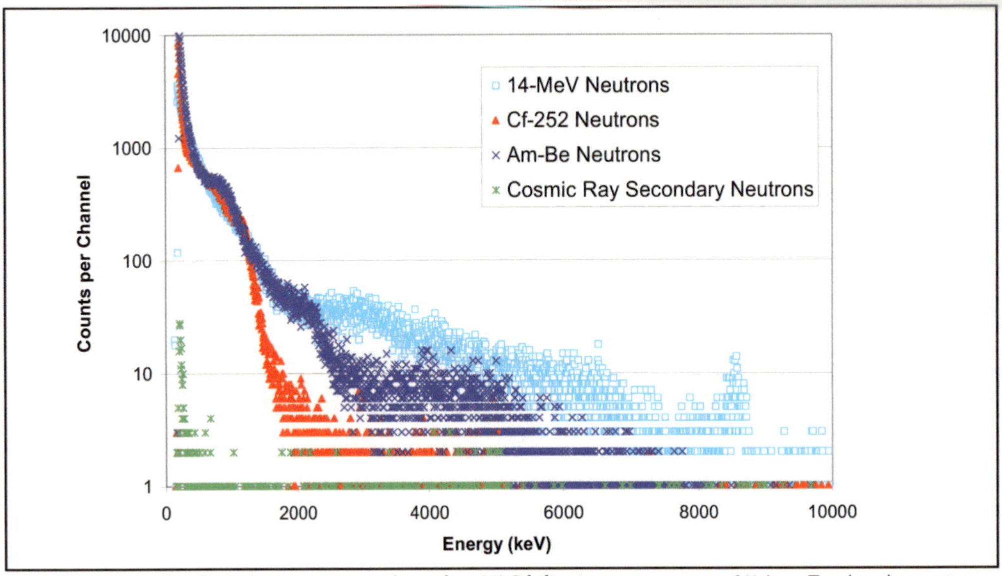

Fig. 10. SiC pulse-height response data for ^{252}Cf fission neutrons, ^{241}Am-Be (α,n) neutrons, 14-MeV neutrons, and cosmic-ray induced background neutrons. (Data reprinted from reference Ruddy, *et al.*, 2003 with permission from the Editorial Department of World Publishing Company Pte. Ltd.)

The pulse-height response spectra clearly shift to higher pulse-heights as a function of incident neutron energy, and structural features corresponding to fast-neutron induced reactions in SiC are visible in the 14-MeV response spectrum. The fast-neutron response measurements were limited by the 10 μm thickness of the SiC detector active volume. Many of the recoil ions that are produced by 14 MeV neutrons have ranges in SiC that are greater than 10 μm and will deposit a variable amount of energy outside of the detector active volume. This lost energy will not contribute to the detector pulse height spectrum and the recovered energy will show in the form of a continuum.

A more detailed examination of the SiC detector response to 14-MeV neutrons is reported in Ruddy *et al.*, 2009a. A 28.3 mm^2 x 100 μm Schottky diode was used without a proton-recoil converter layer. The 100 μm active layer thickness allows much more of the neutron-induced recoil ion energy to be deposited within the active volume of the detector. The resulting 14 MeV pulse-height response data are shown in Figure 11. With the thicker active volume, many more response peaks from fast-neutron reactions become apparent. A listing of the expected nuclear reactions and threshold neutron energies is contained in Table 2.

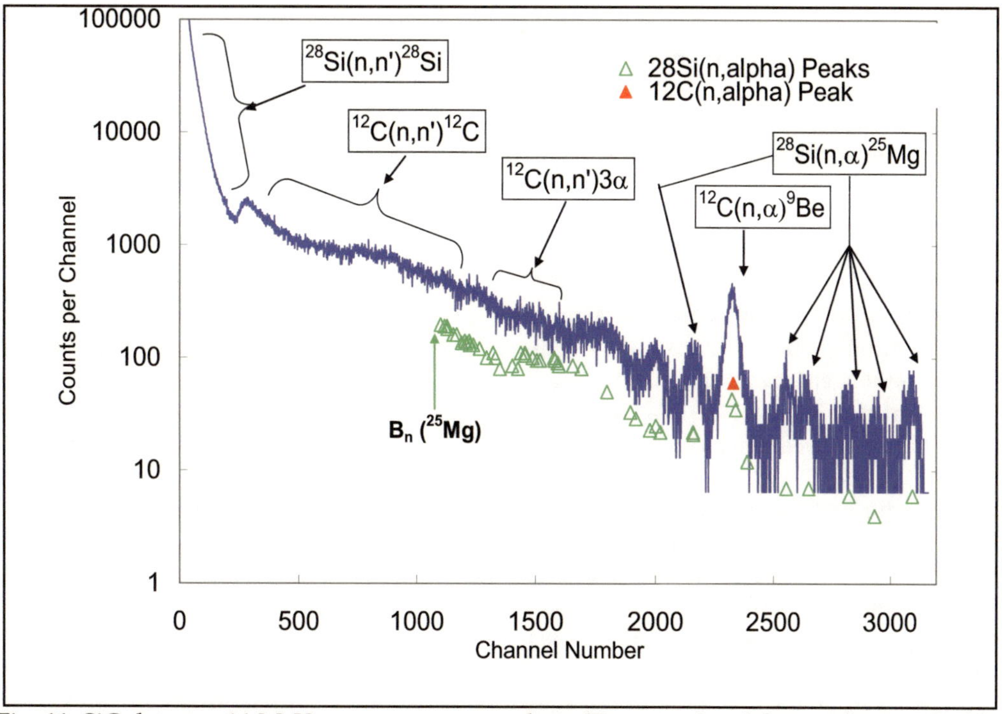

Fig. 11. SiC detector 14 MeV neutron response data. (Data reprinted from reference Ruddy, *et al.*, 2009 with permission from the Editorial Department of World Publishing Company Pte. Ltd.) Channel number is directly proportional to energy deposited in the SiC active volume.

Reaction	Neutron Energy Threshold (MeV)
$^{28}Si(n,n')^{28}Si$	0
$^{28}Si(n,n')^{28*}Si$ (first excited state)	1.843
$^{28}Si(n,\alpha)^{25}Mg$	2.749
$^{28}Si(n,p)^{28}Al$	3.999
$^{12}C(n,n')^{12}C$	0
$^{12}C(n,n')^{12*}C$ (first excited state)	4.809
$^{12}C(n,n')^{12*}C$ (second excited state)	8.292
$^{12}C(n,n')^{12*}C$ (third excited state)	11.158
$^{12}C(n,n')^{12*}C$ (fourth excited state)	13.769
$^{12}C(n,\alpha)^{9}Be$	6.180
$^{12}C(n,n')3\alpha$	7.886
$^{12}C(n,p)^{12}B$	13.643

Table 2. Fast Neutron Reactions in Silicon Carbide.

Only the neutron reactions possible with 14 MeV neutrons are shown in Table 2. Inelastic neutron scattering reactions are shown only for excited states that are bound with respect to particle emission.

At the low-energy portion of the spectrum, the continua for ^{28}Si and ^{12}C elastic and inelastic scattering dominate the detector response shown in Figure 11. At higher energies, specific reaction peaks dominate. The most prominent of these is for the $^{12}C(n,\alpha)^{9}Be$ reaction, which produces a total of 8.3 MeV in recoil-ion energy.

Several peaks corresponding to the $^{28}Si(n,\alpha)^{25}Mg$ reaction are observed. The highest-energy (channel number) peak corresponds to the production of ^{25}Mg in its ground state with a total recoil-ion energy of 11.3 MeV. Satellite peaks, corresponding to the production of excited states of ^{25}Mg can be seen at lower energies. The expected positions of these peaks are indicated by green diamonds. Peaks for the first four excited states are clearly visible.

Peaks for the 5th, 6th and 7th excited states are obscured by the $^{12}C(n,\alpha)^{9}Be$ peak. Evidence for the 8th through 12th excited states is present in the form of unresolved energy peaks. Higher excited states are more closely spaced in energy and blend into a continuum. Eventually, secondary neutron emission becomes possible in ^{25}Mg, which reduces the possibility of observing higher-energy ^{25}Mg excited states.

A comparison of the SiC 14 MeV response with that of a Si passivized ion implanted detector with the same active volume thickness is shown in Figure 12.

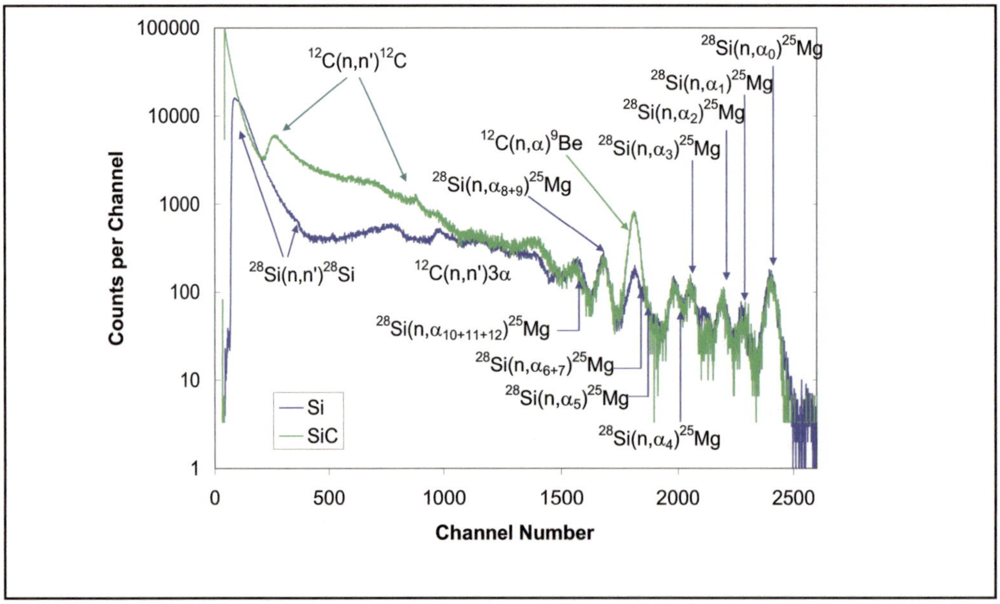

Fig. 12. Comparison of the neutron responses of a 28.3 mm^2 x 100 µm SiC detector and 450 mm^2 x 100 µm Si detector. (Data reprinted from reference Ruddy, *et al.* 2009a with permission from the Editorial Department of World Publishing Company Pte. Ltd.)

The major differences between the two spectra result from the fact that the neutron-induced reactions in carbon are of course absent in the Si detector spectrum. A more detailed comparison of the high-energy peaks is shown in Figure 13. It can be seen that energy positions, peak heights and peak widths are closely matched for the Si and SiC detectors.

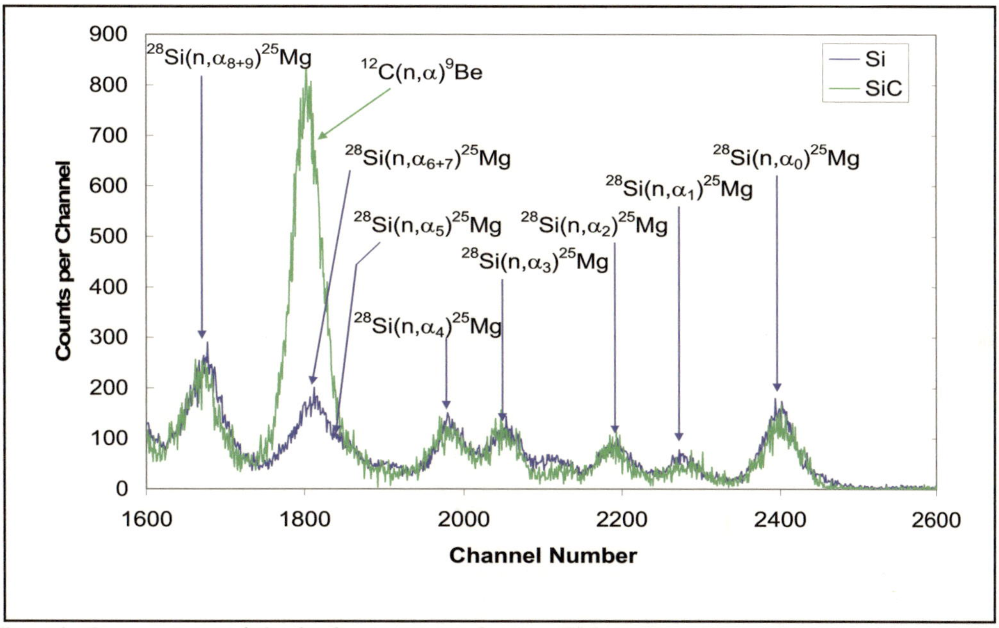

Fig. 13. Comparison of the high-energy peaks in a 28.3 mm² x 100 µm SiC detector and 450 mm² x 100 µm Si detector. (Data reprinted from reference Ruddy, *et al.*, 2009a with permission from the Editorial Department of World Publishing Company Pte. Ltd.)

The response to carbon reactions for the SiC detector can be derived by subtracting the silicon spectrum from the SiC spectrum as shown in Figure 14 (Ruddy, *et al.*, 2009a). The carbon spectrum contains primarily the $^{12}C(n,\alpha)^{9}Be$ peak and continua from neutron elastic and inelastic scattering and multi-particle breakup.

The fast neutron response of SiC detectors to fission neutrons in a reactor was measured by Ruddy, *et al.*, (2006). Three 500 µm diameter x 3 µm SiC Schottky diodes were used to monitor both thermal and fast fission neutron response as a function of reactor power. Two diodes equipped with 24.2 µm and 2.5 µm ^{6}LiF convertor layers were used to monitor thermal neutron response, and the third detector with no convertor layer was used to monitor fast neutrons. The detectors were placed in a beam port at the Ohio State University Research Reactor. Measurements were carried out in the power range from 100 watts to 2000 watts. The SiC fast fission-neutron response compared to the reactor instrumentation was linear over the entire range with a relative standard deviation of $\pm$0.6%.

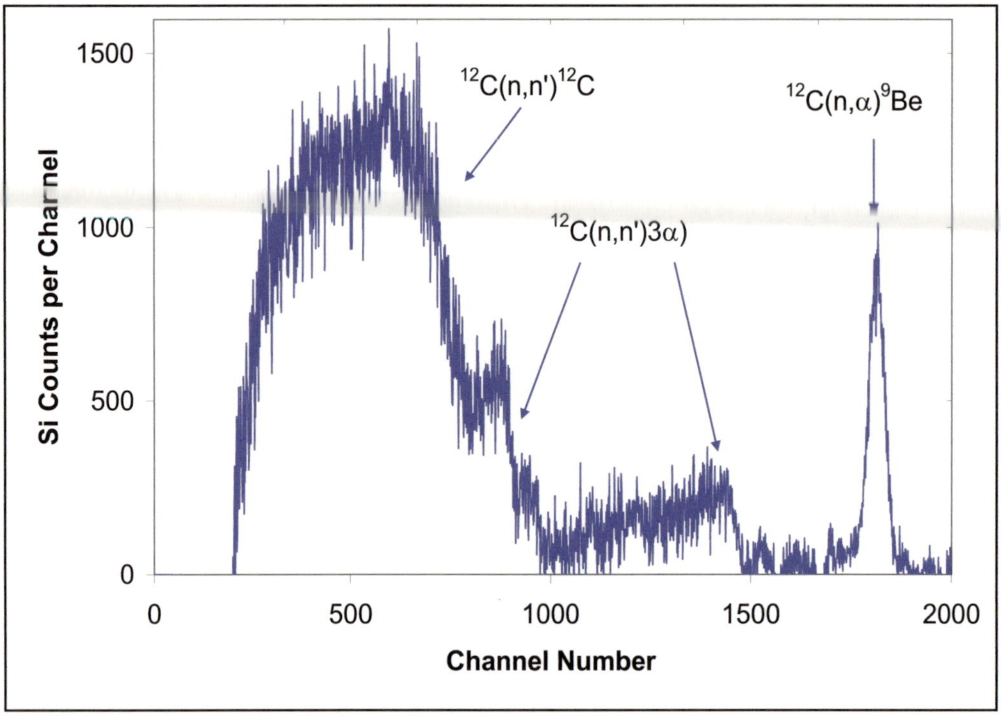

Fig. 14. Response spectrum for 14-MeV reactions in SiC derived by subtracting the response in a Si detector. (Data reprinted from reference Ruddy, *et al.* 2009a with permission from the Editorial Department of World Publishing Company Pte. Ltd.)

The thermal and fast SiC responses were also compared. The relative standard deviation for the measurements at 1000 watts and 2000 watts was $\pm$0.18%.

In a limited set of measurements, SiC detector current was shown to be proportional to reactor power (Ruddy, *et al.*, [2006]).

5. Modeling of the Fast-Neutron Response in Silicon Carbide Neutron Detectors

Modeling of the fast-neutron response of SiC detectors was carried out by Franceschini *et al.*, 2009. A computer code, Particle Generator for SiC (PGSC) was developed to model fast-neutron interactions with SiC and linked to SRIM to streamline the ensuing radiation deposition analysis of the outgoing charged particles. The PGSC code employs a Monte Carlo approach to simulate the particle generation from fast-neutron reactions in the SiC detector active volume, with nuclear cross-section and angular distribution data processed from the ENDF/B-VII.0 data file (ENDF/B-VII.0, 2008). As a result, the collection of possible reactions undergone by source neutrons within SiC is properly sampled and energy, direction and position of the outgoing reaction products can be assigned. The energy deposition of the charged particles is then calculated using the SRIM range-energy code (Ziegler & Biersack, 2006) executed within the PGSC code.

Initial PGSC calculations simulated the response to 14-MeV neutrons for comparison to the measurements by Ruddy, *et al.*, 2009. The (n,p) and (n,α) reactions with C and Si nuclei have been singled out in the prediction by the PGSC code and fitted with Gaussian curves, as shown in Figure 15. Peaks from the (n,α) reactions account for the bulk of the response with the $^{12}C(n,\alpha)^9Be$ ground-state reaction peak most prominent. The $^{28}Si(n,\alpha)^{25}Mg$ ground state peak and fifteen ^{25}Mg excited states peaks are also visible. Peaks from the ground state and thirteen ^{28}Al excited states for the $^{28}Si(n,p)^{28}Al$ reaction are also apparent. The (n,p) peaks are limited by the fact that much of the proton energy is deposited outside of the detector active volume.

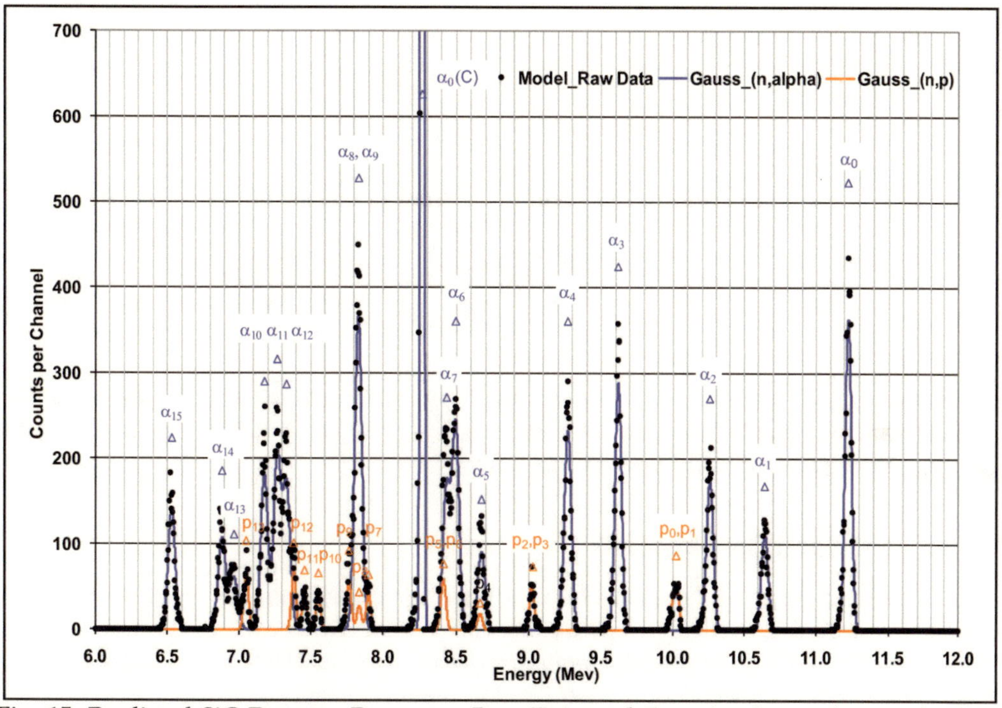

Fig. 15. Predicted SiC Detector Response. Raw Data and Gaussian Interpolations for (n,p) and (n,α) reactions. (Data reprinted from Franceschini *et al.*, 2009 with permission from the Editorial Department of World Publishing Company Pte. Ltd.)

A comparison of the predicted and measured (Ruddy, *et al.*, 2009a) detector responses is shown in Figure 16. A Gaussian representation of the calculated response is compared to the measured data points. The peak positions and intensities match well. The peak widths are narrower for the predicted response than for the measured response. This is likely a result of limitations in the SRIM code.

The eventual goal for the PGSC code is to develop a neutron response spectrum deconvolution methodology, which will allow neutron energy spectra determination for downscattered fission neutrons in complex nuclear reactor environments.

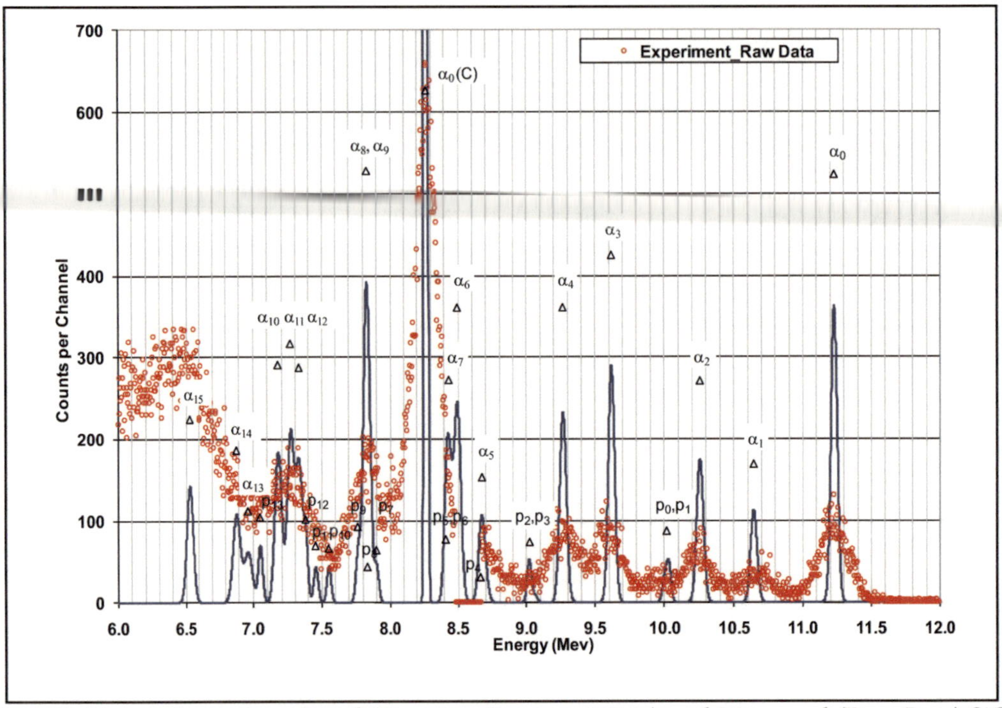

Fig. 16. Comparison of predicted (Gaussian Representation) and Measured (Raw Data) SiC detector responses. (Data reprinted from Franceschini *et al.*, 2009 with permission from the Editorial Department of World Publishing Company Pte. Ltd.)

6. Discussion and Conclusions

Silicon carbide neutron detectors are ideally suited for nuclear reactor applications where high-temperature, high-radiation environments are typically encountered. Among these applications are reactor power-range monitoring (Ruddy, *et al., 2002)*. Fast-neutron fluences at ex-core reactor power-range monitor locations are approximately 10^{17} n cm^{-2}. Semiconductor detectors such as those based on Si or Ge cannot withstand such high fast-neutron fluences and would be unsuitable for this application.

Epitaxial SiC detectors have been shown to operate at temperatures up to 375 °C (Ivanov, *et al.*, 2009). Temperatures do not exceed 350 °C in conventional and advanced pressurized water reactor designs. Therefore, SiC neutron detectors should prove useful for applications in these environments. SiC neutron detectors can potentially be used in reactor monitoring locations with temperatures up to 700 °C (Babcock, *et al.*, 1957; Babcock & Chang, 1963). Such temperatures can be encountered in advanced gas-cooled or liquid-metal cooled reactors. At such temperatures, the long-term integrity of the detector contacts is the key issue rather than the performance of the SiC semiconductor.

Other potential reactor monitoring applications are in-vessel neutron detectors (Ruddy, *et al.*, 2002), monitoring in proposed advanced power reactors (Petrović, *et al.*, 2003) and monitoring of reactors aboard outer space vehicles (Ruddy, *et al.*, 2005).

SiC detectors have also been used to monitor neutron exposures in Boron-Capture Neutron Therapy (Manfreddoti, *et al.*, 2005) as well as the thermal-neutron fluence rates in prompt-gamma neutron activation of waste drums (Dulloo, *et al.*, 2004).

SiC detectors have proven useful for neutron interrogation applications to detect concealed nuclear materials for Homeland Security applications (Ruddy, *et al.*, 2007; Blackburn, *et al.*, 2007; Ruddy, *et al.*, 2009c).

An application that is particularly well suited for SiC detectors is monitoring of spent nuclear fuel. Spent-fuel environments are characterized by very high gamma-ray intensities of the order of 1,000 Gy/hr and very low neutron fluence rates of the order of hundreds per cm^2 per second. Measurements were carried out in simulated spent fuel environments (Dulloo, *et al.*, 2001), which demonstrated the excellent neutron/gamma discrimination capability of SiC detectors. Long-term monitoring measurements were carried out on spent-fuel assemblies over a 2050 hour period, and regardless of the total gamma-ray dose to the detector of over 6000 Gy, the detector successfully monitored both gamma-rays and neutrons with no drift or changes in sensitivity over the entire monitoring period (Natsume, *et al.*, 2006).

SiC detectors have been shown to operate well after a cumulative ^{137}Cs gamma-ray dose of 22.7 MGy (Ruddy & Seidel, 2006; Ruddy & Seidel, 2007). This gamma-ray dose exceeds the total dose that a spent fuel assembly can deliver after discharge from the reactor indicating that cumulative gamma-ray dose to a SiC detector will never be a factor for spent-fuel monitoring applications.

The rapid pace of SiC detector development and the large number of research groups involved worldwide bode well for the future of SiC detector applications.

7. References

Babcock, R. ; Ruby, S. ; Schupp, F. & Sun, K (1957) Miniature Neutron Detectors, Westinghouse Electric Corporation Materials Engineering Report No. 5711-6600-A (November, 1957)

Babcock, R. & Chang, H. (1963) Silicon Carbide Neutron Detectors for High-Temperature Operation, In : *Reactor Dosimetry*, Vol. 1 , p 613 International Atomic Energy Agency, Vienna, Austria.

Bertuccio, G.; Casiraghi, R & Nava, F. (2001) Epitaxial Silicon Carbide for X-Ray Detection, *IEEE Transactions on Nuclear Science*, Vol. 48, pp 232-233.

Bertuccio, G. & Casiraghi, R. (2003) Study of Silicon Carbide for X-Ray Detection and Spectroscopy, *IEEE Transactions on Nuclear Science*, Vol. 50, pp 177-185.

Bertuccio, G.; Casiraghi, R.; Cetronio, A.; Lanzieri, C. & Nava, F. (2004a) A New Generation of X-Ray Detectors Based on Silicon Carbide, *Nuclear Instruments & Methods in Physics Research A*, Vol. 518, pp 433-435.

Bertuccio, G.; Casiraghi, R.; Centronio, A,; Lanzieri, C. & Nava, F. (2004b) Silicon Carbide for High-Resolution X-Ray Detectors Operating Up to 100 °C, *Nuclear Instruments & Methods in Physics Research A*, Vol. 522, pp 413-419.

Bertuccio, G.; Binetti, S.; Caccia, S.; Casiraghi, R.; Castaldini, A.; Cavallini, A.; Lanzieri, C.; Le Donne, A.; Nava, F.; Pizzini, S.; Riquutti, L & Verzellesi, G. (2005) Silicon Carbide for Alpha, Beta, Ion and Soft X-Ray High Performance Detectors. *Silicon Carbide and Related Materials 2004. Materials Science Forum* Vols. 483-485, pp 1015-1020.

Bertuccio, G.; Caccia, S.; Puglisi, D. & Macera, M (2010 Advances in Silicon Carbide X-Ray Detectors, *Nuclear Instruments & Methods in Physics Research A*, In press (available on-line).

Blackburn, B.; Johnson, J..; Watson, S.; Chichester, D.; Jones, J.; Ruddy, F.; Seidel, J. & Flammang, R. (2007) Fast Digitization and Discrimination of Prompt Neutron and Photon Signals Using a Novel Silicon Carbide Detector, *Optics and Photonics in Global Homeland Security* (Saito, T. *et al.* Eds.) *Proceedings of SPIE – The International Society for Optical Engineering*, Vol. 6540, Paper 65401J.

Bruzzi, M.; Lagomarsino, S.; Nava, F. & Sciortino, S. (2003) Characteristics of Epitaxial SiC Schottky Barriers as Particle Detectors, *Diamond and Related Materials*, Vol. 12, pp 1205-1208.

Dulloo, A.; Ruddy, F.; Seidel, J.; Adams, J.; Nico, J. & Gilliam, D (1999a) The Neutron Response of Miniature Silicon Carbide Semiconductor Detectors, *Nuclear Instruments & Methods in Physics Research A*, Vol. 422, pp 47-48.

Dulloo, A.; Ruddy, F.; Seidel, J.; Davison, C.; Flinchbaugh, T. & Daubenspeck, T. (1999b) Simultaneous Measurement of Neutron and Gamma-Ray Radiation Levels from a TRIGA Reactor Core Using Silicon Carbide Semiconductor Detectors, *IEEE Transactions on Nuclear Science,* Vol. 46, pp 275-279.

Dulloo, A.; Ruddy, F.; Seidel, J.; Flinchbaugh, T.; Davison, C. & Daubenspeck, T. (2001) Neutron and Gamma Ray Dosimetry in Spent-Fuel Radiation Environments Using Silicon Carbide Semiconductor Radiation Detectors, In: *Reactor Dosimetry: Radiation Metrology and Assessment* (J. Williams, *et al.*, (Eds.), ASTM STP 1398, American Society for Testing and Materials, West Conshohoken, Pennsylvania, pp 683-690.

Dulloo, A.; Ruddy, F.; Seidel, J.; Adams, J.; Nico, J. & Gilliam, D (2003) The Thermal Neutron Response of Miniature Silicon Carbide Semiconductor Detectors, *Nuclear Instruments & Methods in Physics Research A*, Vol. 498, pp 415-423.

Dulloo, A.; Ruddy, F.; Seidel, J.; Lee, S.; Petrović, B. & McIlwain, M. (2004) Neutron Fluence-Rate Measurements in a PGNAA 208-Liter Drum Assay System Using Silicon Carbide Detectors, *Nuclear Instruments & Methods B*, Vol. 213, pp 400-405.

ENDF/B-VII.0 Nuclear Data File, National Nuclear Data Center, Brookhaven National Laboratory, Upton, NY (on the internet at http://www.nndc.bnl.gov/exfor7/endf00.htm.

Evstropov, V.; Strel'chuk, A.; Syrkin, A. & Chelnokov, V. (1993) The Effect of Neutron Irradiation on Current in SiC pn Structures, *Inst. Physics Conf. Ser.* No.137, Chapter 6, (1993)

Ferber, R. & Hamilton, G. (1965) Silicon Carbide High Temperature Neutron Detectors for Reactor Instrumentation, Westinghouse Research & Development Center Report No. 65-1C2-RDFCT-P3 (June, 1965).

Flammang, R.; Ruddy, F. & Seidel, J. (2007) Fast Neutron Detection With Silicon Carbide Semiconductor Radiation Detectors, *Nuclear Instruments & Methods in Physics Research A,* Vol. 579, pp 177-179.

Franceschini, F.; Ruddy, F. & Petrović, B. (2009) Simulation of the Response of Silicon Carbide Fast Neutron Detectors, In: *Reactor Dosimetry State of the Art 2008*, Voorbraak, W. *et al.* Eds., World Scientific, London, pp 128-135.

Ivanov, A.; Kalinina, E.; Kholuyanov, G.; Strokan, N.; Onushkin, G.; Konstantinov, A.; Hallen, A. & Kuznetsov, A. (2004) High Energy Resolution Detectors Based on 4H-SiC, In: *Silicon Carbide and Related Materials 2004*, R. Nipoti, *et al.* (Eds.), Materials Science Forum Vols. 483-484, pp 1029-1032.

Ivanov, A.; Kalinina, E.; Strokan, N. & Lebedev, A. (2009) 4H-SiC Nuclear Radiation p-n Detectors for Operation Up to Temperature 375 °C, *Materials Science Forum*, Vols. 615-617, pp 849-852.

Lees, J.; Bassford, D.; Fraser, G.; Horsfall, A.; Vassilevski, K.; Wright, N. & Owens, A. (2007) Semi-Transparent SiC Schottky Diodes for X-Ray Spectroscopy, *Nuclear Instruments & Methods in Physics Research A*, Vol. 578, pp 226-234.

Lo Giudice, A.; Fasolo, F.; Durisi, E.; Manfredotti, C.; Vittone, E.; Fizzotti, F.; Zanini, A. & Rosi, G. (2007) Performance of 4H-SiC Schottky Diodes as Neutron Detectors, *Nuclear Instruments & Methods in Physics Research A*, Vol. 583, pp 177-180.

Manfredotti, C.; Lo Giudice, A.; Fasolo, F.; Vittone, E.; Paolini, F.; Fizzotti, F.; Zanini, A.; Wagner, G. & Lanzieri, C. (2005) SiC Detectors for Neutron Monitoring, *Nuclear Instruments & Methods in Physics Research A*, Vol. 552, pp 131-137.

Natsume, T.; Doi, H.; Ruddy, F.; Seidel, J. & Dulloo, A. (2006) Spent Fuel Monitoring with Silicon Carbide Semiconductor Neutron/Gamma Detectors, *Journal of ASTM International*, Online Issue 3, March 2006.

Nava, F.; Bertuccio, G.; Cavallini, A. & Vittone, E. (2008) Silicon Carbide and Its Use as a Radiation Detector Material, *Materials Science Technology*, Vol. 19, pp 1-25.

Nava, F.; Vanni, P.; Lanzieri, C. & Canali, C. (1999) Epitaxial Silicon Carbide Charge Particle Detectors, *Nuclear Instruments and Methods in Physics Research A*, Vol. 437, pp 354-358.

Petrović, B.; Ruddy, F. & Lombardi, C. (2003) Optimum Strategy For Ex-Core Dosimeters/Monitors in the IRIS Reactor, In: *Reactor Deosimetry in the 21st Century*, J. Wagemans, *et al.* (Eds.), World Scientific, London, pp 43-50.

Phlips, B.; Hobart, K.; Kub, F.; Stahlbush, R.; Das, M.; De Geronimo, G. & O'Connor, P. (2006) Silicon Carbide Power Diodes as Radiation Detectors, *Materials Science Forum*, Vols. 527-529, pp 1465-1468.

Ruddy, F.; Dulloo, A.; Seshadri, S.; Brandt, C & Seidel, J. (1997) Development of a Silicon Carbide Semiconductor Neutron Detector for Monitoring Thermal Neutron Fluxes, Westinghouse Science & Technology Center Report No. 96-9TK1-NUSIC-R1, July 24, 1996.

Ruddy, F.; Dulloo, A.; Seidel, J.; Seshadri, S. & Rowland, B. (1999) Development of a Silicon Carbide Radiation Detector, *IEEE Transactions on Nuclear Science*, Vol. 45, p 536-541.

Ruddy, F.; Dulloo, A.; Seidel, J.; Edwards, K.; Hantz, F. & Grobmyer, L. (2000) Reactor Ex-Core Power Monitoring with Silicon Carbide Semiconductor Neutron Detectors, Westinghouse Electric Co. Report WCAP-15662, December 20, 2000, reclassified in October 2010.

Ruddy, F.; Dulloo, A.; Seidel, J.; Hantz, F. & Grobmyer, L. (2002) Nuclear Reactor Power Monitoring Using Silicon Carbide Semiconductor Radiation Detectors, *Nuclear Technology* Vol.140, p 198.

Ruddy, F.; Dulloo, A. & Petrović, B. (2003) Fast Neutron Spectrometry Using Silicon Carbide Detectors, In: *Reactor Dosimetry in the 21st Century*, Wagemans, J., *et al.*, Editors, World Scientific, London, pp 347-355.

Ruddy, F.; Patel, J. & Williams, J. (2005) Power Monitoring in Space Nuclear Reactors Using Silicon Carbide, *Proceedings of the Space Nuclear Conference*, CD ISBN 0-89448-696-9 American Nuclear Society, LaGrange, Illinois, pp 468-475.

Ruddy, F. & Seidel, J. (2006) Effects of Gamma Irradiation on Silicon Carbide Semiconductor Radiation Detectors, *2006 IEEE Nuclear Sciences Symposium*, San Diego, California, Paper N14-221.

Ruddy, F.; Dulloo, A.; Seidel, J.; Blue, T. & Miller, D. (2006) Reactor Power Monitoring Using Silicon Carbide Fast Neutron Detectors, *PHYSOR 2006: Advances in Nuclear Analysis and Simulation*, Vancouver, British Columbia, Canada, 10-14 September 2006, American Nuclear Society, Proceedings available on CD-ROM ISBN: 0-89448-697-7.

Ruddy, F.; Seidel, J. & Flammang, R. (2007) Special Nuclear Material Detection Using Pulsed Neutron Interrogation, *Optics and Photonics in Global Homeland Security* (Saito, T. *et al.* Eds.) *Proceedings of SPIE – The International Society for Optical Engineering*, Vol. 6540, Paper 65401I.

Ruddy, F. & Seidel, J. (2007) The Effects of Intense Gamma Irradiation on the Alpha-Particle Respone of Silicon Carbide Semiconductor Radiation Detectors, *Nuclear Instruments & Methods in Physics Research B*, Vol. 263, pp 163-168.

Ruddy, F.; Seidel, J. & Franceschini, F. (2009a) Measurements of the Recoil-Ion Response of Silicon Carbide Detectors to Fast Neutrons, In: *Reactor Dosimetry State of the Art 2008*, Voorbraak, W. e*t al.* Eds., World Scientific, London, pp 77-84.

Ruddy, F.; Seidel, J. & Sellin, P. (2009b) High-Resolution Alpha Spectrometry with a Thin-Window Silicon Carbide Semiconductor Detector, *2009 IEEE Nuclear Science Symposium Conference Record*, Paper N41-1, pp 2201-2206.

Ruddy, F.; Flammang, R. & Seidel, J. (2009c) Low-Background Detection of Fission Neutrons Produced by Pulsed Neutron Interrogation, *Nuclear Instruments & Methods in Physics Research A*, Vol. 598, pp 518-525.

Strokan, N.; Ivanov, A. & Lebedev, A. (2009) Silicon Carbide Nuclear-Radiation Detectors, *SiC Power Materials: Devices and Applications*, (Feng, Z. Ed.) Chapter 11, Springer-Verlag, New York, pp 411-442.

Tikhomirova, V.; Fedoseeva, O. & Kholuyanov, G. (1972) Properties of Ionizing-Radiation Counters Made of Silicon Carbide Doped by Diffusion of Beryllium, *Soviet Physics – Semiconductors* Vol.6, No. 5 (November, 1972)

Tikhomirova, V.; Fedoseeva, O. & Kholuyanov, G. (1973a) Detector Characteristics of a Silicon Carbide Detector Prepared by Diffusion of Beryllium, *Atomnaya Energiya* Vol. 34, No. 2, (February, 1973) pp 122-124.

Tikhomirova, V.; Fedoseeva, O. & Bol'shakov, V. (1973b) Silicon Carbide Detectors as Fission-Fragment Counters in Reactors, *Izmeritel'naya Tekhnika* Vol. 6 (June, 1973) pp 67-68.

Ziegler, J & Biersack, J (1996) SRIM-96: The Stopping and Range of Ions in Matter, IBM Research, Yorktown, New York.

Ziegler, J & Biersack, J (2006) The Stopping and Range of Ions in Solids, 2006 edition, Yorktown, New York (on the Internet at http://www.srim.org)

14

Fundamentals of biomedical applications of biomorphic SiC

Mahboobeh Mahmoodi[1,2] and Lida Ghazanfari[2]
[1]*Material Group, Faculty of Engineering, Islamic Azad University of Yazd, Yazd, Iran*
[2]*Biomaterial Group, Faculty of Biomedical Engineering, Amirkabir University of Technology Tehran, Iran*

1. Introduction

In recent years, silicon carbide (SiC) has become an increasingly important material in numerous applications including high frequency, high power, high voltages, and high temperature devices. It is used as a structure material in applications which require hardness, stiffness, high temperature strength (over 1000° C), high thermal conductivity, a low coefficient of thermal expansion, good oxidation and corrosion resistance, some of which are characteristic of typical covalently bonded materials. It seems that SiC can create many opportunities for chemists, physicists, engineers, health professional and biomedical researches (Presas et al., 2006; Greil, 2002; Feng et al.2003). Silicon carbides are emerging as an important class of materials for a variety of biomedical applications. Examples of biomedical applications discussed in this chapter include bioceramic scaffolds for tissue engineering, biosensors, biomembranes, drug delivery, SiC-based quantum dots and etc. Although several journals exist that cover selective clinical applications of SiC, there is a void for a monograph that provides a unified synthesis of this subject. The main objective of this chapter is to provide a basic knowledge of the biomedical applications of SiC so that individuals in all disciplines can rapidly acquire the minimal necessary background for research. A description of future directions of research and development is also provided.

2. Properties of Biomorphic SiC

Structural ceramics play a key role in modern technology because of their excellent density, strength relationship and outstanding thermo-mechanical properties. Crystalline silicon carbide is well known as a chemically inert material that is suitable for worst chemical environments even under high temperatures. The same is true for the amorphous modification although the thermal stability is limited to 250 °C. Corrosion resistance under normal biological conditions (neutral pH, body temperature) is excellent. The dissolution rate is well below 30 nm per year (Bolz, 1995; Harder et al., 1999). The properties that make this material particularly promising for biomedical applications are: 1) the wide band gap that increases the sensing capabilities of a semiconductor; 2) the chemical inertness that suggests the material resistance to corrosion in harsh environments such as body; 3) the high hardness (5.8 GPa), high elastic modulus (424 GPa), and low friction coefficient (0.17) that make it an ideal material for smart-implants (Coletti et al., 2007).

Mechanical properties of SiC are altered by changing the sintering additives. At elevated temperature, SiC ceramics with boron and carbon additions, which are free from oxide grain-boundary phases, exhibit high-strength and relatively high-creep resistance. These properties of boron- and carbon-doped SiC originate from the absence of grain-boundary phases and existence of covalent bonds between SiC grains (Zawrah & Gazery, 2007).

Biomimetics is one such novel approach, the purpose of which is to advance man-made engineering materials through the guidance of nature. Following biomimetic approach, synthesis of ceramic composites from biologically derived materials like wood or organic fibres has recently attained particular interest.

Plants often possess natural composite structures and exhibit high mechanical strength, low density, high stiffness, elasticity and damage tolerance. These advantages are because of their genetically built anatomy, developed and matured during different hierarchical stages of a long-term evolutionary process. Development of novel SiC materials by replication of plant morphologies, with tailored physical and chemical properties has a tremendous potential (Chakrabarti, 2004). Biological performs from various soft woods and hard woods can be used for making different varieties of porous SiC ceramics. A wide variety of non-wood ingredients of plant origin commonly used in pulp and paper making can also be employed for producing porous SiC ceramics by replication of plant morphologies (Sieber, 2000).

Wood-based biomorphic SiC has been a matter of consideration in the last decade. There has been a great deal of interest in utilizing biomimetic approaches to fabricate a wide variety of silicon-based materials (Gutierrez-Mora et al., 2005; Greil, 2001; Martinzer et al., 2001; Sieber et al., 2001; Varela-Feria et al., 2002). A number of these fabrication approaches have utilized natural wood or cellulosic fiber to produce carbon performs. Biomorphic SiC is manufactured by a two step process: a controlled pyrolyzation of the wood followed by a rapid controlled reactive infiltration of the carbon preform with molten Si. The result is a Si/SiC composite that replicates the highly interconnected microstructure of the wood with SiC, while the remaining unreacted Si fills most of the wood channels. The diversity of wood species, including soft and hard, provides a wide choice of materials, in which the density and the anisotropy are the critical factors of the final microstructure and hence of the mechanical properties of the porous SiC ceramics (Presas et al., 2006; Galderon et al., 2009). Ceramics mimicking the biological structure of natural developed tissue has attracted increasing interest. The mechanical properties of this material not only depend on the component and porosity, but are also highly dependent on the sizes, shapes, and orientation of the pores as well as grains. The lightweight, cytocompatible for human fibroblasts and osteoblasts (Naji and Harmand, 1991) and open porosity of these materials make them great candidates for biomedical applications.

3. Biomedical applications of SiC

Silicon carbides are emerging as an important class of materials for a variety of biomedical applications, including the development of stents, membranes, orthopedic implant, imaging agents, surface modification of biomaterials, biosensors, drug delivery, and tissue engineering. In the coming chapter, we will discuss our experimental studies and some practical issues in developing SiC for biomedical applications. Hence, we will review some proof-of-concept studies that highlight the unique advantages of SiC in biomedical research.

4. Biocompatibility

Biocompatibility is related to the behavior of biomaterials in various contexts. The term may refer to specific properties of a material without specifying where or how the material is used, or the more empirical clinical success of a whole device in which the material or materials feature. The ambiguity of the term reflects the ongoing development of insights into how biomaterials interact with the human body and eventually how those interactions determine the clinical success of a medical device (such as pacemaker, hip replacement or stent). Modern medical devices and prostheses are often made of more than one material so it might not always be sufficient to talk about the biocompatibility of a specific material. Cell-semiconductor hybrid systems represent an emerging topic of research in the biotechnological area with intriguing possible applications. To date, very little has been known about the main processes that govern the communication between cells and the surfaces they adhere to. When cells adhere to an external surface, an eterophilic binding is generated between the cell adhesion proteins and the surface molecules. After they adhere, the interface between them and the substrate becomes a dynamic environment where surface chemistry, topology, and electronic properties have been shown to play important roles. (Maitz et al., 2003). Coletti et al. studied single-crystal SiC biocompatibility by culturing mammalian cells directly on SiC substrates and by evaluating the resulting cell adhesion quality and proliferation (Coletti et al., 2006). The crystalline SiC is indeed a very promising material for bio-applications, with better bio-performance than crystalline Si. 3C-SiC, which can be directly grown on Si substrates, appears to be an especially promising bio-material. The Si substrate used for the epi-growth would in fact allow for cost-effective and straightforward electronic integration, while the SiC surface would constitute a more biocompatible and versatile interface between the electronic and biological world. The main factors that have been shown to define SiC biocompatibility are its hydrophilicity and surface chemistry. The identification of the organic chemical groups that bind to the SiC surface, together with the calculation of SiC zeta potential in media, could be used to better understand the electronic interaction between cell and SiC surfaces. Using an appropriate cleaning procedure for the SiC samples before their use as substrates for cell cultures is also important. The cleaning chemistry may affect cell proliferation and emphasize the importance of the selection of an appropriate cleaning procedure for biosubstrates. SiC has been shown to be significantly better than Si as a substrate for cell culture, with a noticeably reduced toxic effect and enhanced cell proliferation. One of the possible drawbacks that may be associated with the use of SiC in vivo is related to the unclear and highly debated cytotoxic level of SiC particles. Nonetheless, the potential cytotoxicity of SiC particles does not represent a dramatic issue as much as it does for Si, since the great tribological properties of SiC make it less likely to generate debris.

Several studies have discussed testing SiC in vitro. In one study, the researchers tested SiC deposited from radiofrequency sputtering using alveolar bone osteoblasts and gingival fibroblasts for 27 days (Kotzara et al., 2002). The investigators reported that "Silicon carbide looks cytocompatible both on basal and specific cytocompatibility levels. However, fibroblast and osteoblast attachment is not highly satisfactory, and during the second phase of osteoblast growth, osteoblast proliferation is very significantly reduced by 30%" (Naji et al., 1991). According to another paper, in a 48 h study using human monocytes, SiC had a stimulatory effect comparable to polymethacrylate (Nordsletten et al., 1996). Cytotoxicity and mutagenicity has been performed on SiC-coated tantalum stents. Amorphous SiC did not show any cytotoxic reaction using mice fibroblasts L929 cell cultures when incubated for 24 h or mutagenic potential when investigated using Salmonella typhimurium mutants TA98, TA100, TA1535, and TA1537 (Amon et al., 1996). An earlier study by the same authors of a SiC-coated tantalum stent reported similar results (Amon et al., 1995).

Cogan et al. (Cogan et al., 2003) utilized silicon carbide as an implantable dielectric coating. a-SiC films, deposited by plasma-enhanced chemical vapour deposition, have been evaluated as insulating coatings for implantable microelectrodes. Biocompatibility was assessed by implanting a-SiC-coated quartz discs in animals. Histological evaluation showed no chronic inflammatory response and capsule thickness was comparable to silicone or uncoated quartz controls. The a-SiC was more stable in physiological saline than silicon nitride (Si_3N_4) and well tolerated in the cortex.

Kotzar et al. (Kotzar et al, 2002) evaluated materials used in microelectromechanical devices for biocompatibility. These included single crystal silicon, polysilicon (coating, chemical vapor deposition, CVD), single crystal cubic SiC (3CSiC or β-SiC, CVD), and titanium (physical vapor deposition). They concluded that the tested Si, SiC and titanium were biocompatible. Other studies have also confirmed the good tissue biocompatibility of SiC, usually tested as a coating made by CVD (Bolz & Schaldach, 1990; Naji & Harmand, 1991; Santavirta et al., 1998). Even though crystalline SiC biocompatibility has not been investigated in the past, information exists concerning the biocompatibility of the amorphous phase of this material (a-SiC).

5. Haemocompatibility

The interaction between blood proteins and the material is regarded as an important source of thrombogenesis. The adsorption of proteins is explained, from the thermodynamic point of view, in terms of the systems free energy or surface energy. However, adsorption itself does not induce thrombosis. Theories regarding correlations between thrombogenicity of a material and its surface charge or its binding properties proved not to be useful (Bolz, 1993).

Thrombus formation on implant materials is one of the first reactions after deployment and may lead to acute failure due to occlusion as well as a trigger for neointimal formation. Next to the direct activation by the intrinsic or extrinsic coagulation cascade, thrombus formation can also be initiated directly by an electron transfer process, while fibrinogen is close to the surface. The electronic nature of a molecule can be defined as either a metal , a semiconductor, or an insulator. Contact activation is possible in the case of a metal since electrons in the fibrinogen molecule are able to occupy empty electronic states with the same energy (Rzany et al., 2000). Therefore, the obvious way to avoid this transfer is to use a material with a significantly reduced density of empty electronic states within the range of the valence band of the fibrinogen. This is the case for the used silicon carbide coating (Schmehl, 2008).

Haemocompatibility leads to the following physical requirements (Bolz, 1995): (1) to prevent the electron transfer the solid must have no empty electronic states at the transfer level, i.e., deeper than 0.9 eV below Fermi's level. This requirement is met by a semiconductor with a sufficiently large band gap (precisely, its valence band edge must be deeper than 1.4 eV below Fermi´s level) and a low density of states inside the band gap. (2) To prevent electrostatic charging of the interface (which may interfere with requirement 1) the electric conductivity must be higher than 10^{-3} S/cm. A material that meets these electronic requirements is silicon carbide in an amorphous, heavily n-doped, hydrogen-rich modification (a-SiC:H). The amorphous structure is required in order to avoid any point of increased density of electronic states, especially at grain boundaries (Harder, 1999).

At present, a-SiC:H is known for its high thromboresistance induced by the optimal barrier that this material presents for protein (and therefore platelet) adhesion(Starke et al., 2006).

These properties may translate into less protein biofouling and better compatibility for intravascular applications rather than Si. SiC has relatively low levels of fibrinogen and fibrin deposition when contacting blood (Takami et al., 1998). These proteins promote local clot formation; thus, the tendency not to adsorb them will resist blood clotting. It is now well established that SiC coatings are resistant to platelet adhesion and clotting both in vitro and in vivo. In a study by Bolz et al. (Bolz & Schaldach, 1993), the a-SiC:H films were deposited using the glow discharge technique or plasma-enhanced chemical vapour deposition (PECVD), because it provides the most suitable coating process owing to the high inherent hydrogen concentration which satisfies the electronically active defects in the amorphous layers. They used fibrinogen as an example model for thrombogenesis at implants although most haemoproteins are organic semiconductors. a-SiC:H coatings showed no time-dependent increase in the remaining protein concentration, confirming that no fibrinogen activation and polymerisation had taken place. These results support the electrochemical model for thrombogenesis at artificial surfaces and prove that a proper tailoring of the electronic properties leads to a material with superior haemocompatibility. The in vitro test showed that the morphology of the cells was regular. The a-SiC:H samples showed the same behaviour as the control samples. Blood and membrane proteins have similar band-gaps because the electronic properties depend mainly on the periodicity of the amino acids, and the proteins differ only in the acid sequence, not in their structural periodicity.

A-SiC: H has a superior haemocompatibility; its clotting time is 200 percent longer compared with the results of titanium and pyrolytic carbon. Furthermore, it has been shown that small variations in the preparation conditions cause a significant change in haemocompatibility. Therefore, it is of paramount importance to know exactly the physical properties of the material in use, not only the name. Amorphous silicon carbide can be deposited on any substrate material which is resistant to temperatures of about 250 °C. This property makes amorphous silicon carbide a suitable coating material for all hybrid designs of biomedical devices. The substrate material can be fitted to the mechanical needs, disregarding its haemocompatibility, whereas the coating ensures the haemocompatibility of the device. Possible applications are catheters or sensors in blood contact and implants, especially artificial heart valves.

Bolz and Schaldach (Bolz & Schaldach, 1990) evaluated PECVD amorphous SiC for use on prosthetic heart valves. They showed a decreased thrombogenicity of an amorphous layer of SiC compared to titanium. Several other studies showed that hydrogen-rich amorphous SiC coating on coronary artery stents is anti-thrombogenic (Bolz et al., 1996; Bolz & Schaldach, 1990; Carrie et al., 2001; Monnink et al., 1999). Three studies (on 2,125 patients) showed a benefit that was attributed to the SiC-coated stent (Elbaz et al., 2002; Hamm et al., 2003;

Kalnins et al., 2002). In a direct comparison of silicon wafers and SiC-coated (PECVD) silicon wafers for blood compatibility, both appeared to provoke clot formation to a greater extent than diamond-like coated silicon wafers; silicon was worse than SiC-coated silicon (Nurdin et al., 2003). In conclusion, the haemocompatibility of SiC was demonstrated.

6. Biosensors

In the last decade, there has been a tremendous development in the field of miniaturization of chemical and biochemical sensor devices (Berthold et al., 2002). This is because it is expected that miniaturization will improve the speed and reliability of the measurements and will dramatically reduce the sample volume and the system costs. There is a need for the introduction of a semiconducting material that displays both biocompatibility and great sensing potentiality. Most of the studies conducted in the past on single-crystal SiC provide evidence of the attractive bio-potentialities of this material and hence suggest similar properties for crystalline SiC. The availability of SiC single crystal substrates and epitaxial layers with different dopings and conductivities (n-type, p-type and semi-insulating) makes it possible to fully explore the impressive properties of this semiconductor. In the past, the fact that cells could be directly cultured on Si crystalline substrates led to a widespread use of these materials for biosensing applications. The studies report the significant finding that SiC surfaces are a better substrate for mammalian cell culture than Si in terms of both cell adhesion and proliferation (Coletti et al., 2007). In (bio)-chemical sensor applications, the establishment of a stable organic layer covalently attached to the semiconductor surface is of central importance (Yakimova et al., 2007; Botsoa et al., 2008; Frewin et al., 2009).

Recent interest has arisen in employing these materials, tools and technologies for the fabrication of miniature sensors and actuators and their integration with electronic circuits to produce smart devices and systems. This effort offers the promise of: (1) increasing the performance and manufacturability of both sensors and actuators by exploiting new batch fabrication processes developed including micro stereo lithographic and micro molding techniques; (2) developing novel classes of materials and mechanical structures not possible previously, such as diamond-like carbon, silicon carbide and carbon nanotubes, micro-turbines and micro-engines; (3) development of technologies for the system level and wafer level integration of micro components at the nanometer precision, such as self-assembly techniques and robotic manipulation; (4) development of control and communication systems for microelectromechanical systems (MEMS), such as optical and radio frequency wireless, and power delivery systems, etc. The integration of MEMS, nanoelectromechanical systems, interdigital transducers and required microelectronics and conformal antenna in the multifunctional smart materials and composites results in a smart system suitable for sending and controlling a variety of functions in automobile, aerospace, marine and civil strutures and food and medical industries (Varadan, 2003).

The emerging field of monitoring biological signals generated during nerve excitation, synaptic transmission, quantal release of molecules and cell-to-cell communication, stimulates the development of new methodologies and materials for novel applications of bio-devices in basic science, laboratory analysis and therapeutic treatments. The electrochemical gradient results in a membrane potential that can be measured directly with an intracellular electrode. Extracellular signals are smaller than transmembrane potentials, depending on the distance of the signal source to the electrode. Over the last 30 years, non-invasive extracellular recording from multiple electrodes has developed into a widely-used standard method. A microelectrode array is an arrangement of several (typically more than 60) electrodes allowing the targeting of several sites for stimulation and extracellular recording at once. One can plan the realisation of four activities with the following tasks:

Task 1. Development of new biocompatible substrates favoring neuronal growth along specific pathways.

Task 2. Monitoring of electrical activity from neuronal networks.

Task 3. Resolution of cellular excitability over membrane micro areas.

Task 4. Detection of quantal released molecules by means of newly designed biosensors.

Task number 1 can be realized by means of SiC substrates, by plating the cells directly on the substrate or eventually with an additional proteic layer. For this purpose, 3C-SiC films with controlled stoichiometry, different thickness and crystalline quality can be grown directly on silicon substrates or on silicon substrates previously 'carbonised'.

The main objective of task number 2 is the realization of SiC microelectrode arrays whose dimensions will be compatible with the cellular soma (10-20 μm). In this structure, every element of the array is constituted by a doped 3C-SiC region, with metallic interconnections coated with amorphous silicon carbide, so that silicon carbide represents the only material interfaced to the biological environment. For the realization of task number 3, the SiC array will be improved by constructing microelectrodes in the submicrometric range, in order to reveal electrical signals from different areas of the same cell. The objective of task number 4 is the construction of a prototype of SiC-electrodes array as a chemical detector for oxidizable molecules released during cell activity triggered by chemical substances (KCl or acetylcholine) on chromaffin cells of the adrenal gland. With respect to classical electrochemical methods, requiring polarized carbon fibers with rough dimensions of 10 micrometers in diameter, the SiC multielectrode array should greatly improve the monitoring of secretory vesicles fusion to the plasma-membrane, allowing the spatial localization and temporal resolution of the event.

To date, the majority of the development efforts in the MEMS field has focused on sophisticated devices to meet the requirements of industrial applications. However, MEMS devices for medical applications represent a potential multi-billion dollar market, primarily consisting of microminiature devices with high functionality that are suitable for implantation. These implanted systems could revolutionize medical diagnostics and treatment modalities. Implantable muscle microstimulators for disabled individuals have already been developed. Precision sensors combined with integrated processing and telemetry circuitry can remotely monitor any number of physical or chemical parameters within the human body and thereby allow evaluation of an individual's medical condition. Kotzar et al. selected the following materials as MEMS materials of construction for implantable medical devices: (1) single crystal silicon (Si), (2) polycrystalline silicon, (3) silicon oxide (SiO_2), (4) Si_3N_4, (5) single crystal cubic silicon carbide (3C-SiC or b-SiC), (6) titanium (Ti), and (7) SU-8 epoxy photoresist. The Kotzara et al. study results for SiC showed that when the material was generated using MEMS fabrication techniques, it elicited no significant non-biocompatible responses (Kotzara et al., 2002). Iliescu et al. presented an original fabrication process of a microfluidic device for identification and characterization of cells in suspensions using impedance spectroscopy (Iliescu et al., 2007). The fabrication process of this device consists of three major steps. The steps are shown in Fig. 1.

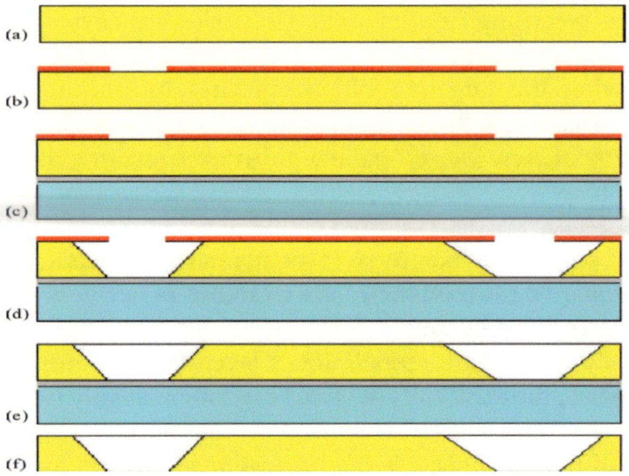

Fig. 1. Main steps of the fabrication process for the etch-through holes in the top glass wafer: (a) starting blank glass wafer, (b) deposition and patterning of the α:Si/SiC/photoresist masking layer, (c) wax bonding of the glass wafer on a dummy silicon wafer, (d) wet etching of glass in HF 49%, (e) strip off the masking layer in an RIE system, (f) debonding from the dummy silicon wafer and cleaning (Iliescu et al., 2007)

Finally, devices with three different electrode geometries (interdigitated; parallel; circular) have been successfully tested. When the introduced cell suspension reaches the measurement region with the electrode structure, it will cause an impedance change between these electrodes depending on the number of cells, their characteristics (complex permittivity) and the applied frequency. Clear differences between dead and live cells have been observed. Therefore, this device can be efficiently used for cell identification and electrical characterization.

Singh and Buchanan (Singh & Buchanan, 2007) studied silicon carbide carbon (SiC-C) composite fiber as an electrode material for neuronal activity sensing and for biochemical detection of electroactive neurotransmitters. Highly adherent SiC insulation near the carbon tip provides highly localized charge transfer, stiffness and protection by inhibition of oxygen, H_2O and ionic diffusion, thereby preventing carbon deterioration. These properties make it a better electrode material than single carbon fiber microelectrodes. Surface morphology plays an important role in the electrode's charge carrying capabilities. For a microelectrode, size is a limiting factor; Hence, there should be ways to increase the real surface area. The SiC-C electrode surface has nanosized pores which significantly increase the real surface area for higher charge densities for a given geometrical area.

For a stimulating neural electrode, the cyclic voltammogram loop and thus the charge density should be as large as possible to provide adequate stimulation of the nervous system while allowing for miniaturization of the electrode. Neurotransmitters including dopamine and vitamin C were successfully detected using SiC-C composite electrodes. Action potentials spikes were successfully recorded from a rat's brain using SiC-C, and a very high signal to noise ratio (20–25) was obtained as compared to (4–5) from commercial electrodes.

In many clinical settings, a decrease of the blood supply to body organs or tissues can have fatal consequences if it is not properly addressed promptly (e.g. mesenteric or myocardial ischemia). Sustained ischemia leads to hypoxia, a stressful condition for cells that is able to induce cell lysis (necrosis) and also to trigger programmed cell death (apoptosis) and, consequently, lead to organ failure. Aside from ischemic diseases, ischemia underlies other natural and clinically induced conditions, like tumor growth, cold-preservation of grafts for transplantation or induced heart-arrest during open heart surgery. Therefore, the ability to monitor ischemia in clinical and experimental settings is becoming increasingly necessary in order to predict its irreversibility (e.g. in the transplantation setting), to develop drugs to prevent and revert its effects, and to treat growing tumors via vascular-targeting drugs. Recently, a minimally invasive system for the continuous and simultaneous monitoring of tissue impedance has been developed (Ivorra et al., 2003), and experimental results have shown its reliability for early ischemia detection and accurate measurement of ischemic effects. This minimally invasive system consists of a small micro-machined silicon needle with deposited platinum electrodes for impedance measurement that can be inserted in biological tissues with minimal damage (Ivorra et al., 2003). High frequency impedance monitoring, based on both the phase and modulus components of impedance, has been correlated to the combined dielectric properties of the extracellular and intracellular compartments and insulating cell membranes and can give complementary information on other effects of sustained ischemia. Moreover, multi-frequency monitoring of impedance has the advantage of yielding to more comprehensive empirical mathematical characterizations (i.e. the Cole model; Cole, 1940) that can provide additional information through the analysis of derived parameters and improve the reproducibility of results (Raicu et al., 2000). Gomez et al. (Gomez et al., 2006) examined the feasibility of producing SiC-based needle-shaped impedance probes for continuous monitoring of impedance and temperature in living tissues. SiC needle-shaped impedance probes (see Fig. 2B) were produced in standard clean room conditions.

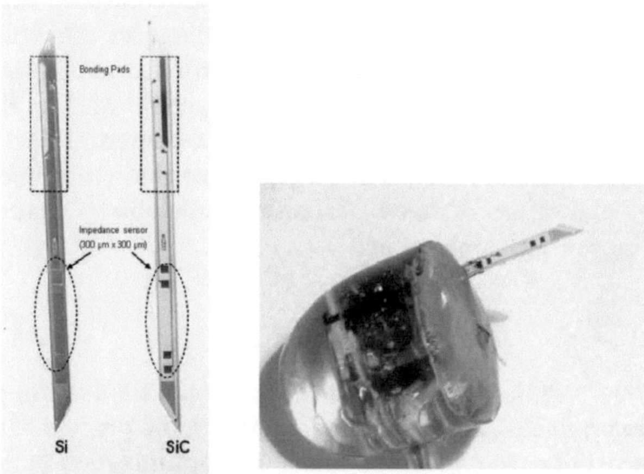

(A) (B) (C)

Fig. 2. (A) Needle-shaped Si probe for impedance; (B) Needle-shaped SiC probe for impedance; (C) Needle-shaped with packaging (Gomez et al., 2006)

In-vitro results obtained with SiC based impedance probes were compared with those obtained with Si-based probes, and they demonstrated that the use of SiC substrates was mandatory to extend the effective operation range of impedance probes beyond the 1 kHz range. In-vivo evaluation of SiC-based impedance probes was conducted on rat kidneys undergoing warm ischemia by dissecting and clamping of the renal pedicles. A substantial rise in impedance modulus was shown throughout the ischemic period (5 to 50 min). This increase can be attributed to the occurrence of hypoxic edema as the result of cell swelling, which leads to a reduction of extracellular space, an increase in extracellular resistance, and cell-to-cell uncoupling (Gersing, 1998). Upon unclamping of the renal artery (50 min), impedance modulus can be seen to return to its basal value, a fact that can be attributed in this experimental setting to a reversion from a short period of ischemia without substantial structural damage to the tissue. A fall in impedance modulus at low frequencies, however, has also been reported as a consequence of membrane breakdown and cell lysis due to the sustained ischemia (Haemmerich et al., 2002). It is in this respect that the multifrequency analysis of the phase component of impedance made possible by the use of SiC-based probes conveys useful complementary information.

Researchers (Godignon, 2005) fabricated impedance and temperature sensors on bulk SiC for a biomedical needle that can be used for open heart surgery monitoring or graft monitoring of organs during transportation and transplantation. According to Godignon (Godignon, 2005) other applications can be foreseen, such as DNA polymerase chain reaction (PCR), electrophoresis chips and cell culture micro-arrays. In DNA electrophoresis devices, the high critical electric field and high resistivity of semi-insulating SiC would be beneficial. In DNA PCR, it is the high thermal conductivity which could improve the device behaviour. In addition, in most of these cases, the transparency of semi-insulating SiC can be used for optical monitoring of the biological process, as for example for the DNA reaction or the cell culture activity.

Caputo et al. (Caputo et al., 2008) reported on biomolecule detection based on a two-color amorphous silicon photosensor. The revealed biomolecules were DNA strands labeled with two fluorochromes (Alexa Fluor 350 or Cy5) with different spectral properties and the device is a p-i-n-i-p amorphous silicon/amorphous silicon carbide stacked structure, that was able to detect different spectral regions depending on the voltage applied to its electrodes. The device design has been optimized in order to maximize the spectral match between the sensor responses and the emission spectra of the fluorochromes. This optimization process has been carried out by means of a numerical device simulator, taking into account the optical and the electrical properties of the amorphous silicon materials. Therefore, according to these set of materials, one can conclude SiC could be considered as a good candidate for biosensing applications.

7. Stent coating

In recent years, coronary stenting has become a well established therapy of coronary artery disease. However, in up to 30% of all stent implantations, the process of restenosis leads to a re-narrowing of the vessel within several months. The optimization of the stent design with regard to mechanical properties only resulted in limited success in reduction of the restenosis rate, and a hybrid concept for stent design was proposed; on the one hand, the mechanical requirements for an optimized geometrical design are met by using 316L stainless steel as bulk material. On the other hand, unwanted interactions of the implant's metal surface with surrounding tissue and blood diminishing biocompatibility and inducing the process of restenosis are reduced by a suitable coating working as a "magic hat" (Harder, 1999; Rzany & Schaldach, 2001). The surface properties of a stent determine the interactions with the surrounding physiologic environment, while properties such as the mechanical performance are determined by the bulk material, the design of which is shown

in Fig.3. The hybrid design of a stent, i.e., a bulk material with a surface coating, allows for optimization with regard to all of the demands (Rzany & Schaldach, 2001).
There are three major stent-related factors influencing the degree of intima proferation:

1- Stent design
2- Stent material
3- Degree of vascular injury

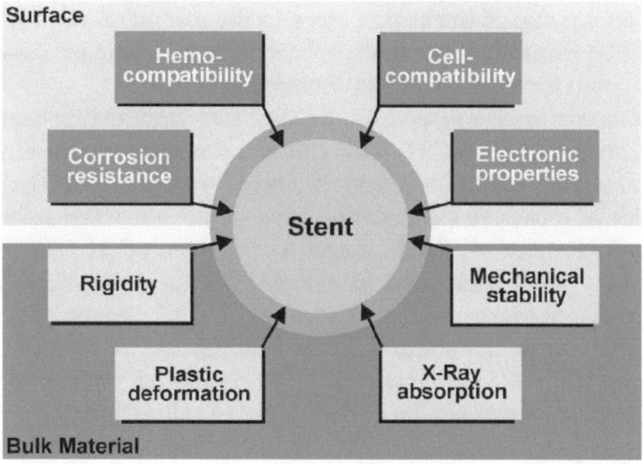

Fig. 3. The surface properties of a stent determine the interactions with the surrounding physiologic environment, while properties such as the mechanical performance are determined by the bulk material and the design. The hybrid design of a stent, i.e., a bulk material with a surface coating, allows for optimization with regard to all of the demands (Rzany & Schaldach, 2001)

Some materials exhibit excellent mechanical properties but have an unfavorable biocompatibility, while other compounds with good biocompatibility are not suitable for stent production, stent coating aims to combine the best desirable characteristics of different materials. Stent coating can be applied as passive and active coatings. While passive coatings serve good biocompatibility, active coatings directly influence the intima proliferation. Active coatings are generally based on the effect of drugs. They are either directly bonded to the surface of the stent or trapped in three- dimensional polymers, which act like a sponge (Karoussos et al., 2002; Wieneke, 2002).
Reduced blood activation and reduced adhesion of blood elements in contact with stent material increase the chances for uncomplicated implantation by minimizing early occlusion became of thrombosis and late occlusion became of the release of growth factors and granulocyte activation. Amorphous silicon carbide, which is known to have antithrombotic effects, can be applied as a coating onto existing stent materials (Van Oeveren, 1999).
After stent implantation, adhesion and thrombocyte formations aggregate at the stent struts, and the injury site can be observed. Consequently, thrombocyte derived factors like platelet derived growth factor serve as chemoattractants for smooth muscle cells and stimulate production of extracellular matrix. Furthermore, stented vessels show reactive inflammatory infiltrates composed of lymphocytes, histiocytes and eosinophiles surrounding the stent struts (Karas et al., 1992). It is assumed that this inflammatory reaction is a mixed response

to vessel injury on the one hand, and non-specific activation mediated through metal ions released from the alloy of the stent on the other.

Cytokines released by inflammatory cells not only serve as smooth muscle mitogens, but also regulate the production of extracellular matrix. Although the detailed mechanisms of inflammation are not completely understood, the correlation between the degree of inflammatory reaction and the extent of neointimal thickness suggest a central role for inflammation in the process of restenosis. It is well accepted that platelet activation and thrombus formation are one of the critical steps in the formation of restenosis. Since it has long been known that thrombus formation is based on electronic processes, semiconductor surfaces have been used for stent coatings (Wieneke, 2002).

The prototype of this coating is a hypothrombogenic semi-conducting coating of amorphous hydrogenated silicon carbide (a-SiC:H). This material can suppress the electron transfer that is crucial in the transformation of fibrinogen to fibrin (Wieneke, 2002). Experimental studies using silicon carbide as a passive stent coating have shown a marked reduction in fibrin and thrombus deposits (Rzany et al., 2000). Based on this theoretical background, stents with silicon carbide coating have been used in patients with acute myocardial infarction with promising short- and long- term results (Scheller et al., 2001; Rzany & Schaldach, 2001). In one randomized study with the silicon carbide coating, the major adverse cardiac events rate after 6 months has reduced significantly as compared with a 316L stainless steel; however, the restenosis rate was similar (Unverdorben et al., 2000).

The deposition of this particular modification of amorphous silicon carbide is performed by means of the PECVD. Since amorphous SiC is a ceramic material, its mechanical properties are significantly different from the metallic substrate. Especially during the dilatation of the stent, enormous mechanical stresses are created at the interface between coating and substrate, while deformations up to 30% are taking place. Therefore, the coating must have strong adhesion to the substrate. There are four steps in the coating process which have to be optimized in sequence to fit both the required electronic properties as well as strong adhesion: the cleaning process, surface activation, deposition of a thin intermediate film and finally coating the surface with a-SiC:H. The specific requirements for the electronic properties of the surface need a careful selection of process parameters.

The electronic band gap is mainly influenced by two physical effects; on the one hand, the band gap of all semiconductors is a property of the material's chemical composition. On the other hand, the band gap of amorphous semiconductors is affected by the density of unsaturated bonds. To achieve a large band gap as well as a low density of states within the gap, the dangling bonds have to be saturated by hydrogen atoms. The most important benefit of the coating with regard to corrosion is that it acts as a diffusion barrier. The uncoated stents may cause cell reactions or reactions of the immune system. However, when coated, the ions must diffuse through the coating before they can get into the patient. Due to the internal structure of amorphous silicon carbide, this diffusion is so slow that the ion release is negligible (Harder, 1999).

The amorphous silicon carbide has been reported to reduce fibrin deposition, which may result in reduced platelet and leukocyte adherence as well (Bolz, 1995; Van Oeveren, 1999). The a-SiC:H surface with multiple clean areas or a loose cell deposit without the fibrin network is shown in Fig. 4. Van Oeveren concluded that the acute response of stainless steel on blood activation can be quenched by a-SiC:H coating.

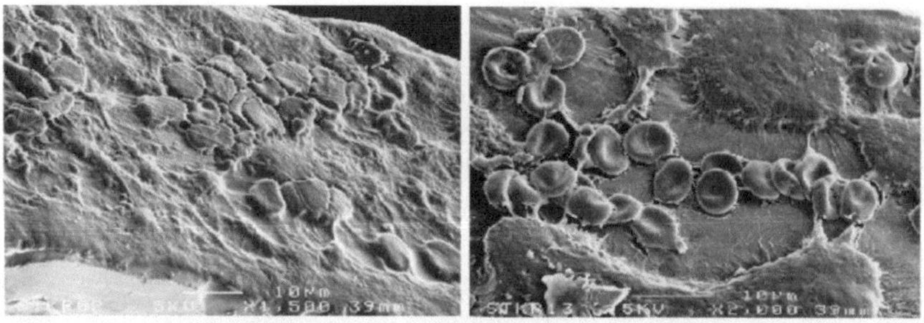

(a) (b)

Fig. 4. (a) Scanning electron micrograph (1500x) of the stainless steel surface showing areas with a dense layer of blood proteins and formed elements, covered by fibrin strands; (b) Scanning electron micrograph (2000x) of the a-SiC:H coated surface showing areas with thrombi and erythrocytes, but not densely packed and not extensively covered with fibrin (Van Oeveren, 1999).

Hanekamp and Koolen reported implantation of a silicon carbide coated stent in coronary arteries < 3.0 mm is a safe and effective treatment when compared to percutaneous transluminal coronary angioplasty alone (Hanekamp & Koolen, 2000). Monnink et al. (Monnink et al., 1999) reported a lower activation of leukocytes expressed by a significantly lower CD11b receptor-mediated adhesion at the SiC-coated stent.

The coating of nitinol stents with a-SiC:H showed an improvement regarding thrombogenic properties as reflected in the platelet count, levels of β-TG and TAT III complex in a short-term in vitro setting. Although the currently used nitinol stents show good results in the iliac vasculature and have shown to be superior to percutaneous transluminal angioplasty alone in the superficial femoral artery, they still tend to restenose in a nonnegligible number (Schillinger, 2006).The coating with a-SiC:H might result in a further decrease of thrombogenicity especially for longer lesions and a conservation of the stent material by decreasing the removal of nickel ions from the alloy (Schmehl, 2008). Therefore, it can be concluded from these reports that the adhesion and activation of human platelets is significantly reduced at silicon carbide coated surfaces.

8. Tissue engineering

Tissue engineering was once categorized as a sub-field of bio materials, but having grown in scope and importance, it can be considered as a field in its own right. It is the use of a combination of cells, engineering and materials methods, and suitable biochemical and physio-chemical factors to improve or replace biological functions. While most definitions of tissue engineering cover a broad range of applications, the term is, in practice, closely associated with applications that repair or replace portions of or whole tissues (i.e., bone, cartilage, blood vessels, bladder, skin etc.). Often, the tissues involved require certain mechanical and structural properties for proper functioning. The term has also been applied to efforts to perform specific biochemical functions using cells within an artificially-created support system (e.g. an pancreas or a bio artificial liver).

Cells are often implanted or 'seeded' into an artificial structure capable of supporting three-dimensional tissue formation. To achieve the goal of tissue reconstruction, scaffolds must meet some specific requirements. A high porosity and an adequate pore size are necessary to facilitate cell seeding and diffusion throughout the whole structure of both cells and nutrients. The scaffold is able to provide structural integrity within the body, and eventually it will break down leaving the neotissue, newly formed tissue which will take over the mechanical load.

Implantation of bone autograft or allograft is a known strategy for the treatment of large bone defects. However, limited supply, donor site morbidity and the risk of transmission of pathological organisms impose major limits to their widespread use. Tissue engineering is trying to address this problem by development of bone substitutes using cells and bioscaffolds. Collagen is the main organic and hydroxyapatite (HA) the main mineral component of the bone extracellular matrix, which determines the mechanical properties of bone and the behavior of cells. Therefore, these components are used, either alone or in combination, for manufacturing most bone substitutes. To date, several bone substitutes have been approved for clinical applications using a wide range of scaffold materials. However, most of them have relatively poor mechanical strength and they cannot meet the requirements for many applications (Ghannam, 2005). Hence, there is a need to fabricate new scaffolds with improved mechanical properties and biocompatibility. Silica-based ceramics are a group of bioactive products, which exhibit better biodegradability in comparison to HA ceramics, promote apatite nucleation and enhance bone bonding in vivo (Hing et al., 2006). In addition, silica-based materials encourage deposition of extracellular matrix, which facilitates cell adhesion and other cellular activities (Thian et al., 2006).

Silicon carbide ceramic is one of the members of this group which is light weight and has excellent mechanical properties. It has been used in manufacturing composite bone scaffolds, for example with a coating of bioactive glass (Gonzal et al., 2003). Silicon carbide supports human osteoblast attachment and growth (Thian et al., 2005; Rokusek et al., 2005).

The concept of using bioscaffolds as one of the strategies for tissue repair has been widely accepted, as they can provide structural stability and a 3D system onto which cells can migrate and grow. Bioscaffolds have been synthesized not only for the repair of bone but also for the repair of various other tissues such as cartilage (Sohier et al., 2008), tendon (Liu et al., 2008), skin (Samadikuchaksaraei, 2008), blood vessels (Zhang et al., 2007), the central nervous system (Samadikuchaksaraei, 2007), and several commercial bioscaffold products are available on the market (Samadikuchaksaraei, 2007).

Fig. 5. SEM micrographs of the constructed scaffold (A) Low magnification (B) Higher magnification of open interconnected micropores (C) grain morphology (Saki et al., 2009)

Saki et al. synthesized a hydroxyapatite-alumina and silicon carbide composite scaffold for bone tissue engineering (Saki et al., 2009). SEM captured images of scaffold morphology show fairly uniform pores, which are suitable for growth of bone tissue cells (Fig 5A, B). Fig. 5C shows the grain morphology of the constructed scaffold. The grains are of uniform morphology and their size ranges between 2.5-5 μm. Cell growth and viability studies show that the scaffold does not significantly change the behaviour of osteoblasts. They show many cells attached to the scaffold (Fig 6).

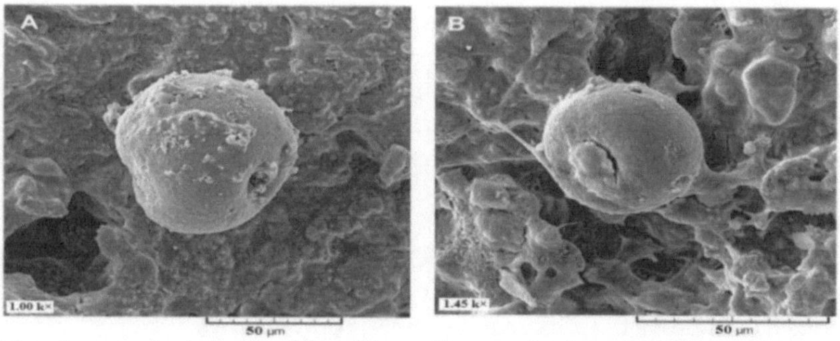

Fig.6. SEM micrographs of osteoblast-like cells attached to HA-alumina-SiC composite surfaces; the cell filaments attaching them to the surface are observed in Fig. 6B (Saki et al., 2009)

Saki et al. used a recently developed method involving impregnation of an organic foam for synthesis of a three dimensional scaffold. One of the advantages of this method of construction is the synthesis of scaffolds with fairly uniform pore morphologies, which are interconnected and in the range suitable for penetration by osteoblasts and vascular tissue.
Silicon-containing HA surfaces have also been shown to be suitable for osteoblast adhesion, migration and proliferation.
Biological materials with complex composite forms and microstructures often display outstanding mechanical properties, which have inspired material scientists in the design of novel materials (Meyers et al., 2008). In the last decade, biotemplating has been widely used to fabricate biomorphic porous materials with various components, such as zeolite, metals, oxides, carbides, nitrides, and with different microstructures, such as fibrilla macroscopic structures, laminated ceramics, or complex micro/macro-pore structures, which might be suitable for technical applications, for example as filters, catalyst carriers or biomedical materials (Zampieri et al., 2006; Liu et al., 2007, Fan et al., 2006; Luo et al., 2007). Due to their hierarchical structure and uniaxial pore morphologies with anisotropic mechanical properties, woods have become the most commonly used biotemplates.
Several investigations were carried out in the recent years to exploit the biomorphic ceramics as new scaffold for bone implants. Porous Si_3N_4/SiC composites have attracted increasing attention because of their excellent physical and chemical properties, such as their strength, resistance to oxidation and corrosion. The mechanical properties of porous ceramics not only depend on the component and porosity, but are also highly dependent on the size, shape, and orientation of the pores as well as grains (Ohji et al., 2008). Fibrous bioceramic scaffolds are favorable candidates since they offer a large specific surface area, giving rise to a high bioactivity for bone tissue engineering. Interconnected pores provide a

framework for tissue in-growth and ensure the nutrition and blood supply for the growing bone (Rambo et al., 2006). Recently, the manufacturing of the Si_3N_4 reinforced biomorphic microcellular SiC composites for potential medical implants for bone substitutions with good biocompatibility and physicochemical properties have been produced (Luo et al., 2009). The open porous SiC reinforced with β-Si_3N_4 are of particular interest for load bearing applications as bioceramic scaffold in bone tissue engineering or as porous support for catalysts.

The remarkable biomechanical properties of human tissues to be replaced or repaired using implants stem from their hierarchic structure. The tissues are an organized assembly of structural units at increasing size levels, which provides optimum fluid transfer and self-healing. Gonzal et al. produced bioSiC by Si-melt infiltration of carbonaceous scaffolds derived from cellulose templates. The natural template selected to develop ceramics for medical implant was sapelli wood. BioSiC presents a biological response similar to titanium controls, but it incorporates the unique property of interconnected porosity, which is colonized by bone tissue, together with lightweight and high strength for optimum biomechanical performance. Bio-derived SiC stands as a new material for biomedical applications (Gonzal et al., 2008). In conclusion, the surface of the fabricated scaffold needs to be optimized to improve the attachment of cells.

9. Dental and orthopaedic implant

In the last decades, many materials have been produced and improved for specific medical applications, such as metals (stainless steel, cobalt–chromium, titanium and alloys), ceramics (alumina, zirconia, graphite), polymers (epoxy, Teflons) and composites.

Moreover, to further improve the fixation and osteointegration performance, different approaches leading to the formation of a bond across the interface between the implant and the tissue via chemical reactions have been attempted. For this purpose, various kinds of bioactive materials have been developed and successfully applied as coatings to artificial bones, such as hydroxyapatite, glass ceramics and glasses.

In this way, the interfacial bond prevents motion between the implant and the host tissue and mimics the type of interface that is formed when natural tissues repair themselves (Gonzalez et al., 2003). The main challenge of the implant technology is the development of a new generation of light implant materials with enhanced mechanical properties, wear resistant and with better biological response. With this aim, biomorphic silicon carbide ceramics are very promising as a base material for dental and orthopaedic implants due to their excellent mechanical properties (Martinzer-Fernandez, 2000).

A new generation of light, tough and high-strength material for medical implants for bone substitutions with a good biological response is reported. The innovative product that fulfills all these requirements is based on biomorphic silicon carbide ceramics coated with a bioactive glass layer. The combination of the excellent mechanical properties and low density of the biomorphic SiC ceramics, used as a base material for implants, with the osteoconducting properties of the bioactive glass materials opens new possibilities for the development of alternative dental and orthopedic implants with enhanced mechanical and biochemical properties ensuring optimum fixation to living tissue. The SiC ceramics have been successfully coated with a uniform and adherent bioactive glass film by pulsed laser ablation using an excimer ArF laser (Gonzalez et al., 2003).

Titanium implants made of commercially pure titanium, medical grade, have a reference tensile yield strength of between 280–345 MPa (Mangonon, 1999). Higher strength implants made of Ti–6Al–4V (alpha-beta alloys) have a reference tensile yield strength of 830–924 MPa. It is then possible to conclude that beech-based SiC biomorphic implants appear as a quite interesting alternative to Ti implants, by showing higher strength and less than 40% of its density. Moreover, taking into account the biomechanical requirements (density, elastic modulus, strain to failure, etc.) of a particular type of bone in the body that should be repaired, biomorphic SiC ceramics can be tailored by an appropriate wood precursor selection. An alternative material for medical implants for bone substitutions based on high-strength and low density biomorphic SiC ceramics coated with bioactive glass is reported, combining the characteristics of both materials into a new product with enhanced mechanical and biochemical properties (Gonzalez et al., 2003).

Several reports (Gonzalez, et al., 2003, 2004; Mayor et al., 1998; LeGeros et al., 1967) showed biomorphic silicon carbide coated with bioactive glass has been proposed as an alternative to titanium and titanium alloy devices due to its low density, bio-inertness, interconnected porosity and improved mechanical properties. Hydroxylapatite coatings was produced by pulsed laser deposition (PLD) on biomorphic silicon carbide ceramics by ablation of non-sintered HA discs with an ArF excimer laser (193 nm, 25 ns, 4.2 J cm^{-2}) at different conditions of water vapour pressure and substrate temperature for dental and orthopaedic applications, reported by Barrajo et al. (Borrajo et al., 2005).

Bioactive silica-based glasses are good candidates to be applied as coatings, thereby improving the physiological response of the ceramic substrate because they promote the intimate bonding of living tissues through the formation of a calcium phosphate layer similar to the apatite found in bone (Fujibayashi et al., 2003), thus preventing the formation of a fibrous capsule around the implant.

Recently (Carlos et al., 2006), in vitro cytotoxicity of wood-based biomorphic silicon carbide ceramics coated with bioactive glass, using MG-63 human osteoblast-like cells, and their application in bone implantology have been reported. The MG-63 osteoblast-like cell monolayer time course formation. A, 1 hr; B, 6 hrs and C, 24 hrs after seeding on a representative beech-based SiC ceramic was coated with bioactive glass is shown in Fig. 7. One hour after seeding (A), rounded cells have involved in cellular division events and can be seen attached to the outer surface inside the pores. Cells begin to penetrate and colonise the inner surface of the existing pores. At 6 hrs after seeding (B), cells were attached and had spread out, displaying a flat configuration and a normal morphology. Neighbouring cells maintained physical contact with one another through extensions of the cytoplasm.

At 24 hrs (C), the bioactive glass coated surface is almost completely covered by the MG-63 cells. No evidence of major deleterious or cytotoxic responses has been observed. The biomorphic beech-based SiC ceramics coated with bioactive glass supports the cellular monolayer formation and the colonisation of the surface of the material. The same results have been obtained for the eucalyptus and the sapelli-based coated ceramics.

SiC ceramics coated with bioactive glass showed the same biological response as the reference materials Ti6Al4V and bulk bioactive glass. The biomorphic SiC ceramics coated with bioactive glass by PLD did not produce a cytotoxic response on the MG-63 osteoblast-like cells. The same behavior was observed for uncoated ceramics. The cellular activity on coated and uncoated SiC ceramics was similar to well known implant materials like Ti6Al4V and bulk bioactive glass (Carlos et al., 2006).

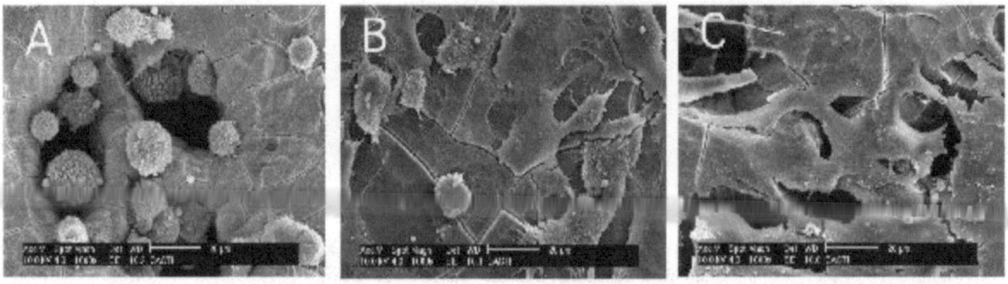

Fig. 7. Scanning electron microscopy images showing the MG-63 osteoblast-like cell monolayer time course formation. A, 1 hour; B, 6 hrs and C, 24 hrs after seeding on a representative beech-based SiC ceramic coated with bioactive glass. All magnifications are 1000×(Carlos et al., 2006)

When using the combination of 316L and gold in liquid, electrolytic media such as blood, the corrosion resistance of the system comes into question. The main reasons for this are contact and crevice corrosion. The corrosive attack in particular could lead to the destruction of the implant or to a release of metallic ions. Along with the potential problem of corrosion, there is the question of bio- and hemocompatibility of gold that is in direct contact with the biological environment. While some authors do not see any deleterious interactions between gold and the vessel wall (Tanigawa et al., 1991, 1995), a significant thrombogenic effect has been attributed to this material in other publications (Sawyer et al., 1965; Schomig et al., 1999). The corrosion behavior of gold-covered stents that have been coated with an additional, amorphous silicon carbide film that is known to be antithrombogenic (a-SiC:H), and which prevents direct contact between the gold coating and the biological environment (Wendler et al., 2000).

For improved osseointegration, titanium-based total hip replacement (THR) components were coated by some manufacturers with hydroxyapatite, which also prevents titanium metallosis. However, hydroxyapatite has a tendency to dissolve from the implant surface, and thus a hydroxyapatite coating does not prevent periprosthetic metallosis permanently. Wear particles from THR prosthetic components cause a local host response directly and indirectly, which leads to component loosening (Santavirta et al., 1990).

Silicon carbide as a ceramic coating material of titanium-based THR implants. The idea is to prevent wear debris formation from the soft titanium surface. SiC is a hard and tightly bonding ceramic surface material, and because of these physical properties it is not easily degradable, as is the case with hydroxyapatite. SiC can be deposited on a titanium implant surface, e.g., by sputtering techniques, and such coatings bond very well to the substrate. In human monocyte cultures phagocytized SiC particles cause a similar stimulation to hydroxyapatite (Santavirta et al., 1998).

Recently titanium-based implant manufacturers have begun to consider modifying surfaces, so that bonds can form a mechanical interlock. The approach is the creation of meshwork through sintered beads or threads that have 350–500 μm pores to promote osteoconduction. The obtained mechanical properties, i.e. elastic modulus ≥335 GPa are favorable as compared with human cancellous bone and titanium. Ceramic on ceramic total hip prostheses are developed to apply to young patients because lifetime of polyethylene joint prostheses is limited by loosening due to biological response. As mating faces of all-ceramic joint must be

highly conformed to reduce stress concentration, wear properties of SiC surface were investigated by Ikeuchi et al. (Ikeuchi et al., 2000). The conclusions are as follows:
(1) Among the four ceramics, alumina and silicon carbide can be applied to ceramic on ceramic joint prostheses because they keep low wear rate, smooth surface and high hardness during sliding in water environments.
(2) Surface film formed on the ceramic surface may contribute to boundary lubrication in a ceramic on ceramic joint prostheses. Therefore, SiC is a candidate material that can be applied for dental and orthopaedic implants.

10. Surface functionalization

Surface functionalization introduces chemical functional groups to a surface. This way, materials with functional groups on their surfaces can be designed from substrates with standard bulk material properties. Prominent examples can be found in the semiconductor industry and biomaterial research. In the recent survey by Stutzmann et al. (Stutzmann et al., 2006) a particular emphasis on the direct covalent attachment of biomolecules on semiconductor surfaces and the resulting electronic properties was given. In that context SiC was suggested as a suitable material for biofunctionalization of H-terminated surfaces. It was also emphasized (Stutzmann et al., 2006) that the different polytypes of SiC were quite well matched to organic systems in terms of band gap and band alignment. Therefore, SiC should be a very interesting substrate material for semiconductor/organic heterostructures. Attachment of covalently bound organic monolayers onto SiC (vide infra) required a pre-treatment that provided the surface of this material with a reproducible reactivity. This pre-treatment involved cleaning of as-received SiC with organic solvents, subsequent oxidation by air plasma, and wet etching in 2.5% aqueous HF solution. Direct, covalent attachment of organic layers to a semiconductor interface provides for the incorporation of many new properties, including lubrication, optical response, chemical sensing, or biocompatibility. In combination with a biocompatible semiconductor material, the hybrid system could be the basis for implantable biosensors or other electrical components inside the human body. One of the major challenges in this area is the stable surface functionalization of mechanically and physicochemically robust materials. Compared to silicon, both diamond and SiC have the same advantages, like stability and biocompatibility, but SiC processing is easier. The development of methods to tune the surface properties of two robust high bandgap materials, silicon-rich silicon nitride (SixN4, $3.5 < x < 4.5$) and SiC would significantly increase the possible use of these materials. Si_xN_4 is widely used, for example, as waveguide material in refractometric (McDonagh et al., 2008) or fluorescence (Anderson et al., 2008) detection, and as coating material for sensors based on electrical impedance (Tlili et al., 2005) or vibrating microcantilevers (Goeders et al., 2008) SiC has a high potential for similar applications (Yakimova et al., 2007). For such sensing and biomedical applications, both materials would benefit from specific surface modification. For the realization, a strategy for the covalent immobilization of the active molecules, e.g. enzymes, on SiC has to be developed. The general method is shown in Fig. 8. Usually the functionalization is carried out in three or more steps.

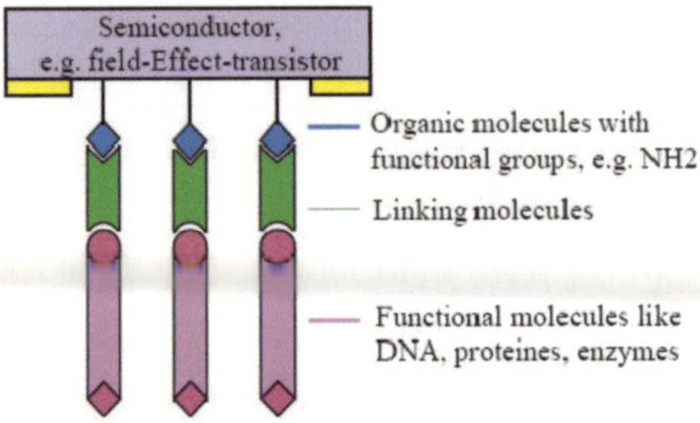

Fig. 8. Schematic representation of the general method for the covalent attachment of biomolecules to semiconductor surfaces (Seino et al., 2002)

Because there are some difficulties with the usual wet chemical processing of SiC, the alternative way has to be followed, which means that small organic molecules with functional groups are immobilised on the clean surface in a vacuum chamber. There is little knowledge about the adsorption of organic molecules on SiC surfaces and hence some questions arise: Molecules with which functional groups can bond covalently to SiC? Can it be achieved to get free functional groups after the bonding process? How do the polarity and the composition of the SiC surface influence the bonding and the structure of the organic layer? In order to answer these questions, a symmetric and an asymmetric molecule, see Fig. 9, with functional groups that bind covalently with silicon surfaces as reported in literature (Seino et al., 2002).

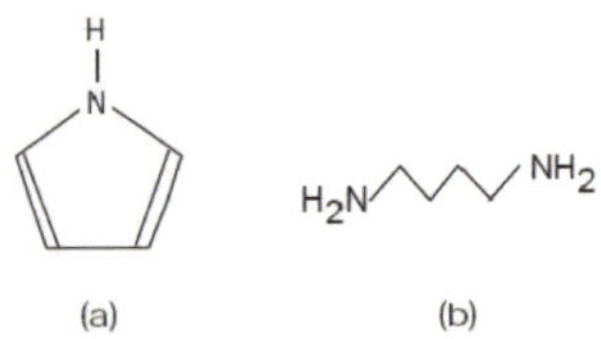

Fig. 9. The organic molecules (a) pyrrole and (b)1,4-diaminobutane (Seino et al., 2002)

The comparison with theoretical calculations (Seino et al., 2002) and other experimental data leads to the conclusion that the molecules adsorb on the surface via N-H-dissociation and the formation of covalent N-Si-bonds. No signs of the formation of ordered adsorbate layers were observed (Seino et al., 2002). Rosso et al. (Rosso et al., 2008) reported the thermal formation of alkyl monolayers from alkenes on hydroxyl-terminated surfaces as depicted in Fig. 10.

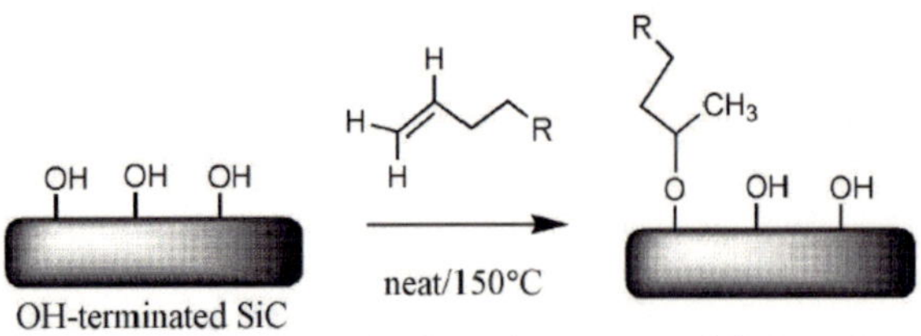

Fig. 10. Attachment of alkyl monolayers on hydroxyl-terminated SiC surfaces (Rosso et al., 2006)

The binding of 1-alkenes via the second carbon of the chain, instead of the first one as was the case for reaction with hydrogen-terminated silicon surfaces, is also probably the

explanation for the formation of somewhat less ordered monolayers. Apart from the stability of methyl-terminated surfaces, the integration of organic functionalities onto inorganic materials opens an extended field of research.
Rosso et al. (Rosso et al., 2009) investigated the UV-assisted formation of organic monolayers on 3C-SiC and SixN4, using only wet chemistry under ambient conditions (temperature and pressure), as this would significantly increase the range of monolayer functionalities that can be attached. After hydrolysis of ester monolayers and subsequent N-hydrosuccinimide activation of the obtained carboxylic acid-terminated surfaces, amide formation with m-(trifluoromethyl) benzylamine (TFBA) showed the possibility of easy surface (bio-) functionalization of SiC. Beside alkyl monolayers, functional coatings from covalently attached ω-functionalized 1-alkenes can be successfully formed onto both SiC surface. Methyl undec-10-enoate and 2,2,2-trifluoroethyl undec-10-enoate (TFE) were successfully attached onto this material using UV light. TFE monolayers were subsequently subjected to further reactions at room temperature, including hydrolysis with a 0.25 M solution of potassium tert-butoxide in DMSO, followed by a NHS activation of the obtained carboxylic acids and subsequent reaction with TFBA, as depicted in Fig. 11.

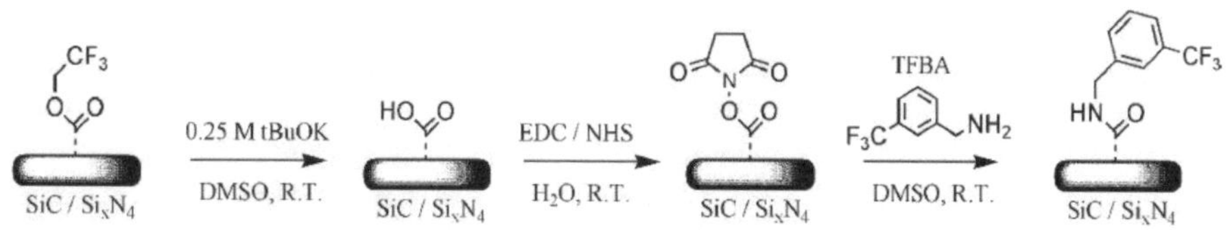

Fig. 11. Functionalization of TFE monolayers on SiC or $SixN_4$ surfaces (Rosso et al., 2009)

Robust functionalization of SiC surface with a wide variety of (bio-)functional moieties is possible using a protecting trifluoethyl ester. A rich chemistry including surface patterning can then be performed. Protein-repellent monolayers can also be formed with these methods, using oligoethylene glycol compounds.
Cicero and Catellani (Cicero and Catellani, 2005) used first-principle predictions of the chemical processes that eventually led to adsorption of organic molecules on the Si-terminated SiC in the search of feasible ways to surface functionalization. Their results suggest a larger stability of the functionalization for the studied groups on SiC rather than on Si. Moreover, the preferential chemisorption of thiolates can lead to the realization of stable self-assembled monolayers (SAMs), with no need of preliminary metallic deposition. The interaction of water with polar SiC surfaces was first investigated. The researchers found that irrespective of coverage, water dissociates on the Si-terminated surface, substantially modifying the clean surface reconstruction. In contrast, the C-terminated surface is non-reactive and hydrophobic. This suggests that by growing SiC substrates with adjacent islands which are either Si- or C-terminated, an atomic control of hydrophilic and hydrophobic environments may be attained. Such surfaces are interesting for DNA and protein attachment. Kanai et al (Kanai et al., 2005) studied theoretically biotin chemisorption on clean and hydrohylated Si-terminated SiC (001) surfaces. These results gained preliminary insight into the understanding of experimental observations showing that biotin molecules directly adsorbed on the hydrohylated surfaces bind strongly to some proteins. Preuss et al (Preuss et al., 2006) made a combined theoretical and experimental study of pyrrole-functionalized Si- and C-terminated SiC surfaces. Recent experimental

studies on surface biofunctionalization of the semiconductors have been focused on the formation of covalently bond SAMs of organic molecules that can serve as a first step for immobilization of biomolecules with the ultimate goal to construct high performance transducers for biosensor devices.

Recently, it has been reported that the biocompatibility of 6H-SiC (0001) surfaces was increased by more than a factor of six through the covalent grafting of NH_2 terminated SAM using aminopropyldiethoxymethylsilane (APDEMS) and aminopropyltriethoxysilane (APTES) molecules. Surface functionalization began with a hydroxyl, OH, surface termination. The study included two NH_2 terminated surfaces obtained through silanization with APDEMS and APTES molecules (hydrophilic surfaces) and a CH_3 terminated surface produced via alkylation with 1-octadecene (hydrophobic surface). H4 human neuroglioma and PC12 rat pheochromocytoma cells were seeded on the functionalized surfaces and the cell morphology was evaluated with atomic force microscopy. The biocompatibility was enhanced with a 2 fold (171-240%) increase with 1-octadecene, 3-6 fold (320-670%) increase with APDEMS and 5-8 fold (476-850%) increase with APTES with respect to untreated 6H-SiC surfaces. H-etching was shown to be a successful technique able to produce atomically flat and repeatable SiC surfaces. Exciting possibilities such as functionalization and nano-patterning of SiC surfaces with biomolecules become more feasible thanks to the morphological atomic order revealed by H-etching treatment. H-etching represents an additional technique, besides wet chemical treatment, that can be used to modify the electronic properties of a surface and therefore to define its electronic behavioral pattern.

Before chemical treatment all the samples were in the 'steady state'. The HF treatment depleted the Si and SiC surfaces of majority carriers. The presence of a depletion layer is ideal for sensing charges added on the surface and therefore is a good starting surface condition for performing semiconductor-cell-electrolyte measurements. From these studies they have obtained an accurate description of the response of SiC surfaces to added charges and the novel result that H-etching electronically passivates 3C-SiC (001) surfaces, that the cell charge effect on semiconducting surfaces is smaller in magnitude than initially estimated and cannot be detected with a measuring system which presents a maximum accuracy of 15 mV (Coletti et al., 2006). The multifunctional molecules at SiC surface has become a useful technique to design hybrid interfaces for the biosensor field, model surfaces for cell-biological studies and drug carrier surfaces for medical application.

11. Biomembrane

The growing capability of microsystems to incorporate sophisticated electronics with mechanical parts and microfluidics is enabling multiple new applications for microdevices. Microsystem interfaces with liquid environments allow environmental monitoring, chemical and biological process monitoring, and various medical applications. Considerable effort has been devoted to developing porous silicon membranes permeable to liquids that can act as such an interface. Microfabricated porous silicon membranes (Chu et al., 1999) have been applied to an implantable artificial pancreas (Desai et al., 1998) and kidney (Fissell et al., 2003) and oral drug delivery systems (Tao & Desai, 2003). The SiC samples used in Rosenbloom et al. (Rosenbloom et al., 2004) study are *n*-type and *p*-type 300–400 μm thick 6H SiC crystals. The porous structures were obtained by electrochemical etching, described for SiC by Shor and Kurtz (Shor & Kurtz, 1994). For measuring protein permeability, free-standing nanoporous SiC membranes were glued onto circular polystyrene plastic supports and placed into a small fluidic chamber. Molecules within the source solution below the membrane diffused through the membrane into the receiving solution above. Biofouling is a significant problem with any nanoporous membrane exposed to high protein concentrations. The two porous SiC morphologies shown in Fig. 12 were used by Rosenbloom et al. (Rosenbloom et al., 2004). Despite considerably different pore structure, membranes of both *n*-type and *p*-type SiC exclude proteins in the same size range. The *n*-type material allowed much more protein to diffuse through, by a factor of as much as four times (for myoglobin). Each membrane type passed proteins of up to 29000 Da molecular weight but excluded larger proteins with a molecular weight of 45000 Da and higher. These molecular weights correspond to molecular diameters of less than 4.7 nm (permeable) and greater than 5.0 nm (excluded). The discrimination between the 29 kD and 44 kD proteins is likely complex, and not simply based on size (4.7 and 5.0 nm, respectively). For example, charge, hydrophilic or hydrophobic interactions, as well as dimerization in solution or differential adsorption to the SiC, to the glue, to the plastic tube or membrane backing, etc., could all play roles. This same behaviour was observed when they tested the same 6-protein mix with commercial membranes, i.e., ovalbumin is excluded to a much greater degree than carbonic anhydrase.

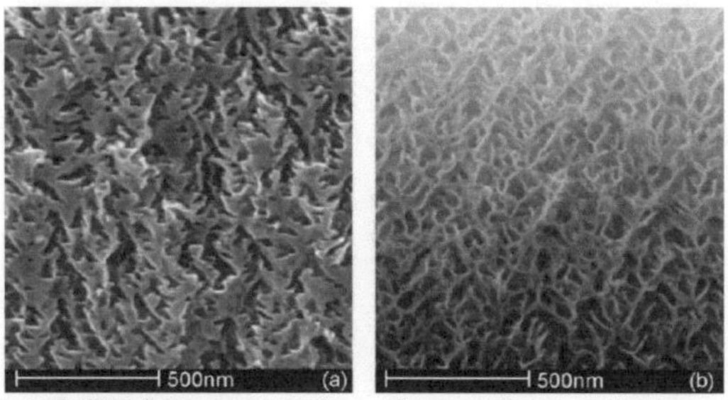

Fig. 12. The cross-sectional SEM images taken from pieces of porous membranes used for the tests: (a) n-type 6 H SiC and (b) p-type 6 H SiC. The p-type porous structure has a more feathery appearance, corresponding to more material removal. These membranes are more fragile than the n-type membranes (Rosenbloom et al., 2004)

It has been shown (Bai et al., 2003), that nanoporous SiC membranes can be made with a wide variety of pore sizes, shapes and densities by varying the conditions of pore formation. The size range of proteins able to cross the membranes includes many important cell signaling molecules, which tend to be small. Nanoporous SiC has a low protein adsorption comparable to the best commercially available polymeric membranes specifically designed for low protein adsorption. This resistance to protein fouling is another desirable property of this material. Early membranes that were tested had nonuniform pores. Having demonstrated good resistance to protein adhesion for SiC, and the ability to make free-standing semi-permeable membranes, it was decided to attempt to improve the membranes by making more uniform and consistent pore structures. The goal was to improve the protein throughput. These membranes had areas of large, relatively straight pores with a high density in some areas. The protein permeability of these membranes was very good. Although these membranes are usable for probe interfaces, it is obvious that the pore structures are nonuniform. Most of the diffusion likely occurs in a few areas of high porosity with large areas of nonpermeable membrane in between. This material was not permeable to protein, likely because the pore sizes are too small. It was also not permeable to small molecules, such as fluorescent dye (Cy3, 767 Da) for unclear reason. The combination of these important properties of nanoporous SiC suggests that this material could be very useful in medical microsystem applications.

12. Quantum dots

A quantum dot (QD) is a semiconductor whose excitons are confined in all three spatial dimensions. As a result, they have properties that are between those of bulk semiconductors and those of discrete molecules. Quantum dots have quickly filled in the role, being found to be superior to traditional organic dyes on several counts, one of the most immediately obvious being brightness (owing to the high extinction co-efficient combined with a comparable quantum yield to fluorescent dyes) as well as their stability (allowing much less photobleaching). Researchers in France have made the first chemically inert, biocompatible silicon carbide quantum dots for fluorescence imaging of living cells. The result is a major advance since all QDs used for imaging so far have been toxic to cells. QDs based on II-IV and III-IV group semiconductors are used for in vitro imaging of biological cells because of their remarkable luminescent properties. They can be tagged onto cells and their fluorescence measured. The only problem is that these QDs are highly toxic to cells, which means that they have to be coated with a protective layer, such as a polymer, before they can be employed. However, no protective layer is really good enough to completely shield cells.

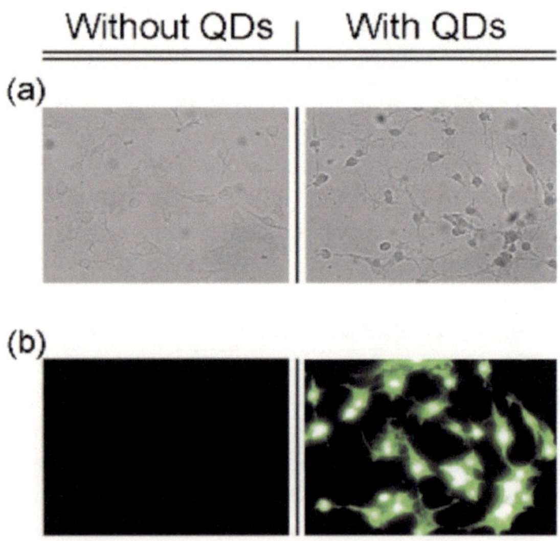

Fig. 13. (Color online) (a) White light microscopy images (100× magnification) of biological cells not having been exposed to QDs (left column) and having been exposed to QDs (right column); (b) Corresponding fluorescencephotos (100× magnification) obtained under UV/violet excitation with the same accumulation times of 2 s (Botsoa et al., 2008)

Jacques Botsoa of the Institut des Nanotechnologies in Lyon and colleagues (Botsoa et al., 2008) found a solution to this problem. The SiC QDs that the researchers used are highly luminescent. They could be used to bio-image and label living cells, as well as study the mechanisms of quantum transport through cells and nuclear membranes. Botsoa and co-workers made their QDs by electrochemical anodization etching of low resistivity grade bulk 3C-SiC polycrystalline wafer. These QDs have dimensions smaller than the Bohr diameter of the exciton (around 5.4 nm) and exhibit highly efficient "above-gap" luminescence thanks to quantum confinement. To test their QDs, the researchers added the suspension to cell cultures of 3T3-L1 fibroblasts (Fig. 13).

Rossi et al (Rossi et al., 2008) prepared SiC nanocrystals by single crystal silicon carbide wafer electrochemical etching. Particles in the size range of 1-3 nm exhibit stable UV-VIS photoluminescence in aqueous media, acids, and organic solvents. Given that the SiC nanostructures below ~ 3 nm would acquire a bandgap above 3 eV that effectively extends their luminescence into the UV range. Moreover, due to the exceptional SiC thermal and chemical stability, such nanostructures could be considered for the UV emitter development and serve as an alternative to the Group-III nitride based semiconductors. SiC based QDs are water-soluble and they may be ideal as biological labels. The nanoparticles show high resistance against photobleaching with no significant cytotoxicity (Fig. 14). In conclusion, SiC-based QDs for biomedical application sounds to be a promising approach for future research.

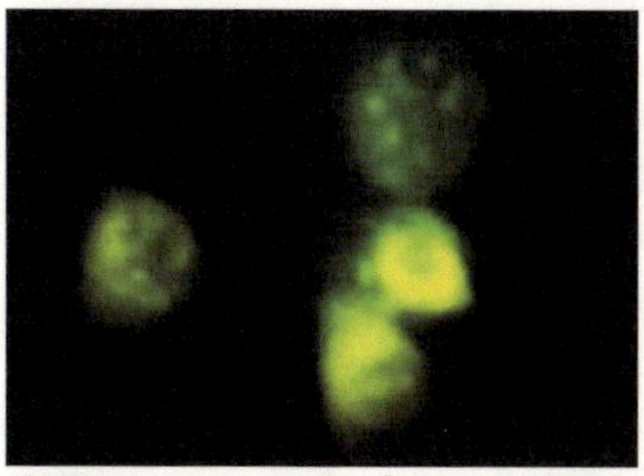

Fig. 14. After the uptake of 3C-SiC nanocrystals, human fetal osteoblast (hFOB) cells exhibit brightgreen-yellow fluorescence (Fan et al., 2008)

13. Nanosilicon carbide

Silicon carbide nanomaterials are widely investigated due to high strength, good creep, oxidation resistance at elevated temperature, chemical inertness, thermal stability, and resistance to corrosion (Kamakaran et al., 2004). The size, shape, and surface composition of these nanostructured materials are some of the few factors that contribute to these unique and fascinating properties (Xia et al., 2003). Efforts have been made to develop nanosize hollow spheres of silicon carbide for technical applications in optical, electronic, acoustic, and sensing devices (Sun and Xia, 2002, Kidambi et al., 2004). Recently, the formation of silicon carbide nanowires from inorganic–organic hybrid of sol–gel-derived silica and the commercially available lignin that is naturally abundant amorphous biopolymer was reported by Mishra et al. (Mishra et al., 2009).

The biocompatibility of the SiC nanomesh renders it a candidate template for biomolecular assembly, opening many unexplored and stimulating possibilities for fundamental and applied studies. As a consequence the sophisticated tools available to surface scientists can shed new light onto nanoscale aspects of biocompatibility (Oliveria & Nanci, 2004), which are related mainly to surface and interface properties, thus guiding the exploration of novel potential biomaterials. Another crucial point for a thorough understanding of the SiC

template is the determination of the chemical nature of the carbonaceous species at the surface (Cicoira & Rosei, 2006).
Bioceramic nanocomposites were synthesized by sintering compacted bodies of hydroxyapatite mixed with 5 or 15 wt% nanosilicon carbide at 1100 or 1200 °C in a reducing atmosphere. The results indicate that the composite of 95 wt% hydroxyapatite and 5 wt% SiC exhibited better mechanical and biological properties than pure hydroxyapatite and further addition of SiC failed strength and toughness (Hesaraki et al., 2010). The preparation of nano-sized silicon carbide has received considerable attention, because it allows the preparation of bulk materials with increased plasticity (Stobierski & Gubernat, 2003) or nanocomposites with enhanced mechanical and tribological properties. In conclusion, it opens up exciting possibilities in the area of template-assisted growth at the nanoscale.

14. Drug delivery

Drug delivery systems (DDS) are an area of study in which researchers from almost every scientific discipline can make a significant contribution. Understanding the fate of drugs inside the human body is a high standard classical endeavor, where basic and mathematical analysis can be used to achieve an important practical end. No doubt the effectiveness of drug therapy is closely related to biophysics and physiology of drug movement through tissue. Therefore, DDS requires an understanding of the characteristics of the system, the molecular mechanisms of drug transport and elimination, particularly at the site of delivery. In the last decade DDS have received much attention since they can significantly improve the therapeutic effects of the drug while minimizing its side effects.In recent years, Poly (D,L-Lactide) (PLA) and Poly (D,L-Lactide-co-Glycolide) (PLGA) have been extensively investigated for use as implantable biodegradable carriers for controlled release of drugs. Silicon carbide coated stents have been coated with a layer of PLA or PLGA containing the drug by dip coating or spray coating techniques. Several drugs have been considered as candidates for stent coatings preventing instent restenosis. SiC is used as a basis for drug delivery systems or bioactive coatings in order to modulate vascular cell growth. For a sufficient polymer-drug coating of a silicon carbide stent and a long-term release of the desired agent, PLA and PLGA are biocompatible materials useful for a variety of applications, including the design and properties of the controlled-release systems for pharmaceutical agents.
Despite the phenomenal pace of stent design technology and the improvements in biocompatibility that have been achieved with the SiC coating, the incidence of in-stent restenosis remains unacceptably high. To address this problem, intense research is being conducted in order to find new stent coatings. Coatings with specific polymer-drug composites or with specific glycosaminoglycans showed promising results in modulating the proliferation of vascular smooth muscle cells and endothelial cells (Bayer, 2001). Using an existing technology for dip coating, glycosaminoglycans can be covalently bonded to the silicon carbide surface via a spacer molecule (Hildebrandt, 2001). Crosslinking the network of coated glycosaminoglycans should result in a stable bioactive layer with long-term anti-proliferative effects (Bayer, 2001). Finally, it is suggested that more detailed experiments are required and would be useful to distinguish and clarify SiC-based materials application in drug delivery.

15. Surface modification of Ti-6Al-4V alloy by SiC paper for orthopaedic applications

It is possible to change localized areas of metals in order to obtain both compositions and microstructures with improved properties. Titanium and titanium alloys are the most frequently used material for load-bearing orthopaedic implants, due to their specific properties such as high corrosion resistance, surface oxidation layer, high strength and high-temperature resistance (Feng et al., 2003). Titanium and its alloys' application like any other biomaterials involve the creation of at least one interface between the material and biological tissues. Biocompatibility and bioactivity of biomaterials rely on the interactions that take place between the interface of the biomaterials and the biological system (Wang & Zheng, 2009). It is generally believed that proteins adsorbed on implant surface can play an important role in cell-surface response. Different proteins such as collagen, fibronectin and vitronectin which are acting as ligands are particularly important in osteoblast interaction with surface. Ligands are the junctions which facilitate adhesion of bone cells to implant surface. In another word, more ligand formation implies a better cell-surface interaction (Tirrell et al., 2002). In vitro studies can be used to study the influence of surface properties on processes such as cell attachment, cell proliferation and cell differentiation. However, in vivo studies must be performed to achieve a complete understanding of the healing process around implants. Previous studies have shown that surface characteristics named above have a significant influence on adhesion, morphology and maturation of cultured osteoblasts (Masuda et al., 1998). Also, it has been demonstrated that for primary bovine osteoblasts, the wettability is one of the key factors. In our studies (Khosroshahi et al., 2007; Khosroshahi), 2007; Khosroshahi et al., 2008; Khosroshahi et al., 2008; Khosroshahi et al., 2009), it is shown that the wettability of the surface can provide a better spreading condition for osteoblast cells due to reduced contact angle. Bearing in mind that the adhesion of bone cells to implant surface consists of two stages. In primary stage the cells must get close enough to surface at an appropriate distance known as focal distance over which the cells can easily be spread over it. In this respect, the wettability can be effective in providing a preferred accessability to surface and thus reaching the focal distance. The secondary stage includes cell-cell attachment which obeys the regular biological facts.

Interface reactions between metallic implants and the surrounding tissues play a crucial role in the success of osseointegration. The titanium and its alloys like some other medical grade metals are the materials of choice for long-term implants. The effect of implant surface characteristics on bone reactions has thus attracted much attention and is still considered to be an important issue (Buchter et al., 2006).

So far as the surface characteristics of the implants are concerned, two main features that can influence the establishment of the osseointegration are the physico-chemical properties and the surface morphology. Cell adhesion is involved in various phenomena such as embryogenesis, wound healing, immune response and metastasis as well as tissue integration of biomaterial. Thus, attachment, adhesion and spreading will depend on the cell-material interaction and the cell's capacity to proliferate and to differentiate itself on contact with the implant (Bigerelle et al., 2005).

Cell behavior, such as adhesion, morphologic change and functional alteration are greatly influenced by surface properties including texture, roughness, hydrophilicity and morphology. In extensive investigations of tissue response to implant surfaces, it has been shown that surface treatment of implant materials significantly influences the attachment of

cells (Heinrich et al., 2008). Additionally, these modified surfaces must resist both the mechanical wear and the corrosion (Sighvi et al., 1998). It is therefore important to evaluate systematically the role of different surface properties and to assess the biological performance of different implant materials.
The surface morphology, as well as manipulation with the physical state and chemical composition of implant surfaces may be significant for bone-implant integration. Surfaces are treated to facilitate an intimate contact between bone and implant. So, the tissue response to an implant involves physical factors, depending on implant design, surface topography, surface charge density, surface free energy and chemical factors associated with the composition of the materials. These substrate characteristics may directly influence cell adhesion, spreading and signaling, events that regulate a wide variety of biological functions (Ronold et al., 2003). Numerous surface treatments including Ion implantation, coating, shot blast, machining, plasma spray, plasma nitrid, nitrogen diffusion hardening are some of the relatively older techniques in the field of material processing which can be used to change implant's surface topography. Thus, the main intention of this work is to extend the earlier research by carrying out some detailed In vitro and In vivo experiments using a 300 and 800 grit SiC papers on surface physico-chemical changes, surface wettability, corrosion resistance, microhardness and osteoblast cells adhesivity of Ti6Al4V with respect to possible orthopaedic applications.

16. Materials and methods

Rectangular-shaped specimens with 20×10 mm dimensions and the thickness of 2 mm, were made from a medical grade Ti6Al4V (ASTM F136, Friadent, Mannheim- Germany- GmbH) with chemical formulation Ti(91.63%)Al(5.12% V(3.25%). The samples were divided into three groups of untreated, 300 and 800 grit SiC paper. Prior to treatment, all samples were cleaned with 97% ethanol and were subsequently washed twice by distilled water in an ultrasonic bath (Mattachanna, Barcelona-Spain). A final rinse was done by de-ionized water at a neutral pH to ensure a clean surface was obtained. They were polished using 300 and 800 grit SiC paper. Finally, an optical microscope with magnification of ×20 was used to ensure that no particles were left on the sample surface.

Surface roughness

The surface micro roughness (Ra) measurements were carried out using a non-contact laser profilemeter (NCLP) (Messtechnik, Germany) equipped with a micro focus sensor based on an auto focusing system. Ra is the arithmetical mean of the absolute values of the profile deviations from the mean line. Five two-dimensional NCLP profiles were obtained for each surface over a distance of 3.094 mm with a lateral resolution of 1μm using a Gaussian filter and an attenuation factor of 60% at a cut-off wavelength of 0.59 mm . The roughness parameters were calculated with the NCLP software similar to that described by Wieland et al. (Wieland et al., 2001).

Surface hardness

Surface microhardness test was carried out with 50 gram load in 10 seconds by a diamond squared pyramid tip (Celemx CMT, Automatic). Each related test was considered at 5 points and reported as an average. The Vickers diamond pyramid hardness number is the applied load divided by the surface area of the indentation (mm^2) which could be calculated from equation bellow:

$$VHN = \{2FSin\ (136°/2)\}/d^2 \tag{1}$$

This equation could be re-written approximately as:

$$VHN = 1.854(F/d^2) \tag{2}$$

Corrosion tests

The standard Tafel photodynamic polarization tests (EG&G, PARC 273) were carried out to study the corrosion behavior of specimens in Hank's salt balanced physiological solution at 37ºC. The metal corrosion behavior was studied by measuring the current and plotting the E-logI (Voltage - Current) diagram. The corrosion rate (milli per year (mpy)) was determined using equation:

$$C.R. = 0.129\ (M/n)\ (I_{corr}/\rho) \tag{3}$$

Where M is the molecular weight, n is the charge, I_{corr} is the corrosion current and ρ is the density.

Surface tension

The surface energy of the samples were determined by measuring the contact angle (θ) of test liquids (diiodo-Methane and water; Busscher) on the titanium plates using Kruss-G40-instrument (Germany).The geometric mean equation divides the surface energy in to two components of dispersive and polar and when combined with Young's equation it yields:

$$\gamma_{lv}\ (1+\cos\theta)=2(\gamma_l^d \cdot \gamma_s^d)^{0.5} +2(\gamma_l^p \cdot \gamma_s^p)^{0.5} \tag{4}$$

Equation (4) can be rearranged as by Ownes-Wendt-Kaeble's equation:

$$\gamma_{lv}\ (1+\cos\theta)/\ (\gamma_l^d)^{0.5} = (\gamma_s^p)^{0.5}\ ((\gamma_l^p)^{0.5}\ /\ (\gamma_l^d)^{0.5}) + (\gamma_s^d)^{0.5} \tag{5}$$

Where s and l represent solid and liquid surfaces respectively, γ_d stands for the dispersion component of the total surface energy (γ) and γ_p is the polar component.

In vitro test

Mice connective tissue fibroblasts (L-929) with 4×10^5 ml were provided and maintained in culture medium (RPMI-1640) consisting of 100U/ml Penicillin, 100U/ml Streptomicine, and 10% fetal calf serum (FCS) .The untreated sample, and SiC treated samples along with a negative control (ie. fibroblast cells only in the cell culture medium) were then placed inside the culture medium in a polystyrene dish. All the samples were incubated at 37°C in 5% CO_2 atmosphere and 90% humidity for 24h. Then the samples were washed with the de-ionized water and sterilized by water steam for 20 min at 120 °C. Subsequently, the samples were then fixed by using 50%, 65%, 75%, 85%, 96% ethanol and stained by Gimsa. Finally, they were evaluated, without extracting the samples from cell culture dish, with an optical microscope

(Nikon TE 2000-U) for cell growth and cytotoxicity. It is worth mentioning that the biocompatibility of the samples was investigated In vitro by L-929 fibroblast cell counting on samples through methyl thiazole tetrazolium (MTT) assay. For this purpose an enzymic method ie.1ml of Trypsin/EDTA was used and the cells were then left to trypsinize in the flask at 37° in the incubator for 3 minutes and were monitored by the same optical microscope.

In vivo test

Anesthetization

Before depilation of the operation site, the animal was completely anesthetized with midazolam (Dormicum®, Roche, Switzerland) 2.5 mg/Kg intravenously (IV). With any sign of recovery during operation, diluted fluanisone/fentanyl (Hypnorm®, India) was injected slowly until adequate effect was achieved, usually 0.2 ml at a time.

Animal implantation

Untreated sample and SiC treated samples were implanted on femur bone of an eight months male goat weighing 30 Kg. Specimens were steam sterilized before implantation in an autoclave (Mattachnna, Barcelona-Spain). The steam sterilization was conducted under 132 °C, 2 bar and in 45 minutes. All the specimens were labeled by separate codes for further studies. The operation site was shaved and depilated with soft soap and ethanol before surgery; the site was also disinfected with 70% ethanol and was covered with a sterile blanket. In order to proceed with implantation, cortex bone was scraped by osteotom (Mattachnna, Barcelona-Spain) after cutting the limb from one-third end in lateral side and elevating it by a self - retaining retractor. Copious physiological saline solution irrigation was used during the implantation to prevent from overheating. To ensure a stable passive fixation of implants during the healing period, they were stabilized by size 4 and 8 titanium wires (Atila ortoped®, Tehran-Iran) without any external compression forces (Fig.15).

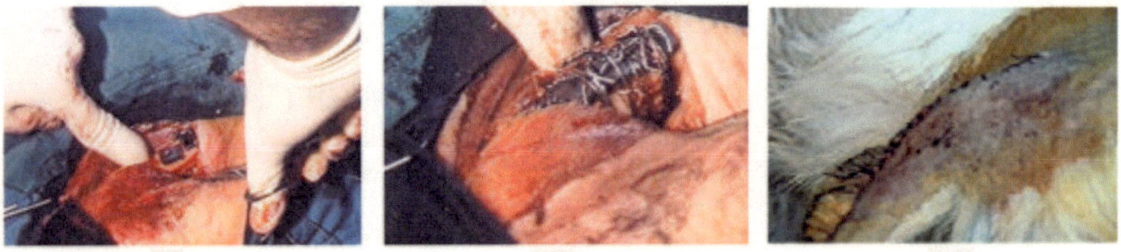

(a) (b) (c)

Fig. 15. Placement of implants in the femur bone of the goat

After the operation the animal was protected from infection by proper prescribed uptake of Penicillin for first four days and Gentamicine for second four days. During the eight days of recovery, the goat was administrated with multi-vitamins to help to regain its strength. During this period, the goat was kept in an isolated space under room temperature, ordinary humidity, lighting and air conditioning, and before it returns to its natural life environment, X-ray radiographs (Fig. 16) were taken in order to ensure that the implant has not been displaced during the maintenance period. It was observed that calus bone had grown in the vicinity of the implant. After five months the animal was sacrificed and the specimens were removed (Fig. 17).

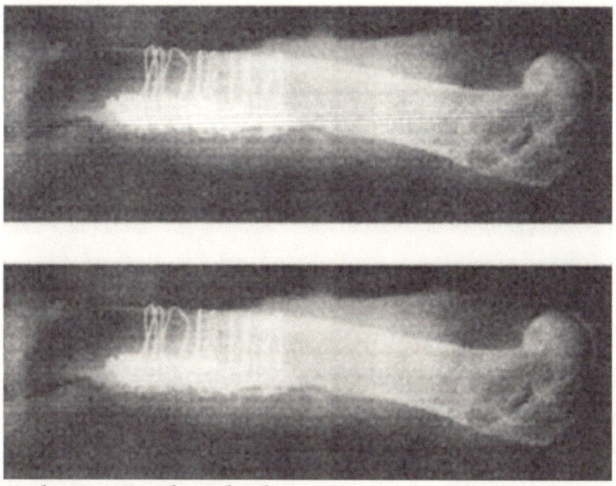

Fig. 16. The X-ray of implants wired to the bone

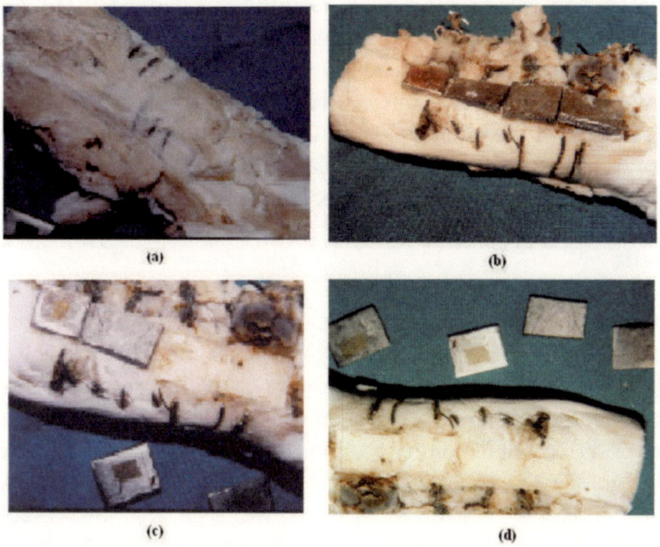

Fig. 17. Implant removal from the femur bone of the goat: (a) before detachment of the wires, (b) after detachmented (c, d) the foot-print of the implants on the bone

The experiments had been approved by the Yazd School of Veterinary Science (Iran) and its animal research authority and conducted in accordance with the Animal Welfare Act of December 20th 1974 and the Regulation on Animal Experimentation of January 15th 1996. The explantation procedure was performed by first cutting the upper and lower section of femur bone using an electric saw and then the implant together with its surrounding tissues was placed in 4% formalin solution for pathological assessment and SEM.

Cell analysis

Osteoblast cells spreading (ie. lateral growth) on the implants was analyzed after removal by SEM (stero scan 360-cambridge) and their spreading condition in a specific area was studied using Image J Program software in three separate regions of each specimen at a frequency of 10 cells per each region. The number of attached cells in 1 cm^2 area of each specimen was calculated by a Coulter counter (Eppendorf, Germany) using enzyme detachment method and Trypsin-EDTA (0.025 V/V) in PBS media at pH = 7.5. The final amount of attached cell can be studied by plotting cell detachment rate versus time.

Histopathology

Surrounding tissues of specimens were retrieved and prepared for histological evaluation. They were fixed in 4% formalin solution (pH = 7.3), dehydrated in a graded series of ethanol (10%, 30%, 50%, 70% and 90%) and embedded in paraffin after decalcification. Then, 10 μm thick slices were prepared per specimen using sawing microtome technique. A qualitative evaluation of macrophage, osteoblast, osteoclast, PMN, giant cells, fibroblast, lymphocyte was carried out by Hematoxylin and Eosin stain and light microscopy (Zeiss, Gottingen-Germany). The light microscopy assessment consisted of a complete morphological description of the tissue response to the implants with different surface topography. Osteoblasts can be in two states; (a) active, forming bone matrix; (b) resting or bone-maintaining. Those make collagen, glycoproteins and proteoglycans of bone the matrix and control the deposition of mineral crystals on the fibrils. Osteoblast becomes an osteocyte by forming a matrix around itself and is buried. Lacunae empty of osteocytes indicate dead bone. Osteoclast, a large and multinucleated cell, with a pale acidophilic cytoplasm lies on the surface of bone, often an eaten-out hollow-Howship's lacuna. Macrophages, are irregularly shaped cells that participate in phagocytosis.

SEM of adhered cells

After implants removal, all three group implants were rinsed twice with phosphate buffer saline (PBS) and then fixed with 2.5% glutaraldehyde for 60 minutes. After a final rinse with PBS, a contrast treatment in 1% osmium tetroxide (Merck) was performed for 1 hour, followed by an extensive rinsing in PBS and dehydration through a graded series of ethanol from 30% to 90% as described in histology section. After free air drying, surfaces were thinly sputter coated with gold (CSD 050, with 40 mA about 7 min). Cell growth on implanted specimens and their spreading condition in a specific area was analyzed using Image J Program software in three separate regions of each specimen for 10 cells per each region.

Statistical analysis

All calculated data were analyzed by using a software program SPSS (SPSS Inc., version 9.0). The results of variance analysis were used to identify the differences between the cells spread area of the treated and cleaned un-treated samples ($p \leq 0.05$).

17. Results and discussion

Characterization of surface topography

SiC paper effect

Figure 18 indicates that SiC treated surfaces have some unevenly distributed microgrooves with occasional scratch and pitting made on it by SiC paper. More directionally defined track lines were produced by 800 than 300.

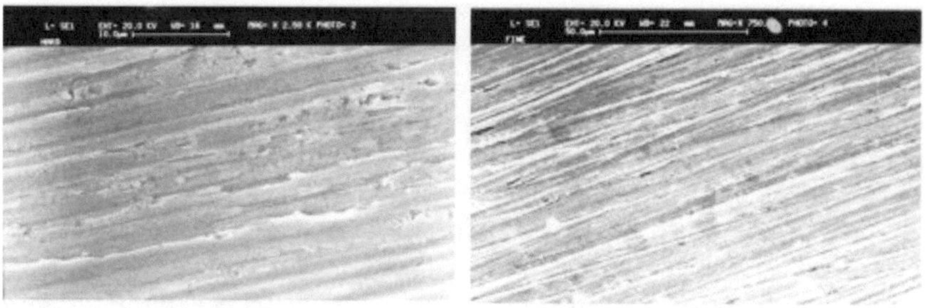

(a) (b)

Fig. 18. SEM of SiC paper treated surface by: (a) 300 grit, (b) 800grit

Surface roughness

In order to obtain a quantitative comparison between the original and treated surface, the arithmetic average of the absolute values of all points of profile (Ra) was calculated for all samples. The Ra values for untreated, 800, and 300 SiC paper were 12.3±0.03, 16.6±0.15, and 21.8 ±0.05 respectively. All the calculations were performed for n=5 and reported as a mean value of standard deviation (SD).

Surface hardness

The surface hardness measurements presented in table 1 clearly indicate that micro hardness of the metal decreases with SiC paper. The surface hardness was found to vary from 377 VHN for SiC treated to 394 VHN for untreated.

Sample	Microhardness (HVN)
Untreated	394
SiC paper (300 grit)	377
SiC paper (800 grit)	378

Table 1. Surface hardness tests before and after treatment

EDX analysis

The experimental results of EDX spectroscopy of the untreated and SiC treated samples in the ambient condition is given in table 2. The analysis exhibited K-α lines for aluminium and titanium for both samples, though it was expected that carbon would be detected too.

Element / Sample	% Al	%V	%Ti
Untreated	5.15	3.25	91.6
SiC paper (300 grit)	5.19	3.37	91.4
SiC paper (800 grit)	6.05	3.35	90.6

Table 2. Surface elements composition before and after treatment

Corrosion test

The comparison of these curves indicates a few important points: 1-a value of 1.77×10^{-3} mpy for untreated sample (Fig. 19a), 2- the corresponding corrosion rates for 300 and 800 grit SiC paper were measured as 1.8×10^{-3} and 1.79×10^{-3} mpy respectively (Figs. 19 b,c) 4- E_{corr} varied from -0.36 V to -0.21 V after the treatment at SiC paper 300 grit. This means that the SiC treated samples are placed at a higher position in the cathodic section of the curve hence releasing hydrogen easier and acts as an electron donor to the electrolyte. Therefore, by smoothly reaching the passivation region, a more noble metal is expected to be achieved. The corrosion current (I_{corr}) was decreased from 2.59 μAcm^{-2} to 0.66 μAcm^{-2} after surface treatment with SiC paper 300 grit and the corrosion current (I_{corr}) for 800 grit was measured 2.51 μAcm^{-2}. A better corrosion resistance was achieved by SiC paper.

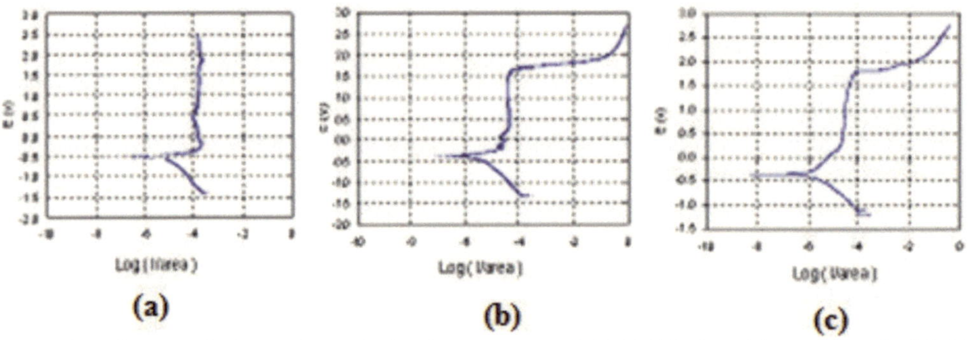

Fig. 19. Tafel potentiodynamic polarization curves of Ti6Al4V for: (a) untreated, (b) SiC paper (300 grit), and (c) SiC paper (800 grit)

Surface tension

The change in surface wettability was studied by contact angle measurement for all specimens treated and untreated (Fig.20). Thus a decrease of contact angle occurred from 70° to 50° indicating a higher degree of wettability. Following the SiC treatment at 800 grit the contact angle reduced to 45 ° showing still a more acceptable hydrophilic behaviour.
Also, variation of surface tension for all specimens was calculated by measured contact angle. It is known that as contact angle decreases, the related surface tension will be increased. Therefore, a value of 46 mN/m was obtained for γ at 300 grit which is considerably higher than 39mN/m of the untreated sample. The corresponding value of γ for 800 grit was found as 50mN/m (Fig. 20b).

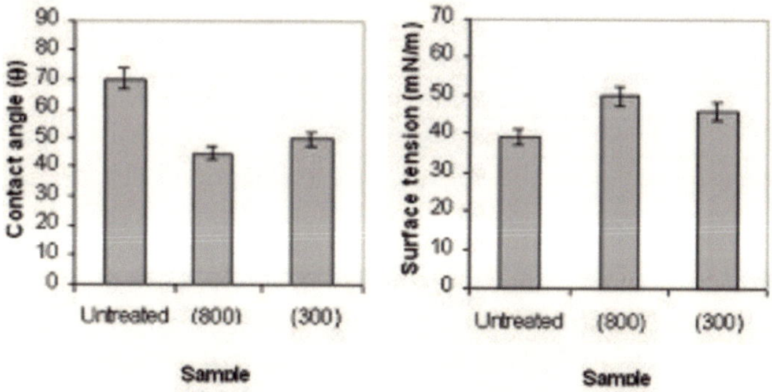

(a) (b)

Fig. 20. Variation of contact angle: (a) and surface tension, (b) with sample surface texture

In vitro

Figures 21 a-c illustrate the morphology and the spreading of cells on the negative control, the untreated and SiC treatment respectively. As it is observed in all cases, some of the attached cells spread radially from the centre and developed a filopodia type shape. The surface of cells which are not spread, were convoluted in to micro ridges and the neighboring cells maintain a physical contact with one another through multiple extensions. Cell spreading is an essential function of cell adhesivity to any surface and it proceeds the proliferation until the surface is fully covered by the cellular network. The number of cells attached to the surface was evaluated by SiC treated samples assay. More cells are attached to the surface for 300 and 800 grits of SiC paper, 9×10^5 and 10×10^5 respectively, which are higher than 8×10^5 for untreated sample.

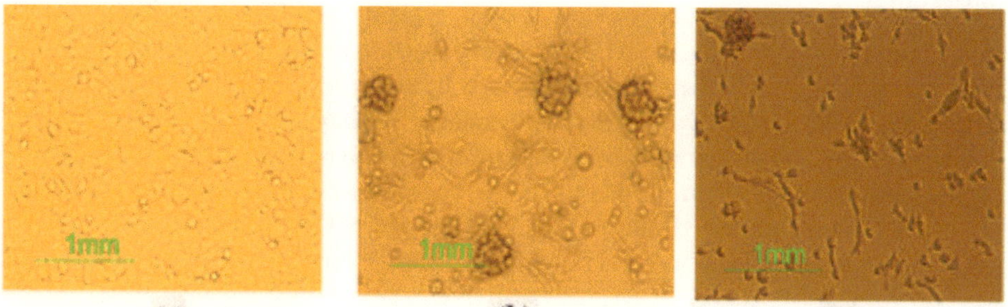

(a) (b) (c)

Fig. 21. Light microscopy of cell culture evaluation (a) negative control, (b) untreated sample, (c) SiC paper (800 grit).

In vivo

Cell spreading analysis

The experimental results of bone cell growth are given in table 3. As it can be seen, cells spreading over the specimen surface are related to surface texture which was measured by Image J program software (IJP). The highest spreading area (383 μm^2) belongs to SiC treated sample (800 grit).

Row	Specimens	Spread cell area (μm²)
1	untreated	352 ± 6
2	SiC paper (800 grit)	383 ± 5
3	SiC paper (300 grit)	367± 3

Table 3. Bone cells spread over the surface of the implanted specimens (average of ten measurements in three separate regions)

The SEM analysis of attached cells morphology (Fig. 22) indicates that the density of cell network is directly dependent on the surface topography. In SiC treated surfaces, the orientation of cells was longitudinal and parallel to the lines made by SiC paper. It is observed from Fig. 22 that SiC treated surfaces have more fibroblast cells compared with the untreated sample.

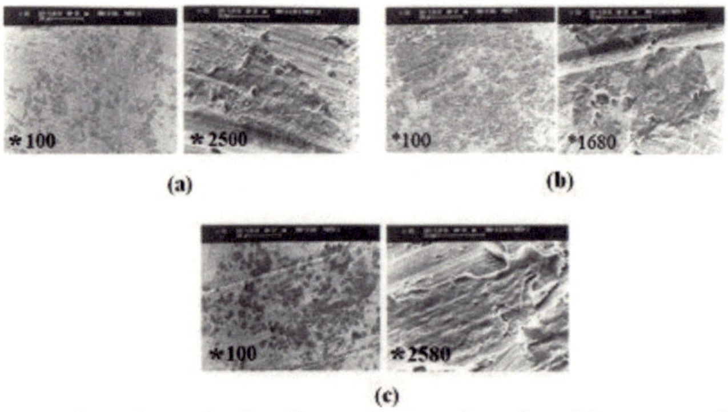

Fig. 22. SEM micrographs of attached cells on the surface for: (a) untreated, (b) 800 grit, (c) 300 grit

Histopathology

When the implants were retrieved, no inflammatory reaction was observed inside or around the implants. Mineralized matrix deposition and bone cells were observed on the surface of implants which are formed during the five months implantation. This deposition was found all around of SiC treated samples (Fig. 23a) and bone formation was characterized by the occurrence of osteocyte embedded in the matrix. Also the above samples were surrounded by fibroblast and osteoblast cells and the untreated sample (Fig. 23b) showed not only fewer number of fibroblast cells, but it also contained osteoclast and polymorpho nuclear leukocytes (PMN).

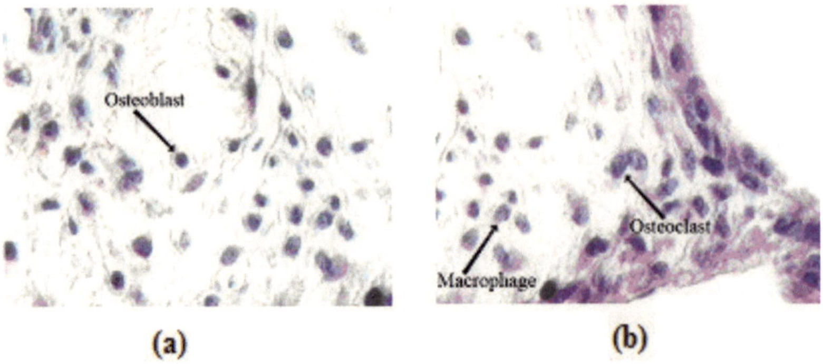

Fig. 23. Light microscopy evaluation of bone tissue for: (a) 800 grit, and (b) untreated

In table 4, the symbols indicate the presence of 2-3 cells (+), 3-5 cells (++), more than 5 cells (+++) and lack of cells (-) respectively. No PMN, giant cells and osteoclast were seen in SiC treated samples. Also tissue healing was better conducted near mentioned implant. Fibroblast and osteoblast cells were seen in samples.
The successful incorporation of bone implants strongly depends on a firm longstanding adhesion of the tissue surrounding the implants. The cellular reaction is influenced by the properties of the bulk materials as well as the specifications of the surface, that is, the chemical composition and the topography (Birte et al., 2003, Siikavitsas et al., 2003). When one is considering materials for application of orthopaedic implants, it is important to consider a number of factors, such as biocompatibility and surface wettability. The interaction of living cells with foreign materials is complicated matter, but fundamental for biology medicine and is a key for understanding the biocompatibility. The initial cellular events which take place at the biomaterials interface mimic to a certain extent the natural adhesive interaction of cells with the extra cellular matrix (ECM).

Sample Cell	SiC paper (800 grit)	SiC paper (300 grit)	untreated
Fibroblast	++	++	++
Osteoblast	+	+	+
Giant cell	-	-	-
Osteoclast	-	-	+
PMN	-	-	+
Lymphocyte	++	++	++
Macrophage	+++	+++	++
Healing	+	+	+

Table 4. Qualitative evaluation of histology results of bone tissue around the implants

The osteoblasts, which play a principal role in bone formation, readily attach to the material surfaces via adsorbed protein layer consisting or RGD containing ligands like fibronectin, vitronectin or fibrinogen. Family of cell surface receptors that provide trans- membrane links between the ECM and the cytoskeleton. Our study showed that surface micro grooves can affect the orientation guidance of bone cells i.e. the deeper grooves were more effective in guiding the cells as it was evaluated by SEM. However, we did not conduct or evaluate systematically the exact effects of grooves depth and size on cell orientation, but our preliminary results were similar to those reported by Xiong et.al (Xiong et al., 2003).
This study was focused on the topographic effects of Ti-6Al-4V produced by SiC paper on goat bone cell adhesion. The results showed a common feature reported in the previous studies on a variety of cell types and substrates ie, topographic features strongly affects the cell guidance. Micro grooved surfaces increase of surface tension and reduction of contact angle. The test confirmed that the highest number of cells is attached to SiC paper modified surface. It is also concluded from the SEM, contact angle measurements and preliminary in vitro and in vivo tests that SiC paper can induce a desirable surface modification on Ti-6Al-4V alloy for cell adhesivity and that a noble and biocompatible. Finally, it is suggested that more detailed experiments are required and would be useful to distinguish and clarify the relation between the grooves size and their orientation must be studied more carefully with respect to cell attachment and their reliability as well as endurance.

18. Future considerations in biomedical applications of SiC

The next decade will see a great increase in scientific research into the biomedical applications of SiC. Many analysis techniques may be used to analyze SiC biocompatibility. In particular, primary cell lines could be cultured on SiC surfaces in the future since their behavior would be a closer description of the in vivo performance of the material. While proof-of concept studies in research laboratories have demonstrated great promise in the use of SiC for scaffold of tissue engineering, several issues will need to be addressed before SiC find way to large-scale clinical application. In particular, researches will need to study toxic and pharmacokinetic effects of SiC in vivo. In addition, research will focuse on the synthesis SiC nanopartiicles that may facilitate the development of multifunctional nanostructures for use in drug delivery and tissue engineering applications. More experiments are required to clarify the relation between SiC and cell attachment in scaffold of tissue engineering. The different polytypes of SiC were quite well matched to organic systems in terms of band gap and band alignment. Therefore, SiC should be a very interesting substrate material for future semiconductor/organic heterostructures. Finally, the feasibility of surface functionalization of SiC leaving free functional groups has been shown while deeper understanding of the chemisorptions of various organic molecules is still needed in order to optimize surface functionalization processes. The preparation and complete characterization of atomically ordered SiC surfaces may lead to the successful implementation of a large variety of biotechnological applications. It is suggested that more investigations are required and would be useful to distinguish SiC biomedical applications.

19. References

Amon, M.; Bolz, A. & Schaldach, M. (1996). Improvement of stenting therapy with a silicon carbide coated tantalum stent. *J Mater Sci: Mater Med*, 7, 5, 273–8

Amon, M.; Winkler, S.; Dekker, A.; Bolz, A.; Mittermayer, C. & Schaldach, M. (1995). Introduction of a new coronary stent with enhanced radiopacity and hemocompatibility. pp. 107–8 *Proceedings of the Annual International Conference of the IEEE Engineers in Medicine and Biology Society* 95CB35746, vol. 17, 1. New York: IEEE Press.

Anderson, A.S.; Dattelbaum, A.M.; Montano, G.A.; Price, D.N.; Schmidt, J.G.; Martinez, J.S.; Grace, W.K.; Grace, K.M.; Swanson, B.I. (2008). Functioal PEG-modified thin films for biological detection. *Langmuir*, *24*, 2240-2247

Bai, S.; Ke, Y.; Shishkin, Y.; Shigiltchoff, O.; Devaty, R.P.; Choyke, W.J.; Strauch, D.; Stojetz, B.; Dorner, B.; Hobgood, D.; Serrano, J.; Cardona, M.; Nagasawa, H.; Kimoto, T. & Porter L.M. (2003). Four Current Examples of Characterization of Silicon Carbide, Mat. Res. Soc. Symp. Proc. 742, K3.1.1

Bayer, G.; Hartwig, S.; Nagel, M.; Tilttelbach, M.; Rzany, A. & Schaldach, M. (2001). Future Strategies for Antiproliferative Stent Coatings. *Progress in Biomedical Research*, 222-225

Berthold, A.; Laugere F.; Schellevis, H.; De Boer, Ch. R.; Laros, M.; Guijt, R. M.; Sarro, P. M. & Vellekoop, M. J. (2002). Fabrication of a glass-implemented microcapillary electrophoresis device with integrated contactless conductivity detection. *Electrophoresis*, 23, 3511–3519

Bigerelle, M. & Anselme, K. (2005). Bootstrap analysis of the relation between initial adhesive events and long-term cellular functions of human osteoblasts cultured on biocompatible metallic substrates. *Acta biomaterialia*, 1, 499-510

Birte, G.S.; Neubert, A.; Hopp, M.; Griepentrog, M. & Lange, K.P. (2003). Fibroblast growth on surface modified dental implants: An in vitro study. *J. Biomed. Mater. Res.*, 64A, 591-599

Bolz, A. & Schaldach, M., (1993). Heamocompatibility optimization of implants by hybrid structuring, S123-S130, World Congress Supplement, Biomaterials, *Koyoto*

Bolz, A. (1995). Applications of Thin-Film Technology in Biomedical Engineering. Bolz, A. & Schaldach, M. (1990). Artificial heart valves: improved blood compatibility by PECVD a-SiC:H coating. *Artificial organs*, 14(4), 260-9

Bolz, A. & Schaldach, M. (1993). Biomaterials haemocompatibility optimization of implants by hybrid structuring. *Med. & Biol. Eng. & Comput.*, 31, 123-130

Bolz, A; Amon, M; Ozbek, C; Heublein, B; Schaldach, M. (1996). Coating of cardiovascular stents with a semiconductor to improve their hemocompatibility. *Texas Heart Institute J.* 23, 2, 162-6

D.L. , Trantolo, D.J. & et al. Encyclopedic Handbook of Biomaterials and Bioengineering. Part A: Materials, In: Wise, 2, 1287-1330

Borrajo, J.P.; Serra, J.; Liste, S.; Gonza´lez, P.; Chiussi, S.; Leo´n, B. & Pe´rez-Amor, M. (2005). Pulsed laser deposition of hydroxylapatite thin films on biomorphic silicon carbide ceramics. *Applied Surface Science*, 248, 2005, 355–359

Botsoa, J., Lysenko, V., G_eloen, A., Marty, O., Bluet, J. M. & Guillot, G. (2008). Application of 3C-SiC quantum dots for living cell imaging. *Appl. Phys. Lett.*, 92, 173902

Buchter, A.; Joos, U.; Wiessman, H.P.; Seper, L. & Meyer, U. (2006). Biological and biomechanical evaluation of interface reaction at conical screw-type implant. *Head and Face Med.*, 2, 5-18

Carlos, A. D.; Borrajo, J.P.; Serra, J.; Gonz´alez, P. & Le´on, B. (2006). Behaviour of MG-63 osteoblast-like cells on wood-based biomorphic SiC ceramics coated with bioactive glass. *J Mater Sci: Mater Med* , 17, 523–529

Carrie, K.; Khalife, M.; Hamon, B.; Citron, J.P.; Monassier, R.; Sabatier, J.; Lipiecky, S.; Mourali, L.; Sarfaty,M.; Elbaz, J.; Fourcade & Puel, J. (2001). Initial and Follow-Up Results of the Tenax Coronary Stent. *J. Interventional Cardiology* 14(1), 1-5

Calderon, N.R.; Martinez-Escandell, M.; Narciso, J. & Rodríguez-Reinoso, F. (2009). The role of carbon biotemplate density in mechanical properties of biomorphic SiC. *J. of the European Ceramic Society*, 29, 465–472

Caputo, D.; de Cesare, G.; Nascetti, A.; Scipinotti, R., (2008). Two-Color Sensor for Biomolecule Detection. ***Sensor Letters***, 6, 4, 542-547

Chakrabarti, O.P.; Maiti, H.S., & Majumdar, R. (2004). Biomimetic synthesis of cellular SiC based ceramics from plant precursor. *Bull. Mater. Sci.*, 27, 5, 467–470

Chu, W.H.; Chin ,R.; Huen, T. & Ferrari, M. (1999). Silicon Membrane Nanofilters from Sacrificial Oxide Removal. *J. Microelectromech. Syst.* 8, 34-42

Cicero, G. & Catellani, A. (2005). Towards SiC surface functionalization: An ab initio study. *J. Chem. Phys.*, 122, 214716, 1-5

Cicoira, f. & Rosei, F. (2006). Playing Tetris at the nanoscale. *Surface Science*, 600, 1–5

Cogan, S.F.; Edell, D.J.; Guzellan, A.A.; Ying, L.P. & Edell, R. (2003). Plasma-enhanced chemical vapor deposited silicon carbide as an implantable dielectric coating. *J. Biomed. Mater. Res.* A, 67, 3, 856-67

Cole, K.S. (1940). Permiability and impermiability of cell membranes for ions. *Sympos Quant.Biol.*, 8, 110-122

Coletti, C.; Jaroszeski, M.; Hoff, A.M. & Saddow, S.E. (2006). Culture of mammalian cells on single crystal SiC substrates, *Mater. Res. Soc. Symp. Proc.* 950

Coletti, C.; Jaroszeski, M.J.; Pallaoro, A.; Hoff, A.M.; Iannotta, S. & Saddow, S.E. (2007). Biocompatibility and wettability of crystalline SiC and Si surfaces. *IEEE EMBS Proceedings*, 5849-5852

Desai, T.A.; Chu, W.H.; Tu, J.K.; Beattie, G.M.; Hayek, A. & Ferrari, M. (1997). Microfabricated Immunoisolating Biocapsules. *Biotech. and Bioeng.*, 57(1), 118-120

Elbaz, M.; El Mokhtar, E.; Fourcade, J.; Mourali, S.; Hobeika, R.; Carrie, D. & Puel, J. (2002). Dose stent design affect the long-term outcome after coronary stenting? *Catheterization & Cardiovascular Interventions,* 56, 305-311

El Ghannam, A. (2005). Bone reconstruction: from bioceramics to tissue engineering. *Expert Rev Med Devices*. 2, 1, 87-101

Fan, T.X.; Li, X.F.; Liu, Z.T.; Gu, J.J.; Zhang, D.; Guo, Q.X. & Am, J. (2006). Microstructure and infrared absorption of biomorphic chromium oxides templated by wood tissues. *Ceram. Soc.* 89, 3511-3515

Fan, J.; Li, H.; Jiang, J., So L. K. Y.; Lam Y. W. & Chu, P. K. (2008). 3C-SiC Nanocrystals as Fluorescent Biological Labels. *Small*, 4, 8, 1058–1062

Feng, B.; Weng, J.; Yang, B.C.; Qu, S.X. & Zhang, X.D. (2003). Characterization of surface oxide films on titanium and adhesion of osteoblast. *Biomaterials*, 24, 4663-4670

Feng, X.; Yang, Q. & De Jonghe, L.C. (2003). Microstructure development of hot-pressed silicon carbide: effects of aluminum, boron, and carbon additives. *Acta Mater*, 51, 3849-60

Fissell, W.H. (2003). Nanoporous Hemofiltration Membranes for Bioartificial Kidneys. oral presentation at 4th Annual BioMEMS and Nanotech World,Washington, DC

Frewin, C.L.; Jaroszeski, M.; Weeber, E.; Muffly, K.E.; Kumar, A.; Peters, M.; Oliveros, A.& Saddow, S.E.J. (2009). Atomic force microscopy analysis of central nervous system cell morphology on silicon carbide and diamond substrates. *Mol. Recognit.*, 22, 5, 380

Fujibayashi, S.; Neo, M.; Kim, H.; Kokubo, T. & Nakamura, T. (2003). A comparative study between in vivo bone in growth and in vitro apatite formation on Na_2O-CaO-SiO_2 glasses. *Biomaterials*, 24, 1349-1356

Gabriel, G.; Erill, I.; Caro, J.; Gomez, R.; Riera, D.; Villa, R. & Godignon, Ph. (2007). Manufacturing and full characterization of silicon carbide-based multi-sensor micro-probes for biomedical applications. *Microelec. J.*, 38, 406–415

Gersing, E. (1998). Impedance spectroscopy on living tissue for determination of the state of organs. *Bioelectrochem. Bioenerg*. 45, 145–149

Goeders, K. M.; Colton, J. S.; Bottomley, L. A. (2008). Microcantilevers: Sensing Chemical Interactions via Mechanical Motion. *Chem. Rev.*, 108, 522-542

Godignon, P. (2005). New Generation of SiC Based Biodevices Implemented on 4" Wafers. *Mater. Sci. Forum* 483–485. 1097-1100

Gomez, R.; Ivorra, A.; Villa, R.; Godignon, Ph.; Mill´an, J.; Erill, I.; Sol`a, A.; Hotter, G. & Palacios, L. (2006). A SiC microdevice for the minimally invasive monitoring of ischemia in living tissues. *Biomed. Microdevices,* 8, 43–49

Gonzalez, P.; Serra, J.; Liste, S.; Chiussi, S.; Leon, B.; Perez-Amor, M.; Martinez-Fernandez, J.; de Arellano-Lopez, A.R. & Varela-Feria, F.M. (2003).New biomorphic SiC ceramics coated with bioactive glass for biomedical applications. *Biomaterials,* 24, 4827–4832

Gonzalez, P.; Borrajo, J.P.; Serra, J.; Liste, S.; Chiussi, S.; Leo´n, B.; Semmelmann, K.; De Carlos, A.; Varela-Feria, F.M.; Martı´nez-Ferna´ndez, J. & De Arellano-Lo´pez, A.R. (2004). *Key Eng. Mater.,* 1029, 254–256

Gonzalez, P.; Serra, J.; Liste, S.; Chiussi, S.; Leon, B.; Perez-Amor, M. & et al. (2003). New biomorphic SiC ceramics coated with bioactive glass for biomedical applications. *Biomaterials,* 24, 26, 4827-432

Gonzalez, P.; Borrajo, J.P.; Serra, J.; Chiussi, J.; Leo, B.; Martı nez-Ferna ndez, J.; Varela-Feria, F.M.; de Arellano-Lopez, A.R.; de Carlos, A.; Mun oz, F.M.; Lo pez, M. & Singh, M. (2008). A new generation of bio-derived ceramic materials for medical applications. *2008 Wiley Periodicals, Inc.*

Greil, P. (2002). Advanced engineering ceramics. *Adv. Eng. Mater,* 4, 5, 247–254

Gutierrez-Mora, F.; Goretta, K.C.; Varela-Feria, F.M.; Arellano Lo pez, A.R. & Martınez Ferna ndez, J. (2005). Indentation hardness of biomorphic SiC. *Inter J. of Refractory Metals & Hard Materials,* 23, 369–374

Haemmerich, D.; Ozkan, R.; Tungjitkusolmun, S.; Tsai, J.Z.; Mahvi, D.M.; Staelin, S.T.; Webster, J.G.; (2002). Changes in electrical resistivity of swine liver after occlusion and postmortem, *Medical and Biological Engineering and Computing,* 40, 1, 29-33

Hamm, C.W.; Hugenholtz, P.G. & Trust, I. (2003). Silicon carbide-coated stents in patients with acute coronary syndrome. Catheterization & Cardiovascular Interventions, 60, 3, 375-381

Hanekamp, C.E.E. & Koolen, J.J. (2000). Coated Stents in Small Coronary Vessels - A Successful Strategy? *Progress in Biomedical Research,* 221-223

Harder, C.; Rzany,A. & Schaldach, M. (1999). Coating of Vascular Stents with Antithrombogenic Amorphous Silicon Carbide, *Progress in Biomedical Research,* 71-77

Heinrich, A.; Dengler, K.; Koerner, T.; Haczek, C.; Deppe, H. & Stritzker, B. (2008). Laser-modified titanium implants for improved cell adhesion. *Laser in Medical science,* 23, 55-58

Hesaraki, S.; Ebadzadeh, T. & Ahmadzadeh-Asl, S. (2010). Nanosilicon carbide/hydroxyapatite nanocomposites: structural, mechanical and in vitro cellular properties. *J Mater Sci: Mater Med,* DOI 10.1007/s10856-010-4068-7

Hildebrandt, P.; Sayyad, M.; Rzany, A. & et al. (2001). Prevention of surface encrustation of urological implants by coating with inhibitors. *Biomaterials,* 22, 503-507

Hing, K.A.; Revell, P.A.; Smith, N. & Buckland, T. (2006). Effect of silicon level on rate, quality and progression of bone healing within silicate-substituted porous hydroxyapatite scaffolds. *Biomaterials,* 27, 29, 5014-5026

Ikeuchi, K.; Kusaka J. & Yoshida, H. (2000). Tribology for all-ceramic joint prostheses. *Journal of Ceramic Processing Research,* 1, 1, 53-56

Iliescu, C.; Poenar, D. P.; Carp, M.; Loe, F. C. (2007). A microfluidic device for impedance spectroscopy analysis of biological samples. *Sensors and Actuators B*, 123, 168–176

Ivorra, A.; Gómez, R.; Noguera, N.; Villa, R.; Sola, A.; Palacios, L.; Hotter, G. & Aguiló, J. (2003). Minimally invasive silicon probe for electrical impedance measurements in small animals. Biosensors & Biolectronics, 19, 4, 391-399

Kamakaran, R.; Lupo, F.F.; Grobert, N.; Scheu, T.; Phillipp, N.Y.G. & Ruhle, M. (2004). Microstructural Characterization of C-SiC Carbon Nanotube Composite Flakes. *Carbon*, 42, 1–4

Kanai, Y.; Cicero, G.; Selloni, A.; Car, R. & Galli, G. (2005). A Theoretical Study of Biotin Chemisorption on Si-SiC(001) Surfaces. J. *Phys. Chem.* B 109, 13656

Kalnins, U.; Erglis, A.; Dinne, I.; Kumsars, I.; & Jegere, S. (2002). *Medical Science Monitor* 8, I16

Karas, S.P.; Gravanis, M.B.; Santoian, E.C.; Robinson, K.A.; Andernerg, K.A. & King, S.B. (1992). Coronary intimal proliferation after balloon injury and stenting in swine: an animal model of restenosis. *J. Am. Coll .Cardiol.*, 20, 467–74

Karoussos, A.; Wieneke, H.; Sawitowski, T.; Wnendt, S.; Fischer, O.; Dirsch, O.; Dahmen, U. & Erbel, R. (2002). Inorganic materials as drug delivery systems in coronary artery stenting. *Mat. Wiss.U. Werkstofftech.*, 33, 738-746

Khosroshahi, M.E.; Mahmoodi, M. & Tavakoli, J. (2007). Characterization of Ti-6Al-4V implant surface treated by Nd:YAG laser and emery paper for orthopaedic applications. *Applied surface science*, 253, 8772-8781

Khosroshahi, M.E.; Tavakoli, J. & Mahmoodi, M. (2007). Analysis of bioadhesivity of osteoblast on titanium by Nd:YAG laser. *J. of Adhesion*, 83, 151-172

Khosroshahi, M.E.; Mahmoodi, M.; Saedinasab, H. & Tahriri, M. (2008). Evaluation of mechanical and electrochemical properties of laser surface modified Ti-6AL-4V for biomedical application: in vitro study. *Surface Engineering*, 24, 209-218

Khosroshahi, M.E.; Mahmoodi, M.; Tavakoli, J. & Tahriri, M. (2008). Effect of Nd: YAG laser radiation on Ti6AL4V alloy properties for biomedical application. *J. of Laser Applications*, 20, 209-217

Khosroshahi, M.E.; Mahmoodi, M. & Saedinasab, H. (2009). In vitro and in vivo studies of osteoblast cell response to a Ti6A14V surface modified by Nd:YAG laser and silicon carbide paper. *Laser Med. Sci*, 24, 925-939

Kidambi, S.; Dai, J.H. & Bruening, M.L. (2004). Selective Hydrogenation of Pd Nanoparticles Embedded in Polyelectrolyte Multilayers. *J. Am. Chem. Soc.*, 126, 2658–9

Kotzara, G.; Freasa, M.; Abelb, Ph.; Fleischmanc, A.; Royc, Sh.; Zormand, Ch.; Morane, J. M. & Melzak, J. (2002). Evaluation of MEMS materials of construction for implantable medical devices. *Biomaterials*, 23, 2737–2750

LeGeros, R.Z.; LeGeros, J.P.; Klein, E. & Shirra, W.P. (1967). Apatite crystallites. Effect of carbonate on mor- phology. *Science*, 155, 1409-1411

Liu, Z.; Fan, T.; Gu, J.; Zhang, D.; Gong, X.; Gu, Q. & Xu, J. (2007). Preparation of Porous Fe from Biomorphic Fe_2O_3 Precursors with Wood Templates. *Mater. Trans.*, 48, 878-881

Liu, Y.; Ramanath, H.S. & Wang, D.A. (2008). Tendon tissue engineering using scaffold enhancing strategies. *TrendsBiotechnol.*, 26, 4, 201-209

Luo, M.; Gao, J.Q.; Yang, J.F. & Am, J. (2007). Biomorphic silicon nitride ceramics with and reduction-nitridation. *Ceram. Soc.* 90, 4036-4039

Luo, M.; Hou, G.Y.; Yang, J.F.; Fang, J.Z.; Gao, J.Q.; Zhao, L. & Li, X. (2009). Manufacture of fibrous β-Si3N4-reinforced biomorphic SiC matrix composites for bioceramic scaffold applications. *Materials Science and Engineering C*, 29, 1422–1427

Maitz, M.F.; Pham, M. & Wieser, E. (2003). Blood compatibility of titanium oxides with various crystal structure and element doping. *J. biomaterials applications*, 17, 4, 303-319

Mangonon, P.L. (1999). The principles of materials selection for engineering design, 1st Ed.*Upper Saddle River*, NJ: Prentice-Hall

Mart nez-Fern!andez, J.; Varela-Feria, F.M. & Singh, M. (2000). Microstructure and thermomechanical characterizacion of biomorphic silicon carbide-based ceramics. *Scr Mater*, 43, 813–8

Martı nez Ferna ndez, J.; de Arellano-Lo pez, A.R.; Varela-Feria, F.M. & Singh, M. (2001). Procedimiento para la Fabricacio´n de Carburo de Silicio a Partir de Precursores Vegetales. *Spanish Patent*, P200102278, submitted by Universidad de Sevilla

Masuda, T.; Yliheikkila, P.K.; Felton, D.A. & Cooper, L.F. (1998). Generolization regarding the process and phenomena of osseointegration: in vivo studies, *part I. Int. J. Oral Maxillofac. Implants*, 13, 17-29

Mayor, B.; Arias, J.; Chiussi, S.; Garcia, F.; Pou, J.; Leo´n, B. & Pe´rez-Amor, M. (1998). Calcium phosphate coatings grown at different substrate temperatures by pulsed ArF-laser deposition. *Thin Solid Films*, 317, 363-366

McDonagh, C.; Burke, C.S. & MacCraith, B.D. (2008). Optical chemical sensors. *Chem. Rev.*, 108, 400-422

Meyers, M.A.; Chen, P.Y.; Lin, A.Y.M. & Seki, Y. (2008). Biological materials: structure and mechanical properties. *Prog. Mater. Sci.*, 53, 1-206

Monnink, S.H.; van Boven, A.J.; Peels, H.O.; Tigchelaar, I.; de Kam, P.J.; Crijns, H.J. & van Oeveren,W. (1999). Silicon-carbide coated coronary stents have low platelet and leukocyte adhesion during platelet activation. *J. Investig Med*, 47, 304–10

Naji, A. & Harmand, M.F. (1991). Cyto compatibilio Qf two coating materials, amorphous alumina and silicon carbide, using human differentiated cell cultures. *Biomaterials*, 12, 690-694

Nordsletten, L.; Hogasen, A.K.; Konttinen, Y.T.; Santavirta, S.; Aspenberg, P. & Aasen, A.O. (1996). Human monocytes stimulation by particles of hydroxyapatite, silicon carbide and diamond: in vitro studies of new prosthesis coatings. *Biomaterials*, 17, 15, 1521-7

Nurdin, N.; François, P.; Mugnier, Y.; Moret, M.; Aronsson, B.O.; Krumeich, J. & Descouts, P. (2003). Haemocompatibility evaluation of DLC and SiC-coated surfaces. *European Cells and Materials*, 5, 17-28

Presas, M.; Pastor, J.Y.; Liorca, J.; Arellano Lo pez, A.R.; Martı nez Ferna ndez, J. & Sepu lveda, R. (2006). Microstructure and fracture properties of biomorphic SiC. *Inter J. of Refractory Metals & Hard Materials*, 24, 49–54

Preuss, M.; Bechstedt, F.; Schmidt, W.G.; Sochos, J.; Schroter Band Richter W (2006). Clean and pyrrole-functionalized Si- and C-terminated SiC surfaces: First-principles calculations of geometry and energetics compared with LEED and XPS. *Phys. Rev. B*, 74, 235406

Ohji, T. (2008). Microstructural design and mechanical properties of porous silicon nitride Ceramics. *Mater. Sci. Eng. A*, 498, 5-11

Oliveira, T.D. & Nanci, A. (2004). Nanotexturing of titanium-based surfaces upregulates expression of bone sialoprotein and osteopontin by cultured osteogenic cells. *Biomaterials*, 25, 403-413

Raicu, V.; Saibara, T. & Irimajiri, A. (2000). Phys. Multifrequency method for dielectric monitoring of cold-preserved organs. *Phys. Med. Biol.*, 45, 1397-1407

Rambo, C.R.; Muller F.A.; Muller, L.; Sieber, H.; Hofmann, I. & Greil, P. (2006). Biomimetic apatite coating on biomorphous alumina scaffolds.*Mater. Sci. Eng. C*, 26, 92-99

Rokusek, D.; Davitt, C.; Bandyopadhyay, A.; Bose, S. & Hosick, H.L. (2005) Interaction of human osteoblasts with bioinert and bioactive ceramic substrates. *J Biomed Mater Res Part A*, 75, 3, 588-594

Ronold, H.J.; Lyngstadaas, S.P. & Ellingsen, J.E. (2003). A study on the effect of dual blasting with TiO_2 on titanium implant surfaces on functional attachment in bone. *J. Biomed. Mater. Res.*, 67A, 524-530

Rosenbloom, A.J.; Sipe, D.M.; Shishkin, Y.; Ke, Y.; Devaty, R.P. & Choyke, W.J. (2004). Nanoporous SiC: A Candidate Semi-Permeable Material for Biomedical Applications. *Biomed.l Microdevices*, 6, 4, 261–267

Rossi, A.M.; Reipa, V. & Murphy, T.E. (2008). Luminescence emission from Silicon Carbide Quantum Dot. *Nanotech Conference Program Abstract* NIST, US

Rosso, M.; Arafat, A.; Schroën, K.; Giesbers, M.; Roper, C.S.; Maboudian, R.; Zuilhof, H. (2008). Covalent attachment of organic monolayers to silicon carbide surfaces. *Langmuir*, 24, 4007-4012

Rosso, M.; Giesbers, M.; Arafat, A.; Schroën, K. & Zuilhof, H. (2009). Covalently Attached Organic Monolayers on SiC and Si*x*N4 Surfaces: Formation Using UV Light at Room Temperature. *Langmuir*, 25, 2172-2180

Rzany, A.; Harder, C. & Schaldach, M. (2000). Silicon carbide as an anti-thrombogenic stent coating; an example of a science-based development strategy. *Prog Biomed Res*, 5, 168-178

Rzany, A. & Schaldach, M. (2001). Smart Material Silicon Carbide: Reduced Activation of Cells and Proteins on a-SiC:H-coated Stainless Steel. *Progress in Biomedical Research*, May 182-194

Saki, M.; Kazemzadeh, M.N.M.; Samadikuchaksaraei, A.; Basir, H.G. & Gorjipour, F. (2009). *Yakhteh Medical Journal*, 11, 1, 55-60

Samadikuchaksaraei, A. (2007). An overview of tissue engineering approaches for management of spinal cord injuries. *J Neuroeng Rehabil*. 4, 15-30

Samadikuchaksaraei, A. (2007). Scientific and industrial status of tissue engineering. *Afr J Biotechnol.*, 6, 25, 897-2909

Samadikuchaksaraei, A. (2008). Engineering of skin substitutes: current methods and products. *Tissue Engineering Research Trends*. In: Greco, G.N. editor. 1st ed, NY: Nova Science Publishers Inc., 251-266, Hauppauge

Santavirta, S.; Konttinen, Y.T.; Bergroth, V.; Eskola, A.; Tallroth, K. & Lindholm, T.S. (1990). Aggressive granulomatous lesions associated with hip arthroplasty. Immunopathological studies. *J Bone Joint Surg (Am)*, 72, 252–258

Santavirta, S.; Takagi, M.; Nordsletten, L.; Anttila, A.; Lappalainen, R. & Konttinen, Y.T. (1998). Biocompatibility of silicon carbide in colony formation test in vitro A promising new ceramic THR implant coating material. *Arch Orthop Trauma Surg*, 118, 89–91

Sawyer, P.N.; Brattain, W.H. & Boddy, P.J. (1965).Electrochemical criteria in the choice of materials used in vascular prostheses, *In: Biophysical mechanism in vascular hemostasis and intravascular thrombosis*, Sawyer, P.N. (Ed.), 337-348, Appleton-Century-Crofts, New York

Seino, K.; Schmidt, W.G.; Furthmüller, J. & Bechstedt, F. (2002). Chemisorption of pyrrole and polypyrrole on Si(001), *Phys. Rev. B*, 66, 235323

Scheller, B.; Hennen, B.; Severin-Kneib, S.; zbek, C.; Schieffer, H. & Markwirth, T. (2001). Long-term follow-up of a randomized study of primary stenting versus angioplasty in acute myocardial infarction. *Am J Med*, 110, 1–6

Schillinger, M.; Sabeti, S.; Loewe, C.; Dick, P.; Amighi, J.; Mlekusch, W.; Schlager, O.; Cejna, M.; Lammer, J. & Minar, E. (2006). Balloon angioplasty versus implantation of nitinol stents in the superficial femoral artery. *N Engl J Med*, 354, 1879–88

Schmehla, J.M.; Harderb, C.; Wendelc, H.P.; Claussena, C.D. & Tepea, G. (2008). Silicon carbide coating of nitinol stents to increase antithrombogenic properties and reduce nickel release. *Cardiovascular Revascularization Medicine*, 9, 255–262

Schömig, A.; Kastrati, A.; Dirschinger; J. & et al. (1999). Randomized comparison of gold-plated steel stent with conventional steel stent: Results of the angiographic follow-up. *JACC. ACCIS*, 1217-54 (abstr.)

Shivani, B.; Mishra, w.; Ajay, K. & et al. (2009). Synthesis of Silicon Carbide Nanowires from a Hybrid of Amorphous Biopolymer and Sol-Gel-Derived Silica. *J. Am. Ceram. Soc.*, 92, 12, 3052–3058

Shor, J.S. & Kurtz, A.D. (1994). Photoelectrochemical Etching of 6H-SiC. J. Electrochemical Society 141(3), 778-781

Sieber, H.; Vogli, E. & Greil P. (2001). Biomorphic SiC-ceramic manufactured by gas-phase infiltration of pine wood. *Ceram Eng Sci Proc*, 22, 4, 109–16

Sighvi, R. & Wang, D.I. (1998). Review: Effects of substratum morphology on cell physiology. *Biotech. Bioeng.*, 43, 764-771

Sikavitsas, V.I.; Dolder, J.; Bancroft, G. & Jansen, J. (2003). Influence of the in vitro culture period on the in vivo performance of cell/titanium bone tissue-engineered constructs using a rat cranial size defect model. *J. Biomed. Mater. Res.*, 67A, 944-951.

Singh, Sh. & Buchanan R.C. (2007). SiC-C fiber electrode for biological sensing. Mater. Sci. and Eng. C, 27, 551–557

Sohier, J.; Moroni, L.; van Blitterswijk, C.; de Groot, K. & Bezemer, J.M. (2008). Critical factors in the design of growth factor releasing scaffolds for cartilage tissueengineering. *Expert Opin Drug Deliv.*, 5, 5, 543-566

Starke, U.; Lee, W.Y.; Coletti, C.; Saddow, S.E.; Devaty, R.P. & Choyke, W.J. (2006). SiC Pore Surfaces: Surface Studies of 4H-SiC (-1102) and 4H-SiC (-110 -2). *Applied Physics Letters*, 88, (3), 031915

Steenackers,M.; Sharp, I.D.; Larsson, K.; Hutter, N.A.; Stutzmann, M. & Jordan, R. (2010). Structured Polymer Brushes on Silicon Carbide. *Chem. Mater.*, 22, 272–278

Stobierski, L. & Gubernat, A. (2003) Sintering of silicon carbide Effect of carbon. *Ceram Int.*, 29, 287–92

Stutzmann, M.; Garrido, J.A.; Eickhoff, M. & Bandt, M.S. (2006) Direct biofunctionalization of semiconductors: A survey. *Phys. Status Solidi*, 203(14), 3424-3437

Sun, Y. & Xia, Y. (2002). Shape-Controlled Synthesis of Gold and Silver Nanoparticles. *Science*, 298, 2176–9

Tanigawa, N.; Sawada, S.; Koyama, T. & et al. (1991). An animal experiment on arterial wall reaction to stents coated with gold, silver and copper. *Nippon Igaku Hoshasen Gakkai Zasshi*, 51, 10, 1195-1200

Takami, Y.; Yamane, S.; Makinouchi, K.; Otsuka, G.; Glueck, J.; Benkowski, R. & Nose Y. (1998). Protein adsorption onto ceramic surfaces. J. Biomed. Mater. Res., 40(1), 24-30

Tanigawa, N.; Sawada, S. & Kobayashi, M. (1995). Reaction of the aortic wall to six metallic stent materials. *Acad Radiol.*, 2, 5, 379-384

Thian, E.S.; Huang, J.; Best, S.M.; Barber, Z.H. & Bonfield, W. (2005). A new way ofincorporating silicon in hydroxyapatite (Si-HA) as thin films. *J Mater Sci Mater Med.*, 16, 5, 411-415

Thian, E.S.; Huang, J.; Best, S.M.; Barber, Z.H.; Brooks, R.A.; Rushton, N. & et al. (2006). The response of osteoblasts to nanocrystalline silicon-substituted hydroxyapatite thin films. *Biomaterials*, 27, 13, 2692-2698

Tlili, C.; Korri-Youssoufi, H.; Ponsonnet, L.; Martelet, C.; Jaffrezic-Renault, N. J. (2005). Electrochemical impedance probing of DNA hybridization on oligonucleotide-functionalised polypyrrole. *Talanta*, 68, 131-137

Tirrell, M.; Kokkoli, E. & Biesalski, M. (2002). The role of surface science in bioengineered materials. *Surf. Sci.*, 500, 61-63 Unverdorben, M.; Sippe, B.; Degenhardt, R.; Sattler, K.; Fries, R.; Abt, B.; Wagner, E.; Scholz, M.; Koehler, H.; Ibrahim, H.; Tews, K.H.; Hennen, B.; Schieffer, H.; Berthold, H.K. & Vallbracht, C. (2000). Langzeitvergleich des Siliziumkarbid-beschichteten Stents mit einem Stahlstent: Die Tenax'- vs. NIR'-Stent Studienresultate (RENISS-L). *Z Kardiol* ,90, 18

Van Oeveren, W. (1999). Reduced Deposition of Blood Formed Elements and Fibrin onto Amorphous Silicon Carbide Coated Stainless Steel. *Progress in Biomedical Research*, February: 78-83

Varadan, V.K. (2003). Nanotechnology: MEMS and NEMS and their applications to smart systems and devices. Conference Title: Smart Materials, Structures, and Systems, Proc. SPIE, Vol. 5062, 20

Varela-Feria, F.M.; Lo pez-Pombero, S.; de Arellano-Lo pez, A.R. & Martı nez-Ferna ndez, J. (2002). Maderas Cera micas: Fabricacio´n y Propiedades del Carburo de Silicio Biomo´rfico. *Bol Soc Esp Ceram Vidrio*, 41, 4, 377–384

Wang, Y.B. & Zheng, Y.F. (2009). Corrosion behaviour and biocompatibility evalution of low modulus Ti-16Nb shape memory alloy as potential biomaterial. *Material Letters*, 63, 1293-1295

Wendler-Kalsch, E.; Mueller, H. & Bonner, S. (2000). Corrosion Behavior of Stents Coated with Gold and a-SiC:H. *Progress in Biomedical Research*. 179-183

Wieneke, H.; Sawitowski, T.; Wnendt, S.; Fischer, A.; Dirsch, O.; Karoussos, I.A. & Erbel, R. (2002). Stent Coating: A New Approach in Interventional Cardiology. *Herz*, 27, 518–26

Wieland, M.; Textor, M.; Spencer, N.D. & Brunette, D.M. (2001). Wavelength-roughness: a quantitative approach to characterizing the topography of rough titanium surfaces. *Int J Oral Maxilloface Impl.*, 16, 2, 163-181

Xia, Y.; Yang, P.; Sun, Y.; Wu, Y.; Mayers, B.; Gates, B.; Yin, Y.; Kim, F. & Yan, H. (2003). One-Dimensional Nanostructures: Synthesis, Characterization, and Applications. *Adv. Mater.*, 15, 353–89

Xiong, L. & Yang, L. (2003). Quantitative analysis of osteoblast behavior on microgrooved hydroxyapatite and titanium substrata. *J. Biomed. Mater. Res.*, 66A, 677-687

Yakimova, R.; Petoral, R. M.; Yazdi, G.R.; Vahlberg, C.; Spetz, L. & Uvdal, K. (2007). Surface functionalization and biomedical applications based on SiC. *J. Phys. D: Appl. Phys.*, 40, 6435-6442

Zampieri, A.; Mabande, G.T.P.; Selvam, T.; Schwieger, W.; Rudolph, A.; Hermann, R.; Sieber, H. & Greil, P. (2006). Biotemplating of Luffa Cylindrica Sponges to ... Catalytic Reactors. *Mater. Sci. Eng. C*, 26 , 130-135

Zhang, W.J.; Liu, W.; Cui, L. & Cao, Y. (2007). Tissue engineering of blood vessel. *J Cell Mol Med.* 11, 5, 945-957

Zawrah, M.F. & El-Gazery, M. (2007). Mechanical properties of SiC ceramics by ultrasonic nondestructive technique and its bioactivity. *Materials Chemistry and Physics*, 106, 330–337

15

Silicon Carbide Whisker-mediated Plant Transformation

Shaheen Asad and Muhammad Arshad
Gene Transformation Lab. Agricultural Biotechnology Division, NIBGE, P.O. Box 577, Jhang Road, Faisalabad, Pakistan

Abstract

With the advancement in molecular biology, several metabolic and physiological processes have been elucidated at molecular levels discovering the involvement of different genes. Since the advent of plant transformation 33 years ago, use of plant transformation techniques sparked an interest in fundamental and applied research leading to the development of biological and physical methods of foreign DNA delivery into 130 plant species. Modern molecular biology tools have developed rich gene sources which are waiting to be transformed into plant species. But unavailability of efficient transformation methods is a major hurdle to expedite the delivery of these genes into plants.

Ever-expanding available gene pools in the era of third generation transgenic plants stressing the delivery of multiple genes for different traits; development and application of new transformation methods is the big need of the time to meet the future challenges for plant improvement. In recent years, silicon-carbide whiskers have proven valuable and effective alternative in which silicon carbide fibers are mixed with plant cells and plasmid DNA, followed by vortexing/oscillation. Cell penetration appears to occur thus whiskers function as numerous fine needles, facilitating DNA entry into cells during the mixing process. This technique is simple, easy and an inexpensive transformation method to deliver the DNA into monocot and dicot plant species. Whiskers, cells and plasmid DNA are combined in a small tube and mixed on a vortex or oscillating mixer. In this chapter we will discuss the use of silicon carbide fibers/whiskers to transform and produce different transgenic plants. This chapter will help the reader to know about emerging applications of silicon carbide and other fibers in the delivery of foreign DNA into plants, and critical parameters affecting DNA delivery efficiency will also be discussed.

1. Introduction

Plant Cell wall is commonly found as the non-living barrier in the ways of DNA deliver technologies being attempted for plant genetic engineering. In case of biological systems, the cell wall is dissolved by cell wall degrading enzymes secreted by donor host for contact of donor cell with recipient cell allowing exchange of biological materials along with net DNA delivery into recipient cells. But this limitation cannot be overcome in monocots which are non-host for prokaryotic mediated deliveries. The cell wall problem in monocots was ruled out by use of protoplasts, ballistic particles penetration opening the way for DNA entry into plants whether monocots or dicots. Although protoplasts are attractive as a transformation target because a large number of treated cells receive, regeneration from protoplasts of many crops remains technically difficult. Particle bombardment is currently considered as the most efficient delivery system. However, it requires sophisticated, expensive equipment and supplies. The same problem also exists for microinjection in which the transformation operation is extremely labor-intensive and technically demanding particularly bacterial cells electro-transformation, but could not be widely applied in plant transformation. Electroporation can be applied to many tissue types, but the efficiency remains poor at this stage.

Genetic improvement of plants has been done by plant breeders for years with great success. Different schemes have been developed by plant breeders for crossing plants in order to transmit and concentrate desirable traits in specific lines. However, in the process of classical plant breeding, success rate of attempts are uncertain and slow due to complex quantitative nature of genes governing agronomic traits. To transmit a required gene of interest by classical methods requires a series of sexual crosses between two lines and then repeated back crossing between the hybrid offspring and one of the parents until a plant with the desired characteristic is obtained. Plant breeding is a lengthy process, taking ten to fifteen years to produce and to release a new variety. This process, however, is limited to those plants which can sexually hybridize, and undesired genes in addition to the desired gene will also be co-transferred.

Recombinant DNA technologies circumvent these limitations by enabling plant geneticists to identify and clone specific genes for desirable traits. Different recombinant techniques can be used to explore biological features like gene transfer by *Agrobacterium tumefaciens*, microprojectile bombardment, electroporation of protoplast, polyethylene glycol method, microinjection, silicon carbide mediated transformation, liposome mediated gene transfer and sonication assisted *Agrobacterium*-mediated transformation.

Among these techniques, silicon carbide mediated gene deliveries hold good promise to expedise the delivery of differet genes into plant species.

2. Silicon whiskers and Transformation mechanism

Silicon carbide (SiC) whiskers have been extensively studied for high tech applications because of several advantages such as high tensile strength, high elastic modulus, excellent shock and degradation resistance (Choi et al., 1997). SiC whiskers have great intrinsic hardness and fracture readily to give sharp cutting edges (Greenwood & Earnshaw, 1984). They are obtained by the thermal reduction of silica in a reducing atmosphere, one source of silica being rice husks (Mutsuddy, 1990). Industrially, silicon carbide whiskers are used as abrasives in the manufacture of cutting tools and in the production of composite materials. Silicon carbide and other whiskers from different sources have been utilized in the transformation of monocot and dicot plant species embryo and cell suspension cultures (Table 1).

The exact mechanism for whisker-mediated transformation is based on different mechanisms. Scanning electron microscopy work on whisker-treated BMS cells described by

Kaeppler et al. (1990) suggested that a SiC whisker may have penetrated the wall of a maize cell. Unlike asbestos fibers (Appel et al., 1988), the surface of SiC whiskers is negatively charged. This negative surface charge probably results in there being little affinity between DNA molecules (which are also negatively charged) and whiskers in neutral pH medium. In these experiments they observed that premixing whiskers and DNA, as reported by Kaeppler et al. (1990, 1992), was not required to achieve efficient DNA delivery. This could suggest that whiskers do not "carry" DNA into the treated cells but function as numerous needles facilitating DNA delivery by cell perforation and abrasion during the mixing process (Fig 1 a and b). We have also found that materials with characteristics similar to SiC whiskers, such as nitride whiskers, can deliver DNA into plant cells.

Whisker/fiber type	Plant species	Genes / Selection strategy	Reference
Silicon carbide	Tobbacco	GUS nptII and BAR gene	Kaepler et al., 1992
Silicon carbide	Maize	GUS nptII and BAR	Kaepler et al.,1992 Petolino et al., 1994 Frame et al., 1994 Frame et al., 2000
Silicon carbide Al-borate K-titanate	Rice	GUS nptII	Nagatani et al., 1993 Matsushita et al., 1999 Mizuno et al., 2004 Terakawa et al., 2005
Silicon carbide	Ryegrass	GUS nptII	Dalton et al., 1998
Silicon carbide	Wheat	GUS nptII	Omirulleha et al., 1996
Silicon carbide	Cotton	GUS, AVP1 nptII	Asad et al., 2008
K-titanate	Soya bean	GFP nptII	Khallafalla et al., 2006

Table 1. Whisker mediated Gene Delivery in different transgenic model and commercial plants

3. Critical parameters for silicon carbide whisker-mediated plant transformation

3.1 Explant types

Transgenic plants have been successfully regenerated from a variety of tissues including leaves, cotyledons, hypocotyls, shoot tips, seed embryos and embryonic meristem transformed by different techniques. However, somatic embryogenesis occurrence is most critical parameter in whisker mediated plant transformation. Plant cell calli with a high density of embryogenic cells are extremely important and remain by far, the best tissue for explants and should be prepared from healthy seedlings/seeds for optimal results.

Among the few reports showing the DNA delivery via whisker method, callus suspension cultures have been used as explants for tobacco, maize, rice, soyabean and cotton. In each plant species, callus suspension cultures were derived from different plant parts. In rice, embryos for whisker treatment were prepared by trimming off of shoots and remaining scutellar tissues were used as explant (Matsushita et al., 1999). Immature maize zygotic embryos of A188 x B73 from greenhouse grown ears 10-12 days post pollination were aseptically excised and plated scutellum surface uppermost on modified N6 medium (Armstrong and Green, 1985; Chu et al., 1975; Mezuno et al., 2004). In case of wheat, mature embryos were used (Omirrulah et al., 1996). To initiate suspension culture, 3 g of proliferating type II callus from a single embryo and suspension cultures were maintained in medium in a 125 ml Erlenmeyer flask at 28°C in darkness on a rotary shaker at 125 r.p.m. Suspensions were subcultured every 3.5 days by addition of 3 ml packed volume of cells and 7 ml of conditioned culture medium to 20 ml of fresh culture medium. Cells on the filter paper disc were placed on solid N6 medium and subcultured weekly for 3 weeks after which the suspension cultures were reinitiated (Frame et al 1994). In cotton, embryogenic cell suspension cultures were prepared from two month old hypocotyls derived calli (Asad et al., 2008).

Unlike particle bombardment and PEG treatment of protoplasts, whiskers deliver plasmid DNA in a non-precipitated form, making it possible to control the quantity of DNA available for transformation. Stable transformation of BMS maize cells with as little as 0.1 pg of input plasmid DNA, the transformation efficiency being 50% as high as, that was achieved using 25 pg. Direct gene transfer tends to introduce multiple gene copies into cells and these gene copies often display complex integration patterns (Spencer et al., 1992). It remains to be shown definitively whether these integration characteristics have any functional significance but the ability to control input DNA quantity with whisker transformation may make it feasible to influence copy number.

3.2 Whisker types

Different types of whiskers have been used in the delivery of foreign DNA into the targert tissues. Transformation efficiency has been largely determined by type of whiskers and even their manufacturing processes. In most cases SiC whiskers have been used to transform different plant species but the drawback associated with them is their persistence and toxicity to lungs if inhaled. Up until now differet types of silicon carbide whiskers including Sc-9, A.A, TW(s), TW(m) and alfa Aesar have been reported with different efficiencies (Table 1).

The transformation efficiency of other materials (glass beads, carborundum a spherical form of SiC and silicon nitride) was much lower than that of silicon carbide. Even using SiC whiskers, transient GUS activities with different sources of whiskers varied significantly. In maize using A X B suspension cells, where transient numbers achieved from SC-9 whiskers (Advanced Composite Materials Co., Greer, SC) were at least 5 times higher than that from other sources, TWS100 (Tokai Carbon Co., Ltd., Tokyo, Japan) and Alfa Aesar (Johnson Matthey, Ward Hill, MA). In a separate experiment, SC-9 whiskers were compared to SiC whiskers supplied by Good fellow (Cambridge, England) using BMS maize suspension cells (Wang et al., 1995). Whiskers of potassium titanate fibers having average diameter 0.5um and length 5-30µm were used for rice transformation. PTW is a physically and chemically stable material and has similar characteristics to those of SCW (0.6µm diameter, 10-80µm length) and ABW with 0.5 µm diameter and length 25µm (Terakawa et al., 2005). They all require careful handling in the fumehood while wearing mask to prevent the mishap of inhaling while using. The difference in transformation efficiency between the two whisker-types was again significant. Moreover other whiskers from other sources such as ABW (alborax Y; alboraxYS3A) and potassium titanate have been developed and utilized to deliver DNA into the crop plant species. Different concentrations of whiskers delivered DNA in different efficiencies.

3.3 Mixing method

In whisker mediated plant transformation, mixing is the most critical step in the successful DNA delivery. It is the mixing method, duration and speed which significantly affect the DNA deliveries in the tissues. In addition different mixing materials have been used to enhance the DNA delivery in silicon carbide mediated plant transformation. Initially most reports involved the use of vortex mixing (Asano et al., 1990; Dunahay et al., 1993; Kaepler et al., 1990; 92). Rapid oscillation of cell and whiskers can result in stable transformation provided that cells survive in sufficient number for adequate regrowth (Frame et al., 1994). DNA delivery in the whisker method is increased by increasing the duration of agitation, agitation speed and amount of whiskers. However cell damage occurs when agitation is severe, extra prolonged and too large an amount of whiskers resulting in severe damage losing cell viability (Petolino et al., 2000; Mizuno et al., 2004). GUS expression units were maximum when mixing speed was 8 for 30 min using 20mg of ABW where as most of silicon carbide whisker preparations were made at the rate of 5% (Asad et al., 2008).

3.4 Pretreatment of explant

Explant serves key role in the efficient transformation of plants by whisker methods. Pretreatment of plant cells with osmotic agents has markedly increased the recovery of stable clones in maize. AXB suspension of cells 1 day after sub-culture were transferred in 1.5ml eppendorf tubes and incubated for 30 min with 1ml of liquid N6 medium containing equal molarities of sorbitol and manitol. 0.5M S/M increased 3-5 fold DNA delivery compared with non-treated control (Vain et al., 1993). No obvious effect on subsequent cell growth was seen following incubation in this medium. Similarly, calli subcultured at different intervals behaved differently for getting kanamycin resistant calli colonies when embryogenic calli were utilized as explants in cotton (Asad et al., 2008). Different transformation results have been obtained for calli from different plants while in case of cotton embryogenic calli, calli at 14 days after subculture were found to be the best explant for silicon carbide whisker mediated plant transformation. Asano et al. (1991) used the fiber technique to transform a suspension culture of *Agrostis alba* L. (Redtop). The genus *Agrostis* includes economically important temperate grass species used for forage and tuff. The frequency of GUS expression units tended to be the highest when plasmid DNA, was delivered to cells 6 days rather than 3 or 10 days after subculture. In Chlaymydomonas *(C. reinhardtii)* High-frequency, stable nuclear transformation can be achieved by agitating wall-less cells in the presence of plasmid DNA, glass beads and PEG (Kindle, 1990). In the whisker-mediated transformation protocol (Dunahay, 1993), plasmid DNA whiskers and PEG were added to -10^8 cells. The mixture was agitated by vortexing for 1 to 3 min. Cell viability after whisker treatment was greater than 80% whereas the viability of cells treated with glass beads was less than 10%. Typically, 10 to 100 stable transformants/10^7 cells were obtained using the whisker protocol.

3.5 Selection strategy

The production of transgenic plants requires efficient selection systems to isolate transformed cells, calluses, embryos, shoots from explants subjected to transformation. The existing selection systems can be divided into two basically different groups, where the conventional systems constitute by far the largest group. They are based on a selective agent (herbicide or antibiotic) being converted to a less toxic substance (detoxification) by the selective gene product. The most widely used selectable marker gene is the neomycin phosphotransferase gene which confers resistance to aminoglycoside type antibiotics such as kanamycin, neomycin and G-418 (geneticin). A number of other selection systems have been developed based on resistance to bleomycin, bromoxynil, chlorsulfuron, 2,4-dichlorophenoxy-acetic acid, glyphosate, hygromycin or phosphinothricin. The other group comprises the so-called positive selection systems where the selective agent is converted into a fully metabolisable compound by the selective gene product, providing the transgenic shoots with a metabolic advantage over non-transgenic shoots which are retarded on such selection media (Penna et al., 2002).

A gene for antibiotic resistance is usually fused with the reporter gene and gene of interest. Antibiotic resistance genes are typically chimeric bacterial genes that can be inserted into the T-DNA portion of the plant transformation plasmids. The resistance gene usually codes for an enzyme that inactivates the antibiotic using phosphorylation. Two frequently used selective markers that employ antibiotics as the selective agent are neomycin phosphotransferase II (*NPT* II) and hygromycin phosphotransferase *(HPT).* When antibiotics such as kanamycin or hygromycin are present in the tissue culture media, transformed cells will continue to divide while untransformed cells will slowly die. The *npt II and hpt* gens are a good selectable marker to use in dicot and monocot transformation work with efficient selection of putative transgenic tissues. The antibiotic hygromycin is very toxic to plant cells and expression of the *hpt* gene effectively causes insensivity to hygromycin in transformed cells (Larkin, 2001). Aragão et al. (2000) reported a system to select transgenic meristematic cells after the physical introduction of a mutant *ahas* gene by using a selection system based on the use of imazapy, a herbicidal molecule of the imidazolinone class capable of systemically translocating and concentrating in the apical meristematic region of the plant. The mechanism of action of imazapyr is the inhibition of the enzymatic activity of acetohydroxyacid synthase [AHAS; acetolactate synthase, acetolactate pyruvate-lyase (carboxilating)], which catalyses the initial step in the biosynthesis of isoleucine, leucine and valine (Shaner et al. 1984). This selectable marker system, combined with an improved multiple-shooting induction protocol, can result in a significant increase in the recovery of fertile, transgenic material after gene delivery process. Zhang et al. (2000) reported other herbicide resistance marker gene for soyabean; gene *bar* encodes a phosphinothricin acetyl-transferase enzyme (PAT), which inactivates the herbicide phosphinothricin (PPT). Joersbo et al., (2003) reported galactose as selective agent and a UDP-glucose: galactose-1-phosphate uridyltransferase gene as selective gene for efficient transformation of potato.

The better selection strategy used in transformation experiments allows rapid growth of transformed tissue as a result of selective growth of transgenic tissues. A fair efficient selection system allows very few non-transformed callus lines to grow through the reasonable selection scheme. Although, in most of the cases, hygromycin and kanamycin have been used as plant tissue selection marker, phosphoinithricin have also been used in the transformation of plant tissues. Kanamycin antibiotic concentration was optimized for cotton embryogenic calli transformation with minimum escapes. Different selection agents have been employed for the whisker-mediated transformed plant calli (Table 1). As the new and efficient plant transformation methods are being developed, use of alternative and more efficient selection methods are required.

3.6 Use of reporter genes

Optimization and success of transformation can be facilitated and checked out by the use of selectable marker and reporter genes, which are excellent for tracking transformation events and are frequently used in transformation protocols. Using genetic engineering, these genes can be placed under regulatory control of specific promoters, making their expression associated to the transformation process.

Gene expression dynamics can often be monitored through the use of reporter genes. These genes are designed to visually reveal the expression of transgenes through destructive or non-destructive procedures. The *Escherichia coli lacZ* gene was one of the first to be used as a reporter. The *lacZ* gene encodes a galactosidase protein, which has been widely characterized genetically and biochemically. This reporter gene was initially used to analyze gene expression in yeast and later introduced into tobacco. The main problem with this reporter gene is the background level of the ®-galactosidase enzyme in plants. Another gene that was widely used for the analysis of transgene expression encodes the chloramphenicol acetyltransferase (CAT) protein. The *cat* gene was successfully introduced into different plants, such as tobacco and several *Brassica* species. The CAT reporter system was a suitable marker for some but not all plant species. It was found that some *Brassica* species had high levels of endogenous CAT activity and the response of unknown inhibitors, drastically reduced the levels of expression of the bacterial CAT in transgenic plants. These problems and the lack of a histochemical detection method limited the applications of *cat* as a reporter gene for plant transformation. Some of the problems encountered with the *cat* and *lacZ* genes were solved by the reporter gene system GUS. The GUS reporter system utilizes a bacterial gene from *Escherichia coli* (*uidA*) coding for a ®-glucuronidase (GUS) and consists in placing this gene in the Ti-plasmid, which is transferred to plant cells during gene delivey. When the plant tissue is assayed, transformation events are indicated by blue spots, which are a result of the enzymatic cleaving of an artificial substrate to give a blue product. The addition of a plant intron to the GUS gene directs expression only in plant cells with no problem of background expression in *Agrobacterium*. Although the GUS system is commonly used, the GUS histochemical assay utilizes cyanide, which kills the tissue, and requires overnight incubation for visualization. Due to the low content or absence of endogenous GUS activity in most plants, the *uidA* gene became one of the most widely-used reporter genes in eukaryotic systems. Unlike the *lacZ* and *cat* reporter genes, the *uidA* gene expression could be quantified relatively easily since it is an inexpensive and sensitive method to assay gene expression; however, the detection assay is destructive, making it impossible to analyze the same piece of tissue over time. For all these reasons the GUS reporter system was successfully used with different plant species (Jefferson et al., 1987; Higuchi et al., 2001). Transformation experiments with soybean mostly use the GUS reporter system for determining gene expression and integration.

A better reporter system would not require killing the tissue and would permit rapid or immediate evaluation of gene expression directly in living tissues. So a new and simpler reporter system was developed which uses the green fluorescent protein (GFP) from jellyfish (*Aquorea victoria*). This reporter gene does not require a destructive staining procedure and allows direct viewing of gene expression in living plant tissue. Similar to the GUS reporter system, *gfp* can be introduced into plants using the Ti-plasmid. Following T-DNA transfer, GFP can be viewed directly in living tissues with blue light excitation. The GFP reporter system permits detection of labeled protein within cells and monitoring plant cells expressing *gfp* directly within growing plant tissue (Haseloff & Siemering, 1998). Since the *gfp* gene was first reported as a useful marker for gene expression in *Escherichia coli* and *Caenorhabditis elegans*, it has been modified by several laboratories to suit different purposes to include elimination of a cryptic intron, alteration in codon usage, changes in the chromophore leading to different excitation and emission spectra, targeting to the endoplasmic reticulum (ER) and mitochondria and understanding the morphology and dynamics of the plant secretory pathway (Brandizzi et al., 2004). GFP has been used as a reporter system for identifying transformation events in *Arabidopsis thaliana,* apple, rice, sugarcane, maize, lettuce, tobacco, soybean, oat, onion, wheat, leek and garlic (Eady et al., 2005). The GFP reporter system has also been used for identifying successful plastid transformation events in potato. The *gfp* gene has successfully been used as a scorable marker to evaluate plant transformation efficiency using *Agrobacterium tumefaciens,* particle bombardment and whisker mediated gene transfer. The gene could be expressed as early as 1.5 h following introduction and, since its detection is nondestructive, *gfp* expression could be followed over extended periods of time. GFP has also been used as a reporter to analyze the compartmentalization and movement of proteins over time in living plant cells using confocal microscopy (Benichou et al., 2003).

As the original *gfp* gene comes from the jellyfish, the coding region was modified to permit expression in plant cells. Codon usage of the gene was altered to stop splicing of a cryptic intron from the coding sequence. The unmodified *gfp* contains an 84 nucleotide sequence that plants recognize as an intron and is efficiently spliced from the RNA transcript, resulting in little or no expression of *gfp*. Using a modified *gfp*, *mgfp4*, expression problems resulting from cryptic intron processing were eliminated for many plants. Although the *mgfp4* gene is clearly an effective reporter gene, brightly fluorescing transformants containing high levels of GFP were difficult to regenerate into fertile plants. GFP in plants accumulates in the cytoplasm and nucleoplasm, while in jellyfish, GFP is compartmentalized in cytoplasmic granules. GFP in plants may have a mildly toxic effect due to fluorescent properties of the protein and accumulation in the nucleoplasm. In order to overcome this problem *mgfp5-ER* was produced, which has targeting peptides fused to GFP to direct the protein to endoplasmic reticulum (ER). With this modification, fertile plants have been regenerated more consistently (Haseloff & Siemering, 1998). Unlike *mgfp4*, *mgfp5-ER* lacks temperature sensitivity found in the wild-type GFP. Wild-type GFP must undergo proper folding with specific temperature requirements to maintain its fluorescent properties.. In addition to better protein folding, *mgfp5-ER* has excitation peaks of 395 and 473 nm. A broad excitation spectrum allows better GFP viewing with UV and blue light sources. The *mgfp5-ER* has shown to be an excellent reporter gene for lettuce and tobacco transformed by *Agrobacterium. Mgfp5-ER* has also been used with success for transient expression in soybean embryogenic suspension cultures via particle bombardment (Ponappa et al., 1999). The gene *gfp* has been modified numerous times and there are several *gfp* versions for plants.

Modified versions, other than *mgfp4* and *mgfp5-ER*, include: *SGFP-TYG* which produces a protein with a single excitation peak in blue light, *smgfp* which is a soluble modified *mgfp4*, *pgfp* which is a modified wild type GFP and *sGFP65T* which is a modified *pgfp* containing a Ser-to-Thr mutation at amino acid 65. Different versions of *gfp* have varying levels of fluorescence. These differences may be dependent upon the transformed species, promoter and termination sequences, or gene insertion sites. In the future, selective markers may not be needed, but while the intricacies of GFP expression need more understanding, selective markers are helpful in providing an advantage to identifying successful transformation event (Wachter, 2005).

In another reporter system, the luciferase reaction occurs in the peroxisomes of a specialized light organ in fireflies (*Photinus pyralis*). The luciferase reaction emits a yellow-green light (560nm) and requires the co-factors ATP, Mg^{2+}, O_2 and the substrate luciferin (Konz et al., 1997). The glow is widely used as an assay for luciferase activity to monitor regulatory elements that control its expression. *Luc* is particularly useful as a reporter gene since it can be introduced into living cells and into whole organisms such as plants, insects, and even mammals. *Luc* expression does not adversely affect the metabolism of transgenic cells or organisms. In addition, the *luc* substrate luciferin is not toxic to mammalian cells, but it is water-soluble and readily transported into cells. Since *luc* is not naturally present in target cells the assay is virtually background-free. Hence, the *luc* reporter gene is ideal for detecting low-level gene expression. A second reporter system based on luciferase expressed by the *ruc* gene from Renilla (*Renilla reniformis*) has also become available. The activities of firefly and *Renilla luciferase* can be combined into a dual reporter gene assay.
Despite the availability of a number of reporter genes, only two reporter genes (GUS and GFP) have been reported in transgenic plants developed through silicon carbide/whisker mediated plant transformation (Khalafalla et al., 2006; Asad et al., 2008).

3.7 Transgene integration and expression improvement

The perfect transformant resulting from any method of transgene delivery, would contain a single copy of the transgene that would segregate as a mendelian trait, with uniform expression from one generation to the next. Ideal transformants can be found with difficulty, depending upon the plant material to be transformed and to some extent on the nature and the transgene complexity. As gene integrations are essentially random in the genome, variability is often observed from one transgenic plant to another, a phenomenon ascribed to 'position effect variation' (Chitaranjan et al., 2010).The general strategy to 'fix' this situation is to generate, probably at a high cost, enough transgenic plants to find some with the desired level of expression.

Efforts are being directed toward achieving stable expression of the transgene with an expected level of expression rather than that imparted by the random site of integration. Scaffold Matrix Attachment Regions (MARs) could potentially eliminate such variability by shielding the transgene from surrounding influence. MARs are A/T rich elements that attach chromatin to the nuclear matrix and organize it into topologically isolated loops. A number of highly expressed endogenous plant genes have been shown to be flanked by matrix attachment regions and reduce the variability in transgene expression (Chitaranjan et al., 2010). Several experiments have been carried out in which a reporter gene like GUS has been flanked by MARS and introduced into transgenic plants and compared to populations containing the same reporter gene without MARs (Mlynarora et al., 2003). Other ways to avoid variation in gene expression due to position effect are plastid transformation and minichromosome transformation. Some guidance might come from genome sequencing, which might provide the necessary DNA ingredients to control gene expression. The ability to target integration could also lead to some control of transgene expression . It is foreseen that site-specific recombinases could assist in this endeavor. All these areas of research, which are primed for breakthroughs, should be carefully monitored for immediate implementation in the design of suitable vectors equally useful for use in different plant transformation methods. In the longer term, it is less expensive and ultimately more desirable to produce higher quality and fewer quantities of transgenic plants.

Prospects

Currently most of the reports on gene deliveries by SCW are limited to model systems and few agronomic plants have been transformed which are largely concerned with transgene delivery and analyses of reporter genes. But no report is available describing the stability and pattern of inheritance in subsequent generations proving the authenticity of this relatively new physical method of plant transformation. So being an emerging transformation method, research on gene delivery with viable markers like GFP and luc genes having uniform integration and expression levels are worth pursuing future tasks. There is also a practical need for a method of transformation that will decrease the complexity of the pattern of transgene integration and expression. Presently, most commercial transgenics are altered in single gene traits. The challenge for the genetic engineers is to introduce large pieces of DNA-encoding pathways and to have these multigene traits function beneficially in the transgenic plants.

Although a clearer understanding of the events surrounding the integration and expression of foreign DNA is emerging, there are many questions that remain unanswered. Are there target cells or tissues not previously attempted that are more amenable to transformation? Is there a physiological stage that allows greater transformation? Can it be manipulated to achieve higher transformation efficiency? Does the tissue chosen as a target affect the level of expression? It is becoming increasingly clear that plants transformed by *Agrobacterium* express their transgene more frequently. Can this be partly attributed to the fact that T-DNAs frequently integrate in telomeric regions (Hoopen et al 1996)? Transformation technologies have advanced to the point of commercialization of transgenic crops. The introduction of transgenic varieties in the market is a multi-step process that begins with registration of the new varieties followed by field trials and ultimately delivery of the seed to the farmer. Technical improvements and employments of new efficient plant transformation methods that have the greatest opportunities for new approaches, probably in the realm of *in planta* transformation, will further increase transformation efficiency by extending transformation to elite commercial germplasm and lower transgenic production costs, ultimately leading to lower costs for the consumer.

4. Conclusion

It is quite clear that whisker-mediated transformation of any species where regenerable suspension cultures exist should be possible once DNA delivery parameters have been established. Up until now most of the work has been focused on the demonstration of the viability of this method by use of reporter genes such as GUS and GFP. Routine transformation protocols are limited in most agriocultural plants. The low success has been attributed to poor regeneration ability (especially via callus) and lack of compatible gene delivery methods, although some success has been achieved by introducing innovative gene delivery technology like silicon/whisker mediated plant transformation. One of the limitations for efficient plant transformation is the lack of understanding of gene expression during the selection and regeneration processes. Therefore, optimization of the transformation efficiency and reproducibility in different laboratories still represents a major goal of investigators. We believe this is because transformation methods have not yet been properly quantified and established for each and every crop plants species. To improve the efficiency of transformation, more appropriate and precise methods need to be developed. For monitoring the efficiency of each step, the jellyfish green fluorescent protein (GFP) perfectly qualifies, because frequent evaluation of transgene expression could provide detailed information about regulation of gene expression *in vitro*. Nowadays, GFP is a useful reporter gene in plant transformation and is also used as a tool to study gene expression dynamics in stably transformed clones. GFP can play an important role in the evaluation of transformation systems and in the avoidance of gene silencing. Progress in soybean transformation suggests that some systems will achieve the transformation efficiency required for functional genomics applications in the near future.

Recently, we have obtained stably transformed lines from silicon carbide whisker treatment of embryogenic callus derived from cotton coker-312, indicating that the method can be extended to target tissues other than suspension cells. In addition to these genes, other genes of agronomic importance have been transformed into commercial crops like cotton and have obtained fertile transgenic AVP1 cotton with significant salt tolerance.

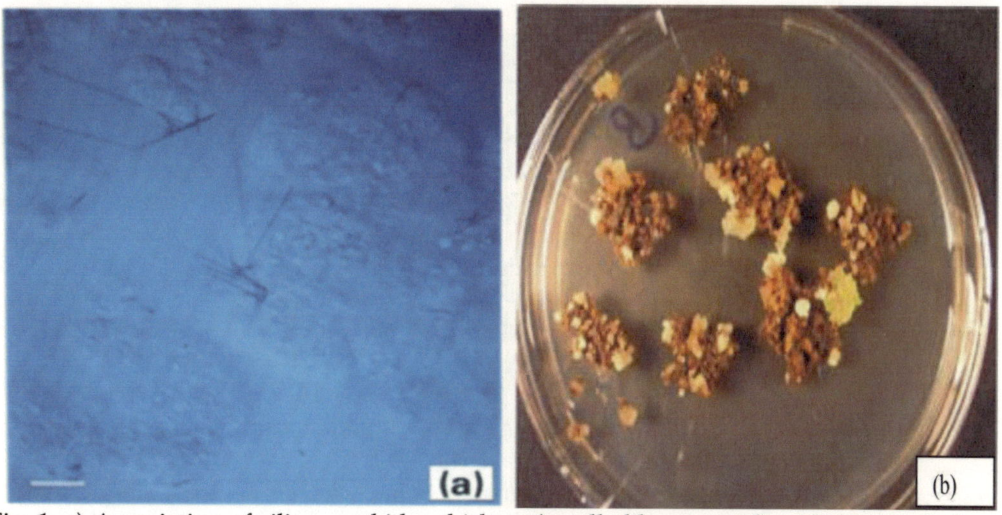

Fig. 1. a) Association of silicon carbide whiskers (needle-like material) with (a) A x B plant suspension cells visualized under light microscopy in maize (Frame et al., 1994); (b) induction of kanamycin resistant cotton calli from embryogenic calli transformed with silicon carbide whiskers (Asad et al., 2008)

5. References

Appel, JD.; Fasy, T. M.; Kohtz, D. S. (1988) Asbestos fibers mediate transformation of monkey cells by exogenous plasmid DNA. Proc. Natl. Acad. Sci. USA. 85 pp.7670-7674; 1988

Aragão, FJL.; Sarokin, L.; Vianna, GR. & Rech, E.L. (2000). Selection of transgenic meristematic cells utilizing a herbicidal molecule results in the recovery of fertile transgenic soybean (*Glycine max* (L.) Merril) plants at high frequency. Theor. Appl. Genet. 101, pp. 1–6

Armstrong, CL. & Green, CE. (1985) Establishment and maintenance of friable, embryogenic maize callus and the nvolvement of L-proline. *Planta,* 164, pp. 207-214

Asad, S.; Mukhtar, Z.; Nazir, F.; Hashmi, AJ.; Mansoor, S.; Zafar, Y. & Arshad, M. (2008) Silicon carbide whisker-mediated embryogenic callus transformation of cotton (*Gossypium hirsutum* L.) and regeneration of salt tolerant plants. Mol. Biotech 40, pp. 161-169

Asano, Y.; Otsuki, Y, & Ugaki, M. (1991) Electroporation-mediated and silicon carbide whisker-mediated DNA delivery in Agrosris alba L. (Redtop). Plant Sci. 9, pp. 247-252.

Benichou, M.; Li, Z.; Tournier, B.; Chaves, AlS.; Zegzouti, H.; Jauneau, A.; Delalande, C.; Latché, A.; Bouzayen, M.; Spremulli, L.L. & Pech, J.-C. (2003) Tomato ef-ts (mt), a functional translation elongation factor from higher plants. Plant mol. biol. 53, pp.411-422

Brandizzi, F.; Irons, SI.; Johansen, J.; Kotzer, A. & Neumann, U. (2004) GFP is the way to glow: bioimaging of the plant endomembrane system. J. of Microsco. 4, pp. 138-158

Chataranjan, K.; Michler, CH.; Abbot, AG. & Hal, TC. (2010) Transgenic crop plants, Volume I, Principles and development, Springer Hieldelberg Dordrecht London New York (PP 145-187)

Choi, GJ. (1997) Silicon carbide fibers from copolymers of commercial polycarbosilane and silazane. Journal of Ind. And Eng, Chem. 3, pp. 223-228

Chu, CC.; Wang, CC.; Sun, CS.; Hsu, C.; Yin, KC.; Chu, Y. & Bi, FY. (1975) Establishment of an efficient medium for another culture of rice through comparative experiments on the nitrogen sources. Sci. Sinica. 18, pp. 659-668

Dalton, SJ.; Bettany, AJE.; Timms, E.; & Morris, P. (1998) Transgenic plants of *Lolium multiflorum, Lolium perenne, Festuca arundiacea, and Agrostis stolonifera* by silicon carbide fibre-mediated transformation of cell suspension cultures . Plant Sci. 132, pp. 31 - 43

Dunahay, TG. (1993) Transformation of Chlamydomonas reinhardtii with silicon carbide whiskers . BioTechn. 15, pp. 452-460.

Eady, C.; Davis, S.; Catanach, A.; Kenel, F. & Hunger, S. (2005) *Agrobacterium tumefaciens*-mediated transformation of leek (*Allium porrum*) and garlic (*Allium sativum*). Plant Cell Rep. 4, pp. 209-215

Frame, BR.; Drayton, PR.; Bagnall, SV.; Lewnau, CJ.; Bullock, WP.; Wilson, HM.; Dunwell, JM.; Thompson, JA. & Wang, K. (1994) Production of fertile transgenic maize plants by silicon carbide whisker-mediated transformation. Plant J. 6, pp. 941 - 948

Gelvin, SB. (2003) *Agrobacterium*-mediated plant transformation: the biology behind the "gene-jockeying" tool. Microbiol. Mol. Biol. Rev. 67, pp. 16–37

Greenwood, NN.; Earnshaw, A. (1984) Silicon carbide, SiC: chemistry of elements. Oxford: Pergamon Press; 386

Haseloff, J. & Siemering, KR. (1998) *In* Green Fluorescent Protein: Properties, applications, and Protocols. Eds. M Chalfie and S Kain. pp. 191–220. Wiley-Liss, New York

Higuchi, M.; Pischke, MS. & Mähönen, AP. (2004) In planta functions of the Arabidopsis cytokinin receptor family. Proceedings of National Academy of Sciences, USA 101, pp. 8821-8826

Jefferson, RA.; (1987) Assaying chimeric genes in plants: the GUS gene fusion system. Plant Mol. Biol. Report. 5, pp. 387-405

Joersbo, M.; Jorgensen, K. & Brunstedt, J. (2003) A selection system for transgenic plants based on galactose as selective agent and a UDP-glucose: galactose-1-phosphate uridyltransferase gene as selective gene. Mol. Breed. 111, pp. 315-328.

Kaeppler, HF; Gu, W; Somers, DA; Rines HW. & Cockburn, AF. (1990) Silicon carbide fiber-mediated DNA delivery into plant cells . Plant Cell Rep. 9, pp. 415-418

Kaeppler, HF; Somers, DA; Rines HW. & Cockbum, AF. (1992) Silicon carbide fiber-mediated stable transformation of plant cells. Theor. Appl. Genet. 84, pp. 560-566

Kaeppler, HF. & Somers, DA. (1994) DNA delivery to maize cell cultures using silicon carbide fibers . In : M. Freeling & V. Walbot (Eds) . The Maize Handbook, pp. 610-613 Springer-Verlag, New York.

Khalafalla, M.; El-Sheny, HA.; Rahman, SM.; Teraishi, M.; Hasegawa, H.; Terakawa, T.; & Ishimoto, M. (2006) Efficient production of transgenic soybean (*Glycine max* [L] Merrill) plants mediated via whisker-supersonic (WSS) method. Afr. J. of Biotech. 5 (18), pp. 1594-1599

Kunz, RE. (1997) Miniature integrated optical modules for chemical and biochemical sensing. Sens. Actuators B 38, pp. 13–28

Larkin, KM. (2001) Optimization of soybean transformation using SAAT and GFP. Wooster: OARDC/OSU, 126p. (Thesis -Master).

Matsushita, M.; Otani, M.; Wakita, M.; Tanaki, O. & Shimida, T. (1999) Transgenic plant regeneration through silicon carbide mediated transformation of rice (Oryza sativa L.). Breed. Sci. 49, pp. 21-26

Mizuno, K.; Takahashi, W.; Ohyama, T.; Shimada, T. and Tanaka, O. (2004) Improvement of the aluminum borate-whisker mediated method of DNA delivery into rice callus. Plant Prod. SCi. 7(1), pp. 45-49

Mlynárová, L.; Hricova, A.; Loonen, A. & Nap, JP. (2003) The Presence of a Chromatin Boundary Appears to Shield a Transgene in Tobacco from RNA Silencing. Plant cell 15(9), pp. 2203-2217

Mutsuddy, BC. (1990). Electrokinetic behaviour of aqueous silicon carbide whisker suspensions . J. Am. Ceram . Soc. 9, pp. 2747-2749

Nagatani, N.; Honda, H.; Shimada, H. & Kobayashi, T. (1997) DNA delivery into rice cells and transformation of cell suspension cultures. Biotechnol. Tech . 11, pp. 471 - 473

Omirrullah, S.; Ismagulava, A.; Karabaev, M.; Meshi, T. and Iwabuchi, M. (1996). Silicon carbide fiber-mediated DNA delivery into cells of wheat (Triticum aestivum L.) mature embryos. Plant Cell Rep. 16, pp. 133-136

Penna, s; Sagi, L.; & Swennen R. (2002) Positive selectable marker genes for routine plant transformation. In vitro Cell Dev. Biol. Plant 38, pp. 125-128

Petolino, JF.; Hopkins, NL.; Kosegi, BD. and Skokut, M. (2000) Whisker-mediated transformation of embryogenic callus of maize. Plant Cell Rep. 19, pp. 781 – 786

Ponapa, T.; Brzozowski, AE; & Finer, JJ. (1999) Transient expression and stable transformation of soyabean using jellyfish green fluorescent protein. Plant Cell Reports 19, pp. 6-12

Spencer, TM.; Gordon-Kamm, WJ.; Daines, RJ.; Start, WG. & Lemaux, PG. (1990) Bialaphos selection of stable transformants from maize cell culture. Theor. Appl. Genet. 79, pp. 625-631

Terakawa, T; Hisakazu, H; & Masanori, Y. (2005) Efficient whisker mediated gene transformation in a combination with supersonic treatment. Breed. sci. 55, pp. 456-358

Vain, P; McMullen, M D. & Finer, JJ. (1993) Osmotic treatment enhances particle bombardment-mediated transient and stable transformation of maize. Plant Cell Rep. 12, pp. 84-88

Wachter, R. (2005) The Family of GFP-Like Proteins: Structure, Function, Photophysics and Biosensor Applications. Photochem. and Photobio. 82, pp. 339-344

Wang, K.; Drayton, P.; Frame, B.; Dunwell, J. & Thompson, J. (1995) Whisker-mediated plant transformation: An alternative technology. In Vitro Cell. Dev. Biol. 31, pp. 101-104

Zhang, J.; Klueva, NY; Wang, Z; Wu, R; T.-H. D. Ho, THD, & Nguyen, HT. (2000) Genetic engineering for abiotic stress resistance in crop plants. In Vitro Cell Dev. Biol. Plant 36, pp. 108-114

Part 3

Bulk Processing, Phase Equilibria and Machining

Silicon Carbide: Synthesis and Properties

Houyem Abderrazak[1] and Emna Selmane Bel Hadj Hmida[2]
[1] *Institut National de Recherche et d'Analyse Physico-Chimique, Pole Technologique Sidi Thabet, 2020, Tunisia*
[2] *Institut Préparatoire Aux Etudes d'Ingénieurs El Manar 2092, Tunisia*

1. Introduction

Silicon carbide is an important non-oxide ceramic which has diverse industrial applications. In fact, it has exclusive properties such as high hardness and strength, chemical and thermal stability, high melting point, oxidation resistance, high erosion resistance, etc. All of these qualities make SiC a perfect candidate for high power, high temperature electronic devices as well as abrasion and cutting applications. Quite a lot of works were reported on SiC synthesis since the manufacturing process initiated by Acheson in 1892. In this chapter, a brief summary is given for the different SiC crystal structures and the most common encountered polytypes will be cited. We focus then on the various fabrication routes of SiC starting from the traditional Acheson process which led to a large extent into commercialization of silicon carbide. This process is based on a conventional carbothermal reduction method for the synthesis of SiC powders. Nevertheless, this process involves numerous steps, has an excessive demand for energy and provides rather poor quality materials. Several alternative methods have been previously reported for the SiC production. An overview of the most common used methods for SiC elaboration such as physical vapour deposition (PVT), chemical vapour deposition (CVD), sol-gel, liquid phase sintering (LPS) or mechanical alloying (MA) will be detailed. The resulting mechanical, structural and electrical properties of the fabricated SiC will be discussed as a function of the synthesis methods.

2. SiC structures

More than 200 SiC polytypes have been found up to date (Pensl, Choyke, 1993). Many authors proved that these polytypes were dependent on the seed orientation. For a long time, (Stein et al, 1992; Stein, Lanig, 1993) had attributed this phenomenon to the different surface energies of Si and C faces which influenced the formation of different polytypes nuclei. A listing of the most common polytypes includes 3C, 2H, 4H, 6H, 8H, 9R, 10H, 14H, 15R,19R, 2OH, 21H, and 24R, where (C), (H) and (R) are the three basic cubic, hexagonal and rhombohedral crystallographic categories. In the cubic zinc-blende structure, labelled as 3C-SiC or β-SiC, Si and C occupy ordered sites in a diamond framework. In hexagonal polytypes *n*H-SiC and rhombohedral polytypes *n*R-SiC, generally referred to as α-SiC, *n*Si-C bilayers consisting of C and Si layers stack in the primitive unit cell (Muranaka et al, 2008).

SiC polytypes are differentiated by the stacking sequence of each tetrahedrally bonded Si-C bilayer. In fact the distinct polytypes differ in both band gap energies and electronic properties. So the band gap varies with the polytype from 2.3 eV for 3C-SiC to over 3.0 eV for 6H-SiC to 3.2 eV for 4H-SiC. Due to its smaller band gap, 3C-SiC has many advantages compared to the other polytypes, that permits inversion at lower electric field strength. Moreover, the electron Hall mobility is isotropic and higher compared to those of 4H and 6H-polytypes (Polychroniadis et al, 2004). Alpha silicon carbide (α-SiC) is the most commonly encountered polymorph; it is the stable form at elevated temperature as high as 1700°C and has a hexagonal crystal structure (similar to Wurtzite). Among all the hexagonal structures, 6H-SiC and 4H-SiC are the only SiC polytypes currently available in bulk wafer form.
The β-SiC (3C-SiC) with a zinc blende crystal structure (similar to diamond), is formed at temperatures below 1700°C (Muranaka et al, 2008). The number 3 refers to the number of layers needed for periodicity. 3C-SiC possesses the smallest band gap (~2.4eV) (Humphreys et al,1981), and one of the largest electron mobilities (~800 $cm^2V^{-1}s^{-1}$) in low-doped material (Tachibana et al, 1990) of all the known SiC polytypes. It is not currently available in bulk form, despite bulk growth of 3C-SiC having been demonstrated in a research environment (Shields et al 1994). Nevertheless, the beta form has relatively few commercial uses, although there is now increasing interest in its use as a support for heterogeneous catalysts, owing to its higher surface area compared to the alpha form.

3. Opto-electronic properties of SiC

Silicon carbide has been known since 1991 as a wide band gap semiconductor and as a material well-suited for high temperature operation, high-power, and/or high-radiation conditions in which conventional semiconductors like silicon (Si) cannot perform adequately or reliably (Barrett et al, 1991). Additionally, SiC exhibits a high thermal conductivity (about 3.3 times that of Si at 300 K for 6H-SiC) (Barrett et al, 1993). Moreover it possesses high breakdown electric-field strength about 10 times that of Si for the polytype 6H-SiC.
In table 1, a comparison of fundamental properties of the main encountered SiC polytypes with the conventional Si semiconductor is depicted (Casady, Johonson, 1996).

Quantity	3C-SiC	4H-SiC	6H-SiC	Silicon
Thermal conductivity ($W\ cm^{-1}K^{-1}$) at 300K doped at ~$10^{17}cm^{-3}$	3.2	3.7	4.9	1.5
Intrinsic carrier concentration at 300K (cm^{-3})	$1.5x10^{-1}$	$5x10^{-9}$	$1.6x10^{-6}$	$1.0x10^{10}$
Saturation velocity ($cm\ s^{-1}$) parallel to c-axis	-	$2.0x10^{7}$	$2.0x10^{7}$	$1.0x10^{7}$
Electron mobility ($cm^2V^{-1}s^{-1}$)	800	1000	400	1400
Hole mobility ($cm^2V^{-1}\ s^{-1}$)	40	115	101	471
Schottky structures ε_s	9.72		9.66	11.7

Table 1. Comparison of some silicon carbide polytypes and silicon material properties (Casady, Johonson, 1996)

However, SiC possesses a much higher thermal conductivity than the semi-conductor GaAs at a temperature as high as 300 K as well as a band gap of approximately twice the band gap of GaAs. Moreover, it has a saturation velocity (v_{sat}) at high electric fields which is superior to that of GaAs and a saturated carrier velocity equal to GaAs at the high field power (Barrett et al, 1993).
The band gap of Si, GaAs and of 6H-SiC are about to 1.1 eV, 1.4 eV and 2.86 respectively.
We found a compilation of properties of: Silicon, GaAs, 3C-SiC (cubic) and 6H-SiC (alpha) with repeating hexagonal stacking order every 6 layers. SiC has a unique combination of electronic and physical properties which have been recognized for several decades (O'Connor, Smiltens, 1960).
In the following table a comparison of several important semiconductor material properties is given (Han et al, 2003).

Properties	Si	GaAs	3C-SiC	6H-SiC	4H-SiC
Band gap (eV) (T<5K)	1.12	1.43	2.40	3.02	3.26
Saturated electron drift velocity (10^7 cm s^{-1})	1.0	2.0	2.5	2.0	2.0
Breakdown field (MV cm^{-1})	0.25	0.3	2.12	2.5	2.2
Thermal conductivity (W cm^{-1} K^{-1})	1.5	0.5	3.2	4.9	3.7
Dielectric constant	11.8	12.8	9.7	9.7	9.7
Physical stability	Good	Fair	Excellent	Excellent	Excellent

Table 2. Comparison of several important semiconductor material properties (Han et al, 2003).

Silicon carbide also has a good match of chemical, mechanical and thermal properties. It demonstrates high chemical inertness, making it more suitable for use in sensor applications where the operating environments are chemically harsh (Noh et al, 2007).

4. Methods to grow SiC single crystals

Naturally silicon carbide occurs as moissanite and is found merely in very little quantities in certain types of meteorites. The most encountered SiC material is thus man made. Traditionally, SiC material has been produced through the Acheson process, in an Acheson graphite electric resistance furnace, which is still used for production of poly-crystalline SiC that is suitable for grinding and cutting applications.
In this process a solid-state reaction between silica sand and petroleum coke at very high temperature (more than 2500°C) leads to the formation of silicon carbide under the general reaction (1) (Fend, 2004):

$$SiO_2(s) + 3C(s) \longrightarrow SiC(s) + 2CO(g) \qquad (1)$$

Crystalline SiC obtained by the Acheson-Process occurs in different polytypes and varies in purity. In fact during the heating process and according to the distance from the graphite resistor heat source of the Acheson furnace, different coloured products could be formed. Thus, colourless, transparent or variously coloured SiC materials could be found (Schwetk et al, 2003). Additionally, the manufactured product has a large grain size and is invariably contaminated with oxygen. Moreover Nitrogen and aluminium are common impurities, and they affect the electrical conductivity of SiC. Thus the as, obtained SiC ceramic, often known by the trademark carborundum, is adequate for use as abrasive and cutting tools.
The conventional carbothermal reduction method for the synthesis of SiC powders is an excessive demanding energy process and leads to a rather poor quality material. Several alternative methods have been reported in the literature for the synthesis of pure SiC.

4.1 Physical vapor transport (PVT)

Physical vapor transport (PVT), also known as the seeded sublimation growth, has been the most popular and successful method to grow large sized SiC single crystals (Augustin et al, 2000; Semmelroth et al 2004). The first method of sublimation technique, known as the Lely method (Lely, Keram, 1955) was carried out in argon ambient at about 2500°C in a graphite container, leading to a limited SiC crystal size. Nevertheless, although the Lely platelets presented good quality (micropipe densities of 1-3 cm^{-2} and dislocation densities of 10^2-10^3 cm^{-2}), this technique has presented major drawbacks which are the uncontrollable nucleation and dendrite-like growth.
Given the fact that the control of SiC growth by the PVT method is difficult and the adjustment of the gas phase composition between C and Si complements and/or dopant species concentration is also limited, (Tairov, Tsvetkov, 1978) have developed a modified-Lelly method also called physical vapor transport method or seed sublimation method. In fact, this latter was perfected by placing the source and the seed of SiC in close proximity to each other, where a gradient of temperature was established making possible the transport of the material vapor in the seed at a low argon pressure.
The conventional PVT method has been refined by (Straubinger et al, 2002) through a gas pipe conveying between the source and the crucible into the growth chamber (M-PVT setup). Considering this new approach, high quality 4H and 6H-SiC, for wafer diameters up to 100 mm, were grow. In addition 15R-SiC and 3C-SiC were also developed.
to control the gas phase composition, (Wellmann et al, 2005) have developed the conventional configuration by the Modified PVT technique (M-PVT) for preparation of SiC crystal. They have also, using an additional gas pipe for introduction of doping gases and/or small amounts of C- and Si- containing gases (silane: SiH_4:H_2-1:10 and propane: C_3H_8).

$v = 10v_0$	$v = 5v_0$	$v = v_0$
No growth	Growth observed	The quality of crystal growth was found improved to the conventional setup configuration without a gas pipe.

v_0 and v are the PVT and the gas fluxes, respectively.
Table 3. The impact of the gas fluxes on the crystal growth.

The modified PVT system showed the improvement of the conventional PVT system of SiC. In fact, (Wellmann et al, 2005) have demonstrated that small additional gas fluxes in the modified- PVT configuration have a stabilizing effect on the gas flow in the growth cell interior compared to the conditional PVT configuration without the gas pipe. Table 3 shows the impact of the gas fluxes on the SiC crystal growth.

In the case of doping, using nitrogen as n-type doping, the gas was supplied directly in front of the growth interface so this modified growth presented an advantage for the high purity doping source. Phosphorus has been either used as n-type doping because it has a higher solid solubility (10 times higher than the state-of-the-art donor nitrogen) (M. Laube et al, 2002) and (Wellmann et al, 2005) have achieved phosphorus incorporation of approximately $2x10^{17}cm^{-3}$ but this hasn't reached the kinetic limitation value.

In contrast, aluminum has been used for p-type doping, the axial aluminum incorporation was improved, the conductivity reached 0.2 $\Omega^{-1}cm^{-1}$ in aluminum doped 4H-SiC which meets the requirement for bipolar high-power devices.

However, many factors can influence the crystal polytype. (Li et al, 2007) reported the effect of the seed (root-mean-square: RMS) on the crystal polytype. (Stein et al, 1992; Stein, Lanig, 1993) attributed this phenomenon for a long time to the different surface energies of Si and C faces which influenced the formation of different polytypes nuclei. In this case the sublimation of physical vapor transport system was used to grow SiC single crystal. In order to do so, the SiC powder with high purity was placed in the bottom of the crucible at the temperature range (2000-2300°C). Whereas the seed wafer was maintained at the top of the graphite crucible at the temperature range (2000-2200°C) in argon atmosphere and the pressure in the reaction chamber was kept at 1000-4000 Pa. After about ten hours of growth, three crystal slices of yellow, mixed (yellow and green) and green zone were obtained.

According to (Li et al, 2007), it was found that polytypes are seed RMS roughness dependent. In fact, the crystal color is more and more uniform with the decreasing seed RMS roughness. The yellow coloured zone corresponds to the 4H-SiC polytype, while the green zone is attributed to 6H-SiC and the mixed zones correspond to the mixture of 6H and 4H-SiC polytypes.

The X-Ray direction funder showed that the two zones were grown in different directions (<0001> and $<11\bar{2}0>$ for yellow and green zone respectively).

The following table summarizes the obtained results of the as synthesized crystals:

Properties	A: Yellow Zone 4H-SiC	B: Green zone 6H-SiC	C: Mixed zone of 4H and 6H-SiC
Raman spectra peaks (cm^{-1})	776-767-966	766-788-796-966	766-776-796-788-966
Hall Effect measurement carrier densities ($10^{16}cm^{-3}$)	6.84	0.82	–
X-Ray direction funder	<0001>	$<11\bar{2}0>$	–
The distance values between the two adjacent faces (nm) by HRTEM micrographs	0.253	0.155	–

Table 4. Raman spectra peaks, Hall Effect and HRTEM of A, B and C (Li et al, 2007).

According to (Ohtani et al, 2009), SiC power diodes and transistors are mainly used in high efficiency power system such as DC/AC and DC/DC converters. For these applications, to obtain a sufficiently low uniform electrical resistivity's and to prevent unnecessary substrate resistance, nitrogen can be easily introduced into the crystals (the growth rates employed were relatively low (0.2-0.5 mm/h)) during physical vapor transport in terms of growth temperature dependence.

4H-SiC crystals were grown on well-and off oriented 4H-SiC $(000\bar{1})$C seed crystals by the PVT growth where the doping resulted in n^+4H-SiC single crystals having bulk resistivities less than 0.01Ωcm.

The resistivity decreased while the nitrogen concentration in SiC crystals increased as the growth temperature was lowered. A 4H-SiC grown at 2125°C exhibited an extremely low bulk resistivity of 0.0015 Ωcm (lowest ever reported). The resistivity change was attributed to the formation of extremely low density of 3C-SiC inclusions and double shockly stacking faults in the substrates. Their presence depended on the surface preparation conditions of substrates then we can say that the primary nucleation sites of the stacking faults exist in the near-surface-damaged layers of substrates.

At low-growth temperatures, nitrogen incorporation was enhanced. This could be attributed to the increased nitrogen coverage of the growing surface at low temperature, or to the low-growth rates resulting from the low-growth temperatures.

On the other hand, authors have showed that the lower growth temperatures cause macrostep formation on the $(000\bar{1})$C facet of heavily nitrogen doped 4H-SiC single crystals. The crystals showed a relatively sharp X-ray rocking curve even when the crystals were doped with a nitrogen concentration of $5.3\text{x}10^{19}$-$1.3\text{x}10^{20}$ cm^{-3} and almost no indication of 3C-SiC inclusions or stacking faults was detected in the crystals. It was also shown that the peaks were quite symmetrical suggesting a good crystalinity of the crystal.

4.2 Chemical Vapor Deposition (CVD)

Chemical vapor deposition (CVD) techniques have the largest variability of deposition parameters. The chemical reactions implicated in the exchange of precursor-to-material can include thermolysis, hydrolysis, oxidation, reduction, nitration and carboration, depending on the precursor species used. During this process, when the gaseous species are in proximity to the substrate or the surface itself, they can either adsorb directly on the catalyst particle or on the surface. Thus the diffusion processes as well as the concentration of the adsorbates (supersaturation) leads to a solid phase growth at the catalyst–surface interface. (Barth et al, 2010)

CVD is one of the suitable used methods to produce SiC in various shapes of thin films powders, whiskers and nanorods using Si-C-HCl system. Amorphous fine silicon carbide powders have been prepared by CVD method in the SiH_4-C_2H_2 system under nitrogen as a carrying gas (Kavcký et al, 2000). Reaction (1) was expected to take place in the reaction zone:

$$2\,SiH_{4(g)} + C_2H_{2(g)} \longrightarrow 2SiC\,(g) + 5H_{2(g)} \qquad (2)$$

The influence of the given reaction conditions are summarized in table 5:

Sample no	Temperature (°C)	Ratio $C_2H_2:SiH_4$	Ratio C:Si	Flow rate ($cm^3\ min^{-1}$)	Colour
1	900	1.2:1	2.4:1	130	Brown
2	1100	1.2:1	2.4:1	130	Black
3	1200	1.2:1	2.4:1	130	Black
4	1250	1.2:1	2.4:1	130	Black
5	1100	1.6:1	3.1:1	178	Black
6	1100	0.9:1	1.8:1	163	Brown-Black
7	1100	2.1:1	4.2:1	143	Brown-Black

Table 5. Reaction condition (Kavcký et al, 2000)

The presence of crystallized form of SiC was proven by infrared spectroscopy (IR) and X-ray Diffraction (XRD) investigations. Whereas, chemical analysis did not detect formation of Si_3N_4 under the indicated reaction conditions. (Kavcký et al, 2000).

From the morphology examinations it was revealed that the powder products were related to the $C_2H_2:SiH_4$ ratio in the initial gas mixture. The temperature of 1100°C was an optimum to $C_2H_2:SiH_4$ ratio 1.2:1. The particle size of SiC powder was narrow in the range of 0.1-0.5 μm and consists of quasi-spherical particles. In addition the growth increased faster than the nucleation rate with temperature.

(Fu et al 2006) obtained SiC using CH_3SiCl_3 (MTS) and H_2 as precursors by a simple CVD process, without using metallic catalyst, and under atmospheric pressure. Thus, SiC nanowires with high purity and homogenous diameter were formed.

High purity argon gas was fed into the furnace to maintain an inert atmosphere. H_2 gas was used as carrier gas, which transfers MTS through a bubbler to the reactor gas as well as diluent's gas. This latter gas regulates the concentration of the mixture containing MTS vapor and carrier gas. During growth, the furnace was maintained at 1050-1150°C for 2 h under normal atmosphere pressure. The substrates were C/C composites.

The following chemical equations can be written to describe the steps leading to SiC crystals:

$$CH_3SiCl_3 + H_2 \longrightarrow CH_4 + SiCl_2 + HCl \qquad (3)$$

$$CH_4 + SiCl_2 \longrightarrow SiC + H_2 \qquad (4)$$

For CVD process using CH_3SiCl_3 (MTS) (Fu et al 2006) the obtained micrographs revealed that the nanowires generally display smooth surface and homogenous diameter of about 70 nm. The AEM image of an individual nanowire and its corresponding selected-area electron diffraction (SAED) pattern, indicated a single crystalline structure of the nanowire. The XRD investigations of nanowires revealed diffracting peaks of both graphite and cubical β-SiC corresponding to the substrate and the nanowires.

Authors reported that the width and length of nanowires were primarily controlled by the deposition temperature. At higher deposition temperature, the growth rate of SiC grains became larger and the crystal grains became bigger. In addition, a special structure with an amorphous SiO_2 wrapping layer on the surface of SiC nanowire was also found.

4.3 Sol-gel processing technique for synthesizing SiC

Sol-gel processing received extensive attention in the 1970s and early 1980s as hundreds of researchers sought after novel, low temperature methods of producing common oxide ceramics such as silica, alumina, zirconia and titania in fully dense monolithic form (Brinker et al, 1984).

The basic advantages of using the sol-gel synthesis approach have been the production of high pure product with extremely uniform and disperse microstructures not achievable using conventional processing techniques because of problems associated with volatilization, high melting temperatures, or crystallization. In addition, the production of glasses by the sol-gel method permits preparation of glasses at far lower temperatures than is possible by using conventional melting. Moreover, the sol-gel process has proved to be an effective way for synthesis of nanopowders. Finally, sol-gel approach is adaptable to producing films and fibers as well as bulk pieces.

The composites produced have included both metal-ceramic and ceramic-ceramic materials, some carefully doped with additional phases (easily performed with sol-gel process). It was found that the as obtained materials have exhibited favourable physical and mechanical properties, some of which can be attributed to their synthesis process (Rodeghiero et al, 1998).

The sol-gel process comprises two main steps that are hydrolysis and polycondensation. The first one starts by the preparation of a Silica-glass by mixing an appropriate alkoxide as precursor, with water and a mutual solvent to form a solution. Hydrolysis leads to the formation of silanol groups (SiOH) subsequently condensed to produce siloxane bonds (SiOSi). The silica gel formed by this process leads to a rigid, interconnected three-dimensional network consisting of submicrometer pores and polymeric chains. The basic structure or morphology of the solid phase can range anywhere from discrete colloidal particles to continuous chain-like polymer networks (Klein et al, 1980; Brinker et al, 1982).

After solvent removal, that requires a drying process, a xerogel is obtained accompanied by a significant shrinkage and densification. This phase of processing affects deeply the ultimate microstructure of the final component. Conversely, the network does not shrink when solvent removal occurs under hypercritical (supercritical) conditions, an aerogel is consequently produced, a highly porous and low-density material.

However, for industrial applications, this process is rather expensive, as it requires costly precursors, especially compared to the Acheson classical one, starting from sand and coke. Furthermore it is not practical to handle important liquid quantities of reagents. Additionally, the carbothermal reduction of silica is performed at temperatures around 1600°C.

As a way to improve the interest of sol-gel process, it would be interesting to prepare high reactivity precursors, and then to decrease the silica carbothermal reduction temperature and/or to increase the SiC production yield. One approach to accelerate the carbothermal reduction of silica and to increase the yield was performed by (**A**. Julbe et al, 1990), who obtained crystalline β-SiC, after pyrolysis at about 1580°C with a 3h hold, starting from colloidal silica sol and saccharose ($C_{12}H_{22}O_{11}$) as silicon and carbon sources respectively. Boric acid (H_3BO_3), soluble in aqueous solutions, was directly introduced in the sol.

The SiC powders had grain size of 100 nm in diameter and were easily sinterable (87% of theoretical density) thanks to boron containing additives resulting in submicron and homogeneous product. Moreover, boric acid added in the original colloidal sols improved

significantly the carbothermal reduction yield as well as the conversion rate and the powder crystallinity.
Another approach to enhance the carbothermal reduction yield for the SiC production consists in increasing the reactivity of the precursors. In this context, and by using also boric acid as additive, (Lj. Čerović et al, 1995) synthesised β-SiC at 1550°C by the reductive heating of gel precursors prepared from silica sol and saccharose or activated carbon as carbon sources. It was proved that for SiC formation when starting from silica and with saccharose being the carbon source, the formation of SiC started hardly at 1300°C and became intensive at 1400°C. In contrast, in the case of gels prepared from activated carbon, the crystallization of β-SiC started at 1400°C and progressed via carbothermal reduction of SiO_2 with a high crystallinity. These differences are due to the close contact between SiO_2 and C molecules obtained only if the gels are prepared using saccharose as carbon source. The same behaviour was also observed by (a White et al, 1987; b White et al, 1987) for SiC powder synthesis starting from organosilicon polymers as silicon and carbon precursors. The molecular intimacy of the SiO_2/C mixture resulted in lower temperatures of synthesis and higher surface areas of the produced SiC powders.
It was also established that even though the synthesis of SiC from gels with activated carbon progressed with greater conversion rate than when using saccharose, the boric acid addition was found to be advantageous.
(V. Raman et al, 1995) synthesizing SiC via the sol-gel process from silicon alkoxides and various carbon sources. Tetraethoxysilane (TEOS), methyltriethoxysilane (MTES) and a mixture of TEOS and MTES were used as silicon precursors whereas, phenolic resin, ethylcellulose, polyacrylonitrile (PAN) and starch were used as carbon sources. After hydrolysis the sol was kept at 40 °C for gelling, ageing and drying. The as obtained gels were then heat treated in order to synthesize silicon carbide by carbothermal reduction of silica, largely a used process at industrial scale for its relatively low coat (Schaffe et al, 1987). It was found that all the products obtained from all the precursors are β-SiC. The colour of the products ranged from light-green to greyish-black depending upon the amount of free carbon in the final product.
Since ceramic nanopowders have demonstrated enhanced or distinctive characteristics as compared to conventional ceramic material, considerable attention was devoted to the synthesis methods for nanoscale particles thanks to their potential for new materials fabrication possessing unique properties (Bouchard et al, 2006).
Among methods used to synthesize ceramic nanoparticles, sol-gel processing is considered to be one of the most common and effective used technique. (V. Raman et al, 1995) for synthesizing β-SiC with crystallite size ranging from 9 to 53 nm. This variation is attributed to the difference in the nature of carbon obtained from the various sources. Whereas, the difference observed with the same carbon source is probably due to Si-C of Si-CH_3 linkage present in MTES which is retained during carbonization and carbothermal reduction.
The materials produced by the sol-gel present rather low density compared to other synthesis methods. Indeed, samples prepared by the sol-gel technique are highly porous in nature due to the evolution of gases during carbonization and carbothermal reduction of gel precursors and thus exhibit lower densities compared with the theoretical value for SiC (3.21g /cm^3). (V. Raman et al, 1995) reported the measured samples densities for different silicon and carbon precursors and they presented a maximum density of 1.86 g /cm^3.

The porosity of these gels is the result of solvents, hydrogen, oxygen and nitrogen loss during the carbonisation of precursors resulting in the formation of porous product (silica and carbon). Table 6 summarizes the characteristics of the obtained SiC materials as a function of the different silicon and carbon sources.

Mixture	Silicon Source	Carbon Source	Colour	Crystallite size (nm)	Density (g/cm^3)
1	TEOS	Phenolic resin	Greyish-black	52.5	1.64
2	MTES	Phenolic resin	Grey	32.6	1.60
3	TEOS + MTES	Phenolic resin	Grey	52.5	1.86
4	TEOS	Ethylcellulose	Light-green	23.3	-
5	MTES	Ethylcellulose	Light-green	9	1.76
6	TEOS	PAN	Greyish-black	< 32.6	1.38
7	TEOS	Starch	Greyish-black	21.3	1.80

Table 6. Properties of SiC prepared by sol-gel process from different silicon and carbon Precursors. (V. Raman et al, 1995)

(J. Li et al, 2000) used two step sol-gel process for the synthesis of SiC precursors. The authors synthesized phenolic resin-SiO_2 hybrid gels by sol-gel technique that was used as silicon source in the presence of hexamethylenetetramine (HMTA) as catalysts.
In the first step for prehydrolysis oxalic acid (OA) was added as catalyst and the ratio OA/TEOS was investigated. OA was considered promoting hydrolysis of TEOS. Moreover it was established that the OA content as well as the prehydrolysis time determined whether gel instead of precipitate could form.
For the second step of the sol-gel process, that is gelation, HMTA was added as catalyst that resulted in a considerable reduction of the gelation time and condensation promoting.
It was considered that the hydrolysis and condensation rates of TEOS were greatly dependent upon the catalyst and the pH value (Brinker, Scherer, 1985). Thus, for pH values below 7, hydrolysis rate increased with decreasing pH, but condensation rate decreased and reached its lowest point at pH=2, the isoelectric point for silica. In both steps in the previous work the pH was bellow 7 and subsequently decreased with increasing OA content.
Given the fact that SiC, is a refractory material which shows a high thermal conductivity, and because of its properties of particle strength and attrition resistance, mesoporous SiC is expected to have extensive application in harsh environments such as catalyst, sorbent or membrane support (Methivier et al, 1998; Keller et al, 1999; Pesant et al, 2004).
Nevertheless, SiC applications as catalyst carrier was limited due to the fact that the specific surface area reachable for this material was rather low. It was shown that the sol-gel process is a promising route to prepare high surface area SiC materials. (G. Q. Jin, X. Y. Guo, 2003) have investigated a modified sol-gel method to obtain mesoporous silicon carbide. As a first step, a binary sol was prepared starting from TEOS, phenolic resin and oxalic acid. Nickel nitrate was used in the sol-gel process as a pore-adjusting reagent. Secondly, a carbonaceous silicon xerogel was formed by the sol condensation with a small amount addition of

hexamethylenetetramine in order to accelerate the condensation of the sol. Finally, mesoporous SiC was obtained with a surface area of 112 m^2/g and an average pore diameter of about 10 nm by carbothermal reduction of the xerogel at 1250°C in an argon flow for 20h. Even though, the mechanism of formation of mesoporous SiC is not yet well defined, interestingly, it was found that the surface areas and pore size distributions are nickel nitrate content dependent.

However, this process is very time consuming and necessitates special conditions, such as flowing argon ($40cm^3/min$). Additionally, the carbothermal reduction of the xerogel is carried out at 1250°C for a relatively long holding time.

(Y. Zheng et al, 2006) prepared a carbonaceous silicon xerogel starting from TEOS and saccharose as silicon and carbon sources respectively. Then the xerogel was treated by the carbothermal reaction at 1450°C for 12h to prepare a novel kind of β-SiC. The characterization of the purified sample revealed mesoporous material nature with a thorn-ball-like structure and a higher surface area of 141 m^2/g. Moreover, the as obtained mesoporous SiC material, revealed two different kinds of pores, with 2-12 nm sized mesopores in the thorn-like SiC crystalloids and 12–30 nm sized textural mesopores in the thorn-ball-like SiC.

Even though the mechanism formation of mesoporous β-SiC is not yet understood, SiC formation could be attributed to either the reaction of SiO with C or SiO with CO in the carbothermal reduction. Nevertheless, it was assumed that the thorn-ball-like β-SiC was probably produced by both reactions. Indeed, it is believed that as a first step, the reaction of gaseous silicon monoxide with carbon results in the formation of thorn-like β -SiC crystalloids whereas, the second reaction of gaseous silicon monoxide with carbon monoxide contributes to the thorn-like β -SiC crystalloids connections to form the thorn-ball-like β -SiC.

(R. Sharma et all, 2008) reported a new and simple sol-gel approach to produce a simultaneous growth of nanocrystalline SiC nanoparticles with the nanocrystalline silicon oxide using TEOS, citric acid and ethylene glycol. After the gel development, a black powder was obtained after drying at 300°C. The powder was subsequently heat treated at 1400°C in hydrogen atmosphere. Interestingly, it was found that under these working conditions, crystalline silicon oxide was formed instead of amorphous silicon oxide which is normally found to grow during the gel growth technique.

On the other hand, SiC is considered to be one of the important microwave absorbing materials due to its good dielectric loss to microwave (Zou et al. 2006). In microwave processing, SiC can absorb electromagnetic energy and be heated easily. It has a loss factor of 1.71 at 2.45 GHz at room temperature. And the loss factor at 695°C for the same frequency is increased to 27.99. This ability for microwave absorption is due to the semiconductivity of this ceramic material (Zhang et al, 2002).

Moreover, SiC can be used as microwave absorbing materials with lightweight, thin thickness and broad absorbing frequency. Since pure SiC posses low dielectric properties that gives barely the capacity to dissipate microwave by dielectric loss, therefore, doped SiC was used in order to enhance the aimed properties. The most applied technique for N-doped SiC powder consists in laser-induced gas-phase reaction (D. Zhao et al, 2001; D.-L. Zhao et al, 2010).

(D. Zhao et al, 2001) prepared nano-SiC/N solid solution powders by laser method. The dielectric properties were measured at a frequency range of 8.2-12.4 GHz.

The interaction between electromagnetic waves and condensed matter can be described by using complex permittivity, ε^* ($\varepsilon^* = \varepsilon' + i\varepsilon''$, where ε' being the real part and ε'' the imaginary part).
Table 7 gives the dielectric constants and the dissipation factors tg δ of Si/C/N compared to bulk SiO_2, hot-pressed SiC, Si_3N_4 and nano SiC particles (D. Zhao et al, 2001).

	Dielectric constants (ε)	Dissipation factors (tg δ)	Frequency (GHz)
Nano-SiC/N solid solution powder	ε' = 46.46-27.69 ε'' = 57.89-38.556	1.25-1.71	8.2-12.4
Nano-SiC/N solid solution powder embedded in paraffin wax matrix	ε' = 5.79-6.33 ε'' = 3.51-4.31	0.61-0.68	8.2-12.4
Nano SiC particles embedded in paraffin wax matrix	ε' = 1. 97-2.06 ε'' = 0.09-0.19	0.045-0.094	8.2-12.4
Bulk SiO_2	8.32	0.12	10
Hot-pressed SiC	9.47	0.003	10
Bulk Si_3N_4	3.8	0.002	10

Table 7. Dielectric properties of nano-SiC/N solid solution powders compared to bulk SiO2, hot-pressed SiC, Si_3N_4 and nano SiC particles (D. Zhao et al, 2001).

The dielectric properties of the nano-SiC/N solid solution powder are very different from those of bulk SiC, Si_3N_4, SiO_2 and nano SiC. The ε', ε'' and tg $\boldsymbol{\delta}$ of nano-SiC/N solid solution powder are much higher than those of nano SiC powder and bulk SiC, Si_3N_4 and SiO_2, particularly the tg $\boldsymbol{\delta}$. The promising features of nano-SiC/N solid solution powder would be attributed to more complicated Si, C, and N atomic chemical environment than in a mixture of pure SiC and Si_3N_4 phase. In fact, in the as obtained solid solution powder, the Si, C, and N atoms were intimately mixed. Even though, the amount of dissolved nitrogen in SiC-N solid solution has not been studied efficiently, where (Komath, 1969) has reported that the nitrogen content of the solid solution is at most about 0:3 wt%, it is supposed that the amount of dissolved nitrogen is larger than that reported. Consequently the charged defects and quasi-free electrons moved in response to the electric field, and a diffusion or polarization current resulted from the field propagation. Since there exists graphite in the nano Si/C/N composite powder, some charge carries are related to the sp^3 dangling bonds (of silicon and carbon) and unsaturated sp^2 carbons. Whereas, the high ε'' and loss factor tg δ were due to the dielectric relaxation.
Owing to the above mentioned sol-gel process advantages, nanocomposite materials are good candidates for sol-gel processing. Moreover, the doping of SiC can be carried out by the homogeneous sol system derived from initial liquid components, and that the sol-gel processing is rather simple.
(B. Zhang et al, 2002), synthesized nano-sized SiC powders by carbothermal reduction of SiO_2 and SiO_2–Al_2O_3 xerogels. This latter was prepared by mixing TEOS, saccharose and some Al_2O_3 powders. The xerogels were subsequently heated at 1550°C for 1h in argon or nitrogen atmosphere to synthesize SiC. It was found that aluminum and nitrogen have

important effects on the polytypes of SiC powders. In the presence of aluminium, the polytype of 12H SiC powders were obtained, whereas, 21R SiC was synthesized under the nitrogen atmospheres (table 8). During the synthesis of silicon carbide, Al_2O_3 is reduced by carbon and forms carbide. At the same time, aluminium dopes into SiC and forms solid solution. Thus, aluminium atoms replace atoms of silicon in the solid solution and induce vacancies of carbon. The lattice parameters were decreased with the increasing of aluminium content. On the contrary, when SiC powders are synthesized in nitrogen atmosphere, nitrogen atoms replace some carbon atoms and form silicon vacancies. The synthesized β-SiC powder has much higher relative permittivity (ε'_r = 30~50) and loss tangent (tg δ = 0.7~0.9) than all of the α-SiC powders, though the α-SiC powders with 5.26 mol% aluminium possess higher conductivities. In fact it was establish that for Al-doped SiC powders, the relative permittivities and loss tangents are in an opposite measure of the aluminium content. For the powders with the same aluminium content, the samples synthesized in nitrogen atmosphere have smaller values for ε'_r and tg δ than those obtained in argon atmosphere in the frequency range of 8.2-12.4 GHz. The fundamental factor on these dielectric behaviours is ion jump and dipole relaxation, namely the reorientation of lattice defect pairs (V_{Si}–V_C , Si_C–C_{Si}). In fact, aluminium and nitrogen decrease the defect pairs that contribute to polarization. With the increase of the aluminium content and the doping of nitrogen, the conductivity of SiC rises, but the relative dielectric constant and loss tangent decrease.

Sample	Atom ratio of Al-Si	Reaction atmosphere	SiC polytype	DC resistivity (Ω cm)	Calculated loss tangent for 10 GHz
1	0	Argon	3C	557.9	8.00×10^{-2}
2	2.63:100	Argon	12H	1803.5	8.78×10^{-2}
3	5.26:100	Argon	12H	511.3	3.57×10^{-1}
4	2.63:100	Nitrogen	21R	1181.7	1.84×10^{-1}
5	5.26:100	Nitrogen	21R	77.5	4.06×10^{-1}

Table 8. DC resistivities and calculated loss tangent of SiC powders as a function of the alumina content (B. Zhang et al, 2002).

Besides the p-type doping by Al, boron atoms can substitute preferably the silicon atoms of SiC lattice. (Z. Li et al, 2009) investigated the effects of different temperatures on the doping of SiC with B. The authors synthesized B-doped SiC powders by sol-gel process starting from the mixture sol of TEOS and saccharose as silicon and carbon sources, respectively, and tributyl borate as dopant at 1500 °C, 1600 °C, 1700 °C and 1800 °C. It was proved that C-enriched β-SiC is completely generated when the temperature is 1700°C and SiC(B) solid solution is generated when the temperature is 1800 °C. The powders synthesized at 1700 °C had fine spherical particles with mean size of 70 nm and narrow particle size distribution. On the contrary, few needle-like particles were generated in the powders synthesized at 1800 °C which is believed to be caused by the doping of B. Thus it is considered that the formed SiC(B) solid solution suppresses the anisotropic growth of SiC whiskers.
The electric permittivities of SiC samples were determined in the frequency range of 8.2–12.4 GHz. Results showed that the SiC(B) sample has higher values in real part ε' and imaginary part ε" of permittivity. The average values of ε' and ε" for the sample synthesized

at 1700 °C were 2.23 and 0.10, respectively. It was also noticed that both ε′ and ε″ have increased for the sample synthesized at 1800 °C, and particularly the ε″ was nearly 2.5 times greater than that of the sample synthesized at 1700 °C. This can suggest an improved capacity of dielectric loss in microwave range.
The basic factor on these dielectric behaviours is that for a temperature heat of 1800°C which generates the SiC(B) solid solution, there exist bound holes in SiC with acceptor doping. Under the alternating electromagnetic field, these bound holes will migrate to and fro to form relaxation polarization and loss, thus leading to higher ε′ and ε″ of the sample at 1800 °C.

4.4 Liquid phase sintering SiC technique

Since SiC possesses a strongly covalent bonding (87%), which is the source of intrinsically high strength of SiC sintered bodies, thus, it is difficult to obtain a fully dense bulk material without any sintering additives.
The sintering of SiC is usually performed at very high temperatures which could reach 2200°C in the solid state with the addition of small amounts of B and C. However, SiC based ceramics boron- and carbon-doped have poor or, at the best, moderate mechanical properties (flexural strength of 300-450 MPa and fracture toughness of 2.5-4 $MPa.m^{1/2}$) (Izhevsky et al, 2000).
Over the last three decades considerable effort has been spent in order to decrease the sintering temperature and to enhance the mechanical properties of silicon carbide ceramics. This effort was based on using sub-micron size, highly sinterable SiC powders and sintering additives able to advance the density of the bulk material and more over to improve or at least to preserve the relevant mechanical properties.
The most effective sintering aids in lowering the sintering temperature and providing the micostructure resistant to crack propagation have been completed by adding the metal oxides Al_20_3 and Y_2O_3 (Omori, Takei, 1982).
Thus the addition of suitable sintering additives leads to dense, fine grained microstructures that result in improved sintered material strength. Nevertheless, these additives are subject to secondary phases' formation at the grain boundaries that frequently cause loss of high temperature strength (Biswas, 2009).
The innovative approach of SiC sintering in the presence of a liquid phase was introduced by (Omori, Takei, 1988) in the early 1980s by pressureless sintering of SiC with Al_20_3 combined with rare earth oxides. Since this initiation several works have been done and have demonstrated that whatever is the starting silicon carbide phase (α or β), SiC was successfully highly densified by pressureless sintering with the addition of Al_20_3 and Y_2O_3 at a relatively low temperature of 1850 ~ 2000°C.
Since its initiation, liquid-phase-sintered (LPS) silicon carbide with metal oxide additives such as Y_2O_3 and Al_20_3 has attracted much attention because it possesses a remarkable combination of desirable mechanical, thermal and chemical properties making it a promising structural ceramic.
One approach to enhance the mechanical properties of β-SiC is to control the gas atmosphere during the sintering. For instance it has been found that sintering LPS SiC in N_2 atmosphere suppresses the α→ β phase transformation and their grain growth, while Ar atmosphere enhances this phase transformation by the formation of elongated grains (Nader et al, 1999; Ortiz et al, 2004).

In this context, (Ortiz et al, 2004) studied the effect of sintering atmosphere (Ar or N_2) on the room- and high-temperature properties of liquid-phase-sintered SiC. It was shown that LPS SiC sintered in N_2 atmosphere possesses equiaxed microstructures and interestingly nitrogen was incorporated in the intergranular phase making the LPS SiC ceramic highly refractory. Moreover, this results in coarsening-resistant microstructures that have very high internal friction (Ortiz et al, 2002).

Two individual batches were prepared, each one containing a mixture of 73.86 wt.% β-SiC , 14.92 wt.% Al_2O_3 and 11.22wt.% Y_2O_3 in order to result in 20 vol.% yttrium aluminium garnet (YAG) in the LPS SiC bodies. Several pellets were prepared and then cold-isostatically pressed under a pressure of 350 Pa before being sintered at 1950°C for 1h in either flowing Ar or N_2 gas atmospheres. Table 9 summarizes the studied mechanical properties at room temperature of the LPS SiC ceramics in Ar atmosphere (Ar-LPS SiC) and in N_2 atmosphere (N_2-LPS SiC) where both materials have densities in excess of 98% of the theoretical limit of 3.484 g cm^{-3}.

At room temperature the microstructure of the LPS SiC sintered in N_2 atmosphere, was characterized by equiaxed grains, as compared with the LPS SiC in Ar-atmosphere which presented rather highly elongated SiC grains.

	Vickers hardness H (GPa)	Vickers indentation toughness K_{IC} (MPa.$m^{1/2}$)	Hertzian indentation
Ar-LPS SiC	17.6±0.3	3.3±0.1	**"quasi-ductile" material**
N2-LPS SiC	**20.7±0.6**	**2.2±0.1**	**Less "quasi-ductile" material**

Table 9. Mechanical properties of the SiC ceramics sintered in flowing Ar and N_2 gas atmospheres (Ortiz et al, 2004).

At high temperature (1400°C) the LPS SiC specimens presented equiaxed-grained structures with a highly viscous N_2-LPS SiC intergranular phase. This microstructure resulted in high resistance to high temperature deformation to a greater extent than Ar-LPS SiC, before the ultimate compressive strength is reached. Table 10 reports the mechanical properties of LPS SiC ceramics in Ar and N_2 gas atmospheres, measured at 1400°C.

	Elasticity limit deformation % ε_e	Ultimate compression strength σ_{UCS} (MPa)	Total strain at catastrophic failure % ε_F
Ar-LPS SiC	~ 1.8	630	**~ 11.4**
N_2-LPS SiC	**~ 1.8**	**870**	**~ 6**

Table 10. Mechanical properties of LPS SiC ceramics in Ar and N_2 gas atmospheres measured at 1400°C (Ortiz et al, 2004).

The relevant contrast in the mechanical properties of the LPS SiC in different atmospheres (N_2 and Ar) was argued to be due to the elongated-grained microstructure and the less viscous intergranular phase devoid of nitrogen.

In another approach (Padture, 1994) succeeded in enhancing the fracture toughness, by seeding the β-SiC powder with 3~5 wt. % of α-SiC through the LPS SiC technique. Additionally, the temperatures were maintained higher than 1900°C for different durations in order to allow the growth of elongated α-SiC grains. Nevertheless, this process despite the fact provides SiC ceramics with improved fracture toughness, it involves high temperatures (> 1900°C) and consequently beats the principle of the LPS process for being low temperature as well as low energy consumption.

(Wang, Krstic, 2003) studied the effect of Y_2O_3 addition and the total oxide volume fraction (Y_2O_3 + Al_2O_3) on mechanical properties of pressureless sintered β-SiC ceramics at low temperature. It was demonstrated that the increase in strength with yttria (Y_2O_3) content is directly related to the increase in relative density of the sintered specimens, which, in turn is related to the level of Y_2O_3 addition. The study of the fracture toughness of SiC ceramics with oxide addition revealed that with increasing the oxide content, the fracture toughness increases and reached a maximum of about 4.3 MPa. $m^{1/2}$ achieved at ~ 14 vol.% of oxide added for samples sintered at 1850°C. This is mostly related to crack deflection mechanism's toughening which occurs when the crack changes its direction as it encounters the SiC grains and the grain boundary phase. Moreover, the mechanism of toughening is further enhanced by thermal mismatch between SiC and the intergranular YAG phase at the grain boundaries this leads to the crack's progress along the grain boundaries (Kim D. Kim C., 1990).

An alternative promising strategy to improve the mechanical properties of β-SiC is to adjust the volume fraction and composition of the boundary phase so as to generate the microstructure with high density and resistance to crack propagation. The variation of the α-SiC and β-SiC proportions of the starting powders mixture is considered to be an efficient way of adjusting both the microstructure and the mechanical properties of the SiC ceramic. (Wang, Krstic, 2003).

According to this approach, (Lee et al, 2002) investigated the effect of the starting phase of the raw material on the microstructure and fracture toughness of SiC ceramic by varying the mixed ratio of α-SiC and β-SiC powders. The authors prepared several samples by altering the β/α phase ratio of SiC starting powder from 0 to 100 vol. % and adding 2 mol % of yttrium aluminium garnet and 2.5 wt. % polyethyleneglycol (PEG). The mixtures were then compacted and hot pressed at 1850°C for 30 min at a pressure of 50 MPa and subsequently sintered at 1950°C for 5h. The densities of the resulting compacted powders were higher than 95 % of theoretical density regardless of the starting phase. Most sintered SiC specimens enclosed elongated grains with rod-like type. However it was noticed that the amount of elongated grain and its aspect ratio was changed with the ratio of α- anf β-SiC in the starting powder. Elongated grains were formed by the β→α phase transformation with a 4H polytype and anisotropic grain growth during heat treatment. It was concluded from this study, that specimen containing 50 vol.% β-SiC in the starting powder showed the highest values of volume fraction, maximum length and aspect ratio for elongated grains. This specimen revealed also the highest fracture toughness of 6.0 MPa. $m^{1/2}$ which is due to the elongated grains induced crack deflection during crack propagation.

In table 11 are reported some mechanical properties of LPS SiS as a function of different sintering aids.

References	Technique	Sintering additives	Hardness H (GPa)	Fracture Toughness (MPa. $m^{1/2}$)	Density
(Chen, Zeng, 1995)	Pressureless sintering SiC	Al_2O_3 + HoO3 (eutectic composition)	17.47	3.68	Up to 3.764 g/cm^3
(Chen, 1993)	Pressureless sintering SiC	Al_2O_3 + SmO3 (eutectic composition)	17.1	4.6	92.6 %
(Hidaka et al,2004)	Hot pressing at 1950°C and P=39 MPa	(Al_2O_3 + Y^{3+} ions) /Polytitanocarbosilane by infiltration	19 - 21	5.9	95 – 98 %
(Scitti et al, 2001)	Hot pressing at 1850-1950°C Annealing (1900°C/3 or 2h)	Al_2O_3 + Y_2O_3	22 24.5-25	2.95 – 3.17 Up to 5.5	Up to 99.4 % (3.24 g/cm^3) 3.22 g/cm^3
(Wang, Krstic, 2003)	Pressureless sintering β-SiC at 1850°C	Y_2O_3 in (Al_2O_3 + Y_2O_3)	22	4.3	~ 98 %
(Mulla, Krstic, 1994)	Pressureless sintering β-SiC at 2050°C	Al_2O_3	_	6	97-98 %
(Hirata et al, 2010)	Hot pressing at 1900-1950°C and P=39 MPa	Al_2O_3 + Y_2O_3	_	6.2	97.3-99.2

Table 11. Mechanical properties of LPS-SiC as a function of different sintering aids.

It could be concluded that, densities in excess of 99% of the theoretical limit can be easily achieved by carefully choosing the composition of the liquid phase and the packing powder configuration (Jensen et al, 2000).

On one hand, the sintering mechanism in the pressureless liquid-phase sintering to full dense SiC with Al_2O_3 and Y_2O_3 additions is considered to be attributed to liquid-phase-sintering via the formation of an eutectic liquid between Al_2O_3 and Y_2O_3 to yield yttrium aluminium garnet, or YAG, based liquid phase. On the other hand, the hardness was mainly related to reduction of secondary phases which generally decrease such property.

Finally, the increase in toughness is related not only to grain morphology but also to second phase chemistry, thus a suitable choice of thermal treatment parameters that modify second phase chemistry without excessive grain growth can theoretically lead to a reinforced microstructure with slight or no strength decrease.

4.5 Mechanical alloying process for SiC synthesis

Nanostructured silicon carbide attracted considerable attentions from the materials and device communities and have been studied intensively in the past decade due to its unique properties and wide applications in microelectronics and optoelectronics (Li et al, 2007).

In fact, it was reported that inorganic structures confined in several dimensions within the nanometer range, exhibit peculiar and unique properties superior to their bulk counterparts. These unique properties can be attributed to the limited motion of electrons in the confined dimensions of the nanomaterial (Barth et al, 2010). However the transition from fundamental science to industrial application requires an even deeper understanding and control of morphology and composition at the nanoscale. Size reduction of well known materials into the nanometre regime or the realization of novel nanostructures can improve device performance and lead to novel discoveries (Barth et al, 2010).

Among the techniques that produce nanostructured materials the mechanical alloying is considered as a powerful process for producing nanomaterials at room temperature with low cost and at a large scale (Basset et al, 1993). However, this process also exhibits disadvantages such as the contamination of both the milling media and the mill atmosphere. During mechanical alloying, two essential processes are involved: cold welding and fracturing of powder particles. Thus after a certain activation time during this process, while the particles size decreases during milling, the number of chemically active defect sites increases. Moreover, towards the milling process which results in repeated fracturing and rewelding of particles the activated zones are continuously increased leading to a considerable temperature decrease for reactions occurring in a planetary mill. For more details see (D. Chaira et al, 2007; Benjamin, 1970). Thus, the normally proceeded reactions at high temperature become possible at low temperature with this technique which constitutes the major advantage of the MA. Then, the restriction caused by thermodynamic phase diagram is overcome by mechanical alloying.

Mechanical alloying as a complex process, it involves therefore, the optimization of a number of variables to achieve the desired product phase and/or microstructure.

The most important parameters that have an effect on the final constitution of the powder are:

. type of mill,
. milling container,
. milling speed,
. milling time,
. type, size, and size distribution of the grinding medium,
. ball-to-powder weight ratio,
. extent of filling the vial,
. milling atmosphere,
. process control agent,
. number and diameter of balls,
. temperature of milling.

Actually these parameters are not completely independent. For more details see (Suryanarayana, 2001).

The applications of MA are numerous, for instance it was used for the synthesis of nanostructured carbides (El Eskandarany et al, 1995; Ye, Quan, 1995; Razavi et al, 2007), nitrides (Calka et al, 1992), composites (Shen et al, 1997) and solid solutions (Li et al, 2002).

Since the work initiated by (Benjamin, 1970; Benjamin, Volin, 1974), MA has attracted more and more attentions due to the all above cited advantages. (Le Caer et al, 1990) synthesized a series of metal carbides and silicides by ball milling mixtures of elemental powders at room temperature with a vibratory mill at fixed miling durations. Whereas, (D. Chaira et al, 2007) used a specially built dual-drive planetary mill to synthesize nanostructured SiC starting from elemental silicon and graphite mixed in 1:1 atomic ratio. The used mill design follows closely the patent design of (Rajamani et al, 2000). The weighed ball-to-powder ratio was 20 to 1 and the mechanical alloying was performed in a planetary ball mill with a critical speed of 63% by using two different balls' diameter (6 and 12 mm). It was shown that in both cases and for the end product milled for 40h, peaks corresponding to Si_5C_3, SiC and β-SiC were detected. Nevertheless, the rate of SiC formation during milling as well as the impact's intensity to dislodge any coating that forms on the ball's surface was found to be better with 12 mm diameter ball. The mechanism of SiC formation under the experimental conditions indicated above corresponds to the following: during the early stage of milling (10h), the powder particles were mechanically activated by Si and C particles mixing forming thus composite particles. Later on, after about 20h of milling, the solid-state reaction started and significant amounts of SiC are produced. Based on the X-ray line broadening it was concluded that the product compound has reduced crystallite size as well as accumulation of lattice strain. The crystallite size was found to be decreased to 10 nm after 40h of milling. Whereas, (El Eskandarany et al, 1995), while using a commercially available vibrating ball mill obtained pure stoichiometric β-SiC powders after 300h of mechanical alloying duration. They started from a mixture of equiatomic Si and C with a ball-to-powder weight ratio of 6 to 1. Under the selected mechanical alloying conditions and at the early stage of MA (after about 6h), the TEM observation revealed that the powders are still a mixture of polycrystalline Si and C. After about 12h of MA the Si and C powders particles were blended together to form composite particles of Si/C. For the alloy MA for 48h a solid-state reaction took place resulting in the β-SiC phase formation. Finally the complete β-SiC phase powder was obtained after 300h of MA which exhibited a homogeneous, smooth spherical shape with an average particle diameter of less than 0.5 μm. A fine cell-like structure with nanoscale dimensions of about 7 nm was also obtained. Table 12 summarizes the mechanism of β-SiC formation with the selected MA parameters in (El Eskandarany et al, 1995).

MA Durations (ks)	**Powder constitution**	**Morphology**	**Particle size**
0	Si and C elemental powders	bulky	300-350 mesh
86	Si/C composite particles	Rod or ellipsoid	6 μm
360	β-SiC and unreacted Si and C	Uniform equiaxed particles	< 1 μm
1080	Pure β-SiC	Spherical with smooth surface	< 0.5 μm

Table 12. The mechanism of β -SiC formation during MA (El Eskandarany et al, 1995).

It is worth to note that, as in this study when a sapphire vial and balls were used there were non contaminants detected coming from the milling media except 0.8 at. % content of oxygen detected in the as milled product, coming probably from the atmosphere medium. In order to study the stability of the MA β-SiC at elevated temperatures, the as obtained end product was annealed at 1773 K and it was demonstrated that there was no transformation to any other phases up to this temperature.
(Lu, Li, 2005) investigated the effect of different Si powders for the synthesis of $Si_{50}C_{50}$ by MA using a conventional planetary ball mill. In order to do so, both Si wafers previously milled for 5h into small particles (particle size of about 1 μm) and commercially available Si powder were separately ball milled with graphite powder under the same experimental conditions. It was shown that the end products obtained from different Si powders were almost similar. Under the selected milling conditions and with a powder-to-ball mass ratio of about 1:20, the SiC phase starts to form after about 100h of milling duration. The reaction between C and Si was gradual and completed after 180h of MA where pure and fine SiC powders were obtained.
(Aberrazak, Abdellaoui, 2008) obtained equiatomic nanostructured SiC starting from elemental Si and C powders, using a commercialized planetary ball mill with milling conditions corresponding to 5.19 W/g shock power (M. Abdellaoui, E. Gaffet, 1994; M. Abdellaoui, E. Gaffet, 1995). Under the optimized milling conditions a pure SiC phase was formed after just 15h. The reaction between C and Si was proved to be gradual and started after about 5h of MA with a SiC weight content around 1 %. This proportion increased to about 66 % after 10h alloying and to about 98 % after 15 h alloying. Contrary to the previous cited works a steady state was reached by (Aberrazak, Abdellaoui, 2008) after only 20 h of alloying duration and was characterized by a crystallite size of about 4nm. The SEM characterization revealed that the powder exhibited homogeneous distribution of the particles with 0.3 μm in size.
(Ghosh, Pradhan, 2009) reproduced the same milling duration for the synthesis of a pure β-SiC phase by high-energy ball-milling. It was revealed from this study that a development of thin graphite layer and amorphous Si were detected after 3 h of milling. Whereas, the formation of a nanocrystalline SiC phase started, by re-welding mechanism of amorphous Si and graphite layers, within 5 h of milling. Thanks to the microstructural characterization of the mechanically alloyed sample by Rietveld's analysis, it was shown that at the early stage of milling, graphite layers are distributed on the Si nano-grain boundaries as very thin layers. Finally, the complete formation of nanocrystalline SiC phase was accomplished after 15 h of milling. For the Rietveld analysis investigation, conversely to the fitting done without considering the amorphous phase contribution, the fitting profile presented better quality when amorphous silicon was taken in account.
Same calculated lattice parameters of β-SiC are reported in table 13.

β-SiC Lattice constant	**(El-Eskandarany et al, 1995) calculated**	**(Aberrazak, Abdellaoui, 2008) calculated**	**(Li, Bradt, 1986) Reported literature**	**(Liu et all, 2009) Reported literature**
a = b= c (nm)	0.4357	0.4349	0.4358	0.4348

Table 13. β-SiC lattice parameter as calculated for materials obtained by MA and as reported in the literature.

The lattice parameters variations revealed in table 2 is mainly related to the severe deformation on powder particles which occurs during the mechanical alloying. Moreover, with increasing the milling time, the crystalline defects, such as point defects and dislocations, increase too.

Mechanically alloyed SiC can also be used as reinforcement in the Al matrix. In fact, Al/SiC has received particular interests during recent years due to their high specific modulus, high strength and high thermal stability. One of the largest applications of this composite is the automobiles industry such as electronic heat sinks, automotive drive shafts, ground vehicle brake rotors, jet fighter aircraft fins or explosion engine components (Clyne, Withers, 1993).

Due to Al/SiC remarkable properties, considerable effort has been developed for its fabrication. The most used technique for composites fabrication is casting. However, even though this technique is the cheapest one, it is difficult to apply for the synthesis of Al/SiC composites owing to the extreme gap difference in the thermal expansion coefficients between Al and SiC. In addition there is a poor wettability between molten Al (or Al alloys) and SiC. Moreover, undesirable reaction between SiC and molten Al, might occurs resulting in brittle phases' formation of Al_4C_3 and Si. In order to circumvent to these main drawbacks, solid state process such as mechanical alloying (MA) is considered as a promising way to avoid brittle phases and particle agglomerations during fabrication of Al/SiC composite (Saberi et al, 2009). (Chaira et al, 2007) prepared Al/SiC composite by mixing nanostructured SiC powder, obtained by MA, with Al particles. The mixture was then sintered at 600°C for 1h to obtain Al-10 vol.% SiC composite. The studied SEM micrographs revealed a good compatibility particles as well as a good interfacial bond between Al matrix and SiC respectively. Moreover, no cracks or voids were present. It is worth to know that the interface between the metal and the particle is decisive for the determination of the overall properties of metal matrix composite. In fact, it was reported that a well-bonded interface facilitates the efficient transfer and distribution of load from matrix to the reinforcing phase, which results in composite strengths enhancement. For the two studied specimens differing in SiC proportion in the prepared composite (10 and 20 vol. %), a maximum of 90 % of theoretical density was achieved.

The co-effect of varying process parameters of milling on the evolution of powder's properties was not well established up to now. (Kollo et al, 2010) investigated the effect of milling parameters on the hardness of Al/SiC composite obtained by milling aluminium powders with 1 vol. % of nanostructured silicon carbide using a planetary ball mill. The varied milling parameters were such like rotation's disc speed, milling time, ball diameter, ball-to-powder ratio and processing control agent (stearic acid or heptane).

The studied parameters revealed that the hardness of the compacted materials when milled with heptane, as processing control agent, was not greatly influenced by the input energy. Besides, there was no sticking of aluminium on the vessel and the balls. Whereas, when using stearic acid, a remarkable difference of hardness response was found. Indeed, it is well established that stearic acid has a tendency to react during high-energy milling, in introducing carbon to the powder mixture. Consequently, it is supposed that reaction kinetics in addition to mixing is contributing to the hardening of the densified materials where a density range of 92-95 % of the theoretical was achieved. On the other hand, it was shown that the milling speed has a higher influence when the milling is performed with smaller balls, yielding thus to higher hardness values. It was also found that the hardness generally increases with the time and the speed of milling, while, milling speed has a higher influence when the milling is performed with smaller balls. Whereas, for vial filling volume, depending on the ball size, a local minimum in filling parameter was found.

Table 14 reports Vickers hardness of different Al/SiC composites prepared by MA.

Al/SiC composition	Technique	Vickers hardness (Hv)	Reference
Al-20 vol.% SiC	(Al + nanoSiC) Sintered at 600°C for 1h	40	Chaira et al, 2007
Al/SiC composite	SiC incorporated by mechanically stirring the fully molten Al	36±2 - 39±1	Tham et al, 2001
Al–1 vol.% nano SiC	(Al + nanaoSiC) hot pressed	163	Kolloa et all, 2010

Table 14. Microhardness of different Al/SiC composites obtained by MA.

(Chaira et al, 2007), demonstrated that with increasing sintering temperature, the hardness of Al-SiC composites increased too due to good compatibility of Al and SiC particles. However, the hardness values of the obtained composite remained by far lower than the one given by (Kolloa et all, 2010) who had studied and optimized the milling's parameters on the hardness of the material. Moreover, a better density was also achieved, a property which is also related to the hardness of the material.

5. Conclusion

Silicon carbide can occur in more than 250 crystalline forms called polytypes. The most common ones are: 3C, 4H, 6H and 15R. Silicon carbide has attracted much attention a few decades ago because it has a good match of chemical, mechanical and thermal properties that makes it a semiconductor of choice for harsh environment applications. These applications include high radiation exposure, operation in high temperature and corrosive media. To obtain high-performance SiC ceramics, fine powder with narrow particles-size distribution as well as high purity are required. For this purpose, many effective methods have been developed.

The simplest manufacturing process of SiC is to combine silica sand and carbon in an Acheson graphite electric resistance furnace at temperatures higher than 2500 °C. The poor quality of the obtained product has limited its use for abrasive.

Sol-gel process has proved to be a unique method for synthesis of nanopowder, having several outstanding features such as high purity, high chemical activity besides improvement of powder sinterability. Nevertheless, this process suffers time consuming and high cost of the raw materials. On the other hand, mechanical alloying is a solid state process capable to obtain nanocrystalline silicon carbide with very fine particles homogeneously distributed at room temperature and with a low coast. Moreover this process has a potential for industrial applications.

Liquid-phase-sintered ceramics represent a new class of microstructurally toughened structural materials. Liquid phase sintering technique, for instance, is an effective way to lower the sinterability temperature of SiC by adding adequate additives in the appropriate amounts. In fact, as the main factors affecting the improvements of the mechanical properties of the LPS-SiC, depend on the type and amount of sintering aids these latter have to be efficiently chosen. Whereas, physical vapor transport technique is versatile for film depositions and crystals growth. One of the large applications of PVT technique is crystalline materials production like semi-conductors. Indeed this method was considered to be the most popular and successful for growing large sized SiC single crystals.

6. References

a. Abdellaoui M., Gaffet E., (July, 1994), A mathematical and experimental dynamical phase diagram for ball-milled $Ni_{10}Zr_7$, *Journal of Alloys and Compounds*, 209, 1-2, pp: 351-361.

b. Abdellaoui M., Gaffet E., (March 1995), The physics of mechanical alloying in a planetary ball mill: Mathematical treatment , *Acta Metallurgica et Materialia*, 43, (3), pp: 1087-1098.

Abderrazak H., Abdellaoui M., (2008), Synthesis and characterisation of nanostructured silicon carbide, *Materials Letters*, 62, pp: 3839-3841.

Augustin G., Balakrishna V., Brandt C.D., Growth and characterization of high-purity SiC single crystals, *Journal of Crystal Growth,* 211, (2000), pp: 339-342.

Barrett D.L., McHugh J.P., Hobgood H.M., Hobkins R.H., McMullin P.G., Clarke R.C., (1993), Growth of large SiC single crystals, *Journal Crystal Growth*, 128, pp: 358-362.

Barth S., Ramirez F. H., Holmes J. D., Rodriguez A. R., (2010), Synthesis and applications of one-dimensional semiconductors, *Progress in Materials Science*, 55,pp: 563-627.

Basset D., Mattieazzi P., Miani F., (August, 1993), Designing a high energy ball-mill for synthesis of nanophase materials in large quantities, *Materials Science and Engineering: A*, 168, 2, pp: 149-152.

Benjamin J. S., (1970), Dispersion strengthened superalloys by mechanical alloying, *Metallurgical transactions*, 1, 10, pp: 2943-2951.

Benjamin J. S., Volin T. E., (1974), The mechanism of mechanical alloying, Metallurgical and Materials Transactions B, 5, 8, pp: 1929-1934

Biswas K., (2009), Liquid phase sintering of SiC-Ceramic, *Materials science Forum*, 624, pp: 91-108.

Brinker C.J., Clark D.E., Ulrich D.R. (1984) (Eds.), Better Ceramics Through Chemistry, North-Holland, New York.

Brinker C.J., Keefer K.D., Schaefer D.W., Ashley C.S., (1982), Sol-Gel Transition in Simple Silicates, *Journal of Non-Crystalline Solids*, 48, pp.47-64

Brinker C. J., Scherer G. W., (1985), Sol→gel→glass: I. Gelation and gel structure. *Journal of Non-Crystalline Solids*, 70, pp: 301-322.

Bouchard D., Sun L., Gitzhofer F., Brisard G. M., (2006), Synthesis and characterization of $La_{0,8}Sr_{0,2}MO_{3-\delta}$ (M = Mn, Fe or Co) cathode materials by induction plasma technology, *Journal of thermal spray and technology*, 15(1), pp: 37-45.

Calka A, Williams J. S., Millet P., (1992), Synthesis of silicon nitride by mechanical alloying, *Sripta Metallurgica and Materiala*, 27, pp: 1853-1857

Casady J.B., Johonson R.W., (1996), Status of silicon carbide (SiC) as a wide-bangap semiconductor for high-temperature applications, A Review, *Solid State Electronics*, 39, pp: 1409-1422.

Čerović Lj., Milonjić S. K., Zec S. P, (1995), A comparison of sol-gel derived silicon carbide powders from saccharose and activated carbon, *Ceramics International*, 21, 27 l-276.

Chaira D., Mishra B.K., Sangal S., (2007), Synthesis and characterization of silicon carbide by reaction milling in a dual-drive planetary mill, *Materials Science and Engineering A,* 460–461, pp: 111–120.

Chen Z., (1993), Pressureless sintering of silicon carbide with additives of samarium oxide and alumina, *Materials Letters*, 17, pp: 27-30.

Chen Z., Zeng L., (1995), Pressurelessly sintering silicon carbide with additives of holmium oxide and alumina, *Materials Research Bulletin*, 30(3), pp. 256-70.

Clyne T. W., Withers P. J., An introduction to metal matrix composites, Cambridge University Press, Cambridge, ISBN 0521418089.

El Eskandarany M. S., Sumiyama K., Suzuki K., (1995), Mechanical solid state reaction for synthesis of β-SiC powders, *Journal of Materials Research*, 10, 3, pp: 659-667.

Ellison A., Magnusson B., Sundqvist B., Pozina G., Bergman J.P., Janzén E., Vehanen A., (2004), SiC crystal growth by HTCVD, *Materials Science Forum*, 457-460, pp: 9- 14.

Fend Z. C., (2004), SiC power materials: devices and applications. Ed. Springer series in material science, Springer-Verlag Berlin Heidelberg, ISBN: 3-540-20666-3.

Fu Q-G., Li H. J., Shi X. H., Li K. Z., Wei J., Hu Z. B., (2006), Synthesis of silicon carbide by CVD without using a metallic catalyst, *Matetrials Chemistry and Physics,* 100, pp: 108-111.

Ghosh B., Pradhan S.K., (July, 2009), Microstructural characterization of nanocrystalline SiC synthesized by high-energy ball-milling, *Journal of Alloys and Compounds,* 486, pp: 480–485.

Han R., Xu X., Hu X., Yu N., Wang J. Tan Y. Huang W., (2003), Development of bulk SiC single crystal grown by physical vapor transport method, *Optical materials*, 23, pp: 415-420.

Hidaka N., Hirata Y., Sameshima S., Sueyoshi H., (2004), Hot pressing and mechanical properties of SiC ceramics with polytitanocarbosilane, *Journal of Ceramic Processing Research,* 5, 4, pp: 331-336.

Hirata Y., Suzue N., Matsunaga N., Sameshima S., (2010), Particle size effect of starting SiC on processing, microstructures and mechanical properties of liquid phase-sintered SiC, *Journal of European Ceramic Society,* 30 pp: 1945-1954.

Humphreys R.G., Bimberg D. , Choyke W .J., Wavelength modulated absorption in SiC, *Solid State Communications,* 39, (1981), pp:163-167.

Izhevsky V. A., Genova L. A., Bressiani A. H. A., Bressiani J. C., (2000), Liquid-phase-sintered SiC. Processing and transformation controlled microstructure tailoring, *Materials Research,* 3(4) pp: 131-138.

Jensen R. P., Luecke W. E., Padture N. P., Wiederhorn S. M., (2000), High temperature properties of liquid-phase-sintered α-SiC, *Materials Science and Engineering,* A282, pp. 109-114.

Jin G. Q., Guo X. Y., (2003), Synthesis and characterization of mesoporous silicon carbide, *Microporous and Mesoporous Materials,* 60 (203), pp: 207-212.

Julbe A., Larbot A., Guizard C., Cot L., Charpin J., Bergez P., (1990), Effect of boric acid addition in colloidal sol-gel derived SiC precursors, *Materials and Research Bulletin,* 25, pp. 601-609.

Kamath G.S., (1968), International Conference on Silicon Carbide, Pennsylvania, USA, (1969), Special Issue to *Material Research Bulletin*, 4, S1-371, pp. S57–S66.

Kavecký Š., Aneková B., Madejová J., Šajgalík P., (2000), Silicon carbide powder synthesis by chemical vapor deposition from siliane/acetylene reaction system, *Journal of the European Ceramic Society,* 20, pp: 1939-1946.

Keller N. , Huu C. P., Crouzet C., Ledoux M. J., Poncet S. S., Nougayrede J-B., Bousquet J., (1999), Direct oxidation of H_2S into S. New catalysts and processes based on SiC support, *Catalyst Today*, 53, 535-542.

Klein L.C., Garvey G.J., (1980), Kinetics of the Sol-Gel Transition, *Journal of Non-Crystalline Solids*, (38-39), pp:45-50.

Kleiner S., Bertocco F., Khalid F.A., Beffort O., (2005), *Materials Chemistry and Physics*, 89, 2-3, pp: 362-366.

Kim D. H., Kim C. H., (1990), Toughening behavior of silicon carbide with addition of yttria and alumina, Journal of American Ceramic Society, 73, 5, pp. 1431-1434..

Kollo L., Leparoux M., Bradbury C. R., Jäggi C., Morelli E. C., (2010), Arbaizar M. R., Investigation of planetary milling for nano-silicon carbide reinforced aluminium metal matrix composites, *Journal of Alloys and Compounds*, 489, pp: 394-400.

Laube M., Schmid F., Pensl G., Wagner G., (2002), Codoping of 4H-SiC with N- and P-Donors by Ion Implantation, *Materials Science Forum*, 389-393, pp: 791-794..

Le Caer G., Bauer-Grosse E., Pianelli A., Bouzy E., Matteazzi P., (1990), Mechanically driven synthesis of carbides and sillicides, *Journal of Materials Science*, 25, 11, pp: 4726-4731.

Lee J-K., Park J-G., Lee E-G., Seo D-S., Hwang Y., (2002), Effect of starting phase on microstructure and fracture toughness of hot-pressed silicon carbide, *Materials Letters*, 57 pp: 203-208.

Lely J.A., Keram B.D., (1955), Darstellung von Einkristallen von Silizium Karbide und Beherrschung von Art und Menge der eingebauten Verunreinigungen, *Ber. Deut. Keram. Ges* 32, pp: 229-231.

Li J. L., Li F., Hu K., (December, 2002), Formation of SiC-AlN solid solution via high energy ball milling and subsequent heat treatment, *Materarials Science and Technology*, 18, pp: 1589-1592.

Li J., Tian J., Dong L., Synthesis of SiC precursors by a two-step sol-gel process and their conversion to SiC powders, (2000), Journal of the European Ceramic Society 77 pp: 1853-1857.

Li K. Z., Wei J., Li H. J., Li Z. J., Hou D. S., Zhang Y. L., (2007), Photoluminescence of hexagonal-shaped SiC nanowires prepared by sol-gel process, *Materials Science and Engineering*, A 460-461, pp: 233-237.

Li Z., Zhou W., Lei T., Luo F., Huang Y., Cao Q., (2009), Microwave dielectric properties of SiC(β) solid solution powder prepared by sol-gel, *Journal of Alloys and Compounds*, 475, pp: 506–509.

Li X. B., Shi E. W., Chen Z. Z., Xiao B., Polytype formation in silicon carbide single crystals, *Diamond & Related Materials*, 16, (2007), pp: 654-657.

Liu H. S., Fang X. Y., Song W. L., Hou Z. L., Lu R., Yuan J., Cao M. S., (2009), Modification Modification of Band Gap of β-SiC by N-Doping, *Chinese Physics Letters*, 26, 6, 067101-1-067101-4

Lu C. J., Li Z. Q., (2005), Structural evolution of the Ti-Si-C system during mechanical alloying, *Journal of Aloys and Compounds*, 395, pp: 88-92

Methivier Ch., Beguin B., Brun M., Massardier J., Bertolini J., (1998), Pd/SiC catalysts: characterisation and catalytic activity for the methane total oxidation , *Journal of Catalyst*, 173, pp: 374, 382.

Moore J. J., Feng H. J., (1995), Combustion synthesis of advanced materials: Part I. Reaction parameters, *Progress in Materials Science*, 39, (4-5), pp: 243-273.

Mulla M. A., Krstic V. D., (1994), Mechanical properties of β-SiC pressureless sintered with Al_2O_3 additions, *Acta metallurgica et materiala,* 42, 1, pp. 303-308.

Muranaka T., Kikuchi Y., Yoshizawa T., (2008), Akimitsu J., Superconductivity in carrier-doped silicon carbide, Science and Technolology of Advanced Materials, *9, 044204, pp: 1-8.*

Nader M., Aldinger F., Hoffman M. J., (1999), Influence of the α/β-SiC phase transformation on microstructural development and mechanical properties of liquid phase sintered silicon carbide, *Journal of Materials Science,* 31, pp: 1197-1204.

Noh S., Fu X., Chen L., Mehregany M., (2007), A study of electrical properties and microstructure of nitrogen-doped poly-SiC films deposited by LPCVD, *Sensors and Actuators,* A 136, pp: 613–617.

O'Connor J.R., Smiltens J., Eds, Silicone Carbide, a High Temperature Semiconductor, Pergamon, Oxford, 1960.

Omori M., Takei H., (1988), Preparation of pressureless-sintered SiC--- Y_2O_3--- Al_20_3, Journal of Materials Science, 23, pp: 3744-3749.

Omori M., Takei H., (1982), Pressureless sintering of SiC, *Journal of American Ceramic Society,* 65(6), pp: C92.

a. Ohtani N., Katsuno M., Nakabayachi M., Fujimoto T., Tsuge H., Yaschiro H., Aigo T., Hirano H., Hoshino T., Tatsumi K., (2009), Investigation of heavily nitrogen-doped n^+ 4H-SiC crystals grown by physical vapor transport, *Journal of Crystal Growth,* 311, 6, pp: 1475-1481.

b. Ohtani N., Fujimoto T., Katsuno M., Yshiro H., in: Feng Z.C. (Ed), SiC Power Materials-Devices and Applications, Springer Series in Materials, 73, Springer, Berlin, 2004, p. 89.

Ortiz A. L., Bhatia T., Padture N. P., Pezzotti G., (2002), Microstructural evolution in liquid-phase-sintered SiC: III, effect of nitrogen-gas sintering atmosphere, *Journal of American Ceramic Society,* 88, pp: 1835-1840.

Ortiz A. L., M-Bernabé A., Lopez O. B., Rodriguez A. D., Guiberteau F., Padture N. P., (2004), Effect of sintering atmosphere on the mechanical properties of liquid-phase-sintered SiC, *Journal of European Ceramic Society,* 24, pp: 3245-3249.

Padture N. P., (1994), In-situ toughened silicon carbide, *Journal of American Ceramic Society,* 77(2), pp: 519-523.

Pensl G., Choyke W.J., Electrical and optical characterization of SiC, *Physics B,*185, (1993), 264-283.

Pesant L., Matta J., Garin F., Ledoux M.J., Bernhard P., Pham C., Huu C. P., (2004), A high-performance Pt/ß-SiC catalyst for catalytic combustion of model carbon particles (CPs), *Applied Catalysis A,* 266, pp: 21-27.

Polychroniadis E. K., Andreadou A., Mantzari A., (2004), Some recent progress in 3C-SiC growth. A TEM characterization, *Journal of Optoelectronics and Advanced Materials,* 6,1, pp: 47-52.

Rodeghiero E.D., Moore B.C., Wolkenberg B.S., Wuthenow M., Tse O.K., Giannelis E.P., (1998) Sol-gel synthesis of ceramic matrix composites, *Materials Science and Engineering A24,* pp: 11–21.

Raman V., Bahl O. P., Dhawan U., (1995), Synthesis of silicon carbide through the sol-gel process from different precursors, *Journal of Materials Science,* 30, pp: 2686-2693.

Rajamani, R.K., Milin L., Howell G., (2000), United States Patent no. 6,086,242.

Razavi M, Rahimipour M. R., Rajabi-Zamani A. H., (2007), Synthesis of nanocrystalline TiC powder from impure Ti chips via mechanical alloying, *Journal of Alloys and Compounds*, 436, pp: 142-145.

Rost H.-J, Doerschel J., Irmscher K., Robberg M., Schulz D., Siche D., (2005), Polytype stability in nitrogen-doped PVT—grown 2″—4H-SiC crystals, *Journal of Crystal Growth*, 275, pp: e451e-454.

Saberi Y., Zebarjad S.M., Akbari G.H., (may, 2009), On the role of nano-size SiC on lattice strain and grain size of Al/SiC nanocomposite, *Journal of Alloys and Compounds*, 484, pp: 637–640.

Scitti D., Guicciardi S., Bellosi A., (2001), Effect of annealing treatments on microstructure and mechanical properties of liquid-phase-sintrerd silicon carbide, *Journal of European Ceramic Society*, 21, pp: 621-632.

Shaffer P. T. B., Blakely K. A., Janney M. A., (1987), Production of fine, high-purity, beta SiC powder, *Advances in Ceramics*, 21, *Ceramic Powder Science*, ed. G. L. Messing, K. S. Mazdiyasni, J. W. Mazdiyasni and R. A. Haber. The American Ceramic Society, Westerville, OH, pp: 257-263.

Semmelroth K., Schulze N., Pensl G. , Growth of SiC polytypes by the physical vapour transport technique, *Journal of Physics: Condensed Matter*, 16, (2004), pp: S1597-S1610.

Schwetk K. A., Werheit H., Nold E., (2003), Sintered and monocrystalline black and green silicon carbide: Chemical compositions and optical properties, *Ceramic Forum International*, 80 (12).

Sharma R., Rao D.V. S., Vankar V.D., (2008), Growth of nanocrystalline β-silicon carbide and nanocrystalline silicon oxide nanoparticles by sol gel technique, *Materials Letters*, 62, pp: 3174-3177.

Shen T. D., Koch C. C., Wang K. Y., Quan M. X., Wang J. T., (1997), Solid-state reaction in nanocrystalline Fe/SiC composites prepared by mechanical alloying, *Journal of Materials Science*, 32, 14, pp: 3835-3839.

a. Stein R.A., lanig P, (1993) Control of polytype formation by surface energy effects during the growth of SiC monocrystals by the sublimation method, *Journal of Crystal Growth*, 131, pp: 71-74.

b. Stein R.A., Lanig P., Leibenzeder S., (1992), Influence of surface energy on the growth of 6H- and 4H-SiC polytypes by sublimation, *Materials Science and Engeneering* B,11, pp: 69-71.

Straubinger T.L., Bickermann M., Weingaertner R., Wellmann P.J., Winnacker A., Aluminum p-type doping of silicon carbide crystals using a modified physical vapor transport growth method, *Journal of Crystal Growth*, 240, (2002), pp: 117-123.

Suryanarayana C., (2001), Mechanical alloying and milling, *Progress in Materials Science*, 46, pp: 1-184.

Tachibana T., Kong H.S., Wang Y.C, Davis R.F., (1990), Hall measurements as a function of temperature on monocrystalline SiC thin films, *Journal of Applied Physics*, 67, pp: 6375-6381.

Tairov M Yu., Tsvetkov V. F., (1978), Investigation of growth processes of ingots of silicon carbide single crystals, *Journal of Crystal Growth*, 43, pp: 209-212.

Tham M. L., Gupta M., Cheng L., (2001), Effect of limited matrix-reinforcement interfacial reaction on enhancing the mechanical properties of aluminium-silicon carbide composites, *Acta Materiala*, 49, pp: 3243-3253.

Vadakov Y.A., Mokhov E.N, M.G. Ramm, A.D. Roenkov, (1992), Amorphous and crystalline silicon carbide III, in: Harris G.L., Spencer M.G., C.Y.- W. Yang (Eds.), Springer, New York, , p. 329.

Wang G., Krstic V., (2003), Effect of Y_2O_3 and total oxide addition on mechanical properties of pressureless sintered β-SiC, *Journal of Materials Science and Technolology*, 19(3), pp: 193-196

Wellmann P., Desperrier P., Müller R., Straubinger T., Winnacker A., Baillet F., Blanquet E., Dedulle J.M., Pons M., SiC single crystal growth by a modified physical vapor transport technique, *Journal of Crystal Growth*, 275, (2005), pp: e555-e560.

White A. D., Oleff M. S., Boyer D. R., Budinger A. P., Fox R. J., (1987), Preparation of silicon carbide from organosilicon gels: I. Synthesis and characterization of precursor gels. *Advanced Ceramic Materials*, 2(l), pp: 45-52.

White A. D., Oleff M. S., Boyer D. R., Budinger A. P., Fox R. J., (1987), Preparation of silicon carbide from organosilicon gels: II. Gel pyrolysis and SiC characterization. *Advanced Ceramic Material*, 2(l), pp: 53-59.

Ye LL, Quan MX. (1995), Synthesis of nanocrystalline TiC powders by mechanical alloying, *Nanostructured Materials*,5, 1, pp :25-31.

Zhang B., Li J., Sun J., Zhang S., Zhai H., Du Z., (2002), Nanometer silicon carbide powder synthesis and its dielectric behavior in the GHz range, *Journal of the European Ceramic Society*, 22, pp: 93-99.

Zhao D.L., Luo F., Zhou W.C., (2010), Microwave absorbing property and complex permittivity of nano SiC particles doped with nitrogen, *Journal of Alloys and Compounds*, 490, pp: 190–194.

Zhao D., Zhao H., Zhou W., (2001), Dielectric properties of nano Si/C/N composite powder and nano SiC powder at high frequencies, *Physica E*, 9, pp: 679-685.

Zou G., Cao M., Lin H., Jin H., Kang Y., Chen Y., (2006), Nickel layer deposition on SiC nanoparticles by simple electroless plating and its dielectric behaviours, *Powder Technology*, 168, 2, pp:84-88.

Zheng Yo., Zheng Yi., Lin L. X., Ni J., Wei K. M., (2006), Synthesis of a novel mesoporous silicon carbide with a thorn-ball-like shape, *Scripta Materialia*, 55, pp: 883–886.

Combustion Synthesis of Silicon Carbide

Alexander S. Mukasyan
University of Notre Dame
USA

1. Introduction

Combustion synthesis (CS) is an effective technique to produce a wide variety of advanced materials that include powders and net shape products of ceramics, intermetallics, composites and functionally graded materials. This method was discovered in the beginning of 1970's in the former Soviet Union (Merzhanov & Borovinskaya, 1972), and the development of this technique has led to the appearance of a new material science related scientific direction. There are two modes by which combustion synthesis can occur: *self-propagating high-temperature synthesis* (SHS) and *volume combustion synthesis* (VCS). A schematic diagram of these modes is shown in Figure 1. In both cases, reactants may be in the form of loose powder mixture or be pressed into a pellet. The samples are then heated by an external source (e.g. tungsten coil, laser) either locally (SHS) or uniformly (VCS) to initiate an exothermic reaction.

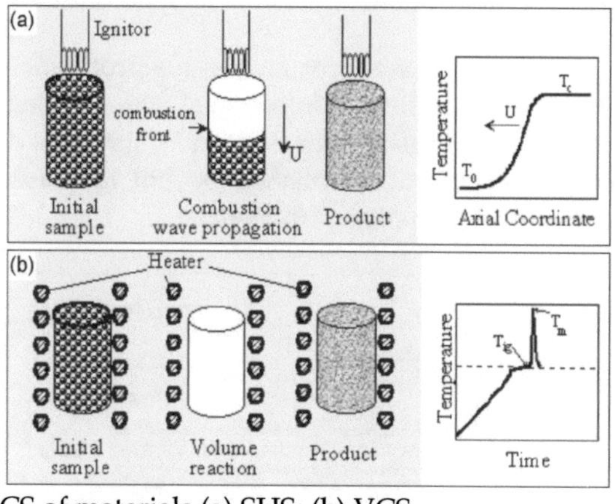

Fig. 1. Two modes for CS of materials (a) SHS; (b) VCS

The characteristic feature of the SHS mode (Fig.1a) is that locally initiated, the self-sustained reaction rapidly propagates in the form of a reaction wave through the heterogeneous mixture of reactants. The temperature of the wave front typically has quite high values (2000-4000 K). If the physico-chemical parameters of the medium, along with the chemical kinetics in the considered system are known, one may calculate the combustion velocity and reaction rate throughout the mixture. Thus, the SHS mode can be considered as a well-organized wave-like propagation of the exothermic chemical reaction through a heterogeneous medium, which leads to synthesis of desired materials.

During volume combustion synthesis (VCS) mode (Fig.1b), the entire sample is heated uniformly in a controlled manner until the reaction occurs simultaneously throughout the volume. This mode of synthesis is more appropriate for weakly exothermic reactions that require preheating prior to ignition, and is sometimes referred to as the *thermal explosion* mode. However, the term *"explosion"* used in this context refers to the rapid rise in temperature (see insert in Fig.1b) after the reaction has been initiated, and not the destructive process usually associated with detonation or shock waves. For this reason, volume combustion synthesis is perhaps a more appropriate name for this mode of synthesis (Varma et.al, 1998).

Figure 2 represents the sequence of operations necessary for CS technology. The dried powders of required reactants (e.g. silicon and carbon) in the appropriate ratio are wet mixed for several hours to reach the highly homogeneous condition. Thus prepared green mixture is loaded inside the reactor, which is then sealed and evacuated by a vacuum pump. After this, the reactor is filled with inert or reactive gas (Ar, N_2, air). A constant flow of gas can also be supplied at a rate such that it permeates through the porous reactant mixture.

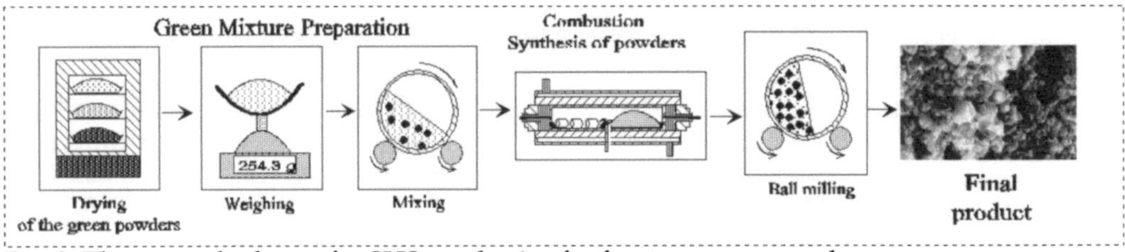

Fig. 2. The general scheme for SHS synthesis of refractory compounds

The design of a typical commercial reactor for large-scale production of materials is shown in Figure 3. Typically, it is a thick-walled stainless-steel water-cooled cylinder with volume up to 30 liters. The inner surface of the reactor is lined by graphite during SHS of carbides. Local reaction initiation is typically accomplished by hot tungsten wire. After synthesis product can be milled and sieved for desired fractions.

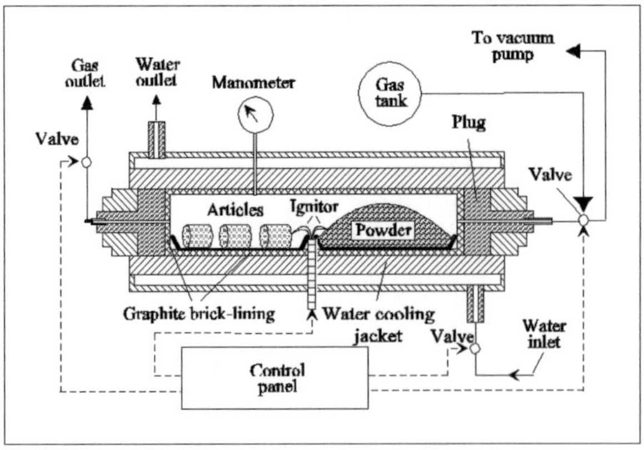

Fig. 3. Schematics of the SHS - reactor

The CS method has several advantages over traditional powder metallurgical technologies (Merzhanov, 2004). These advantages include (i) short (~minutes) synthesis time; (ii) energy saving, since the internal system chemical energy is primarily used for material production; (iii) simple technological equipment; (iv) ability to produce high purity products, since the extremely high-temperature conditions (up to 4000 K), which take place in the combustion wave, burn off most of the impurities. This approach also offers the possibilities for nanomaterials production (Merzhanov et.al, 2005; Aruna & Mukasyan, 2008). The number and variety of products produced by CS has increased rapidly during recent years and currently exceeds several thousands of different compounds. Specifically, these materials include carbides (TiC, ZrC, B_4C, etc.), borides (TiB_2, ZrB_2, MoB_2, etc.), silicides (Ti_5Si_3,TiSi2, $MoSi_2$, etc.), nitrides (TiN, ZrN, Si_3N_4, BN, AlN), oxides (ferrites, perovskites, zirconia, etc.), intermetallics (NiAl, TiNi, TiAl, CoAl, etc.) as well as their composites. The principles and prospects of CS as a technique for advanced materials production are presented in various reviews and books (Munir & Anselmi-Tamburini, 1989; Moore & Feng, 1995; Varma et.al, 1998; Merzhanov, 2004; Merzhanov & Mukasyan 2007, Mukasyan & Martirosyan, 2007). In this chapter the focus is on the combustion synthesis of silicon carbide (SiC), which due to its unique properties is an attractive material for variety of applications, including advanced high temperature ceramics, microelectronics, and abrasive industry.

2. Combustion Synthesis of Silicon Carbide from the Elements

From the viewpoint of chemical nature, *gasless combustion synthesis from elements* is described by the general equation:

$$\sum_{i=1}^{n} X_i^{(s)} = \sum_{j=1}^{m} P_j^{(s,l)} + Q \tag{1}$$

where $X_i^{(s)}$ are elemental reactant powders (metals or nonmetals), $P_j^{(s,l)}$ are products, Q is the heat of reaction, and the superscripts (s) and (l) indicate solid and liquid states, respectively. In the case of SiC synthesis from elements the reaction can be written as follows:

$$Si + C = SiC + 73 \text{ kJ/mol} \tag{2}$$

The reaction (2) has a moderate enthalpy of product formation (compared to ΔH_{273} = -230 kJ/mol for Ti-C system) and thus has relatively low adiabatic combustion temperature (T_{ad}=1860 K; compared with 3290 K for Ti-C reaction). Thus it is not easy to accomplish a self-sustained SHS process in this system. However, almost all available literature on CS of silicon carbide is related to this chemical pathway. Several approaches have been developed to enhance the reactivity of Si-C system. They can be sub-divided in five major groups:

(a) CS with preliminary preheating of the reactive media;
(b) CS with additional electrical field;
(c) chemical activation of CS process;
(d) SHS synthesis in Si-C-air/nitrogen systems;
(e) mechanical activation of the initial mixture

The employment of one or another approach depends on the desired product properties, e.g. purity, particle size distribution and morphology, yield and cost considerations. To understand these specifics, including advantages and disadvantages of different technologies, let us discuss them in more details.

2.1 CS with preliminary preheating of the reaction media

The obvious way to increase reaction temperature is a preliminary preheating of the reactive mixture to some initial temperature (T_0). The dependence of T_{ad} as a function of T_0 for stoichiometric (1:1 mol) mixture is shown in Figure 4. It can be seen that increase of T_0 above 900 K allows increasing T_{ad} to ~2300 K. The first publication on SHS of SiC from elements, describes the optimization of the preheating procedure of the reactive media to produce pure silicon carbide powder (Martynenko & Borovinskaya, 1978). It was shown that initial temperature of 900K and synthesis conducted in argon flow leads to the stable combustion wave propagation in stoichiometric Si + C mixture with formation of β-SiC powder with grain size of ~ 3 μm. Later this general approach, i.e. to increase the combustion temperature by preliminary preheating of the reaction media, has been used by many other researchers leading to the development of effective technologies for CS of SiC powder.

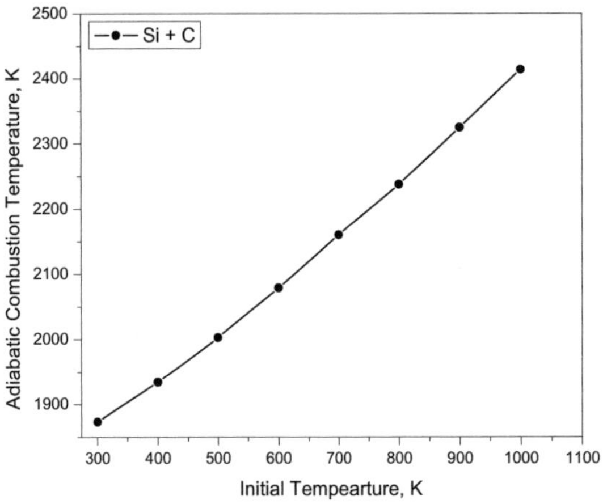

Fig. 4. Adiabatic combustion temperature in Si+C system as a function of initial temperature of the reaction mixture

For example, Pampuch, et al, 1987, showed that uniform preheating of the stoichiometric Si+C mixture in the flow of argon gas, leads to the self-ignition (VCS mode) of the heterogeneous media at temperature ~1300C with formation of β-SiC powders, which has a morphology similar to that of initial carbon as it is demonstrated in Figure 5. Two types of carbon precursors were used: carbon black (Fig.5a) and charcoal (Fig.5c). The BET surface area of the β-SiC obtained by using carbon black and charcoal, was 5.8 and 6.2 m^3/g, respectively. The crystallite size, determined from the broadening of the (111) X-ray peak, was 200 nm in the former and 145 nm in the latter case.

It was further demonstrated that suggested VCS approach allows effective synthesis of pure SiC powders, containing 99,6% of β–phase, <0.05wt% of free Si; 0.1 wt% of free carbon; and 0.3 wt% oxygen. It was also outlined (Yamada et al., 1985; Pampuch et.al, 1989) that self-purification effect is a characteristic feature of SC-based methods. Indeed, it was shown that

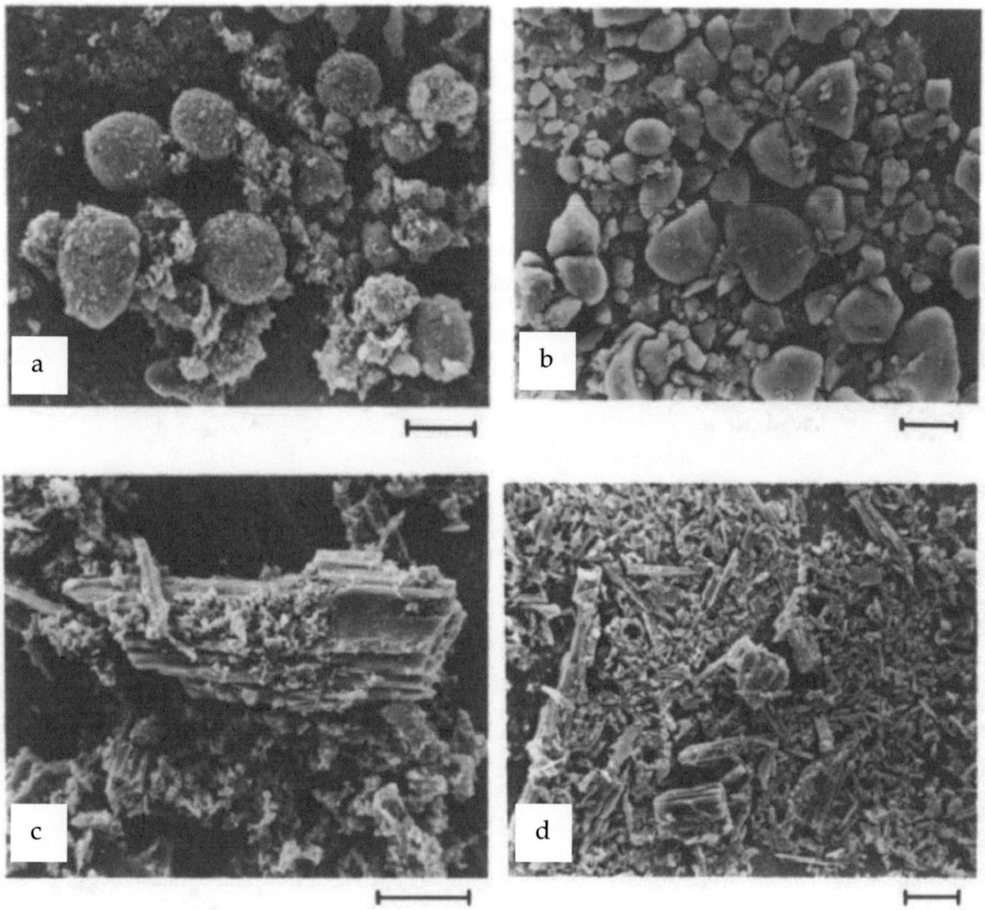

Fig. 5. Micrographs of initial carbon precursors (a) - carbon black; (c) - charcoal and corresponding products (β–SiC) (b) and (d); the bar scale = 50 μm.

50% of metallic impurities presented in the initial precursors were eliminated in the high temperature conditions of CS wave. Also it was shown that addition of Al to the initial mixture leads to the formation of ~10wt.% of α–SiC phase. It is important to outline the difference between approaches suggested by Martinenko and Pampuch. While in both cases the preheating was used as a tool to enhance the reaction rate, in the former case the mixture was preheated to ~ 900 K, followed by local mixture initiation, i.e. SHS mode was used, and in the later case, the self-ignition conditions ~1500 K was reached to promote the VCS mode. More recently another approach for preheating of the Si+C carbon mixture to produce SiC powder by SC was suggested by Chinese scientist (Wu & Chen, 1999; Chen et.al 2002). This method suggests using of a custom-built oxy-acetylene torch, which is moving along the surface of reactive mixture in air with speed (~3 mm/s) of the propagation of the combustion wave, leading to the relatively high yield (~94%) of desired product. From the view point of energy consumption this method is more affected as compared to the discussed above and allows synthesis to be accomplished in air. While the purity of thus obtained product is not so high, the microstructure of the powder is attractive, involving high surface area agglomerates with sub-micron grains (see Figure 6).

It is important that CS +preheating approach allows one-step production of the SiC ceramics. It was for the first time demonstrated by Japanese scientists in 1985 (Yamada et al., 1985), who used a high pressure self-propagating sintering method. In this case Si+C mixture was encapsulated into the high-pressure heating cell, on which pressure of 3GPa was applied in a cubic anvil device. Reaction was initiated by preheating the cell by carbon heater. Ceramics, which was synthesized under optimum conditions, contains ~96% of β-SiC phase, has density 2.9 g/cm^3 and micro hardness 23 GN/m^2.

Fig. 6. Microstructure of the SiC powders synthesized by torch-related CS method under different combustion temperatures: (a) higher; (b) lower.

2.2 CS with additional electric field

The other way to use preheating to provide conditions for CS self-sustained regime is to pass the current through the initial reactive medium. This approach was for the first time suggested by Yamada et al., 1986, followed by works of Steinberg's (Gorovenko et al., 1993; Knyazik, et al.,1993) and Munir's groups (Feng & Munir, 1995; Xue & Munir, 1996; Munir, 1997; Gedevanishvili & Munir, 1998).

The direct passing of the electric current through the sample, i.e. Joule preheating (Figure 7a) reaching, self-ignition VCS mode was used by Yamada and Shteinberg. It was shown that the process involves three stages; (i) the first stage is just inert preheating of the media to excitation of the SHS reaction. Heat, generated by the resistivity of the reactant, preheats the sample and raises the temperature. If the applied electric power is cut off on this stage, SiC product is not detected; (ii) the second stage—the SHS reaction self-initiated, typically in the middle part of the sample, where the heat losses are minimal. As SiC is produced, the electric resistivity increases rapidly and the current drops suddenly, as seen in Figure 7b; (iii) the third stage—the spontaneous reaction propagates toward both ends of the sample producing stoichiometric β–SiC phase. The duration of the reaction is on the order of 0.1 s. It was shown that decreasing particle size of the initial precursor one may synthesize sub-micron SiC powders by using this method. Figure 8 shows morphology of silicon carbide powders obtained by using 5 μm (a) and 0.1 μm (b) silicon particles.

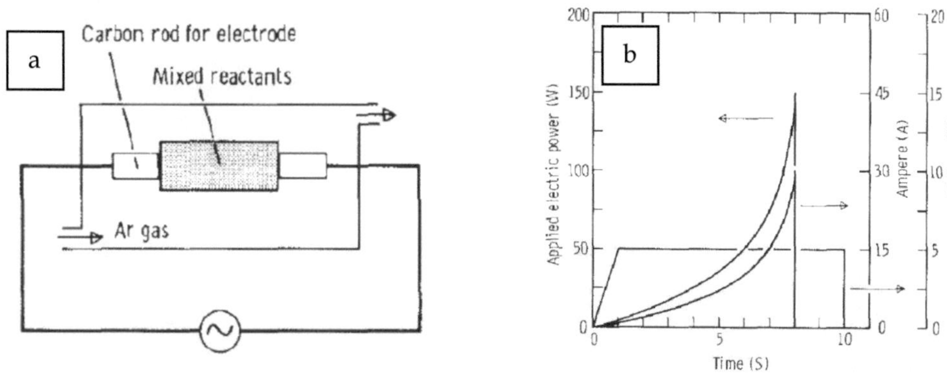

Fig. 7. Schematics of the of the set-up for CS of silicon carbide with Joule media preheating (a) and characteristic I-U diagram of the process (b)

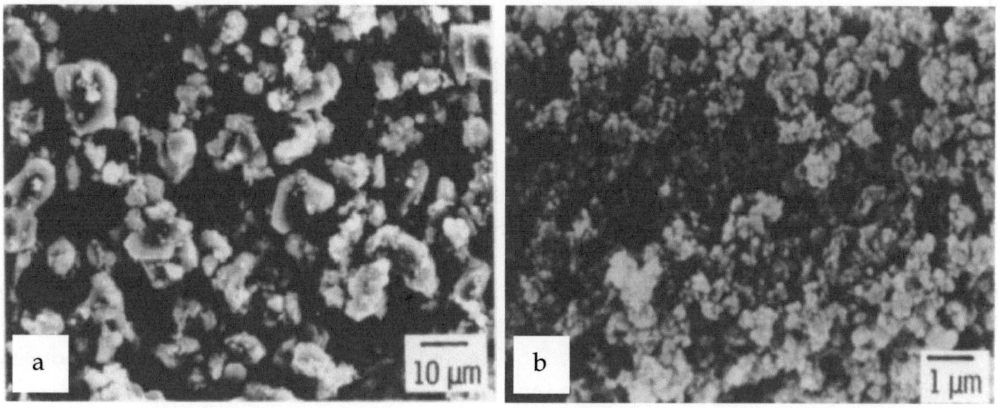

Fig. 8. Microstructure of the SiC powder obtained by CS with Joule preheating by using silicon powders of different size: (a) 5 µm; (b) 0.1 µm.

It was also confirmed that CS method leads to the self-cleaning of the powders during combustion process. For example, it was shown that oxygen content in the final products (0.2-0.3wt.%) was much less than that (0.5-0.7 wt.%) in the corresponding initial mixtures.
Different scheme for using of electrical field for synthesis of materials was suggested by Munir (cf. Munir 1997). The approach involves the imposition of a voltage across (not along) the reactant compact and the reaction is initiated by a heating coil as it is shown in Figure 9a. With the imposition of an ignition source and a field, it is possible to accomplish the self-sustaining reaction wave propagation in the powder mixtures of Si and C in inert atmosphere. Dynamics of the electric voltage and current during field-assisted SHS of silicon carbide, is shown in Figure 9b. It can be seen that as the reaction front starts to propagate (indicated by 'S' in the figure), the voltage drops and then remains relatively constant until the wave reaches the end of the sample (indicated by 'E' in the figure). The behavior of the current is consistent with that of the voltage, so is the behavior of the calculated resistance. The steadiness of the electrical parameters during SHS may indicate

that the current is primarily confined to the narrow reaction zone. Such localization of the current to the reaction zone is a typical phenomenon for all reaction systems with resistivity of products much higher than that for initial mixture. It was shown that the critical value for the applied electrical field (E~6 V/cm) exist, below which the reaction front cannot not propagate in self-sustained manner. Increase of E above critical leads to almost

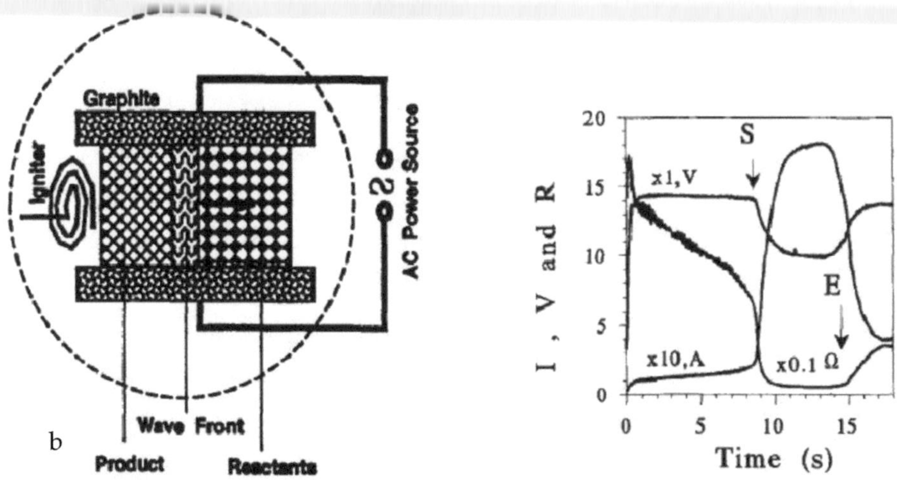

Fig. 9. Schematics of field-assisted SHS process (a) and typical current (I), voltage (V) and resistivity (R) behaviour during field-assisted SHS in Si+C sysetm (b)

linear increase of combustion front velocity in the range 0.1-0.8 cm/s followed by the thermal explosion conditions that take place at E~20 V/cm. As a result of field-assisted self-propagating high-temperature synthesis β-SiC powder (Feng & Munir, 1995) is produced with well crystallized morphology in the form of plates having thickness ~ 2 μm and size of ~ 20 μm (Figure 10). This approach was widely used for synthesis of different complex ceramics (AlN-SiC; $MoSi_2$-SiC etc), which cannot be produced by conventional means (Xue & Munir, 1996; Munir, 1997; Gedevanishvili & Munir, 1998).

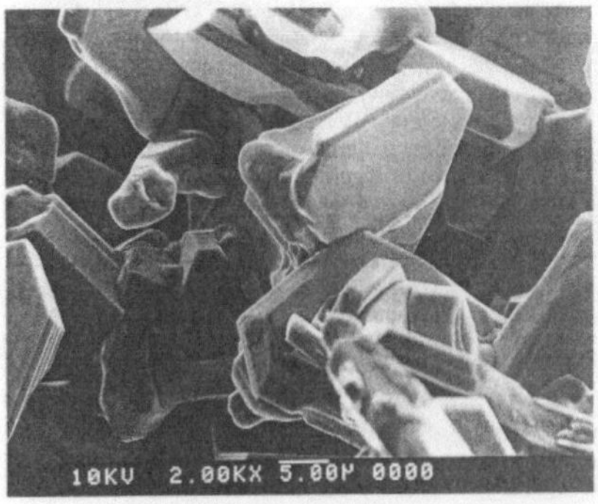

Fig. 10. Microstructure of SiC powder produced by field-assisted SHS method

2.3 Chemically activated SHS

Another universal approach to enhance reactivity of the low exothermic Si-C system is, so-called, chemically activated SHS, which was for the first time suggested by Nersisyan et al., in 1991. It was proposed to use $(-CF_2-CF_2-)_n$ polytetrafluoroethylene (PTFE) as an additive for the Si+C powder mixture. The following set of equations represents the main chemical reactions take place in the combustion front:

Low temperature reaction $Si(s) + (-CF_2-CF_2-)_n \rightarrow SiF_2\ (g\uparrow) + C(s) + Q_1$ (3)

Intermediate reaction $SiF_2(g) + C(s) \rightarrow SiF_4(g\uparrow) + SiC(s) + Q_2$ (4)

High-temperature reaction $Si(s,l) + C(s) \rightarrow SiC\ (s) + Q_3$ (5)

It was proved that in Si + C + PTFE system combustion wave consists of two main zones. The first one involves mainly reaction (3) between silicon and PTFE which on one hand leads to the preheating of the Si+C reaction media and on the other hand to the formation of gaseous silicon containing species (SiF_2, SiF_4). The second, carbidization stage proceeds owing to reactions (4) and (5). Reaction (5) may occur in self-sustain manner partially owing to preheating of the media by heat of reaction (3) and additional heat released in reaction (4). Note, that gas phase reaction (4) and condensed phase reaction (5) should lead to different morphologies of the SiC product. Indeed it was shown that two different types of particles can be synthesized (Nersisyan et al., 1991). With the certain amount of PTFE additive the cube-shape particles with size on the order of 10 μm can be produced, while for the other composition the formation of the long ~1 mm thin (0.5 μm) fibers were observed.

Next, the set of additives including KNO_3, $NaNO_3$,NH_4NO_3 and $BaNO_3$ was investigated (Kharatuan & Nersisysn, 1994). It was demonstrated that among these promoters (KNO_3 - Si) is the best one leading to effective synthesis of relatively small (~1 μm) SiC particles with the amount of free carbon less than 0.5 ew.% as a major impurity. Also it was shown that SHS process consists of two main stages:

Stage I - oxidation

$$KNO_3 + Si(s) \rightarrow SiO_2 + K_2O(SiO_2) + Si_3N_4 + Q_1 \quad (6)$$

$$KNO_3 + C(s) \rightarrow K_2O + CO_2 + N_2 + Q_2 \quad (7)$$

Stage II - carbidization

$$Si + C \rightarrow SiC\ (s) + Q_3 \quad (8)$$

With importance of the additional reactions

$$SiO_2 + Si \rightarrow 2SiO\ (g\uparrow) \quad (9)$$

$$SiO\ (g) + 2C \rightarrow SiC\ (s) + CO\ (g) \quad (10)$$

As in the case of PTFE additive the reactions (6) and (7) provide enough additional heat to support the main reaction (8), which in this case may self-propagate in the inert atmosphere. It was also shown that because the envolvement of gas phase reactions the argon gas pressure is a critical parameter to control the process. The typical temperature time pofiles obtained under different synthesis conditions are shown in Figure 11 and illustrate the mentioned above important conclusions. It can be seen, that two stage combustion front may propagate only if argon pressure is above some critical value.

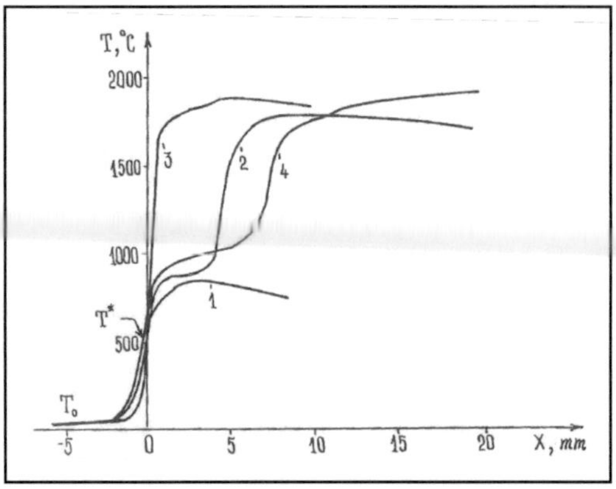

Fig. 11. Typical temperature profiles in SHS wave for Si+C+15 %(KNO3+Si) mixture at different argon pressure (MPa): (1) 0.1; (2) 0.2; (3) 1.5; (4) 5.5

Later more complex [Mg + (-CF_2-CF_2-)$_n$] additive was used to synthesize SiC powder n SHS mode (Zhang et al., 2002). The following reaction scheme takes place in the combustion front:

$$2Mg + (\text{-}CF_2\text{-}CF_2\text{-})_n = 2MgF2 + 2C + Q(-1518 kJ/mol) \quad (11)$$

$$Si + C = SiC + Q\ (-69\ kJ/mol) \quad (12)$$

The pure β-SiC powder was obtained after leaching of the as-synthesized product in 10% vol. % (HNO3 + HF) solution for 4 h. The characteristic microstructure is shown in Figure 12. It can be seen that this approach allows production of powders with average particle size of ~ 5 μm .

Chemical activation method was also widely used in combination with other approaches to enhance reactivity of Si-C system, i.e. reaction in the atmosphere of the reactive gases (e.g. nitrogen and air)

Fig. 12. Typical microstructure of as-synthesized product (a) and after leaching (b)

2.4 SHS in Si C-nitrogen/air system

SHS reaction between silicon and nitrogen (Mukasyan et al., 1986), as well as carbon burning in air are much more exothermic as compared to Si+C reaction. First, it was suggested to use air as an atmosphere to carry CS of silicon carbide (Martynenko, 1982). It was shown that optimization of synthesis conditions, which include the usage of initial Si-C mixture with slight excess of carbon, air pressure above critical (~3PMa) and clever change of the air content in the reactor, allows production of silicon carbide powder with 5-7 wt.% of silicon nitride and relatively high specific surface area up to 10 m^2/g. Microstructure of this powder is shown in Figure 13. Later conducting experiments in pure nitrogen (Yamada et al., 1989) it was shown that combustion wave consists of two stages (see Figure 14). First is reaction of nitrogen with silicon to produce some amount of silicon nitride which leads to preheating of the rest of the media, followed by carbidization reaction with maximum temperature around 2100 K.

This approach was further widely used for synthesis of silicon carbide - based powders (Agrafiotis, et al., 1990; Kata et al., 1997; Puszynski & Miao, 1998; Kata & Liz, 2005; Kharatyan et al., 2006; Yang et al., 2009, etc). First, it is worth noting the idea of possible reduction and decomposition of the silicon nitride in the combustion front, which may lead to the synthesis of silicon carbide powder with minimum (<1wt%) amount of silicon nitride. Indeed, as it was suggested by Kata, the following reactions may occur under certain conditions:

$$Si_3N_4 + 3C = 3SiC + 2N_2 \qquad (13)$$

$$Si_3N_4 \rightarrow Si + N_2 \qquad (14)$$

leading to the complete elimination of the undesired silicon nitride phase in the synthesized product.

Fig. 13. Microstructure of SiC powder obtained By SHS in Si-C-air system

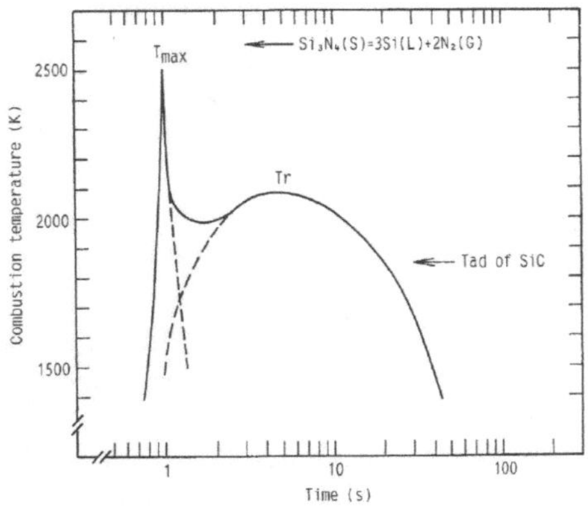

Fig. 14. Typical temperature profile of SHS wave in Si-C-N2 system

Method for combustion synthesis of SiC/Si_3N_4 powders in $Si\text{-}C\text{-}(\text{-}CF_2\text{-}CF_2\text{-})_n$ system was also developed (Kharatyan et al., 2006). It was shown that the use of fluoroplastic as an activating component allows a single-stage synthesis of Si_3N_4–SiC composite with contents of individual components varying from 0 to 100%. More important, that using chemical activation plus combustion in reactive atmosphere allows one to produce SiC whiskers and fibers (cf. Puszynski & Miao, 1999; Chen et al., 2001). For example, long (3mm) thin (0.2-1μm) SiC fibers were synthesized by using carbamide (urea) as a promoter, while combustion process took place in nitrogen atmosphere. Microstructure and results of XRD analysis of these fibers are shown in Figure 15, correspondingly.

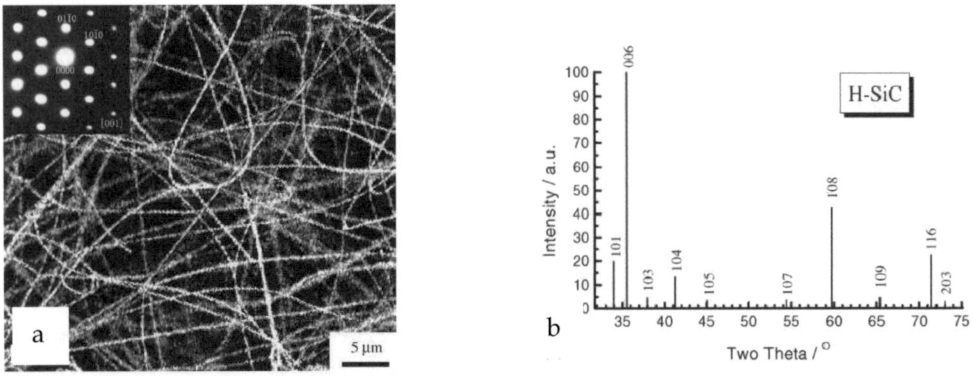

Fig. 15. Typical microstructure (a) and XRD data (b) of silicon carbide fibers synthesized in Si-C-urea-N2 system

Combination of different activation approaches to enhance reaction in silicon-carbon system is currently widely used for developing of novel effective CS methods. The most recent ones are related to the so-called mechanical activation approach.

2.5 Mechanical activation of CS reaction in Si - C system

High energy ball milling (HEBM) is the processing of powder mixture in high-speed planetary ball mills and other devices, where the particles of the mixture are subjected to significant mechanical impacts with a force sufficient to break the brittle and plastically deform the ductile components (Suryanarayana, 2005). Brittle particles are milled to finer grains, whereas ductile particles (usually, metals) are subjected to multiple deformations, all together forming layered composites particles with the layer thickness decreasing as the milling time is increased. Thus such mechanical treatment not only decreases the particle size of reactants, but also increases their contact surface area, which is typically free from oxide films. Moreover, the defects of the crystalline structure are accumulated in the media during HEBM. All these factors enhance the chemical activity of the combustible mixture and thus, are called mechanical activation (MA). The MA may include partial or complete dissolution of one reactant in the other (mechanical doping or mechanical fusion); otherwise, the components of the mixture can be involved into a chemical reaction with formation of a new compound during HEBM (mechanical synthesis). The analysis of literature allows one to conclude that mechanical activation (i) decreases the self-ignition temperature of various combustible systems, (ii) expands the flammability limits, (iii) favors a more complete reaction, and (iv) typically increases the combustion wave front velocity (cf. Rogachev & Mukasyan, 2010). In the above context, it looks logical to apply this approach for reaction enhancement in low exothermic carbon-silicon systems.

Recently the process was developed (Yang et al., 2007a), which involves the following steps: (i) adding a small amount (1-3wt.%) of NH4Cl and PVC into the Si-C powder mixture; (ii) mechanically activating of the mixtures through HEBM (2-12 h); mixture preheating to temperature (950-1200°C) and keeping it at this temperature until the reaction self-initiates. It was demonstrated, that MA allows one to initiate reaction at relatively low temperature 1050°C and reach full conversion by using small amounts of additives. Fine β-SiC powders with specific surface area 4.4 m^2/g, and the particle size < 5 μm was synthesized. However, it was shown that optimum MA time should be applied, because long HEBM leads to synthesis of larger SiC particles (compare Fig. 16 a and b).

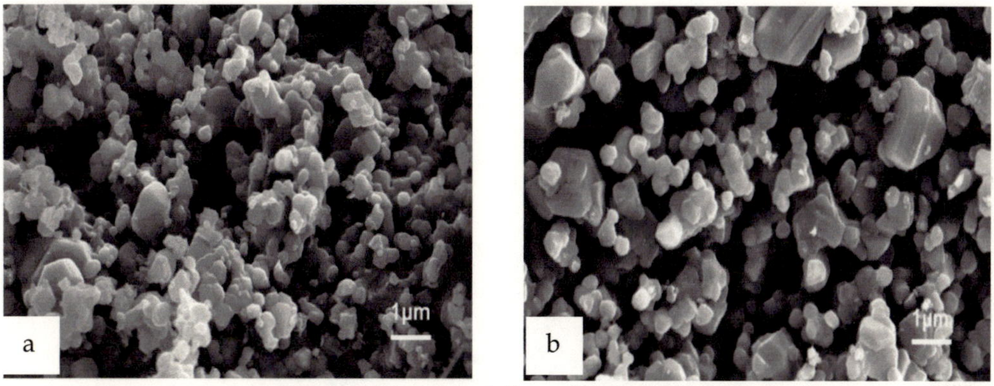

Fig. 16. Microstructure of SiC powders after different time of MA: (a) 2 h; (b) 12 h.

Similar approach but with synthesis in SHS mode was also developed (Yang et al., 2007b). While slightly larger amount (~6wt.%) of PTFE should be used as compared to VCS mode, but still it is almost three times less than critical promoter concentration for not activated mixture. It is also important that much finer particles (Figure 17a) can be synthesized in this mode which also has narrow particle size distribution (Figure 17b).

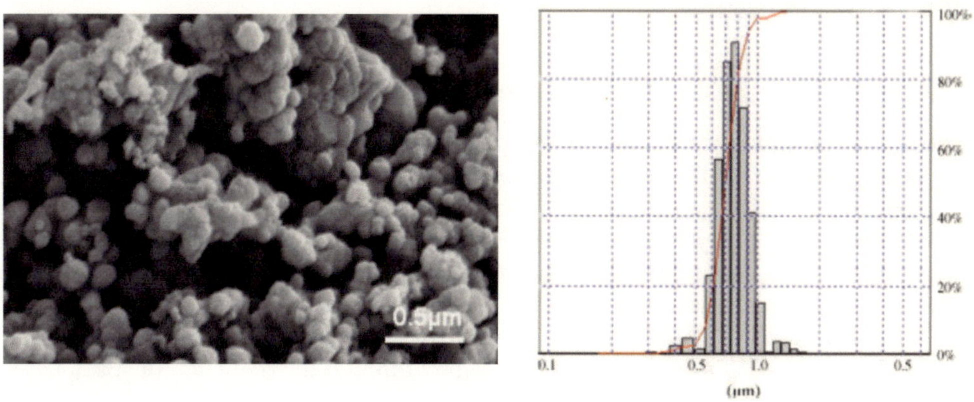

Fig. 17. Microstructure (a) and particle size distribution (b) for SiC powder synthesized in SHS mode after MA of the initial mixture

Both mechanical and chemical activations (PTFE) with synthesis in nitrogen atmosphere were used for large scale synthesis of nanopowder of β- SiC (Liu, et al., 2008). It was shown that if MA takes place with 1:16 ball to mixture mass ratio during 2 h than under 4 MPa of nitrogen only 1.5 wt.% of PTFE is required to reach full conversion of mixture to SiC phase with amount of other impurities less than 1 %.

Mechanically activated pure (without any additives) Si+C mixture was burned in air to synthesized SiC powder in SHS mode (Yang et al., 2009). It was demonstrated that after 4 h of HEBM the mixture can be ignited in air at 1 atm. First surface reaction between oxygen and mainly silicon leads to the formation of the relatively thin (~0.5 mm) layer of SiO_2 phase, simultaneously preheating bulk media. Second combustion wave due to reaction between silicon and carbide results in synthesis of β-SiC with small amount of Si_2N_2O phase. Typical microstructure of the SiC powder is shown in Figure 18. It can be seen that about 100 nm particles can be produced by this approach.

Fig. 18. Microstructure of SiC powder synthesized in air after 4h of HEBM of Si+C mixture

The most widely used method in conventional powder metallurgy is based on the Acheson idea on thermal reduction of silica by carbon, where different types of the silicon containing precursors are used (cf. Choyke &. Matsun, 2004). Similar approach, but taking place in SHS mode has also been developed.

3. Combustion Synthesis of Silicon Carbide by Reduction Reactions

The other way to produce SiC powder by using SHS method involves sequence of two reactions that take place in the combustion front: reduction of silica by a metal to make pure silicon, followed by silicon reaction with carbon. The SiO_2 + Me(Mg, Al) + C system is much more exothermic, as compared to binary Si + C composition (see below). Thus, it is relatively easy to initiate the SHS mode in such reduction-type mixture without using any special enhancing means. However, only few patents (Merzhanov et.al, 1992; Merzhanov et al., 1994) and scientific publications (cf. Yermekova et al., 2010) may be found that are related to the combustion synthesis of silicon carbon through the reduction of silica. Let us discuss this approach in detail.

3.1 Thermodynamic considerations

The overall combustion reaction for reduction synthesis of SiC, when magnesium (Mg) is used as a reducing element, can be written as follows:

$$SiO_2 + 2Mg + C = SiC + 2MgO \tag{15}$$

The thermodynamic analysis (Shirayev, 1993; Mamyan, 2002) allows calculating the adiabatic combustion temperature (T_{ad}) and equilibrium products composition for reaction (15) as a function of the inert gas (argon) pressure (P) in the reaction chamber (Fig.19). It can be seen (Fig.19a) that T_{ad} increases and the amount of gas phase products decreases, with increase of inert gas pressure. Also, the absolute value of T_{ad} is > 2000 K, which is above melting (m.p.) and boiling points of magnesium (922 K and 1363 K, correspondingly), m.p. of Si (1683 K) and SiO_2 (1923K), but well below m.p. of MgO (3073 K) and carbon (4093K). It is clear that the amount of gas phase products (which includes Mg, CO and SiO) can be decreased by increasing inert gas pressure in the reaction chamber, since higher P suppresses the gasification processes.

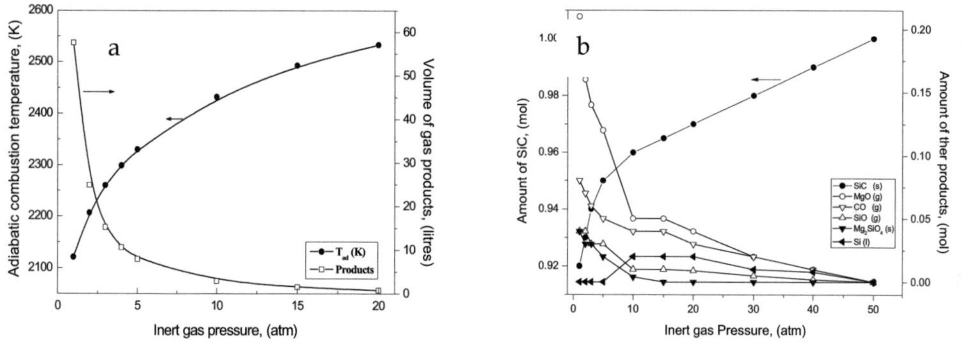

Fig. 19. Thermodynamic characteristics of the SiO_2:Mg:C=1:2:1 system as a function of inert gas pressure: (a) T_{ad} and Volume of gas phase products; (b) equilibrium products

In addition to main solid products, i.e. SiC and MgO, two other phases, i.e. Mg_2SiO_4 and Si, can be produced (Fig.19b). The last two phases are undesirable, because it is not easy to leach them out from the as-synthesized product. Thermodynamics suggests how one can reduce the amount of these phases. It appears that increasing of P leads to a significant decrease of the Mg_2SiO_4 and Si quantities. Moreover, at P=50atm only SiC and MgO are the equilibrium combustion products in the considered system (Fig.19b). Thus 100% of SiO_2 conversion to silicon carbide for stoichiometric 1:2:1 composition can be reached under high argon pressure.

3.2 Silicon carbide powder by SHS in SiO_2-C-Mg system

The thermodynamic calculation reveals that by conducting experiments under optimum gas pressure and adjusting the composition of the initial mixture one can expect the synthesis of product which involves only two solid phases SiC and MgO. The simple chemical treatment of such mixture (see details in Amosov et al., 2007) allows complete leaching of the MgO phase and obtaining pure silicon carbide powder. However, the thermodynamics cannot suggest how one may control the microstructure of the product synthesized in the SHS. The last issue was recently investigated by Ermekova et al, 2010.

Three different types of the silicon oxide (SiO_2) powder were used: (i) from Yerken-deposit, Kazakhstan, (KZ) 98,8% purity, particle size d≤100μm, (ii) Laboratory Cerac (LC), WI, USA, 99.5% purity, d<10 μm (iii) nano Untreated Fumed Silica (UFS), Cabot Corporation, MA, USA, 99.9% purity. The typical microstructures of these powders are shown in Figure 20. It can be seen that KZ powder has a wide range of particle size distribution with average size about 20 μm and thus low specific surface area (1 m^2/g). In contrast the UFS powder possesses extremely uniform and fine microstructure, as well as high BET ~ 390m^2/g. Laboratory Cerac (LC) powder has properties which are somewhere in between KZ and UFS silica with surface area ~7 m^2/g. Carbon black powder and Mg (Alfa Aesar, MA, USA, 99.8% purity, d<44mm) were used as precursors for initial reactive mixture.

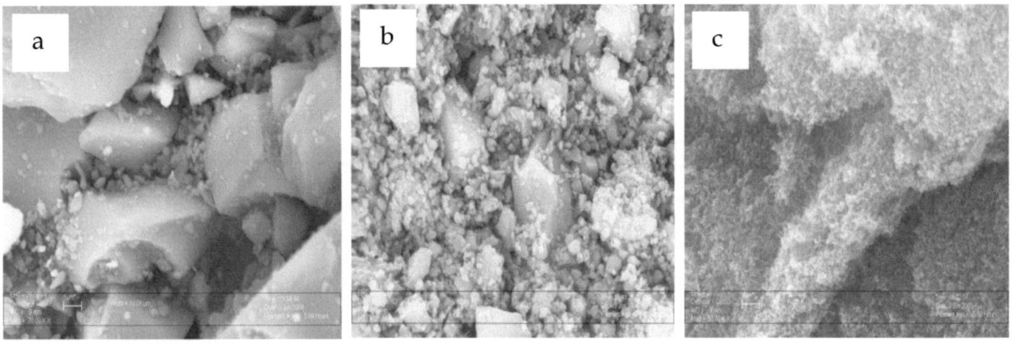

Fig. 20. Microstructure of the initial SiO_2 powders: (a) KZ; (b) Cerac; (c) UFS

Correspondingly, three SiO_2-Mg-C mixtures (named below KZ, LC and UFC) with different silica, but with the same optimized composition were prepared by 2 hours of dry mixing. All thus prepared mixtures were placed on graphite tray and inserted into stainless steel cylindrical reactor. The reactor was evacuated up to the 10^{-3} atm and filled with argon up to the desired pressure, in the range of 1-20 atm. The ignition was carried out by passing short electrical current impulse (I=10A, U=20V) through the coil of tungsten wire. All as-synthesized products were chemically treated for 3 hours in 36% solution of hydrochloric acid at room temperature. The acid was taken into amounts according to the predicted mass of magnesium oxide to be leached out. This process was followed by thorough powder washing in ionized water and drying at 100° C for about 2 hours.

Microstructures of as-leached powders are shown in Figure 21. It can be seen that all powders have relatively uniform particle size distribution with sub-micron average values: d_{KZ} =280 nm; d_{Cerac} =150 nm; d_{UFS}=90 nm. More close inspection (Ermekova et al.,2010) reveals that in all cases particles involve set of extremely thin hexagonal plates sintered to each other. The diagonal of the hexagon is about average particle size, while thickness is as small as 10 nm for KZ-SiC; 3-5 nm for Cerac-SiC ; and 1 nm UFC-SiC

Fig. 21. Microstructure of the β–SiC powders synthesized from different SiO_2 precursors: (a) KZ; (b) Cerac; (c) UFS ▬▬ 200 nm

It is shown, that silicon carbon nanopowder can be directly synthesized by combustion reaction in the silica-magnesium-carbon system. While even micro-size precursors (KZ) lead to formation of nano-particles, the scale of heterogeneity of used silica is a parameter which allows one to control the size of SiC grains. It was also demonstrated that under optimum conditions pure β–Sic powder with surface above 100 m^2/g can be synthesized in SHS mode.

4. Concluding remarks

Various combustion synthesis methods, which lead to production of SiC powders, with different morphologies and particle size distributions, are discussed in this chapter. Almost all of them can be easily scaled-up to produce tons of materials per year (Merzhanov & Mukasyan, 2007). Because of the energy saving nature of combustion - based technologies, the total production cost for SHS methods primarily involves the cost of the initial precursors and milling cost (Golubjatnikov, 1993). It is important that "self-purifying" feature of the CS approaches, mentioned in this chapter (see also Bloshenko, et al, 1992), allows to use not high purity raw powders to produce high purity products. The latter contributes to the low production cost of the powders. Simple technological equipment results in relatively low cost capital investment to organize the production site. All above makes CS technology very attractive for industrial production of advanced ceramic materials.

However, it is also well recognized that world market of such advanced materials is not large enough and already distributed among well known manufacturers. Thus being attractive SHS technologies cannot compete with traditional approaches, if even slightly better and cheaper product will be suggested. The cases which may lead to success are to synthesize the materials which cannot be produced by traditional technologies or provide much cheaper product. In the author's opinion, the sub-micron silicon carbide powder is such a product. Indeed, the current conventional technologies, which are based on Acheson idea, can produce low cost silicon carbide powders in the micron range (>5 μm) and impurity of 96-98 wt%. Production of smaller particles requires long term milling processes which typically decrease powder purity and increase cost. Not-traditional approaches require high temperature equipment and are characterized by relatively long synthesis time, otherwise using same precursors as in SHS method and thus they have higher production costs as that for combustion-based technologies. Search of market prices for nano-sized SiC powders confirmed this conclusion.

As shown in this chapter, almost all CS-based methods lead to production of sub-micron β-SiC powders. In many cases, nano-powders (<100 nm) as well as nano-fibers can be routinely synthesized in effective energy saving way. The purity issue, i.e. amount of SiC phase, oxygen, free carbon, iron contents in the final product, is very important. Thus, selection of the specific CS-method is dictated by specific requirements on the powder properties (particle size, morphology, impurity, cost, etc.).

High pure SiC (>99 wt.%) powders can be produced by CS in Si+C system and using ***preliminary preheating*** approach. Note that in this case some external energy source is required to preheat the media and the typical range of synthesized particles size is 1-10 micron. Also nano-sized (average particle size 80 nm) pure product can be achieved by using ***reduction reaction*** in SiO_2-Mg-C system. In this case the purity will be defined not only by optimum synthesis conditions, but also by optimization of leaching stage. The latter may lead to increase of the production cost. The other aspect to be considered is the utilization of magnesium salts, which are formed during as-synthesized product leaching. However, this problem allows elegant solution. All approaches with using different ***chemical activators,*** including nitrogen and air environments, being very simple for technical realization and allowing synthesis of extremely fine particles, have one obstacle, which is related to the product purity. If small amount (3-7wt.%) of such refractory phases as Si_3N_4 is not the concern for the powder application then ***combustion in nitrogen*** is the best method to synthesized pure ceramic powder. It looks like that optimum combustion in ***argon*** of Si+C mixture with ***PTFE additive*** allows to produce nanopowders with small amount of impurities. Note, that clever evacuation of fluorides compounds from the volume of the reactor is required for this approach. ***Combustion at 1atm in a***ir is very attractive approach. But this technology requires relatively long term high energy ball milling of initial mixtures. The estimation of the production costs is required to make conclusion regarding the effectiveness of this method.

Finally, note that CS-based methods allow direct synthesis of materials in combustion wave. Indeed, it is logical to use high temperatures formed in the reaction front not only to synthesize desired phase but also to sinter the produced powders to make pore-free materials. The list of such methods is relatively long (Merzhanov & Mukasyan, 2007) and it can be a topic for the next book.

5. References

Agrafiotis, Ch.C., Lis, J,. Puszynski, J.A., Hlavacek, V., Combustion synthesis of silicon carbide in nitrogen atmosphere, *J . Amer. Ceram. Soc.*, Vol.72, No.9, 1735-1738

Amosov, A.P.; Borovinskayaand,I.P.; Merzhanov, A.G., (2007) *Poroshkovaia tehnologia SVS materialov* (Powder Technology of SHS Materials), Moscow: Mashinostroenie, p.136.

Aruna, S.T.& Mukasyan, A.S. (2008), Combustion synthesis and nanomaterials, *Current Op. Sol. State & Mater. Sci.*, Vol.12, 44-50

Bloshenko, V.N.; Bokii, V.A.; Borovinskaya, I.P.; Merzkanov, A.G., (1992) Self-purification of SHS products from oxygen impurities, *Int. J. of Self-Propagating high-temperature synthesis,* Vol. 1, No.2,257-265

Chen, C-C.; Li, C.-L.; Liao, K.-Y., (2002), A cost-effective process for large-scale production of submicron SiC by combustion synthesis, *Mat. Chem. & Phys.,* Vol. 73, 198-205

Choyke, W. J. Matsunami, H. (2004) in a book: *Silicon Carbide: Recent Major Advances.* G. Pensl (eds), Springer-Verlag, Berlin, Heidelberg, New York.

Ermekova, Z.S.; Mansurov, Z.A.; Mukasyan A.S. (2010) Influence of Precursor Morphology on the Microstructure of Silicon Carbide Nanopowder produced by Combustion Syntheses, *Ceramics International,* Vol 36, No.8, 2297-2305.

Feng, A & Munir Z.A., (1995), Effect of an electric field on self-propagating combustion synthesis: Part II. Field-assisted synthesis of b-SiC, *Metallurgical and Materials Transactions B: Process Metallurgy & Mat. Proces. Sci.,* Vol. 26, No. 3, 587-593

Gedevanishvili, S. & Munir, Z.A., (1998) An investigation of the combustion synthesis of $MoSi_2$ b-SiC composites through electric-field activation, *Mat. Sci.& Eng.,*A242, 1-6

Golubjatnikov, K. A.; Stangle, G. C.; Spriggs, R. M. (1993) Economics of advanced self-propagating, high-temperature synthesis materials fabrication, *American Ceramic Society Bulletin,* Vol. 72, No. 12, 96-102

Gorovenko, V.I., Knyazik, V.A., Shteinberg, A.S., (1993) High-temperature interaction between silicon and carbon, *Ceram. Inter.,* Vol.19, no.2, 129-132

Kata D., Lis, R., Pampuch (1997) Combustion synthesis of muptiphase powders in Si-C-N system, *Solid State Ionics,* 101-103, 65-70

Kata, D.; Lis, J. (2005) Silicon nitride rapid decomposition for ceramic nanopowder manufacturing, *Glass Physics and Chemistry,* Vol. 31, No. 3, 364-369

Kharatyan, S.L & Nersisyan, H.H., (1994), *International Journal of Self-Propagating high-Temperature Synthesis, Vol.3, No.1, 17-25*

Khachatryan, G.L., Arutyunyan, A.B.; Kharatyan, S.L. (2006) Activated combustion of a silicon-carbon mixture in nitrogen and SHS of Si_3N_4-SiC composite ceramic powders and silicon carbide, *Combus., Explos.&Shock Waves,* Vol. 42, No. 5, 543-548

Knyazik, V.A., Shteinberg, A.S., Gorovenko, V.I., Thermal analysis of high-speed high-temperature reactions of refractory carbide synthesis, (1993), *J. Thermal. Anal.,* Vol.40, No.1, 363-371

Liu, G.; Yang, K.; Li, J., Combustion synthesis of nanosized β-SiC powder on a large scale, (2008) *J. Phys. Chem,* Vol. 112, 6285-6292

Martynenko, V.M. & Borovinskaya, I.P. (1978), Thermodynamic analyses for silicon carbide synthesis in combustion regime, *Proc. II All-Union Conf. on Combustion Technology,* Chernogolovka, 180-182

Martynenko, V.V., Self-Propagating high temperature synthesis of silicon carbide, (1984) *Ph.D Thesis,* Branch of Institute of Chemical Physics, USSR Academy of Sciences, Chernogolovka, Russia.

Merzhanov, A.G. & Borovinskaya, I.P. (1972), Self-propagating high-temperature synthesis of refractory inorganic compounds, *Dokl. Chem.,* Vol. 204, No.2, 429-431

Merzhanov, A.G..; Borovinskaia, I.P.; Mahonin, M.S.; Popov, L.S. (1992) Patent № 4409571/26, C01B31/36, Russia

Merzhanov, A.G.; Borovinskaia, I.P.; Mamian, S.S.; Mikabidze,G.V.; Vershinikov, V.I.; Tavadze, G.F., (1994). Patent No. № 4445557/26, C01B31/36. Russia.

Merzhanov, A.G., (2004), The chemistry of self-propagating high-temperature synthesis, *J. Mat. Chem.* Vol.14, No.12, 1779-1786

Merzhanov, A.G.; Borovinskaya, I.P.; Sytchev,A.E., (2005), SHS of nano-powders, in: Lessons in nanotechnology from traditional materials to advanced ceramics, Baumard, J.F.(ed); Dijon, France : Techna Group Srl., 1-27

Merzhanov, A.G. & Mukasyan A.S., (2007), book: Solid Combustion, Moscow, Torus Press, 336

Moore, J.J.,& Feng, H.J., Combustion synthesis of advanced materials, (1995) *Prog. Mater. Sci.,* Vol. 39, 243- .

Mukasyan A.S. & Martirosyan K.S. (ed.), (2007), Combustion of heterogeneous systems: fundamnetas and applications for materials sythesis, Transworld Res. Network, Kerala, India, 245

Mukasyan A.S., Merzhanov A.G., Martinenko V.M., Borovinskaya I.P., and Blinov M.Y., (1986) "Mechanism and Principles of Silicon Combustion in Nitrogen," *Combust. Explos. Shock Waves*, Vol. 22, No. 5, 534-540.

Mamyan,S.S., (2002), Thermodynamic analysis of SHS processes, *Progress In Self-Propagating High-Temperature Synthesis*, book Series: *Key Eng. Materials,* Vol.217 1-8

Munir, Z.A., & Anselmi-Tamburini, U., (1989), Self-propagating exothermic reactions: the synthesis of high-temperature materials by combustion, *Mater. Sci. Reports*, Vol. 3, No.7-8, 277-365

Munir, Z.A., (1997), Field effects in Self-propagating solid-state synthesis reactions, *Sol.State Ionics,* Vol.101-103, 991-1001

Nersisyan, G.A.; Nikogosov, V.N.; Kharatyan, S. L.; Merzhanov, A.G., (1991) Chemical transformation mechanism and combustion regimes in the system silicon-carbon-fluoroplast, *Combustion explosion and Shock Wave*, Vol.27, No.6, 729-724

Pampuch, R ; Stobierski, L.; Liz, J.; Raczka, M, (1987) Solid Combustion Synthesis of β-SiC powder, *Mat. Res. Bull.*, Vol. 22, 1225-1231

Pampuch, R ; Stobierski, L.; Liz, J.; (1989) Synthesis of siterable B-SiC powders by solid combustion method, *J. Amer. Cer. Soc.,* Vol.72, No.8 1434-1435

Puszynski, J.A.; Miao, S. (1998) Chemically-assisted combustion synthesis of silicon carbide from elemental powders, in: Sigh, J.P. (ed,) *Innovative Process/Synthesis;Ceramics, Glasses, composites II,* Amer. Cer. Soc, Westerville 13-28

Rogachev, A.S. & Mukasyan, A.S, (2010), *Combust. Explos. Shock Waves,* Vol 46, No.3, 243-266

Shiryaev, A.A., Distinctive features of thermodynamic analysis in SHS investigations, (1993) *J. of Eng. Phys. and Thermophys.* Vol. 65, 957 -96.

Suryanarayana, C., (2005), book: *Mechanical Alloying and Milling,* Marcel Dekker, New York, 457

Varma, A.; Rogachev, A.S.; Mukasyan, A.S. & Hwang, S. (1998), Combustion synthesis of advanced materials: principles and applications, *Adv. Chem. Eng.,* Vol.24, Academic Press, 79-226

Wu, Ch.-Ch. & Chen, Ch.-Ch., (1999), Direct combustion synthesis of SiC powders, *J. Mat. Sci.,* Vol.34, 4357-4363

Xue, H., & Munir, Z.A., (1996) The synthesis of composites and solid solutions of a-SiC-AlN by field -activated combustion, *Scr. Mater.,* Vol.35, no.8, 97 9-982

Yamada, O.; Miyamoto, Y.; Koizumi, M., (1985) High-pressure self-sintering of silicon carbide, *Amer. Cer. Soc. Bul.,* Vol. 64, No.2, 319-321

Yamada, O.; Miyamoto, Y.; Koizumi, M., (1986), Self-propagating high-temperature synthesis of SiC, *J. Mater. Res.*, Vol.1, No.2, 275-279

Yamada, O.; Hirao,K; Koizumi, M., Miyamoto, Y.; (1989), Combustion synthesis of silicon carbide in nitrogen atmosphere, *J. Amer. Ceram. Soc,* Vol 72, No.9 1735-1738

Zhang, J; Jeong, J.C.; Lee, J.H., Won, C.W.; Kim, D.J.; Kim, C.O. (2002), The effect of carbon sources and activative additive on the formation of SiC powder in combustion reaction, *Mat. Res. Bull.,* Vol.37, 319-329

Yang, Y; Lin, Zh.-M., Li, J.-T., (2009) Synthesis of SiC by silicon and carbon combustion in air, *J. Europ. Cer. Soc.,* Vol. 29, 175-180

Yang, Y.; Yang, K.; Lin, Z-M.; Li, J-T., (2007a) Mechanical-activation-assisted combustion synthesis of SiC, *Mat. Lett.,* Vol.61, 671-676

Yang, Y.; Yang, K.; Lin, Z-M.; Li, J-T.,(2007b), Mechanical-activation-assisted combustion synthesis of SiC powders with polytetrafluoroethylene as promoter, (2007) *Mat. Res. Bull,* Vol.42 1625-1632

Yang, Y.; Lin, Z-M.; Li, J-T.,(2009),Synthesis of SiC by silicon and carbon combustion in air, *J. Europ. Ceram. Soc,* Vol 29, 175-180.

18

In Situ Synthesis of Silicon-Silicon Carbide Composites from SiO_2-C-Mg System via Self-Propagating High-Temperature Synthesis

Sutham Niyomwas

*Ceramic and Composite Materials Engineering Research Group (CMERG),
Department of Mechanical Engineering, Faculty of Engineering,
NANOTEC Center of Excellence at Prince of Songkla University,
Prince of Songkla University, Hat Yai, Songkla, Thailand 90112*

1. Introduction

Silicon carbide is one of the most important non-oxide ceramic materials which are produced on a large scale in the form of powders, molded shapes and thin film (Boulos et al., 1994). It has wide industrial application due to its excellent mechanical properties, high thermal and electrical conductivity, excellent chemical oxidation resistance and it has potential application as a functional ceramic or a high temperature semiconductor. The main synthesis method of SiC is a carbothermal reduction known as the Acheson process. The general reaction (Pierson, 1996) is:

$$SiO_{2(s)} + 3C_{(s)} \rightarrow SiC_{(s)} + 2CO_{(g)} \quad (1)$$

A conventional carbothermal reduction method for the synthesis of pure SiC powders involves many steps and is an energy intensive process. Several alternate methods such as sol-gel (Meng et al., 2009), thermal plasma (Tong and Reddy, 2006), carbothermal reduction (Gao et al., 2001), microwave (Satapathy et al., 2005) and SHS (Morancais et al., 2003; Gadzira et al., 1998; Feng et al., 1994; Niyomwas 2008; and Niyomwas 2009) have been reported in the literature for the synthesis of SiC powders. Sol-gel process requires expensive precursor's solutions and complicated process while the thermal plasma synthesis, laser synthesis and microwave synthesis have very high operating cost with expensive equipments; on the other hand SHS is considered as less expensive to produce SiC powders.

SHS process can be used to prepare a fine powder of high temperature materials at 1800 to 4000 ^{o}C, using their high exothermic heats of reaction. It is well known that the SHS process is a very energy-efficient method because a high-temperature furnace is not required and the process is relatively simple. Many researchers reported of using elemental silicon and carbon to synthesis SiC via SHS. However, this reaction is not strong enough and without constant maintenance of temperature at a certain level, SHS for SiC does not take place.

Due to the high reaction temperature and long reaction time, the powders produced have a large particle size. It is well known that materials with fine microstructures exhibit markedly improved properties without change in their physical appearance. These characteristics include improved strength, stiffness, wear resistance, fatigue resistance and lower ductility and toughness. Nanoscale SiC particles have important potential application in nanoelectronics, field emission devices and nanocomposite therefore, efforts have been made by many research groups to fabricate SiC nanoparticles by method like dual-beam implantation of C+ and Si+ ions (Kogler et. al., 2003), solid state reaction of Si-C (Meng et. al., 2000; Larpkiattaworn et. al., 2006) and thermal plasma synthesis (Tong and Reddy, 2006). These methods involve multi-steps and energy intensive process that have difficulty in establishing commercial viability.

In this report, Si-SiC composite and nanocomposites particles were synthesized by SHS from powder mixtures of SiO_2-Mg-C and SiO_2-Mg-C-NaCl without preheating the precursors. Thermodynamics model for SHS reaction was developed. The experimental results of the synthesis of Si-SiC composite particles were compared with the model calculation. The effects of silica sources, carbon mole ratio in precursor mixture and the amount of added NaCl in precursor on the Si-SiC conversion were investigated. An excellent agreement between model results and experimental data from this study was obtained

2. Experimental

The raw materials used in this report were Mg, C, NaCl and different sources of SiO_2 powders; i.e. precipitated SiO_2, fumed silica, rice husk ash (RHA), and natural sand whose properties are listed in Table 1 and 2 and shown in Figure 1 and 2.

Reactant	Vendor	Size	Purity (%)
Mg	Riedel-deHaen	-	99
C	Ajax Finechem	-	99
SiO_2 (pp)	Aldrich	-325 mesh	99.6
SiO_2 (Fumed)	Aldrich	-	99.8
SiO_2 (RHA)	-	-325 mesh	90.0
SiO_2 (Sand)	-	-325 mesh	98.5
NaCl	Lab-Scan Analytical Science	-	99.0

Table 1. Properties of the reactant powders

Compound	wt%
SiO_2	89.96
K_2O	7.68
CaO	1.64
P_2O_5	0.7

Table 2. Composition of RHA from XRF analysis

Fig. 1. SEM image of (a) RHA, (b) fumed silica (c) precipitated SiO_2 and (d) natural sand

Fig. 2. SEM image of (a) Mg and (b) activated carbon

The experimental setup used in this work is schematically represented in Figure 3. It consisted of a SHS reactor with controlled atmospheric reaction chamber and tungsten filament connected to power source through current controller which provides the energy required for the ignition of the reaction.

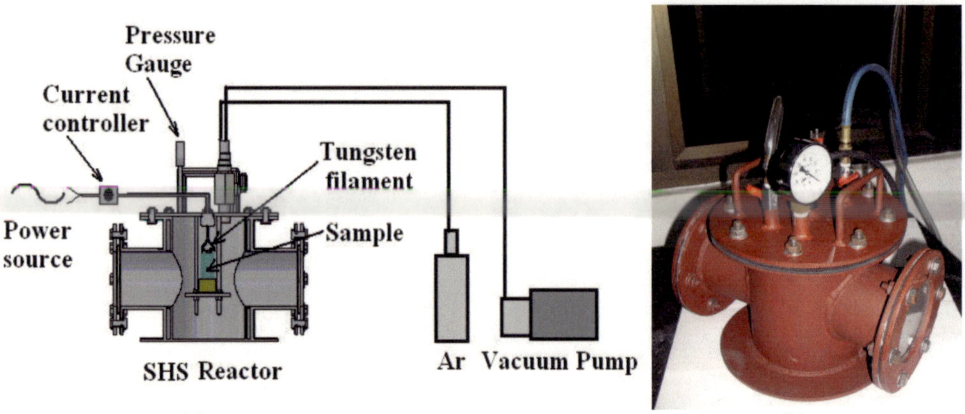

(a) (b)

Fig. 3. (a) Schematic of the experimental setup and (b) SHS reactor

Reactant powders were weighted as stoichiometric ratio and milled in planetary ball-milled with the speed of 250 rpm for 30 minute. The obtained mixture was uniaxially pressed to form cylindrical pellets (25.4 mm. diameter and about 25 mm high) with green density in the range of 50-60% of the theoretical value. Green sample was then loaded into reaction chamber of SHS reactor. The reaction chamber was evacuated and filled with argon. This operation was repeated at least twice in order to ensure an inert environment during reaction revolution. The combustion front was generated at the top end of a sample by using a heated tungsten filament. Then, under self-propagating conditions, the reaction front travels until it reaches the opposite end of the sample.

The obtained products were characterized in terms of chemical composition and microstructure by XRD (PHILIPS with Cu Kα radiation), TEM (JEM-2010, EOL), and SEM (JEOL, JSM-5800 LV) analyses.

3. Thermodynamic Analysis

Calculations for equilibrium concentration of stable species produced by SHS reaction were performed based on the Gibbs energy minimization method (Gokcen and Reddy, 1996). The evolution of species was calculated for a reducing atmosphere and as a function of temperature in the temperature range of 0-3000°C. Calculations assume that the evolved gases are ideal and form an ideal gas mixture, and condensed phases are pure. The total Gibbs energy of the system can be expressed by the following equation:

$$G = \sum_{gas} n_i\left(g_i^o + RT\ln P_i\right) + \sum_{condensed} n_i g_i^o + \sum_{solution} n_i\left(g_i^o + RT\ln x_i + RT\ln\gamma_i\right) \quad (2)$$

where G is the total Gibbs energy of the system; g_i^o is the standard molar Gibbs energy of species i at P and T; n_i is the molar number of species i; P_i is the partial pressure of species i; x_i is the mole fraction of species i; and γ_i is the activity coefficient of species i. The exercise is to calculate n_i in a way G is a minimized subject to mass balance constraints.

The equilibrium composition of SiO_2-Mg-C system at different temperatures was calculated using Gibbs energy minimization method and the results were shown in Figure 4. The overall chemical reactions can be expressed as:

$$2SiO_{2(s)} + C_{(s)} + 4Mg_{(s)} = Si_{(s)} + SiC_{(s)} + 4MgO_{(s)} \tag{3}$$

During the process of SHS, the mixture of SiO_2, Mg and C may have interacted to form some possible compounds as the following intermediate chemical reactions.

$$SiO_{2(s)} + 2Mg_{(s)} = Si_{(s)} + 2MgO_{(s)} \tag{4}$$

$$Si_{(l)} + C_{(s)} = SiC_{(s)} \tag{5}$$

$$SiO_{2(s)} + C_{(s)} = SiO_{(g)} + CO_{(g)} \tag{6}$$

$$SiO_{(g)} + 2C_{(s)} = SiC_{(s)} + CO_{(g)} \tag{7}$$

$$2MgO_{(s)} + SiO_{2\,(l)} = Mg_2SiO_{4\,(s)} \tag{8}$$

The adiabatic temperature of the SHS process can be calculated from the enthalpy of reaction (Moore and Feng, 1995). This is the maximum theoretical temperature that the reactants can be reached, and determined from Equation (9). This equation applies to a phase change occurring between initial temperature and T_{ad}. The calculated result of overall reaction from Equation (3) is 2162.7 °C.

$$\Delta H = \int_{298}^{T_m} C_{p,solid}\, dT + \Delta H_f + \int_{T_m}^{T_{ad}} C_{p,liquid}\, dT \tag{9}$$

where, ΔH is the enthalpy of reaction, ΔH_f is the enthalpy of transformation, C_p is specific heat capacity, T_m is the melting temperature and T_{ad} is adiabatic temperature.

In order to synthesize nanocomposites particles, NaCl was added to the precursor as a diluents and the reaction can be expressed as:

$$2SiO_{2(s)} + C_{(s)} + 4Mg_{(s)} + nNaCl_{(s)} = Si_{(s)} + SiC_{(s)} + 4MgO_{(s)} + nNaCl_{(s)} \tag{10}$$

Adiabatic temperature of each sample with the difference amount of NaCl was calculated from Equation (3) and shown in Table 3 .

NaCl (mole)	0	0.25	0.5	0.75	1.0
T_{ad} (°C)	2162.7	2036.9	1924.2	1816.6	1725.4

Table 3. Adiabatic temperature of the chemical reaction of TiO_2-C-Mg-NaCl system

It can be seen that T_{ad} of all reactions which added 0 - 0.75 mole of NaCl were higher than 1800 °C thus those reactions were feasible for SHS process. The reduction of T_{ad} with adding NaCl resulted from the melting of NaCl and absorbed by molten salt.

It has been accepted that the reaction can be self-sustained combustion when adiabatic temperature of the reaction higher than 1800°C (Moore and Feng, 1995). From calculated adiabatic temperature of the reactions, they are higher than 1800°C, thus the using of SHS is feasible for these systems.

It can be seen from Figure 4 that it is thermodynamically feasible to synthesize Si-SiC-MgO by igniting the reactant of reaction (3) and (10). Due to a highly exothermic reaction at room temperature (ΔH =-655.15 kJ) and thermodynamic instability at room temperature, the reactant phases of SiO_2, C and Mg were not shown in the calculated stable phases in Figure 4. After ignition the reaction (4) took place and has followed by reaction (5) to form Si-SiC-MgO phases. At temperatures higher than 1500°C, the system was unstable and formed an intermediate phase of Mg_2SiO_4 and gas phases. When the reaction front moved further away the products cooled down and rearranged phases in such a way that was shown in Figure 4.

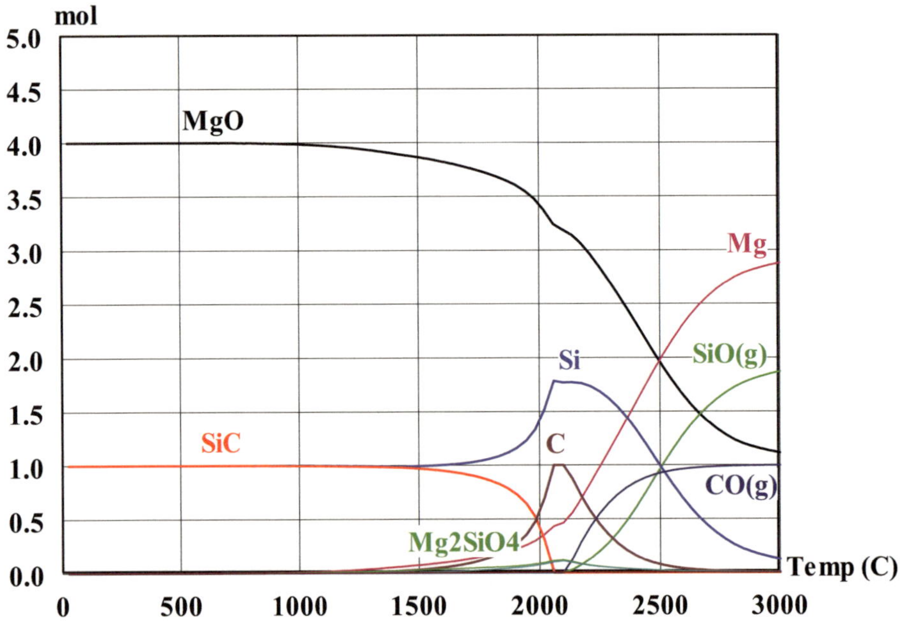

Fig. 4. Equilibrium composition of SiO_2-C-Mg systems in Ar gas atmosphere

4. Synthesis of Si-SiC Composites

4.1 The effects of carbon molar ratio in the precursor mixture

Figure 5 shows XRD patterns of reaction products with different relative amounts of carbon at a constant molar ratio of SiO_2:Mg = 2:4. All three conditions can produce resulting products as in Equation (3) and intermediate phase of Mg_2SiO_4. As the relative amount of carbon to SiO_2:Mg increased, the Si content in the final products changed accordingly. The Si content was highest when using 1 mole of carbon in the precursor mixture which is the stoichiometric ratio of precursors of Equation (3). When mixing 2 mole of carbon into the reaction, it is more favorable to form SiC (Niyomwas 2008), as shown in Equation (11), that results in less Si in the product. The other evidence of higher reaction temperature is the relative higher intensity of α-SiC phase which forms at higher temperature than β-SiC phase appeared at the condition of 2 mole carbon reaction. For adding 3 mole of carbon into the precursor mixture, Si content in the product increases while SiC decreases and some free carbon is left in the system. This may be due to the excess carbon absorbed heat released from exothermic reaction that causes the decrease of overall heat of reaction and less favor for forming SiC.

$$2SiO_{2(s)} + 2C_{(s)} + 4Mg_{(s)} = 2SiC_{(s)} + 4MgO_{(s)} \quad ; T_{ad} = 2393.2^{\circ}C \tag{11}$$

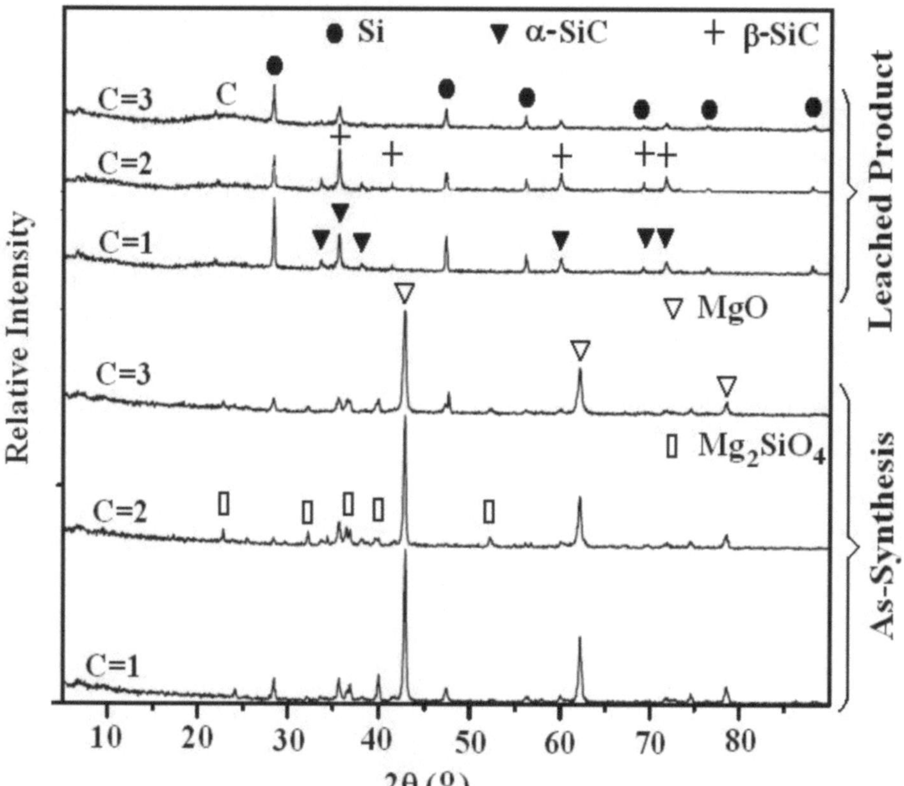

Fig. 5. XRD patterns of as-synthesized products and leached products of different carbon mole ration in the precursor mixture

Figure 6 (a) to (f) shows low and high magnification SEM micrographs of the products from SHS reaction after leaching process from the precursor mixture of 1, 2, and 3 mole of carbon, respectively. The morphology of products reveals an agglomerated particle of Si-SiC as identified by XRD patterns in Figure 5. Figure 6 (a), (c) and (e) shows the free carbon left in the products which is more apparent when 3 mole of carbon was added into the precursor mixture.

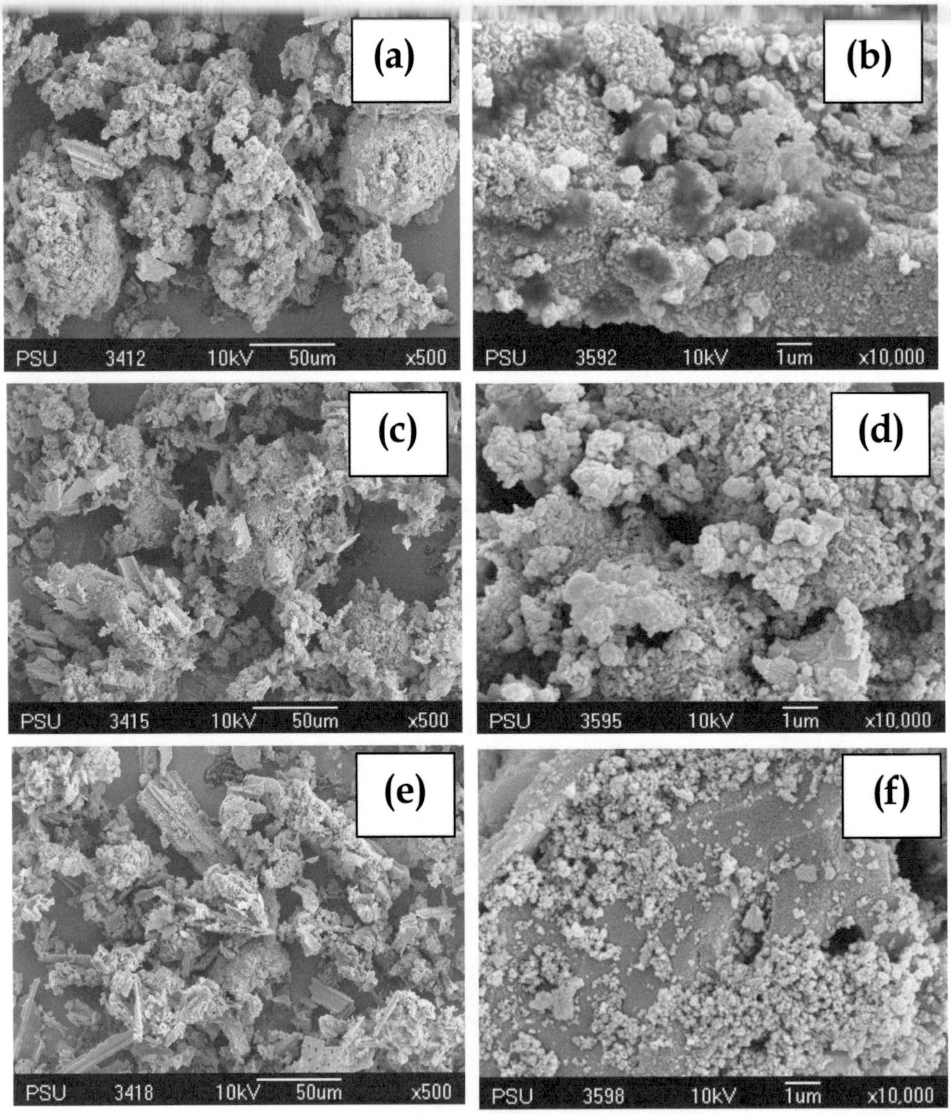

Fig. 6. SEM image of leached products from different carbon mole ratio on mixture of SiO_2 (pp)as silica sources: (a) and (b) for 1 mole C, (c) and (d) for 2 mole C, and (e) and (f) for 3 mole C

It is believed that the agglomeration of fine SiC particle occurred because the Si-SiC-MgO composite powder was synthesized by melting of the reactants followed by recrystallization. The reactants are in the solid state at the early stage of the reaction. As the reaction temperature increase to 650°C, the Mg particles start to melt. At the higher temperature than 650°C, the carbon and SiO_2 particles are surrounded by the Mg melt, and the SiO_2 particles are reduced by the Mg melt. It is assumed that the theoretical adiabatic temperature, 2162.7 °C, is reached, the Si is completely melted; hence, the diffusion of the carbon, Si, and oxygen is rapid, and it is believed that the SiC and MgO particles are synthesized simultaneously based on the thermodynamic calculation resulted (Figure 4).

4.2 The effects of silica sources in the precursor mixture

Figure 7 shows XRD patterns of reaction products with different silica sources at a constant molar ratio of C:SiO_2:Mg = 2:2:4. It shows the presence of Si, SiC, MgO and Mg_2SiO_4, and no Mg peak appeared for the as-synthesized product. After leaching with dilute HCl, only Si-SiC phases were found. Not much difference was found in the as-synthesized products from different silica sources, only the products from fumed silica that showed relatively higher intensity of Si. The resulting SiC were in both α-SiC phase and β-SiC phase. The peak at 2θ = 33.82° in XRD patterns of leached product was detected near the peak of cubic structure (β-SiC phase) of silicon carbide at 2θ = 35.64° which is characteristic of hexagonal polytypes (α-SiC phase). Peaks due to Si-SiC phases are observed together with the broad peak of amorphous carbon, indicating that the formation of SiC has taken placed and that residual free carbon still exists. To get rid of free carbon in the final products, the more studies on lower molar ratio of carbon in the precursor mixtures are planned in the future.

Figure 8 shows SEM micrograph of the products from SHS reaction after leaching process from the precursor mixture of different silica sources, i.e. (a) RHA, (b) fumed silica, (c) sand and (d) precipitated silica. The morphology of products reveals an agglomerated particle of Si-SiC as identified by XRD patterns in Figure 7. The existence of Mg_2SiO_4 in the product may be due to incomplete reaction from the rapid nature of SHS reaction. Free carbon left in the products is also evidently found in every condition.

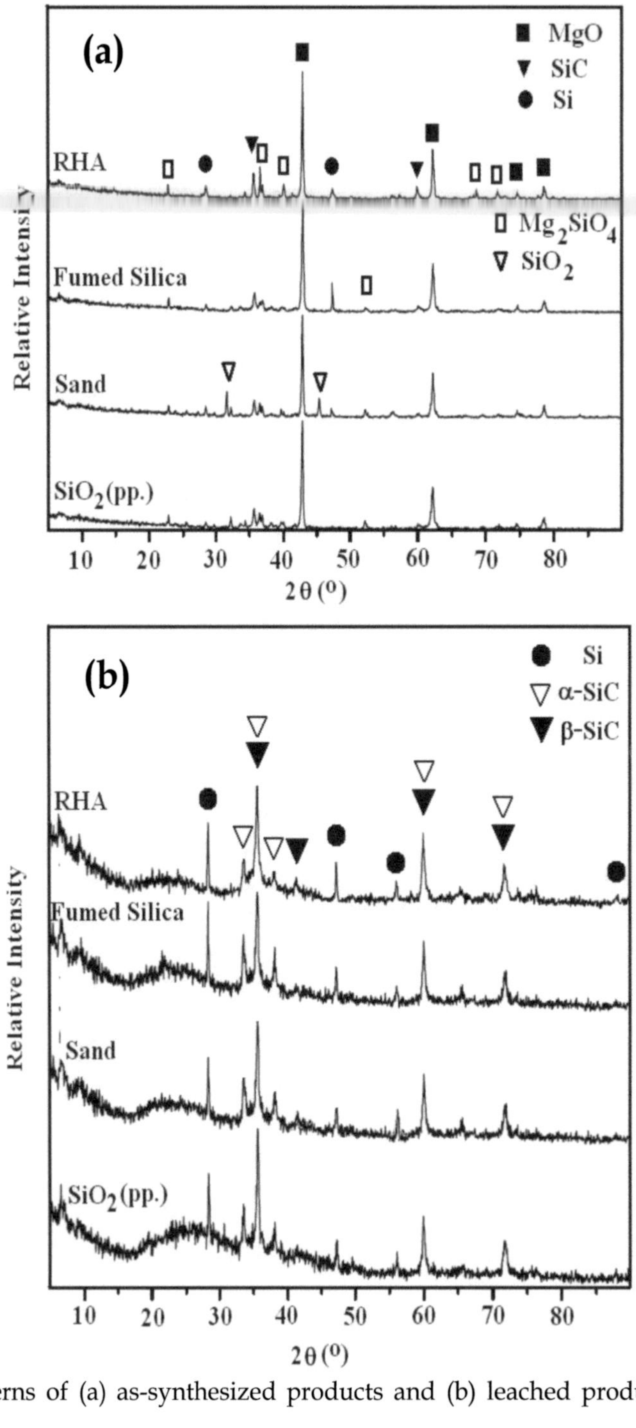

Fig. 7. XRD patterns of (a) as-synthesized products and (b) leached products of different silica sources

Fig. 8. SEM image of leached products from difference silica (a) RHA, (b) fumed silica, (c) sand and (d) precipitated silica

4.3 The effects of adding NaCl in the precursor mixture

The final product Si-SiC nanoparticle composites were obtained from washing out NaCl from synthesized product by distilled water and leaching the rest with 0.1M HCl for 12 hours and identifying with SEM, XRD and TEM images, shown in Figure 9, 10 and 11, respectively.

The XRD patterns of the samples after reaction showed the presence of MgO, Mg_2SiO_4 in the as-synthesized powder and no Mg peak appeared. After leaching with dilute HCl, only Si-SiC phase were found. When NaCl was added with the reaction mixture, diffraction peak due to NaCl were also detected. As the NaCl content was increased, the intensity of the Si peaks were increased, at the same time α-SiC peaks were decreased. This is due to the lower reaction temperature caused by NaCl diluents to the reactions. As the reaction temperature goes down, the reaction in Equation 3 was incompleted and left more Mg_2SiO_4 phase and also have less energy to form SiC from element Si and carbon causing higher peak of Si.

The morphology of leached and washed Si-SiC powders were shown in Figure 9. They appeared as lumpys of agglomerated powders with finer powders for higher NaCl content. The microstructure observations of the leached samples revealed the generation of fine powders. Transmission electron micrographs of 0, 0.25 and 0.75 mole of NaCl samples are given in Figure 11 (a), (b) and (c), respectively. The micrographs revealed that the powders remained agglomerated even after leaching and drying. With the addition of NaCl, the particles were observed to be very fine and less agglomerated. During exothermic reactions, very high temperature (>1800 °C) were achieved within a fraction of minutes. Since the combustion temperature was much higher than the boiling point of NaCl (1465 °C) evaporation of NaCl was expected. The presence of NaCl in the XRD pattern indicated that all of the NaCl did not escape out from the mixture after the synthesis. This NaCl may have partially melted and vaporized which give a coating on the SiC particle. This coating could reduce the grain growth of particles and finally decrease the particle sizes. The decrease of adiabatic temperatures with the addition of NaCl may also have helped in reducing the grain growth.

Fig. 9. SEM micrographs of synthesized products after leaching when added (a) 0.25 and (b) 0.75 mole of NaCl

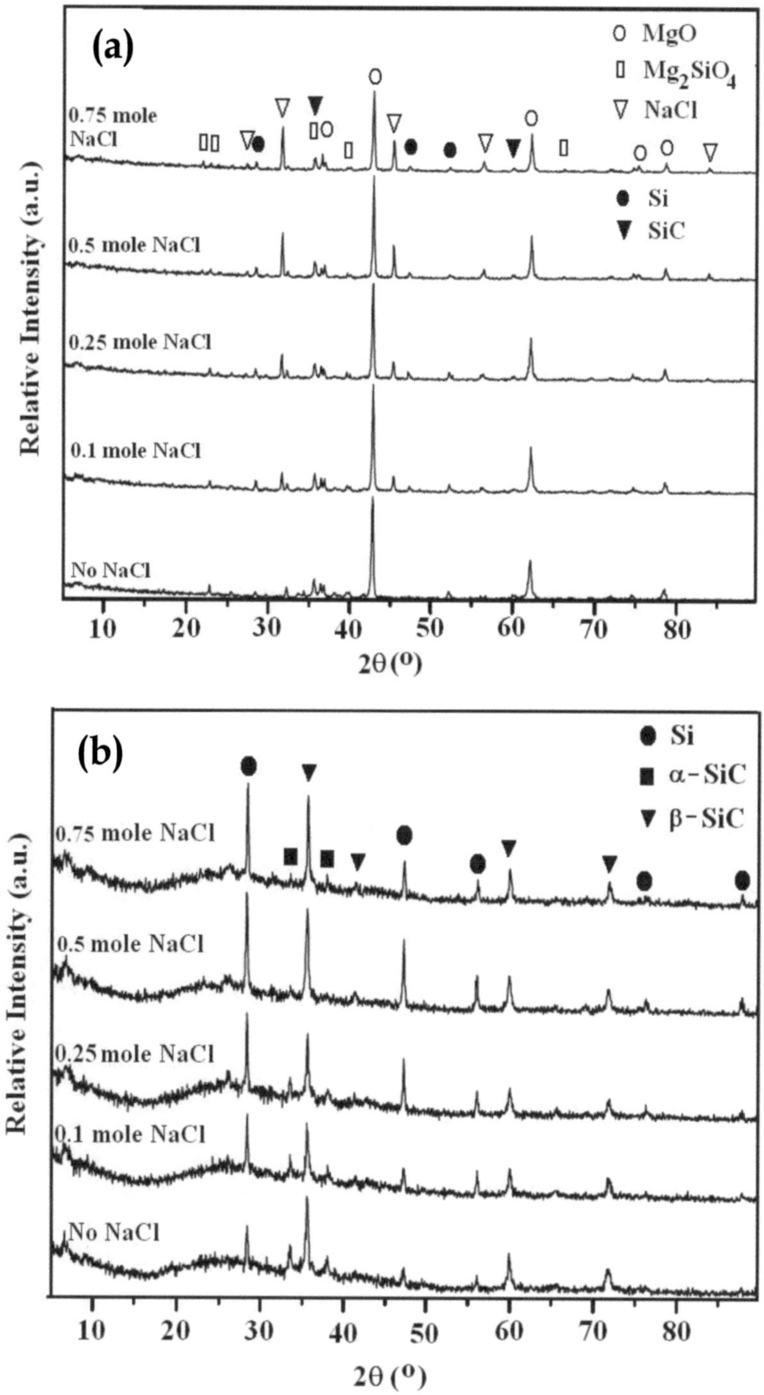

Fig. 10. XRD patterns of (a) as-synthesized and unleached products and (b) leached and washed products

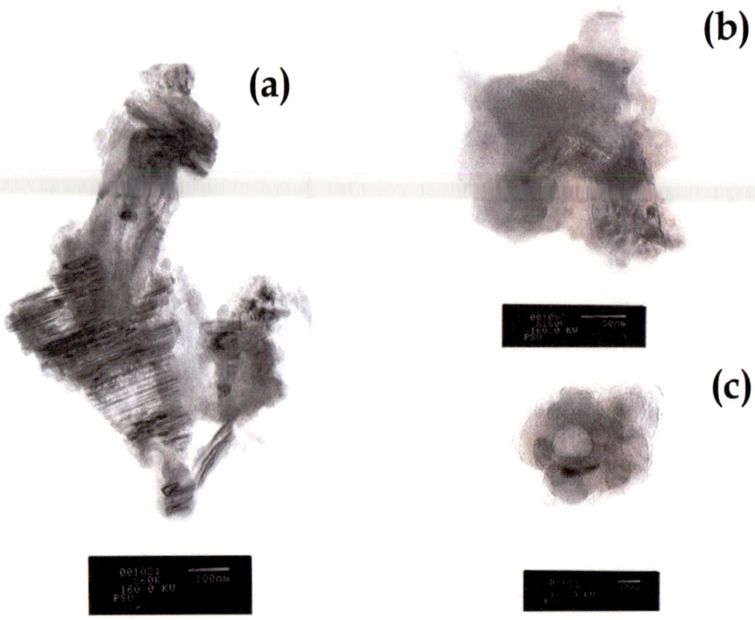

Fig. 11. TEM micrographs of synthesized products after leaching when added (a) 0, (b) 0.25 and (c) 0.75 mole of NaCl

5. Conclusion

The Si-SiC composite powders were produced from leaching out MgO and Mg_2SiO_4 from reaction products that were *in-situ* synthesized via self-propagating high temperature synthesis reaction from precursors of SiO_2–Mg–C. The highest Si to SiC was found when use mixture molar ratio of C:SiO_2:Mg = 1:2:4, while the lowest one was found on the result of mixture molar ratio of C:SiO_2:Mg = 2:2:4. As the relative molar ratio of carbon to SiO_2:Mg increase (C:SiO_2:Mg = 3:2:4), the free carbon left in the final product was relatively high. The resulting products from different silica sources of RHA, fumed silica, sand and precipitated silica, show no significant different in Si-SiC composite. In all conditions, the resulting SiC were in both α-SiC phase and β-SiC phase.

The nano-sized Si-SiC nanoparticle composites were successfully synthesized via self-propagating high temperature synthesis (SHS) from the reactants of SiO_2, C, and Mg with the mole ratio 2:1:4 and nNaCl (n = 0.1-0.75 mole). As the amount of NaCl in precursor mixture was increased, the resultant product showed decreasing particle size to nanometer range.

6. Acknowledgements

This work has been financial supported from the National Nanotechnology Center (NANOTEC), NSTDA, Ministry of Science and Technology, Thailand, through its “Program of Center of Excellence Network" (NANOTEC Center of Excellence at Prince of Songkla University), and partially financial support from the Ceramic and Composite Materials Engineering Research Group (CMERG) of the Faculty of Engineering, Prince of Songkla University, Thailand

7. References

Boulos, M.I.; Fauchais, P. & Pfender, E.(1994). *Thermal Plasmas Fundamentals and Applications, Vol. 1*, Plenum Press, New York and London, USA, 37.

Feng, A. & Munir, Z.A. (1994). Field-assisted Self-propagating Synthesis of β-SiC, *J. Appl. Phys.*, Vol. 76, 1927-1928.

Gadzira, M.; Gnesin, G.; Mykhaylyk, O. & Andreyev, O. (1998). Synthesis and Structural Peculiarities pf Nonstoichiometric β-SiC, *Diamond and Related Mater.*, Vol. 7, 1466-1470.

Gokcen, N.A. & Reddy, R.G.(1996). *Thermodynamics*, Plenum Press, New York, NY, USA, 291.

Gao, Y.H.; Bando. Y.; Kurashima K. & Sato, T. (2001). The Microstructural Analysis of SiC Nanorods Synthesized Through Carbothermal Reduction, *Scripta Mater.*, Vol. 44, 1941-1944.

Kogler, R.; Eichhorn, F.; Kaschny, J. R.; Mucklich, A.; Reuther, H.; Skorupa, W.; Serre, C. & Perez-Rodriguez, A. (2003). Synthesis of nano-sized SiC precipitates in Si by simultaneous dual-beam implantation of C_+ and Si_+ ions, *Applied physics. A, Materials science & processing*, Vol. 76, No. 5, 827-835.

Larpkiattaworn, S.; Ngernchuklin, P.; Khongwong, W.; Pankurddee, N. & Wada, S. (2006) . The influence of reaction parameters on the free Si and C contents in the synthesis of nano-sized SiC, *Ceramic Internatonal*, Vol. 32, 899-904.

Meng, G.W.; Cui, Z.; Zhang, L.D. & Phillipp, F. (2000). Growth and Characterization of Nanostructured β-SiC via Carbothermal Reduction of SiO_2 Xerogels Containing Carbon Nanoparticles, *J. Cryst. Growth*, Vol. 209, 801-806.

Moore, J. & Feng, H. (1995). Combustion synthesis of advanced materials: Part I Reaction parameters, *Prog. In Mat. Sci.*, Vol. 39, 243-273.

Morancais, A.; Louvet, F.; Smith, D.S. & Bonnet, J.P. (2003). High Porosity SiC Ceramics Prepared Via a Process Involving an SHS stage, *J. Euro. Ceram Soc.*, Vol. 23, 1949-1956.

Niyomwas, S. (2008). The effect of carbon mole ratio on the fabrication of silicon carbide from SiO_2-C-Mg system via self-propagating high temperature synthesis, *Songklanakarin J. Sci. Technol.*, Vol. 30, No. 2, 227-231.

Niyomwas, S. (2009). Synthesis and Characterization of Silicon-Silicon Carbide from Rice Husk Ash via Self-Propagating High Temperature Synthesis, *Journal of Metals, Materials and Minerals*, Vol. 19, No. 2, 21-25.

Pierson, H.O.(1996). *Handbook of Refractory Carbides and Nitrides*, William Andrew, Noyes, 137.

Satapathy, L.N.; Ramesh, P.D.; Agrawal, D. & Roy, R. (2005). Microwave synthesis of phase-pure, fine silicon carbide powder, *Mater. Res. Bull.*, Vol. 40, 1871-1882.

Tong, L. & Reddy, R.G. (2006). Thermal Plasma Synthesis of SiC nano-powders/nano-fibers, *Mater. Res. Bull.*, Vol. 41, No.12, 2303-2310.

19

High Reliability Alumina-Silicon Carbide Laminated Composites by Spark Plasma Sintering

Vincenzo M. Sglavo and Francesca De Genua
University of Trento
Italy

1. Introduction

High-melting temperature oxides, carbides and nitrides (Al_2O_3, ZrO_2, SiC, B_4C, Si_3N_4, BN) are superior in hardness and strength to metals, especially at high temperature and in severe conditions. However, the extensive use of such ceramics in structural engineering applications often encountered critical problems due to their lack of damage tolerance ability and to the limited reliability of the mechanical properties.

Several ceramic composites and, in particular, laminated structures were developed in recent years to enhance strength, toughness and to improve flaw tolerance. Fracture resistance and R-curve behaviour were achieved in laminated composites with (i) thin layers in residual compression alternated to thick layers in tension able to arrest surface and internal cracks and produce a threshold strength (Rao et al., 1999; Orlovskaya et al., 2005; Bermejo et al., 2006; Bermejo & Danzer, 2010; Náhlík et al., 2010), (ii) weak or porous interlayers that generate graceful and tortuous crack propagation paths (Davis et al., 2000; She et al., 2000), (iii) high strength surface layers and high toughness core, able to arrest starting surface flaws and to limit long cracks propagation, respectively, (Cho et al., 2001) and (iv) metallic layers that promote ductile bridging effects (Mekky & Nicholson, 2007).

The limitations of such laminates are related to processing difficulties and to the fact that they can be used only when the tensile load is applied parallel to the layers, thus being not easily suitable to produce real components such as shells or tubes. Nevertheless the improvement in toughness is often achieved at the expense of strength and still leaves limited relief from the variability in fracture stress.

Both significant strength increase and improved mechanical reliability, in terms of Weibull modulus or minimum threshold failure stress, can be achieved by the "engineering" of the critical surface region in the ceramic component. Such effect can be realized by using a laminated composite structure with tailored sub-surface insertion of layers with different composition. Such laminate is able to develop, upon co-sintering, a spatial variation of residual stress with maximum compression at specific depth from the surface due to the differences in thermal expansion coefficient of the constituting layers. The arising apparent fracture toughness curve (Sglavo et al., 2001; Sglavo & Bertoldi, 2006 a; Sglavo & Bertoldi, 2006 b) is responsible for the stable growth of surface flaws before final catastrophic failure

and, consequently, for a certain surface damage insensitivity that leads to outstanding mechanical reliability. Using this approach, some oxide laminated structures have been produced by tape casting, a convenient and well-developed technique for the preparation of thin green sheets, and pressureless sintering and in most cases very high reliability has been observed experimentally. (Sglavo et al., 2005; Costabile & Sglavo, 2006; Sglavo & Bertoldi, 2006 a; Sglavo & Bertoldi, 2006 b; Leoni et al, 2008).
In the present work silicon carbide has been selected as second phase to graduate the thermal expansion coefficient of alumina due to its relatively low specific density that could allow the production of lighter components with improved mechanical performance and reliability. Silicon carbide, in the shape of particle, nanoparticle, platelet or whisker, has been used many times in the past to improve hardness, strength, toughness, thermal shock resistance and creep resistance of alumina ceramics (Lee & Rainforth, 1994; Peters, 1998) Nevertheless, it has been shown that pressureless sintering to full density of alumina/silicon carbide composites can occur only at temperatures above 1700°C and usually involves considerable weight loss due to the formation of volatile compounds (Gadalla et al., 1992; Hue et al., 1997). In order to overcome such problems the alumina/silicon carbide composites have been usually prepared by pressure assisted sintering techniques such as Hot Pressing (HP).
In the present work, Spark Plasma Sintering (SPS) technology has been chosen for the consolidation of the laminates. SPS is a new promising solid compressive and pulsed electric current energizing sintering technology whose consolidation mechanisms are still under debate. During the SPS process a mechanical pressure is applied to the powder compact within the die while heating is provided internally thanks to a pulsed direct electric current passing through the die and, depending on the electrical conductivity of the material to be sintered, also through the powder compact. Such electrical current propagation significantly improves heat-transfer resulting in a current activated sintering. Therefore, SPS has several advantages over conventional sintering methods including higher heating rates, lower processing temperatures and shorter holding times, this allowing the production of highly dense materials with good control of grain coarsening (Chae et al., 2006; Munir et al., 2006).
The purpose of this work was to produce by tape casting and SPS alumina/silicon carbide monolithic composite laminates, made by stacking together laminae of identical composition, and an engineered laminate, with layers of different composition, specifically designed to manifest high mechanical reliability.

2. Experimental

2.1 Sample Processing

Commercially available high-purity α-Al_2O_3 (A-16SG, Alcoa Inc., Pittsburgh, PA, USA) and α-SiC (Sika ABR I F1500S, Saint-Gobain, Courbevoie, France) powders with average particle size of 0.4 μm and 1.8 μm and specific surface area of 8.6 m^2/g and 4.9 m^2/g, respectively, were used in this work.
Alumina/silicon carbide laminae with 0 ÷ 30 vol% SiC loads were produced and labelled as ASx where x is the volume content of the second phase (SiC). The green laminae were obtained by tape casting of water-based slurries prepared using a two-stage milling/mixing procedure necessary to break up powder agglomerates and to maintain the mechanical and chemical stability of the used binders. The alumina slurry was firstly prepared by mixing

the powder to deionized water with ammonium polymethacrylate as dispersant (Darvan C®, R. T. Vanderbilt Inc., Norwalk, CT, USA) - 0.4 mg/m^2 of active matter per unit area of alumina powder - for 24 h with alumina milling media. Then SiC powder was added and the dispersion was mixed for additional 24 h. The SiC powder was electrostatically dispersed so that no further amount of a dispersant was required.
The slurry was filtered through a 41 µm filter and de-aired for 30 min using a low-vacuum Venturi pump to remove air entrapped during the milling stage. An acrylic emulsion (Duramax® B-1014, Rohm & Hass, Philadelphia, PA, USA) as binder and a low-Tg acrylic emulsion (Duramax ® B-1000, Rohm & Hass, Philadelphia, PA, USA) as plasticizer were added to the dispersions and mixed for 30 min, taking care to avoid the production of new air bubbles. Three drops of 10 wt% wetting agent water solution (NH_4-lauryl sulphate, code 09887, Fluka Chemie AG, Buchs, Switzerland) were also added after 20 min of mixing to improve the tape spread onto the casting substrate. All suspensions were produced with a powder content equal to 39 vol%. Binder content was always equal to 15.3 vol% and plasticizer was added in a volume ratio equal to 1:2 with respect to the binder.
Tape casting was carried out using a double doctor-blade assembly (DDB-1-6, 6 in wide, Richard E. Mistler Inc., Yardley, PA, USA) with a gap height of 250 µm at a speed of 1 m/min. The tapes were casted on the polyethylene hydrophobic surface side of a composite film substrate (PET12/Al7/LDPE60, BP Europack, Vicenza, Italy) fixed on a rigid float glass plate. The relative humidity of the over-standing environment was set to 80%, after casting, in order to control the tapes drying. Green tapes with smooth surfaces and a uniform thickness of about 120 µm were therefore produced. Disks of 20 mm diameter were cut from the tapes by a hollow punch.
Green laminates were produced by thermo-compressing in a 20 mm nominal diameter aluminium die with a predefined stacking of 60 disks at 80°C under 30 MPa for 15 min.
According to the results of the thermogravimetric analysis (STA409 Netzsch GmbH, Bayern, Germany), performed at 5°C/min in air on binders and SiC powder, the adopted debinding heating cycle to allow slow and complete burn-out of organic additives was 1°C/min up to 600°C; the temperature was then raised to 4°C/min up to 1000°C for the successive pre-sintering treatment (dwell time = 12 min) useful to produce samples that could be easily handled without noticeable oxidation of the carbide.
Each pre-sintered laminate measuring 20 mm in diameter and 5.8-6.1 mm in height was wrapped into a graphite foil and carefully placed into the graphite mold. This latter was then closed and placed in the SPS apparatus (Dr. Sinter 1050, Sumimoto Coal Mining Co., Tokyo, Japan). A preliminary uniaxial pressure equal to 6.4 MPa was applied at room temperature. Vacuum level of 10^{-2} mbar was reached and pulsed current (12 impulses of 3 ms on and 2 impulses of 3 ms off) was applied. The temperature was raised up to 600°C in 5 min and then monitored and regulated by means of an optical pyrometer. Then the uniaxial pressure was risen to 28.7 MPa and the temperature was increased at first up to 1650°C (heating rate = 100°C/min) and then to 1700°C (heating rate = 50°C/min). After a holding time of 5 min the pressure was released, the current was turned off and the sample was cooled down.
Graphite foil residues were eliminated from the 3.2-3.4 mm thick sintered laminates by grinding on 250 µm grit diamond disks. A graphite spray was also used to protect the surfaces of the engineered laminates (that realized with laminae of different composition, as

reported below) during SPS, and no grinding procedure was used to remove graphite residues from such surfaces.

2.2 Design of the Engineered Laminate

Layers with different composition were stacked to obtain, after SPS, the architecture of the engineered laminate shown in Fig. 1 and labelled as AS-1. Such architecture was selected on the basis of the results presented in previous works to guarantee the stable growth of surface cracks before final failure (Sglavo et al., 2005; Costabile & Sglavo, 2006; Sglavo & Bertoldi, 2006 a; Sglavo & Bertoldi, 2006 b).

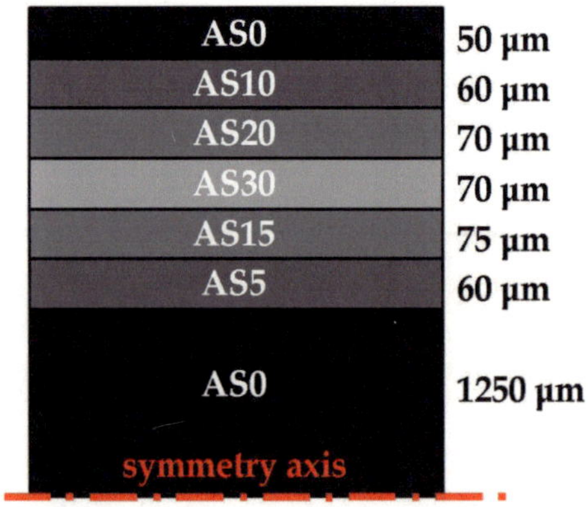

Fig. 1. Architecture (i.e. composition, stacking order and layers thickness) of the engineered alumina/silicon carbide multilayer. Dimensions are not in scale.

Sintering of a ceramic composite laminate made of layers of different composition is responsible for the formation of biaxial residual stresses in the laminate. During cooling, the difference in elastic modulus and, most of all, thermal expansion coefficient of the constituent layers is accommodated by diffusion mechanisms as long as the temperature is high enough (Rao et al., 2001). Below a certain temperature, T_{SF}, which is called the "stress free" temperature, the different components behave like a perfect elastic solid and internal stresses appear. The residual stresses $\sigma_{res,i}$ within each single layer (rank = i) under the hypotheses of symmetrical laminate geometry and perfect adhesion between layers, can be calculated using the following equation (Sglavo et al., 2005; Sglavo & Bertoldi, 2006 a; Sglavo & Bertoldi, 2006 b):

$$\sigma_{res,i} = E_i^*(\overline{\alpha} - \alpha_i)(T_{SF} - T_{RT}) \tag{1}$$

where T_{RT} is the room temperature and:

$$E_i^* = \frac{E_i}{1 - \upsilon_i} \tag{2}$$

E_i being the elastic modulus and υ_i the Poisson's ratio.
The average thermal expansion coefficient $\overline{\alpha}$ of the whole laminate, is:

$$\overline{\alpha} = \frac{\sum_{i=1}^{n} E_i^* t_i \alpha_i}{\sum_{i=1}^{n} E_i^* t_i} \quad (3)$$

t_i being the layer thickness and n the total number (odd) of the layers.
The effect of residual stresses on fracture behaviour can be combined with the intrinsic fracture toughness of each layer K_c^i thus obtaining the so-called apparent fracture toughness as a function of the distance from the surface, i.e. for increasing crack depths, (Sglavo et al., 2005; Sglavo & Bertoldi, 2006 a; Sglavo & Bertoldi, 2006 b):

$$T = K_C^i - \sum_{j=1}^{i} \left[2Y \left(\frac{c}{\pi} \right)^{0.5} \Delta\sigma_{res,j} \left[\frac{\pi}{2} - \text{arcsen} \left(\frac{x_{j-1}}{c} \right) \right] \right] \quad x_{i-1} < x < x_i \quad (4)$$

where $Y \approx 1.1215$, $\Delta\sigma_{res,j}$ being the stress increase of layer i with respect to the previous one, x_j the starting depth of layer j.
Equation (4) represents a short notation of n different equations, the sum being calculated for different number of terms for each i. On the basis of such approach, once the crack model is defined, the material failure is determined by comparing T with the stress intensity factor associated to the external loads only.
The properties of alumina and silicon carbide required for the design calculations can be obtained from NIST Structural Ceramics Database (www.nist.gov). Young modulus (E) and Poisson's ratio (υ) values for monolithic alumina/silicon carbide composites are calculated as the Hashin-Shtrikman bounds (Kingery et al., 1976) and thermal expansion coefficient (α) is evaluated by the Kerner equation (Kingery et al., 1976). Hardness (H) and fracture toughness (K_C) were determined by the conventional indentation fracture method (Anstis et al., 1972). Vickers indentations were produced using a load of 100 N for 15 s.
The properties of the composite materials used in the design procedure are summarized in Table 1. The reported thickness corresponds to the average actual value measured on optical micrographs of polished fracture surface of the sintered engineered laminate.

	AS0	AS5	AS10	AS15	AS20	AS30
K_C [MPa $m^{0,5}$]	3.73	2.61	2.80	2.87	3.04	2.99
E [GPa]	415.9	415.9	416.0	416.1	416.1	416.2
υ	0.23	0.227	0.224	0.22	0.217	0.21
α (0÷1000°C) [10^{-6} °C^{-1}]	8.1	7.98	7.85	7.72	7.59	7.32
thickness [μm]	52	57	57	75	67	66

Table 1. Materials properties and layers thickness used to estimate the stress distribution and the apparent fracture toughness of the engineered laminate AS-1. The thickness corresponds to the average actual value measured on the sintered samples.

The resulting residual stress profile calculated according to Eq. (1) for the engineered laminate AS-1 is shown in Fig 2. A stress free temperature equal to 1200°C was used for the calculations. Similarly to previous situations where the material was tailored to manifest high mechanical reliability (Green et al., 1999; Sglavo & Green, 2001; Sglavo et al., 2005; Costabile & Sglavo, 2006; Sglavo & Bertoldi, 2006 a; Sglavo & Bertoldi, 2006 b; Leoni et al, 2008), the stress state is slightly tensile (around 50 MPa) on the surface while a considerable residual compression (in excess to 400 MPa) occurs at a certain depth in the fourth layer.

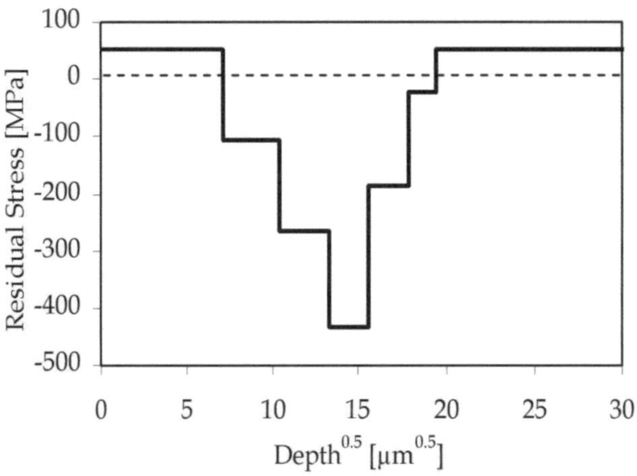

Fig. 2. Calculated residual stress profile in the surface region of the engineered laminate AS-1.

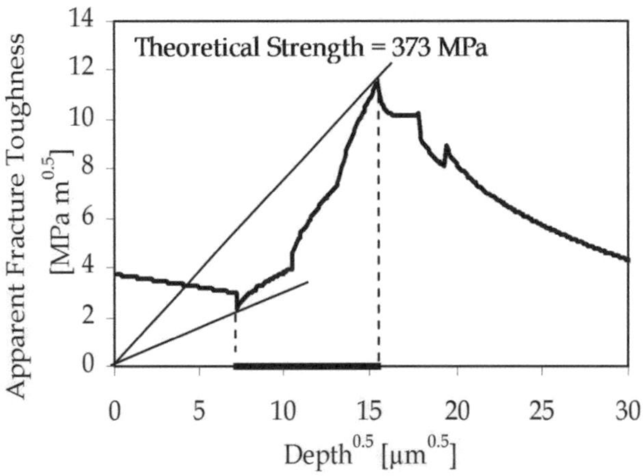

Fig. 3. Apparent fracture toughness for the engineered laminate AS-1. The straight line corresponding to the design failure stress and the minimum stress required for crack propagation are also shown and used for the determination of the stable growth interval (horizontal bar).

The apparent fracture toughness, T, of the laminate due to this stress profile and calculated according to Eq. (4) is shown in Fig. 3 as a function of the distance from the surface. A wide stable crack growth interval (from ≈ 12 μm to ≈ 250 μm) before final failure, that could be responsible for limited strength variability, can be observed. The failure stress associated to the maximum apparent fracture toughness, graphically determined by the line intersecting T at the maximum, is equal to ≈ 375 MPa.

2.3 Sample characterization

Green compact porosity and pore size distribution were measured by mercury porosimetry on the pre-sintered laminates (Pascal 140 and Porosimeter 2000, CE Instruments, Milano, Italy). The relative densities of the sintered samples were measured by the Archimedes technique using distilled water as immersion medium. The theoretical densities of the composites were calculated by the rule of mixtures assuming a value of 3.984 g/cm^3 for Al_2O_3 and 3.16 g/cm^3 for SiC.

X-ray diffraction (XRD) analyses (Geiger Flex Dmax III, Rigaku Inc., Tokyo, Japan) were carried out on monolithic pre-sintered and sintered laminates.

Fracture strength of all the laminates was measured by piston-on-three-balls biaxial flexural test (Shetty et al., 1980) using a universal mechanical testing machine (mod. 810, MTS Systems, Minneapolis, MN, USA). The disks were supported by three steel balls (3.2 mm diameter) lying on a circle (12.4 mm diameter) 120° apart. The load was applied at the centre by a hardened steel cylinder (1.5 mm diameter). A certain number of AS-1 engineered laminates were also pre-cracked in the center of the prospective tensile face by Vickers indentation using loads ranging from 10 N to 100 N before flexural test.

Fracture surfaces were finally examined by optical (BH2-UMA, Olympus Optical Co. GmbH, Hamburg, Germany) and scanning electron microscopy (JSM-5500, Jeol Inc., Tokyo, Japan). The average grain size was measured by the linear intercept method (Wurst & Nelson, 1972) on optical and scanning electron micrographs. A minimum of 200 grains from 5 randomly selected areas were counted for each sample. To reveal the microstructure specimens were diamond polished to 1 μm to obtain a mirror finish and etched: alumina was thermally etched at 1500°C for 20 min in air, while the composites were chemically etched in phosphoric acid at 250°C for 3 min.

3. Results and Discussion

The pre-sintered monolithic laminates exhibited a total porosity of approximately 40% (Fig. 4) with narrow pore size distributions from 0.14 μm to 0.16 μm. The packing density and quality of green samples is high enough to help the sintering of the compacts. Successive SPS treatment allowed to produce dense composite materials (97-99% relative density). The addition of non sinterable inclusions of SiC to alumina determined a reduction of the bulk density without noticeable influence on the sintering of the composites that always reach almost theoretical density as shown in Fig. 4.

XRD analysis performed on AS0 and AS30 monolithic laminates did not point out the presence of any impurity and confirmed the expected composition by showing only α-Al_2O_3 peaks for the first one and α-Al_2O_3 and α-SiC peaks for the second one. The formation of mullite is very limited considering the amount of silica normally present as surface layer on SiC powder or formed during the pre-sintering step. The relevant change in colour of the final samples from white, in the case of pure alumina, or light grey to black revealed that a non-controlled small carbon amount is introduced during the SPS process probably from the graphitic foil and apparatus. It is not a surface contamination but it is included deep inside within the samples.

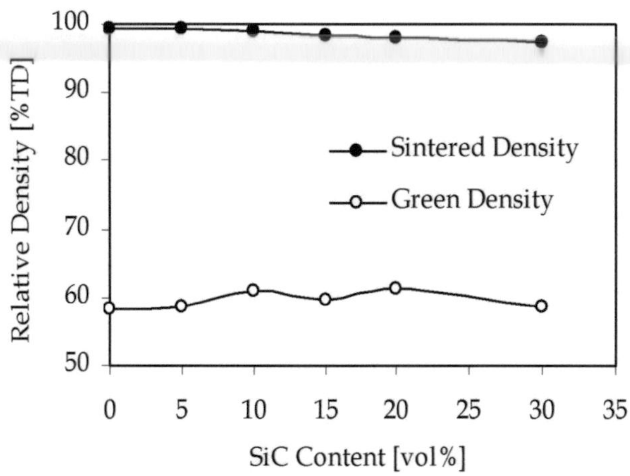

Fig. 4. Pre-sintered and sintered relative density as a function of SiC content in the monolithic laminates. The theoretical densities (TD) of the composites were calculated by the rule of mixtures assuming a value of 3.98 g/cm^3 for Al_2O_3 and 3.16 g/cm^3 for SiC.

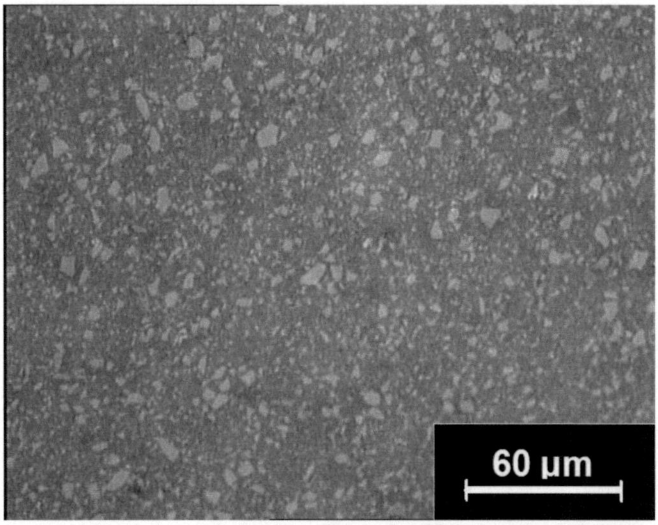

Fig. 5. Optical micrograph of the polished surface of AS30 sample. White particles represent silicon carbide.

Microscopical observations of the laminates polished surface confirmed the presence of limited porosity and revealed a homogeneous distribution of the two phases as shown in Fig. 5. Perfect adhesion between the initial layers was also observed both in monolithic and engineered laminates (Fig. 6) where layers of different composition were assembled; this confirms the goodness and the reliability of the processing procedure used in the present work.

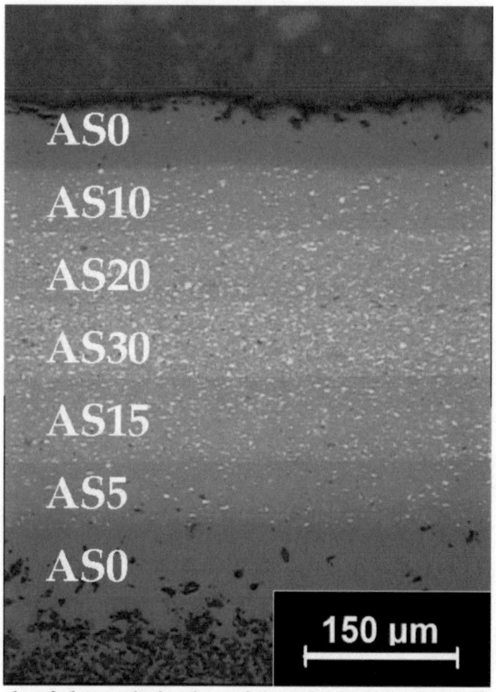

Fig. 6. Optical micrograph of the polished surface of AS-1 sample. Perfect adhesion between the layers of different composition can be observed.

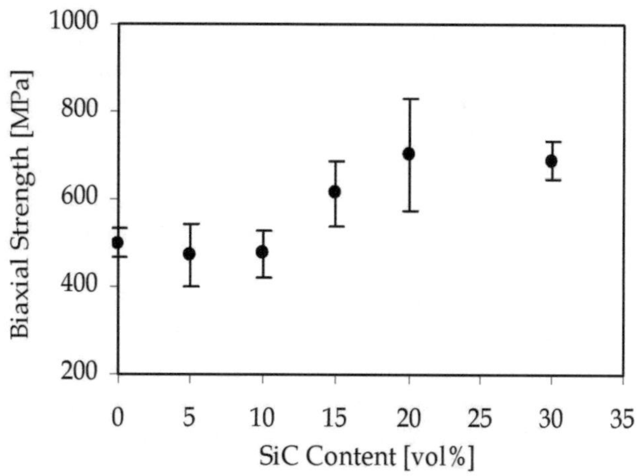

Fig. 7. Strength of monolithic laminates.

Fig. 8. Fracture surface of AS0 (a) and AS30 (b) monolithic laminates.

Grain size distribution is not homogeneous in monolithic alumina and 1 µm grains were observed as well as some larger 10 µm grains. However, the alumina mean grain size ranges from 6.9±0.5 µm in the pure alumina monolithic laminate to approximately 1 µm in the monolithic 30 vol% SiC containing composite. On the other hand, silicon carbide grain size is equal to 4.2±0.3 µm in all monolithic composites. The alumina grain refinement in the composites has been already reported in literature and has been attributed to a decreased grain boundary mobility, diffusivity and rate of densification due to SiC particles presence (Chae et al., 2006).

The strength measured on monolithic laminates is shown in Fig. 7. The addition of limited amount of SiC, up to 10 vol%, does not affect the strength of the composite which exhibits the resistance of pure alumina, around 500 MPa. For silicon carbide loads exceeding 10 vol% an appreciable increase of the biaxial failure stress is observed and values as high as 700 MPa are obtained for SiC content of 20 and 30 vol%. The standard deviation shown for all monolithic samples is always quite large due to the limited number of samples, four to five, used for the mechanical tests. Only for pure alumina ten samples were used. The observed increase in failure stress with SiC content can be accounted for by the transformation of the fracture propagation from intergranular in pure alumina samples to transgranular in composites containing larger amount of silicon carbide as shown in Fig. 8; such effect of matrix weakening and grain boundary strengthening is in good agreement with findings reported in previous works (Levin et al., 1995; Sternitzke, 1997; Pérez-Riguero et al., 1998) and is a consequence of the tensile residual stresses field which develops upon cooling in the matrix around SiC particles because of the thermo-elastic mismatch between Al_2O_3 and SiC. The reduction of critical defect size associated to smaller grain size obtained with SiC additions, that inhibit the grain growth of Al_2O_3, can also be considered for the observed increase of strength.

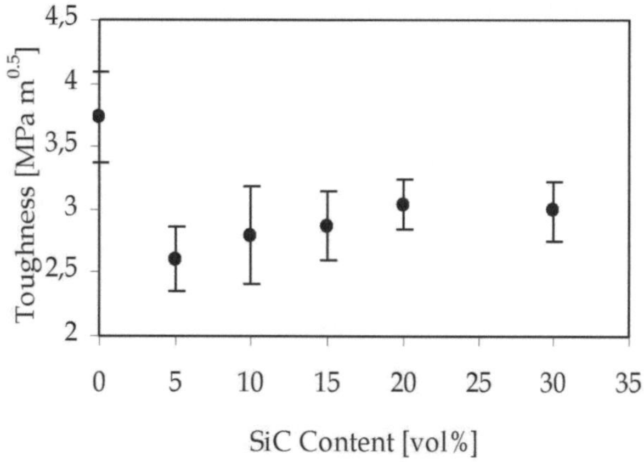

Fig. 9. Toughness of monolithic laminates.

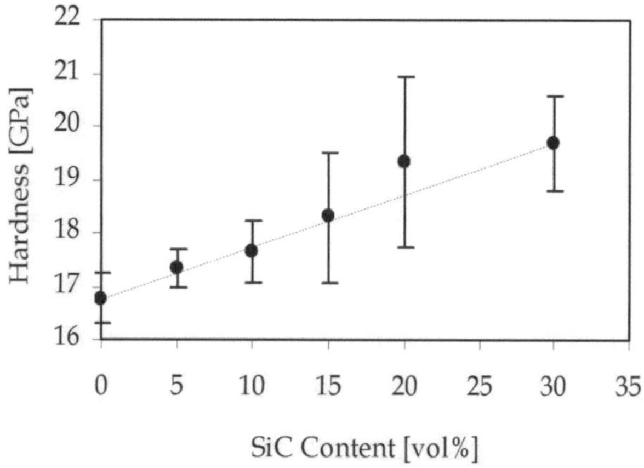

Fig. 10. Hardness of monolithic laminates. The straight line represents the values calculated by the rule of mixture.

As shown in Fig. 9 all monolithic composites show toughness values lower than those of alumina independently on the added SiC volumetric fraction and despite the change in fracture mode. The fracture mode change from intergranular in pure alumina to

transgranular in the composites (Fig. 8) seems to be not effective in the present case. Conflicting results and theories are reported in literature about the grain boundary strengthening and toughening, especially in the case of Al_2O_3/SiC nanocomposites (Pérez-Riguero et al., 1998). Through a simple comparison of the experimental values, one can concludes that the average tensile stress field in the alumina matrix due to the thermal expansion mismatch of alumina and SiC reduces the fracture toughness more than the increase due to the change in the fracture mode. A net increase in toughness can be achieved only in the case the average internal stresses are small (Carrol et al., 1996).
All composites show an increase of hardness over monolithic alumina (Fig. 10). The experimental trend is comparable to a linear rule of mixture assuming values of 16.8 GPa and 26.5 GPa for pure alumina and silicon carbide respectively.

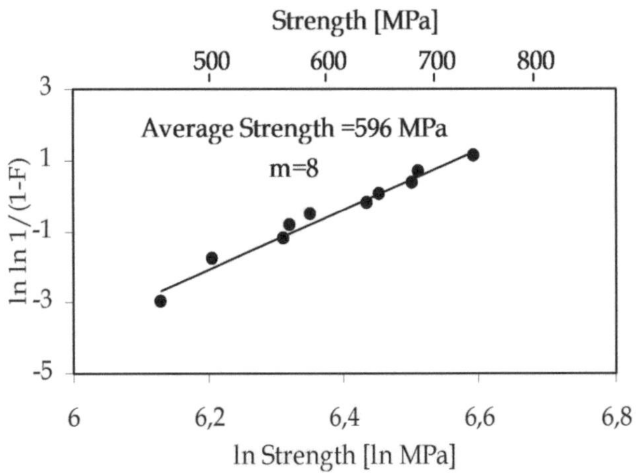

Fig. 11. Weibull diagram for the engineered laminate AS-1. The straight line represents best linear fitting of the experimental data.

The biaxial bending strength measured on the engineered laminate AS-1 is graphically represented on the Weibull plot in Fig. 11. The fracture probability, F, was calculated as F = (j-0.5)/N where N is the total number of samples and j the rank in the ascending ordered distribution. The linear fitting of the experimental data allows the determination of the Weibull modulus, m, equal to 8.3. The average strength is equal to 596 ± 85 MPa.
It is now useful to compare the obtained results with the design strength that can be evaluated on the basis of the chosen laminate architecture. There is a substantial difference between the experimental average strength (596 MPa) and the design failure stress (373 MPa); even the weakest sample in the Weibull distribution shown in Fig. 11 possesses a strength of 460 MPa, definitely higher than the design stregth. These observations require some explanation. The outer layer of AS-1 laminate is made of pure alumina. Therefore, the dimension of surface cracks can be estimated from the fracture toughness and the strength measured on alumina specimens that range from 26 to 38 μm if semi-circular cracks are supposed. This means that, in spite of the simplifications made in the model and in the calculations, on the basis of Fig. 3, surface flaws in AS-1 laminate should undergo to stable growth upon bending before final catastrophic failure. Nevertheless, since the T curve is

very steep before its maximum, after an initial growth in the depth, surface flaw could propagate more easily along the surface, evolving at first from semi-circular to semi-elliptical and then to through-thickness crack. Careful observation of fracture surfaces allows to confirm such hypothesis.

Examination of fractographic features of monolithic alumina samples shows typical crack patterns of biaxial moderate strength with bent mirror and main crack branching soon after the mirror boundaries. The samples break into 3÷5 fragments, with a quite clear-cut and planar fracture surface (Fig. 12). The engineered laminate shows a more extensive fragmentation due to more intensive stress state. All samples possess fracture surfaces with more complex crack propagation, involving steps and jogs in the mirror and excursion of the crack front out of the plane in proximity of the surface in tension (Fig. 13). Crack propagation occurs initially orthogonally to the surface in tension but then kinks as it approaches the highly compressed layers.

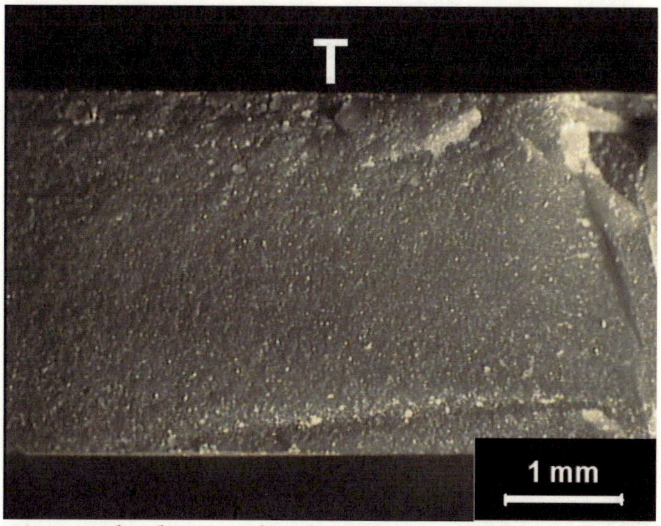

Fig. 12. Optical micrographs showing the planar fracture surface of the alumina monolithic laminate. The surface in tension in the flexure test is marked with T.

This effect is evident in Fig. 14 that corresponds to a polished section of a sample subjected to biaxial bending where a surface crack initially propagates perpendicularly to the surface in tension and then deviates tilting out from the main fracture plane once encountering high intensity biaxial residual compressive stresses.

In addition, SEM micrograph (Fig. 15) taken on the fracture surface in proximity of the surface in tension shows a smooth region extending along ≈ 400 μm from the surface due to a stable crack growth of defects in the residual compressive field. Similar peculiar crack propagation behaviour has already been observed in pre-stresses ion-exchanged glasses (Sglavo et al., 2007) where, similarly to the engineering process used in the present work, a specific residual stress field with maximum compression at a certain depth from the surface was created (Green et al., 1999; Sglavo & Green, 2001).

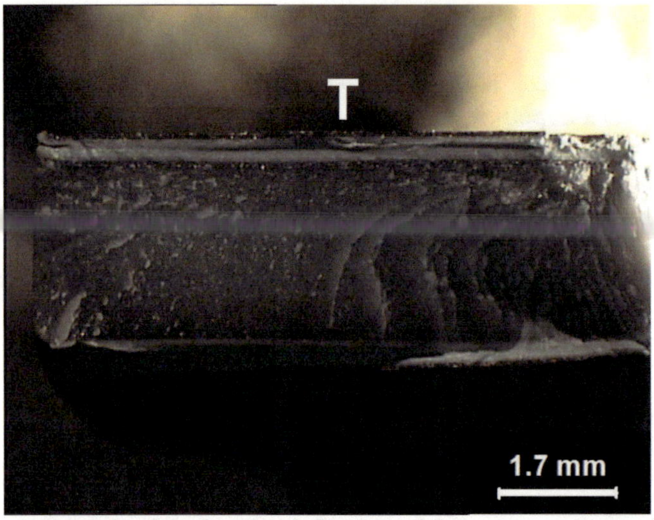

Fig. 13. Optical micrographs showing the deflection of the propagating surface crack on the as broken fracture surface of a fragment of the AS-1 laminate. The surface in tension in the flexure test is marked with T.

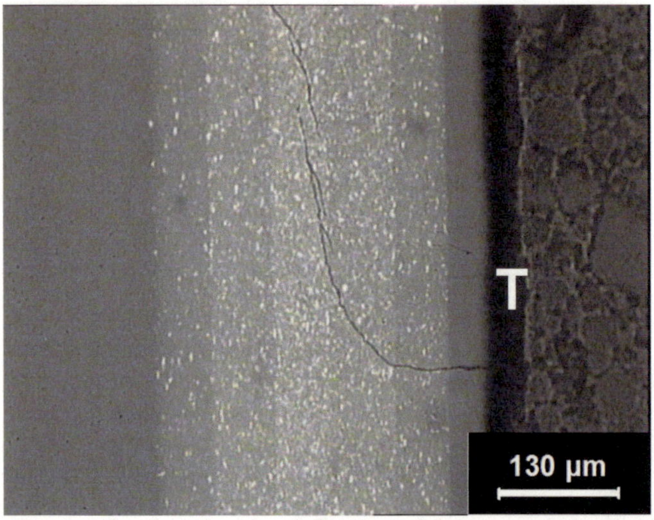

Fig. 14. Optical micrographs showing the deflection of the propagating surface crack on the polished fracture surface of a fragment of the AS-1 laminate. The surface in tension in the flexure test is marked with T.

This phenomenon is responsible for a shielding of the crack tip and consequently for higher failure stress. In addition, due to the randomness of such crack kink, the final strength is no more strictly controlled by the T-curve (Fig. 3) and becomes quite scattered. It is anyway interesting to observe that failure stress remains invariant even when large defects are introduced on the surface by Vickers indentation (Fig. 16). Strength values around 600 MPa are obtained even when indentation loads as high as 100 N, which produces radial cracks of the order of 100 μm, are used.

Fig. 15. SEM micrograph showing the deflection of the propagating surface crack at a certain depth. The surface in tension in the flexure test is marked with T.

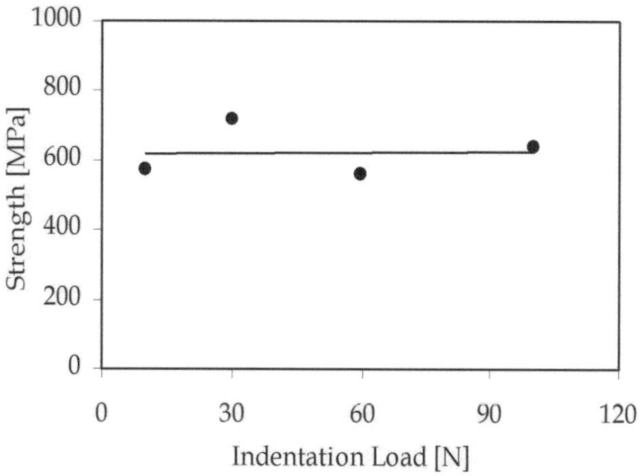

Fig. 16. Failure stress as a function of the indentation load for the engineered laminate.

4. Conclusion

Fully dense SiC-Al_2O_3 composites characterized by perfect adhesions between constituting layers were produced by Spark Plasma Sintering tape casted composite laminae.
Monolithic composite disks with SiC load up to 30 vol% and an engineered laminate with specific layers combination able to promote stable growth of surface defects before final failure were considered. High homogeneity of the two phases is always obtained.

Monolithic composites show an increasing strength with SiC content and biaxial failure stress as high as 700 MPa is obtained for the highest SiC load. A graceful crack propagation, first inward and then parallel to the surface of the laminate, can be observed in the engineered laminate. Such fracture behaviour is shown to be responsible for the high strength (about 600 MPa) and the peculiar surface damage insensitivity.

5. References

Anstis, G. R.; Chantikul, P.; Lawn, B. R & Marshall, D. B. (1981). A critical evaluation of indentation techniques for measuring fracture toughness: I, Direct crack measurements. *J. Am. Ceram. Soc.,* Vol. 64, No. 9, (September 1981) 533-538, ISSN 0002-7820

Bermejo, R.; Torres, Y.; Sanchez-Herencia, A. J.; Baudin, C.; Anglada, M. & Llanes, L. (2006). Residual stresses, strength and toughness of laminates with different layer thickness ratios. *Acta Mater.,* Vol. 54, No. 18, (October 2006) 4745–4757, ISSN 1359-6454

Bermejo, R. & Danzer, R. (2010). High failure resistance layered ceramics using crack bifurcation and interface delamination as reinforcement mechanisms. *Eng. Fract. Mech.,* Vol. 77, No. 11, (July 2010) 2126–2135, ISSN 0013-7944

Carroll, L.; Sternitzke, M. & Derby, B. (1996). Silicon carbide particle size effects in alumina-based nanocomposites. *Acta Mater.,* Vol. 44, No. 11, (November 1996) 4543-4552, ISSN 1359-6454

Chae, J. H.; Kim, K. H.; Choa, Y. H.; Matsushita, J.; Yoon, J.-W. & Shim, K. B. (2006). Microstructural evolution of Al_2O_3-SiC nanocomposites during spark plasma sintering. *J. Alloys Compounds,* Vol. 413, No. 1-2, (March 2006) 259-264, ISSN 0925-8388

Cho, K. S.; Choi, H. J.; Lee, J. G. & Kim, Y. W. (2001). R-curve behaviour of layered silicon carbide ceramics with surface fine microstructure. *J. Mater. Sci.,* Vol. 36, No. 9, (May 2001) 2189-2193, ISSN 0022-2461

Costabile, A. & Sglavo, V. M. (2006). Influence of the architecture on the mechanical performances of alumina-zirconia-mullite ceramic laminates. *Adv. in Science and Technology,* Vol. 45, (October 2006) 1103-1108, ISSN 1662-8969

Davis, J. B.; Kristoffersson, A.; Carlström E. & Clegg, W. J. (2000). Fabrication and Crack Deflection in Ceramic Laminates with Porous Interlayers. *J. Am. Ceram. Soc.,* Vol. 83, No. 10, (October 2000) 2369-2374, ISSN 0002-7820

Gadalla, A.; Elmasry, M. & Kongkachuichay, P. (1992). High temperature reactions within SiC-Al_2O_3 composites. *J. Mater. Res.,* Vol. 7, No. 9, (September 1992) 2585-2592, ISSN 0884-2914

Green, D. J.; Tandon R. & Sglavo, V. M. (1999). Crack arrest and multiple cracking in glass using designed residual stress profiles. *Science,* Vol. 283, No. 5406, (February 1999) 1295-1297, ISSN 0036-8075

Hue, F.; Jorand, Y.; Dubois, J. & Fantozzi, G. (1997). Analysis of the weight loss during sintering of silicon-carbide whisker-reinforced alumina composites. *J. Eu. Ceram. Soc.,* Vol. 17, No. 4, (February 1997) 557-563, ISSN 0955-2219

Kingery, W. D.; Bowen, H. K. & Uhlmann, D. R. (1976). *Introduction to ceramics,* J. Wiley & Sons, ISBN 0471478601, NY, pp. 603-606, pp. 773-777

Lee, W. E. & Rainforth, M. (1994). *Ceramic Microstructures – Property control by processing*, Chapman & Hall, ISBN 0412431408, London, U.K., pp. 509-570

Leoni, M.; Ortolani, M.; Bertoldi, M.; Sglavo, V. M. & Scardi, P. (2008). Nondestructive measurement of the residual stress profile in ceramic laminates. *J. Am. Ceram. Soc.*, Vol. 91, No. 4, (April 2008) 1218-1225, ISSN 0002-7820

Levin, I; Kaplan, W. D.; Brandon, D. G. & Layyous, A. A. (1995). Effect of SiC submicrometer particle size and content on fracture toughness of alumina-SiC "nanocomposites". *J. Am. Ceram. Soc.*, Vol. 78, No. 1, (January 1995) 254-256, ISSN 0002-7820

Mekky, W. & Nicholson, P. S. (2007). R-curve modeling for Ni/Al_2O_3 laminates. *Composites. Part B, Engineering*, Vol. 38, No. 1, (January 2007) 35-43, ISSN 1359-8368

Munir, Z. A.; Anselmi-Tamburini, U. & Ohyanagi, M. (2006). The effect of electric field and pressure on the synthesis and consolidation of materials: A review of the spark plasma sintering method. *J. Mater. Sci.*, Vol. 41, No. 3, (February 2006) 763-777, ISSN 0022-2461

Náhlík, L.; Šestáková, L; Hutar, P. & Bermejo, R. (2010). Prediction of crack propagation in layered ceramics with strong interfaces. *Eng. Fract. Mech.*, Vol. 77, No. 11, (July 2010) 2192–2199, ISSN 0013-7944

Orlovskaya, N.; Kuebler, J.; Subbotin, V. & Lugovy, M. (2005). Design of Si_3N_4-based ceramic laminates by the residual stresses. *J. Mat. Sci.*, Vol. 40, No. 20, (October 2005) 5443–5450, ISSN 0022-2461

Pérez-Riguero, J.; Pastor, J. Y.; Llorca, J.; Elices, M.; Miranzo, P. & Moya, J. S. (1998). Revisiting the mechanical behavior of alumina/silicon carbide nanocomposites. *Acta Mater.*, Vol. 46, No. 15, (September 1998) 5399-5411, ISSN 1359-6454

Peters, S. Y. edt. (1998). *Handbook of composites*, Chapman & Hall, ISBN 0412540207, London, U.K., pp. 307-332

Rao, M. P.; Sánchez-Herencia, A. J.; Beltz, G. E.; McMeeeking, R. M. & Lange, F. F. (1999). Laminar ceramics that exhibit a threshold strength. *Science*, Vol. 286, No. 5437, (October 1999) 102-105, ISSN 0036-8075

Rao, M. P.; Rödel, J. & Lange, F. F. (2001). Residual stress induced R-Curves in laminar ceramics that exhibit a threshold strength. *J. Am. Ceram. Soc.*, Vol. 84, No. 11, (November 2001) 2722-2724, ISSN 0002-7820

Sglavo, V. M.; Larentis, L. & Green, D. J. (2001). Flaw insensitive ion-exchanged glass: I, Theoretical aspects. *J. Am. Ceram. Soc.*, Vol. 84, No. 8, (August 2001) 1827-1831, ISSN 0002-7820

Sglavo, V. M. & Green, D. J. (2001). Flaw insensitive ion-exchanged glass: II, Production and mechanical performance. *J. Am. Ceram. Soc.*, Vol. 84, No. 8, (August 2001) 1832-1838. ISSN 0002-7820

Sglavo, V. M.; Paternoster, M. & Bertoldi, M. (2005). Tailored residual stresses in high reliability alumina-mullite ceramic laminates. *J. Am. Ceram. Soc.*, Vol. 88, No. 10, (October 2005) 2826–2832, ISSN 0002-7820

Sglavo, V. M. & Bertoldi, M. (2006 a). Design and production of ceramic laminates with high mechanical resistance and reliability. *Acta Mater.*, Vol. 54, No. 18, (October 2006) 4929-4937, ISSN 1359-6454

Sglavo, V. M. & Bertoldi, M. (2006 b). Design and production of ceramic laminates with high mechanical reliability. *Composites. Part B, Engineering*, Vol. 37, No. 6, (2006) 481-489, ISSN 1359-8368

Sglavo, V. M.; Prezzi, A. & Green, D. J. (2007). In situ observation of crack propagation in ESP (engineered stress profile) glass. *Eng. Fract. Mech.*, Vol. 74, No. 9, (June 2007) 1383-1398, ISSN 0013-7944

She, J.; Inoue T. & Ueno K. (2000). Damage resistance and R-curve behavior of multilayer Al_2O_3/SiC ceramics. *Ceram. Int.*, Vol. 26, No. 8, (2000) 801-805, ISSN 0272-8842

Shetty, D. K.; Rosenfield, A. R.; McGuire, P.; Bansal, G. K. & Duckworth, W. H. (1980). Biaxial flexure tests for ceramics. *Ceramic Bullettin*, Vol. 59, No. 12., (1980) 1193-1197, ISSN 002-7812

Sternitzke, M. (1997). Review: structural ceramic nanocomposites. *J. Eu. Ceram. Soc.*, Vol. 17, No. 9, (1997) 1061-1082, ISSN 0955-2219

Wurst, J. C. & Nelson, J. A. (1972). Linear intercept technique for measuring grain size in two-phase polycrystalline ceramics. *J. Am. Ceram. Soc.*, Vol. 55, No. 2, (February 1972) 109, ISSN 0002-7820

High Temperature Phase Equilibrium of SiC-Based Ceramic Systems

Yuhong Chen, Laner Wu ,Wenzhou Sun , Youjun Lu and Zhenkun Huang
School of Material Science & Engineering, Beifang University of Nationalities
Ningxia, China

1. Introduction

Silicon carbide (SiC) is one of the promising structure materials for mechanical and thermal applications(Nitin P. ,1994). Although SiC ceramic has been developed for several decades, it is still important to study in some areas, ally the high temperature phase relations in SiC-based ceramic systems. In addition, the SiC/Si_3N_4 composites are of increasing interest because they should have the complement each other in the mechanical properties.(Kim Y. & Mitomo.M, 2000, Lee Y et.al., 2001) SiC and Si_3N_4 are the strong covalent compounds. The self-diffusion coefficient of Si and C, also Si and N, are too low to get the fully dense ceramics without sintering aids. Rare-earth oxides are often used as liquid phase sintering aids for densification. the behaviours of their high temperature reactions and the derived phase relations are still unknown. Becher (Becher et al ,1996) found that the chemical composition of the grain boundary amorphous phase could significantly influence the interfacial debonding behaviour in silicon nitride. Other study (Keeebe H. et.al., 1996)also showed that the secondary phase chemistry could play a key role in toughening Si_3N_4 ceramic due to its influences on the grain morphology formation, secondary-phase crystallization and residual stress distribution at grain boundaries. For SiC ceramics less of reaction behaviour at high temperature was known due to its sluggish diffusion. About phase relations the Si_3N_4-containing systems have been much published (Anna E. McHale. 1994), but either SiC-based ceramic or SiC/ Si_3N_4 composite systems were rarely done. Even so, the compatibility relations of SiC with neighbour phases should be revealed. Doing so is beneficial to practical use in the manufacture of SiC-based ceramics, as well as SiC/ Si_3N_4 composites.

The present work focused on the determination of the phase relations in the quaternary systems of SiC- Si_3N_4-SiO_2-R_2O_3 (R=La,Gd,Y) at high temperatures. Lanthanum which has lower atomic number in 17 rare earth elements, as a typical light rare-earth oxide, Gd_2O_3 as middle and Y_2O_3 as heavy one with similar property as heavy rare earth oxide were chosen to use in this study. Rare earth oxides used as sintering aids retained in intergranular phases after reaction, which cause strength degradation of the material at high temperature. The investigation of phase relations in this quaternary system will be a summary of work from studies of Si-N-O-R(ANNA E. McHale. (1994)) to Si-C-N-O-R systems. Extensive investigation for the phase relations and reactives in high temperature is beneficial to practical use in the manufacture of SiC-based ceramics, as well as SiC/ Si_3N_4 composites.

2. Experimental

The starting powders were α-SiC (H.C.Starck), β- Si_3N_4 (H.C.Starck), La_2O_3, Gd_2O_3 and Y_2O_3 (R_2O_3 with 99.9% purity, from Baotou Rare-earth Institute, China). The rare earth oxides were calcined in air at 1200°C for 2h before use.The compositions investigated were restricted to the region bound by the poins SiC, Si_3N_4 and R_2O_3 (R=La,Gd,Y), but SiO_2 came from in situ oxygen impurity on the surface of powders. Selected compositions were made by mixing the required amounts of the starting powders in agate jar mills with absolute alcohol for 2hr. The dried mixtures were hot-pressed in graphite dies 10 mm in diameter lined with BN in a graphite resistance furnace under a pressure of 30MPa at a subsolidus temperature under a mild flow of Ar, as well as N_2 used for comparison. For the systems SiC-R_2O_3, the melting behaviours of SiC and R_2O_3 (1:1 mole ratio) shown in the table 1. In which the subsolidus temperatures were used as the hot-pressing temperatures for some compositions.

R_2O_3:SiC (1:1)	Temperatures (°C)					
R_2O_3	1600	1700	1750	1800	1850	1900
La_2O_3	not melted	partly melted	melted			
Gd_2O_3		not melted	Little melted	partly melted	melted	
Y_2O_3		not melted	Little melted	Little melted	partly melted	melted

Table 1. Melting behaviors for R_2O_3 : SiC (1:1)

The specimens were hot-pressed for 1 to 2 hr in the high temperature region and then cooled at 200°C/min.. End points of hot-pressing were obtained where no further phase change was observed when specimens were heated for longer times. An automatic recording X-ray diffraction with monochromated CuKα radiation was used to scan the samples at a rate of 2°/min.

3. Phase relation of binary subsystem

3.1 Phase relation of R_2O_3- Si_3N_4 subsystem

Table 2 shows the phase relation for different Si_3N_4-R_2O_3 binary subsystems in Ar or N_2 atmosphere respectively.

	Si_3N_4- La_2O_3	Si_3N_4-Gd_2O_3	Si_3N_4-Y_2O_3
Ar	2:1,K,J	M,J	M
N_2	2:1,K,J,	M,J	M,J

Table 2. phase relation of Si_3N_4-R_2O_3 binary subsystem

In the Y_2O_3- Si_3N_4 subsystem Y_2O_3- Si_3N_4 mililite(M phase) was determined after hot-pressing under Ar and N_2 atmosphere. On the M- Y_2O_3 join a richer-oxygen phase, 2 Y_2O_3 Si_2N_2O (J-phase, monocl.) was determined, The binary phase diagram of Y_2O_3- Si_3N_4 under 1MPa N_2 is presented as Fig 1(Huang Z. K. & Tien T. Y.,1996).

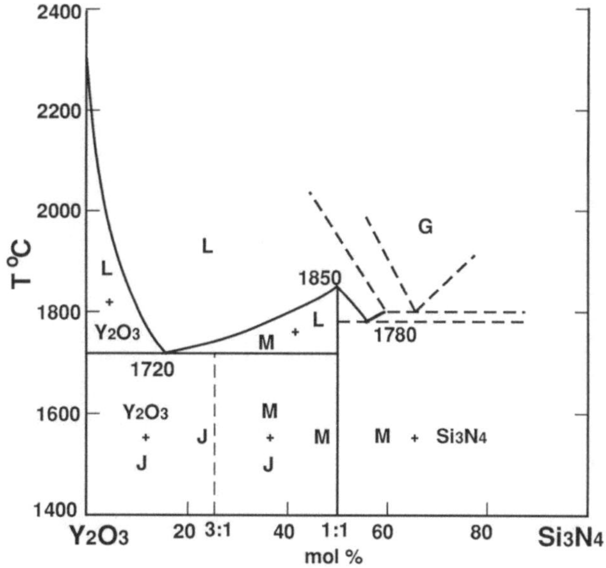

Fig. 1. Phase diagram of Y_2O_3- Si_3N_4 subsystem

The reaction can be written as follows:

Si_3N_4+ Y_2O_3→Si_3N_4 ·Y_2O_3 ($Y_2Si_3N_4O_3$, M)
Si_3N_4+SiO_2+ 2 Y_2O_3→2(2 Y_2O_3 Si_2N_2O) ($Y_4Si_2N_2O_7$, J)
The Gd_2O_3- Si_3N_4 subsystem has similar phase relations and reactions.

SiC + Gd_2O_3+ SiO_2 + 3/2N_2 → Gd_2O_3 Si_3N_4 (M phase) + CO_2↑

Si_3N_4+SiO_2+ 2 Gd_2O_3→2(2 Gd_2O_3 Si_2N_2O) ($Gd_4Si_2N_2O_7$, J)

In the La_2O_3- Si_3N_4 subsystem La_2O_3 ·2 Si_3N_4 (monoclinic 2:1) was determined repeatedly after hot-pressing under either Ar or N_2 atmosphere. A disputed La-melilite (La_2O_3: Si_3N_4) was not found, because of the large radius of La^{3+} ion. It could form only in bigger cell to be La_2O_3. Si_2N_2O. AlN (La_2Si_2 AlO_4N_3, melilite) by Al-N replaced for Si-N(Huang Z. K. & Chen I. W.,1996). $LaSiNO_2$ (K phase, monoclinic) were determined because of the impurty of powder. On the 2:1- La_2O_3 join a richer-oxygen phase, 2 La_2O_3 Si_2N_2O (J-phase, monocl.) was determined, indicating the presence of excess oxygen from SiO_2 impurity in the powder mixtures. M.Mitomo (Mitomo M.,et.al. 1982)found that an equi-molar mixture of and heated to 1800°C showed that there were three temperature regions in which chemical reaction took place.

$$Si_3N_4 + La_2O_3 \xrightarrow{1200 to 1250°C} Si_3N_4 + (La_4Si_2N_2O_7 + LaSiNO_2)$$
$$\xrightarrow{1400 to 1500°C} LaSiNO_2 + Si_3N_4$$
$$\xrightarrow{1650 to 1750°C} La_2O_3 \cdot 2\,Si_3N_4 + liquid$$

3.2 Phase relation of R_2O_3-SiC subsystem

No new phase was detected in SiC- Si_3N_4 and SiC-R_2O_3 (R=La,Gd,Y) systems, it can be due to its very low self-diffusion coefficient of Si and C with very strong covalence of Si-C bond. However , a few of $2R_2O_3 \cdot Si_2N_2O$ (J phase)was observed in SiC-R_2O_3 system. The oxygen content of SiC powder, existing either as surface SiO_2 or as interstitial oxygen is between 0.8 to 1.1wt%. The reduction of SiC (lower X-ray peak intensity of SiC) indicated that a part of SiC could directly react with R_2O_3 after being oxidized/nitrided under N_2. The reaction can be written as follows:

$3SiC + 2N_2 \rightarrow Si_3N_4 + 3C$,

$4R_2O_3 + SiO_2 + Si_3N_4 \rightarrow 2(2R_2O_3.\ Si_2N_2O)$ (J phase)

It should be noted that only a little amount of oxygen content is enough to form much more rare-earth silicon-oxynitrides as shown below: For the examples of La-siliconoxynitrides, one mole of oxygen can cause formation of 2 moles of J phase (La), ($Si_2N_2O.2La_2O_3$). It means that 1 wt% O_2 can cause formation of 47.0 wt% J(La) phase.

In fact, it is difficult to make SiC reaction under N_2, but when rare-earth oxide entered in system, SiC can be reacted even at lower temperature (1550°C for SiC- La_2O_3, 1600°C for SiC-Gd_2O_3 system). The addition of rare-earth oxide benefits the nitride reaction of SiC. Table 3 shows the phase relation in SiC -R_2O_3 binary system in different atmosphere.

	SiC- La_2O_3	SiC-Gd_2O_3	SiC-Y_2O_3
Ar	No reaction	No reaction	No reaction
N_2	J, SiC	J, SiC	J,SiC

Table 3. Formed phase of SiC:R_2O_3=1:1 compositions

4. The phase equilibrium of SiC-Si_3N_4 -R_2O_3

The binary phases of $La_2O_3 \cdot 2Si_3N_4$ and $Si_3N_4.R_2O_3$ (M(Gd),M(Y)) coexist with SiC forming a tie-line which separated every ternary system of SiC- Si_3N_4-R_2O_3 (R=La,Gd,Y) into two triangles, respectively. The $2R_2O_3 \cdot Si_2N_2O$ (J phase) also coexist with SiC forming another tie-line in triangle near R_2O_3 side. Based on the experimental results of binary subsystem, the subsolidus phase diagrams of SiC- Si_3N_4-R_2O_3 (R=La,Gd,Y) systems are presented as Fig. 2.Comparing SiC- Si_3N_4-R_2O_3 with AlN- Si_3N_4-R_2O_3 systems (Cao G.Z., et.al,1989) reported by Cao G.Z. et, the similarity is evident except SiC couldn't participate to form α-Sialon because of its tough Si-C bond with bigger bond length 1.89Å.

The XRD pattern of typical sample after hot-pressed of SiC- Si_3N_4 -Y_2O_3 system in N_2 atmosphere is shown in Fig3, phase analysis indicated that M phase ($Si_3N_4 \cdot Y_2O_3$), K phase ($Si_2N_2O \cdot Y_2O_3$), or J phase ($Si_2N_2O \cdot 2Y_2O_3$) were formed. And in these samples, SiC coexisted with M, K-phase (Fig3-a) , coexisted with Si_3N_4, M-phase(Fig3-b) and with Y_2O_3 ,J phase(Fig3-c). But in sample sintered in Ar atmosphere, K phase had formed instead of J phase(Fig4). The reason is higher oxygen partial pressure in Ar atmosphere. The introduction of Si_2N_2O transformed the ternary system into the quaternary system. In the system, three compatible tetrahedrons, namely, SiC-M-K-J , SiC-M-J-Y_2O_3 , SiC- Si_3N_4-M-K (in N_2) or SiC- Si_3N_4-M-J(in Ar) have been determined. SiC and Si_3N_4 would selectively equilibrate with these three phases in the order of M < K < J < Y_2O_3 with respect to the effects of the oxygen content of SiC and Si_3N_4 powders and the oxygen partial pressure in high temperature. Based on those results, the subsolid phase diagram for the ternary SiC-Si_3N_4-Y_2O_3 system and the quaternary SiC- Si_3N_4-Si_2N_2O-Y_2O_3 system are given in Fig 5.

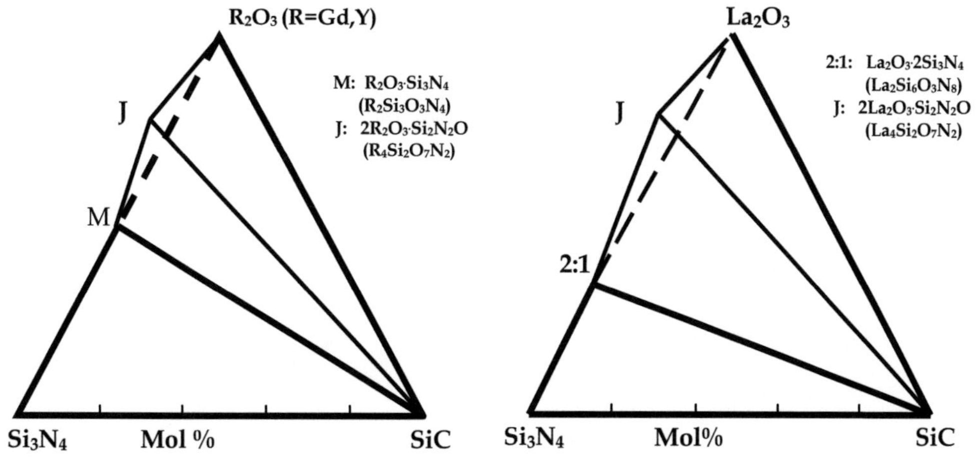

Fig. 2. Subsolidus phase diagram of the system SiC-Si_3N_4-R_2O_3 in Ar or N_2

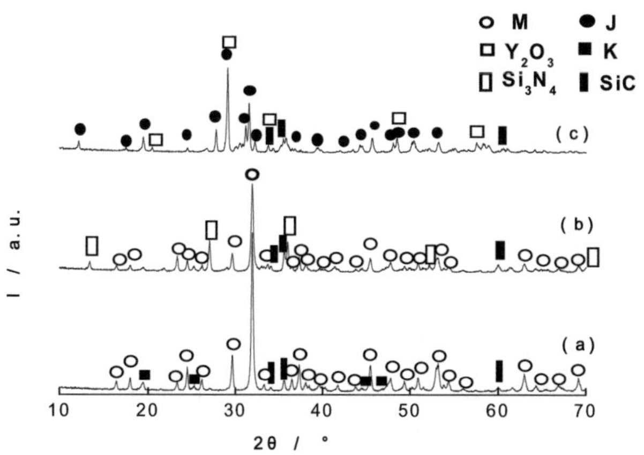

Fig. 3. XRD pattern of SiC-Si_3N_4-Y_2O_3 hot pressed sample in N_2

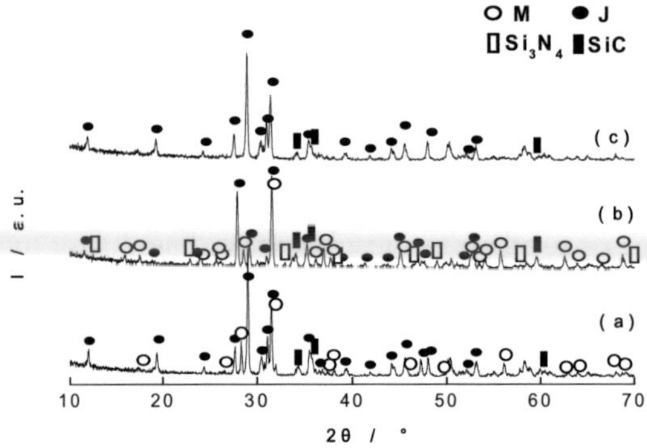

Fig. 4. XRD pattern of $SiC-Si_3N_4-Y_2O_3$ hot pressed sample in Ar

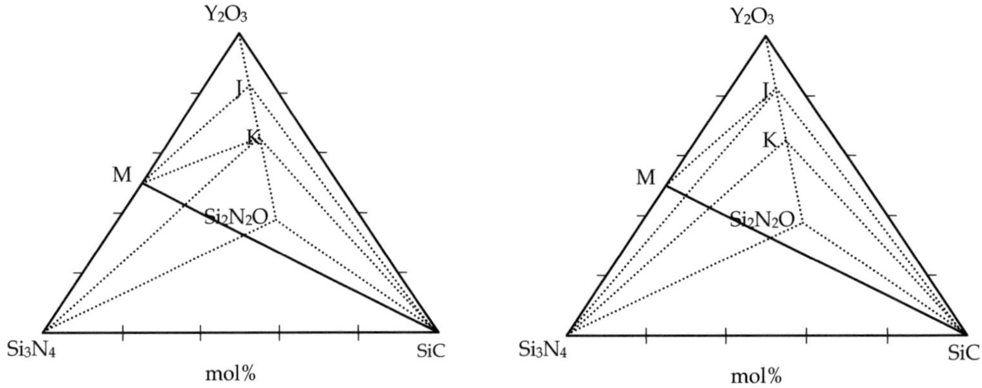

Fig. 5. Subsolidus phase diagram of $SiC- Si_3N_4-Si_2N_2O-Y_2O_3$ system(a: in N_2 ,b:in Ar

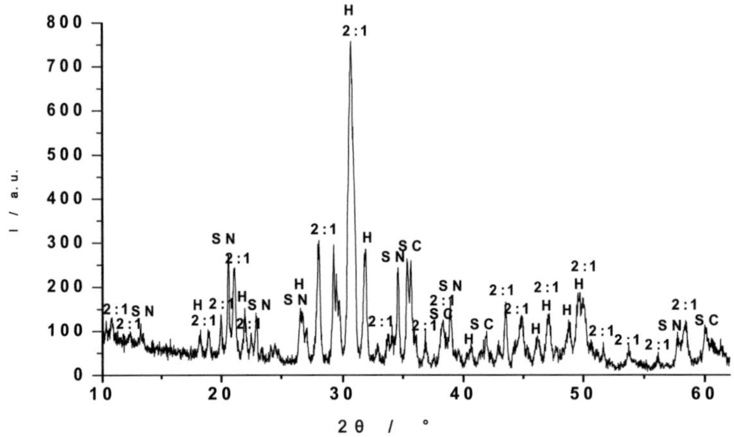

Fig. 6. XRD pattern of $SiC-Si_3N_4$-2:1-H showing coexistence of four phases in the system $SiC-Si_3N_4-La_2O_3-SiO_2$.

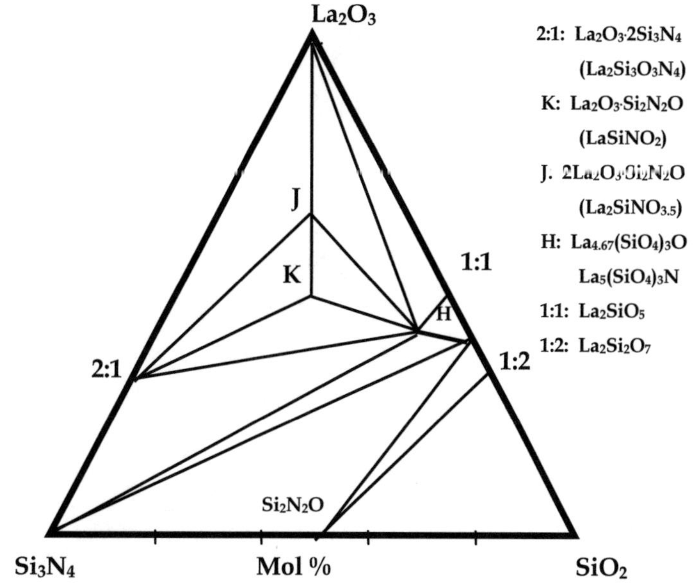

Fig. 7. Subsolidus phase diagram of the system Si_3N_4-SiO_2-La_2O_3 in Ar or N_2[9,13]

As the typical example, Fig 6 showed XRD patterns of four phase coexistence in two typical tetrahedrons respectively in SiC- Si_3N_4 -La_2O_3 system. The oxygen-richer rare-earth silicon-oxynitrides phase $La_5(SiO_4)_3N$ (H phase) had been indicated in this system. K-phase ($Si_2N_2O \cdot La_2O_3$) $2La_2O_3 \cdot Si_2N_2O$ (J-phase) were indicated in this system similar with Si_3N_4 -La_2O_3 system, in which J phase also occurred on the binary composition Si_3N_4:$2La_2O_3$. It indicates that the formation of above oxynitrides was related to the presence of excess oxygen from SiO_2 impurity in the powder mixtures. It should be noted that these oxygen-richer rare-earth silicon-oxynitrides do not lie on the plane SiC- Si_3N_4-La_2O_3 even so synthesized by these three powders, but lie in the Si_3N_4-SiO_2-La_2O_3 system . The isothermal section at 1700°C of Si_3N_4-SiO_2-La_2O_3 system was reported by M.Mitomo(M.Mitomo,1982). Where he obtained J- and K-phase by crystallization from liquid phase, because they lie by a liquid area. In the present work they were obtained directly by solid-state reaction under hot-pressing at 1550°C and led to construct the subsolidus phase relations of Si_3N_4-SiO_2-La_2O_3 system (Fig. 7)(Toropov,et al ,1962, Mitomo,1982) showing some similarity in both.
Above all the oxygen-richer rare-earth silicon-oxynitrides and the three members of ternary systems Si_3N_4-SiO_2-La_2O_3 were compatible with SiC forming ten four-phase compatibility tetrahedrons as follows:
SiC-Si_3N_4-2:1-H, SiC-Si_3N_4-H-Si_2N_2O, SiC-H-Si_2N_2O-1:2, SiC-Si_2N_2O-1:2-SiO_2, SiC-2:1-K-H, SiC-2:1-K-J, SiC-K-J-H, SiC-2:1-J-La_2O_3, SiC-J-La_2O_3-H, SiC-H-La_2O_3-1:1.
The subsolidus phase relationship of this quaternary system with ten four-phase compatibility tetrahedrons is plotted in Fig 8.

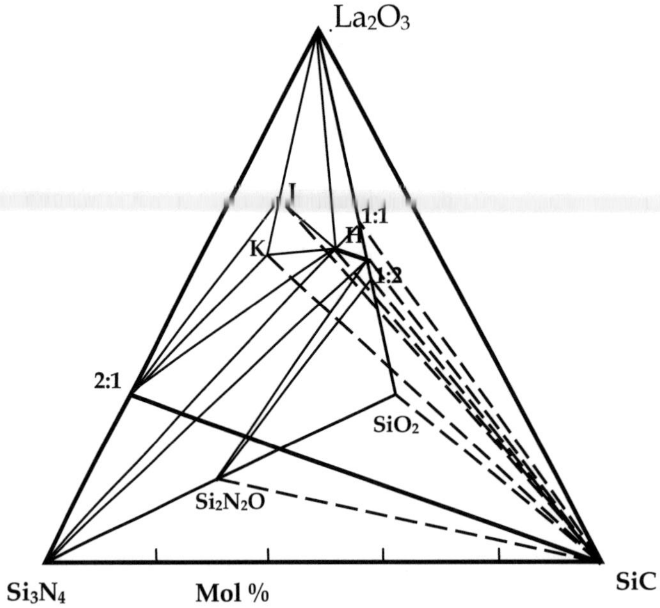

Fig. 8. Subsolidus phase diagram of the system SiC-Si_3N_4-La_2O_3 -SiO_2 in N_2 or Ar

Fig. 8. Subsolidus phase diagram of the system SiC-Si_3N_4-La_2O_3 -SiO_2 in N_2 or Ar

In the Si_3N_4-SiC-Gd_2O_3 system, the M-phase($Si_3N_4 \cdot Gd_2O_3$、 J-phase($Si_2N_2O \cdot 2Gd_2O_3$) and H-phase ($Gd_{10}(SiO_4)_6N_2$) were indicated, a typical XRD pattern of hot-pressure in 1700℃ is shown in Fig 9.

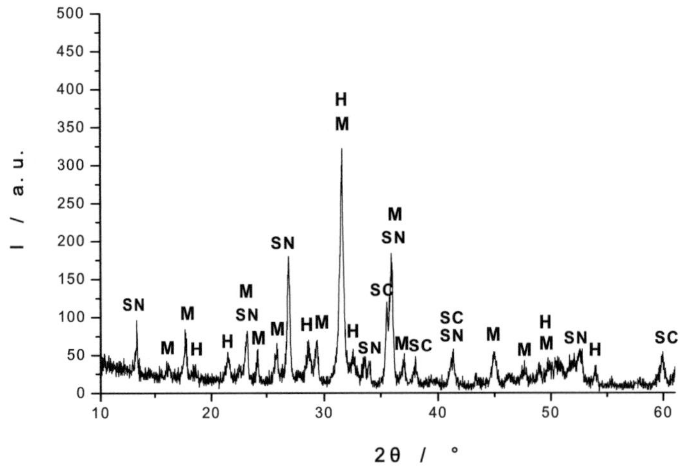

Fig. 9. XRD pattern of SiC-Si_3N_4-M(Gd)-H(Gd) four-phases coexistence in the system SiC-Si_3N_4-Gd_2O_3-SiO_2.

Table 4 shows the phase analysis of different compositions in Si_3N_4-SiC-Gd_2O_3 system. With the increasing of SiC and Si_3N_4, which means the increasing oxygen content in system, M-phase, J-phase and H-phase would be formed. In the Ar atmosphere, H-phase, which is more oxygen-rich inclined to generation than in N_2 since the higher oxygen particle pressure.

No.	the composition of raw material /mol	Phase composition(in Ar)	Phase composition (in N_2)
1#	SiC : Si_3N_4 : Gd_2O_3= 4 : 4 : 1	M(vs),Si_3N_4(s),SiC(m),H(w)	M(vs), Si_3N_4(s), H(m),SiC(w)
2#	SiC : Si_3N_4 : Gd_2O_3= 1 : 1 : 1	M(vs),J(m),SiC(w)	M(vs),J(m),SiC(w)
3#	SiC : Si_3N_4 : Gd_2O_3= 1 : 1 : 2	J(s),H(m),SiC(w)	J(s),SiC(w)
4#	SiC : Si_3N_4 : Gd_2O_3= 1 : 1 : 4	J(s),Gd_2O_3(w)	J(vs),SiC(m),Gd_2O_3(w)

vs: very strong, s: strong, m: middle w: weak

Table 4. The compositions of raw material and phase compositions in ternary systems SiC-Si_3N_4-Gd_2O_3 (in Ar or N_2 , 1700℃)

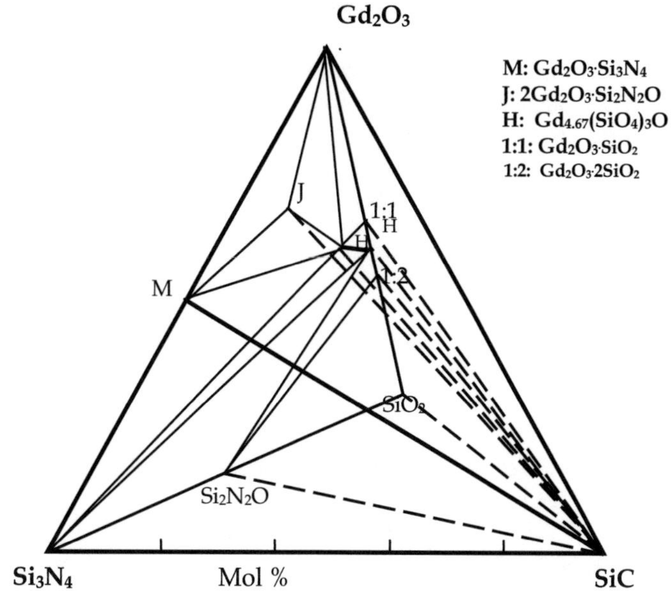

Fig. 10. Subsolidus phase diagram of the system SiC-Si_3N_4-Gd_2O_3-SiO_2 in Ar or N_2

The compositions in the triangles bounded by R-SiC tielines and Gd_2O_3 always led to the formation of rare-earth silicon-oxynitrides, indicating the presences of excess oxygen in the powder mixture, that means SiO_2 in powder also participated in the reaction in the system. Presence of SiO_2 leads to the quasiternary system Si_3N_4-SiC-Gd_2O_3 extend into the quaternary system Si_3N_4-SiC-SiO_2-Gd_2O_3. All rare earth silicon-oxinitrides wrer compatible with SiC, forming eight four-phases compatibility terahedrons as follows:
SiC-Si_3N_4-M-H, SiC-Si_3N_4-H-Si_2N_2O, SiC-H-Si_2N_2O-1:2, SiC-Si_2N_2O-1:2-SiO_2, SiC-M-J-H, SiC-M-J-Gd_2O_3, SiC-J-Gd_2O_3-H, SiC-H-Gd_2O_3-1:1,
Hence the subsolidus phase diagram of this quaternary system is plotted in Fig. 10.

5. The high temperature reaction

Generally, the oxygen content of SiC powder, existing either as surface SiO_2 or as interstitial oxygen is between 0.8 to 1.1wt%. More than 1.5% of oxygen content exists in Si_3N_4 powder. The in-situ SiO_2 coexisting with powder mixture leads to the quasiternary systems SiC-Si_3N_4-R_2O_3 extend into the quaternary systems SiC-Si_3N_4-SiO_2-R_2O_3 (R=La,Gd,Y).Just as discussed, only a little amount of oxygen content is enough to form much more rare-earth siliconoxynitrides. That is the reason for easier and much more formation of oxygen-richer rare-earth siliconoxynitrides in the present systems. Their formations are essentially based on the reactions of SiO_2 and Si_3N_4 with R_2O_3 , but without Si_2N_2O presence as following:

J(R): $4R_2O_3 + SiO_2 + Si_3N_4 \rightarrow 2(Si_2N_2O.{}_2R_2O_3)$,
K(R): $2R_2O_3 + SiO_2 + Si_3N_4 \rightarrow 2(Si_2N_2O.R_2O_3)$,
H(R): $10R_2O_3 + 9SiO_2 + Si_3N_4 \rightarrow 4(R_5(SiO_4)_3N)$,

The formation of oxygen-richer rare-earth siliconoxynitrides are often accompanied with not only consuming Si_3N_4 but also reducing SiC (much lower X-ray peak intensity of SiC) specific when hot-pressing under N_2 atmosphere. This implies that a part of SiC could also directly react with R_2O_3 after being oxidised/nitrided. A few of ${}_2R_2O_3$ Si_2N_2O were observed from SiC-R_2O_3 binary system when firing in N_2 atmosphere. In this case the reactions of SiC and R_2O_3 can be written as follows:

$SiC + O_2 \rightarrow SiO_2 + C$,

$4SiO_2 + 2N_2 \rightarrow 2Si_2N_2O + 3O_2$;

$4SiC + O_2 + 2N_2 \rightarrow 2Si_2N_2O + 4C$;

$3SiC + 2N_2 \rightarrow Si_3N_4 + 3C$,

then $2SiC + 2R_2O_3 + N_2 + 1.5O_2 \rightarrow 2R_2O_3\ Si_2N_2O$ (J phase) + 2CO.
$2SiC + R_2O_3 + N_2 + 1.5O_2 \rightarrow R_2O_3\ Si_2N_2O$ (K phase) + ${}_2CO$.
$6SiC + 5R_2O_3 + N_2 + 7.5O_2 \rightarrow 2(R_5(SiO_4)_3N)$ (H phase) + 6CO.
$3SiC + R_2O_3 + 2N_2 \rightarrow R_2O_3.\ Si_3N_4$ (M phase) + 3C.

Table 5 summarizes the formation of rare-earth silicon-oxynitrides in the present systems, indicating the trend of formation lessens with decreasing bond ionicity from SiO_2 to SiC.

	Ionicity	La_2O_3	Gd_2O_3	Y_2O_3
SiO_2	5	2:1,H*,1:1	2:1,H*,1:1	2:1,H*,1:1
Si_2N_2O	4#	J(1:2),K(1:1),H**	J(1:2),H**	J(1:2) ,K(1:1),H**
Si_3N_4	3	2:1	M(1:1)	M(1:1)
SiC (in Ar)	2	No	No	No
SiC (in N_2)##	2	J	J	J

*H : $R_{4.67}(SiO_4)_3O$.
**H: $R_5(SiO_4)_3N$ or $5R_2O_3.4SiO_2.Si_2N_2O$.
Ionicity of Si_2N_2O : 5 for Si-O bond, 3 for Si-N bond.
##A few of J phase formed.

Table 5. Formation of some rare-earth siliconoxynitrides (mole ratio)

6. Conclusion

Subsolidus phase diagrams of the ternary systems SiC- Si_3N_4-R_2O_3 (R=La,Gd,Y) were determined. The in-situ SiO_2 impurity in the powder mixtures leads to form some oxygen-richer rare-earth siliconoxynitrides and extend the quasiternary systems into quaternary system of SiC-Si_3N_4-SiO_2-R_2O_3. The phase relations of these quaternary systems were established with several SiC-containing four-phase compatibility tetrahedrons. The formation of oxygen-richer rare-earth siliconoxynitrides was discussed. When firing under nitrogen atmosphere a part of SiC could also directly tend to react with R_2O_3 after being oxidised/nitrided forming some rare-earth siliconoxynitrides. They all contributed to construct the phase diagrams of quaternary systems SiC- Si_3N_4-SiO_2-R_2O_3.

Acknowledgements

This study was supported by National Natural Science Foundation of China (50962001). The authors are grateful to Mr. Jiang and Mr. Han for their assistance.

7. References

Nitin P. Padture. (1994)In situ-toughened silicon carbide. J.Am.Ceram.Soc., 1994,77[2]519-523 ISSN :1551-2916

Kim Y. & Mitomo.M . (2000) Fabrication and mechanical properties of silicon carbide-silicon nitride nanocomposites. J. materail Science. 35(2000)5885-5890 ISSN :0022-2461

Lee Y,Kim Y., Choi H., .Lee J.(2001) Effects of additive amount on microstructure and mechanical properties of silicon carbide –silicon nitride composite J. material Science. 36(2001)699-702 ISSN :0022-2461

Becher P.F., Sun Y., Hsueh C., Alexander,K., et..(1996) Debonding of interfaces between beta silicon nitride and Si-Al-Y oxynitride glass Acta Mater., 1996, 44 3881-3893 ISSN :1359-6454

Keeebe H., Pezzotti G., Ziegler G.(1999) Microstructure and fracture toughness of Si_3N_4 ceramics : combined roles of grain morphology and secondary phase chemistry J.Am. Ceram. Soc., 1999, 82,1642-1644 ISSN :1551-2916

Anna E. McHale. (1994) Phase Diagrams for Ceramists[M]. Vol. X. Compiled at National Inst. of Standards and Techn. Edited and Published by The American Ceramic Society, 36-115. ISBN:0-944904-74-2

Huang Z. K. & Tien T. Y.(1996) Soli-liquid reaction in the Si_3N_4-AlN-Y_2O_3 system J. Am. Ceram. Soc., 1996, 79[6], 1717-1719. ISSN :1551-2916

Huang Z. K. & Chen I. W.(1996) Rare-earth melilite solid solution and its phase relations with neigh-boring phase J. Am. Ceram. Soc., 1996, 79 [8] 2091-2097. ISSN :1551-2916

Mitomo M., Izumi F., Horiuchi S., Matsui Y.. (1982)Phase relationships in the system Si_3N_4-SiO_2-La_2O_3 (1982) J. Of Mater. Sci., 1982,17, 2359-2364 ISSN :0022-2461

Jack K.H., (1978)Mater. Sci.Res. 1978,11,561-578

Cao G.Z., Huang Z.K., Yan D.S.. (1989) Phase relationships in Si_3N_4-Y_2O_3-La_2O_3 system. the Science China, Ser.A. 1989 32,4, 429-433 ISSN:1674-7216

Yan D.S., Sun W.Y. (2000) phase relationship and material design in the Ln-Si-Al-O-N system. Science in China ,series B, 2000, 6 ,225-232 ISSN:1674-7224

N.A.Toropov,I.A.Bondar F.Ya.Galakhov. Trans. Inter.Ceram. Congr. 8th, Copenhagen Denmark, 1962. 87-90

Harue Wada, Ming-Jong Wang, and Tseng-Ying Tien, (1988) J. Am. Ceram. Soc.,1998, 71 [10] 837-840.

Liquid Phase Sintering of Silicon Carbide with AlN-Re_2O_3 Additives

Laner Wu, Yuhong Chen ,Yong Jiang, Youjun Lu and Zhenkun Huang
School of Material Science & Engineering, Beifang University of Nationalities
Ningxia, China

1. Introduction

Silicon carbide can be pressureless sintered by a solid stated process with the sintering aids of B and C to near full density at temperatures in excess of 2100℃(Prochazka,1974). However, the lower fracture toughness (3 to 4 Mpa m$^{1/2}$) limit their use in many potential structural applications. It has been known that sintering of SiC can be achieved at relatively lower temperature (1850℃-2000℃) with the addition of oxides (Al_2O_3 and Y_2O_3) via liquid phase sintering(Omori & Takei, 1988; Nitin, 1994). The resulting material obtained with homogeneous and equiaxed fine-grained microstructure. Oxides like SiO_2 and Al_2O_3, which are normally considered as thermodynamically stable, are prone to react with SiC at temperature of about 2000℃, leading to formation of gaseous products such as CO, SiO and Al_2O.

Al_2O_3+SiC→Al_2O(g)+SiO(g)+CO(g)

In order to suppress these reactions, a powder bed is generally required (Tan et al, 1998). Alternatively, the additive system of AlN and rare earth oxides including Y_2O_3, is used where the decomposition of AlN into Al and N_2 can be efficiently controlled by using N_2 atmosphere, leading to lower weight lost (Chia et al, 1994; Ye et al, 20002). The AlN -Y_2O_3 phase diagram indicates that eutectic temperature in this system is about 1850℃ (Kouhik, 2002). It might avoid forming a liquid with rather low melting temperature and a coarse surface of ceramic caused by vaporized gases from the reaction of SiO_2 and Al_2O_3-Y_2O_3. Also in this system the intermediate compositions can offer sufficient amount of liquid with melting temperature higher than 1700℃ as sintering aid of LPS-SiC. Some studies have been carried out by using rare-earth oxide containing densification aids (Koushik et al, 2004; Koushik et al, 2005). Our previous study on melting behaviours of SiC and a series of Re_2O_3 (1 : 1 mol mixture) has shown that melting temperatures raise with increasing the atomic number of rare earth element (from La to Er and Y) (Wu et al, 2008). The aim of this work was to study the sintering behavior of liquid phase sintered SiC with AlN and Re_2O_3 (La_2O_3, Nd_2O_3 , Y_2O_3) additive system and their mechanical property in both pressureless sintering and hot press sintering.

2. Material and Method

2.1 Materials

The submicron α-SiC powder was manufactured by Beifang University of Nationalitie. SiC content >97%(mass fraction, the same below), free C<1%, SiO2<1.2%; median particle size of the powder: D_{50} = 0.7μm. AlN powder (D_{50} < 0.8μm, purity>98%) were provided by Beijing Iron Rcscarch Institute, Y_2O_3, La_2O_3 and Nd_2O_3 (purity>99.9% D_{50} = 2-5 μm) was provided by Baotou Rear Earth Research Institute. The particle size distribution of the powders was measured by Laser Sizer (model Microtrac X–100, Honeywell, USA). The chemical analysis of the SiC powder was carried out according to Abrasive Grains –chemical analysis of silicon carbide(National Standard of China GB/T 3045-2003) .

2.2 Experimental Methods

2.2.1 Preparation of the powder mixtures

SiC powder and additives were mixed in an attrition mill for about 1 hr in alcohol using Si_3N_4 balls as medium. The compositions of various powder mixtures prepared and the nomenclature used to describe the samples are specified in Table 1. All of the powder mixtures have content of 85% SiC and 15% additives(mass fraction) except "Slay -1". The milled slurry was separated from the milling media and possible wear debris by screening through 320 mesh screen. The slurry was dried in a stirring evaporator and completely dried in a drying oven at 80°C. The dried powder mixture was sieved through 100 mesh.

sample code	AlN /mol%	Y_2O_3 /mol%	Nd_2O_3 /mol%	La_2O_3 /mol%	Theoretical Density ρ/g·cm^{-3}
Sly-1	40	60	0	0	3.40
Sly-2	60	40	0	0	3.38
Sly-3	80	20	0	0	3.34
Sln	60	0	40	0	3.50
Slny	60	20	20	0	3.44
Sla	60	0	0	40	3.47
Slay	60	20	0	20	3.43
Slay-1*	66	17		17	3.45

*Sample " Slay - 1 "has 80% SiC and 20% sintering additives (mass fraction)

Table 1. Compositions and Theoretical Density of powder mixtures

2.2.2 Pressureless Sintering

The mixed powder was axial pressed under pressure of 100Mpa and then cold isostatic pressed under 250 MPa. The rectangular shaped green samples of approximately 10×50×50 mm were sintered in a graphite furnace (made by Robert furnace Co. China). The samples were put into a graphite crucible using BN powder as separator. A high purity N_2 gas atmosphere was used during sintering. The gas pressure was maintained at 0.02 Mpa during sintering. The samples were sintered at 1800, 1850, 1900, 1950, 2000°C and 2050°C for

1 hr separately. Heating rates of 20°C/min from ambient temperature to 1600°C and 10°C /min from 1600°C to final sintering temperature were used.

2.2.3 Hot press sintering

The powder mixtures were put in a 40 mm ×40 mm graphite mould (lined with BN powder as separator), hot press sintered under an axial pressure of 30 MPa in N_2 protected atmosphere with a sintering temperature of 1 850 °C, held for 0.5 h (the furnace made by Shanghai Chenrong Co., China).

2.3 Characterization

The weight loss and linear shrinkage of both green body and sintered specimen of all samples were measured. Bulk density were measured by Archemede's principle by a water displacement method. The hardness was determined by using a load of 98 N in a micro-hardness test fitted with a Vickers square indenter (Wolpert U.S.A). The fracture toughness was calculated by the length of the cracks originating from the edges.: K_{1C}=0.016 (E/Hv)0.5×(p/c-1.5) where K_{1C} is the fracture toughness of the material, Hv is the Vickers hardness, E is the Young's modulus (for LPS-SiC a value of 400 was assumed) c is the crack length(μm) and a is indentation diagonal (Anstis et al, 1981). The specimens were cut into rectangular beams with dimensions of 3×4×36 mm to test three point bending strength. The tensile edges were bevelled to remove stress concentrations and edge flaws caused by sectioning. Observation of the microstructure has been performed by SEM (SSX-550 Shimadzu Japan) on fracture surfaces and also on finished surface polished by 1 μm diamond paste. The phase composition of samples was determined by X-ray diffraction using Cu-Kα radiation (XRD-6000 Shimadzu Japan) , a step width of 0.2 with an exposure time of 2 degree/min per position.

3. Results and discussion

3.1 Sinterability of SiC-AlN-Y_2O_3 system

Similar to other works (Rixecker et al, 2000; Magnani & Beaulardi 2005), the sintering temperature for completed densification is a function of the additive composition and the best densification behavior does not coincide with the eutectic composition in the AlN-Y_2O_3 system(See Fig 1) (Kouhik, 2002), which is about 40 mol% AlN as shown in Fig 2. Sample Sly-2 with 60 mol%AlN reached full density at much lower temperature compared with the other two samples. Further more the temprature range of dentification is much wider than others also. It can be seen obviously that the sample with less AlN content as Sly-1(with 40%AlN) need higher sintering temperature and has very limited adaptive temperature range. Sample Sly-3(with 80%AlN) need even higher sintering temperature, its adaptive sintering temperature range is also very limted. It is well known that an important requirement of liquid phase sintering is that there must be good wetting of the solid phase (SiC) by the liquid phase (additive) and there must be a small contact angle θ between the solid SiC and the liquid drops formed by the additive. R.M.Balestra's work showed that at this additive system with 60%mol% AlN had good wettability (θ_{min}≌6°)(Balestra, et al, 2006).The viscosity of silicate melts increases with their nitrogen content, in analogy to the glass transition temperatures of oxynitride glasses.

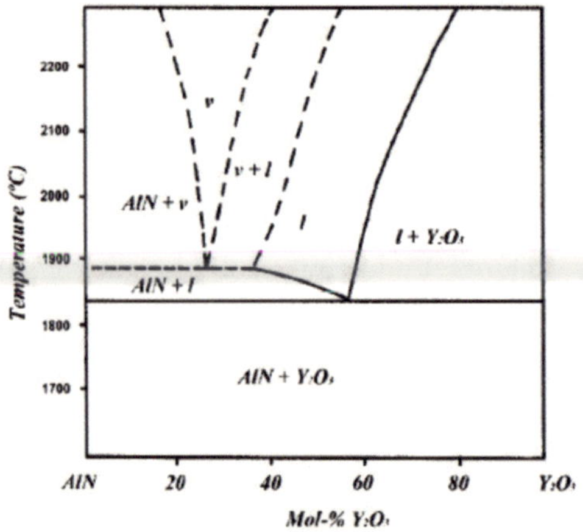

Fig. 1. Phase diagram of the Y_2O_3/ AlN system(Kouhik, 2002)

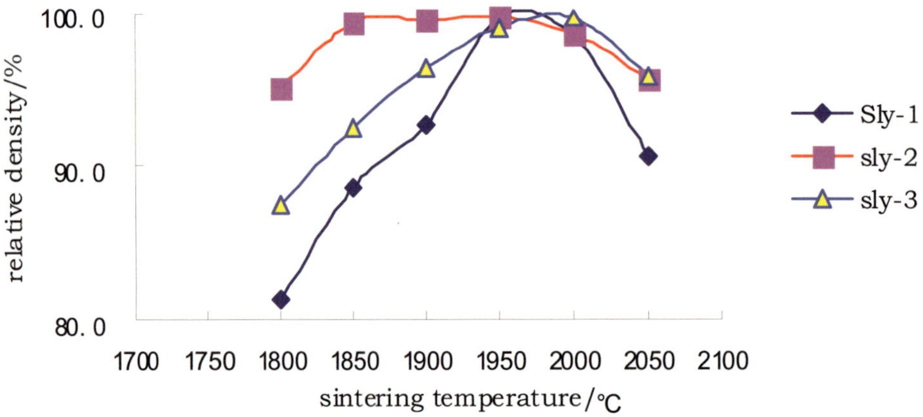

Fig. 2. Sinterable behavior as a function of nitrogen content in the additive

The weight loss of all full density specimens kept about 2%, as shown in Fig 3. When the sintering temperature was raised higher than 2000℃, the weight loss of all specimens increased rapidly up to more than 5%, and the diametric linear shrinkage was less than those in full density temperature. Hence at that temperature, additive decomposition made the density of specimens decrease. It can be seen from Fig. 3 that Sly-1 has less shrinkage than others in the whole temperature range, and less weight loss at lower temperature. Among all samples, Sly-3 has the most even curve both in shrinkage and weight loss . It will bring more convenient sintering process design for densification of SiC. Experimental results showed that SiC-AlN-Y_2O_3 could be fully densified in wide temperature range (1850℃-2000℃), and keep low weight loss around 2% in this range. The surface of specimens remains smooth, indicating that sintering could be done without powder bed.

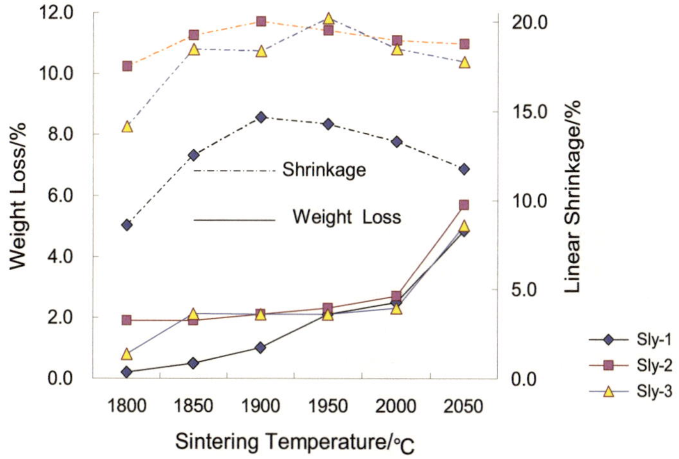

Fig. 3. Weight loss and linear shrinkage of samples VS sintering temperature

It can be seen in the phase diagram of the Y_2O_3/AlN system shown in Fig. 1 that there is a liquid region with sharp lines between the gas(v)/liquid(l) phase region and liquid /Y_2O_3 region. Its eutectic point is near 1 830 ℃(Kouhik, 2002) . The actual sintering temperature is close to the eutectic temperature in order to prevent unfavorable influence of volatilization. The material with the liquid region composition showed less mass loss during high temperature sintering. Experimental results show that SiC-AlN-Y_2O_3 can be fully densified over a wide temperature range (1850℃-2000℃), and keep low weight loss around 2%. The surface of specimens remains smoothly, indicating that sintering could be done without powder bed.

3.2 Sinterability of SiC-AlN-R_2O_3 (R=Nd, La) systems

The best sintered density and corresponding weight loss data of specimens of all test using AlN-Re_2O_3 additive system by using pressureless sintering are shown in Table 2. These test results indicated that the specimens wouldn't been fully densified by using AlN-Nd_2O_3 or AlN-La_2O_3 additive system, all these systems showed much higher weight loss than those results reported in gas pressure sintering (Izhevskyi et al, 2003) which indicated much decomposition reaction occurred without N_2 gas protection.

sample code	Sintering temperature/℃	Weight loss / %	RD ρ/ %
Sln	1900	5.9	96.5
Slny	1950	3.1	99.2
Sla	1900	6.9	92.4
Slay	1950	5.1	98.1
Slay-1	1930		97.0
Slay-1 (H P)*	1850		99.3

*Slay-1(HP) was sintered by hot press

Table 2. sintering density and weight loss of AlN- R_2O_3 systems

Interestingly, AlN-Re_2O_3-Y_2O_3 additive system showed much better sintering behaviours than AlN-Re_2O_3 system. Although more weight loss occured than in the AlN- Y_2O_3 system did, and higher sintering temperature was needed for densification.

3.3 Mechanical properties

Mechanical properties of all densified specimens are summarized in Table 3. For AlN-Y_2O_3 system specimens, the hardness (Hv) increased with increasing AlN content. AlN-Nd_2O_3-Y_2O_3 additive specimen show higher hardness than that of all other specimens, which has same hardness as SSSiC(21-25 GPa)(Wu A.,et al,2001). All specimens have bending strength in range of 350-500MPa. All specimens have relative higher fracture toughness compared to SSSiC which is in range of 3-5 MPa·$m^{1/2}$. The SEM picture of crack and the fracture surface are shown in Fig 4. The indicated fracture mode was intergranular fracture. Grain refinement and inter-crystal deflection are the main reasons for the toughness increasing.

sample code	Hardness /GPa	Bending strength/Mpa	Fracture toughness /MPa·$m^{1/2}$
Sly-1	18.7±0.7	410±4.8	6.8±0.4
Sly-2	19.4±0.8	435±42	8.0±0.7
Sly-3	20.8±0.2	481±57	6.1±0.2
Slny	22.2±0.2	--	6.9±0.3
Slay	18.9±1.1	367±13	6.5±0.3
Slay-1	20.5±1.2	434±52	4.8±1.0
Slay-1(HP)	19.0±1.0	828±55	8.6±1.9

Table 3. mechanical properties of best densified specimens

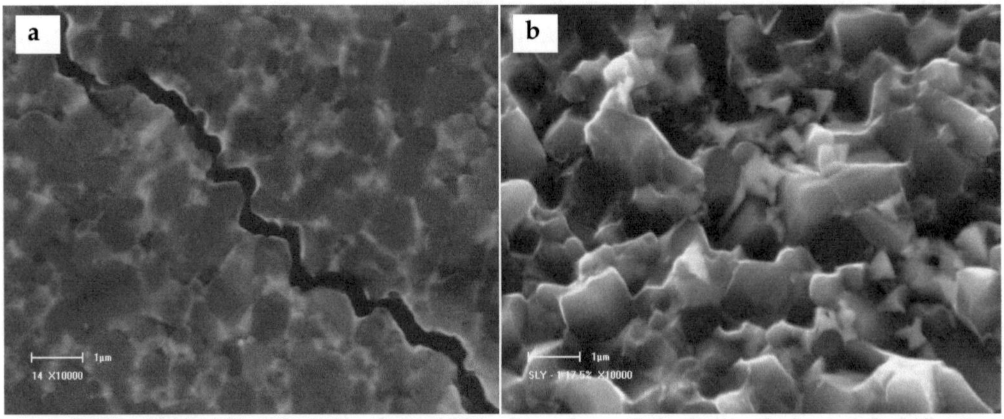

Fig. 4. SEM picture of crack deflection and break surface of sly-2 sample (a. crack deflection, b. fracture surface)

3.4 Microstructure and phase composition

3.4.1 SiC-AlN-Y_2O_3 system

Typical microstructure of AlN-Y_2O_3 system are shown in Fig 5, similar to the microstructure described in previous report(Rixecker G., et al, 2001, Koushik B.,et al, 2005, Wu L., et al, 2008, L.S.Sigl ,2003) . The SiC grains are predominantly equiaxed with a mean grain size of 1-2μm. Relatively little grain growth occurred during densification, indicating that the atomic transport through the melt is sluggish. The core-rim structure is found more clearly in higher AlN content samples.

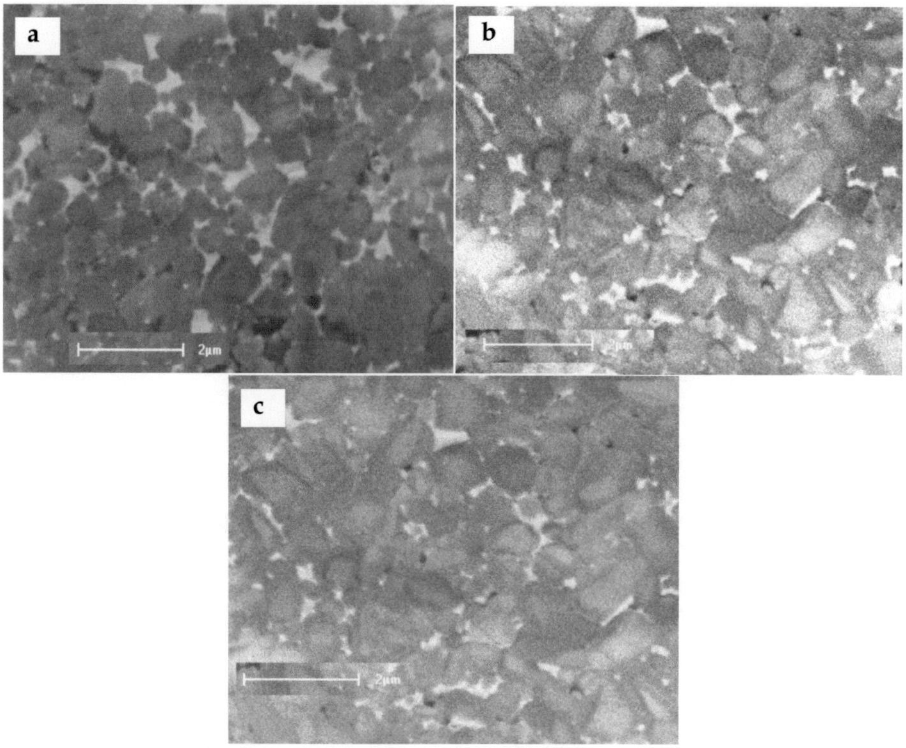

Fig. 5. Microstructure of LPS-SiC with AlN-Y_2O_3 additive
a) sly-1, b) sly-2, c) sly-3

The XRD pattern of the sample is shown in Fig 6. The major phase is 6H - SiC, the minor phases are AlN, Y_2O_3 and $Y_{0.54}Si_{9.57}Al_{2.43}O_{0.81}N_{15.19}$ (α-Sialon). The work of Haihui Ye described that for sample sintered in 1 MPa N_2 atmosphere the AlN, $Y_{10}Al_2Si_3O_{18}N_4$, and $Y_2Si_3N_4O_3$ phase (melilite) were identified; but in Ar , Y_2O_3 and $Y_{10}Al_2Si_3O_{18}N_4$ phase were identified (YE. et al, 2002). Formation of minor $Y_2Si_3N_4O_3$ (melilite) means that a little SiC has been reduced/nitrided to be Si_3N_4 during firing in 1 MPa N_2 atmosphere. In this experiment the nitridation of partial SiC to Si_3N_4 also happened in N_2, 0.02 atm. leading to the formation of $Y_{0.54}Si_{9.57}Al_{2.43}O_{0.81}N_{15.19}$ (α-Sialon), which was from the reaction of the compositions on the one dimension α-Sialon line of Si_3N_4 -Y_2O_3:9AlN with the formula of $YxSi_{12-(m+n)}Al_{(m+n)}O_nN_{16-n}$, x=0.33-0.67(Sigl, 2003). It has been shown that the core-shell structure which can be seen clearly in Fig 5 formed mainly by solution-reprecipitation of oxynitride or α-Sialon during matter transport.. The subsolidus phase diagram of SiC-AlN-Y_2O_3 system in N_2 is shown in Fig 7.

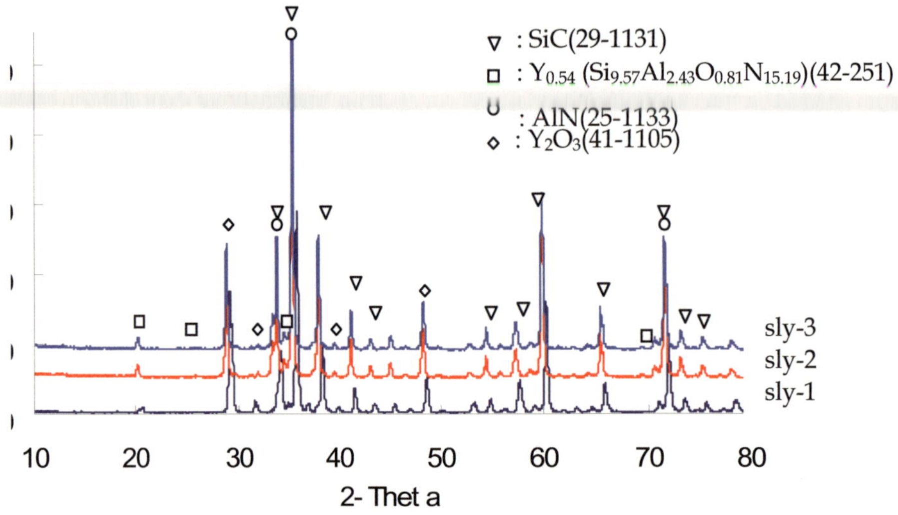

Fig. 6. XRD analysis of sintered sample with AlN-Y_2O_3 additive

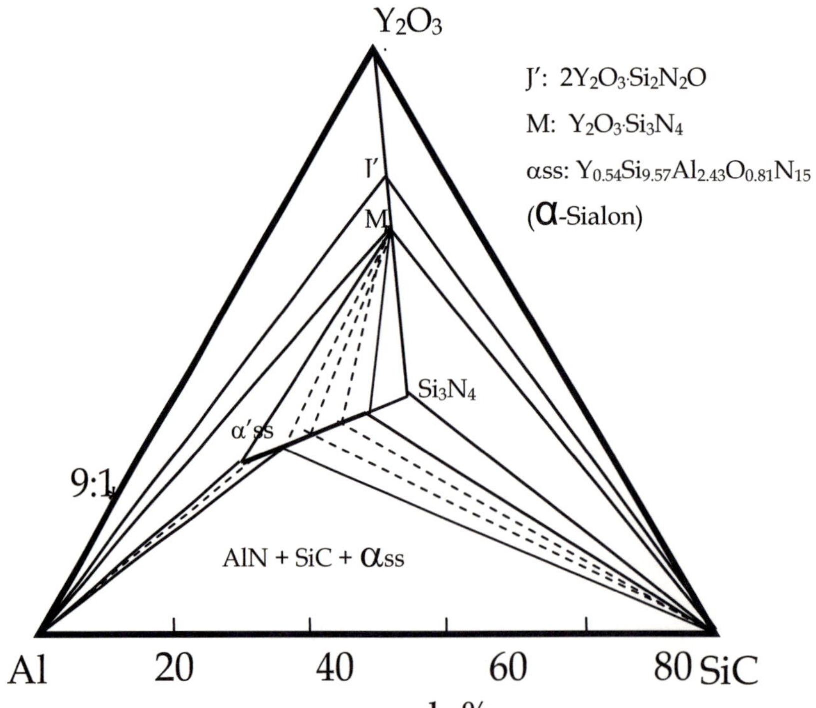

Fig. 7. Subsolidus phase diagram of SiC-AlN-Y_2O_3 system in N_2
* See Huang 1983

3.4.2 SiC-AlN-R_2O_3 (R=Nd, La) systems

For the AlN-Re_2O_3-Y_2O_3 additive system, the microstructure of LPS- SiC is similar to the AlN-Y_2O_3 system, but core-rim structure are hardly found in SEM(Fig 8 a),b),c)). Only in the hot-pressed samples (Fig 8 d)), "core-shell" could be observed obviously. Although the SEM images shown in Fig 8 c) and d) came from the samples with exactly the same composition. The different sintering process bring unlike microstructure of the ceramics. Certainly hot press sintering gains better results. It can be further explained by their mechanical properties.

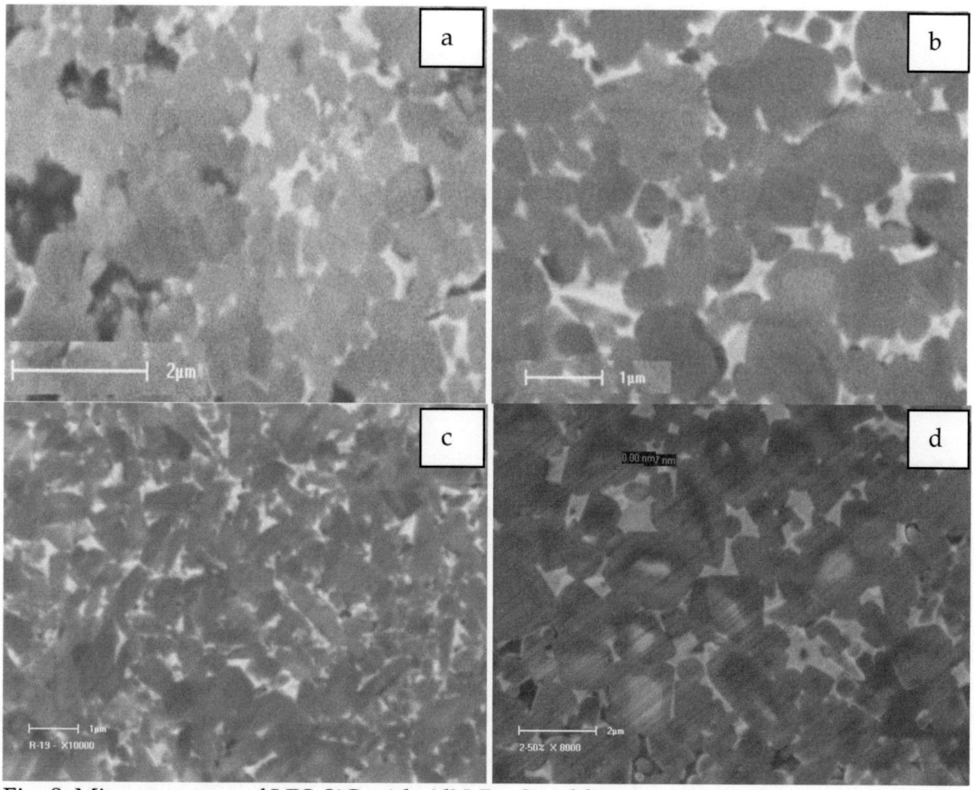

Fig. 8. Microstructure of LPS-SiC with AlN-Re_2O_3 additive
a) slny, b) slay, c) Slay-1, d) Slay-1(HP)

The XRD pattern of the sample with AlN-Nd_2O_3-Y_2O_3 additive is shown in Fig 9, two nitrogen-richer phases of $Y_{0.54}Si_{9.57}Al_{2.43}O_{0.81}N_{15.19}$ (α-Sialon) and $Nd_4Si_2O_7N_2$ (NdAM') were found.

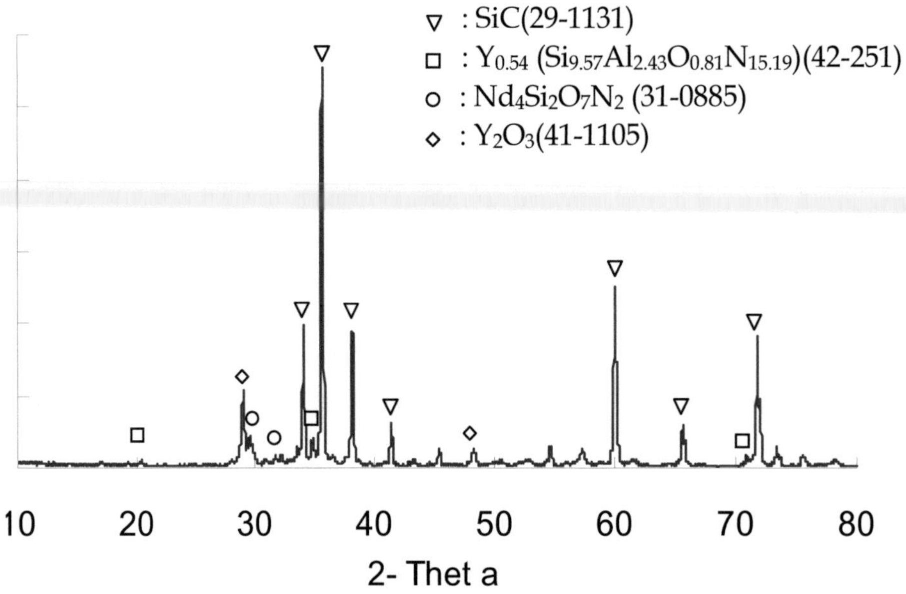

Fig. 9. XRD analysis of sintered sample with $AlN-Nd_2O_3-Y_2O_3$ additive

4. Conclusion

1. Fully dense SiC ceramics were obtained by liquid phase sintering with $AlN-Y_2O_3$ and $AlN-R_2O_3-Y_2O_3$ additives. Materials with 60mol% AlN in $AlN-Y_2O_3$ additives show that SiC can be sintered over a wide temperature range (1850℃-2000℃), and keeps low weight loss around 2%. The surface of specimens remains smooth, indicating that sintering could be done without powder bed. The specimens made using the $AlN-R_2O_3-Y_2O_3$ additives systems show higher weight loss, around 5%.
2. The materials obtained have fine-grained and homogeneous microstructure. The core-rim structure can be found in high AlN content specimen. All specimens have higher fracture toughness in the range of 6-8 MPa·$m^{1/2}$. Grain refinement and inter-crystal deflection are the main reasons for toughness increasing. The specimens for $AlN-Nd_2O_3-Y_2O_3$ additive system showed higher hardness.
3. XRD analysis identified that nitrogen-richer phases of $Y_{0.54}Si_{9.57}Al_{2.43}O_{0.81}N_{15.19}$ (α-Sialon) in $AlN-Y_2O_3$ additive system and $Nd_2Si_4O_7N_2$ (NdAM′) in $AlN-Nd_2O_3-Y_2O_3$ additive system formed in the present study, indicating partial SiC was reduced/nitrided to Si_3N_4 or Si_2N_2O.

Acknowledgements

This study was supported by National High Technology Research and Development program of China (2002AA332110) and National Natural Science Foundation of China

(50962001). The authors are grateful to Mr. Jiang and Mr. Han for their assistance with experiments.

5. References

Huiwen Xu, et al. Microstructural evolution in Liquid-Phase-Sintered SiC : Part 1, Effect of starting powder. J. Am. Ceram. Soc., 2001, 84[7]1578-1584. ISSN :1551-2996

G.R.Anstis, P.Chantikul, B.R.Lawn and D.B.Marshall. A critical evaluation of indentation techniques for measureing fracture toughness I Direct crack measurements. J. Am. Ceram. Soc., 1981, 64[9]533-543. ISSN :1551-2996

K.Y.Chia, W.D.G.Boecker and R.S.Storm.U.S.Pat., 5, 298, 470(1994)

Kouhik Biswas. Liquid phase sintering of SiC ceramics with rare earth sesquioxides [D]. Stuttgart: University of Stuttgart, 2002.

Koushik Biswas Rixecker.G, Aldinger F.. Effect of rare-earth cation additions on the high temperature oxidation behaviour of LPS-SiC. Materials Science and Engineering A 2004, 374 56–63. ISSN :0921-5093

Koushik Biswas, Georg Rixecker, Fritz Aldinger. Gas pressure sintering of SiC sintered with rare-earth-(III)-oxides and their mechanical properties. Ceramics International 2005, 31 703–711. ISSN : 0272-8842

L.S.Sigl. Thermal conductivity of liquid phase sintered silicon carbide. J. Euro. Ceramic Soc., 2003, 23, 1115-112 ISSN : 0955-2219

M. Balog, P. Šajgal´ık, M. Hnatko, Z. Len˘céš, F. Monteverde, J. Ke˘ckéš, J.-L. Huangd Nanoversus macro-hardness of liquid phase sintered SiC. J. Euro. Ceramic Soc., 2005, 25, 529–534.. ISSN : 0955-2219

Magnani, G. and Beaulardi, L.. Properties of liquid phase pressureless sintered SiC-based materials 83 obtained without powder bed. J.Aus.Ceram.Soc., 2005, 41(1), 31-36. ISSN :0004-881X

Nitin P. Padture. In situ-toughened silicon carbide. J.Am.Ceram.Soc., 1994, 77[2]519-23 ISSN :1551-2996

Omori M, Takei H. Preparation of pressureless sintering of SiC–Y_2O_3–Al_2O_3. J Mater Sci, 1988, 23(10): 3 744–3 747. ISSN :

Prochazka S. Sintering of silicon carbide// Proceedings of the Conference on Ceramics for High Performance Applications (Hyannis, MA). Hyamnis: Brook Hill Publishing Co, 1975: 7–13.

Rixecker G, Biswas.K, Wiedmann I and Aldinger.F.. Liquid-phase sintered SiC ceramic with oxynitride additives. J.Ceramic processing research 2000, 1:1 12-19. ISSN :1229-9162

Rixecker G, Wiedmann I, Rosinus A, et al. High-temperature effects in the fracture mechanical behaviour of silicon carbide liquid-phase sintered with AlN-Y_2O_3 additives. J.Euro. Ceram Soc., 2001, 21: 1 013–1 019. ISSN :0955-2219

R.M. Balestra, S. Ribeiro, S.P. Taguchi, F.V. Motta, C. Bormio-Nunes Wetting behaviour of Y_2O_3/AlN additive on SiC ceramics. J. Euro. Ceram. Soc., 2006, 26(16) 3881-3886 ISSN :0955-2219

TAN Shouhong, CHEN Zhongming, JIANG Dongliang. Liquid phase sintering SiC Ceramics. J. China Ceram Soc., (in Chinese), 1998, 26(2): 191–197. ISSN :0454-5648

V.A.Izhevskyi, L.A.Genova, A.H.ABressiani , J.C.Bressiani Liquid phase sintered SiC processing and Transformation controlled microstructure tailoring. Material research, 2000, 3[4]131-138. ISSN :1516-1439

Wu Laner, Chen Yuhong, Jiang Yong, Huang Zhenkun. Liquid sintering of SiC with AlN–Re_2O_3 Additives. J.of the Chinese ceramic society 2008, 36, [5] 593-596 .ISSN : 0454-5648

Wu Anhua, Cao Wenbin, Li Jiangtao, Ge Changchun, Solid State Sintered SiC Ceramics, J. of Materials Engineering(China), 2001, 4, 3-5. ISSN :1001-4381

Ye Haihui, Rixecker.G, Siglinde. H, et al. Compositional identification of the intergranular phase in liquid phase sintered SiC. J Euro Ceramic Soc., 2002, 22: 2 379–2 387. ISSN :0955-2219

Z. K. Huang, P. Greil, G. Petzow, Formation of α - Si3N4 solid solution in the system Si3N4-AlN-Y2O3. J. Am. Ceram. Soc., 1983, 66, C96-C97. ISSN :1551-2996

Investigations on Jet Footprint Geometry and its Characteristics for Complex Shape Machining with Abrasive Waterjets in Silicon Carbide Ceramic Material

S. Srinivasu D. and A. Axinte D.
University of Nottingham
United Kingdom

1. Introduction

Due to a combination of desirable mechanical properties, such as high hardness, wear resistance, strength at elevated temperatures in addition to corrosion resistance, chemical inertness, electromagnetic response and bio-compatibility, advanced ceramics are widely used for the manufacture of components for the optical, electronic, mechanical and biological industries (Siores et al., 1996; Chen et al., 1996). Silicon carbide (SiC), a structural ceramic, is increasingly being used for highly demanding engineering applications (bearings, valves, rotors, cutting tools, face seals, textile thread guides) where close dimensional/geometrical tolerances are required (Kahlmana, 2001). Furthermore, due to its low coefficient of thermal expansion, high thermal conductivity, high decomposition temperature, chemical inertness, low wettability by molten metal and low density, SiC is commonly used for heat resistant parts and refractory applications (Lee & Rainforth, 1992; Inasaki, 1987).

Although advanced ceramic components can be fabricated to net shape through hot-isostatic-pressing (HIP), a highly expensive process, final machining cannot be avoided to generate intricate surfaces and to meet the required quality specifications - dimensional accuracy, surface finish, elimination of surface flaws, as HIP is not economical for all applications (Richerson, 2006). In general, machining of ceramic components is usually done in the green state. However, to maintain dimensional control and production optimization, post-sintering machining unavoidable (Hocheng & Chang, 1994). On the other hand, milling of 3D surfaces is challenging with conventional machining approaches due to (i) component high hardness, which lead to excessive tool wear (ii) low fracture toughness of the workpiece materials, which lead to part fracture under the local machining forces (Chen et al., 1996; Hocheng & Chang, 1994). Diamond cutting (e.g. drilling)/grinding are commonly adopted for generating 3D features in sintered ceramics (Samant & Dahotre, 2009; Tuersley et al., 1994). However, conventional machining approaches might not always be suitable as they demand dedicated tooling/fixtures to generate 3D features in such difficult-to-cut materials with the associated high cost and extended lead times.

As the machining often represents a significant portion of the fabrication cost of the final ceramic components, technologies suitable for cost effective machining of brittle materials (e.g. SiC) have to be identified. Various non-conventional approaches for machining of ceramics, such as ultrasonic, abrasive water jet (AWJ), electrical discharge and laser applications are reported and the trade-off between the material removal rate and the level of near-surface damage that these processes induce have been detailed (Samant & Dahotre, 2009; Tuersley et al., 1994). Among the non-conventional machining approaches, AWJ is a well utilized approach for machining hard to machine materials, such as ceramics at high material removal rate (MRR), with the advantage of reducing the above mentioned drawbacks of the conventional machining approaches (Chen et al., 1996; Kahlmana et al., 2001; Hashish, 1987; Chen et al., 1998). Furthermore, AWJs are suitable for machining of a wide range of materials due to its unique features, such as exertion of low cutting forces (<10N) at the machining zone, omni-directional cutting capability and material removal by micro cutting by abrasive particles, which eliminates the problem of chipping and fracture of brittle workpiece and leaving no heat affected zone (Zeng & Kim, 1996). A considerable amount of work has been reported on machining of ceramics with AWJs (Chen et al., 1996; Chen et al., 1998; Zeng et al., 1997; Hocheng & Chang, 1994; Gudimetla et al., 2002; Momber et al., 1996; Momber & Kovacevic, 2003; Zeng & Kim, 1996).

Freist et al. (1989) introduced the concept of 3D machining with AWJs. The concept is to break up the desired geometry into several parts and achieve each by the kerf shape generated by AWJs. The kerf geometry is described mathematically by a cosine function; however, milling is not demonstrated. There have been very limited efforts in the direction of controlled 3D complex shape generation with AWJs (Andrews & Horsfield, 1983). Ramp surface generation has been reported in X5CrNi189 material achieved by multi-pass linear-traverse milling in which the superposition of several kerfs is exploited as a strategy. A multi stage concentric pocket, 3D geometry, in a WC-Co hard metal is milled by the discrete milling approach proposed by Ojmertz. In milling of complex parts in advanced engineering materials (AEM), it is believed by the authors that jet footprints (JFP) with different shapes and dimensional characteristics (e.g. slopes equal to the local slope of the desired 3D complex geometry) have to be generated to approximate the local geometry of the desired complex part with the geometry of the suitable JFP. The JFPs with desired features can be realised by varying the jet feed rates and jet impingement angles, i.e. the angle between the jet axis and the target surface. Hence, complex 3D shapes in AEMs can be realised by AWJs, mainly, in two methods by: (i) varying the jet feed rate, and (ii) varying the jet impingement angle. In 2-axis and 5-axis AWJ milling of pockets, the bulk material in the targeted region is removed by raster scanning the surface by the jet (Ojmertz, 1997; Ojmertz & Amini, 1994; Laurinat et al., 2004). In 2-axis machining, the desired shape of the final component can be achieved by approximating the local geometry of the final complex 3D shape with the JFP characteristics generated at various jet feed rates at normal jet impingement angle. On the other hand, in complex shape generation by 5-axis machining, the jet is continuously manoeuvred, by varying the jet impingement angle, according to the local geometry of the final complex shape - to match the JFP at that specific orientation with the local geometry of the desired part to be manufactured (Srinivasu et al., 2009).

In order to generate the controlled 3D complex shapes in AEMs, the variation in geometrical and dimensional characteristics (erosion depth (***h***), top width (w_t), slope of the side walls) of the jet footprint (JFP), which is the building block of the AWJ milling, with the variation in jet feed rate and jet impingement angle have to be known apriori (Miller, 2004; Srinivasu et al., 2009). Among them, the w_t is a critical measure that dictates the degree of the overlap to be employed, which influences the flatness of the bottom surface, depth of the pocket, surface roughness and longitudinal dimensions of the part, i.e. length and width (Ojmertz, 1997; Miller, 2004). Similarly, in profiling, drilling and slotting of hard to machine materials by macro/micro AWJs, the machined dimensions (e.g. diameter of hole) depend on the top width of the JFP. Freist et al. (1989) emphasised the need to study the JFP and its variation with the change in operating parameters, such as pump pressure (***P***), abrasive flow rate (m_f), jet feed rate (v), focusing nozzle diameter (d_f) and standoff distance (SOD), i.e. the axial distance between the tip of focusing nozzle and target material to be exposed to the jet. The characteristics of the JFP, such as erosion depth (h(α)), slope of the side walls, top width (w_t(α)) depend on the d_f, φ (as in 2-axis machining); and SOD, which is influenced by α, and the jet impingement angle (Shipway, 1997). Hence, the w_t (α) has to be known at different α by considering the φ for controlled milling of complex shapes. Zeng et al. (1997) performed an experimental investigation to identify the optimal operating parameters and their influence on performance measures, such as depth of penetration, surface roughness and MRR in AWJ milling of alumina ceramics. In this study, d_f, lateral feed between passes, and SOD were found to be of critical importance in controlling the kerf geometry. Laurinat et al. (1993) and Freist et al. (1989) described the kerf geometry at both single pass and following multi-passes with different degrees of overlap by employing cosine function in hardened, austenitic and ferritic steels by varying operating parameters, such as ***P***, d_f, m_f and v. The quality of the pockets with flat surfaces in terms of variation in depth of penetration and surface texture are analyzed by considering lateral feed, m_f, v, ***P*** and SOD (Ojmertz et al. (1997), Freist et al. (1989) and Laurinat et al. (1993)). It must be noted that despite different approaches up to now, only the generation of flat pockets have been achieved; the challenge of identifying methods to generate truly 3D surfaces still exists. Hashish (1993) has studied the influence of jet impingement angle in AWJ machining (milling, turning and drilling) of different materials, such as stainless steel 304, Ti-6-4 on volume removal rate, depth of cut and surface finish are studied. However, attention was not given to kerf geometry and its characteristics. From the literature, it is observed that considerable efforts have been reported on empirical investigation on the influence of various operating parameters, such as water pressure, abrasive mass flow rate and jet feed rate on geometrical characteristics of the JFP. However, there are very limited studies on the dependence of w_t on the a and its modelling. In addition to jet impingement angle, in 2-axis/5-axis milling with AWJs, jet feed rate has considerable influence on the top width of kerf (Hascalik et al., 2007). For AWJ cutting applications, the influence of v on top kerf width is studied (Srinivasu et al, 2009, Niu et al., 1995; Gropetti et al., 1998). Furthermore, the change in the SOD results in generation of corners along the edge of the kerf with specific radius of curvature. The SOD along the JFP varies gradually when the jet impinges at an angle in 5-axis milling (Srinivasu et al., 2009). Hence, the SOD has to be considered in determining the w_t. Nevertheless, there have been very limited reports on studying the influence of $\boldsymbol{a}$ in the presence of variation in v for AWJ milling applications. For machining (milling, turning and drilling) of different materials, such as stainless steel 304, Ti-6-4 and ceramics, an improved depth of cut (***h(a)***), MRR and surface finish are observed with the change in jet impingement angle (Wang, 2003; Hashish, 1993). However, there are very limited studies that have considered the influence of a on top width of JFP. Although some empirical models exist for prediction of geometrical characteristics of the JFP, they cannot readily be adoptable for AWJ milling as are developed for cutting applications; most of the models in the literature have assumed the top width of kerf is equal to the $\boldsymbol{d_f}$, which is not true in practice due to the divergence of jet plume (Srinivasu et al., 2009).

From the literature review, it is inferred that the key enabling element for generation of complex geometries in AEMs using AWJ technology is a unified understanding of the influence of the interaction of jet at different feed rates and impingement angles on the JFP generated. Furthermore, there is a need to develop models for prediction of the geometry of the JFP and its dimensional characteristics, such as top width of kerf in 2-axis/5-axis macro/micro milling. In order to address the above issues, in this chapter, the research work done at the University of Nottingham under the NIMRC sponsored research project titled "*Freeform Abrasive WaterJet Machining in Advanced Engineering Materials* (*Freeform_JET*)", under the following headings was presented: (i) comprehensive investigation on the physical phenomenon involved in the formation of JFP, (ii) development of models for (a) prediction of geometry, and (b) top width, of the JFP.

2. Experimentation and methodology

In order to understand the physical phenomenon involved in generation of the geometry of the JFP at various jet impingement angles and jet feed rates, and to generate the data required to develop models for prediction of JFP geometry and top width, experimental trials were conducted and the complete details are as follows: Milling trials were conducted on 5-axis AWJ (Ormond) cutting system with a streamline SL-V100D ultra-high pressure pump capable of providing a maximum pressure of 413.7 MPa at various mass flow rates (0-1 kg/min) while the jet feed rate can be varied in the range of 0-20,000 mm/min. Garnet (80 mesh size, average Φ180μm - GMA Garnet) abrasive media with sub-angular particle shapes was employed throughout the experimentation to mill SiC ceramic plate (100mmX100mmX10mm). The hardness of the SiC was evaluated as 2500VH. Figure 1a shows a photograph of the experimental setup employed in this study. The structure of the SiC consists of two different regions: α-SiC and β-SiC displaying two different wear characteristics; as α-SiC was reported to have increased strength than β-SiC phase and lower fracture toughness (Lee & Rainforth, 1992), it is expected that the first one will be easier to be removed under AWJ impingement. The two constituents of the SiC ceramic have been revealed by fine diamond polishing (# 6μm/5min followed #1μm/5 min) followed by etching with 'Murakami' (aqueous solution of NaoH and $K_3[Fe(CN_6)]$) solution for 10 minutes. Figure 1b explains the notations used in describing the characteristics of the AWJ process and its erosion outcomes (i.e. kerf shape/dimensions).

As the kerf characteristics are influenced by various operating parameters such as ***P***, d_f, m_f, ***a***, *v*, SOD and properties of workpiece material, careful consideration has been taken in selecting their values in relation to material of study. Since, SiC is a hard material, a high ***P*** of 345 MPa was employed. Furthermore, to maintain the optimum ratio of focusing nozzle diameter to orifice diameter of 3-4 for optimum performance (Chalmers, 1991), a d_f of 1.06 mm and d_o of 0.3 mm were employed. Garnet abrasive of 80 mesh size with an m_f of 0.7 kg/min was employed (Hashish, 1989). SOD of 3 mm was employed as it has been demonstrated that the MRR is insensitive to SOD within the range of 2-5 mm and decreases beyond 5 mm (Hashish, 1987; Laurinat et al., 1993; Ojmertz, 1997). The above operating parameters were kept constant throughout the experimental program. In order to study the influence of v and ***a*** on the JFP and its characteristics, the following experimental plan was followed.

- *Examination of the influence of jet feed rate on jet footprint generation*: To understand the influence of jet feed rate on JFP generation, experiments were conducted by varying the *v* in the range of 100-1700 mm/min in steps of 400 mm/min.

- *Examination of the influence of jet impingement angle on jet footprint generation*: To understand the influence of jet impingement angle on JFP generation, experiments were conducted by varying ***a*** in the range of 40^0-90^0 in steps of 10^0. Further, to study the influence of ***a*** on kerf geometry at different jet feed rates, cutting trials were performed at different jet impingement angles for smaller (*v* = 100 mm/min) and higher (*v* = 900 mm/min) levels of feed rate.

- *Examination of the influence of number of passes on jet footprint generation*: To understand the influence of number of passes on erosion depth, the contribution of preceding jet pass on the increase in SOD (SOD_{actual}: $SOD_{n+1} = SOD_n + h_n$) and shape of kerf geometry were analyzed. For this purpose, different kerfs were generated by single and double jet passes at *v* = 900 mm/min and ***a*** = 90^0 at nominal SOD (i.e. 3 mm) and their variation in geometries/characteristics were discussed. Additionally, trials for compensating the increase in SOD at a second jet pass were performed as follows: 1st pass with SOD = 3mm and 2nd pass with a corrected SOD ($SOD_{corrected}$ = SOD-h) have been carried out; where, 'h' represents the erosion depth in single jet pass.

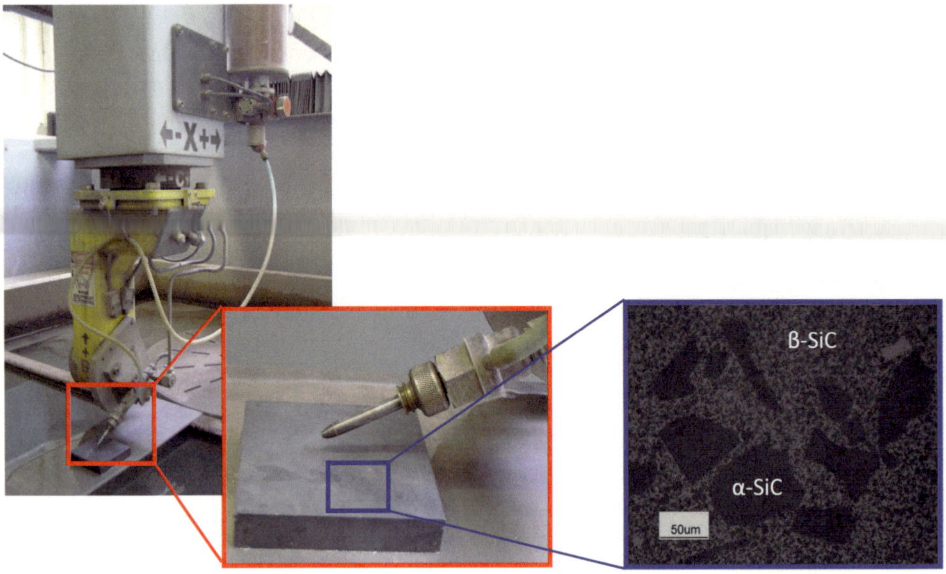

(a)

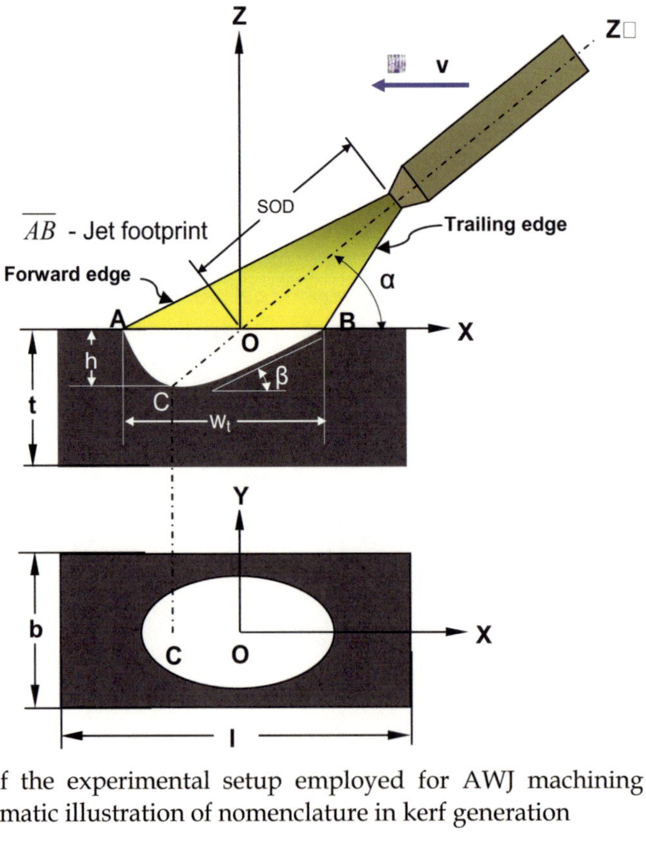

Fig. 1. (a) Photograph of the experimental setup employed for AWJ machining of SiC ceramic material, (b) Schematic illustration of nomenclature in kerf generation

A summary of the testing program is presented in Table 1. To study the influence of jet impingement angle and jet feed rate on the kerf generation in AWJ machining, the cut surfaces were analysed in two stages (i) geometry of the kerf generated at different jet impingement angles; and (ii) dimensional characteristics of the kerf, such as erosion depth, kerf width, slope of the kerf trailing wall. To enable these investigations, sections across the kerfs have been cut, followed by diamond polishing (# 60μm grit / 10min and 15μm grit / 15min.) to ensure their flatness and to allow accurate measurement of geometry of JFP and its geometrical measurements, such as top width, depth, slope of walls using fibre optic digital microscope (Keyence-VHX) and profilometer. Once the jet footprints were generated they have been 3D scanned (Fig. 4) using a Talysurf CLI 1000 from which the ten kerf profiles were extracted at equal spaced intervals (along jet feed direction) to allow the evaluation of the averaged profiles and their variability at various experimental conditions. The average profiles have then been fed into the geometrical models (developed in MATLAB codes) for their calibration and validation.

Constant operating parameters			
d_f(mm) 1.06	P (MPa) 345	d_o (mm)	0.3
m_f (kg/min)	0.7 (Garnet, 80 mesh)	SOD (mm)	3.0
Variable operating parameters			
S. No.	**Objective**	**Operating parameters**	
I	Influence of v on top with of jet footprint	v (mm/min)	100, 500, 900, 1300, 1700
		a (deg)	90
II	Influence of α on top width of jet footprint	v (mm/min)	100, 900
		a (deg)	50, 60, 70, 80 , 90

Table 1. Overview of experimental plan to study the influence of jet impingement angle and jet feed rate on top width of the jet footprint on SiC material

3 Analysis and modelling of abrasive waterjet footprint

3.1 Physical phenomenon involved in the formation of jet footprint (*Srinivasu et al., 2009*)

Understanding the influence of jet footprint at various impingement angles can be done by analyzing the 2D cross-sectional view of the kerf in the plane of the focusing nozzle/jet tilt. Hence, in the following sections, the variation in 2D geometry of the kerf by considering the key kinematic operating parameters (a and v) is discussed with the help of schematic illustrations and the experimental results on kerf geometry and dimensional characteristics, such as erosion depth, top kerf width and slope of kerf walls.

3.1.1 Influence of kinematic operating parameters (α and v) on kerf geometry

a) Influence of jet impingement angle on kerf geometry

For better understanding of the kerf generation phenomena at different jet impingement angles, the experimental results are analysed in two distinct situations: (a) normal jet impingement angle ($\alpha = 90^0$) and (b) shallow jet impingement angle ($40^0 < \alpha < 90^0$)

(i) Normal jet impingement (a = 90⁰)

Figure 2a presents the photographs of the kerf cross sectional geometry generated at normal jet impingement angle at various jet feed rates in the range of 100-1700 mm/min while Fig. 2b shows their measured 2D cross-sectional profiles. The geometry of the kerf generated at $\alpha = 90^0$ is symmetric about the vertical axis, which coincides with the jet axis, in this case. The observations are explained with the help of a schematic illustration of jet-material interaction in kerf generation at normal jet impingement (Fig. 3). The kerf geometry is dictated by: (i) jet energy across the jet-material interaction site ($\overline{AB}$); (ii) local impact angles of abrasive particles (θ) across the JFP. Energy of the jet across the jet footprint varies depending on the jet impingement angle (α) and the jet plume divergence, which in turn influences the velocities of water/abrasive particles.

As the exact energy distribution in the jet is not known clearly, uniform (Leber & Junkar, 2003) and Gaussian distributions (Henning & Westkamper, 2003) have been considered by the researchers. On the other hand, by using flow separation technique (Simpson, 1990) and Laser Doppler Anemometry (Chen & Siores, 2003) these distributions are experimentally determined as double slope distribution. Furthermore, it is found that at higher abrasive flow rates and high water pressures, the abrasive flow increases at the core region and decreases towards walls of the focusing nozzle (Simpson, 1990). As higher water pressure and abrasive flow rates were employed in this study, the velocity of water and abrasive particles were assumed to follow the shape of Gaussian distribution. At any cross-section of jet plume (perpendicular to jet axis), velocity profile of water follows nearly Gaussian distribution (Henning & Westkamper, 2003); Yanaida & Ohashi, 1978; Gropetti & Capello, 1992; Kovacevic & Momber, 1995). On the other hand, with the increase in axial distance from the focusing nozzle, the divergence of jet plume increases which in turn cause decrease in axial velocity (Fig. 3). As the velocity distribution in the radial direction of the jet footprint when $\alpha = 90^0$ is symmetric, the erosion energy which is proportional to the velocity (velocity exponent) of water/abrasive particles also follows the same profile. This leads to maximum erosion at centre of jet axis and gradual decrease on either side. At normal jet impingement angle, due to jet plume divergence (Fig. 3), the local impact angle of abrasive particles (θ) with the target surface decreases gradually on either side of the jet axis across the JFP. Thus, the local impact angle varies from $\theta = 90^0$ at centre of jet axis to a critical angle θ_c (where there is no significant erosion of target material) on either side of the JFP. Furthermore, for brittle materials, the maximum erosion is typically observed at normal impact angle ($\theta = 90^0$) and it reduces gradually with the decreasing in θ (Ruff & Wioderborn, 1979). Hence, the comprehensive effect of reduction in (i) velocity of water/abrasive particles (ii) impact angle of abrasive particles, on either side of jet axis contributes to the symmetric nature of the kerf geometry at $\alpha = 90^0$.

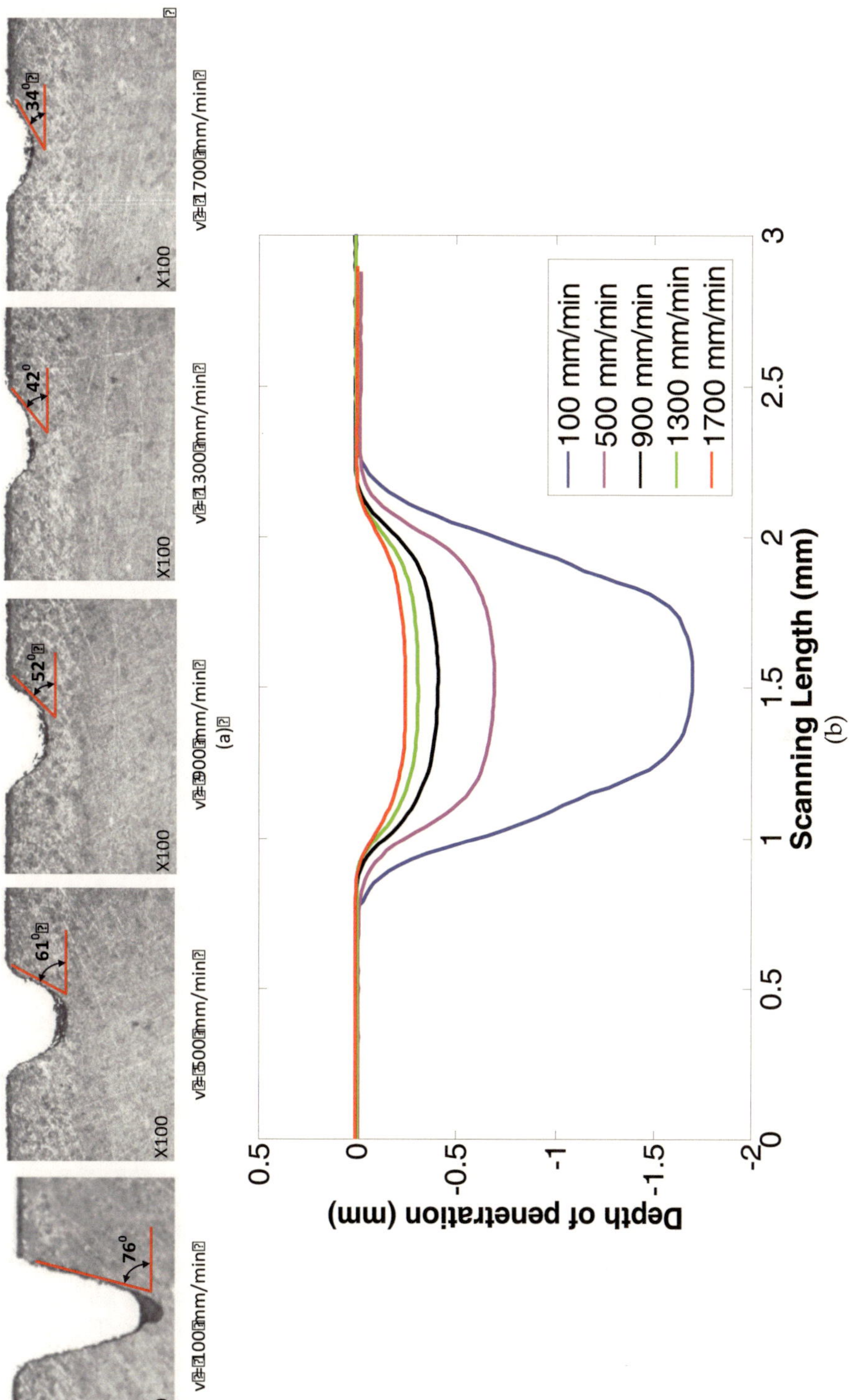

Fig. 2. Kerfs generated at different jet feed rates (α = 900) (a) photograph of cross-section, (b) 2D cross-sectional profile.

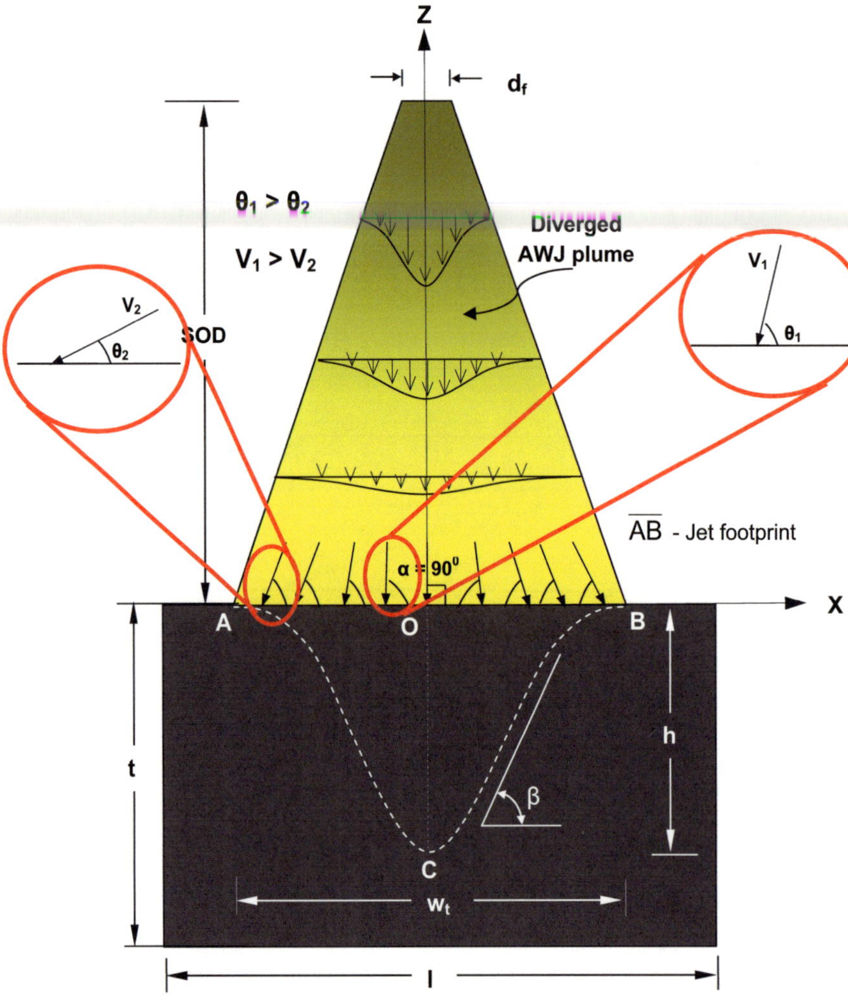

Fig. 3. Schematic illustration of kerf generation at normal jet impingement angle (α = 90°)

(ii) Shallow angle jet impingement ($40^0 < a < 90^0$)

Figure 4 presents the photographs of kerf cross-sections generated at the different jet impingement angles, i.e. 90^0-40^0, in steps of 10^0 at both lower v = 100 mm/min (Fig. 4a (ii)) and higher v = 900 mm/min (Fig. 4a (iii)). From Fig. 4, it can be observed that at α = 90^0, the kerf geometry is symmetric about the vertical axis (which is the same as the jet axis) as discussed earlier (Fig. 3). However, as the jet impingement angle decreases, the kerf geometry becomes asymmetric. This is explained as follows by the use of Figures 5 and 6 that show the schematic illustration of kerf generation at shallow jet impingement angles. The top view of the kerf gradually transforms from circular (at α = 90^0) to elliptical (at $0^0 < a < 90^0$) whereas the side cross-sectional view moves towards the right deviating from the symmetry (Fig. 4(i), Fig. 5). Furthermore, along the jet footprint ($\overline{AB}$), the erosion depth decreases at a slow rate from 'C' to 'B' and at a fast rate from 'C' to 'A'. These issues can be attributed to: (i) the interaction of various zones of the jet plume which are at varying axial distances from the tip of focusing nozzle and radial distances from jet axis, at footprint and (ii) variation in 'effective' impact angle of abrasive particles at jet footprint.

With the decrease in jet impingement angle, the width of footprint increases ($\overline{AB} < \overline{A'B'} < \overline{A''B''}$ in Fig. 5) in the direction of $\overrightarrow{XO}$ due to jet plume divergence. However, as α varies in the XZ plane, the increase in the width of JFP in the direction of the XY plane is not significant compared to that on the XZ plane. Hence, the top-view of the kerf gradually transforms from circle (at $\alpha = 90^0$) to an ellipse (at $0^0 < \alpha < 90^0$) with the decrease in α. Maximum erosion depth, $\overline{OC}$ or $\overline{OC'}$ or $\overline{OC''}$, is observed along the jet axis, $\overline{OZ'}$ or $\overline{OZ''}$ or $\overline{OZ'''}$ (Fig. 5). This is due to high velocity of water/abrasive particles along the jet axis. However, the depth decreased rapidly from point 'C' to point 'A' where the forward edge of the jet in the XZ plane meets the target surface (Figures. 5 and 6) and decreases slowly from point 'C' to 'B' where the trailing edge of the jet meets the target surface and that results in asymmetric geometry of kerf. This is explained in the following way: in contrast to normal jet impingement, the footprint on target surface $\overline{A'B'}$ or $\overline{A''B''}$ (Fig. 5) at shallow jet impingement angle occurs at different axial distances (D5 > D4 > D3 > D2 > D1, etc. (Fig. 6) from the tip of the focusing nozzle. As the distance Di increases, the velocity of jet decreases due to jet plume divergence that can be explained with decrease in height of Gaussian profile which in turn causes the decrease in erosive capability of the abrasive particles. The rapid decrease in depth of penetration across the forward part of the footprint ($\overline{OA}$) from 'C' to 'A' can be attributed to the increase in radial distance from jet axis ($\overline{OZ'}$ or $\overline{OZ''}$ or $\overline{OZ'''}$) and the longitudinal distance (D1, D2, D3 D4, D5 etc.), in the direction of the jet axis, across the jet footprint ($\overline{AB}$) from the tip of focusing nozzle. In addition to this, the impact angle of abrasive particles in the direction of footprint $\overrightarrow{OA}$ decreases due to shallower α (Fig. 6). Hence, the cumulative negative influence, i.e. increase in radial and axial distances as well as reduction in impact angle of abrasive particles, results drastic decreases in the velocity of abrasive particles which in turn cause decrease in erosion depth at higher rate towards 'A'. The decreased rate of erosion depth, in the trailing part of the jet footprint ($\overline{OB}$), can be attributed to decrease in axial distance along the jet axis (D2 < D1) and the increase in impact angle of abrasive particles in the direction $\overrightarrow{OB}$. The impact angle of abrasive particles increases gradually in the $\overrightarrow{OB}$ direction that increases the erosion capability of the abrasive particles in brittle materials. Further, the axial distance across the trailing part of the jet footprint ($\overline{OB}$) from the tip of the focusing nozzle decreases which in turn increases the erosion capability of the abrasive particles. However, the increase in radial distance in the direction of $\overrightarrow{OB}$ due to divergence of jet plume reduces the velocity of abrasive particles. Moreover, the divergence along the trailing part of jet plume is geometrically less compared to that in the forward edge of the jet. Hence, the slow rate of decrease in depth of erosion is due to the comprehensive result of positive effect of increase in θ, decrease in axial distance and the negative effect of increase in radial distance from jet axis. The rate of decrease of depth of penetration in forward part and trailing part depends on α. This is in contrast to the case of normal jet impingement where, across the footprint, the distance from the tip of the focusing nozzle is the same (= SOD) which results in symmetric geometry.

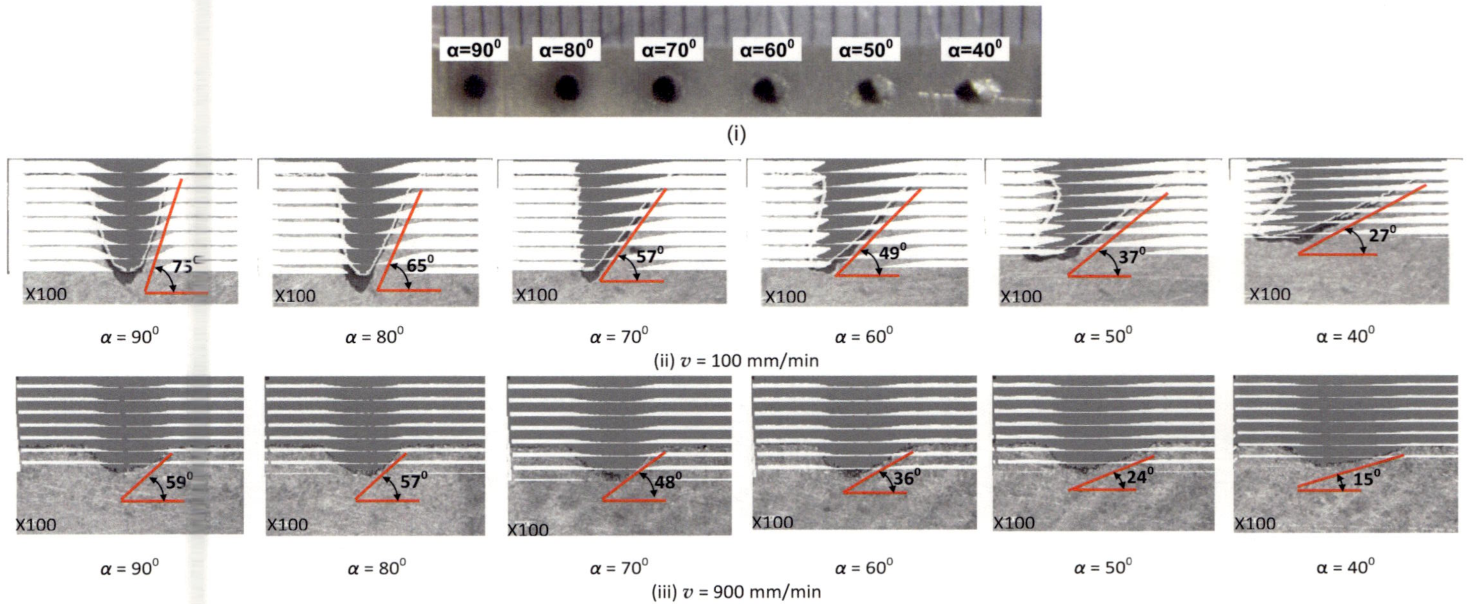

Fig. 4. Photographs of the 3D jet footprints generated at various jet impingement angles ($40^0 < \alpha < 90^0$) (i) top view, and 2D cross sections at (ii) v = 100 mm/min, (iii) v = 900 mm/min

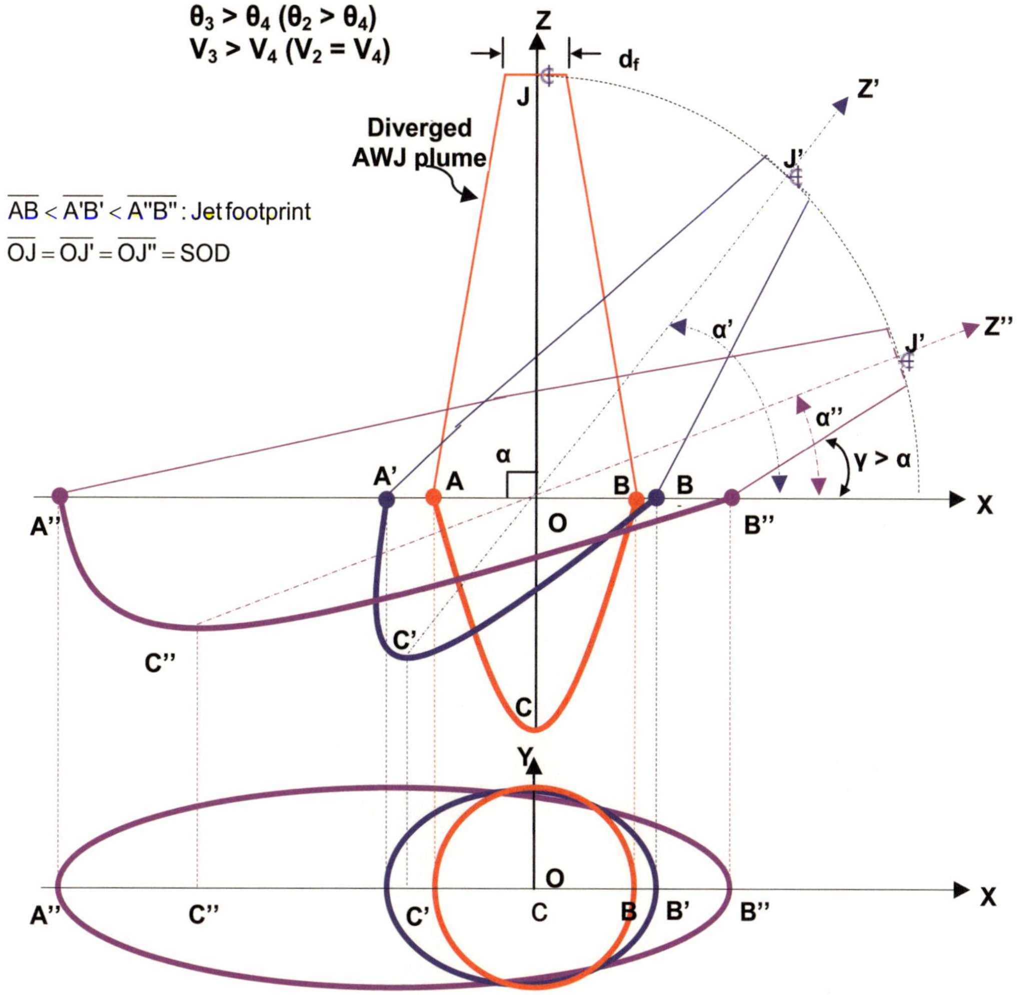

Fig. 5. Schematic illustration of variation in jet structure at various jet impingement angles (0 < α < 90⁰) on kerf generation

In addition to the change in geometry, the following changes in dimensional characteristics were observed which influence the geometry of the kerf significantly. From Fig. 4a, it can be observed that the slope of the kerf trailing edge is decreasing with the decrease in ***α***. This can be attributed to the shift of jet axis towards the workpiece surface at shallower ***α***. In addition to this, with the decrease in α, the depth of erosion was decreased and the top kerf width was increased (Fig. 7) which results in decrease in slope of kerf wall. Further, the slope (***β***) of kerf trailing wall is less than the jet impingement angle (***α***) employed. This can be attributed to the velocity profile that is similar to Gaussian distribution across the jet cross-section. When the jet impinges at a shallow angle, the maximum erosion is along the jet axis $\overline{OZ'}$ (Fig. 6) and the erosion depth in the direction of jet axis across $\overline{OB}$ decreases as the velocity of water/abrasive particle decreases due to its Gaussian nature. This makes the slope of the kerf trailing edge less than the jet impingement angle.

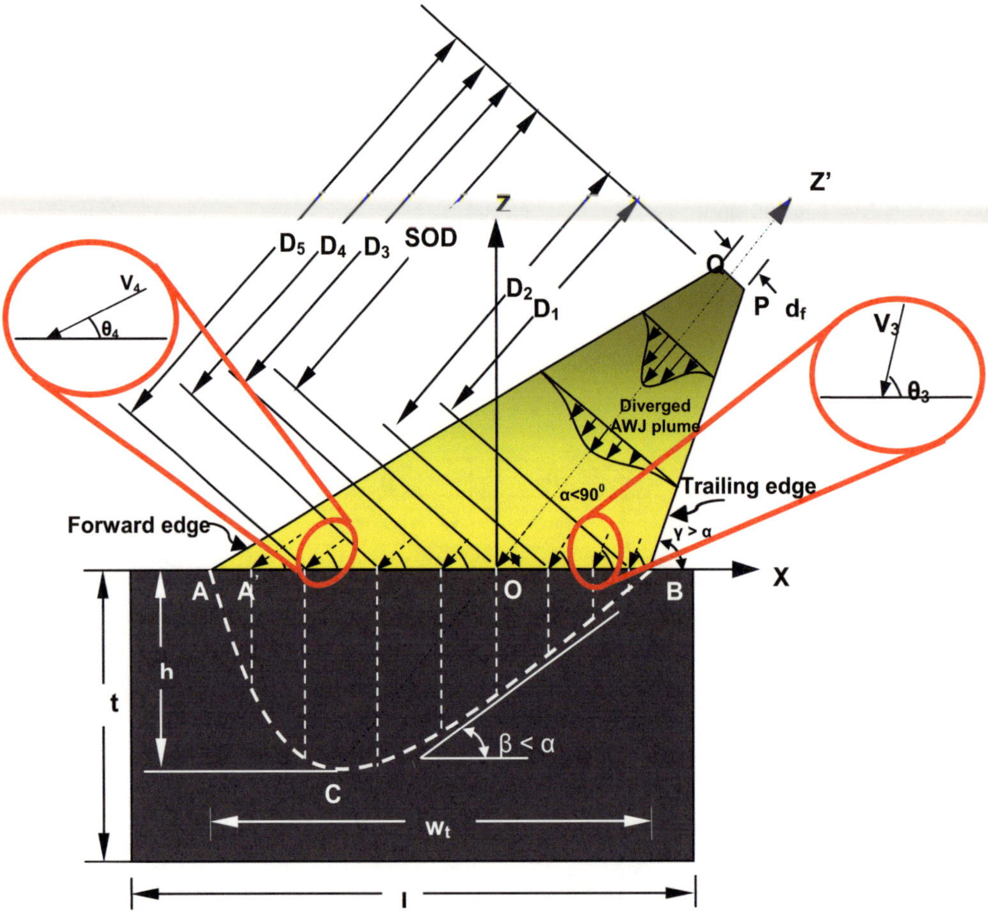

Fig. 6. Schematic illustration of local impact angles of abrasive particles and standoff distance at shallow ($40^0 < \alpha < 90^0$) jet impingement angle on kerf geometry

Maximum erosion depth was observed in the range of 70^0-80^0 jet impingement angle, i.e. for v = 100 mm/min erosion is maximum at $\alpha = 80^0$ and v = 900 mm/min erosion is maximum at $\alpha = 70^0$ (Fig. 7). This is a slightly different observation compared to the well known observation of maximum erosion at normal impingement ($\alpha = 90^0$) in gas-solid particle erosion for brittle materials (Ruff & Wioderborn, 1979). Similar shift of peak in erosion rate has been reported previously for sodalime glass w11x and WC-Co alloys w12x in certain erosion conditions (Chen et al., 1998; Kim & Park, 1998; Anand et al., 1986; Konig et al., 1988). This can be attributed to the effective average impact angle of abrasive particles and hardness of workpiece (Oka et al., 1997). The effective average impact angle (θ) of particles cannot be 90^0 at $\alpha = 90^0$ and approaches 90^0 at $\alpha < 90^0$. Furthermore, with the increase in hardness, the maximum erosion occurs at higher impact angles. Hence, the shift in maximum erosion was in the range of 70^0-80^0 ($\alpha < 90^0$). The width of the kerf increased with the decrease in α (Fig. 7). This is due to the combined effect of jet impingement angle and jet plume divergence. At lower α, the abrasive particles along the forward edge of the jet plume impinges on workpiece at a farther distance compared to higher jet impingement angle, due to divergence, which results in increase in width of jet footprint ($\overline{A''B''} > \overline{A'B'} > \overline{AB}$) as was shown in Fig. 5.

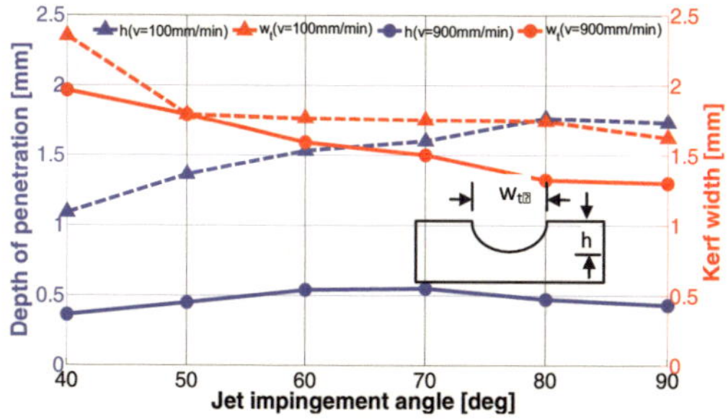

Fig. 7. Influence of jet impingement angle ($40^0 < a < 90^0$) on (a) erosion depth and (b) top kerf width

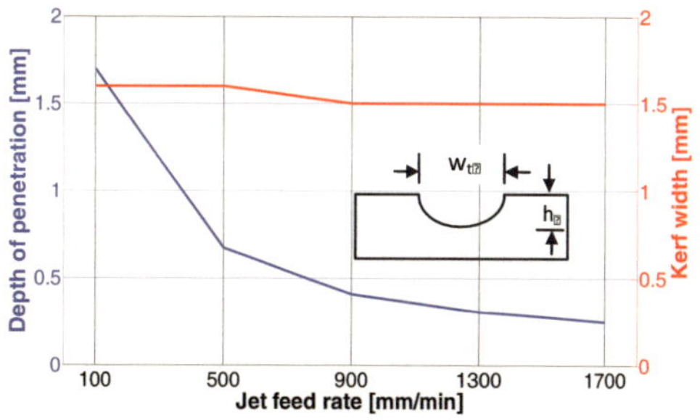

Fig. 8. Influence of jet feed rate ($a = 90^0$) on (a) erosion depth and (b) top kerf width

b) Influence of jet feed rate on kerf geometry

(i) Normal jet impingement ($a = 90^0$)

From the Fig. 2a (v = 100-1700 mm/min), it can be observed that the symmetric nature of the kerf geometry is maintained at different v when $a = 90^0$. However, there is a significant variation in the geometry of kerf at different jet feed rates. This can be explained with the change in dimensional characteristics of the kerf geometry, such as depth of penetration (***h***), top kerf width (w_t) (Fig. 8) and slope of kerf walls (β) (Fig. 2a) with the variation in v. The well known decreasing trend of h with the increase in v can be attributed to the increased exposure time of the material to the jet at lower v (Fig. 8). As the exposure time increases, more abrasive particles participate in erosion and penetrate more into the material which result in increased erosion depth. However, it can be observed that the ***h*** is not uniformly increased along the kerf geometry with the decrease in v as the increase in erosion along the kerf corner/walls is smaller than the increase in erosion along jet axis (Fig. 2b). This is explained in the following: As the abrasive particles along the trailing edge of jet plume are at shallower impact angle and the abrasive particles along the jet axis are nearly normal, the scaling of erosion is less for the same time. Furthermore, water/particle velocity along the jet axis is higher than jet plume edges. Moreover, at lower v, at an instantaneous time of 't+1', the abrasive particles interacts with the kerf generated at an instantaneous time 't' which is not a flat surface and cause decrease in 'effective' abrasive particle impact angle from the bottom of the kerf towards the edges of the kerf which results in decreased erosion in this direction. Hence the kerf geometry deviates from the sinusoidal curve and be approximated using simple 'cosine function' approximation. Further, rounding of edges on right side of kerf can be seen from Fig. 4a. This effect was significant at lower feed rates. This may be due to passage of rebounded jet along the left edge ($\overrightarrow{CA}$) of the kerf from the bottom as the jet enters from the left side ($\overrightarrow{BC}$) of the kerf. The kerf width decreased with the increase in jet feed rate, although the difference is insignificant (Fig. 8). This is explained in the following: when a cut is made, at an instantaneous time of 't' sec, the jet footprint, $\overline{AB}$ (Fig. 3), first pass through the material and generates a kerf with top width, which is nearly equal to the width of the JFP. Following that (at infinitesimally small incremental time, (t+Δt), the jet that has lower width than the footprint (due to divergence of jet plume) passes through the kerf already formed at an instantaneous time of 't' sec and cannot result in any further increase in kerf top width. However, at lower v, the abrasive particles along the boundary of jet, which have low erosion capability, gets enough time to interact with the material and enhance the erosion which results in slight increase in kerf width where these particles cannot make significant erosion at higher v. Hence, a slight decrease in kerf width was observed at higher v (Fig. 8). As a comprehensive view, with the increase in v, the erosion depth of the kerf is decreased and the width of kerf is nearly constant which results in a decrease in the slope of the kerf wall (Fig. 8 and Fig. 4a(i)). The slope of the kerf walls has direct influence on the geometry of the kerf generated. Hence, the jet feed rate plays a significant role in generating the desired kerf geometrical characteristics.

(ii) Shallow angle jet impingement ($40^0 < a < 90^0$)
It can be observed from Fig. 4a that, for the same jet impingement angles, the cross sectional geometry of the kerf generated at higher jet feed rates (v = 900 mm/min) is considerably different in terms of erosion depth, top kerf width and slope of kerf trailing edge from the same generated at lower v (= 100 mm/min). This is also due to the fact, that was observed for normal jet impingement angle at lower v, i.e. interaction of the jet at an instantaneous time of 't+1' on the surface generated at time 't' which is a non-flat surface; and increase in exposure time with the decrease in v. Furthermore, the slope of the kerf trailing wall (β) is decreased at lower v for the same a (Fig. 4a). This can be attributed to the increase in erosion capability of abrasive particles along the jet plume trailing edge $\overline{PQ}$ (Fig. 6) at lower v. The water/abrasive particles have low energy along the trailing edge of diverged jet plume (Fig. 6). At higher v, due to low exposure time of material to the low energy abrasive particles, material cannot be eroded in the direction of jet plume trailing edge $\overline{PQ}$ (Fig. 6). The water/abrasive particles along the jet axis $\overline{OZ'}$, which is less steep than jet plume trailing edge, i.e. $\gamma > a$, are responsible for material removal. Hence, β is smaller. In contrast to this, at lower v, the exposure time of material to the low energy particles increase which enhances the erosion of material in the direction of jet trailing edge which is steeper than jet axis. Hence, at lower jet feed rate, the slope of kerf trailing edge is higher than higher v.

3.1.2 Variation in depth of penetration along jet feed direction

From the bottom of the kerf cross sectional views presented in Figs. 2a and 4a, it is clear that the h along the jet feed direction is not uniform. Figure 9a presents an example of a 3D axonometric plot of the kerf generated at a = 900, P = 345 MPa (= 50,000 psi), v = 500 mm/min and m_f = 0.7 kg/min, where it can be noted the variation in the direction of jet feed. The same behaviour was observed at all jet impingement angles. In order to analyze further, the kerf generated at $a = 90^0$ was considered. Kerf profiles taken at five equally distanced sections along the jet feed direction are presented in Fig. 9b. Figure 9c presents the errorbar graph (1 standard deviation) of the 3D kerf that presents the variation of kerf profile around the mean profile. From the errorbar graph, it is evaluated that the depth of erosion along the jet feed direction was varying with a standard deviation of 0.015 mm around the mean erosion depth of 0.704 mm. The variation in kerf profile can be attributed to the fluctuations in the pump pressure, jet feed rate employed, abrasive particle mass flow, transverse feed of milling etc. (Hashish, 1989; Oka et al., 1997; Ansari & Hashish, 1993;

Hashish, 1989; Paul et al., 1998). From the previous studies it is observed that the low energy jet (at low P and at higher v levels) generates uniform kerf (Hashish, 1989; Ansari & Hashish, 1993).

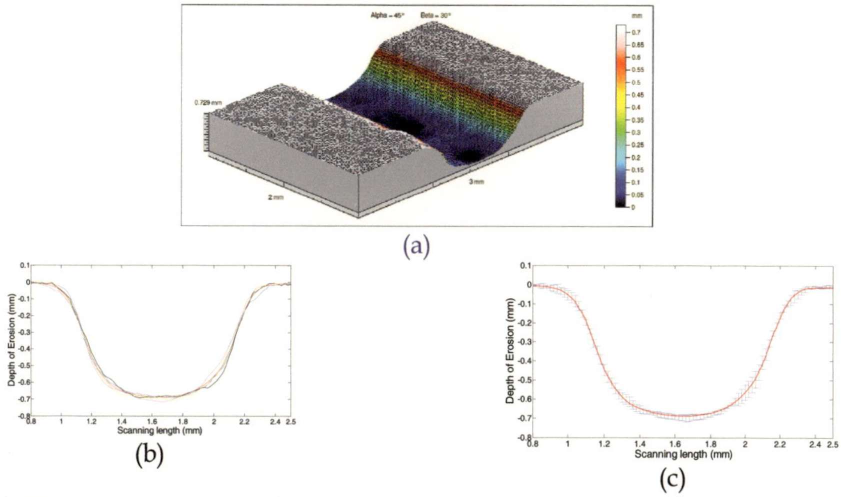

Fig. 9. (a) 3D axonometric plot of kerf generated by AWJ (v=500mm/min), (b) kerf profiles at different regions in the direction of jet feed, (c) Error bar graph of the kerf

3.1.3 Influence of multi-pass on kerf generation

Figure 10 presents the experimental kerf profiles obtained in single (blue profile) and double pass (red profile) operations by keeping all the other operating parameters constant. Intuitively, the double pass is expected to generate the kerf with erosion depth of H ='2xh' (green profile) whereas in reality the generated depth is less than '2xh'.The decrease in depth of penetration in double pass can be attributed to the combined effect of (i) change in local impact angles (θ) of the abrasive particle due to non-flat kerf geometry generated in the first pass and (ii) increase in SOD due to kerf generated in the first pass. The kerf formation in double pass approach is schematically illustrated in Fig. 11. ACB is the kerf geometry generated in single pass with an erosion depth of '*h*' and A'C'B' is the kerf geometry generated in double pass operation by considering all the other operating parameters constant.

Influence of kerf geometry generated on the following pass

In a second (or subsequent) pass, erosion is taking place on the previously generated kerf which is a non-flat surface (ACB) (Fig. 11). This differs from a single pass where erosion starts on a flat surface (AB). As explained earlier, at $\alpha = 90^0$, the impact angle of the abrasive particles (θ) is 90^0 on the jet axis ($\overline{OZ}$) and decreases in value on either side of jet axis across the footprint ($\overline{AB}$). However, in erosion by subsequent (e.g. second) passes, the abrasive particles interact with a (non-flat) kerf surface formed in the previous passes. Hence, for subsequent passes, the impact angle of abrasive particles is the angle between the kerf surface formed by the previous passes and abrasive particles impact direction (Fig. 11) which decreases away from the centreline and causes decrease in erosion rate, since erosion is lower at shallower θ for brittle materials.

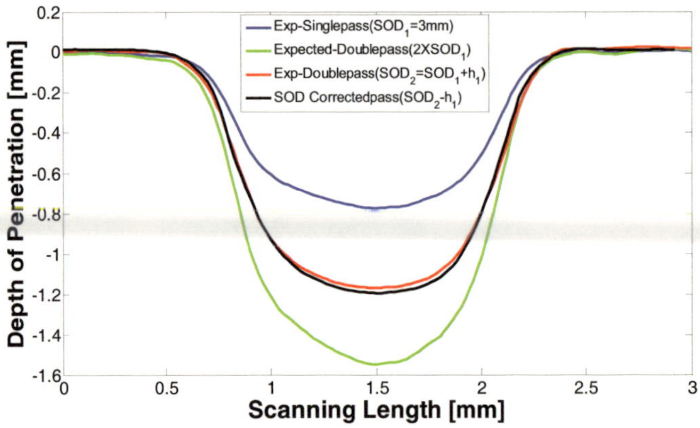

Fig. 10. Influence of number of passes and standoff distance on kerf generation ($\alpha = 90^0$)

b) Influence of SOD on characteristics of kerf generated in double pass

Furthermore, the actual standoff distance (SOD_{actual}) in a second (or subsequent) pass is the sum of SOD set in the first pass and the erosion depth (h), i.e. SOD_{actual} = SOD + h. In order to support the intuition experimentally, SOD correction experiments were conducted: 1st pass achieved a maximum penetration depth of 0.45 mm ($\boldsymbol{P}$ = 50,000 PSI, $\boldsymbol{m_f}$ = 0.7 kg/min, v = 500 mm/min, $\boldsymbol{d_f}$ = 1.06 mm, $\boldsymbol{d_o}$ = 0.3 mm, SOD = 3 mm). After the first pass, the SOD becomes 3.45 mm as the depth of penetration achieved in single pass was 0.45 mm, i.e. SOD_{actual} = 3+0.45. Hence, the focusing nozzle was moved down by 0.45 mm which makes the SOD again to 3mm that was employed for a single pass cut. Figure 10 presents the kerf profiles of (a) single pass, (b) intuitive/expected double pass (two folds depth of penetration in single pass cut), (c) experimentally achieved double pass and (d) corrected double pass (cut made after moving down the focusing nozzle by 0.45 mm). All the other operating parameters were kept constant throughout the study. From the results, it can be observed that there is very little change in erosion depth due to the movement of focusing nozzle down by 0.45 mm. However, when the number of passes increases, the small effect of SOD becomes cumulative. From this, it can be concluded that although the increased SOD has influence on the reduction of depth of penetration in following passes, it is not significant. However, it has to be considered for tight dimensional control. Hence, in a comprehensive view, to control the erosion depth in multi-passes, the influence of kerf generated in previous pass and the SOD have to be considered for control over the surface to be generated.

With the knowledge gained from the experimental investigation on the jet footprints, in the following section, development and validation of an analytical model for prediction of JFP at normal jet impingement angle is presented.

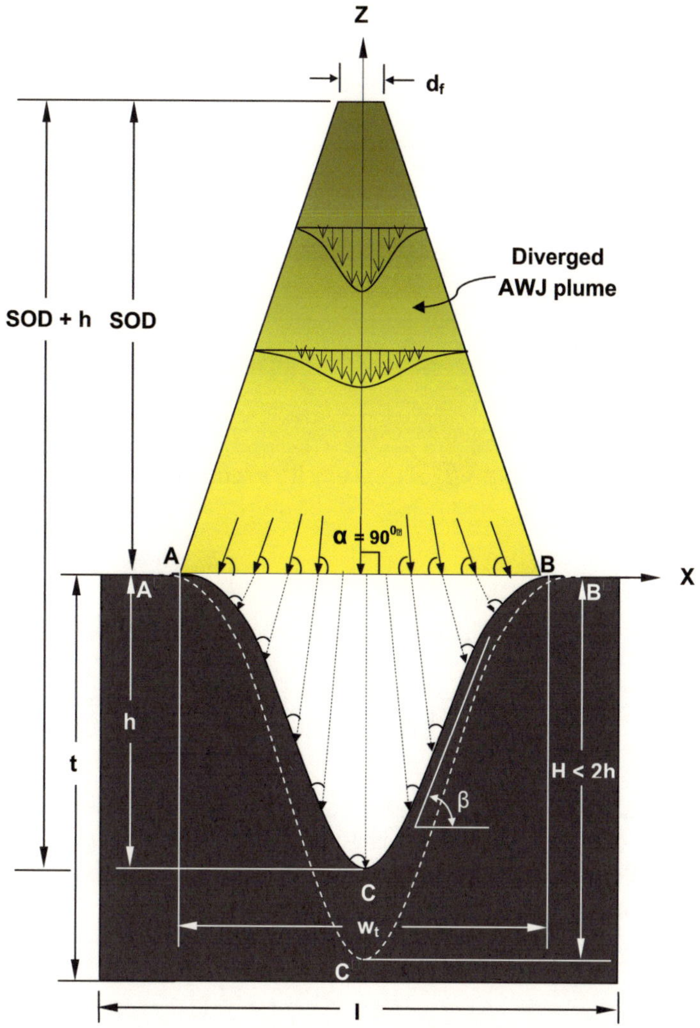

Fig. 11. Schematic illustration of kerf generation in multi-pass operation ($\alpha = 90^0$)

3.2 Model for the jet footprint geometry (*Axinte et al., 2010*)

For this model it was assumed that an abrasive waterjet jet, with radius a, impacts at 90° a flat workpiece surface while moving with a constant feed speed v_f in the y-direction. This means that, for the time being, the proposed model is more applicable to brittle target workpiece materials on which normal jet impingement angle has been found to give higher material removal rates (Hashish, 1993). The jet footprint (Fig. 12a) is a function z= Z(x, y, t) in which x - distance from the jet axis, y - the direction in which the jet moves and t - dwell (exposure) time. For brittle materials, maximum erosion occurs when the jet is perpendicular to the workpiece so this model assumes that the etching rate is proportional to some power (k) of the impingement velocity (Slikkerveer, 1999; Getu et al., 2008; Ghobeity et al., 2008) Eq. (1), of the jet in the direction of the inwards unit normal, Eq. (2), of the surface being etched.

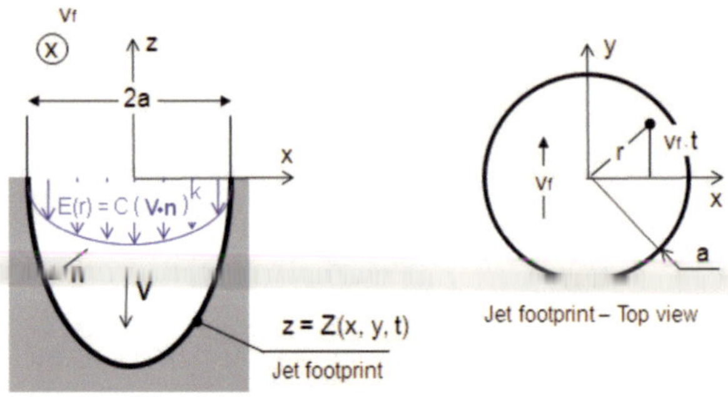

(a) (b)

Fig. 12. Schematic of the jet footprint: cross section (a) and top view (b)

$$\mathrm{V} = V(r)\begin{pmatrix} 0 \\ 0 \\ -1 \end{pmatrix} \text{ where V(r) is the velocity profile of the impinging jet} \tag{1}$$

$$\mathrm{n} = \frac{1}{\left(1+\left(\frac{\partial Z}{\partial x}\right)^2+\left(\frac{\partial Z}{\partial y}\right)^2\right)^{\frac{1}{2}}}\begin{pmatrix} \frac{\partial Z}{\partial x} \\ \frac{\partial Z}{\partial y} \\ -1 \end{pmatrix} \tag{2}$$

The component of the jet's velocity in the direction of the inwards unit normal of the surface is now given by the dot product of V and n (Slikkerveer, 1999; Getu et al., 2008; Ghobeity et al., 2008; Ten Thije Boonkamp & Jansen, 2002; Hashish, 1993) as shown in Eq. 3. where, n is normal to the kerf surface.

$$\mathrm{V{\cdot}n} = \frac{\mathrm{V(r)}}{\left(1+\left(\frac{\partial Z}{\partial x}\right)^2+\left(\frac{\partial Z}{\partial y}\right)^2\right)^{\frac{1}{2}}} \tag{3}$$

As the jet moves along the y axis (Fig. 12b), a point situated at distance x from the jet axis ($-a \le x \le a$) will be impinged by the abrasive jet when $-\sqrt{a^2-x^2} \le v_f t \le \sqrt{a^2-x^2}$ at which times the jet footprint follows Eq. (4).

$$\frac{\partial \mathrm{Z}}{\partial \mathrm{t}} \propto (\mathrm{V{\cdot}n})^k \tag{4}$$

Considering that the etching (material specific erosion) rate, E, is dependent only on the radial position $r \le a$ and defining it as $E(r) = C(V{\cdot}n)^k$ (Slikkerveer, 1999; Getu et al., 2008; Ghobeity et al., 2008), C and k positive constants, and then substituting Eq. (3), the jet footprint becomes Eq. (5).

$$\frac{\partial Z}{\partial t}=\frac{-E(r)}{\left(1+\left(\frac{\partial Z}{\partial x}\right)^2+\left(\frac{\partial Z}{\partial y}\right)^2\right)^{\frac{k}{2}}}=\begin{cases}\dfrac{-E\left(\sqrt{x^2+v_f{}^2t^2}\right)}{\left(1+\left(\frac{\partial Z}{\partial x}\right)^2+\left(\frac{\partial Z}{\partial y}\right)^2\right)^{\frac{k}{2}}} & \text{for } -\sqrt{a^2-x^2}\le v_f t\le\sqrt{a^2-x^2}\\ 0 \text{ for } -a\le v_f t\le-\sqrt{a^2-x^2} \text{ and } \sqrt{a^2-x^2}\le v_f t\le a\end{cases}\qquad(5)$$

As the etching rate E is dependent on the properties of the workpiece material and other process parameters (e.g. pressure, mass flow of abrasives) it is treated as a function that needs to be calibrated using experimental data; note also that for this simple geometry and jet motion, the final surface will be a trench, with no variation in the y-direction. The problem can be made dimensionless by defining: $x=a\bar{x};\ Z=a\bar{Z};\ t=\frac{a}{v_f}\bar{t};\ E=\varepsilon v_f\bar{E};\ y=v_f\bar{t}$; note that in the following 'barred notations' (e.g. $\bar{x}$) represent dimensionless measures. Here εvf is the maximum etching rate, so that the dimensionless etching rate function $\bar{E}$ has a maximum value of one. In dimensionless terms the geometrical model, Eq. (5), of the kerf becomes as described in Eq. (6).

$$\frac{\partial\bar{Z}}{\partial\bar{t}}=\begin{cases}\dfrac{-\varepsilon\bar{E}\left(\sqrt{\bar{x}^2+\bar{t}^2}\right)}{\left(1+\left(\frac{\partial\bar{Z}}{\partial\bar{x}}\right)^2+\left(\frac{\partial\bar{Z}}{\partial\bar{t}}\right)^2\right)^{\frac{k}{2}}} & \text{for } -\sqrt{1-\bar{x}^2}\le\bar{t}\le\sqrt{1-\bar{x}^2}\\ 0 \text{ for } -1\le\bar{t}\le-\sqrt{1-\bar{x}^2} \text{ and } \sqrt{1-\bar{x}^2}\le\bar{t}\le1\end{cases}\qquad(6)$$

This is a nonlinear partial differential equation, and has no obvious analytical solution. However, if ε is small (obtained at high jet feed speeds), the kerf slope is small and the equation can be linearized Eq. (7).

$$\frac{\partial\bar{Z}}{\partial\bar{t}}=\begin{cases}-\varepsilon\bar{E}\left(\sqrt{\bar{x}^2+\bar{t}^2}\right) & \text{for } -\sqrt{1-\bar{x}^2}\le\bar{t}\le\sqrt{1-\bar{x}^2}\\ 0 \text{ for } -1\le\bar{t}\le-\sqrt{1-\bar{x}^2} \text{ and } \sqrt{1-\bar{x}^2}\le\bar{t}\le1\end{cases}\qquad \text{with } -1\le\bar{x}\le1\qquad(7)$$

If now this simple equation is integrated, the shape of the eroded shallow kerf can be obtained as in Eq. (8).

$$\bar{Z}_o(\bar{x})=-2\varepsilon\int_{\bar{x}}^{1}\frac{\bar{r}\cdot\bar{E}(\bar{r})}{\sqrt{\bar{r}^2+\bar{x}^2}}d\bar{r}\qquad(8)$$

After some manipulation, this integral equation can be inverted to obtain Eq. (9).

$$\bar{E}(\bar{r})=\frac{1}{\varepsilon\pi}\left[\int_{\bar{r}}^{1}\frac{R\left(\bar{Z}_o(R)-\bar{Z}_o(\bar{r})\right)}{\left(R^2-\bar{r}^2\right)^{3/2}}dR-\frac{\bar{Z}_o(\bar{r})}{\sqrt{1-\bar{r}^2}}\right]\qquad(9)$$

It is now possible to calibrate the model using an experimentally obtained shallow kerf shape $\overline{Z}_0(\overline{x})$ in Eq. (9) to find the specific etching rate $\overline{E}(\overline{r})$. With $\overline{E}(\overline{r})$ known, the Eq. (6) can be integrated numerically, using Newton method to find $\frac{\partial \overline{Z}}{\partial \overline{t}}$ at each time step and ultimately to determine Z(x) for any value of ε, and hence predict the profile of any kerf, including those whose slope is not small (obtained at various jet feed speeds). Therefore, to find the kerf profile at any jet feed speed the following successive steps should be followed:

1. Generate a shallow profile (i.e. $\overline{Z}_0(\overline{x})$) with small kerf slopes.
2. Knowing $\overline{Z}_0(\overline{x})$, the specific etching rate $\overline{E}(\overline{r})$ can be numerically evaluated using Eq. (9).
3. Once $\overline{E}(\overline{r})$ is known, the kerf profile, Z(x), can be obtained for any jet feed speed by solving Eq. (5).

Using process parameters (mf= 0.7 kg/min; P= 345 MPa; D0= 3 mm) mentioned in the previous section at a high value of jet feed speed (vf = 1700 mm/min) a shallow kerf on SiC target workpiece was generated and its scanned profile (i.e. $\overline{Z}_0(\overline{x})$) has been used as calibrated (Fig. 13a) input into Eq. (9) from where employing a MATLAB code the specific etching rate $\overline{E}(\overline{r})$ was evaluated. Then, with $\overline{E}(\overline{r})$ known, predictions of the kerf profiles for smaller jet feed speeds (vf = 100, 500, 900 and 1300 mm/min) have been generated and compared with those experimentally obtained; this revealed that only for small jet feed speed (i.e. vf = 100 mm/min), the model predicts noticeably deeper (approx. 30 µm) kerf than the experimental one (see Model 1 - Fig. 13e). This is somehow an expected finding since with deeper kerf profiles, the real stand-off distance (D) increases considerably (40%) with the increase of jet dwell time (decrease of vf). Therefore, the bottom of the kerf is situated at larger values of stand-off distance from the jet nozzle and consequently less erosion than the theoretical one occurs here. To account for this phenomenon, it has been thought that the distance between the jet and the workpiece should be included in the model.

If D0 is the initial SOD between the centre of the nozzle and the point (x, z) = (0, 0), then the distance from the opening of the jet to any point on the kerf surface is given by D(x, t) = D0 – Z(x, t). Preliminary experimental trials on SiC target workpiece showed that the maximum jet penetration occurs at a stand-off distance of 3 mm but if the jet is any closer then cracks or breaks may occur on the target surface. This suggests that:

$$\frac{\partial Z}{\partial t} \propto F(D(x,t)) \tag{10}$$

where F is the linear function F(D(x, t)) = 1 – b·[D(x, t) – 3], and b ≥ 0 should be determined by fitting the model to experimental data. Combining this with Eq. (4), the new kerf profile dependence it is obtained, Eq. (11).

$$\frac{\partial Z}{\partial t} \propto \frac{-E(r) \cdot F(D(x,t))}{\left(1+\left(\frac{\partial Z}{\partial x}\right)^2+\left(\frac{\partial Z}{\partial y}\right)^2\right)^{\frac{k}{2}}} \tag{11}$$

Hence, after applying the same non-dimensionalization procedure, Eq. (6) now becomes the corrected jet footprint dependence as shown in Eq. (12).

$$\frac{\partial \overline{Z}}{\partial \overline{t}} = \begin{cases} \dfrac{-\varepsilon \overline{E}\left(\sqrt{\overline{x}^2+\overline{t}^2}\right)\cdot F\left(D(\overline{x},\overline{t})\right)}{\left(1+\left(\dfrac{\partial \overline{Z}}{\partial \overline{x}}\right)^2+\left(\dfrac{\partial \overline{Z}}{\partial \overline{t}}\right)^2\right)^{\frac{k}{2}}} \text{ for } -\sqrt{1-\overline{x}^2} \le \overline{t} \le \sqrt{1-\overline{x}^2} \\ 0 \text{ for } -1 \le \overline{t} \le -\sqrt{1-\overline{x}^2} \text{ and } \sqrt{1-\overline{x}^2} \le \overline{t} \le 1 \end{cases} \tag{12}$$

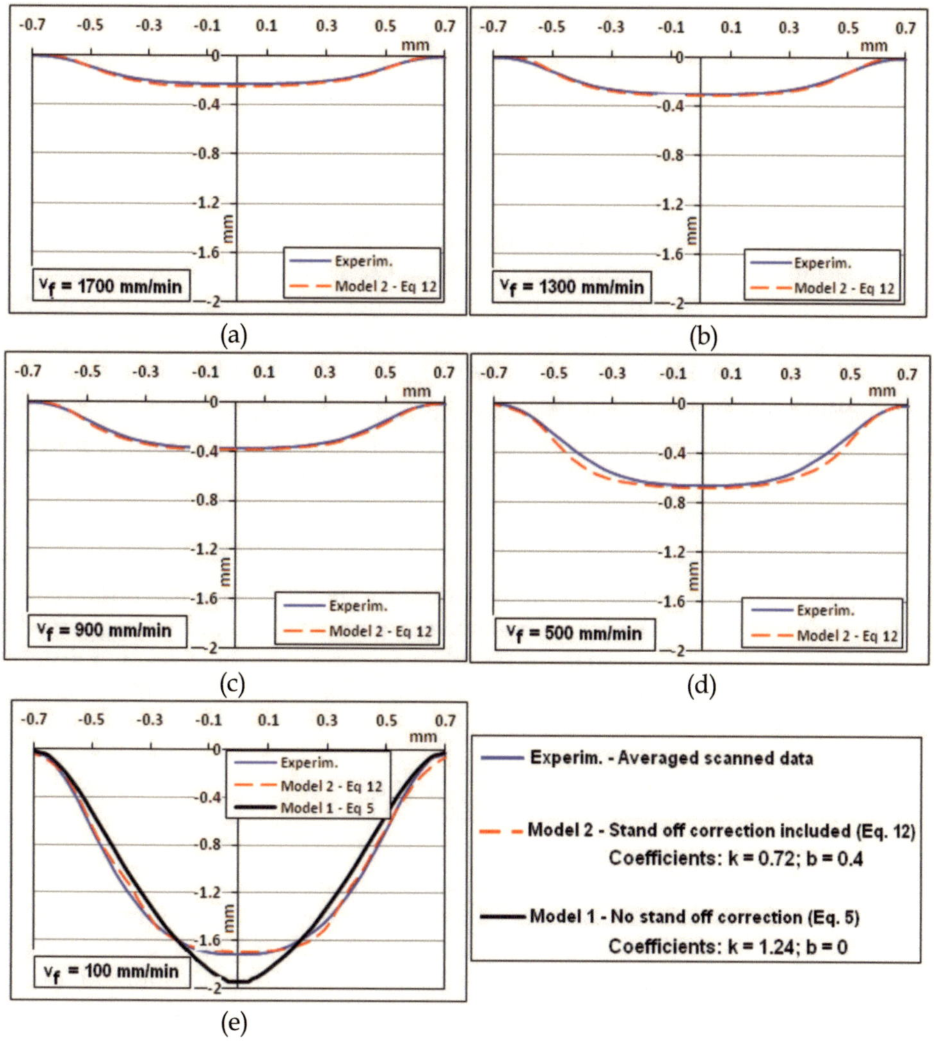

Fig. 13. Experimental vs. predicted kerfs at various jet feed speeds

This can again be solved as before by linearizing the problem which approximates D(x, t) = D0 when vf is large (vf= 1700 mm/min). Therefore, F(D(x, t)) is just a constant and $\overline{E}(\overline{r})$ is the same as in Eq. (9). Utilising the same calibration method as before and the corrected jet footprint model, Eq. (12), the comparison between the predicted (modelled) kerf profiles and the experimental data is presented in Fig. 13 where the notations Model 1 and Model 2 refer to footprint mathematical predictions on which the correction of stand-off distance was omitted (Eq. (5) for k=1.24) and included (Eq. (12) for k=0.72, b=0.1) respectively. The summarised results in Fig. 13 clearly show that the corrected jet footprint model is capable to predict at high degree of accuracy (error <5%) the kerf profile at various jet feed speeds.
The model presented, in this section, can only approximate the geometry of the JFP and cannot effectively predict the jet footprint's top width, which is one of the critical parameters in controlling the dimensions of the final component milled as mentioned in earlier sections. Hence, in the following section, a model developed exclusively for prediction of top width of the JFP is presented.

3.3 Model for top width of jet footprint (*Srinivasu, D.S. & Axinte, D.A.*)

3.3.1 Development of model for top width of jet footprint

In 2-axis machining, the top width (w_t) of the JFP can be determined by considering the diameter of focusing nozzle (d_f), jet plume divergence in air (φ) and SOD. The development of model for the w_t of the JFP involves two stages: Firstly, an expression for the jet plume divergence (φ) at normal jet impingement angle is derived by considering the d_f, SOD and the w_t generated at a jet feed rate of 1700mm/min. Secondly, an expressions for the $w_t(a)$ at shallow jet impingement angles is derived based on the geometry of the tilted structure of the jet plume at shallow a (Fig. 14b), by considering the previously evaluated φ.

a) Expression for jet plume divergence, φ

The jet plume diverges in air with an angle of φ after emerging from the focusing nozzle of bore diameter, d_f, before interacting with the target workpiece situated at a distance of SOD (Fig. 14a). Due to the divergence of the jet plume at normal jet impingement angle, the width of the jet increases gradually from d_f (at the tip of focusing nozzle) to w_t (at a distance of SOD) depending on the jet plume divergence angle. The expression for the φ can be derived as follows:

From Δ^{le} AEC (Fig. 14a), $$\varphi = \tan^{-1}\left(\frac{w_t - d_f}{2(SOD)}\right) \quad (13)$$

b) Expression for width of jet footprint at normal jet impingement, w_t

The top width of the JFP at a distance x, at normal jet impingement, can be approximated by the width (diameter) of jet plume at that distance (Fig. 14a). As the width of the jet depends on the distance between workpiece and the tip of focusing nozzle, initial diameter of focusing nozzle and jet plume divergence, the w_t can be derived as follows:

From Δ^{le} MNC (Fig. 14a), w_t at a distance of $X = d_x = d_f + 2(\overline{MN}) = d_x = d_f + 2x\tan(\varphi)$ (14)

c) Expression for width of jet footprint at shallow jet impingement, $w_t(a)$

The jet plume divergence angle is constant and will not change with the change in jet impingement angles. Hence, at shallower jet impingement angles ($a < 90^0$), the width of the JFP can be approximated geometrically from the consistent structure of the jet plume as the summation of three parts on the target workpiece surface (Fig. 14b): (i) leading part of the top width of JFP generated by leading portion of jet plume (a - $\overline{A'E'}$), (ii) right part of the top width of JFP generated by trailing portion of jet plume (b - $\overline{F'B'}$) and (iii) middle part generated by the diameter of focusing nozzle ($\overline{E'O'}$ + $\overline{O'F'}$).

$$w_t(a) = \overline{A'B'} = \overline{A'E'} + \overline{E'O'} + \overline{O'F'} + \overline{F'B'} \tag{15}$$

Evaluation of $\overline{A'E'}$: (a)

From Δ^{le} C'A'E' (Fig. 14b),
$$\overline{A'E'} = \left(SOD + \frac{d_f}{2\tan(\alpha)}\right)\frac{\sin(\varphi)}{\sin(\alpha-\varphi)} \tag{15.1}$$

Evaluation of $\overline{E'O'}$:

From Δ^{le} J'E'O' (Fig. 14b),
$$\overline{E'O'} = \frac{d_f}{2\sin(\alpha)} \tag{15.2}$$

Evaluation of $\overline{O'F'}$:

From Δ^{le} I'O'F'(Fig. 14b),
$$\overline{O'F'} = \frac{\overline{I'F'}}{\sin(\alpha)} = \frac{d_f}{2\sin(\alpha)} \tag{15.3}$$

Evaluation of $\overline{F'B'}$:(b)

From Δ^{le} D'F'B' (Fig. 14b),
$$\overline{F'B'} = \overline{F'D'}\frac{\sin(\varphi)}{\sin(\alpha+\varphi)} = \left(SOD - \frac{d_f}{2\tan(\alpha)}\right)\frac{\sin(\varphi)}{\sin(\alpha-\varphi)} \tag{15.4}$$

Finally, from Eq. (3.1), Eq. (3.2), Eq. (3.3) and Eq. (3.4)

$$w_t(a) = \overline{A'B'} = \overline{A'E'} + \overline{E'O'} + \overline{O'F'} + \overline{F'B'}$$
$$= \left(SOD + \frac{d_f}{2\tan(\alpha)}\right)\frac{\sin(\varphi)}{\sin(\alpha-\varphi)} + \frac{d_f}{\sin(\alpha)} + \left(SOD - \frac{d_f}{2\tan(\alpha)}\right)\frac{\sin(\varphi)}{\sin(\alpha+\varphi)} \tag{16}$$

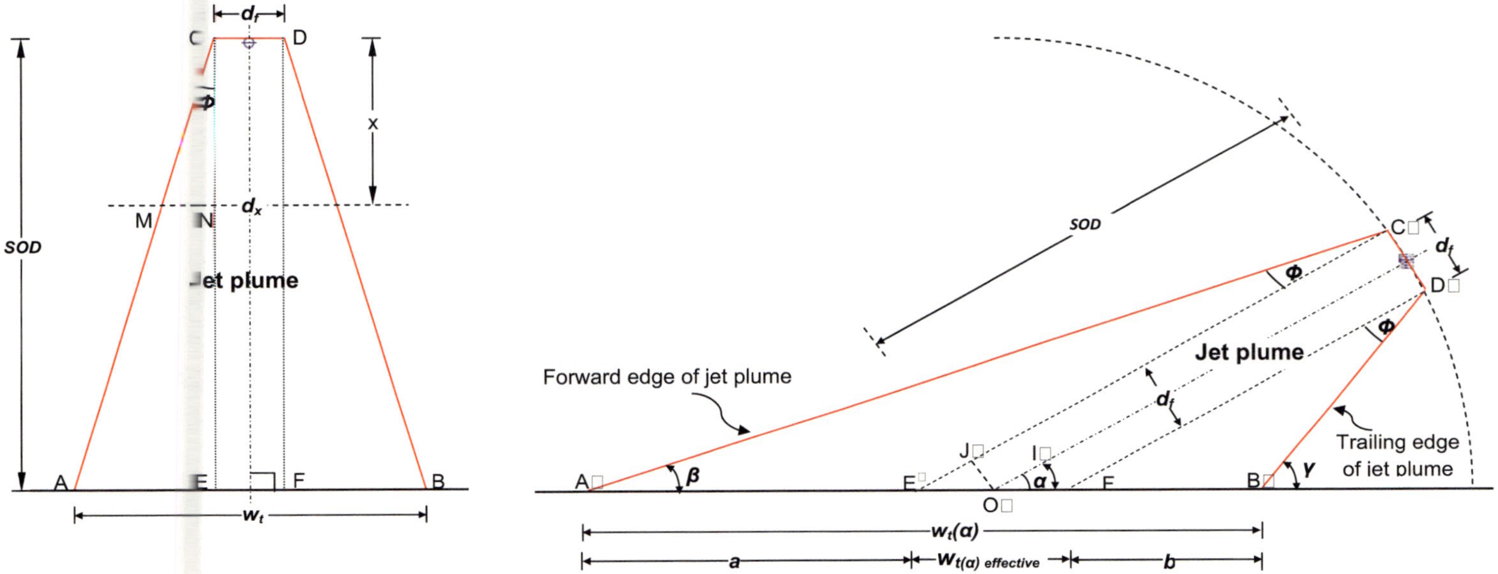

(a) (b)

Fig. 14. Schematic illustration of jet plume structure in air before impingement onto the target surface (a) normal impingement, (b) shallow angle impingement

However, this model cannot be employed at jet feed rates higher than 100 mm/min as the effective divergence of jet plume, that is responsible for erosion, decreases at higher jet feed rates, although the jet plume divergence angle does not change with v as discussed earlier. Hence, the prediction of $w_t(a)$ at desired jet feed rates, higher than 100 mm/min, needs the evaluation of φ at that desired v, and is termed in this modelling study as effective divergence angle, φ_V. The φ_V replaces the jet plume divergence angle (φ) in the above model for evaluation of top width of JFP at any jet impingement angle at desired v.

3.3.2 Assessment of the proposed model for the top width of jet footprint

a) Without considering the 'effective jet plume divergence'

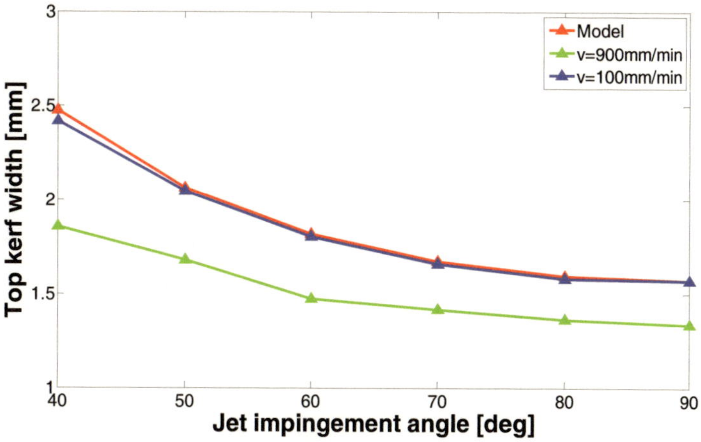

Fig. 15. Variation in top width of jet footprint with the jet impingement angle without considering the 'effective jet plume divergence'

Figure 15 presents the w_t predicted by the proposed model without considering the parameter: *'effective jet plume divergence'*, along with the experimentally achieved values at lower (v = 100mm/min) and higher (v = 900mm/min) jet feed rates. The predicted values are in well agreement with the experimental values at v = 100 mm/min. However, the difference is considerable at higher jet feed rate (v = 900 mm/min). Furthermore, this difference increased at shallower jet impingement angles. This may be explained as follows:

(i) With the increase in v, the exposure time of the material to the jet plume decreases. Furthermore, the erosion capability of the abrasive particles along the jet plume edges is less due to low impact angle of abrasive particles (Fig. 6 and Fig. 3). Because of the combined effect of exposure time and low impact angle, the erosion capability of the abrasive particles along the jet plume edges decreases with the increase in jet feed rate. Hence, the *'effective jet plume divergence'* is, indirectly, a function of jet feed rate. However, the proposed model evaluated the erosion at higher jet feed rate (v = 900mm/min) according to the jet plume divergence angle (φ) that was evaluated at lower jet feed rate (v = 100mm/min) which results in increase in difference between predicted and experimental values.

(ii) On the other hand, as the *a* decreases, the width of spray region at jet-material interaction site increases which cannot generate considerable erosion due to lower impact angle of abrasive particles and lesser velocity of water along the forward edge of the jet plume ($\overline{A'C'}$ - Fig. 6). However, in this region, the model assumes considerable erosion, according to geometrical structure of the jet plume divergence, without considering the influence of abrasive particle impact angle. Hence, there is significant difference between predicted and experimental w_t.

Hence, in the following section, the model is modified by including the influence of jet feed rate by a parameter called *'effective jet plume divergence'*, φ_V, to improve its predictive capability at any v.

b) Considering the 'effective jet plume divergence', φ_V

Figure 16a and 16b presents the top width of the JFP predicted by the proposed model with the consideration of the influence of jet feed rate with the introduction of the parameter - *'effective jet plume divergence'* φ_V, along with the experimental results at lower (100mm/min) and higher (900mm/min) v. The difference between the predicted and experimental w_t values at higher jet feed rate was reduced significantly compared to the earlier case. This can be attributed to the inclusion of φ_V parameter that has taken care of the reduction in effective jet plume divergence at higher jet feed rates. For this purpose, φ evaluated at desired jet feed rate (v = 900 mm/min), i.e. φ_V, which is 3^0, was substituted in the model (Eq. 16) and becomes Eq. 16′.

$$w_t(\alpha) = \left(SOD + \frac{d_f}{2\tan(\alpha)}\right)\frac{\sin(\varphi_V)}{\sin(\alpha-\varphi_V)} + \frac{d_f}{\sin(\alpha)} + \left(SOD - \frac{d_f}{2\tan(\alpha)}\right)\frac{\sin(\varphi_V)}{\sin(\alpha+\varphi_V)} \qquad (16')$$

The reduction in effective jet plume divergence can be observed as it is reduced from 5.24^0 at lower jet feed rate ($\varphi_{v=100\text{ mm/min}}$) to 3^0 at higher v ($\varphi_{v=900\text{ mm/min}}$).

Although, the predicted w_t of the JFP at different jet impingement angles is in well agreement with the experimental values, there exists an error at shallower jet impingement angles (Fig. 16). Furthermore, this error increased at higher jet impingement angles and higher v (= 900 mm/min) (Fig. 16b). This can be explained as follows: Figure 2b presents the schematic illustration of the variation in leading ($\overline{A'E'}$ - caused by the leading edge - $\overline{A'C'}$) and trailing ($\overline{F'B'}$ - caused by the trailing edge - $\overline{D'B'}$) parts of w_t at shallow *a*. Figure 17 present the trends of variation in $\overline{A'E'}'$ (Eq. 3.1) and $\overline{F'B'}'$ (Eq. 3.2) at different jet impingement angles according to the proposed model. Although, the length of the leading and trailing parts of the JFP increases with the decrease in *a*, the rate at which $\overline{A'E'}$ increases is higher (Fig. 17). This is due to the fact that the increased distance of points on the target surface from the tip of focusing nozzle along the forward part is higher than the trailing part and is responsible for the error in prediction by the model. At larger SOD, the jet structure becomes very unstable and do not follow exactly the variation as assumed in

the modelling. However, the model assumes considerable erosion at this point. Hence, the actual jet plume divergence that was evaluated at desired v cannot be applied exactly and differs slightly at higher distances from the tip of focusing nozzle and this is the case that arises at shallower jet impingement angles. This leads to the over prediction of top width of JFP by the proposed model at shallower jet impingement angles.
Furthermore, as discussed earlier, at shallower $\boldsymbol{a}$, the particles along the leading edge ($\overline{A'C'}$) of the diverged plume cannot initiate considerable erosion due to shallower particle impact angles and is also not stable (Fig. 6). The scale of erosion reduces further at higher jet feed rates as the exposure time of the workpiece material to the abrasive particles reduces. Hence, the combined effect of unstable portion of the jet along the edges of the plume at larger distances, smaller impact angles along jet plume edges contribute to the error. Finally, the error can also be attributed to the non consideration of (i) quantitative measure of erosion caused at different impact angles of the abrasive particles and (ii) the erosion capability of the deflected jet from the target surface.

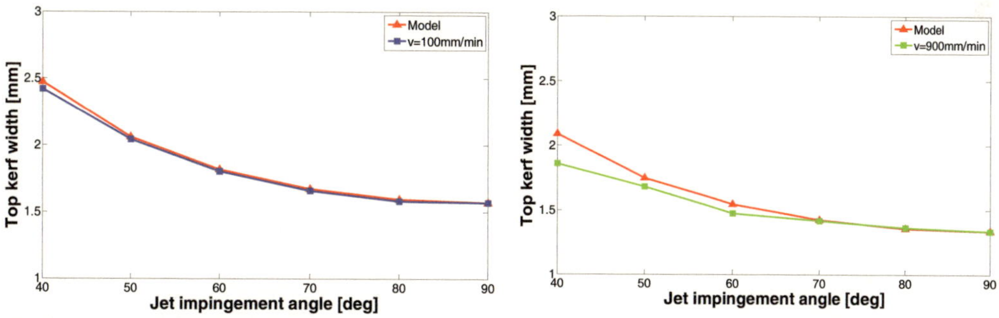

Fig. 16. Variation of top width of jet footprint with the jet impingement angle by considering the effective jet plume divergence (a) φv=100 mm/min = 5.24^0, (b) φv=90^0 mm/min = 3^0

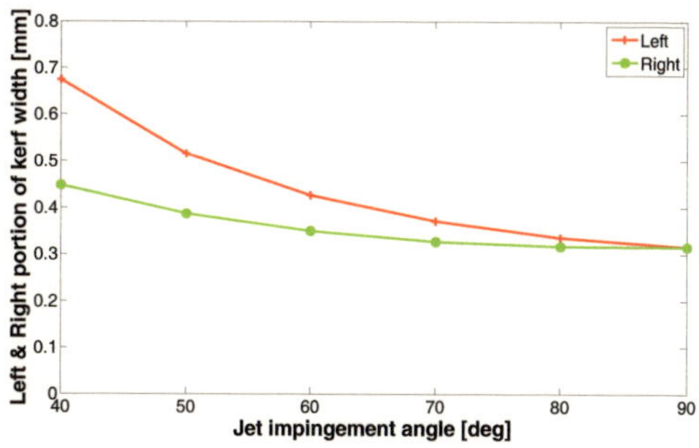

Fig. 17. Variation of leading (A'E') and trailing (F'B') portions of the top width of jet footprint at different jet impingement angles by the proposed model

4. Conclusions

Jet footprint (JFP) is the building block in fabrication of complex shaped components by 2-axis/5-axis abrasive waterjet milling in which the jet feed rate and impact angle are varied continuously according to the shape to be produced. Hence, in this chapter, the research work done at the University of Nottingham under the NIMRC sponsored research project titled "*Freeform Abrasive WaterJet Machining in Advanced Materials* (Freeform_JET)", under the following headings: (i) investigations on the physical phenomenon involved on the formation of JFP, (ii) development of models for prediction of geometry of JFP, and (iii) development of model for top width of the JFP were presented. The findings of this research work was summarised as follows:

4.1 Physical phenomenon involved in generation of JFP

For the first time, a detailed analysis of the geometry of the JFP by considering the key kinematic operating parameters (jet impingement angle, jet feed rate) on abrasive waterjet machining of advanced engineered SiC ceramic, which is difficult to machine with conventional approaches due to its high hardness and low fracture toughness, is presented. In-depth analysis of the mechanism of kerf geometry formation under the influence of key kinematic operating parameters of AWJ: emphasizing the dependence of JFP on both jet impingement angle and erosion time (given by jet feed rate), the variation in the geometry of the kerf was explained, with the help of schematic illustrations and dimensional characteristics obtained from experiments, based on two key elements: (i) variation of axial distance from tip of focusing nozzle and radial distance from jet axis, and (ii) impact angle of the individual abrasive particles across JFP. While the variation of impact angle of individual abrasives across the JFP accounts for the "cosine type" symmetrical kerf at normal jet impingement angles, the variation of SOD and divergence of jet plume, when jet impinges at shallower angles, are the main factors in obtaining asymmetrical kerf geometries. Amplitudes of these non/symmetrical kerf geometries are very much dependent on the jet exposure time (jet feed rate) due to increase in erosion along jet axis and erosion capability of abrasive particles at shallower impact angle. At 90^0 jet impingement angle, as the effect of local impact angle of abrasives is cumulative, with the decrease of the jet feed rate, the symmetrical kerfs become more elongated along their axis and therefore simple cosine function approximation is less appropriate for their geometry. Further, the kerf generated at shallower jet impingement angles cannot be approximated by simple 'cosine function' approximation.

Relationship between the key kinematic parameters of AWJ and dimensional characteristics of kerf: The influence of a and v on dimensional characteristics, such as depth of penetration, top kerf width, slopes of kerf walls was analyzed. Maximum erosion depth was observed at jet impingement angle in the range of 70^0–80^0 as the effective average impact angle of abrasive particles approaches 90^0 in this range and also this shift can be attributed to the material hardness. In contrast to nearly constant width at normal jet impingement, an increase in width was observed at a shallower jet impingement angle which is due to the increase in width of jet footprint.

Study of the influence of successive AWJ passes on the "resultant" kerf geometry: Careful analysis of the kerf amplitude (penetration) along jet impingement axis showed that once a footprint is generated after the first jet pass, the next one "inherits" a non-flat target surface that indirectly leads to change in local standoff distances and impact angles of abrasives that ultimately results in lesser erosion on the lateral walls of the kerfs. Based on these aspects, an in-depth analysis of the kerf geometry after two consecutive jet passes has shown that real jet penetration does not follow a pure factorial rule (penetration depth number of passes); this observation is of key importance for generating multi-pass AWJ milled surfaces at controlled depth. Apart from investigating the dependencies between the kinematic parameters of AWJ machining with the geometrical and dimensional characteristics of the jet footprint, this work flags-up to the machining community the technical challenges to be addressed when proposing machining strategies for milling complex geometry surfaces.

4.2 Analytical model for prediction o f the geometry of the JFP

A geometrical model was developed that enables the prediction of the geometry of the JFP for maskless controlled-depth AWJ milling when the jet impinges normally on target workpiece. The model relies on a nonlinear partial differential equation in which a specific response (etching rate) of the workpiece material to the waterjet conditions is included. By taking limiting conditions (i.e. high jet feed speeds) on the proposed geometrical model, shallow kerf profiles are generated that enable the calibration/identification of the specific etching rate of the workpiece materials to be identified. Then, knowing this key characteristic of the jet-workpiece interaction, the full geometry of the jet footprint can be obtained for any (technologically required) jet feed speed by employing numerical methods to solve the non-linear differential equation which the proposed model is based on. An example of model validation was presented for 90° jet impingement angle against a SiC ceramic as target material, on which, the specific etching rate was determined (at constant m_f= 0.7 kg/min; P= 345 MPa) by the use of analytical models first, to which a scanned shallow kerf profile, obtained for high jet feed (v_f= 1700 mm/min), was input. Then, the JFP at various jet feed speeds (v_f= 100, 500, 900, 1300 mm/min) were obtained and compared with the experimental ones. This revealed that only at low values of jet feed speeds (high jet dwell times) the predicted kerf profile was deeper (by 30 μm) than the real ones; however, the gradient of the kerf was found to be very similar. This finding revealed the need to introduce into the model a linear correction on the SOD of the nozzle to the real target surface that develops as eroded kerf profile. Such addition to the geometrical model enabled accurate (errors < 5%) prediction of the JFP over a wide range of jet feed speed values. The proposed geometrical JFP model, initially developed for normal jet impingement angle, finds its use in developing innovative jet paths capable to generate complex geometry surfaces by continuously altering the jet feed speed in such a way that the succession of JFPs can envelope the desired AWJ milled surface.

4.3 Analytical model for prediction of top width of jet footprint

In this section, the physical phenomenon involved in formation of the top width of the JFP at various jet impingement angles was analysed and the model for the top width was developed by considering the jet impingement angle, jet plume divergence, SOD and jet feed rate. The findings of this work are as follows:

- *Physical phenomena involved in generation of top width*: The change in the cross-section of the jet plume, parallel to the surface of the target workpiece, with the orientation of the jet dictates the shape of the JFP generated and its geometrical characteristics. The jet plume cross-section at the machining zone is nearly circular in shape at the normal jet impingement and gradually transforms to elliptical with the decrease in jet impingement angles. Hence, the top width of the footprint is the diameter of the circle at normal impingement and is length of major axis at shallower jet impingement angles. The top width of the JFP generated is influenced by the diameter of focusing nozzle, jet plume divergence in air and SOD.

- *Edge quality*: The edge quality of the kerf deteriorated with the increase in jet feed rate at normal jet impingement. At shallower jet impingement angles, chipping is significant along the leading edges of the JFP due to unstable nature of the part of the forward portion of the jet plume.
- *Top width variation*: An increasing trend of the top width of the JFP was observed with the (i) decrease in jet impingement angle, and (ii) decrease in jet feed rates. For the same jet impingement angle, the top width of the JFP is decreased with the increase in jet feed rate. The influence of jet feed rate is considered in the model by a parameter defined as 'effective jet plume divergence'.
- *Assessment of the model predictions*: The inclusion of the 'effective jet plume divergence' parameter into the model has enabled the prediction of top width of JFP at any jet feed rates. The current model does a correct prediction at lower jet feed rates if the divergence of the water jet is taken into account, however, it over predicts at higher velocities and smaller jet impingement angles because the effect of particle velocity variation along top width of JFP and impingement angle on the abrasion rate is not taken into account.

Although the deflected jet is capable of doing secondary machining and influences the characteristics of the JFP, this aspect was not considered in the above models to avoid mathematical complications. Furthermore, these models have not included the effect of particle impact angle on the erosion. By taking into consideration, the effect of abrasive particle impact angle, erosion theories and the secondary machining by the deflected jet plume, the predictive capability of the models in prediction of JFP geometry, top width can be improved further at shallower impact angles. In addition to considering the above aspects, the researchers are developing models for prediction of JFPs at shallower jet impingement angles and resulting kerfs at various degrees of overlap, which are the building blocks in generation of complex shaped parts in advanced engineering materials.

5. References

Ahmet Hascalik; Ulas Caydas & Hakan Gurun (2007). Effect of traverse speed on abrasive waterjet of Ti–6Al–4V alloy, *Materials and Design*, 28, 1953-1957.

Anand, K.; Morrison, C.; Scattergood, R.O.; Conrad, H.; Routbort, J.L. & Warren, R. (1986). Erosion of multi phase materials, in: Institute of Physics Conference Series, Adam Higler, London, pp. 949–961.

Andrews, D.R. & Horsfield, N. (1983). Particle collisions in the vicinity of an eroding surface, Journal of Physics D: *Applied Physics*, vol. 16, pp. 525-538.

Ansari, A.I. & Hashish, M. (1993). Volume removal trends in abrasive water jets, Transactions of ASME, *Manufacturing Science and Engineering*, 64, 629-641.

Axinte, D.A.; Srinivasu, D.S.; Billingham, J. & Cooper, M. (2010). Geometrical modelling of abrasive waterjet footprints: a study for 90° jet impact angle, *Annals of CIRP - Manufacturing Technology*, 59, 341-346.

Chalmers, E.J. (1991). Effect of parameter selection on abrasive waterjet performance, in *Proceedings of 6th American Waterjet Conference*, Houston, Texas, pp. 345-354.

Chen, F.L. & Siores, E. (2003). The effect of cutting jet variation in surface striation formation in abrasive waterjet cutting, *Journal of Materials Processing Technology*, 135, 1-5.

Chen, L.; Siores, E. & Wong, W.C.K. (1996). Kerf characteristics in abrasive waterjet cutting of ceramic materials, *International Journal of Machine Tools and Manufacture*, 36(11), 1201-1206.

Chen, L.; Siores, E. & Wong, W.C.K. (1998). Optimising abrasive waterjet cutting of ceramic materials, *Journal of Materials Processing Technology*, 74, 251–254.

Freist, B; Haferkamp, H; Laurinat, A. & Louis, H. (1989). Abrasive waterjet machining of ceramic products, in: *Proceedings of 5th American Waterjet Conference*, Toronto, Canada, pp. 191-204.

Gropetti, R. & Capello, E. (1992). On an energetic semi-empirical model of hydro-abrasive jet material removal mechanism for control and optimization in: *Proceedings of 11th International Conference on Jet Cutting Technology*, St-Andrews, Scotland, pp. 101-122.

Groppetti, R; Gutema, T & Dilucchio, A. (1998). A contribution to the analysis of some kerf quality attributes for precision abrasive waterjet cutting, in *14th International Conference on Jet Cutting Technology*, Belgium, pp. 253-269.

Gudimetla, P.; Wang, J. & Wong, W. (2002). Kerf formation analysis in the abrasive waterjet cutting of industrial ceramics, *Journal of Material Processing Technology*, 128(1-3), 123-129.

Hashish, M. (1987). Milling with abrasive waterjets: A preliminary investigation, in *Proceedings of 4th U.S. Waterjet Conference*, Berkeley, C.A., pp. 1-10.

Hashish, M. (1989). An investigation of milling with abrasive waterjet, Transactions of ASME, *Journal of Engineering for Industry*, 111, 158-166.

Hashish, M. (1989). Machining of advanced composites with abrasive waterjets, *Manufacturing Review*, 2(2), 142-150.

Hashish, M. (1993). The effect of beam angle in abrasive waterjet machining, Transactions of ASME, *Journal of Engineering for Industry*, 115, 51–56.

Henning, A. & Westkamper, E. (2003). Modelling of wear mechanisms at the abrasive waterjet cutting front, in: *Proceedings of WJTA American Waterjet Conference*, Houston, Texas, USA, pp. Paper 3-A.

Hocheng, H. & Chang, K.R. (1994). Material removal analysis in abrasive waterjet cutting of ceramic plates, *Journal of Materials Processing Technology*, 40(3-4), 287-304.

Inasaki, I. (1987) Grinding of hard and brittle materials, *Annals of the CIRP- Manufacturing Technology*, 36, 463-471.

Kahlmana, L.; Öjmertz, K.M.C. & Falk, L.K.L. (2001). Abrasive-waterjet testing of thermo-mechanical wear of ceramics, *Wear*, 248, 16–28.

Kim, J.J. & Park, S.K. (1998). Solid particle erosion of SiC and SiC–TiB composite hot-pressed with Y2 O3, *Wear*, 222, 114–119.

Konig, W.; Dauw, D.F.; Levy, G. & Panten, U. (1988). EDM-future steps towards the machining of ceramics, *Annals of CIRP - Manufacturing Technology*, 37, 623–631.

Kovacevic, R. & Momber, A.W. (1995). Statistical character of the failure of multiphase materials due to high pressure waterjet impingement, *International Journal of Fracture*, 71, 1-14.

Laurinat, A.; Louis, H. & Wiechert, G. M. (1993). A model for milling with abrasive water jets, *Proceedings of 7th American Water Jet Conference*, Seattle, Washington, pp. 119-139.

Lebar, A. & Junkar, M. (2003). Simulation of abrasive waterjet machining based on unit event features, Proceedings of Institution of Mechanical Engineering-Part B: *Journal of Engineering Manufacture*, 217(B5), 699 - 703.

Lee, W.E. & Rainforth, W.M. (1992). Ceramics Microstructures: property control and processing. London: Chapman & Hall.

Miller, D.S. (2004). Micromachining with abrasive waterjets, *Journal of Materials Processing Technology*, 149, 37–42.

Momber, A.W. & Kovacevic, R. (2003). Hydro abrasive erosion of refractory ceramics, *Journal of Materials Science*, 38, 2861-2874.

Momber, A.W.; Eusch, I. & Kovacevic, R. (1996). Machining refractory ceramics with abrasive water jets, *Journal of Materials Science*, 31(24), 6485-6493.

Niu, M.S.; Kobayashi, R. & Yamaguchi, T. (1995). Kerf width in abrasive waterjet machining, in *Proceedings of 4th Pacific Rim Interenational Conference on Waterjet Technology*, Shimizu, Japan.

Ojmertz, K.M.C. & Amini, N. (1994). A discrete approach to the abrasive waterjet milling process, *Proceedings of 12th International Conference on Jet Cutting Technology*, pp. 425-434.

Ojmertz, K.M.C. (1997). *A study on abrasive waterjet milling*, Ph.D. Thesis, Chalmers University of Technology.

Oka, Y.I.; Ohnogi, H.; Hosokawa, T. & Matsumura, M. (1997). The impact angle dependence of erosion damage caused by solid particle impact, *Wear*, 203-204, 573-579.

Paul, S.; Hoogstrate, A.M.; van Luttervelt, C.A. & Kals, H.J.J. (1998). An experimental investigation of rectangular pocket milling with abrasive water jet, *Journal of Material Processing Technology*, 73, 179 -188.

Richerson, D.W. (2006). Modern Ceramic Engineering: properties, processing and use in design: CRC, Taylor Francis.

Ruff, A.W. & Wioderborn, S.W. (1979). Erosion by solid particle impact, in: Treatise on Material Science and Technology: Erosion, New York., pp. 69-126.

Samant, A.N. & Dahotre, N.B. (2009). Laser machining of structural ceramics - A review, Journal of European Ceramic Society, 40(3-4), 287-304.

Shipway, P.H. & Hutchings, I.M. (1993). Influence of nozzle roughness on conditions in a gas blast erosion rig, *Wear*, vol. 162-164, pp. 148-158.

Shipway, P.H. (1997). The effect of plume divergence on the spatial distribution and magnitude of wear in gas-blast erosion, *Wear*, Vol. 205, 169-177.

Simpson, M. (1990). Abrasive Particle Study in High Pressure Water jet Cutting, *International Journal of Water Jet Technology*, 1, 17-28.

Siores, E.; Wong, W.C.K.; Chen, L. & Wager, J.G. (1996). Enhancing abrasive waterjet cutting of ceramics by head oscillation techniques, *Annals of CIRP - Manufacturing Technology*, 45(1), 327-330

Srinivasu, D.S. & Axinte, D.A. (in press). An analytical model for top width of jet footprint in abrasive waterjet milling: a case study on SiC ceramics, *Proceedings of the Institution of Mechanical Engineers, Part B: Journal of Engineering Manufacture*.

Srinivasu, D.S.; Axinte, D.A.; Shipway, P.H. & Folkes, J. (2009). Influence of kinematic operating parameters on kerf geometry in abrasive waterjet machining of silicon carbide ceramics, *International Journal of Machine Tools & Manufacture*.

Tuersley, I.P.; Jawaid, A. & Pashby, I.R. (1994). Review: Various methods of machining advanced ceramic materials, *Journal of Materials Processing Technology*, 42(4), 377-390.

Wang, J. (2003). The Effects of the Jet Impact Angle on the Cutting Performance in AWJ Machining of Alumina Ceramics, Key Engineering Materials, *Advances in Abrasive Technology* V, vol. 238-239, 117-122.

Yanaida, K. & Ohashi, A. (1978). Flow characteristics of water jets in air, in: 4th *International Symposium on Jet Cutting Technology*, pp. A3-39.

Zeng, J. & Kim, T.J. (1996). An erosion model of polycrystalline ceramics in abrasive waterjet cutting, *Wear*, 193, 207-217.

Zeng, J.; Munoz, J. & Kain, I. (1997). Milling ceramics with abrasive waterjets - An experimental investigation, in: *Proceedings of 9th American Waterjet Conference*, Dearborn, Michigan, pp. 93-107.

Ductile Mode Micro Laser Assisted Machining of Silicon Carbide (SiC)

Deepak Ravindra, Saurabh Virkar and John Patten
Western Michigan University
USA

1. Introduction

This chapter is divided into three parts: (1) background research, (2) experimental work and (3) simulations on ductile mode micro laser assisted machining. The origin and science behind ductile regime machining will briefly be discussed prior to discussing the experimental and simulated study conducted on SiC. Although the results of both studies (experimental and simulation) are not intended for direct comparison, the main objective of both studies are similar, that is to analyze the effects of laser heating/thermal softening on ductile mode machining of single crystal 4H-SiC.

1.1 Background

Although silicon carbide (SiC) has been around since 1891, it was not until the mid 1990's that this material was introduced into the precision manufacturing industry. SiC is well known for its excellent material properties, high durability, high wear resistance, light weight and extreme hardness. However, SiC is also well known for its low fracture toughness, extreme brittleness and poor machinability. SiC is one of the advanced engineered ceramics designed to operate in extreme environments. This material is pursued as both a coating and structural material due to its unique properties, such as:

- Larger energy bandgap and breakdown field allowing it to be used in high-temperature, high-power and radiation-hard environments
- Mechanical stiffness, expressed by its high Young's modulus (Gao et al., 2003)
- Desirable tribological properties, such as wear resistance and self-lubricating (Ashurst et al., 2004)

SiC is commercially available in various forms/phases (polytypes) such as single crystal, polycrystalline (sintered and CVD) and amorphous. The most common polytypes of SiC are 2H, 3C, 4H, 6H, and 15R. The numbers refer to the number of layers in the unit cell and the letter designates the crystal structure, where C=cubic, H=hexagonal, and R=rhombohedral. In this study, only one polytype will be discussed: 4H. The 4H polytype is a single crystal.

1.2 Ductile Regime Machining

Materials that are hard and brittle, such as semiconductors, ceramics and glasses, are amongst the most challenging to machine. When attempting to machine ceramics, such as SiC, especially to improve the surface finish, it is important to carry out a 'damage free' machining operation. This often can be achieved by ductile mode machining (DMM) or in other words machining a nominally hard and brittle material in the ductile regime. Material removal processes can be considered in terms of fracture dominated mechanisms or localized plastic deformation. A fracture dominant mechanism for ceramics, i.e., brittle fracture, can result in poor surface finish (surface damage) and also compromises on material properties and performance (Ravindra et al., 2007).

The insight into the origins of the ductile regime during single point diamond turning (SPDT) of semiconductors and ceramics was provided by the research done by Morris, et al in 1995 in collaboration with one of the current authors (Patten). A detailed study of machining chips (debris) and the resultant surface was studied (analyzed using a TEM) to evaluate evidence of plastic material deformation. This seminal research concluded that the machining chips were plastically formed and are amorphous (not due to oxidation) due to the back transformation of a pressure induced phase transformation, and the machining debris (chips) contain small amounts of micro-crystalline (brittle) fragments.

According to the grinding research carried out by Bifano et al. in 1991, there are two types of material removal mechanisms associated with the machining process: ductile; plastic flow of material in the form of severely sheared machining chips, and brittle; material removal through crack propagation. This previous research discusses several physical parameters that influence the ductile to brittle transition in grinding of brittle materials. The researchers were successful in performing ductile mode grinding on brittle materials. However, these researchers did not propose or confirm a model or suitable explanation for the origin of this ductile regime. Bifano et al. also proposed a model defining the ductile to brittle transition of a nominally brittle material based on the material's brittle fracture properties and characteristics. A critical depth of cut model was introduced based on the Griffith fracture propagation criteria. The critical depth of cut (d_c) formula is as follows:

$$d_c = (E \cdot R) / H^2 \qquad (1)$$

where E is the elastic modulus, H is the hardness and R is the fracture energy. The value of the fracture energy (R) can be evaluated using the relation:

$$R \sim K_c^2 / H \qquad (2)$$

where K_c is the fracture toughness of the material. The above two equations can be combined to represent the critical depth (d_c) as a measure of the brittle transition depth of cut:

$$d_c \sim (E / H) \cdot (K_c / H)^2 \qquad (3)$$

The researchers were successful in determining a correlation between the calculated critical depth of cut and the measured depth (grinding infeed rate). The constant of proportionality was estimated as to be 0.15 and this is now added into Equation (2.3) to generate a more accurate empirical equation:

$$d_c \sim 0.15 \cdot (E / H) \cdot (K_c / H)^2 \quad (4)$$

1.3 Chip Formation

A critical depth, d_c is determined before any ductile mode machining operation is carried out. Any depth beyond or exceeding the critical depth, which is also known as the Ductile to Brittle Transition (DBT) depth, will result in a brittle cut. Since the equipment used in the current study (Universal Micro-Tribometer by CETR) is a load controlled (and not a depth controlled) machine, thrust force calculations were carried out for corresponding required depths of cuts. The Blake and Scattergood ductile regime machining model (as shown in Fig. 1) was used to predict the required thrust force for a desired depth of cut (Blake & Scattergood, 1991). In this model it is assumed that the undesirable fracture damage (which extends below the final cut surface) will originate at the critical chip thickness (t_c), and will propagate to a depth, y_c. This assumption is consistent with the energy balance theory between the strain energy and surface energy (Bifano et al., 1991).

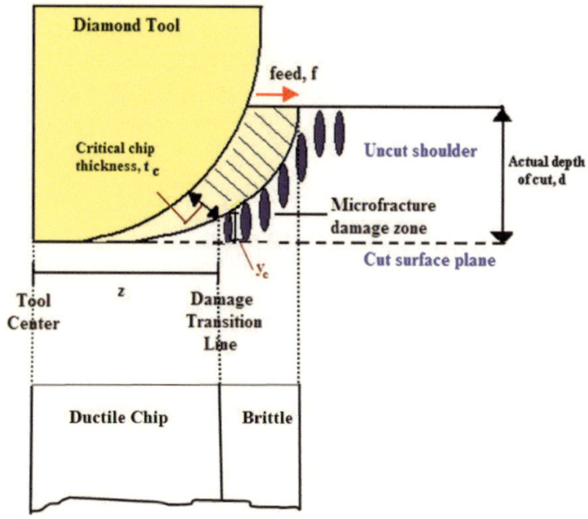

Fig. 1. Model for ductile regime machining

In general, the ductile-to-brittle transition (DBT) is a function of variables such as tool geometry (rake and clearance angle, nose and cutting edge radius), feed rate, and depth of cut.

1.4 High Pressure Phase Transformation (HPPT)

Although SiC is naturally very brittle, micro/nanomachining this material is possible if sufficient compressive stress is generated to cause a ductile mode behavior, in which the material is removed by plastic deformation instead of brittle fracture. This micro-scale phenomenon is also related to the High Pressure Phase Transformation (HPPT) or direct amorphization of the material (Patten et al., 2005). Fig. 2 shows a graphical representation of the highly stressed (hydrostatic and shear) zone that results in ductile regime machining.

Patten and Gao state that ceramics in general undergo a phase transformation to an amorphous phase after a machining process. This transformation is a result of the High Pressure Phase Transformation (HPPT) that occurs when the high pressure and shear caused by the tool (during the chip generation process) is suddenly released after a machining process. This phase transformation is usually characterized by the amorphous remnant that is present on the workpiece surface and within the chip. This amorphous remnant is a result of a back transformation from the high pressure phase to the atmospheric pressure phase due to the rapid release of pressure in the wake of the tool. There are two types of material removal mechanisms during machining: ductile mechanism and the brittle mechanism (Bifano et al., 1991). In the ductile mechanism, plastic flow of material in the form of severely sheared machining chips occur, while material removal is achieved by the intersection and propagation of cracks in the brittle fracture mechanism. Due to the presence of these two competing mechanisms, it is important to know the DBT depths (or critical size) associated with these materials before attempting a machining operation.

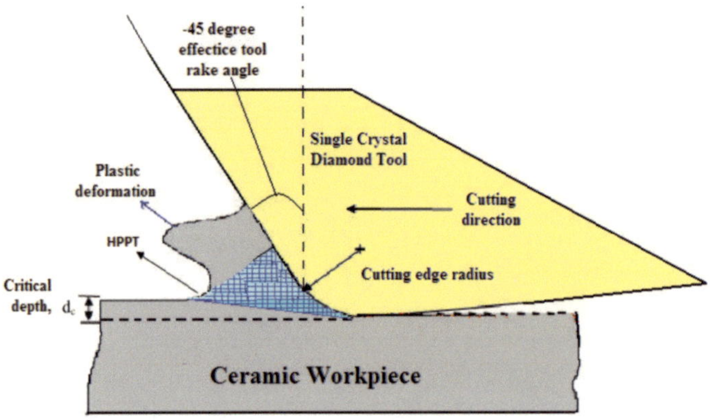

Fig. 2. A ductile machining model of brittle materials

Fig. 2 shows a ductile cutting model showing the high compressive stress and plastically deformed material behavior in brittle materials. A -45° rake angle tool is demonstrated in the above schematic as a negative rake angle tool yields in higher compressive stresses at the tool-workpiece interface.

1.5 Challenges in Ductile Regime machining of Ceramics

Since the hardness of SiC is approximately 30% of the hardness of diamond, machining SiC with a diamond tool is an extremely abrasive process. As a result of the abrasive material removal process, there are several limitations in terms of machining parameters that have to be considered. The primary limitation in the process of ductile mode machining is to not exceed the critical depth of cut or the DBT depth of the material. Exceeding the DBT depth during the machining process will result in fracture and thus leaving a poor surface finish. Another important parameter during machining is the feed. In general, lower feed rates result in a better surface finish however; lower feed rates also result in more tool wear due to the longer track length covered by the tool during machining. Tool wear can be crucial when trying to improve the surface finish of a SiC workpiece. Any wear along the cutting edge radius (rake and flank wear) will directly affect the machined surface finish, possibly causing cracks and fracture. A small chipped area or crack in the tool tip could potentially grow during the machining process, eventually causing the tool to fail. Tool failure at times can be observed in the cutting forces during the machining process. In general, low cutting forces are desired to minimize the diamond tool wear. The micro laser assisted machining (μ-LAM) process, which will be discussed in the next few sections, shows positive results in addressing the challenges faced in conventional ductile regime machining of SiC.

2. Experimental Study on Ductile Mode μ-LAM

2.1 Introduction to μ-LAM

Semiconductors and ceramics share common characteristics of being nominally hard and brittle, which stems from their covalent chemical bonding and crystal structure. These materials are important in many engineering applications, but are particularly difficult to machine in traditional manufacturing processes due to their extreme hardness and brittleness. Ceramics have many desirable properties, such as excellent wear resistance, chemical stability, and high strength even at elevated temperatures. All of these properties make them ideal candidates for tribological, semiconductor, MEMS and optoelectronic applications. In spite of all these characteristics, the difficulty during machining and material removal has been a major obstacle that limited the wider application of these materials (Jahanmir et al., 1992). The plastic deformation of these nominally brittle materials at room temperature is much less than in metals, which means they are more susceptible to fracture during material removal processes. Surface cracks generated during machining are subsequently removed in lapping and polishing processes, which significantly increases the machining time and cost. Machining mirror-like surface finishes contribute significantly to the total cost of a part. In some cases, grinding alone can account for 60-90% of the final product cost (Wobker & Tonshoff, 1993). In this context, developing a cost effective method to achieve a flawless surface in ultra fine surface machining of an optical lens or mirror has become a challenge. In many engineering applications, products require a high quality surface finish and close tolerances to function properly. This is often the case for products made of semiconductor or ceramic materials. The real challenge is to produce an ultra precision surface finish in these nominally brittle materials at low machining cost.

Current limitations for brittle material machining include the high cost of processing and low product reliability. The cost is mainly due to the high tool cost, rapid tool wear, long machining time, low production rate and the manufacturing of satisfactory surface figure and form. The low product reliability is primarily due to the occurrence of surface/subsurface damage, i.e., cracks, and brittle fracture. In order to develop a suitable process, ductile regime machining, considered to be one of the satisfactory precision machining techniques, has been continuously studied over the last two decades (Blake & Scattergood, 1990; Blackley & Scattergood, 1994; Morris et al., 1995; Leung et al., 1998; Sreejith & Ngoi, 2001; Yan et al., 2002; Patten et al., 2003; Patten et al., 2005). Laser assisted micro/nano machining is another important development in this direction (Dong & Patten, 2007; Rebro et al., 2002).

In past research, it has been demonstrated that ductile regime machining of these materials is possible due to the high pressure phase transformation (HPPT) occurring in the material caused by the high compressive and shear stresses induced by the single point diamond tool tip (Ravindra et al., 2009; Ravindra & Patten, 2008). To further augment the ductile response of these materials, traditional scratch/single point diamond turning tests are coupled with a micro-laser assisted machining (μ-LAM) technique (Shayan et al., 2009). A schematic of the basic underlining concept of the μ-LAM process is shown in Fig. 3. This hybrid method could potentially increase the critical depth of cut (DoC) (larger DBT depth) in ductile regime machining, resulting in a higher material removal rate. μ-LAM was previously successfully carried out on single crystal Si yielding a greater DBT (for the scratch performed with laser heating)(Ravindra et al., 2010).

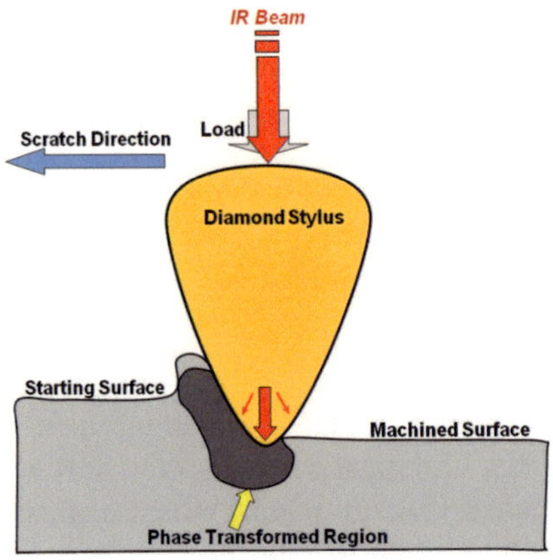

Fig. 3. A schematic cross-section of the μ-LAM process

The objective of the current study is to determine the effect of laser heating (using the μ-LAM process) on the material removal of single crystal 4H-Silicon Carbide (SiC) using scratch testing. The scratch tests were carried out to examine the effect of temperature in thermal softening of the high pressure phases formed under the diamond tip. There were two studies done from these scratch experiments: studying the laser heating effect on the DBT of the material and evaluating the thermal softening and relative hardness as a result of irradiation of the laser beam at a constant cutting speed. The effects of laser heating were studied by verifying the depths of cuts and the nature of the scratches (i.e. ductile, DBT or brittle) for diamond stylus scratch tests carried out on single crystal SiC with increasing loads (thrust force). The load range was selected such that the scratches show both ductile and brittle response (with a DBT region within the scratch). Cutting forces and three-dimensional cutting surface profiles (using a white light interferometer) were investigated.

2.2 Experimental Process

The scratch tests were performed on a Universal Micro-Tribometer (UMT) which is produced by the Center for Tribology Research Inc. (CETR). This equipment was developed to perform comprehensive micro-mechanical tests of coatings and materials at the micro scale. This system facilitates cutting speeds as low as 1μm/sec at nanometric cutting depths. The tribometer is a load controlled device where the required thrust force (Fz) is applied by the user to obtain the desired DoC (based on the tool geometry and workpiece material properties). The unit is equipped with a dual-axis load cell that is capable of constantly monitoring the thrust and cutting forces (Fx); obtained as an output parameter from the cutting experiment. A typical scratch test setup along with the μ-LAM system is shown in Fig. 4. All scratch tests were performed on a single crystal 4H-SiC wafer. All cuts were performed on the {1010} plane along the <1010> direction.

A 90° conical single crystal diamond stylus (with a spherical end tip radius of 5μm) was used as the scratch tool. The details of the diamond tip attachment were depicted in Fig. 5. An infrared (IR) diode fiber laser (λ=1480nm and P_{max}=400mW,) with a Gaussian profile with a beam diameter of ~10μm was used in this study. The laser beam is guided through a 10μm fiber optic cable to the ferrule, which is attached to the diamond stylus. The μ-LAM system is configured in such a way that the laser beam passes through the diamond tip and impinges on the work piece material at the tool work piece interface (contact) (Dong, 2006).

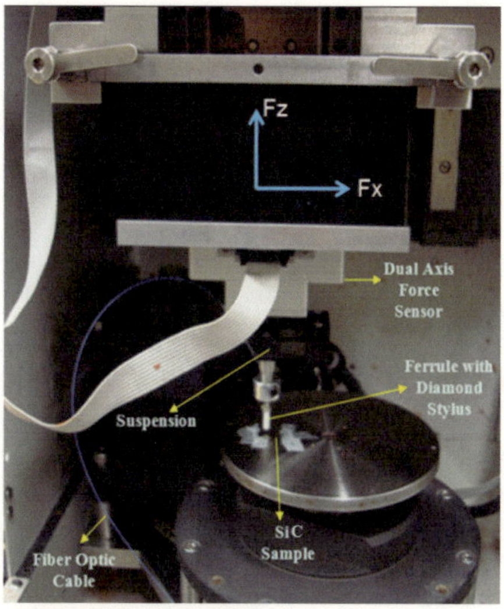

Fig. 4. μ-LAM setup on the Universal Micro Tribometer

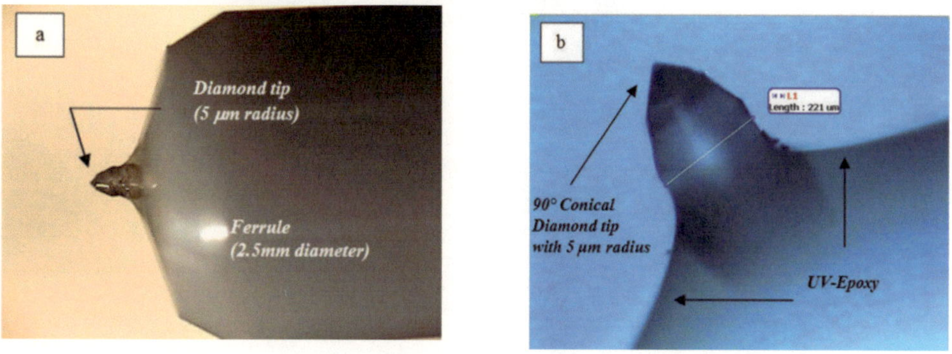

Fig. 5. Diamond tip attachment: (a) 5μm radius diamond tip attached on the end of the ferrule using epoxy, (b) Close up on diamond tip embedded in the solidified epoxy

Scratch tests were chosen to be the principal method of investigation in this study as it is a better candidate for evaluating machining conditions than indenting because the mechanics during scratching are more applicable to the machining process such as single point diamond turning (SPDT). In this study, two conditions of scratches were performed: with and without laser heating. The scratches were carried out at low cutting speeds (1 µm/sec) in order to maximize the thermal softening of the material during the laser heating. Scratch lengths of 500 µm were produced on the SiC wafer specimen. The loads were increased linearly with time from 2 mN to 70 mN along the scratch. The scratch test parameters are summarized in Table 1.

Scratch Number	Load Range, Fz (mN)	Machining Condition	Cutting Speed, (µm)	Laser Power (mW)
1	2 - 70 mN	No Laser	1	0
2	2 - 70 mN	With Laser	1	350

Table 1. Scratch testing parameters
**350mW is the laser power, approximately 150mW is actually delivered to the work piece material, the balance of the laser power is lost due to scattering and reflections.*

2.3 Experimental Results & Discussion

Fig. 6 shows two scratches that represent the two conditions: without (scratch 1) and with laser heating (scratch 2). The load range (2-70 mN) performed on these scratches was ideal for this study as it had both the ductile and brittle regime along the same scratch. The DBT is identified somewhere between the ductile and brittle regime of the scratch using optical microscopy, white light interferometry and force analysis (from variations in cutting forces). It is seen in Fig. 6 that the scratch performed without laser heating exhibits brittle fracture along the cut before, i.e., at a shallower depth, than the scratch performed with laser heating.

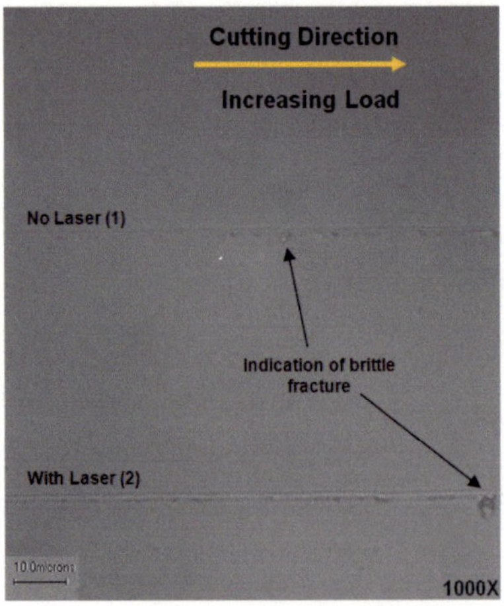

Fig. 6. Micrograph showing brittle fracture along the scratch

In this study, there were two different analyses done based on the results obtained from the scratch tests. The first analysis compares the depth and cutting forces (Fx) for a constant thrust force (Fz) for both cutting conditions (with and without laser heating) to see the effect of thermal softening on the material. For this analysis, scratches analyzed for both conditions were in the ductile regime. The results summarized in Table 2 show that for the same amount of applied thrust force (Fz = 30mN), the scratch performed with laser heating yielded a greater depth of cut (145nm vs. 95nm). It is also evident that cutting forces were equal for both these conditions for an equal applied thrust force (although the scratch performed with laser heating was significantly deeper). A scratch without laser heating done at higher loads to result in a depth of 145nm will most definitely result in higher cutting forces due to the hardness of the material (Shayan et al., 2009).

The second analysis done was to study the effects of laser heating on the DBT of the material. To determine this, two-dimensional scratch/groove profiles obtained using a white light interferometric profilometer were analyzed. Fig. 7 shows the cross-section of the two scratches taken at an equal thrust force of approximately 35mN. It can be seen that the scratch performed with laser heating (left) exhibits a perfectly ductile behavior whereas the scratch done without laser heating (right) indicates slight fracture (brittle behavior) of the material. The DBT depth identified for the scratch performed without laser heating just before the point of fracture is approximately 105nm. The brittle behavior is identified by the imperfect pattern of the groove edge which is a representation of the stylus imprint on the material. It is important to note from Fig. 7, that the scratch performed without laser heating is (apparently) deeper (210nm vs.113nm) as it is difficult to control the depth when the material removal mechanism is brittle (i.e. difficult to predict the depth due to fracture of the material). The clear and defined edges that depict the stylus imprint is a good indication of ductile response of the material (as seen in the scratch performed with laser heating).

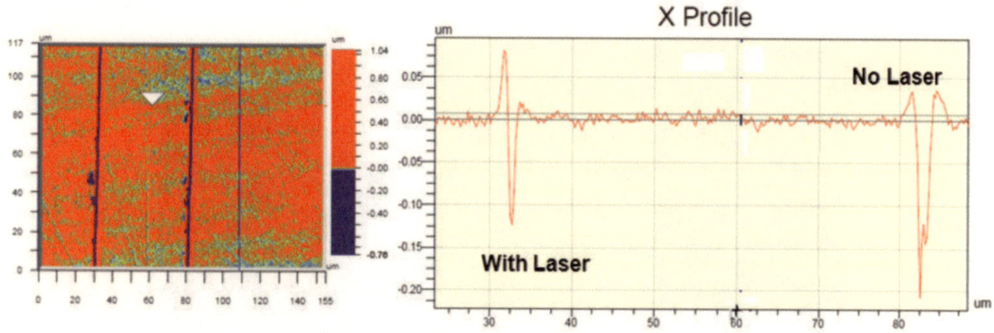

Fig. 7. Cross-section of scratches obtained from a white light interferometric profilometer

Fig. 8 shows the cross-section of the same two scratches (at a different point) taken at an equal thrust force of approximately 40mN. The DBT depth identified for the scratch performed with laser heating just before the point of fracture is approximately 240nm. At this load, the scratch performed with no laser heating shows signs of severe fracture. In comparison, the DBT depth of the scratch performed with laser heating was approximately 135nm greater than the DBT depth of the scratch performed without laser heating.

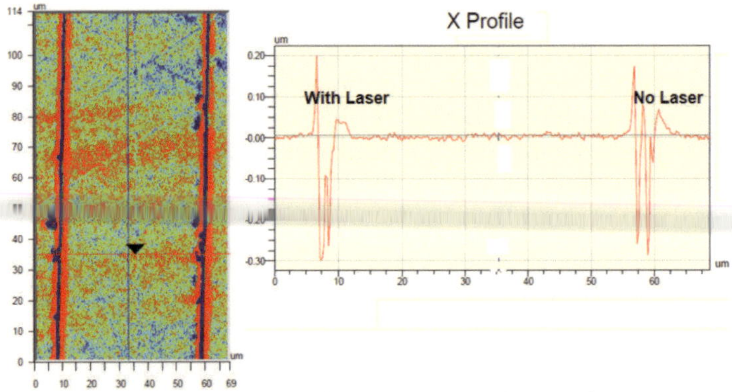

Fig. 8. Cross-section of scratches obtained from a white light interferometric profilometer

From Table 2, it is seen that the cut performed with laser heating yields a slightly higher cutting force at the DBT. This is due to the higher thrust force (40mN vs. 35mN) and larger depth of cut (240nm vs. 105nm).

Machining Condition	Thrust Force (Fz), (mN)	Cutting Force (Fx), (mN)	Depth of Cut(nm)	Scratch Nature
No Laser	30	10	95	Ductile
With Laser	30	10	145	Ductile
No Laser	35	12	105	DBT
With Laser	40	14	240	DBT

Table 2. Scratch test results
Just before the DBT occurs.

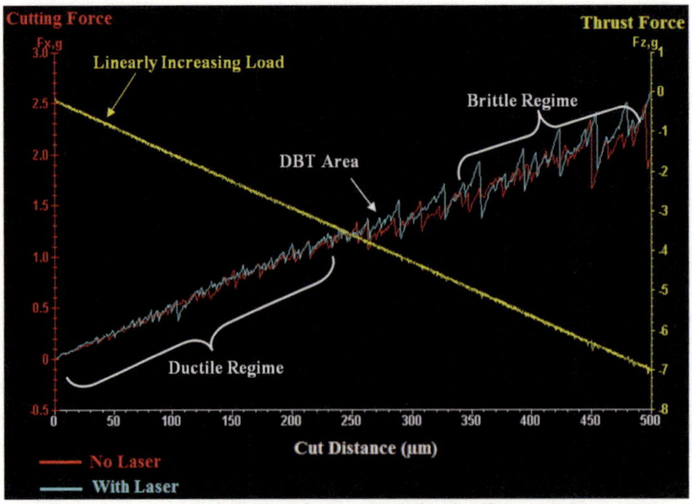

Fig. 8. Plot shows cutting force and thrust force data for both scratches

Analyzing the force data after the scratch experiments helps in correlating the onset of brittle fracture along the scratches. Brittle mode material removal is usually seen in the force data (especially cutting forces as it is more sensitive towards brittle fracture) and can be identified by its unstable behavior (higher standard deviation/ higher peaks-valleys in the force plots). Fig. 8 shows the force data plot obtained from both scratching conditions (with and without laser heating). Monitoring the cutting forces during the material removal process is also an effective in-situ method to detect the onset of brittle regime machining (onset of fracture occurrence).

3. Simulation Study of Thermal Effects for Analysis of Ductile Mode μ-LAM

This section of the chapter describes the numerical simulations of the ductile mode machining process conducted on single crystal 4H Silicon Carbide. The aim of the simulation work was to incorporate the laser heating effects in the simulation model to study the thermal softening behavior of SiC. In μ-LAM, a laser is used for heating the workpiece where the laser passes through the optically transparent diamond tool (Shayan et al., 2009). A laser source cannot be directly simulated in the simulation software hence thermal boundary conditions were defined on the tool and workpiece to mimic the laser heating effect. Initially, an approximate thermal softening curve was used to study the compatibility of the software (AdvantEdge from Third Wave Systems) with the desired laser heating and thermal softening effect (Virkar & Patten, 2009); i.e., a proof of concept. A new and more accurate thermal softening curve was developed based on references to incorporate more realistic thermal behavior (Virkar & Patten, 2010). The simulations were run at various temperatures throughout the thermal softening regime (up to the melting or decomposition point) and the changes in chip formation, cutting forces and pressures were studied.

3.1 FEM for μ-LAM Process

μ-LAM is a ductile mode material removal process developed for machining of nominally brittle materials augmented with thermal softening (provided by laser heating). Ductile mode machining implies plastic deformation and material removal, rather than brittle fracture, resulting in a smooth fracture free machined surface. This ductile mode material removal can be attributed to a High Pressure Phase Transformation (HPPT) at the tool-chip interface and the resultant high pressure phase is metallic or amorphous, and ductile. The HPPT occurs due to contact between the sharp tool and workpiece at or below critical depth of cut, i.e., the ductile to brittle transition. The recent work by (Patten et al., 2005) has determined the critical depth for ductile regime machining of single crystal SiC (Patten et al. 2005). These critical depths are in nano scale (< 1 μm) for SiC (Patten et al., 2004; Patten et al. 2005; Patten & Jacob, 2008).

Due to the metallic and ductile nature of the high pressure metallic phase that occurs at the tool chip interface, the metal machining software 'AdvantEdge' can be used to simulate the μ-LAM process and the scope of the work reported herein is limited to plastic deformation and ductile material removal. The software currently considers only ductile or plastic material removal and does not consider a fracture criterion or brittle material removing mechanisms. In the ductile mode, the software can be used to predict the forces and pressures generated by the tool-chip interaction for a given set of process conditions and tool geometry, assuming an appropriate material model is used (Jacob, 2006). The material properties include elastic and plastic behavior, thermal softening, strain rate and heat transfer parameters.

The material model used in the current work is the Drucker-Prager model, which is used to accommodate the pressure induced phase transformation (pressure sensitivity) and the resultant elastic-plastic behavior (Ajjarapu et al., 2004). The Drucker-Prager yield criterion is given by:

$$\sqrt{3J_2} + I_1\alpha - \kappa = 0 \quad (5)$$

Where I_1 is the first invariant of stress tensor and J_2 is the second invariant of deviatoric stress tensor, α is the Drucker-Prager coefficient and κ is the material constant.

The quantity I_1 is given by

$$I_1 = \sigma_1 + \sigma_2 + \sigma_3 \quad (6)$$

The quantity J_2 is given by

$$J_2 = 1/6\,[(\alpha_1 - \alpha_2)^2 + (\alpha_2 - \alpha_3)^2 + (\alpha_3 - \alpha_1)^2] \quad (7)$$

where σ_i, σ_2, and σ_3 are the principal stresses.

The quantity κ is given by

$$\kappa = \frac{2\sigma_c\sigma_t}{\sigma_c + \sigma_t} \quad (8)$$

which is equal to yield stress in the case when $\sigma_c = \sigma_t$, i.e. no pressure dependency, and where σ_c and σ_t are the yield stress in compression and tension respectively.

The hardness of the SiC material is 26 GPa and the initial yield stress is taken to be 11.82 GPa based on a proposed initial tensile yield (σ_t) value of H/2.2 for brittle materials (Gilman, 1975). The compressive yield (σ_c) is set to equal the hardness (H) of the material, to reflect the effect of the HPPT (Jacob, 2006; Ajjarapu et al., 2004).

For uniaxial stress state (where σ_2 and σ_3 are zero), I_1 and J_2 are evaluated as:

$$I_1 = \sigma_1 \quad (9)$$

$$J_2 = \sigma_1^2/3 \quad (10)$$

Now using equation (5), κ equals 16.25 GPa while α equals -0.375. These parameters are essential to provide a pressure sensitive model.

3.2 Thermal Softening Behavior

The thermal softening function (AdvantEdge Manual, 2009) is defined as

$$\Theta(T) = c_0 + c_1 T + c_2 T^2 + c_3 T^3 + c_4 T^4 + c_5 T^5 \quad if\, T < T_{cutoff} \tag{11}$$

$$\Theta(T) = \Theta(T_{cut})\left(1 - \frac{T - T_{cut}}{T_{melt} - T_{cut}}\right) \quad if\, T \geq T_{cutoff} \tag{12}$$

The polynomial coefficients c_0 through c_5 are used fit to a 5th order polynomial, T is the temperature, T_{cutoff} is the linear cutoff temperature and T_{melt} is the melting temperature.

The initial work was started by using a simplified approximate thermal softening curve given below (Virkar & Patten, 2009).

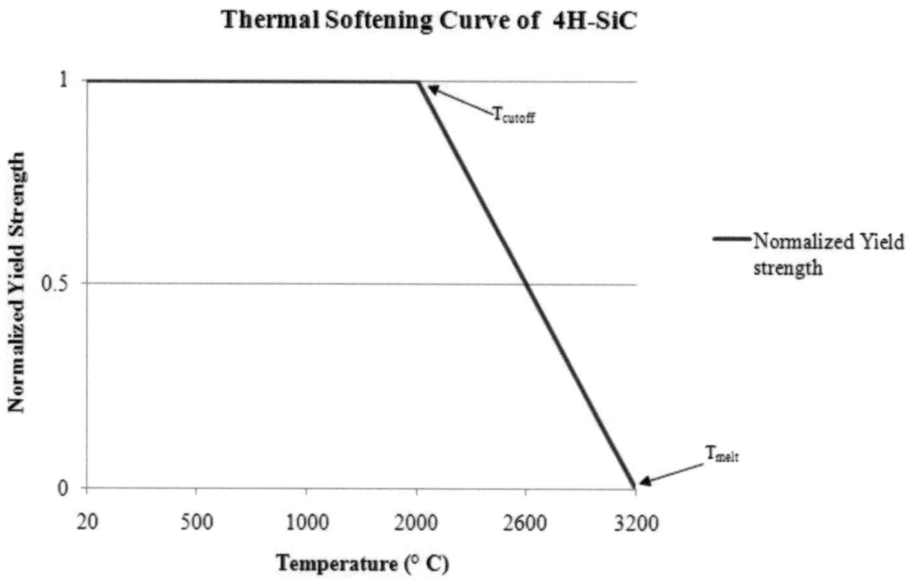

Fig. 9. Approximate thermal softening curve

In this curve, 2000° C and 3200° C were assumed as the thermal cutoff and melting temperature respectively. This model was used to test (evaluate) the thermal softening effect and to establish that it can be adequately incorporated in the simulation model (Virkar & Patten, 2009). For this curve, it is assumed that the yield strength remains constant from 20° C up to 2000° C and then it decreases linearly from 2000° C to the melting point (3200° C). Using polynomial equation (11) (Ajjarapu et al., 2004), the constants c_0 to c_5 are made zero to get a straight line. This thermal softening curve was later modified by using the values of thermal softening and melting temperatures from literature references (Shim et al., 2008; Yonenaga, 2001; Yonenaga et al., 2000; Samant et al., 1998; Tsetkov et al., 1996; CREE material data sheet; Naylor et al., 1979). The new thermal softening curve is given below in Fig. 10, which will be the emphasis of the remainder of this section.

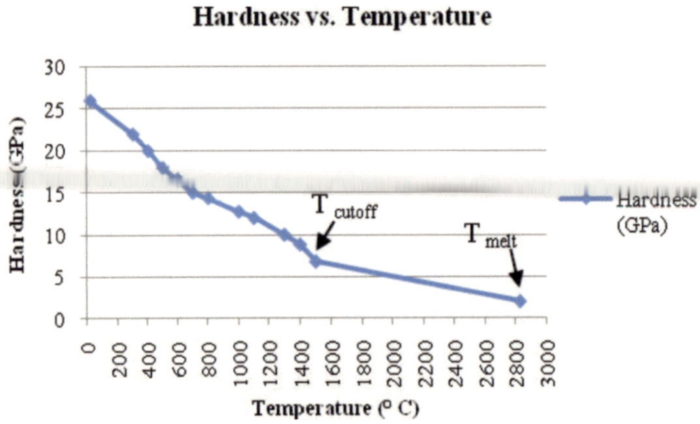

Fig. 10. Thermal Softening Curve based on references

Note: The temperature values T_{cutoff} and T_{melt} are estimated based on different values from various references (Shim et al., 2008; Yonenaga, 2001; Yonenaga et al., 2000; Samant et al., 1998; Tsetkov et al., 1996; CREE material data sheet; Naylor et al., 1979)

In this simulation work, a 3rd order polynomial curve was used to represent the thermal softening behavior. Using the polynomial equation (11), the constants c_0 to c_5 are assigned values based on the decrease in hardness values to get a 3rd order curve. Beyond the cutoff temperature, there is a linear decrease in the yield strength (there was no thermal softening values found in the literature over this range). This thermal softening curve demonstrates the thermal effects on SiC.

3.3 Tool and Workpiece Geometry and Properties

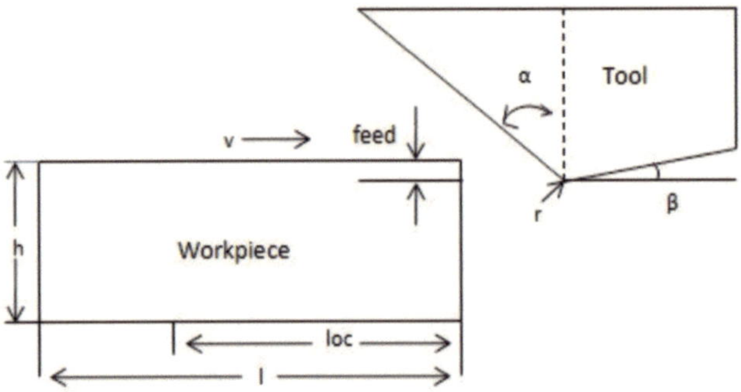

Fig. 11. Tool and Workpiece geometry

The workpiece was made long enough (L = 0.08 mm) to ensure that the length of cut (loc) would allow steady state conditions to be achieved during the simulation. The height (h = 0.02 mm) of the workpiece was much larger (between 10 to 100 times) in comparison to the feed (f) or uncut chip thickness (t) (Virkar & Patten, 2009; Virkar & Patten, 2010), which generally is between 50 and 500 nm. The boundary conditions of the workpiece surface are assumed to be traction free and constrained in the vertical direction. The workpiece material properties (Refer Table 3) used in the simulation model are given below

Material properties	**Value**	**Units**
Elastic Modulus, E	330	GPa
Poisson's ratio	0.212	-
Hardness, H	26	GPa
Initial yield stress, σ_0 or κ	16.25	GPa
Reference plastic strain, ε_0^P	0.049	-
Drucker-Prager coefficient (DPO)	-0.375	-
Thermal Cutoff temperature	1500	° C
Melting temperature	2830	° C

Table 3. Workpiece material properties

The tool used in the simulation is a single point diamond. The tool geometry is given in Table 4. The simulations were conducted in 2-D and as a result a round nose tool geometry could not be simulated, therefore the simulated tool cutting edge is flat, with a width (depth) of 20 μm (0.02 mm). The top and back surfaces of the tool are fixed with adiabatic conditions.

Cutting Edge Radius, r, (nm)	100
Rake angle, α	- 45°
Relief angle, β	5°
Width of tool (μm)	20
Thermal Conductivity, W/m □C	1500
Heat Capacity, J/kg °C	471.5
Density, kg/m^3	3520
Elastic Modulus, GPa	1050
Poisson's ratio	0.2

Table 4. Tool Geometry and material properties

The -45° rake angle creates a high pressure (compression) sufficient to accommodate the HPPT, thus the chip formation zone is conducive for ductile deformation (Patten et al., 2005).

3.4 Simulation Model

The simulation method is based upon a 2-D Lagrangian finite-element machining model assuming plane strain conditions (AdvantEdge Manual, 2009). The simulations were carried out by specifying the material properties for 4H-SiC. The constitutive model does not incorporate crystallographic planes/orientations and treats the material as elastic-plastic and ductile. To reflect the ductile behavior in ceramics, promoted by the HPPT, a pressure sensitive Drucker-Prager constitutive model is used (Patten et al., 2004; Patten et al., 2005).

The objective of this work was to incorporate the thermal effects that would closely replicate the actual heating conditions during the μ-LAM process. The software cannot directly incorporate the laser heating source in the model hence thermal boundary conditions similar to the laser heating effect are defined. The thermal effects were simulated by defining thermal boundary conditions in three different ways to capture the essence of the laser heating effect. In the first case a thermal boundary condition was provided on the tool tip about ~2μm along the rake and clearance faces away from the cutting edge (Refer Fig. 12).

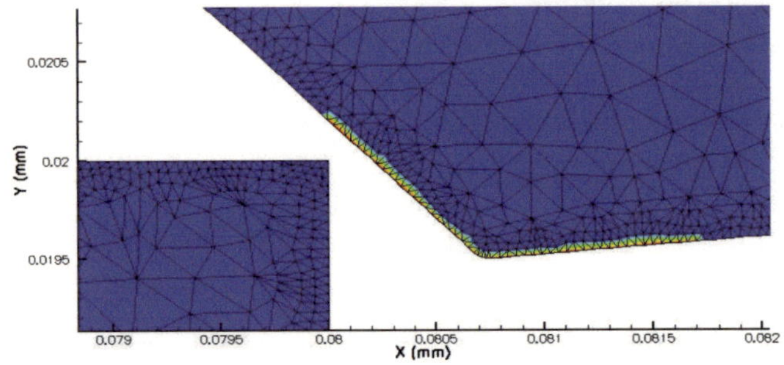

Fig. 12. Tooltip Boundary Condition

The blue color contour identifies the tool and workpiece being at room temperature and at the tool tip, the contour color varies which is due to a thermal boundary defined for heating effect. The temperature of the thermal boundary would be similar to the temperature of laser heating during machining. In the second case, the entire rake and clearance face of the tool was heated. In Fig. 4, the thermal boundary is at the tool tip but for the second case the thermal boundary was defined such that it extended on the entire rake and clearance face. For the third case, a thermal boundary was provided on the workpiece top surface (Refer Fig. 13) to reflect the heating effect of the laser on the workpiece surface. Note in the initial work (Virkar & Patten, 2009), the entire tool and workpiece system was uniformly heated to the elevated temperatures instead of a boundary condition.

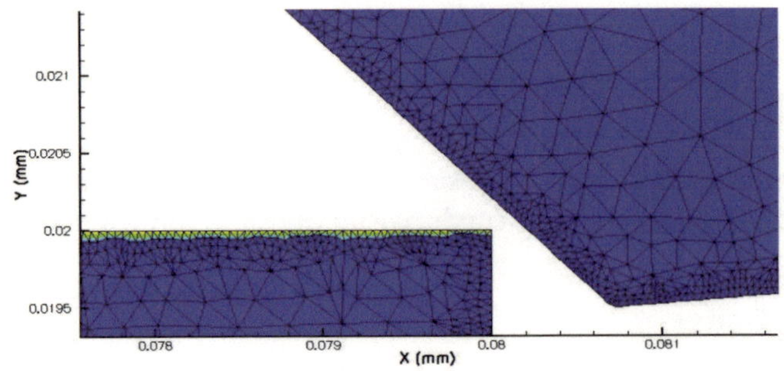

Fig. 13. Workpiece Boundary Condition

The process parameters are given in Table 5

Parameters	Values
Feed (nm)	500
Coefficient of friction	0.3
Cutting speed (m/s)	1
Depth (width) of cut (mm)	0.02

Table 5. Process Parameters

The simulations were conducted with the values given in Tables 3, 4, and 5. The feed in the 2-D simulation relates to the uncut chip thickness, and the depth of cut refers to the workpiece width. With the given values of high negative rake angle and small feed, ductile mode machining is achieved (Jacob, 2006). The coefficient of friction was taken as 0.3, however the results are not very sensitive to this value (AdvantEdge Manual, 2009).

The simulation results include the cutting and thrust forces that are used for comparison to evaluate the various machining conditions (primarily the workpiece temperature and the resultant thermal softening effect). The simulations were carried out at various temperatures: 20° C, 700° C, 1500° C, 2200° C, and 2700° C where 1500° C is the thermal cutoff temperature as shown in Fig. 2. Beyond the thermal cutoff temperature, there is a linear decrease in the material strength until it reaches the melting point (2830° C). The results of the simulations were viewed in the post processing tool Tecplot (AdvantEdge Manual, 2009), which provides temperature, pressure, stress and strain contours along with the cutting and thrust force plots for analysis of the simulation results. At each of the simulated temperatures, the chip formation, force plots and pressure contours are used and compared to evaluate the results.

3.5 Analysis of Force Data

The cutting force data from the simulation output is used for calculating the approximate cutting pressures at the tool chip interface. The other parameters used are width of the tool and feed or uncut chip thickness to determine the uncut chip cross sectional area. The formula used is given below

$$\text{Cutting Pressure} = \frac{\text{Cutting Force } (F_c)}{\text{Chip Area } (A_c)} \quad (8)$$

The chip area is calculated by multiplying the width of the tool (20 μm) with programmed feed (0.5 μm) of the simulation. To generate a ductile cutting environment through applied stresses (hydrostatic and shear) requires that the pressures at the workpiece-chip interface be equal to, or greater than, the hardness of the material, which is taken as 26 GPa (i.e. the compressive yield) for 4H-SiC. The thermal softening effect results in decreasing cutting pressures as the simulation temperature increases. In μ-LAM the cutting pressures are lowered at the tool-chip interface due to thermal softening (Virkar & Patten, 2009). Tool wear is reduced due to the lower cutting forces and pressures.

3.6 Simulation Results

3.6.1 Tool Tip Boundary Condition

A thermal boundary is provided at the tip of the tool, ~2 µm from the cutting edge, along the surface of the rake and clearance faces. During the actual µ-LAM process, the laser beam passes through the tip (cutting edge) of the tool, is absorbed by the workpiece and results in the heating of the workpiece locally at the tool-workpiece interface. Hence, this boundary condition (on the cutting edge, ~2µm along the rake and clearance faces) is a very close approximation to the actual location of the heating source in the actual µ-LAM process (note that it is on the wrong surface however, located on the tool rather than the workpiece). Table 6 summarizes the simulation results, where the heating effect is implemented on the tool rather than on the workpiece and then the heat is conducted to the workpiece at the tool-workpiece interface.

Temperatures (° C)	Cutting Force (mN)	Thrust Force (mN)	Chip formation	Pressure (GPa)
20	500	900	Yes	50
700	460	890	No	46
1500	370	610	No	37
2200	200	300	No	20
2700	80	130	Yes	8

Table 6. Tooltip Boundary Condition Results

In this case, the chip formation is seen at 20° C and 2700 ° C (Refer Table 6) and for the rest of the temperature points there is a smooth ductile material deformation (pile up and plowing in front of the tool), but no chip formation.

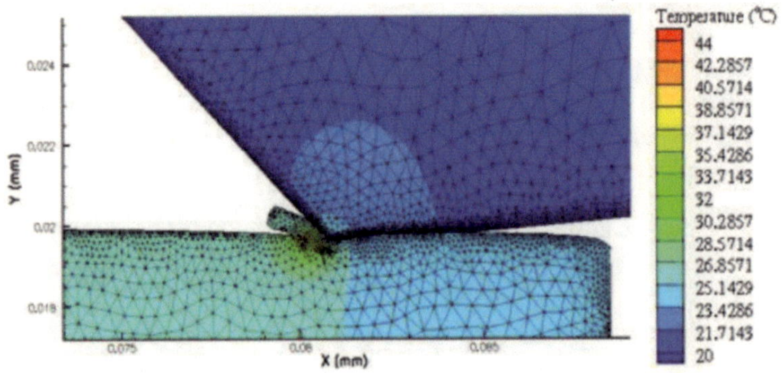

Fig. 14. At 20° C

The chip formation at 2700° C (Fig. 7) is quite different than that at 20° C (Refer Fig. 8) as the chip is very thick at the base and pointed at its end in the former case.

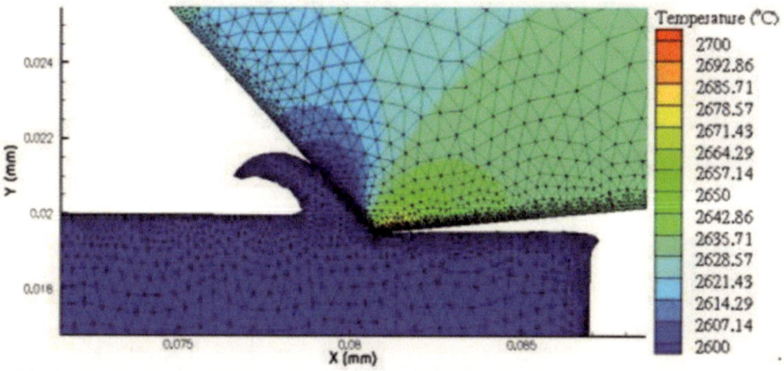

Fig. 15. At 2700° C

Figs. 16 and 17 describe the response of force and cutting pressure with respect to the temperature. The cutting pressure is above the hardness value (26 GPa) at 1500° C which means that the workpiece is still retaining its strength. Above 1500° C the pressure drops below the hardness value (Refer Table 6). The cutting forces and pressures show a decreasing trend with increase in temperature, as expected (Refer Fig. 16 and 17).

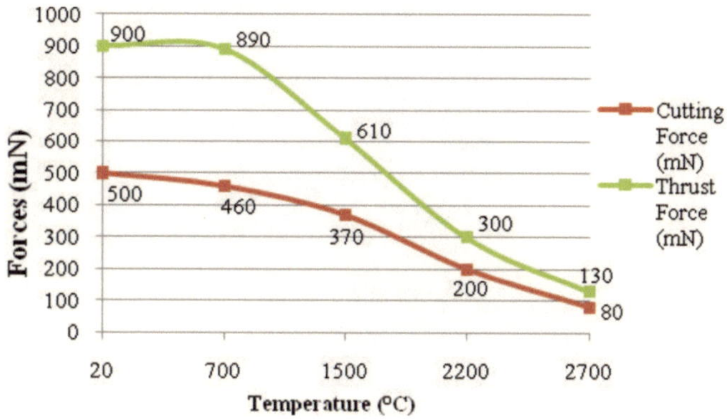

Fig. 16. Forces vs. Temperature curve

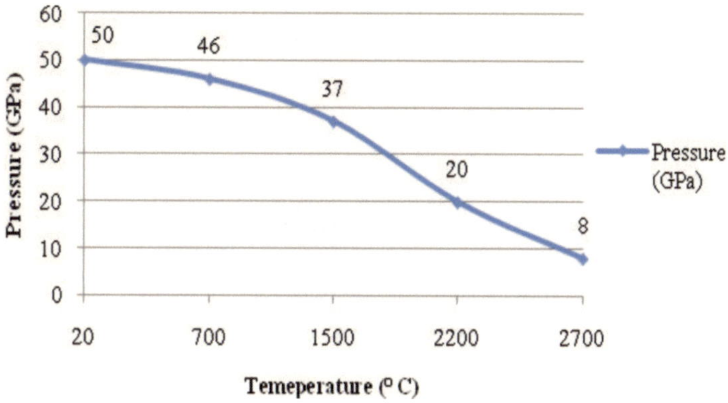

Fig. 17. Cutting Pressure vs. Temperature

3.6.2 Rake and Clearance Face heated Boundary Condition

In this case, a thermal boundary condition was defined on the entire rake and clearance face. This case was the first condition to start with this study and the easiest boundary condition to implement on the tool. Using this boundary condition, the effect on chip formation and cutting forces and pressures was established. Table 7 summarizes the result details:

Temperatures (° C)	Cutting Force (mN)	Thrust Force (mN)	Chip formation	Pressure (GPa)
20	500	1060	Yes	50
700	450	1000	No	45
1500	380	620	No	38
2200	200	300	No	20
2700	60	90	Yes	6

Table 7. Tool Rake and Clearance face Boundary Condition Results

In this case, the chip formation was again seen at 20° C and 2700° C (Refer Table 7). This behavior is also seen in the tooltip boundary condition (Refer Table 4). At 2700° C the chip formed was thicker as compared to the chip at 20° C (Refer Figs. 18 and 19)

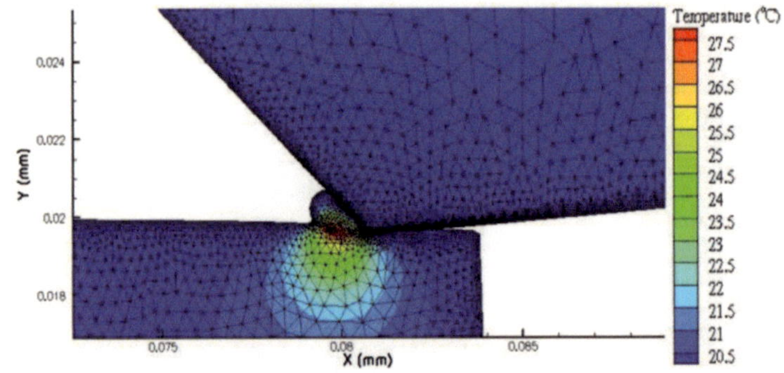

Fig. 18. At 20° C

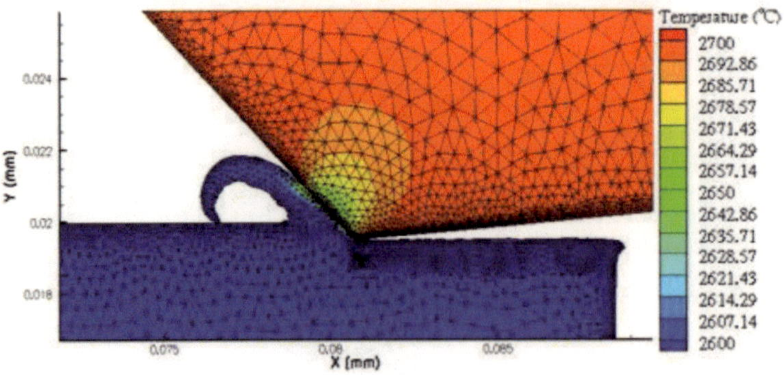

Fig. 19. At 2700° C

Figs. 20 and 21 shows a decreasing trend in the cutting forces and pressures with an increase in temperature.

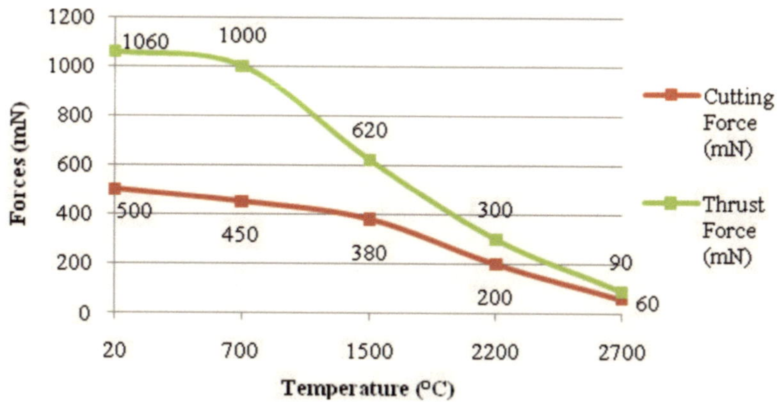

Fig. 20. Forces vs. Temperature curve

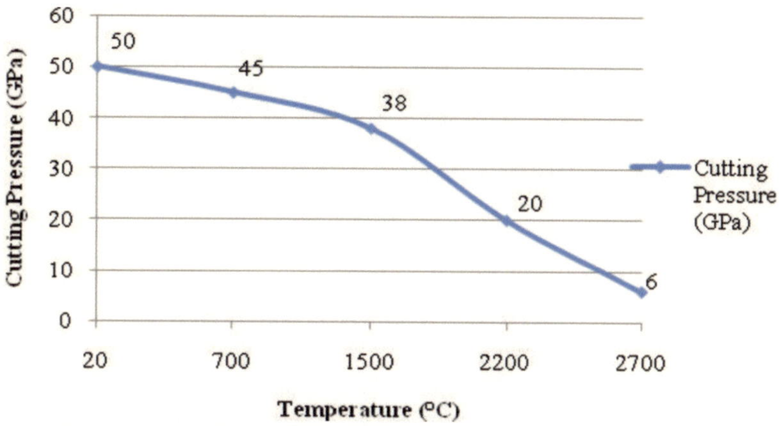

Fig. 21. Cutting Pressure vs. Temperature

3.6.3 Workpiece Boundary Condition

In this case, a thermal boundary is provided on the top surface of the workpiece. In μ-LAM, the workpiece surface is directly heated by a laser (the workpiece absorbs the laser energy and is subsequently heated), which passes through the tool hence the workpiece boundary condition closely replicates the actual process, i.e., the workpiece is directly heated. This phenomena is not quite directly comparable with the experiments as in μ-LAM process just the area under the tool tip (tool workpiece interface) is heated (based upon the laser spot size) Table 8 summarizes the simulation results.

Temperatures (° C)	Cutting Force (mN)	Thrust Force (mN)	Chip formation	Pressure (GPa)
20	470	1040	Yes	47
700	450	950	No	45
1500	390	570	No	39
2200	200	260	No	20
2700	30	40	No	3

Table 8. Workpiece Boundary Condition Results

(Note that the temperature scale is changed in the above figures to more clearly show the thermal effect)

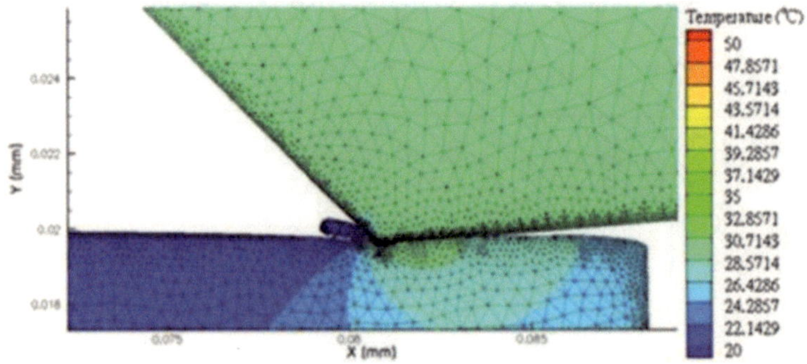

Fig. 22. At 20° C

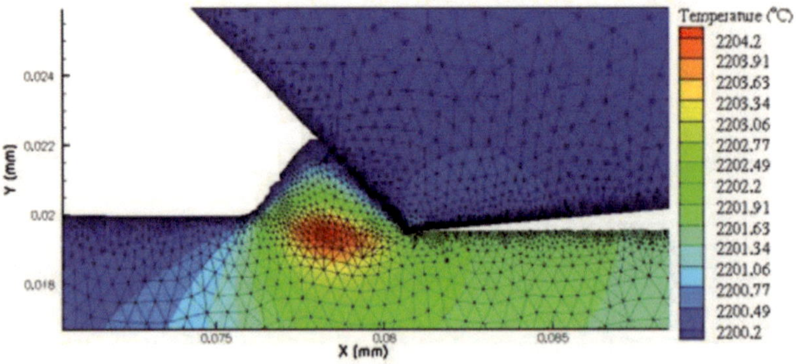

Fig. 23. At 2200° C

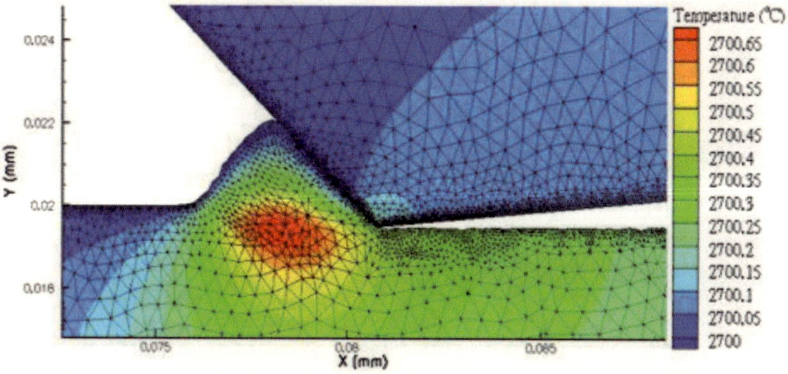

Fig. 24. (At 2700⁰ C)

In this case, the actual chip formation is seen only at 20° C. There is no chip formation seen at other temperature points (Refer to Table 8). The material deformation in this case is seen in Fig. 22, 23 and 24 (Note that the temperature scale is changed in both figures to show and highlight the thermal effect). The cutting forces and pressures show a decreasing trend (Refer Fig. 25 and 26) with increase in temperature which is an indication of thermal softening.

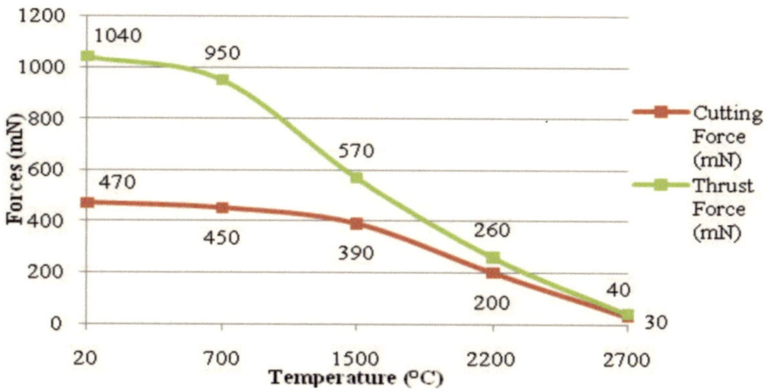

Fig. 25. Forces vs. Temperature curve

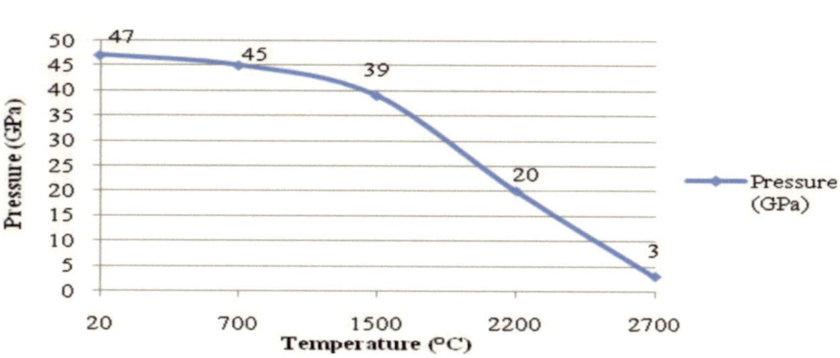

Fig. 26. Cutting Pressure vs. Temperature

3.7 Interaction between Stress and Temperature in μ-LAM

This interaction study between stress and temperature is developed from the simulation results (Virkar & Patten, 2010). The results for workpiece and tool-tip boundary conditions were considered for the analysis. The temperature dependence is determined from the yield strength vs. temperature curve given in Table 9 (this figure is similar for Fig. 10, Hardness vs. Temperature). The yield strength for brittle materials like SiC is calculated using H/2.2 (Gilman, 1975), where H is hardness value of SiC. The H value is derived from the hardness vs. temperature curve given in Fig. 10. The hardness values and the corresponding yield strength values are given in Table 7.

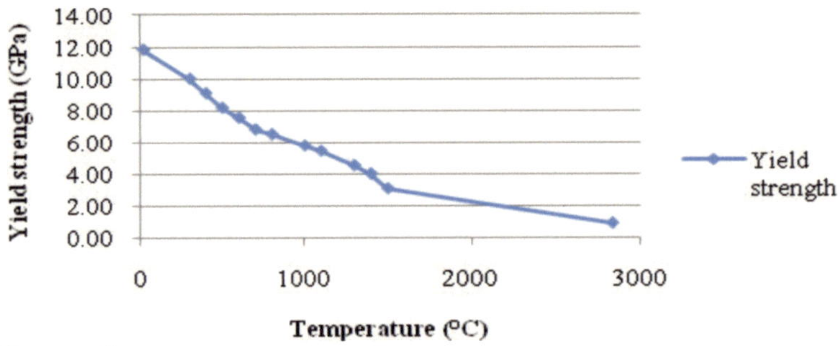

Fig. 27. Yield strength vs. temperature curve

Note: The yield strength in above figure refers to estimated tensile yield strength

Temperature (°C)	Hardness, (σ_c) GPa	Yield strength, (σ_t) GPa
20	26	11.82
300	22	10.00
400	20	9.09
500	18	8.18
600	16.6	7.55
700	15	6.82
800	14.35	6.52
1000	12.75	5.80
1100	12	5.45
1300	10	4.55
1400	8.835	4.02
1500	6.8	3.09
2830	2	0.91

Table 9. Yield strength as a function of temperature

Two approaches were selected for analyzing the interaction between the stress and temperature namely,

1. Normalized Cutting Force Approach
2. Yield strength as a function of pressure and temperature

3.7.1 Normalized Cutting Force approach

An approximate equation for cutting energy was developed using the cutting forces from the simulation at selected temperature points in the thermal softening regime (Virkar & Patten, 2010). The equation is given as

$$\text{Normalized Cutting Force} = \text{X stress} + \text{Y temperature} \tag{9}$$

where X and Y are proportional multipliers and X + Y = 100%.

The Normalized cutting force is obtained from the cutting forces taken directly from the simulation output (Refer Table 6) at each designated temperature point and dividing them by the force at 20° C. The stress factor in the equation is substituted as 1 because in the Drucker-Prager model stress or yield is not a function of temperature. The thermal (temperature) contribution is the tensile yield strength value taken from the yield strength vs. temperature curve (Refer Fig. 27) at the designated simulation temperature. In this way the percentage interaction or contribution of stress and temperature is determined for all boundary condition simulations given in Table 10.

	Temperatures (° C)	Stress %	Temperature %
Tooltip Boundary Condition simulation	20	100	0
	700	80.95	19.05
	1500	64.86	35.14
	2200	26.83	73.17
	2700	6.67	93.33
Workpiece Boundary Condition simulation	20	100	0
	700	76.19	23.81
	1500	59.52	40.48
	2200	26	74
	2700	4	96

Table 10. Percentage interaction of stress and temperature

From this approach it is seen that for both the boundary conditions at 20° C, the stress contribution dominated (no thermal effect). As the temperature of the boundary condition increases the temperature contribution increases reducing the effect due to stress contribution. The cutting and thrust forces decrease as the temperature contribution increases thus showing the thermal softening effect. The apparent crossover in contribution of stress to temperature occurs between 1500° C to 2200° C.

3.7.2 Yield strength as a function of pressure and temperature approach

This approach is based on the Drucker-Prager yield criterion. At first, the tensile yield strength was calculated as a function of temperature (Refer Table 11). In the Drucker-Prager model, the κ value is the calculated yield which is determined using equation (8). The σ_c is taken as the room temperature hardness of SiC, which is kept constant and σ_t is the tensile yield which is substituted from the yield strength vs. temperature curve (Refer Fig. 27). The calculated yield (κ) from 20° C until 2830° C is calculated using the equation (8) and is given in Table 11:

Temperature (°C)	Hardness,(σ_c) GPa	Tensile Yield strength,(σ_t) - GPa	Calculated yield with constant σ_c - GPa
20	26	11.82	16.25
300	22	10.00	14.44
400	20	9.09	13.47
500	18	8.18	12.45
600	16.6	7.55	11.70
700	15	6.82	10.80
800	14.35	6.52	10.43
1000	12.75	5.80	9.48
1100	12	5.45	9.02
1300	10	4.55	7.74
1400	8.835	4.02	6.96
1500	6.8	3.09	5.53
2830	2	0.91	1.76

Table 11. Calculated yield strength as a function of temperature

Note: The initial compressive yield σ_c is kept constant to see the effect due to pressure without any thermal effect as we assume it to be equal to the hardness previously.

From Table 11 we can see that the yield strength goes on decreasing as the temperature of the simulation is increasing. The data for hardness vs. temperature is limited only up until 1500° C based on the references. From 1500° C upto 2830° C, the hardness vs. temperature is approximate and hence there is a linear decrease in the yield strength (Fig. 2). Hence the σ_t value at 2200° C is based on interpolation between 1500° C and 2830° C on the temperature scale.

To calculate the yield strength as a function of pressure, we consider the Drucker-Prager yield criterion from equation (5). In that equation, I_1 and J_2 are the invariants of stress tensors and are dependent on the three dimensional stresses which are taken from the simulation output. By substituting the values of σ_1, σ_2, σ_3 in equations (6) and (7), we get value of I_1 and J_2. Substituting these values back into equation (5) gives us the value of calculated yield κ. The value of Drucker-Prager coefficient (α) is kept constant at 0.375 which is obtained assuming plain strain conditions. Table 12 shows all the calculated values.

From Simulation					Calculated Values			
Boundary Condition	Temp. (° C)	σ_x (GPa)	σ_y (GPa)	σ_z (GPa)	I_1 (GPa)	J_2 (GPa)	K (GPa)	Max. pressure (GPa)
WBC	20	38.74	25.92	32.41	97.06	41.08	47.50	50.00
WBC	700	22.06	21.62	21.97	65.65	0.05	25.02	46.00
WBC	1500	8.86	10.79	9.81	29.46	0.94	12.72	37.00
WBC	2200	4.78	2.58	3.68	11.03	1.21	6.04	20.00
WBC	2700	0.40	0.78	0.59	1.77	0.04	0.99	8.00
Tooltip	20	35.12	24.53	29.80	89.45	28.06	42.71	47.00
Tooltip	700	31.55	14.27	22.86	68.67	74.63	40.71	45.00
Tooltip	1500	16.28	12.65	14.71	43.64	3.31	19.51	39.00
Tooltip	2200	10.07	6.51	8.29	24.87	3.17	12.41	20.00
Tooltip	2700	4.41	4.08	4.25	12.74	0.03	5.06	3.00

Table 12. Yield strength as a function of pressure

From Table 12, we can see that the calculated yield goes down with increase in temperature. This effect is seen as the σ_t factor in Drucker-Prager model is affected by temperature. There is also a column of maximum pressure which is calculated by dividing the cutting forces by cross sectional area of contact surface of tool. Thus, by both approaches the same trend is seen in the behavior of SiC. As the temperature goes on increasing the dominance due to stress reduces.

4. Conclusion

For the experimental study, micro-laser assisted scratch tests were successful in demonstrating the enhanced thermal softening of the material resulting in a greater ductile to brittle transition depth. Laser heating was successfully demonstrated as evidenced by the significant increase in the ductile response of single crystal 4H-Silicon Carbide in the {1010} plane along the <1010> direction. Laser assisted (heating, thermal softening and reduced brittleness) material removal resulted in greater depths of cuts (in the ductile regime) at less applied thrust forces, smaller cutting forces and a larger critical DoC. Force analysis (thrust and cutting), optical microscopy and white light interferometry served as useful analysis methods to detect the enhanced ductile response and reduced brittle fracture as a result of preferential material heating (of the high pressure phase transformed material). Results obtained from this study are promising to further implement micro-laser assisted machining (μ-LAM) in operations such as single point diamond turning. Lower cutting forces obtained from the μ-LAM process are favorable to minimize tool wear while machining abrasive ceramics/semiconductors such as Quartz, Silicon and Silicon Carbide. The results from this study also will benefit the manufacture of brittle materials as laser heating is proven to decrease the brittle response in ceramics and semiconductors which can result in higher productivity rates (i.e. higher material removal rate).

For the simulation study, a set of carefully designed thermal boundary conditions were simulated in 2D using the software TWS AdvantEdge software. The simulations were used to predict the thermal softening behavior of SiC using the Drucker-Prager yield criterion. In the results of each of the cases, there is a decreasing trend in cutting forces and pressures with an increase in temperature. This confirms the thermal softening behavior due to simulated laser heating, which reduces the hardness of the material. The interaction between stress and temperature was also determined from the boundary condition simulation outputs. It showed that the dominance of stress decreases with increase in temperature.

5. Future Research

The results and data obtained will be implemented in establishing machining parameters to perform ductile mode micro-laser assisted single point diamond turning (SPDT) on SiC. Various scratch tests are currently being conducted to establish optimized correlation between laser power and machining parameters such as depth of cut, feed and cutting speed.

High pressure and temperature experiments have been studied (and currently under analysis) to determine the pressure-temperature correlation with the phase transition during the μ-LAM process using laser heated diamond anvil cells (DAC). This analysis however neglects the effect due to shear forces (which is believed to contribute significantly in the phase transformation of SiC). To include shear force analysis, a new DAC is being developed that will account for the missing shear component in a regular DAC cell. Furthermore, the change in lattice volume will be calculated and the cell structure of the newly developed high pressure phase material will be determined. Laser absorption measurements are also being conducted on the current high pressure phase samples to optimize the laser power during the μ-LAM process.

6. Acknowledgements

The authors would like to thank Husyein Bogac Poyraz for his contribution in the experiments and Third Wave Systems (TWS) Inc., for their generous support and technical assistance. The authors are also thankful to the National Science Foundation (NSF) for their grant CMMI-0757339.

7. References

AdvantEdge User Manual, version 5.3, 2009

Ajjarapu, S.K.; Cherukuri, H.; Patten, J.A.; Brand, C.J. (2004). Numerical simulations of ductile regime machining of Silicon Nitride using Drager-Prager model, *Proc. Institute of Mechanical Engineers*, 218(C), pp. 1-6

Ashurst, W.R.; Wijesundara, M.B.J.; Carraro, C. & Maboudian, R. (2004). Tribological impact of SiC coating on released polysilicon structures, *Tribol. Lett.*, Vol.17, No.2, pp. 195-198

Bifano, T.G.; Dow, T.G. & Scattergood, R.O. (1991). Ductile regime grinding- a new technology for machining brittle materials, *Journal of Engineering for Industry*, Vol.113, pp. 184-189

Blake, P.N. & Scattergood, R.O. (1990). Ductile-regime machining of germanium and silicon, *Journal of the America Ceramic Society*, Vol.73, Issue 4, pp 949-957

Blake, P.N. & Scattergood, R.O. (1990). Precision machining of ceramic materials, *Journal of the American Ceramic Society*, Vol.73, No.4, pp. 949-957

Blackley, W.S. & Scattergood, R.O. (1994). Chip topography for ductile-regime machining for germanium, *Journal of Engineering for Industry*, Vol.116, pp. 263-266

CREE material data sheet http://www.cree.com/products/pdf/MAT-CATALOG.pdf

Dong, L. & Patten, J.A. (2007). Real time infrared (IR) thermal imaging of laser-heated high pressure phase of silicon, *Advanced Laser Applications Conference & Expo (ALAC 2007)*, Boston, MA.

Dong, L. (2006). In-situ detection and heating of high pressure metallic phase of silicon during scratching, *PhD dissertation, University of North Carolina at Charlotte*

Gao, D; Wijesundara, M.B.J.; Carraro, C.; Low, C.W.; Howe, R.T. & Maboudian, R. (2003). High modulus polycrystalline 3C-SiC technology for RF MEMS, *Proc. Transducers 12th Int. conf. Solid-State Sensors and Actuators*, pp. 1160-1163

Gilman J.J. (1975). Relationship between impact yield stress and indentation hardness, Journal of Applied Physics, 46(4), pp. 1435-1436

Jahanmir, S; Ives, L.K.; Ruff, A.W. & Peterson, M.B. (1992). Ceramic machining: assessment of current practice and research needs in the United States, *NIST Special Publication*, Vol. 834, p.102

Leung, T.P.; Lee, W.B. & Lu, X.M. (1998). Diamond turning of silicon substrates in ductile-regime, *Journal of Materials Processing Technology*, Vol.73, pp. 42-48

Marusich, T.D.; Askari, E. (2001). Modelling residual stress and workpiece surfaces in machined surfaces, www.thirdwavesys.com

Morris, J.C.; Callahan, D.L.; Kulik, J.; Patten, J.A. & Scattergood, R.O. (1995). Origins of the ductile regime in single-point diamond turning of semiconductors, *Journal of American Ceramic Society*, Vol.78, No.8, pp. 2015-2020

Morris, J.C.; Callaham, D.L.; Kulik; Patten, J.; & Scattergood, R.O. (1995). Origins of the ductile regime in single point diamond turning of semiconductors, *Journal of American Ceramic Society*, Vol.78, No.6, pp. 2015-2020

Naylor, M.G.; Page, T.F. (1979). The effect of temperature and load on the indentation hardness behavior of Silicon Carbide engineering ceramics, Proceedings of International Conference on erosion of soil and impact, pp. 32

Jacob, J. (2006). Numerical simulation on machining of silicon carbide, *Master's Thesis*, Western Michigan University, MI, USA

O'Connor, B.; Marsh, E.; Couey, J. (2005). On the effect of crystallographic orientations for ductile material removal in Silicon, *Precision Engineering*, Vol. 29(1), pp. 124-132

Patten, J.A.; Bhattacharya, B. (2006). Single point diamond turning of CVD coated silicon carbide, *Journal of Manufacturing Science- ASME*, Vol. 127, pp. 522

Patten, J.A.; Cherukuri, H.; Yan, J.W. (2004). Ductile regime machining of semiconductors and ceramics, *Institute of Physics Publishing*, pp. 661

Patten, J.A.; Fesperman, R.; Kumar, S.; McSpadden, S.; Qu, J.; Lance, M.; Nemanich, R. & Huening, J. (2003). "High-Pressure Phase Transformation of Silicon Nitride", *Applied Physics Letters*, 83(23), 4740-4742, 2003.

Patten, J.A.; Gao, W. & Yasuto, K. (2005). Ductile regime nanomachining of single-crystal silicon carbide, *Journal of Manufacturing Science- ASME*, Vol.127, No.3, pp. 522-532

Patten, J.A.; Jacob, J. (2008). Comparison between numerical simulations and experiments for single point diamond turning of single crystal silicon carbide, *Journal of Manufacturing Processes*, Vol. 10, pp. 28-33

Patten, J.A.; Jacob, J.; Bhattacharya, B.; Grevstad, A. (2008). Comparison between numerical simulation and experiments for single point diamond turning of Silicon Carbide, *Society of Manufacturing Engineers NAMRAC conference, pp.2*

Ravindra, D; Patten, J. & Tano, M. (2007). Ductile to brittle transition in a single crystal 4H SiC by performing nanometric machining, *Proc. ISAAT 2007 Precision Grinding and Abrasive Technology at SME International Grinding Conference, Advances in Abrasive Technology*, X, pp 459-465

Rebro, P.A.; Pfefferkorn, F.E.; Shin, Y.C. & Incropera, F.P. (2002). Comparative assessment of laser-assisted machining of various ceramics, *Transactions of NAMRI*, Vol.30, pp. 153-160

Ravindra, D.; Patten, J. & Qu, J. (2009). Single point diamond turning effects on surface quality and subsurface damage in ceramics, *Proceedings of the ASME International Manufacturing Science and Engineering Conference*, West Lafayette, IN.

Ravindra, D. & Patten, J. (2008). Improving the surface roughness of a CVD coated SiC disk by performing ductile regime single point diamond turning, *Proceedings of the ASME International Manufacturing Science and Engineering Conference*, Evanston, IL.

Samant, A.V.; Zhou, W.L.; Pirouz, P. (1998). Effect of test temperature and strain rate on the yield strength of monocrystalline 6H-SiC, *Physica Status Solidi (a)*, Vol. 166, pp. 155

Shayan, A.R.; Poyraz, H.B.; Ravindra, D.; Ghantasala, M. & Patten, J.A. (2009). Force analysis, mechanical energy and laser heating evaluation of scratch tests on silicon carbide (4H-SiC) in micro-laser assisted machining (μ-LAM) process, *Proceedings of the ASME International Manufacturing Science and Engineering Conference*, Evanston, IL.

Shayan, A.R.; Poyraz, H. B.; Ravindra, D. & Patten, J.A. (2009). Pressure and temperature effects in micro-laser assisted machining (μ-LAM) of silicon carbide, *Transactions NAMRI/SME*, Vol.37, pp. 75-80

Shim, S.; Jang, J.I., Pharr, G. M. (2008). Extraction of flow properties of single crystal Silicon Carbide by nanoindentation and Finite Element simulation, *Act. Materialia, Vol. 58, pp. 3824-3832*

Shin, Y.C.; Pfefferkorn, F.E.; Rozzi, J.C. (2000). Experimental evaluation of laser assisted machining of Silicon Nitride ceramics, *Journal of Manufacturing Science- ASME*, Vol.122, pp. 666

Sreejith, P.S.; Ngoi, B.K.A. (2001). Material removal mechanisms in precision machining of new materials, *International Journal of Machine Tools & Manufacture*, Vol.41, pp. 1831-1843

Tsevetkov, V.F.; Allen, S.T.; Kong, H.S.; Carter, C.H. (1996). Recent progress in SiC crystal growth, *Institute of Physics*, Vol no. 142, pp.17

Virkar, S.R.; Patten, J.A. (2009). Numerical simulations and analysis of thermal effects on Silicon Carbide during ductile mode micro-Laser Assisted Machining, *Proceedings of the ASME International Manufacturing Science and Engineering Conference*, West Lafayette, IN

Virkar, S.R.; Patten, J.A. (2010). Simulation of thermal effects for analysis of micro Laser Assisted Machining, Proceedings of ICOMM conference, Madison, WI

Wobker, H.G. & Tonshoff, H.K. (1993). High efficiency grinding of structural ceramics, *International Conference on Machining of Advanced Materials, NIST Special Publication 847*, pp. 455-463, Gaithersburg, MD.

Yan, J.W.; Syoji, K. & Kuriyagawa, T. (2002). Ductile regime turning at large tool feed, *J. Mater. Process Tech.*, Vol. 121, No.2-3, pp. 363-372

Yan, J.W.; Maekawa, K. & Tamaki, J.(2004). Experimental study on the ultra-precision ductile machinability of single-crystal germanium", *JSME International Journal C-Mech Sy.*, Vol.47, No.1, pp. 29-36

Yonenaga, I. (2001). Thermo-Mechanical stability of wide-bandgap semiconductors: High temperature hardness of SiC, AlN, GaN, ZnO and ZnSe, *Physica B.*, 308-310, pp. 1150-1152